WITHDRAWN

AQUATIC TOXICOLOGY

MOLECULAR, BIOCHEMICAL, AND CELLULAR PERSPECTIVES

EDITED BY

DONALD C. MALINS
GARY K. OSTRANDER

LEWIS PUBLISHERS
Boca Raton   Ann Arbor   London   Tokyo

Library of Congress Cataloging-in-Publication Data

Aquatic toxicology : molecular, biochemical, and cellular perspectives
  / edited by Donald C. Malins, Gary K. Ostrander.
     p.   cm.
  Includes bibliographical references (p.   ) and index.
  ISBN 0-87371-545-4 (acid-free paper)
  1. Aquatic organisms—Effect of water pollution on.   2. Fishes—Effect of
water pollution on.   3. Molecular toxicology.
I. Malins, Donald C.   II. Ostrander, Gary Kent.
QH90.8.T68A67   1993
574.2'8—dc20   93-6117
                                                                                                      CIP

This book contains information obtained from authentic and highly regarded sources. Reprinted material is quoted with permission, and sources are indicated. A wide variety of references are listed. Reasonable efforts have been made to publish reliable data and information, but the author and the publisher cannot assume responsibility for the validity of all materials or for the consequences of their use.

Neither this book nor any part may be reproduced or transmitted in any form or by any means, electronic or mechanical, including photocopying, microfilming, and recording, or by any information storage or retrieval system, without prior permission in writing from the publisher.

All rights reserved. Authorization to photocopy items for internal or personal use, or the personal or internal use of specific clients, may be granted by CRC Press, Inc., provided that $.50 per page photocopied is paid directly to Copyright Clearance Center, 27 Congress Street, Salem, MA 01970 USA. The fee code for users of the Transactional Reporting Service is ISBN 0-87371-545-4/94/ $0.00+$.50. The fee is subject to change without notice. For organizations that have been granted a photocopy license by the CCC, a separate system of payment has been arranged.

CRC Press, Inc.'s consent does not extend to copying for general distribution, for promotion, for creating new works, or for resale. Specific permission must be obtained in writing from CRC Press for such copying.

Direct all inquiries to CRC Press, Inc., 2000 Corporate Blvd., N.W., Boca Raton, Florida 33431.

© 1994 by CRC Press, Inc.
Lewis Publishers is an imprint of CRC Press

No claim to original U.S. Government works
International Standard Book Number 0-87371-545-4
Library of Congress Card Number 93-6117
Printed in the United States of America 1 2 3 4 5 6 7 8 9 0
Printed on acid-free paper

To my wife, Mary, with thanks for always being there.
D.C.M.

To my wife, Debbie. Your constant encouragement and support is appreciated.
G.K.O.

INTRODUCTION

Studies of the effects of chemicals on the cellular and subcellular systems of aquatic organisms seemed to have lagged behind in quantity, and sometimes in quality, those of terrestrial animals (e.g., mammals). Fortunately, there have been welcome changes in these circumstances in the last decade or so. The traditional focus on rats, mice, and other laboratory mammals in molecular biology, biochemistry, and toxicology has shifted significantly toward an interest in how a broad variety of cellular and subcellular aquatic systems respond to insults from environmental contaminants.

There are undoubtedly many reasons for this shift in emphasis. Among them is the recognition that an understanding of chronic effects on aquatic biota can only be attained if we can increase our knowledge of stresses on their molecular cellular, and subcellular systems.

Scientists are also becoming increasingly aware that aquatic systems offer many outstanding opportunities to broaden our perspectives of biological processes, many of which are unique and reflect special adaptations to life in the aquatic environment. Moreover, those responsible for managing threatened aquatic environments have long suffered from a paucity of information on chronic effects and would welcome increased knowledge of the processes and mechanisms that govern such effects.

In editing this book, we recognized that very exciting work with aquatic vertebrates and invertebrates is being undertaken at the cellular and subcellular level. Thus, we want to provide a forum for those investigators who, because of their well-recognized expertise and accomplishments, would be able to convey an understanding of the latest developments in aquatic toxicology at the molecular and cellular level of organization.

This book is not intended to be comprehensive. Practical considerations relating to bringing virtually all the excellent research being conducted into one volume would have been an almost impossible task. We wish to make the point, however, that our contributors are among the very best in their respective fields, and their scientific interests are pivotal to understanding toxic chemical stresses on a variety of important aquatic systems and in charting the course of future research.

Our contributors were given considerable latitude. We asked them to prepare a thoughtful essay, including a cogent evaluation of the relevant literature and not just a recapitulation of what has been published. Furthermore, we encouraged them to provide to the extent possible a sense of how their particular area of interest interfaces with mammalian toxicology, to suggest practical applications of the work to environmental problems, to delineate research priorities, and to provide a sense for future research directions.

We hope that we have succeeded in providing the researcher, student, environmental manager, and others with valuable insight into a variety of exciting provocative events occurring in aquatic toxicology — events that form an essential basis for protecting precious and endangered aquatic environments from the long-term effects of pollution.

Donald C. Malins
Gary K. Ostrander
January 1994

THE EDITORS

Donald C. Malins, Ph.D., D.Sc., is Director of the Molecular Epidemiology Program at the Pacific Northwest Research Foundation. He is also Editor-in-Chief of the international journal *Aquatic Toxicology*. Previously, he directed a multidisciplinary environmental research group for the National Oceanic and Atmospheric Administration, U.S. Department of Commerce in Seattle. His present research interests relate to the etiology of cancer and its prediction at the molecular level using human and aquatic model systems. Among his honors are the D.Sc. from his alma mater, the University of Aberdeen, Scotland, for biochemical research of high distinction and the U.S. Department of Commerce's Gold Medal for outstanding contributions to understanding the effects of toxic chemicals on marine ecosystems.

Gary K. Ostrander, Ph.D., is an Associate Professor in the Department of Zoology at Oklahoma State University in Stillwater. He is presently a member of the Environmental Toxicology Program at the university and serves as an editor of *Aquatic Toxicology*. Dr. Ostrander received his Ph.D. degree from the University of Washington. He was an NIH postdoctoral fellow in the Department of Pathology, School of Medicine, University of Washington and also served as a Staff Scientist at the Pacific Northwest Research Foundation in Seattle. His current research interests include comparative studies of mammalian and aquatic animal models to include the function of tumor suppressor genes and carbohydrate antigens during chemical carcinogenesis.

CONTRIBUTORS

Robert S. Anderson, Ph.D.
The University of Maryland System
Center for Environmental and
 Estuarine Studies
Chesapeake Biological Laboratory
Solomons, MD

Stephen G. George, Ph.D.
NERC Unit of Aquatic Biochemistry
University of Stirling
Stirling, Scotland

Mark E. Hahn, Ph.D.
Biology Department
Woods Hole Oceanographic Institution
Woods Hole, MA

David E. Hinton, Ph.D.
Department of Medicine
School of Veterinary Medicine
University of California
Davis, CA

Margaret O. James, Ph.D., D.Sc.
Department of Medicinal Chemistry
University of Florida
Gainesville, FL

Kevin M. Kleinow, D.V.M., Ph.D.
Department of Veterinary Physiology,
 Pharmacology and Toxicology
School of Veterinary Medicine
Louisiana State University
Baton Rouge, LA

Alexander E. Maccubbin, Ph.D
Department of Experimental
 Therapeutics
Grace Cancer Drug Center
Roswell Park Cancer Institute
Buffalo, NY

James M. McKim, Ph.D.
U.S. Environmental Protection Agency
Environmental Research Laboratory
Duluth, MN

Michael J. Moore, Vet. M.B., Ph.D
Biology Department
Woods Hole Oceanographic Institution
Woods Hole, MA

Mark S. Myers, M.A.
Environmental Conservation Division
Northwest Fisheries Science Center
National Marine Fisheries Service
National Oceanic and Atmospheric
 Administration
Seattle, WA

John W. Nichols, Ph.D.
U.S. Environmental Protection Agency
Environmental Research Laboratory
Duluth, MN

Gary K. Ostrander, Ph.D.
Environmental Toxicology Program
Department of Zoology
Oklahoma State University
Stillwater, OK

W. E. Robinson, Ph.D.
University of Massachusetts —
 Boston, Harbor Campus
Environmental Sciences Program
Boston, MA

G. Roesijadi, Ph.D.
The University of Maryland System
Center for Environmental and
 Estuarine Studies
Chesapeake Biological Laboratory
Solomons, MD

Richard E. Spieler, Ph.D.
Oceanographic Center
Nova University
Dania, FL

John J. Stegeman, Ph.D.
Biology Department
Woods Hole Oceanographic
 Institution
Woods Hole, MA

Rebecca J. Van Beneden, Ph.D.
Department of Zoology
University of Maine
Orono, ME

Daniel N. Weber, Ph.D.
Marine and Freshwater Biomedical
 Core Center
University of Wisconsin — Milwaukee
Milwaukee, WI

REVIEWERS

Gary J. Atchison, Ph.D.
Department of Animal Ecology
Iowa State University
Ames, IA

Keith Cheng, M.D., Ph.D.
M.S. Hershey Medical Center
Pennsylvania State University
Hershey, PA

William S. Fisher, Ph.D.
U.S. Environmental Protection Agency
Center for Marine and Estuarine
 Disease Research
Gulf Breeze, FL

John W. Fournie, Ph.D.
U.S. Environmental Protection Agency
Environmental Research Laboratory
Gulf Breeze, FL

John Harshbarger, Ph.D.
Office of Tumor Registry
National Museum of Natural History
Smithsonian Institution
Washington, D.C.

William L. Hayton, Ph.D.
Pharmaceutics and Pharmaceutical
 Chemistry
College of Pharmacy
The Ohio State University
Columbus, OH

Frank M. Hetrick, Ph.D.
Department of Microbiology
University of Maryland
College Park, MD

Margaret O. James, Ph.D., D.Sc.
Department of Medicinal Chemistry
University of Florida
Gainesville, FL

Pam Kloepper-Sams, Ph.D.
Environmental Safety Department
The Proctor & Gamble Co.
Cincinnati, OH

Peter Landrum, Ph.D.
Great Lakes Environmental Research
 Laboratory
Ann Arbor, MI

John Lech, Ph.D.
Medical College of Wisconsin
Pharmacology and Toxicology
Milwaukee, WI

James M. McKim, Ph.D.
U.S. Environmental Protection Agency
Environmental Research Laboratory
Duluth, MN

Michael Miller, Ph.D.
Department of Biochemistry
Health Sciences Center
West Virginia University
Morgantown, WV

Mark S. Myers
Environmental Conservation Division
Northwest Fisheries Science Center
National Marine Fisheries Service
National Oceanic and Atmospheric
 Administration
U.S. Department of Commerce
Seattle, WA

Michael C. Newman, Ph.D.
Savannah River Ecology
 Laboratory
University of Georgia
Aiken, SC

David Noakes, Ph.D.
Department of Zoology
University of Guelph
Guelph, Ontario, Canada

Wolfgang K. Vogelbein, Ph.D.
Virginia Institute of Marine Science
The College of William and Mary
Gloucester Point, VA

Dennis Watson, Ph.D.
Center for Molecular and
 Structural Biology
Hollings Cancer Center
Medical University of South Carolina
Charleston, SC

CONTENTS

Chapter 1
Trophic Transfer of Chemicals in the Aquatic Environment 1
Margaret O. James and Kevin M. Kleinow

Chapter 2
Enzymology and Molecular Biology of Phase II Xenobiotic-
Conjugating Enzymes in Fish 37
Stephen G. George

Chapter 3
Biochemistry and Molecular Biology of Monooxygenases: Current
Perspectives on Forms, Functions, and Regulation of Cytochrome
P450 in Aquatic Species 87
John J. Stegeman and Mark E. Hahn

Chapter 4
Cells, Cellular Responses, and Their Markers in Chronic Toxicity
of Fishes 207
David E. Hinton

Chapter 5
Modulation of Blood Cell-Mediated Oxyradical Production in
Aquatic Species: Implications and Applications 241
Robert S. Anderson

Chapter 6
DNA Adduct Analysis in Fish: Laboratory and Field Studies 267
Alexander E. Maccubbin

Chapter 7
Expression of Oncogenes and Tumor Suppressor Genes in
Teleost Fishes 295
Rebecca J. Van Beneden and Gary K. Ostrander

Chapter 8
Pathobiology of Chemical-Associated Neoplasia in Fish 327
Michael J. Moore and Mark S. Myers

Chapter 9
Metal Regulation in Aquatic Animals: Mechanisms of Uptake,
Accumulation, and Release 387
G. Roesijadi and W. E. Robinson

Chapter 10
Behavioral Mechanisms of Metal Toxicity in Fishes 421
Daniel N. Weber and Richard E. Spieler

Chapter 11
Use of Physiologically Based Toxicokinetic Models in a Mechanistic
Approach to Aquatic Toxicology 469
James M. McKim and John W. Nichols

Index 521

Chapter 1

Trophic Transfer of Chemicals in the Aquatic Environment

Margaret O. James and Kevin M. Kleinow

I. INTRODUCTION

Numerous xenobiotics are found in anthropogenically altered aquatic environments. These chemicals often are available for uptake by aquatic organisms. One of the important issues in aquatic toxicology is our ability to predict the rate and extent of organismic uptake of pollutant chemicals as well as the *in vivo* fate and effects of the absorbed chemicals. It has been well documented that animals living in chemically polluted aquatic environments will acquire body burdens of the chemical pollutants.[1-3] The actual body burden acquired by an individual animal will depend on a number of factors, including the physicochemical properties of the xenobiotic chemical, the routes of exposure, and the physiological and biochemical make-up of the animal. Several routes of exposure are possible. Chemicals dissolved or suspended in water may enter through the gills, the skin, or the gastrointestinal tract. Chemicals present in sediment may be absorbed by direct dermal contact and ingestion, while chemicals present in plants and low trophic-level organisms may be ingested and absorbed through the gastrointestinal tract. The relative importance of each route of entry varies with the intrinsic chemical properties of each xenobiotic and also with the likelihood that the chemical under consideration will be degraded in the aquatic environment or biotransformed by low trophic-level organisms. There is still much to be learned about the impact of preconsumptive biotransformation by plants and animals on the fate of contaminants in the consumer species and about the relative importance of dietary pollutants to the overall body burdens of aquatic species. Knowledge of the bioavailability and biotransformation of chemical contaminants in components of the aquatic food chain may be of aid in defining the risk to animals and humans of consumption of fish and shellfish from polluted waters.

A. Importance of Route of Exposure on the Process of Chemical Uptake

The aquatic environment provides a number of classic avenues for the uptake of xenobiotic chemicals into biota. Respiratory, dermal, and oral routes are important portals of entry for xenobiotics in aquatic species, as they are for their mammalian counterparts. What differs, however, is the medium from which xenobiotics are presented. Xenobiotics appear to be more readily taken up by organisms when presented from water rather than air. This is due in large part to the universal solvent properties of water. Even compounds that are virtually insoluble in water often are soluble in the organic matter suspended in the water column. The colloids, particulates, and sediments of aquatic systems provide a conduit for exchange. A dynamic equilibrium exists between organisms and xenobiotic reservoirs, with water as the interface. Air, while certainly capable of similar particle transfers, is limited primarily to compounds that are volatile or suspended into the air. Overall, contaminant uptake by branchial routes in aquatic species appears to be relatively more important than the pulmonary route of their terrestrial counterparts. It is generally recognized that respiratory routes of uptake of xenobiotics from water are important for many compounds in aquatic species.[4] The exposure potential of organisms to a variety of xenobiotics is enhanced by physiological adaptations of the organisms to the oxygen-poor aqueous environment. Many organisms have adapted to low oxygen by developing countercurrent exchange mechanisms in the gill, which besides being an efficient means to transfer oxygen, facilitates the exchange of xenobiotics. These factors, along with the direct accessibility of the gill to water, provide a minimally encumbered route of uptake.

Exposure to waterborne contaminants also provides uptake potential by dermal contact as well as by ingestion of water (especially for marine species). The role of the dermis in the uptake of xenobiotics has not received extensive attention for aquatic vertebrates. Tovell et al.[5] demonstrated in goldfish that as much as 20% of a sodium lauryl sulfate burden was due to uptake across the dermis. Likewise, Varanasi et al.[6] produced evidence that naphthalene was absorbed across the dermis in the rainbow trout. Other studies have demonstrated the uptake of compounds such as selenium,[7] as well as other xenobiotics, into the mucus layer of fish.[8] While the toxicological implications of these processes are poorly understood, such studies demonstrate that the skin, scales, and mucal layer in fishes may be contributors to the uptake and disposition of aquatic contaminants. Information regarding the uptake of xenobiotics from ingested water is not currently available.

B. The Components of the Aquatic Food Chain and the Potential for Trophic Transfer

Considerable controversy exists regarding the relative contribution and importance of diet in the uptake and transfer of contaminants along the aquatic food chain. Numerous workers have demonstrated and argued that, at least for persistent compounds such as organochlorines, levels in aquatic organisms can be explained by direct partitioning between water and tissues.[9-14] A number of early studies have indicated in laboratory experiments that fish and crabs accumulated greater organochlorine levels from water than did prey organisms.[15-18] These results, as well as similar results with other organisms, were attributed to the relative lipid contents of the different organisms rather than to trophic transfer. Contrasting studies employing short and single-pathway food chains have demonstrated significant bioaccumulation of many lipophilic toxicants by dietary routes. Studies with fish,[19,20] euphausiids,[21] and benthic polychaetes,[22] when considered in natural settings, indicate food as an important contaminant source. Intestinal absorption efficiencies of greater than 50% for a variety of lipophilic toxicants in a number of aquatic species further support the concept of dietary assimilation.[23-26]

Several studies have suggested that experimental methodology is perhaps at the center of the uptake route controversy. Use of solvent carriers and unrealistically high concentrations of organochlorines beyond the chemicals' water solubilities for waterborne exposures are cited as a probable cause of the discrepancies.[27,28]

Mechanistic models considering exchange across the gills and via the food suggest that for contaminants whose assimilation exceeds their elimination (compounds that are lipophilic, environmentally persistent, and resistant to biotransformation), biomagnification is a distinct possibility.[29,30] Such models have suggested that 90 to 99% of contaminants in Lake Michigan lake trout, for example, are a result of concentration via the food chain. Studies by Oliver and Niimi,[31] examining polychlorinated biphenyls (PCBs) in Lake Ontario salmonids, have suggested that as one progresses through the pelagic food chain, each trophic level takes a greater divergence from the line expected by plotting bioconcentration factor against the octanol/water partition coefficient when lipid content and direct uptake from the water are considered. More recently, field studies examining lake trout from PCB-contaminated lakes with varying trophic structures indicated that lake trout PCB concentrations increased with food chain length and lipid content.[32] A 3.5-fold biomagnification factor was correlated for each successive trophic level.

Additional models have examined other food chain structures relative to contaminant transfer. One such model examined PCBs in food chains for which the lobster and winter flounder were the targeted terminal consumers.[33] This model suggested that PCBs in the flounder and to a lesser extent in the lobster are derived from the sediment, with dietary uptake exceeding waterborne uptake for all of the PCBs examined. While dietary uptake was more pronounced for the higher chlorinated derivatives, assimilation efficiency of ingested PCBs declined through a tri-, tetra-, penta-, and hexachlorobiphenyl series. PCB concentrations in lobster and flounder appeared to correlate to differences in the relative importance of the benthic component of the food chains to each organism and to whole body lipid content. Another model examining a pelagic food chain indicated that 18 to 42% of PCB and dieldrin uptake into Lake Michigan alewives can be accounted for by ingestion of organisms such as copepods, cladocerans, and mysids.[34] Although food chain transfer appears to explain data for many studies, other lines of modeling have suggested that the importance of oral routes of exposure are overstated.[35]

Trophic transfer studies with metals have also shown considerable variation in the perceived importance of dietary routes of exposure. Differences appear to exist with not only the contaminant but also the nature of the food chain. Arsenic, for example, appears to be marginally available to organisms such as copepods, barnacles, and oysters when presented as the dissolved form in water. In contrast, when phytoplankton capable of incorporating dissolved arsenic were fed to these organisms, significant trophic transfer of 25 to 50% of the arsenic resulted.[36] Demonstrably different, the uptake of silver by several invertebrates including the grass shrimp, *Paleomonetes pugio,* and oyster, *Crassostrea virginica gigas,* was predominantly from water, whereas absorption from contaminated phytoplankton or particles was minimal.[37–40]

The relative influence and contribution of each trophic level in the progressive transfer of dietary contaminants are not well defined. Direct uptake of contaminants from water has been demonstrated for both phytoplankton[41] and zooplankton.[42–46] Relatively few studies have examined the actual trophic transfer of contaminants at the primary producer and secondary consumer level.[47–50] Likewise, while many studies have established the biotransformation of toxicants such as polycyclic aromatic hydrocarbons (PAH) and p'-dichlorodiphenyl-trichloroethane (DDT) by primary producers[51,52] and lower trophic-level consumers,[53–56] few have examined the effect of this process upon the successive accumulation, disposition, and toxicity of xenobiotics.

Much is to be learned regarding the significance, mechanics, and mechanisms operative in the food chain transfer of contaminants. In particular, the role of the intestine in modulating the systemic uptake and disposition of pollutants present in the diet is a topic that has received relatively little attention in the aquatic toxicology arena. The remainder of this chapter will address specific issues that may prove to be critical to the physiological, molecular, and biochemical understanding of this complex, important, but yet undefined area of research.

II. INTRINSIC FACTORS THAT AFFECT THE BIOAVAILABILITY OF XENOBIOTICS INGESTED IN THE DIET

Of crucial importance in a discussion of trophic transfer of xenobiotics is a consideration of the ability of xenobiotics to cross the intestinal membranes and enter the systemic circulation. There is considerable uncertainty regarding the nature of toxicant transport through the enterocyte brush border and basal membranes. Generally, the movement of lipid-soluble compounds has been thought to occur by diffusion.[57-59] Several types of diffusion have been identified in mammals. Passive diffusion, which appears to be the major transport mechanism for lipophilic xenobiotics, involves the integration of xenobiotics into the lipid layer of the membrane. Movement by lateral membrane flow provides access to the basal lateral membrane. This nonsaturable process is dependent upon the concentration gradient and partition coefficient of the contaminant. Coabsorption of multiple xenobiotics by diffusion does not appear to alter component compound absorption rates. The composition of membrane lipid may affect the efficiency of absorption by passive diffusion by altering the solubility of the xenobiotic in the membrane lipids. It has been shown that acclimation temperature significantly affects the fatty acid composition of fish lipids, especially membrane phospholipids.[60-63] The major difference noted with lower temperature is an increase in the degree of unsaturation. It has been noted even among various phospholipids and the different tissues of an animal that there are differences in the fatty acid composition response to different acclimation temperatures.[64,65] While the essential fatty acid requirement of the animal and its thermal history influence in large part the phospholipid composition, dietary fatty acid composition serves as a model for the make-up of triglycerides or depot lipids in fish.[64-66] Carrier-mediated or facilitated diffusion has been identified in mammals for a number of endogenous and exogenous agents, including short chain fatty acids,[67] cadmium,[68] and lead.[69] These mechanisms, like passive diffusion, follow a downhill concentration gradient in either direction across the membrane. Unlike passive diffusion, carrier-mediated diffusion is saturable, as it is dependent on the amount of carrier protein present.

Other absorptive mechanisms, such as active transport, pinocytosis, and filtration, have been identified in the intestine. While relatively little information exists, these mechanisms currently appear to be more compound-specific and limited in scope than diffusive processes and are unlikely to be of major importance for xenobiotics. Active transport in the intestine is an important mechanism for the normal absorptive process of many dietary solutes, such as sugars, amino acids, electrolytes, and bile acids. Competitive interactions for carriers have been shown to occur for toxicants that are structurally similar to nutrients that are normally transported. Cobalt and manganese are known to compete for the iron transport system,[70,71] as does 5-fluorouracil for the pyrimidine absorptive pathway.[72] Pinocytosis is best known as the transport mechanism responsible for the absorption of macromolecules, such as antigens, in the early life stages of mammals. Relatively few compounds, such as azo dyes,[73] have been shown to be transported by the membrane invagination and vesicle formation typical of this process. Filtration or paracellular transport of toxicants is of unknown importance. Molecular size (small) and charge of the toxicant appear to be the major determinants for transmembrane movement through pores and leaky junctions. This type of movement may be generated by diffusion or by osmotically driven bulk water flow. The extent, importance, and mechanisms of each of these processes are known only in a rudimentary fashion in mammals. Less is known in aquatic species in both absolute and extrapolative terms.

Considering the importance of the passive diffusion pathway for absorption of most xenobiotics, major factors that influence oral bioavailability are the physicochemical properties of the xenobiotic, the properties of the matrix in which the chemical is ingested, and the physiology of the gastrointestinal tract. Van Veld has discussed some of the factors important in the process of absorption.[74]

A. Influence of Physicochemical Properties of the Chemical on the Rate and Extent of Absorption

The physicochemical properties of the xenobiotic, such as lipophilicity, hydrophilicity, presence of acidic or basic groups and their pKa, and the size of the molecule, will separately and together influence the process of uptake of dietary chemicals from the gastrointestinal tract. Since passive diffusion of the unionized xenobiotic or ion pair is the most important uptake process, the solubility of the xenobiotic chemical in water and lipid is of great importance. Molecules that are easily soluble in water and somewhat soluble in lipid are normally absorbed best, as the solubility in water ensures exposure to the entire surface area of the intestinal cells and solubility in lipid ensures passive diffusion through intestinal cell membranes and into the blood. Molecules that are poorly soluble in water but are highly lipid soluble will often exhibit great interindividual and interspecies variability in absorption, and absorption of xenobiotics with these characteristics will be strongly influenced by parameters such as the matrix in which the chemical is ingested and the physiology of the animal (see Section II.B). Molecules that are poorly soluble in both water and lipid are normally poorly absorbed, unless they can be presented to the intestinal tract in more soluble forms by dispersion in solubilizing matrices or, if they have acidic or basic groups, by forming soluble salts or ion pairs.

Molecules with ionizable hydrogens (acidic groups) are normally best absorbed if administered as sodium or potassium salts, as the salts are usually more water soluble than the un-ionized compound, which is usually more lipid soluble. The importance of forming a salt will depend on the pKa of the acidic group, the intrinsic solubility of the compound, and the physiology of the digestive tract. For example, the antibacterial drug sulfadimethoxine, a weak acid of pKa 6.7,[75] is poorly water soluble as the free drug, but freely water soluble as the sodium salt. Free sulfadimethoxine had a bioavailability to trout of 34 ± 9%, whereas sodium sulfadimethoxine had a bioavailability to trout of 63 ± 9%.[76] In catfish, however, no differences in bioavailability were found between the salt, which had a bioavailability of 34% and the free drug, which was 31% bioavailable.[77] Acidic groups most commonly encountered in xenobiotics of toxicological importance are the carboxylate group and the aromatic hydroxy group (phenolic group), although others, such as the sulfonylurea group, are also found in pesticides of environmental importance. In animals with acid-producing stomachs (see next section) acidic xenobiotics with pKa in the range of 3 to 6 will normally be absorbed from the stomach, as well as the more alkaline intestine, as they will exist in the un-ionized form at low pH of the stomach but largely as anions at the intestinal pH. Offsetting this pH effect on the lipid solubility of weak acids is the fact that the surface area of the intestine is larger and presents more opportunity for absorption than the stomach.

Molecules with atoms that can easily gain a proton (basic groups) are also usually more water soluble and therefore better absorbed if they are ingested as hydrochloride salts or as salts with other acids. Again, the pKa of the basic group, the likelihood of the group being protonated at physiological pH, and the water solubility of the ionized and un-ionized forms will markedly influence the absorption from the gastrointestinal tract. Xenobiotics with basic groups of pKa >6 are poorly absorbed from the stomachs of animals with acid-producing stomachs, as they will be largely ionized as the cation at acidic pH. More of the un-ionized and therefore more lipid-soluble form of the xenobiotic will be present at neutral or alkaline pH, as exists in the intestine. Basic groups commonly encountered in xenobiotics of relevance to aquatic toxicology are aromatic nitrogens, such as are found in carbazole and related nitrogen heterocyclic compounds, and amino and substituted amino groups, such as are found in aniline-based compounds and several pesticides.

B. Influence of the Matrix and Physiological Parameters

Toxicants may be presented to the gastrointestinal tract in a multitude of forms, as well as matrices. The most obvious route of uptake via the gastrointestinal tract is in a food matrix. Food, however, may exist as many different complex mixtures, dependent on factors as diverse as trophic level (food source), phylogeny of the consumed organisms, health status, and body condition. In cases where plants are a major food source, nutrient and light availability, as well as other seasonal factors, may be added to this list. All of these factors are dynamic determinants for the composition of the food matrix. The availability of toxicants from these foods is dependent upon the associated matrix, the nature of the association of toxicant with the matrix, the form of the toxicant, and the favorability of the gastrointestinal milieu for contaminant absorption. Besides ingestion in food, contaminants may be delivered with ingested water or sediments. The relative contribution of toxicants via active consumption of water is for the most part unknown, although it is known for select marine species that water is consumed daily at a rate of 5 to 12% of body weight.[78] It appears likely that this mechanism may only serve to concentrate contaminants found at significant waterborne concentrations. Likewise, evidence suggests that toxicants may be incidentally ingested while associated with sediments.[79-81] This phenomenon, documented largely in benthic organisms, appears to be related to the capture of prey or ingestion of waters rich in suspended sediments. Van Veld has suggested that sediment association would probably render contaminants less bioavailable from the gastrointestinal tract than if the xenobiotic were presented with dietary foodstuffs.[74] This supposition appears to be related to the association of contaminants with foodstuffs and their physiologically mandated uptake.

Digestibility of the foodstuff plays an important role in liberating toxicants for absorption and associating toxicants with media that facilitate absorption. Lovell,[82] examining a number of foodstuffs, indicated considerable variation in nutrient digestibility among different fish species. A direct correlation appears to exist between diet and the digestion process. Animals that have dissimilar food in natural settings often have differing digestion rates. It has been suggested that digestion rates have evolved to suit the natural diet,[83] with longer times associated with the progression through microphagous (planktivores), to mesophagous (invertebrates such as mollusks, annelids), to macrophagous fish (vertebrates, large crabs). These differences in digestive rates appear to persist even with unnatural foods.

The enzymes involved in the digestion of lipids, proteins, and carbohydrates appear to be ubiquitously distributed, even among aquatic organisms with diverse dietary habits.[84] The relative activity and distribution of digestive enzymes along the gastrointestinal tract, however, appear to correlate with the diet.[85] Seasonal changes in enzyme activity and proportional changes in dietary composition also appear to alter digestibility.[86] Decreases in protein digestibility in channel catfish have been shown to correlate with high levels of dietary carbohydrate.[87] Lipid digestibility appears to vary according to composition,[88] with carp exhibiting values of 75.6 and 90% for lard and soybean oil, respectively.

Gastrointestinal pH may be biologically significant in regard to absorption of food-related xenobiotics.[89] Differences exist between fish, with approximately 15% of teleostean fish species lacking a true stomach and acid digestion.[90] Gastric juice, when present in other species, may vary substantially in pH. A consistent finding in both elasmobranchs and teleosts is that stomach acidity increases upon ingestion of food, with values changing from near neutrality preingestion to a pH of around 2 postingestion.[91-93] Interestingly, in several species, including *Tilapia*, secretion of gastric acid stops at the end of the daylight feeding hours with an ensuing rise in gastric juice pH.[94] Intestinal pH has been characterized in relatively few species; however, in *Tilapia* a range of 6.8 to 8.8 has been reported.[85,95] Although both trout and catfish possess acid-producing stomachs, their gastrointestinal tracts vary in length and are 60 and 150% of total body length, respectively. Differences in gut length are further amplified with herbivores such as the carp and milkfish, which exhibit gastrointestinal lengths of 300 and 500% of total body length. Such findings may be related to plant digestibility and the nearly continuous feeding behavior of herbivores. Carnivorous fish tend to eat fewer meals less frequently and to retain food for longer periods.

Gastric evacuation rates in aquatic organisms can be a contributing factor to the assimilation efficiency or apparent bioavailability of a toxicant. Many parameters are thought to influence gastrointestinal transit times. The size of the meal relative to the size of the organism in terms of weight, volume, and fullness of the stomach is one such factor. Large meal size, for example, has been shown not only to increase the time of digestion[96] but also to result in decreases in the absorption efficiency.[97,98] Such variances in digestion appear to be common for not only predacious but also herbacious fish species. Large items of food, such as are consumed by predacious fish, are gradually eroded, with the outer layers passed to the intestine while the original meal is still in the gastric phase of digestion. Likewise, studies with herbacious fish have indicated the possibility of only partial retention of phytoplankton in the stomach, with much of the meal bypassing the stomach and entering the intestine.[99,100] Other factors, such as feeding frequency, length of the premeal fast period, type of meal, variations in water temperature, and previous thermal history, may be influential in determining transit time. Evacuation rates tend to decrease with lower temperatures or increase with rising temperatures up to the species temperature tolerance.[101–105] For example, studies with rainbow trout indicate that a three- to fourfold longer time interval was required for either a 50 or 100% gastric clearance at 5°C compared with 20°C. Depending on the diet type, the time required for trout to evacuate the stomach was increased from 16.4 hr at 20°C to 58.5 and 72.4 hr at 5°C for diets of oligochaetes and pellets, respectively.[105] The greatest change in evacuation rate per unit temperature was at the lower temperatures. High fat levels also have been shown to delay gastric emptying in rainbow trout.[106]

Lipid content and form have been shown to influence the absorption of endogenous substrates and xenobiotics from the intestinal lumen. Studies with mammals have indicated that absorption of fat-soluble drugs and vitamins is facilitated by the presence of dietary triglycerides.[59,107,108] Conversely, high amounts of fat can restrict digestibility of available lipids, as well as reduce uptake of fat soluble contaminants.[109] This latter phenomenon appears to result from a competition of dietary triglyceride and digestive transport products, such as micelles, for toxicants. Other studies with fat-soluble vitamins in mammals have demonstrated that even dietary lipid saturation influences the mucosal transport of exogenous agents.[110–112] Unsaturated fatty acids of similar chain length can reduce gut transport by as much as 50% when compared to their saturated counterpart. The mechanism of this action is unknown. In general, lipophilic contaminants with high triglyceride solubility,[74] molecular weights <600,[23,113] low degrees of chlorine substitution,[26] and molecular volumes <0.25 nm^3 are more readily absorbed.[113]

Lipid- and fat-soluble contaminants, as well as other gastric digestion products, are often presented to the intestine as a partially digested chyme. In both mammals and fish, pancreatic lipases in conjunction with bile serve to further digest and emulsify lipids liberated from gastric digestion to form small micelles. These micelles in mammals, and apparently in fish, are composed of triglycerides and hydrolysis products such as diglycerides, monoglycerides, free fatty acids, and glycerol.[114] Where fat-soluble xenobiotics are present, the micelles will include these fat soluble contaminants. Micelles may differ in their association and behavior with xenobiotics, depending upon their composition. Studies in mammals indicate that aqueous solubilization of polycyclic hydrocarbons and PCBs are different, according to micellar composition. While PAH (such as 7,12-dimethylbenzanthracene and 3-methylcholanthrene) association with micelles was favored by long chain monounsaturated fatty acids and their acyl glycerol components, polychlorinated biphenyls' aqueous solubilization was favored by polyunsaturated components of the micelle.[115] Additionally, the lengths of the fatty acids in micelles also appear to influence solubilization of both hydrocarbons[115] and vitamins.[116] Long chain mixtures appear to be much more effective than shorter chain mixtures in this regard. The mechanisms for the saturation and length effects are unclear; however, stereochemical fit and micellar interior size have been postulated, respectively. Once formed, micelles facilitate the transport of xenobiotics through the aqueous phase to the brush border and increase the surface area for mucosal cell-associated lipases. In mammals, constituents of micelles are liberated by lipase action in a 200 to 500 μm unstirred water layer.[117] These molecules must traverse this layer, a mucus layer,[118] and a proton rich layer, referred to as an acid microclimate,[119] before reaching the lipid phase of the brush border membrane.

The absorption of lipophilic xenobiotics into enterocytes appears to be closely associated with lipolytic product uptake.[120] The exact mechanism is unclear. In both mammals and fish, lipolytic products once in enterocytes are reassembled to form triglycerides within the smooth endoplasmic reticulum of the cell.[121–124] Upon association of a phospholipid membrane by action of the smooth endoplasmic reticulum, the vacuoles are further processed by the rough endoplasmic reticulum and packed by the Golgi apparatus. Mammals release these products as chylomicrons and very low density lipoproteins (VLDL). In fish, several contrasting stories have emerged. Studies in rainbow trout have shown that dietary lipids were released into lymphatic routes largely as triglycerides associated with VLDL-like particles, much like their mammalian counterparts.[122,125] (There is controversy regarding the existence of lymphatic systems in fish.) Other investigations have demonstrated direct release of lipid products into the circulation.[126–130] The products released include small amounts of free fatty acids and free fatty acids associated with circulatory high density lipoproteins (HDL).[114] The bulk of dietary lipid, however, appears to be resynthesized to triglycerides and secreted into blood in the form of HDL,[127,128,130] followed by low density lipoproteins (LDL) and VLDL.[127,129,130] The relative importance of lipid transport to contaminant absorption and transport in fish has been demonstrated by microscopic fluorescence and radiotracer studies. These studies with the killifish have indicated that benzo(a)pyrene (BaP) and dietary fat are coprocessed at least through enterocyte vacuole formation.[123]

Relatively few studies other than those with lipids or lipid solubility have examined factors influential to dietary xenobiotic bioavailability in fish. Among these studies are a limited number of investigations with aquacultural drugs, which have examined regional absorptive differences along the intestinal tract and the influence of chemical form, dosage, particle size, diet composition, and temperature on the process of absorption. The results indicate that xenobiotic bioavailability in fish is influenced by factors that are operative in mammals, as well as by other conditions, such as poikilothermy, which are fish specific.

Studies with the proximal intestine of fish indicate many functional similarities to mammals in the face of some morphological differences. In mammals, it is generally recognized that the duodenum and jejunum are responsible for the absorption of the majority of nutrients and drugs. In addition, drug absorption from the colon/rectum as well as bile acid absorption from the ileum have been well documented. Kleinow et al.,[76] in studies with ^{35}S-sulfadimethoxine (SDM), showed that the distal gastrointestinal tract of rainbow trout exhibited the greatest capacity to absorb ^{35}S-SDM, followed by the proximal intestine and the stomach region, when examined in isolated preparations. Differences of 20% between the highest and lowest values were demonstrated as a threefold difference in plasma concentrations. While the mechanism was not elucidated, these results suggest that all gastrointestinal segments were capable of absorbing SDM, and like mammals for at least this compound, the stomach region exhibited the lowest bioavailability. The reason the distal gastrointestinal tract exhibited such high bioavailabilities is unknown in fish; however, it may be speculated that regional differences in metabolism or in circulatory patterns (as in mammals[131]) may explain these differences.

Few studies in fish have examined the effect of dosage form on gastrointestinal bioavailability. Drug particle size has been shown to affect bioavailability.[132] Studies with oxolinic acid in yellowtail indicated that particle size reduction enhanced bioavailability. On a case-by-case basis, dose has been shown to sometimes influence the bioavailability of drugs used in aquaculture. The bioavailability of oxolinic acid in rainbow trout was reported to be 14.3% after an oral dose of 100 mg/kg and 38.1% when the dose was 20 mg/kg.[133] This compares to other studies that found bioavailabilities of 13.6% when administered as a 75 mg/kg dose[134] and 91% for trout dosed with 5 mg/kg.[135] These results suggest that bioavailability of oxolinic acid is dose dependent and higher at the lower doses. Diverse administration techniques, such as carrier vehicle (food vs. aqueous) and ionization (salt vs. free drug) appear as likely modifiers to bioavailability in the cited papers. The effect of dose appears drug and species dependent. Other studies with oxolinic acid in Atlantic salmon indicated no difference in bioavailability when dosed in the feed at 9 and 26 mg/kg.[136] Similarly, studies with sulfadimethoxine showed absolute increases in plasma concentration but only marginally lower bioavailabilities when doses were raised threefold from 42 to 126 mg/kg. Other examples are evident in the literature. In general, bioavailability values in aquatic animals are lower than in humans.[137]

Temperature is known to have a profound effect on the persistence of drug residues in aquatic species, with lower temperatures resulting in greater tissue concentrations and more prolonged retention times.[138] Recent studies with a variety of compounds have suggested that lower temperature may not only reduce drug clearance[135] but may, in concert with diet, alter the biotransformational profile of the existing residue.[139] Few studies have directly measured bioavailability relative to temperature; however, both higher and lower bioavailabilities have been reported with lower temperatures.[135,140] These results suggest that lower trophic levels exposed to xenobiotics at colder temperature may pose a greater residue risk when consumed by higher animals, but clearly more studies are needed in this area to better define the processes involved.

III. THE ROLE OF BIOTRANSFORMATION AS A MODULATOR OF BIOAVAILABILITY AND DISPOSITION

The bioavailability and subsequent fate of a dietary xenobiotic is highly dependent upon enzymatic biotransformation of the xenobiotic, at various stages in the presentation of the xenobiotic to the organism. These stages are biotransformation in the lower trophic-level organism, biotransformation in the intestinal microflora of the consumer species, biotransformation in the gastrointestinal tract of the consumer species, and biotransformation in the liver.

A. Preconsumptive Biotransformation

Biotransformation of a xenobiotic by one organism prior to ingestion of this organism by a second animal may profoundly alter the bioavailability, subsequent disposition, and potential toxicity of the xenobiotic. Preconsumptive biotransformation should usually be expected to reduce the bioavailability and hence the potential toxicity of most xenobiotics, because the biotransformation products should be more polar than the starting material and therefore less readily absorbed, more readily excreted, and perhaps less biologically active.

Exceptions to this general scenario may be envisioned. For example, for molecules such as BaP, which require several metabolic steps for conversion to a reactive metabolite,[141] preconsumptive biotransformation may enhance the likelihood of deleterious effects on higher trophic-level animals. If the first organism transforms the BaP into BaP-7,8-dihydrodiol, a proximate carcinogen, then the consumer may absorb the BaP-7,8-dihydrodiol and convert it by a single P450-dependent biotransformation to the ultimate carcinogen, BaP-7,8-dihydrodiol-9,10-epoxide. The bioactivation may take place in the target organs, such as intestine, liver, and biliary epithelium.[142] It has been shown that orally administered BaP-7,8-dihydrodiol was more readily absorbed by two uninduced flounder species than unchanged BaP, probably because introduction of the hydroxyl groups increased the water solubility of the extremely hydrophobic BaP.[143,144] In uninduced southern flounder, however, binding of reactive metabolites to intestinal and hepatic DNA was similar for animals treated with BaP or BaP-7,8-dihydrodiol (Table 1).[144] This was probably because, in the uninduced southern flounder, conjugation of the BaP-7,8-diol to glucuronide metabolites was more efficient than monooxygenation to the diol-epoxide.[144] We speculate that in fish whose intestinal P4501A was induced by exposure to PAHs (see below), activation of BaP-7,8D by P450 may predominate over detoxication by glucuronide conjugation, resulting in higher DNA binding in the intestines of induced fish exposed to BaP-7,8D relative to those exposed to BaP.

Similar arguments may be made for other PAHs. The contribution of preconsumptive biotransformation to the carcinogenicity of PAH is not known; however, it is known that many invertebrates that are eaten by carcinogen-sensitive fish can convert BaP into BaP-7,8-oxide, which is then rearranged to 7-hydroxyBaP or hydrated to BaP-7,8-dihydrodiol.[55,145] As well as forming BaP-7,8-oxide, 7-OHBaP, and BaP-7,8-dihydrodiol, it is likely that conjugates of these BaP metabolites will be formed in the prey organism (see Figure 1a). Expected conjugates are the glutathione, cysteinyl-glycine, cysteine, and mercapturic acid conjugates of BaP-7,8-oxide, the glucoside conjugates of BaP-7,8-dihydrodiol and 7-OHBaP, and the sulfate conjugates of BaP-7,8-dihydrodiol and 7-OHBaP.[146,147] While these metabolites are more hydrophilic than the parent BaP, some of these metabolites, especially unconjugated metabolites and glucoside conjugates, are still lipophilic enough for absorption by passive diffusion.

If fish are like other vertebrates, intestinal tissue and microflora (see below) will contain hydrolases and lyases that could convert xenobiotic conjugates into forms that may be absorbed across the intestine.[148–150] Intestinal cytosol usually contains several glucosidases and glucuronidases that could convert sugar conjugates to the aglycones. This process may be compared with the second step of enterohepatic cycling of biliary metabolites, in which xenobiotic conjugates formed in liver are excreted into bile; the bile is secreted into the intestine, where conjugates may be hydrolyzed back to the parent xenobiotic or its phase 1 metabolite and reabsorbed.[151] Enterohepatic cycling has been shown to occur in goldfish dosed with phenol; the phenol was converted into phenylglucuronide in liver, excreted into bile, and then hydrolyzed back to phenol in the intestine and subsequently excreted as the sulfate conjugate in urine.[152] Studies in the rat have shown that β-glucoside conjugates of phenolic xenobiotics formed in plants may be hydrolyzed in the rat intestine (the extent of hydrolysis depending on the structure and properties of the aglycone) and the aglycones absorbed and ultimately excreted in urine as sulfate and glucuronide conjugates.[153] Xenobiotic β-glucoside conjugates are also formed in invertebrates[147] and may be expected to undergo hydrolysis in the intestine of consumers. Experiments have not yet been

Table 1. Influence of Biotransformation on the Retention and DNA Binding of Orally Administered BaP 24 Hours after a Dose to Uninduced Southern Flounder

			pmol BaP equiv. bound/mg DNA	
Chemical administered	% Recovered in fish tissues	% Recovered in bile	Liver	Intestine
BaP	12.8	6.2	0.24[a]	0.76
BaP-7,8-dihydro-diol	12.4	21.3	0.33	0.44
BaP metabolite mixture[b]	4.3[c]	2.4[c]	0.006[a,c]	0.026[c]

Source: James, M.O., et al., Chem-Biol. Interact., 79, 305, 1991. With permission.

[a] Significantly lower binding of BaP metabolites to hepatic vs. intestinal DNA.
[b] BaP metabolites formed *in vivo* by the spiny lobster. The mixture included 0.5% unchanged BaP, 9% unconjugated metabolites, and the remainder a mixture of glucoside and glutathione conjugates of BaP metabolites.[144]
[c] The extent of absorption and DNA binding of the metabolite mixture was significantly less than of either BaP or pure BaP-7,8-dihydrodiol.

done in fish species to test the hypothesis that xenobiotic β-glucosides may be hydrolyzed in the intestine and the parent xenobiotic absorbed, but preliminary *in vitro* studies have shown that catfish intestinal cytosol contains β-glucosidase activity.[154] Sulfate conjugates are often more readily hydrolyzed than carbohydrate conjugates due to their inherent instability, especially at pH <5.[155] Although this has not been studied, it may be expected that xenobiotic sulfates could be hydrolyzed in the fish stomach and intestine and the parent xenobiotic absorbed.

It has been shown in mammals that glutathione conjugates are subject to extensive further processing.[156] When glutathione conjugates of styrene oxide were injected into winter flounder, the major urinary metabolite was the cysteine conjugate with small amounts of the mercapturic acid, while bile contained mainly the intact GSH conjugates, with small amounts of cysteinyl-glycine, cysteinyl, and mercapturic acid conjugates.[157] Other unidentified urinary metabolites were also observed, which could have been produced by the cysteine conjugate β-lyase pathway (Figure 1b). The site of processing of the glutathione conjugates was not established but may have been liver, intestine (following biliary secretion), or kidney.[157] Other studies showed that kidney and intestine of the rainbow trout contained γ-glutamyl transpeptidase activity.[158] It has recently been demonstrated that skate liver is capable of forming premercapturic acid and mercapturic acid metabolites of 1-chloro-2,4-dinitrobenzene, as well as the initial glutathione conjugate.[159] Cysteine conjugate β-lyase, which converts xenobiotic cysteinyl conjugates to thiols, is found in intestinal microflora in mammals.[160] It has been shown in mammalian species that metabolites of hexachlorobutadiene and related halogenated compounds that are formed from GSH conjugates by the cysteine conjugate β-lyase pathway can cause renal toxicity.[161] Thiol-containing xenobiotics may be metabolized by the FAD-monooxygenase to sulfoxides and sulfones, some of which may be reactive metabolites.[162] The possibility that

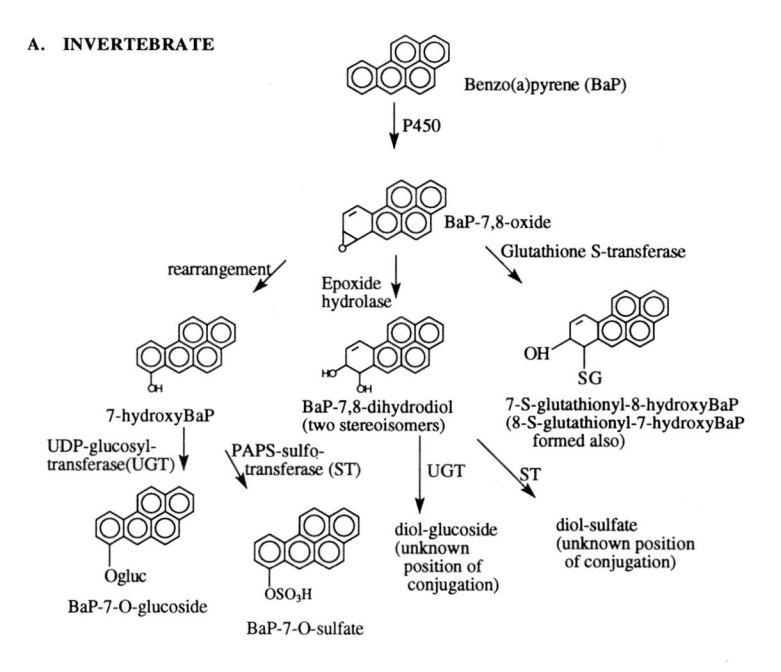

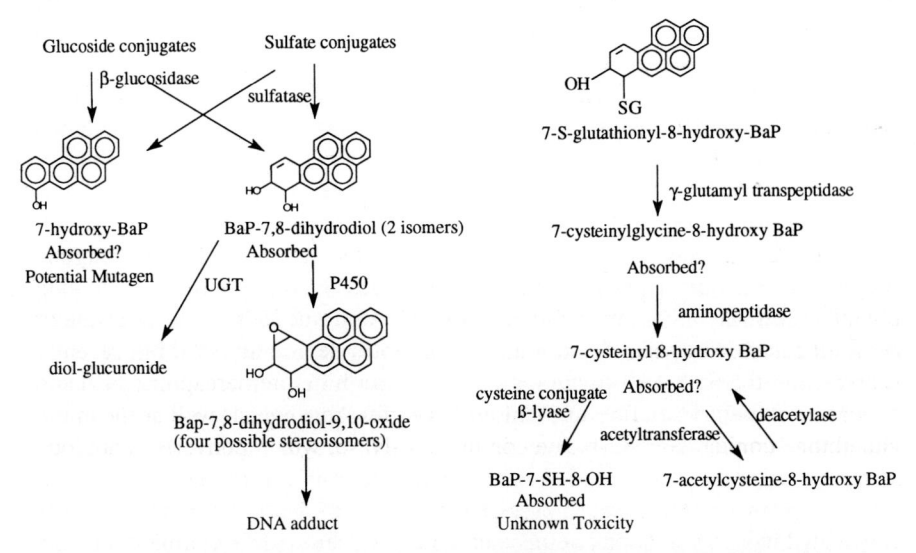

Figure 1. Possible biotransformation pathways of BaP and its metabolites in aquatic organisms. (A) Pathways likely to occur in invertebrates exposed to parent BaP.[146,147,201] (B) Pathways likely to occur in a fish following consumption of an organism that produced the metabolites depicted in (A).[175,177]

dietary glutathione conjugates and their breakdown products may be hydrolyzed in the intestine and reabsorbed has not yet been directly examined in aquatic species. The importance of cysteine conjugate β-lyase in aquatic species is as yet unknown, as is the toxicity of thiol-containing xenobiotics.

These examples illustrate the point that preconsumptive biotransformation of lipophilic molecules does not always reduce bioavailability or abolish their potential toxicity to consumers, because further metabolic processing of the metabolites in intestine or intestinal flora may yield metabolites than can be readily absorbed and converted to toxic species in the consumer organism. Further investigation of this aspect is indicated.

B. Microfloral Biotransformation

Studies in mammalian species have shown that the bacteria inhabiting the intestine (the intestinal microflora) can play an important role in determining the extent of biotransformation of a xenobiotic, and that the composition of the intestinal microflora can be altered by diet.[148,160] Bacterial enzymes that play a major part in xenobiotic biotransformation include hydrolytic enzymes such as β-glucosidases, β-glucuronidases, and sulfatases, as well as reductases.[148] The hydrolytic enzymes are likely to assist in the release of unconjugated xenobiotic, which, being generally more lipophilic than the conjugate, will be more readily absorbed across the gastrointestinal tract (discussed above). Reductases are likely to be of importance in converting nitroaromatics to reactive nitroso compounds, which may cause local tissue damage, or to amines, which may be absorbed and later oxidatively metabolized. As yet, there have been no systematic studies that the authors are aware of concerning the role of the intestinal microflora in fish in xenobiotic biotransformation.

C. Biotransformation in the Gastrointestinal Tract

In aquatic species, as in terrestrial species, the liver contains the highest concentration of most biotransformation enzymes. In many cases, the liver is quantitatively the major site of biotransformation of pollutants entering the body from all routes. Chapters 2 and 3 address the overall importance of the liver in determining the biotransformation products formed. In considering the biotransformation of dietary xenobiotics, however, it is clear that the intestine and, in some cases, the stomach contain biotransformation enzymes at sufficiently high concentrations that their contribution is not negligible. The contribution of intestinal tract enzymes to xenobiotic biotransformation is particularly important when relatively low concentrations of the xenobiotic are present, as is normally the case for environmental chemicals. Recent studies with catfish (Table 2) have shown that the K_m values for intestinal BaP monooxygenase activity and for UDP-glucuronosyltransferase and sulfotransferase activities with 3-, 7-, and 9-hydroxybenzo(a)pyrene were all in the micromolar (μM) range.[163] Thus, low

Table 2. Properties of Xenobiotic-Metabolizing Enzymes in the Uninduced Catfish

Activity	Substrate	Intestine K_m µM	Intestine V_{max} nmol/min/mg	Liver K_m µM	Liver V_{max} nmol/min/mg
Monooxygenase	Benzo(a)pyrene	1.6	0.004	0.5	0.151
Epoxide hydrolase	Styrene oxide	ND[a]	4.8	ND[a]	4.0
Glucuronosyl transferase	3-OHBaP	2.9	0.24	8.6	2.60
	7-OHBaP	1.6	0.36	6.7	1.82
	9-OHBaP	2.3	0.24	2.3	0.61
Sulfotransferase	3-OHBaP	1.7	0.88	1.4	1.35
	7-OHBaP	1.5	0.30	1.8	0.84
	9-OHBaP	0.8	0.36	1.1	1.60
Glutathione S-transferase	1-Chloro-2,4-dinitrobenzene	1700	5800	1300	5700

Source: Adapted from James, M.O., et al.[163] and James, M.O. and Altman, A.H.[154]

Note: The K_m values shown are for the xenobiotic substrate, and the cofactor or cosubstrate was present in saturating concentration. The K_m values for UDP-glucuronic acid were 5 µM and for 3′-phosphoadenosine-5′ phosphosulfate (PAPS) 1.2 µM. All values shown are mean values from 2 to 8 animals.

[a] K_m was not determined for epoxide hydrolase. The styrene oxide concentration used in assays was 0.8 mM.

concentrations of pollutant substrates should be metabolized readily in the intestine. High dietary doses of xenobiotics, such as are encountered when therapeutic drugs are given, will often overwhelm the biotransformational capacity of the intestine, and the xenobiotic will be absorbed and pass to the liver for further biotransformation. The catfish study (Table 2) showed that similar K_m values were found in liver and intestine preparations but that higher V_{max} values were found in liver, indicating that more enzyme was present per mg protein in liver than intestine. The whole liver contains more microsomal and cytosolic protein than all of the intestinal mucosa, underscoring the larger quantities of xenobiotic-metabolizing enzymes and higher total capacity of the liver to biotransform pollutants than the intestinal mucosa.

1. Are the Enzymes Present in the Same Forms that are Found in Liver?

In fish, as in mammals, most of the important enzymes of xenobiotic biotransformation, including cytochrome P450,[164,165] UDP-glucuronosyl transferase,[166,167] sulfotransferase,[168] and glutathione S-transferase,[169–171] exist in different isoforms with differing inducibility and substrate selectivity. The nature of the isozymes present in different tissues is therefore an important determinant of the likelihood of xenobiotic biotransformation in that tissue. Several xenobiotic metabolizing enzymes have been found in the intestine and in the liver of many fish species. Table 2 presents data for biotransformation of typical substrates in a representative freshwater species, the channel catfish. Similar results have been found for other freshwater and marine fish for various xenobiotic-metabolizing enzymes.[54,146,167,172,173] The question of whether or not the same forms are present in liver and intestine has been addressed in a few cases. Western blot analysis of microsomes from intestine and liver of control and induced spot *(Leiostomus xanthurus)* with a monoclonal antibody to scup P450 1A1 (formerly named P450E) showed that P4501A1 was present in intestine and liver of spot, and that this isozyme was inducible in both tissues.[174] Clarke et al.[167] showed that UDP-glucuronosyl-transferase activity for p-nitrophenol and testosterone but not bilirubin was present in plaice intestine, and this was confirmed by Western blotting of microsomes from liver and intestine. The isozyme composition of glutathione S-transferases in intestine has not to our knowledge yet been characterized.

Although sparse, the data so far indicate that isozymes found in intestine are also present in liver, although there is little information on the relative quantities of different isozymes in liver and intestine. There remains the possibility that as yet undiscovered isozymes will be found that are expressed in intestine but not in liver and vice versa.

2. Inducibility of Enzymes by Dietary-Inducing Agents

Many of the induction studies conducted in aquatic species have employed waterborne exposure or injection of the inducing agent. These studies have shown that cytochrome P4501A homologues of many fish species are exquisitely sensitive to induction by the PAH class of inducing agents.[175] Studies in which inducing agents of the PAH type have been administered in the diet have shown that intestinal and hepatic P4501A can be induced by dietary PAH.[74,176] It was shown that, at low concentrations of inducing agent, only intestinal P450 was induced in killifish and spot.[174,176] It was suggested that the reason these low dietary concentrations of BaP and 3-methylcholanthrene induced cytochrome P4501A in intestine but not in liver was because these low concentrations were extensively metabolized in the intestine, especially after induction, such that not enough unchanged PAH reached the liver.[74]

Studies of the responsiveness of glutathione S-transferase and UDP-glucuronosyltransferase to PAH-type inducing agents in fish have shown marked species differences in inducibility in liver.[177] The extent of induction of these conjugating enzymes in liver is usually less than twofold, and whether or not induction is observed is highly variable and dependent on species and induction protocol, whereas increases in P450-dependent monooxygenase activity are typically at least fivefold and have occurred in all fish species tested.[178] To date there have been very few studies of the responsiveness of intestinal conjugating enzymes to dietary PAH. Clarke et al.[167] showed that intraperitoneal injection of 3-methylcholanthrene or Aroclor 1254 did not induce 1-naphthol-UDP-glucuronosyltransferase in intestine, although a 1.5 to 1.8-fold increase was found in liver. If conjugating enzymes are not induced in intestine while monooxygenase activity is induced, it is likely that activation of BaP and related PAH to reactive metabolites in the intestine will predominate over detoxication by conjugation in induced fish. Studies in mammals have shown that incorporation of inducing agents such as 3-methylcholanthrene, β-naphthoflavone and phenobarbital into the diet resulted in induction of intestinal monooxygenase activity,[179] with minimal effects on intestinal UDP-glucuronyltransferase and glutathione S-transferase, even when hepatic conjugating activities were induced.[180,181] Clearly, more studies delineating the responses of intestinal xenobiotic metabolizing enzymes to inducing agents are needed to determine the likely effect of induction on toxicity due to repeated exposure to PAH.

3. Modulation of Biotransformation by Other Dietary Constituents or by Pollutants

A number of common constituents of food, including lipids and organic molecules that occur naturally in plants, have the potential for inhibiting or stimulating xenobiotic biotransformation enzymes. Alteration of the extent of intestinal biotransformation of the xenobiotic may increase or decrease whole animal bioavailability.

The lipid composition of food, as well as influencing absorption of xenobiotics (see above), can also regulate biotransformation and the inducibility of biotransformation enzymes, such as cytochrome P450. The lipid composition (amount and type) of the diet can have direct effects at the level of the enzyme-catalyzed reaction, as well as effects on the responses of fish to inducing agents. A direct effect on intestinal monooxygenase activity was demonstrated by *in vitro* studies of Van Veld et al.[176] They showed that when BaP was presented in artificial fat vacuoles to intestinal microsomes, the amount of triglyceride present was inversely related to the extent of BaP metabolism. The apparent K_m for BaP increased with concentration of triglycerides present, with no change in V_{max}, and the results were interpreted as being an effect on the actual amount of BaP available to microsomal enzymes, an intuitively attractive hypothesis. Several studies have addressed the question of the influence of diet on the expression and basal activities of xenobiotic-metabolizing enzymes, focusing on dietary lipid composition,[182–185] but all of these studies have examined hepatic enzyme activity. The general findings from these studies were that the lipid composition of the diet affected both basal hepatic monooxygenase activities and the inducibility of hepatic monooxygenase activity by PAH-type inducing agents. It may be speculated that the effect of diet on fish intestinal monooxygenase activity would be similar to effects on the hepatic enzymes.

Plant secondary metabolites and natural products, such as oxygen and nitrogen heterocyclic compounds, may inhibit or stimulate xenobiotic metabolism in the intestine. For example, in mammals, flavonoid natural products inhibit some forms of P450, such as 1A, and stimulate other forms of P450, such as 2 and 3 family isozymes.[186] Coadministration of certain flavonoid compounds and BaP to mice results in lower mutagenicity and carcinogenicity.[187] Selected flavonoid compounds can also inhibit β-glucuronidase activity[188] but stimulate epoxide hydrolase.[189] Fish 1A1 is very sensitive to inhibition by α-naphthoflavone.[178] Imidazole derivatives also inhibit monooxygenase activity and stimulate epoxide hydrolase activity in fish.[190] Dietary exposure of fish to BaP together with plants that contain certain oxygen or nitrogen heterocyclic compounds may be expected to result in reduced biotransformation of BaP in the intestine, due to inhibition of monooxygenase activity, with resulting lower formation of mutagenic metabolites. Assuming the dietary modulators of drug metabolism are absorbed along with the BaP, inhibition of mutagenicity may be expected in the liver. The role of diet and presence of dietary components that inhibit or stimulate xenobiotic biotransformation in explaining species differences in carcinogenicity of PAH has not been explored in carcinogen-sensitive and resistant aquatic species.

IV. BINDING OF DIETARY CHEMICALS TO CELLULAR MACROMOLECULES

Dietary xenobiotics can potentially bind reversibly or irreversibly to macromolecules in the stomach, intestinal flora, or intestinal mucosal cells, and, following absorption, to cells of other organs. Reversible or noncovalent binding of dietary xenobiotics in the gastrointestinal tract can have the effect of preventing systemic absorption of the xenobiotic and may facilitate excretion from the fish. On the other hand, covalent binding to macromolecules in fish gastrointestinal cells and cells of other organs, such as liver, may initiate cellular changes leading to cell death or neoplasia.

A. Covalent Binding to DNA — Relationship to GI Tract Cancer Observed in Fish from Polluted Areas

Binding of reactive metabolites of chemical carcinogens to DNA is a necessary but not sufficient step in the development of tumors in animals exposed to carcinogens.[191] Fish from waters polluted with PAH and related compounds often exhibit neoplasms or tumors of the gastrointestinal tract, including mouth, liver, and bile duct.[192,193] A number of studies have demonstrated a correlation between the presence of procarcinogens in water and sediments, DNA adducts in the liver, and tumor incidence in fish,[194–196] although there was no attempt in these studies to determine the route of exposure to the procarcinogens. Laboratory studies of DNA adduct formation in liver after oral or intraperitoneal injections to English sole and starry flounder showed that higher hepatic BaP concentrations and adduct levels were observed after i.p. than after oral doses and that adduct concentrations were higher in the more carcinogen-sensitive English sole.[197] Once formed, the hepatic DNA adducts of the English sole were persistent for at least 2 months after a single exposure.[198] There do not appear to be any studies of the incidence of DNA adducts in intestinal tissues of fish taken from polluted environments or dosed with BaP in the laboratory. Laboratory experiments with the uninduced southern flounder have shown that oral administration of BaP and BaP-7,8-dihydrodiol results in binding of metabolites to intestinal and hepatic DNA (Table 1).[144] The persistence of the intestinal adducts was not studied, but it is possible that adducts to the DNA of epithelial cells will be rapidly removed from the body as the cells turn over and are sloughed off into feces. There is still much to be learned, both in mammals and fish, about the relationship of adduct formation and stability to the carcinogenic process at a particular site in the body.

B. Covalent Binding to Other Macromolecules — Toxicological Implications

It has been shown that BaP metabolites bind to hepatic RNA and protein in the California killifish and speckled sanddab.[199] The significance of this binding as regards toxicity is unclear but may conceivably be linked to disturbances in protein synthesis (RNA binding) or cellular function (protein binding). It is not known if metabolites of dietary BaP that are formed in the intestinal mucosal cells will bind to the RNA and protein of these cells as they do to DNA.[144] Modification of proteins by active metabolites of xenobiotics such as halogenated hydrocarbons has often been linked with toxic outcomes, such as neoantigen formation or cell death. While this phenomenon has largely been investigated in liver, it has been shown to occur in other organs in which the xenobiotics are metabolically activated and may occur in the intestine. More studies are needed to fully understand the implications of covalent binding to intestinal macromolecules.

C. Reversible Binding of Xenobiotics to Components of the Gastrointestinal Tract

It has been discussed in Section II that xenobiotics reversibly associate with lipid components of the diet and the mucosal cell membranes during the process of absorption. Modulation of consumer risk from dietary carcinogens may reside with functions and processes characteristic of the intestine. Since intestinal mucosal epithelial cells turn over rapidly and are sloughed off into feces, this noncovalent binding to mucosal cell components may be a mechanism for preventing absorption of xenobiotics. Recent studies have examined the role of mucosal cell turnover and food transit on intestinal residue retention of dietary BaP. These studies have demonstrated significant regional differences in BaP concentration following intestinal evacuation (at 2, 3, and 4 days) in unfed fish, with the highest concentrations in the rectal segment. Upon feeding, differences between segments were less apparent, and all segments contained significantly less BaP than was in unfed animals.[200] Microscopic autoradiography of cut sections of unfed fish indicated significant BaP concentrations associated with the surface of mucosal cells. These results suggest that food passage accelerates clearance of BaP residues from fish intestine. It is unknown at this time if this is accomplished by a stripping of the mucus layer, by stimulation of mucosal cell turnover, or by some other mechanism.

V. CONCLUSIONS

Many intellectual gaps exist in our knowledge of the trophic transfer of contaminants. While it is evident that the trophic transfer of xenobiotics occurs, it is also evident that we, as scientists, have yet to quantitatively or biochemically define its inner wrappings. In aquatic systems, it has been easier to "black box" the entire process by modeling or simplistically approaching the process by one-dimensional studies. Clearly, the transfer of contaminants along the aquatic food chain is multidimensional. Food chain transfer of contaminants is no less complicated than drug delivery and is perhaps more complicated, as physiological/ ecological variables throughout the transfer are pervasively altering the outcome. The variance in opinion and results regarding the importance of dietary transfer may be clouded by experimental artifacts resulting from our lack of understanding or may be a direct reflection of presently misunderstood processes. It is obvious that definition of all of the operative variables may not be feasible or necessary with an understanding of fundamental processes.

From a risk assessment standpoint, it is of importance to determine if contaminant transfer is either more or less important from waterborne or dietary routes. Studies have approached this question by waterborne exposures, food exposures, and food/water exposures. Often, exposure levels and conditions are not reflective of a natural state. Water with low organic loads, idealized water quality, carrier vehicles, and use of experimentally convenient artificial diets have been the norm. It appears necessary for these types of studies to take into account dose, existing residues, and concentrations, the type of diet, and perhaps most importantly, the concept that uptake for many compounds is a diffusive process. The intestine provides a two-way avenue, with the concentration gradient of the compound determining not only the rate of uptake but also the direction. It has in fact been demonstrated that the intestine can function as an excretory organ in mammals. There are many opportunities within the study design to change the experimental outcome based on the equilibria established between routes of uptake.

Compounds that are recalcitrant to biotransformation, such as PCB, are generally recognized as being trophically transferred. In fact, evidence now suggests that compounds that are biotransformed and their biotransformation products are presented in the diet. This area of research may have been largely overlooked, as often only the parent compound was analyzed and biotransformation products were often less cumulative and persistent. Nevertheless, these products are presented to the consumer. Their disposition and ultimate toxicity is largely unknown. Perhaps they are eliminated with the feces, hydrolyzed to reform the parent, or absorbed as the metabolite.

Preconsumptive metabolism may play an inherent role in the toxicity of consumed contaminants. It is most likely that preconsumptive biotransformation will result in detoxication; however, activation is a possibility. In the case of indirect carcinogens, it is unknown if consumption of the premetabolized compound disposes the consumer to additional risk as further steps in the cascade toward metabolic activation have occurred, or if the risk is reduced by affecting absorption and facilitating elimination. These questions are only now receiving attention.

The nature of contaminant residues in tissues and their association with tissue will probably prove to be an important building block for a true understanding of dietary contaminants and their transfer. This is particularly true in aquatic organisms, where unusual components such as waxy esters and polyunsaturated lipids are more the norm than the exception. It is known from nutritionally related studies that different dietary components offer differing digestibilities and unique properties. The nature of the association of contaminants with these components includes not only chemical-to-chemical association, but also organotaxis due to the ingestive behavior of most higher aquatic organisms. Differences in matrix chemical composition between organisms, contaminant association with the matrix, and feeding interactions between trophic levels may contribute to an understanding of routes of xenobiotic transfer.

Finally, advances in the elucidation of mechanisms involved in contaminant transport, absorption, transcellular processing, and disposition will provide a better understanding of aquatic toxicology. Even basic physiological parameters relevant to xenobiotic uptake are virtually unknown for many aquatic species. Given the diversity of life, lifestyles, and function, these may prove to extend or confirm by extrapolation our understanding of trophic transfer of xenobiotics.

ACKNOWLEDGMENTS

The authors' work discussed in this chapter was supported in part by the U.S. Public Health Service grant ES-05781 and in part by the U.S. Food and Drug Administration grants FD-R-000158 and FD-O-01466.

REFERENCES

1. Koeman, J.H., Ten Noever De Brauw, M.C., and De Vos, R.H., Chlorinated biphenyls in fish, mussels and birds from the river Rhine and the Netherlands coastal area, *Nature,* 221, 1126, 1969.
2. Whittle, K.J., Hardy, R., Holden, A.V., Johnston, R., and Pentreath, R.J., Occurrence and fate of organic and inorganic contaminants in marine animals, *Ann. N.Y. Acad. Sci.,* 298, 47, 1977.
3. Connor, M.S., Fish/sediment concentration ratios for organic compounds, *Environ. Sci. Technol.,* 18, 31, 1984.
4. Spacie, A. and Hamelink, J.L., Bioaccumulation, in *Fundamentals of Aquatic Toxicology,* P., Rand, G.M., and Petrocelli, S.R., Eds., Hemisphere Publishing Corporation, New York, 1985, chap. 17.
5. Tovell, P., Howes, D., and Newsome, C., Absorption, metabolism and excretion by goldfish of the anionic detergent, sodium lauryl sulphate, *Toxicology,* 4, 17, 1975.
6. Varanasi, U., Uhler, M., and Stranahan, S., Uptake and release of napthalene and its metabolites in skin and epidermal mucus of salmonids, *Toxicol. Appl. Pharmacol.,* 44, 277, 1978.
7. Kleinow, K.M. and Brooks, A.S., Selenium compounds in the fathead minnow *(Pimephales promelas).* II. Qualitative approach to gastrointestinal absorption, routes of elimination and influence of dietary pretreatment, *Comp. Biochem. Physiol.,* 83, 71, 1986.
8. Part, P. and Lock, R.A.C., Diffusion of calcium, cadmium and mercury in a mucous solution from rainbow trout, *Comp. Biochem. Physiol.,* 76, 259, 1983.
9. Moriarty, F., *Organochlorine Insecticides: Persistent Organic Pollutants,* Academic Press, New York, 1975.
10. Moriarty, F., *Ecotoxicology,* Academic Press, New York, 1988.
11. Clayton, J.R., Pavlou, S.P., and Brietner, N.F., Polychlorinated biphenyls in coastal marine zooplankton. Bioaccumulation by equilibrium positioning, *Environ. Sci. Technol.,* 11, 676, 1977.
12. Ellgehausen, G., Guth, J.A., and Esser, H.O., Factors determining the bioaccumulation potential of pesticides in the individual compartments of aquatic food chains, *Ecotoxicol. Environ. Safety,* 4, 134, 1980.
13. Bruggeman, W.A., Martron, L.B.J.M., Koolman, D., and Hutzinger, O., Accumulation and elimination kinetics of di-, tri- and tetrachlorobiphenyls by goldfish after dietary and aqueous exposure, *Chemosphere,* 10, 811, 1981.
14. Shaw, G.R. and Connell, D.W., Physiochemical properties controlling PCB concentrations in aquatic organisms, *Environ. Sci. Technol.,* 18, 18, 1984.
15. Chadwick, G.G. and Brocksen, R.W., Accumulation of dieldrin by fish and selected fish food organisms, *J. Wildl. Manage.,* 33, 693, 1969.
16. Epifanio, C.E., Dieldrin uptake by larvae of the crab *(Leptodius floridanus), Mar. Biol.,* 19, 320, 1973.
17. Jarvinen, A.W., Hoffman, M.J., and Thorslund, T.W., Long-term toxic effects of DDT food and water exposure on fathead minnows *(Pimephales promelas), J. Fish. Res. Board Can.,* 34, 2089, 1977.
18. Reinert, R.E., Accumulation of dieldrin in an algae *(Scenedesmus obliquus), (Daphnia magna),* and the guppy *(Pecilia reticulata), J. Fish. Res. Board Can.,* 29, 1413, 1972.

19. Macek, K.J. and Korn, S., Significance of the food chain in DDT accumulation by fish, *J. Fish. Res. Board Can.*, 27, 1496, 1970.
20. Reinert, R.E., Stone, L.J., and Bergman, H.L., Dieldrin and DDT: accumulation from water and food by lake trout *(Salvelinus namaycush)* in the laboratory, *Proc. 17th Conf. Great Lakes Res.*, 1974, 1974.
21. Cox, J.L., Uptake, assimilation, and loss of DDT residues by *(Euphausia pacifica)*, a euphansiid shrimp, *U.S. Fish. Wildl. Serv. Fish. Bull.*, 69, 627, 1971.
22. Fowler, S.W., Polikarpov, G.G., Elder, D.L., Parsi, P., and Villeneuve, J.P., Polychlorinated biphenyls: accumulation from contaminated sediments and water by the polychaete *(Nereis diversicolor)*, *Mar. Biol.*, 48, 303, 1978.
23. Bruggeman, W.A., Opperhuizen, A., Wijbenga, A., and Hutzinger, O., Bioaccumulation of super-lipophilic chemicals in fish, *Toxicol. Environ. Chem.*, 7, 173, 1984.
24. Mitchell, A.I., Plack, P.A., and Thomson, I.M., Relative concentrations of ^{14}C-DDT and two polychlorinated biphenyls in the lipids of cod tissues after a single oral dose, *Arch. Environ. Contam. Toxicol.*, 6, 525, 1977.
25. Niimi, A.J., Biological half-lives of chlorinated diphenyl ethers in rainbow trout *(Salmo gairdneri)*, *Aquat. Toxicol.*, 9, 105, 1986.
26. Tanabe, S., Maruyama, K., and Tatsukawa, R., Absorption efficiency and biological half-life of individual chlorobiphenyls in carp *(Cyprinus carpio)* orally exposed to Kanechlor products, *Agric. Biol. Chem.*, 46, 891, 1982.
27. Harding, G.C., Vass, W.P., and Drinkwater, K.F., Importance of feeding, direct uptake from seawater, and transfer from generation to generation in the accumulation of an organochlorine (p,p'-DDT) by the marine planktonic copepod *(Calanus finmarchicus)*, *Can. J. Fish. Aquat. Sci.*, 38, 101, 1981.
28. Lederman, T.C. and Rhee, G., Bioconcentration of a hexachlorobiphenyl in Great Lake planktonic algae, *Can. J. Fish. Aquat. Sci.*, 39, 380, 1982.
29. Thomann, R.V., Connolly, J.P., and Nelson, A.T., The great lakes ecosystem—modelling the fate of PCBs, in *PCBs and the Environment*, CRC Press, Boca Raton, FL, 1986, 3, 153.
30. Thomann, R.V., Bioaccumulation model of organic chemical distribution in aquatic food chains, *Environ. Sci. Technol.*, 23, 699, 1989.
31. Oliver, B.G. and Niimi, A.J., Trophodynamic analysis of polychlorinated biphenyl congeners and other chlorinated hydrocarbons in the Lake Ontario ecosystem, *Environ. Sci. Technol.*, 22, 338, 1988.
32. Rasmussen, J.B., Rowan, D.J., Lean, D.R.S., and Carey, J.H., Food chain structure in Ontario lakes determines PCB levels in lake trout *(Salvelinus namaycush)* and other pelagic fish, *Can. J. Fish. Aquat. Sci.*, 47, 2030, 1990.
33. Connolly, J.P., Application of a food chain model to polychlorinated biphenyl contamination of the lobster and winter flounder food chains in New Bedford Harbor, *Environ. Sci. Tech.*, 25, 760, 1991.
34. Breck, J.E. and Bartell, S.M., Approaches to modeling the fate and effects of toxicants in pelagic systems, in *Toxic Contaminants and Ecosystem Health*, Evans, M.S., Ed., John Wiley & Sons, New York, 1988, 427.
35. Barber, M.C., Suarez, L.A., and Lassiter, L.A., Modeling bioaccumulation of organic pollutants in fish with an application to PCBs in Lake Ontario salmonids, *Can. J. Fish. Aquat. Sci.*, 48, 318, 1991.

36. Sanders, J.G., Osman, R.W., and Riedel, G.F., Pathways of arsenic uptake and incorporation in estuarine phytoplankton and the filter-feeding invertebrates *(Eurytemora affinis, Balanus improvisus* and *Crassosterea virginica), Mar. Biol.,* 103, 319, 1989.
37. Amiard-Triquet, C., Amiard, J.C., Ballan-Dufrancais, C., Berthet, B., Gouzerh, P., Jeantet, A.Y., Martoja, R., and Truchet, M., Study of the bioaccumulation at the molecular, cellular and organism levels of silver transferred to the oyster, *Crassostrea gigas* Thunberg, directly from water or via food, in *Metals in the Environment,* 2, Lindberg, S.E. and Hutchinson, T.C., Eds., CEP Consultants Ltd., Edinburgh, U.K., 2, 488, 1987.
38. Martoja, R., Ballan-Dufrancais, C., Jeantet, A.Y., Gouzerh, P., Amiard, J.C., Amiard-Triquet, C., Berthet, B., and Baud, J.P., Effects chimiques et cytologiques de la contamination experimentale de l'huitre, *(Crassostrea gigas)* Thunberg par l'argent administre sous forme dissoute et par voie alimentaire, *Can. J. Fish. Aquat. Sci.,* 45, 1827, 1988.
39. Abbe, G. R. and Sanders, J.G., Pathways of silver uptake and accumulation by the American oyster *(Crassostrea virginica)* in Chesapeake Bay, *Estuarine Coastal Shelf Sci.,* 31, 113, 1990.
40. Connell, D.B., Sanders, J.G., Riedel, G.F., and Abbe, G.R., Pathways of silver uptake and trophic transfer in estuarine organisms, *Environ. Sci. Technol.,* 25, 921, 1991.
41. Neudorf, and Khan, M.A.Q., Pick-up and metabolism of DDT, dieldrin and photodieldrin by a freshwater alga *(Ankistrodesmus amalloides)* and a microcrustacean *(Daphnia pulex), Bull. Environ. Contam. Toxicol.,* 13, 443, 1975.
42. Crosby, D.G. and Tucker, R.K., Accumulation of DDT by *(Daphnia magna), Environ. Sci. Technol.,* 5, 714, 1971.
43. Herbes, S.E. and Risi, G.F., Metabolic alteration and excretion of anthracene by *(Daphnia pulex), Bull. Environ. Contam. Toxicol.,* 55, 147, 1978.
44. Southworth, G.R., Beauchamp, J.J., and Schmieder, P.K., Bioaccumulation potential of polycyclic aromatic hydrocarbons in *(Daphnia pulex), Water Res.,* 12, 973, 1978.
45. Harding, G.C.H. and Vass, W., Uptake from seawater and clearance of pp'-DDT by marine planktonic crustacea, *J. Fish. Res. Board Can.,* 36, 247, 1979.
46. Gardner, W.S., Landrum, P.F., and Cavaletto, J.F., Lipid-partitioning and disposition of benzo[a]pyrene and hexachlorobiphenyl in Lake Michigan *(Pontoporeia hoyi* and *Mysis relicta), Environ. Toxicol. Chem.,* 9, 1269, 1990.
47. Dobroski, C.J., Jr. and Epifanio, C.E., Accumulation of benzo[a]pyrene in a larval bivalve via trophic transfer, *Can. J. Fish. Aquat. Sci.,* 37, 2318, 1980.
48. Wyman, K.D. and O'Connors, H.B., Jr., Implications of short term PCB uptake by small estuarine copepods (genus *Acartia)* from PCB contaminated water, inorganic sediments and phytoplankton, *Estuarine Coastal Mar. Sci.,* 11, 121, 1980.
49. Pinkney, A.E., Poje, G.V., Sansurnoi, F.M., Lee, C.C., and O'Connor, J.M., Uptake and retention of ^{14}C-Aroclor 1254 in the amphipod *(Gammarus tigrinus),* fed contaminated fungus *(Fusarium oxyporum), Arch. Environ. Contam. Toxicol.* 14, 59, 1985.
50. Harding, G.C., Organochlorine dynamics between zooplankton and their environment, a reassessment, *Mar. Ecol. Prog. Ser.,* 33, 167, 1986.

51. Rice, C.P. and Sikka, H.C., Uptake and metabolism of DDT by six species of marine algae, *J. Agric. Food Chem.*, 21, 148, 1973.
52. Miyazaki, S. and Thorsteinson, A.J., Metabolism of DDT by freshwater diatoms, *Bull. Environ. Contam. Toxicol.*, 8, 81, 1972.
53. Johnson, B.T., Saunders, C.R., and Sanders, H.O., Biological magnification and degradation of DDT and aldrin by freshwater invertebrates, *J. Fish. Res. Board Can.* 28, 705, 1971.
54. James, M.O., Khan, M.A.Q., and Bend, J.R., Hepatic microsomal mixed-function oxidase activities in several marine species common to coastal Florida, *Comp. Biochem. Pharmacol.*, 62C, 155, 1979.
55. McElroy, A.E., In vivo metabolism of benzo(a)anthracene by the polychaete *Nereis virens*, *Mar. Environ. Res.*, 17, 133, 1985.
56. James, M.O., Biotransformation and disposition of PAH in aquatic invertebrates, in *Metabolism of Polycyclic Aromatic Hydrocarbons in the Aquatic Environment*, Varanasi, U., Ed., CRC Press, Boca Raton, FL, 1989, 69.
57. Schanker, L.S., Tocco, D.J., Brodie, B.B., and Hogben, A.M., Absorption of drugs from the rat small intestine, *J. Pharmacol. Exp. Ther.*, 123, 81, 1958.
58. Thomson, A.B.R. and Dietschy, J.M., Intestinal lipid absorption: major extracellular and intracellular events, in *Histology of the Gastrointestinal Tract*, Johnson, L.R., Ed., Raven Press, New York, 1981, 1147.
59. Kuksis, A., Intestinal digestion and absorption of fat-soluble environmental agents, in *Intestinal Toxicology*, Schiller, C.M., Ed., Raven Press, New York, 1984, 69.
60. Kemp, P. and Smith, M.W., Effect of temperature acclimatization on the fatty acid composition of goldfish intestinal lipids, *Biochem. J.*, 117, 9, 1970.
61. Leslie, J.M. and Buckley, J.T., Phospholipid composition of goldfish *(Carassius auratus* L.) liver and brain and temperature-dependence of phosphatidylcholine synthesis, *Comp. Biochem. Physiol.*, B53, 335, 1976.
62. Leger, C., Specific distribution of fatty acids in the triglycerides of rainbow trout adipose tissue. Influence of temperature, *Lipids*, 12, 538, 1977.
63. Farkas, T., Csengeri, I., Majoros, F., and Olah, J., Metabolism of fatty acids in fish. III. Combined effect of environmental temperature and diet on formation and deposition of fatty acids in the carp *(Cyprinus carpio* Linnaeus 1758), *Aquaculture*, 20, 29, 1980.
64. Wodtke, E., Lipid adaptation in liver mitochondrial membranes of carp acclimated to different environmental temperatures. Phospholipid composition, fatty acid pattern, and cholesterol content, *Biochem. Biophys. Acta*, 529, 280, 1978.
65. Wodtke, E., Temperature adaptation of biological membranes: effects of acclimation temperature on the unsaturation of the main neutral and charged phospholipids in mitochondrial membranes of the carp *(Cyprinus carpio)*, *Biochem. Biophys. Acta*, 640, 698, 1981.
66. Castell, J.D., Lee, D.J., and Sinnhuber, R.O., Essential fatty acids in the diet of rainbow trout *(Salmo gairdneri)*: metabolism and fatty acid composition, *J. Nutr.*, 102, 93, 1972.
67. Stremmel, W., Lotz, G., Strohmeyer, G., and Berk, P.D., Identification, isolation and partial characterization of a fatty acid binding protein from rat jejunal microvillus membranes, *J. Clin. Invest.*, 75, 1068, 1985.

68. Foulkes, E.C., Some determinants of intestinal cadmium transport in the rat, *J. Environ. Pathol. Toxicol.*, 3, 471, 1980.
69. Aungst, B.J. and Fung, H.L., Kinetic characterization of *in vitro* lead transport across the rat small intestine, *Toxicol. Appl. Pharmacol.*, 69, 39, 1981.
70. Thomson, A.B.R., Olatunbosun, D., and Valberg, L.S., Interrelation of intestinal transport system for manganese and iron, *J. Lab. Clin. Med.*, 78, 642, 1971.
71. Schade, S.G., Felsher, B.F., Glades, B.E. and, Conrad, M.E., Effect of cobalt upon iron absorption, *Soc. Exp. Biol. Med.*, 134, 741, 1970.
72. Schanker, L.S. and Jeffery, J.J., Active transport of foreign pyrimidines across the intestinal epithelium, *Nature*, 190, 727, 1961.
73. Benson, J.A., Culver, P.J., Ragland, S., Jones, C.M., Drummey, G.D., and Bougas, E., The D-xylose absorption test in malabsorption syndromes, *N. Engl. J. Med.*, 2567, 335, 1957.
74. Van Veld, P.A., Absorption and metabolism of dietary xenobiotics by the intestine of fish, *Rev. Aquat. Sci.* 2, 185, 1990.
75. Suzuki, A., Higuchi, W.I., and Ho, N.F.H., Theoretical model studies of drug absorption and transport in the gastrointestinal tract II, *J. Pharm. Sci.*, 59, 651, 1970.
76. Kleinow, K.M., Beilfuss, W.L., Jarboe, H.H., Droy, B.F., and Lech, J.J., Pharmacokinetics, bioavailability, distribution and metabolism of sulfadimethoxine in the rainbow trout *(Oncorhynchus mykiss)*, *Can. J. Fish. Aquat. Sci.*, 49, 1070, 1992.
77. Squibb, K.S., Michel, C.M.F., Zelikoff, J.T., and O'Connor, J.M., Sulfadimethoxine pharmacokinetics and metabolism in the channel catfish *(Ictalurus punctatus)*, *Vet. Hum. Toxicol.*, 30, 31, 1988.
78. Motais, R., Rankin, I., and Maetz, J., Water fluxes in marine fishes, *J. Exp. Biol.*, 51, 529, 1969.
79. Knezovich, J.P., Harrison, F.L., and Wilhelm, R.G., The bioavailability of sediment-sorbed organic chemicals: a review, *Water, Air Soil Poll.*, 32, 233, 1987.
80. Maccubbin, A.E., Black, P., Trzeciak, L., and Black, J.J., Evidence for polynuclear aromatic hydrocarbons in the diet of bottom-feeding fish, *Bull. Environ. Contam. Toxicol.*, 34, 876, 1985.
81. Varanasi, U., Stein, J.E., Nishimoto, M., Reichert, W.L., and Collier, T.K., Chemical carcinogenesis in feral fish: uptake, activation and detoxification of organic xenobiotics, *Environ. Health Perspect.*, 71, 155, 1987.
82. Lovell, R.T., Digestibility of nutrients in feedstuffs for catfish, in *Nutrition and Feeding of Channel Catfish*, Stickney, R.R. and Lovell, R.T., Eds., *Southern Cooperative Series Bull.*, Auburn University, Auburn, AL, 218, 33, 1977.
83. Fänge, R. and Grove, D., Digestion, in *Fish Physiology*. Volume VIII. *Bioenergetics and Growth*, Hoar, W.S., Randall, D.J., and Brett, J.R., Eds., Academic Press, Orlando, FL, 1979, 162.
84. Phillips, A.M., Jr., Nutrition, digestion and energy, in *Fish Physiology*, Vol. I. Hoar, W.S. and Randall, D.J., Eds., Academic Press, New York, 1969, 391.
85. Fish, G.R., The comparative activity of some digestive enzymes in the alimentary canal of *Tilapia* and perch, *Hydrobiologia*, 15, 161, 1960.
86. Chesley, L.C., The concentration of proteases, amylase and lipase in certain marine fishes, *Biol. Bull. (Woods Hole, Mass.)*, 60, 133, 1934.
87. Page, J.W. and Andrews, J.W., Interaction of dietary levels of protein and energy on channel catfish *(Ictalurus punctatus)*, *J. Nutr.*, 102, 1339, 1973.

88. Takeuchi, M., Digestibility of dietary lipids in carp, *Bull. Tokai Regional Fish. Res. Lab.,* 99, 55, 1979.
89. Guarino, A.M., Plakas, S.M., Dickey, R.W., and Zeeman, M., Principles of drug absorption and recent studies of bioavailability in aquatic species, *Vet. Hum. Toxicol.,* 30 (Suppl. 1), 41, 1988.
90. Gronel, J.L., Nouws, J.F.M., DeJong, M., Schutte, A.R., and Driessens, F., Pharmacokinetics and tissue distribution of oxytetracycline in carp, *Cyprinus carpio* L., following different routes of administration, *J. Fish Dis.,* 10, 153, 1987.
91. Bernard, F., La digestion chez les poissons, *Trav. Lab. Hydrobiol. Piscicult. Univ. Grenoble,* 44, 61, 1952.
92. Western, J.R.H., Feeding and digestion in two cottid fishes, the freshwater *(Cottus gobio)* and the marine *(Enophrys bubalis), J. Fish Biol.,* 3, 225, 1971.
93. Moriarty, D.J.W., The physiology of digestion of blue-green algae in the cichlid fish *(Tilapia nilotica), J. Zool.,* 171, 25, 1973.
94. Bowen, S.H., Feeding, digestion and growth-qualitative considerations, in *The Biology and Culture of Tilapias,* Pullin, R.S.V. and Lowe-McConnell, R.H., Eds., ICLARM Conf. Proc. 7, Manila: Int. Cent. Living Aquat. Resour. Manage., 141, 1982.
95. Nagase, G., Contribution to the physiology of digestion in *Tilapia mossambica* Peters; digestive enzymes and the effect of diets on their activity, *Z. Vgl. Physiol.,* 49, 270, 1964.
96. Jobling, M., Gwyther, D., and Grove, D.J., Some effects of temperature, meal size and body weight on gastric evacuation time in the dab, *(Limanda limanda* [L.]*), J. Fish Biol.,* 10, 291, 1977.
97. Pandian, T.G., Intake, digestion, absorption and conversion of food in the fishes *(Megalops cyprinoides* and *Ophiocephalus striatus), Mar. Biol.,* 1, 316, 1967.
98. Solomon, D.J. and Brafield, A.G., The energetics of feeding, metabolism, and growth of perch *(Perca fluviatilis* L.*), J. Anim. Ecol.,* 41, 699, 1972.
99. Moriarty, C.M. and Moriarty, D.J.W., Quantitative estimation of the daily ingestion of phytoplankton by *Tilapia nilotica* and *Haplochromis nigripennis* in Lake George, Uganda, *J. Zool.,* 171, 15, 1973.
100. Moriarty, C.M. and Moriarty, D.J.W., The assimilation of carbon from phytoplankton by two herbivorous fishes *(Tilapia nilotica)* and *(Haplochromis nigripennis), J. Zool.,* 171, 41. 1973.
101. Shrable, J.B., Tiemeier, O.W., and Deyoe, L.W., Effects of temperature on the rate of digestion by channel catfish, *Prog. Fish-Cult.,* 31, 131, 1969.
102. Brett, J.R. and Higgs, D.A., Effects of temperature on the rate of gastric digestion in fingerling sockeye salmon, *(Oncorhynchus nerka), J. Fish. Res. Board Can.,* 27, 1767, 1970.
103. Tyler, A.V., Rates of gastric emptying in young cod, *J. Fish. Res. Board Can.,* 27, 1177, 1970.
104. Elliott, J.M., Rates of gastric evacuation in brown trout, *(Salmo trutta* L.*), Freshwater Biol.,* 2,1, 1972.
105. Windell, J.T., Kitchell, J.F., Norris, D.O., Norris, J.S., and Foltz, J.W., Temperature and rate of gastric evacuation by rainbow trout *(Salmo gairdneri), Trans. Am. Fish. Soc.,* 6, 712, 1976.

106. Windell, J.T., Norris, D.O., Kitchell, J.F., and Norris, J.S., Digestive response of rainbow trout *(Salmo gairdneri)* to pellet diets, *J. Fish. Res. Board Can.*, 26, 1801, 1969.
107. Bloedow, D.C. and Hayton, W.L., Effects of lipids on bioavailability of sulfisoxazole acetyl, dicumarol, and griseofulvin in rats, *J. Pharm. Sci.*, 65, 328, 1976.
108. Yoshitomi, H., Nishihata, T., Frederick, G., Dillsaver, M., and Higuchi, T., Effect of triglyceride on the small intestinal absorption of cefoxitin in rats, *J. Pharm. Pharmacol.*, 39, 887, 1987.
109. Andrews, J.W., Murray, M.W., and Davis, J.M., The influence of dietary fat levels and environmental temperature on digestible energy and absorbability of animal fat in catfish diets, *J. Nutr.*, 108, 749, 1978.
110. Hollander, D., Retinol lymphatics and portal transport. Influence of pH, bile and fatty acids on transport, *Am. J. Physiol.*, 239, G210, 1980.
111. Hollander, D., Intestinal absorption of vitamins A, E, D, and K, *J. Lab. Clin. Med.*, 97(4), 449, 1981.
112. Hollander, D., Rim, E., and Ruple, P.E., Vitamin K-2 colonic and ideal *in vivo* absorption. Bile, fatty acids and pH effects on transport, *Am. J. Physiol.*, 233, 3124, 1977.
113. Niimi, A.J. and Oliver, B.G., Influence of molecular weight and molecular volume on dietary absorption efficiency of chemicals by fishes, *Can. J. Fish Aquat. Sci.*, 45, 222, 1988.
114. Iijima, N., Aida, S., and Kayama, M., Intestinal absorption and plasma transport of dietary fatty acids in carp, *Nippon Suisan Gakkaishi*, 56, 1829, 1990.
115. Laher, J.M. and Barrowman, J.A., Polycyclic hydrocarbon and polychlorinated biphenyl solubilization in aqueous solutions of mixed micelles, *Lipids*, 18, 216, 1983.
116. Takahashi, Y.I. and Underwood, B.A., Effect of long and medium chain length lipids upon aqueous solubility of α-tocopherol, *Lipids*, 9, 855, 1974.
117. Wilson, F.A., Sallee, V.L., and Dietschy, J.M., Unstirred waterlayer in the intestine: rate determinant for fatty acid absorption from micellar solutions, *Science*, 174, 1031, 1971.
118. Braybrooks, M., Barry, B.W., and Abbs, E.T., The effect of mucin on the bioavailability of tetracycline from the gastrointestinal tract; *in vivo, in vitro* correlations, *J. Pharmacol.*, 27, 508, 1975.
119. Hogben, C.A., Tocco, D.J., Brodie, B.B., and Schanker, L.S., On the mechanism of intestinal absorption of drugs, *J. Pharmacol. Exp. Ther.*, 125, 275, 1959.
120. Rahman, A. and Barrowman, J.A., The influence of bile on the bioavailability of polynuclear aromatic hydrocarbons from the rat intestine, *Can. J. Physiol. Pharmacol.*, 64, 1214, 1986.
121. Bauermeister, A.E.M., Pirie, B.J.S., and Sargent, J.R., An electron microscopic study of lipid absorption in the pyloric caeca of rainbow trout *(Salmo gairdneri)* fed wax ester-rich zooplankton, *Cell Tissue Res.*, 200, 475, 1979.
122. Sire, M.F., Lutton, C., and Vernier, J.M., New views on intestinal absorption of lipids in teleostean fishes; an ultrastructure and biochemical study in the rainbow trout, *J. Lipid Res.*, 22, 81 1981.
123. Vetter, R.D., Carey, M.C., and Patton, J.S., Coassimilation of dietary fat and benzo(a)pyrene in the small intestine: an absorption model using the killifish, *J. Lipid Res.*, 26, 428, 1985.

124. Van Veld, P.A., Vetter, R.D., Lee, R.F., and Patton, J.S., Dietary fat inhibits the intestinal metabolism of the carcinogen benzo[a]pyrene in fish, *J. Lipid Res.*, 28, 810, 1987.
125. Babin, P.J. and Vernier, J.-M., Plasma lipoproteins in fish, *J. Lipid Res.*, 30, 467, 1989.
126. Robinson, J.S. and Mead, J.F., Lipid absorption and deposition in rainbow trout *(Salmo gairdneri)*, *Can. J. Biochem.*, 51, 1050, 1973.
127. Kayama, M. and Iijima, N., Studies on lipid transport mechanism in the fish, *Nippon Suisan Gakkaishi*, 42, 987, 1976.
128. Mankura, M., Iijima, N., Kayama, M., and Aida, S., Plasma transport form and metabolism of dietary fatty alcohol and wax ester in carp, *Nippon Suisan Gakkaishi*, 53, 1221, 1987.
129. Iijima, N., Kayama, M., Okazaki, M., and Hara, I., Time course changes of lipid distribution in carp plasma lipoprotein after force-feeding with soybean oil, *Nippon Suisan Gakkaishi*, 51, 467, 1985.
130. Iijima, N., Aida, S., Mankura, M., and Kayama, M., Intestinal absorption and plasma transport of dietary triglyceride and phosphatidylcholine in the carp, *Comp. Biochem. Physiol.*, 96A, 45, 1990.
131. Schummer, A., Wilkens, H., Vollmerhaus, B., and Habermehl, K.H., *The Circulatory System, the Skin, and the Cutaneous Organs of Domestic Mammals*, Springer-Verlag, New York, 1981.
132. Endo, T., Onozawa, M., Hamaguchi, M., and Kusuda, R., Enhanced bioavailability of oxolinic acid by ultra-fine size reduction in yellowtail, *Nippon Suisan Gakkaishi*, 53, 1711, 1987.
133. Cravedi, J.P., Choubert, G., and Delous, G., Digestibility of chloramphenicol, oxolinic acid and oxytetracycline in rainbow trout and influence of these antibiotics on lipid digestibility, *Aquaculture*, 60, 133, 1987.
134. Björklund, H.V. and Bylund, G., Comparative pharmacokinetics and bioavailability of oxolinic acid and oxytetracycline in rainbow trout *(Oncorhynchus mykiss)*, *Xenobiotica*, 21, 1511, 1991.
135. Kleinow, K.M., Jarboe, H.H., Shoemaker, K.E., and Greenlees, K.J., *Can. J. Fish. Aquat. Sci.*, submitted, 1993.
136. Hustvedt, S.O., Salte, R., Kvendset, O., and Vassvik, V., Bioavailability of oxolinic acid in Atlantic salmon *(Salmo salar* L.) from medicated feed, *Aquaculture*, 97, 305, 1991.
137. Guarino, A.M., Plakas, S.M., Dickey, R.W., and Zeeman, M., Principles of drug absorption and recent studies of bioavailability in aquatic species, *Vet. Hum. Toxicol.*, 30, 41, 1988.
138. Little, P.J., James, M.O., Pritchard, J.B., and Bend, J.R., Temperature dependent disposition of ^{14}C-benzo(a)pyrene in the spiny lobster, *Panulirus argus.,Toxicol. Appl. Pharmacol.*, 77, 325, 1985.
139. Kleinow, K.M., unpublished data, 1993.
140. Bowser, P.R., Wooster, G.A., St. Leger, J., and Babish, J.G., Pharmacokinetics of enrofloxacin in fingerling rainbow trout *(Oncorhynchus mykiss)*, *J. Vet. Pharmacol. Ther.*, 15, 62, 1992.
141. Dipple, A., Polycyclic aromatic hydrocarbon carcinogenesis, an introduction, in *Polycyclic Hydrocarbons and Carcinogenesis, ACS Symposium Series* 283, Harvey, R.G., Ed., American Chemical Society, Washington, D.C., 1985, 1.

142. Smolowitz, R.M., Hahn M.E., and Stegeman J.J., Immunohistochemical localization of cytochrome P4501A1 induced by 3,3',4,4'-tetrachlorobiphenyl and by 2,3,7,8-tetrachlorodibenzofuran in liver and extrahepatic tissues of the teleost *Stenotomus chrysops* (scup), *Drug Metab. Disp.*, 19, 113, 1991.
143. Kleinow, K.M., Cahill, J.M., and McElroy, A.E., Influence of preconsumptive metabolism upon the toxicokinetics and bioavailability of a model carcinogen in the flounder *(Pseudopleuronectes americanus), Bull. Mt. Desert Isl. Biol. Lab.*, 28, 122, 1989.
144. James, M.O., Schell, J.D., Boyle, S.M., Altman, A., and Cromer, E.A., Southern flounder hepatic and intestinal metabolism and DNA binding of benzo(a)pyrene (BaP) metabolites following dietary administration of low doses of BaP, BaP-7,8-dihydrodiol or a BaP metabolite mixture, *Chem.-Biol. Interact.*, 79, 305, 1991.
145. Sims, P. and Grover P.L., Epoxides in polycyclic aromatic hydrocarbon metabolism and carcinogenesis, *Adv. Cancer Res.*, 20, 165, 1974.
146. James, M.O., Bowen, E.R., Dansette, P.M., and Bend, J.R., Epoxide hydrase and glutathione S-transferase activities with selected alkene and arene oxides in several marine species. *Chem.-Biol. Interact.*, 25, 321, 1979.
147. Li, C.-L.J. and James, M.O., Glucose and sulfate conjugations of phenol, β-naphthol and 3-hydroxybenzo(a)pyrene by the American lobster, *(Homarus americanus), Aquat. Toxicol.*, in press, 1993.
148. Goldman, P., Role of intestinal microflora, in *Metabolic Basis of Detoxication,* Jakoby, W.B., Bend, J.R., and Caldwell, J., Eds., Academic Press, New York, 1982, 323.
149. Larsen, G.L. and Bakke, J.E., Metabolism of mercapturic acid-pathway metabolites of 2-chloro-N-isopropylacetanilide (propachlor) by gastrointestinal bacteria, *Xenobiotica,* 13, 115, 1983.
150. Bakke, J.E., Gustafsson, J.-Å., and Gustafsson, B.E., Metabolism of propachlor by the germfree rat, *Science,* 210, 433, 1980.
151. Smith, R.L., *The Excretory Function of Bile,* Chapman and Hall, London, 1974, chap. 8.
152. Layiwola, P.J., Linnecar, D.F.C., and Knights, B., Hydrolysis of the biliary glucuronic acid conjugate of phenol by the intestinal mucus/flora of goldfish *(Crassius auratus), Xenobiotica,* 13, 27, 1983.
153. Edwards, V.T. and Hutson, D.H., The disposition of plant xenobiotic conjugates in animals, in *Xenobiotic Conjugation Chemistry,* Paulson, G.D., Caldwell, J., Hutson, D.H., and Menn, J.J., Eds., American Chemical Society, Washington, D.C., 1986, 322.
154. James, M.O. and Altman, A.H., Unpublished data, 1993.
155. Roy, A.B., The chemistry of sulfate esters and related compounds, in *Sulfation of Drugs and Related Compounds,* Mulder, G.J., Ed., CRC Press, Boca Raton, FL, 1981, 5.
156. Bakke, J.E., Catabolism of glutathione conjugates, in *Xenobiotic Conjugation Chemistry,* Paulson, G.D., Caldwell, J., Hutson, D.H., and Menn, J.J., Eds., American Chemical Society, Washington, D.C., 1986, 301.
157. Yagen, B., Foureman, G.L., Ben-Zvi, Z., Ryan, A.J., Hernandez, O., Cox, R.H., and Bend, J.R., The metabolism and excretion of ^{14}C-styrene oxide-glutathione adducts administered to the winter flounder, *Pseudopleuronectes americanus,* a marine teleost, *Drug. Metab. Disp.*, 12, 389, 1984.

158. Bauermeister, A., Lewendon, A., Ramage, P.I.N., and Nimmo, I.A., Distribution and some properties of the glutathione S-transferase and γ-glutamyl transpeptidase activities of rainbow trout, *Comp. Biochem. Physiol.,* 74C, 89, 1983.
159. Simmons, T.W., Hinchman, C.A., and Ballatori, N., Polarity of hepatic glutathione and glutathione S-conjugate efflux, and intraorgan mercapturic acid formation in the skate, *Biochem. Pharmacol.,* 42, 2221, 1991.
160. Bakke, J.E. and Gustafsson, J.-Å., Role of intestinal flora in metabolism of agrochemicals conjugated with glutathione, *Xenobiotica,* 16, 1047, 1986.
161. Lock, E.A., The nephrotoxicity of haloalkane and haloalkene glutathione conjugates, in *Selectivity and Molecular Mechanisms of Toxicity,* De Matteis, F. and Lock, E.A., Eds., Macmillan, New York, 1987, chap. 3.
162. Cashman, J.R., Olsen, L.D., Young, G., and Bern, H., S-oxygenation of Eptam in hepatic microsomes from fresh- and saltwater striped bass *(Morone saxatilis), Chem. Res. Toxicol.,* 2, 392, 1989.
163. James, M.O., Altman, A.H., Feistner, H., and Gahne, Å., Intestinal and hepatic glucuronidation and sulfation of 3,7, and 9-hydroxy-benzo(a)pyrenes in the catfish, *Toxicologist,* 13, 63, 1993.
164. Nebert, D.W., Nelson, D.R., Coon, M.J., Estabrook, R.W., Feyereisen, R., Fujii-Kuriyama, Y., Gonzalez, F.J., Guengerich, F.P., Gunsalus, I.C., Johnson, E.F., Loper, J.C., Sato, R., Waterman, M.R., and Waxman, D.J., The P450 superfamily: update on new sequences, gene mapping, and recommended nomenclature, *DNA Cell Biol.,* 10, 397, 1991.
165. Miranda, C.L., Wang, J.-L., Henderson, M.H., and Buhler, D.R., Immunological characterization of constitutive isozymes of cytochrome P450 from rainbow trout. Evidence for homology with phenobarbital-induced rat P450s, *Biochim. Biophys. Acta,* 1037, 155, 1990.
166. Burchell, B., Nebert, D.W., Nelson, D.R., Bock, K.W., Iyanagi, T., Jansen, P.L.M., Lancet, D., Mulder, G.J., Roy-Chowdhury, J., Siest, G., Tephly, T.R., and MacKenzie, P.I., The UDP-glucuronosyltransferase gene superfamily: suggested nomenclature based on evolutionary divergence, *DNA Cell Biol.,* 10, 487, 1991.
167. Clarke, D.J., Burchell, B., and George, S.G., Differential expression and induction of UDP-glucuronosyltransferase isoforms in hepatic and extrahepatic tissues of a fish, *Pleuronectes platessa:* immunochemical and functional characterization, *Toxicol. Appl. Pharmacol.,* 115, 130, 1992.
168. Falany, C.N., Molecular enzymology of human liver cytosolic sulfotransferases, *Trends Pharmacol. Sci.,* 12, 255, 1991.
169. Mannervik, B. and Danielson, U.H., Glutathione transferases — structure and catalytic activity, *CRC Crit. Rev. Biochem.,* 23, 283, 1988.
170. Foureman, G.L., Hernandez, O., Bhatia, A., and Bend, J.R., The stereoselectivity of four hepatic glutathione S-transferases purified from a marine elasmobranch *(Raja erinacea)* with several K-region polycyclic arene oxide substrates, *Biochim. Biophys. Acta,* 914, 127, 1987.
171. George, S.G. and Buchanan, G., Isolation, properties and induction of plaice liver cytosolic glutathione S-transferases, *Fish Physiol. Biochem.,* 8, 437, 1990.
172. Lindström-Seppa, P., Koivusaari, U., and Hänninen, O., Extrahepatic xenobiotic metabolism in north-European freshwater fish, *Comp. Biochem. Physiol.,* 69C, 259, 1981.

173. Stegeman, J.J., Binder, R.L., and Orren, A., Hepatic and extrahepatic microsomal electron transport components and mixed-function oxygenases in the marine fish *Stenotomus versicolor, Biochem. Pharmacol.,* 28, 3431, 1979.
174. Van Veld, P.A., Stegeman, J.J., Woodin, B.R., Patton, J.S., and Lee, R.F., Induction of monooxygenase activity in the intestine of spot *(Leiostomus xanthurus),* a marine teleost, by dietary polycyclic aromatic hydrocarbons, *Drug Metab. Disp.,* 16, 659, 1988.
175. Stegeman, J.J., Polynuclear aromatic hydrocarbons and their metabolism in the marine environment, in *Polycyclic Hydrocarbons and Cancer,* Gelboin, H.V. and Ts'O, P.O.P., Eds., Academic Press, New York, 1981, 1.
176. Van Veld, P.A., Vetter, R.D., Lee, R.F., and Patton, J.S., Dietary fat inhibits the intestinal metabolism of the carcinogen benzo(a)pyrene in fish, *J. Lipid Res.,* 28, 810, 1987.
177. James, M.O., Conjugation and excretion of xenobiotics by fish and aquatic invertebrates, in *Xenobiotic Metabolism and Disposition,* Kato, R., Estabrook, R.W., and Cayen, M.N., Eds., Taylor and Francis, U.K., 1989, 283.
178. Stegeman, J.J., Cytochrome P450 forms in fish: catalytic, immunological and sequence similarities, *Xenobiotica,* 19, 1093, 1989.
179. Combes, R.D., Cytochrome P450, mixed-function oxidases and formation of genotoxic metabolites by the intestinal tract, in *Intestinal Metabolism of Xenobiotics,* Koster, A.Sj., Richter, E., Lauterbach, F., and Hartmann, F., Eds. Progress in Pharmacology and Clinical Pharmacology Vol 7/2, Gustav Fischer, Stuttgart, 1989, chap 2.2.
180. Schwenk, M., Glucuronidation and sulfation in the gastrointestinal tract, in *Intestinal Metabolism of Xenobiotics,* Koster, A.Sj., Richter, E., Lauterbach, F., and Hartmann, F., Eds. Progress in Pharmacology and Clinical Pharmacology Vol 7/2, Gustav Fischer, Stuttgart, 1989, chap 2.4.
181. Siegers, C.-P., Glutathione and GSH-dependent enzymes, in *Intestinal Metabolism of Xenobiotics,* Koster, A.Sj., Richter, E., Lauterbach, F., and Hartmann, F., Eds. Progress in Pharmacology and Clinical Pharmacology Vol 7/2, Gustav Fischer, Stuttgart, 1989, chap. 2.5.
182. LeMaire, P., Mathieu, A., Guidicelli, J., and LaFaurie, M., Effect of diet on the responses of hepatic biotransformation enzymes to benzo(a)pyrene in the European sea bass *(Dicentrarchus labrax), Comp. Biochem. Physiol.,* 102C, 413, 1992.
183. Ankley, G.T. and Blazer, V.S., Effect of diet on PCB-induced changes in xenobiotic metabolism in the liver of channel catfish *(Ictalurus punctatus), Can. J. Fish. Aquat. Sci.,* 45, 132, 1988.
184. Ankley, G.T., Blazer, V.S., Plakas, S.M., and Reinert, R.E., Dietary lipid as a factor modulating xenobiotic metabolism in channel catfish *(Ictalurus punctatus), Can. J. Fish. Aquat. Sci.,* 46, 1141, 1989.
185. Eisele, T.A., Coulombe, J.L., Williams, J.L., Shelton, D.W.W., and Nixon, J.E., Time and dose-dependent effects of cyclopropenoid fatty acids on the hepatic microsomal mixed function oxidase system of rainbow trout, *Aquat. Toxicol.,* 4, 139, 1983.
186. Huang, M.-T., Johnson, E.F., Muller-Eberhard, U., Koop, D.R., Coon, M.J., and Conney, A.H., Specificity in the activation and inhibition by flavonoids of benzo(a)pyrene hydroxylation by cytochrome P450 isozymes from rabbit liver microsomes, *J. Biol. Chem.,* 256, 10897, 1981.

187. Slaga, T.J., Thompson, S., Berry, D.L., DiGiovanni, J., Juchau, M.R., and Viaje, A., The effects of benzoflavones on polycyclic hydrocarbon metabolism and skin tumor initiation, *Chem.-Biol. Interact.,* 17, 297, 1977.
188. Havsteen, B., Flavonoids, a class of natural products of high pharmacological potency, *Biochem. Pharmacol.,* 32, 1141, 1983.
189. Alworth, W.L., Dang, C.C., Ching, L.M., and Viswanathan, T., Stimulation of mammalian epoxide hydrase activity by flavones, *Xenobiotica,* 10, 395, 1980.
190. Little, P.J., James, M.O., Bend, J.R., and Ryan, A.J., Imidazole derivatives as inhibitors of cytochrome P-450 dependent oxidation and activators of epoxide hydrolase in hepatic microsomes from a marine fish, *Biochem. Pharmacol.,* 30, 2876, 1981.
191. Miller, J.A. and Miller, E.C., Ultimate chemical carcinogens as reactive mutagenic electrophiles, in *Origins of Human Cancer,* Hiatt, H.H., Watson, J.D., and Winsten, J.A., Eds., Cold Spring Harbor Conferences on Cell Proliferation, Vol. 4, Cold Spring Harbor Laboratory, Cold Spring Harbor, NY, 1977, 605.
192. Myers, M.S., Landahl, J.T., Krahn, M.M., and McCain, B.B., Relationships between hepatic neoplasms and related lesions and exposure to toxic chemicals in marine fish from the U.S. West Coast. *Environ. Health Perspect.,* 90, 7, 1991.
193. Black, J.J. and Baumann, P.C., Carcinogens and cancers in freshwater fishes. *Environ. Health Perspect.,* 90, 27, 1991.
194. Reichert, W.L. and Varanasi, U., Detection of damage by ^{32}P-analysis of hepatic DNA in English sole from an urban area and in fish exposed to chemicals extracted from the urban sediment, *Proc. Am. Assoc. Cancer Res.,* 28, 95, 1987.
195. Dunn, B.P., Black, J.J., and Maccubbin, A., ^{32}P-postlabeling analysis of aromatic DNA adducts in fish from polluted areas, *Cancer Res.,* 47, 6543, 1987.
196. Dunn, B.P., Carcinogen adducts as an indicator for the public health risks of consuming carcinogen-exposed fish and shellfish, *Environ. Health Perspect.,* 90, 111, 1991.
197. Varanasi, U., Nishimoto, M., Reichert, W.L., and Le Eberhart, B.-T., Comparative metabolism of benzo(a)pyrene and covalent binding to hepatic DNA in English sole, starry flounder and rat, *Cancer Res.,* 46, 3817, 1986.
198. Varanasi, U., Reichert, W.L., Le Eberhart, B.-T., and Stein, J.E., Formation and persistence of benzo(a)pyrene-diolepoxide-DNA adducts in liver of English sole *(Parophrys vetulus), Chem.-Biol. Interact.,* 69, 203, 1989.
199. von Hofe, E. and Puffer, H.W., *In vitro* metabolism and *in vivo* binding of benzo(a)pyrene in the California killifish *(Fundulus parvipinnis)* and speckled sanddab *(Citharichthys stigmaeuos), Arch. Environ. Contam. Toxicol.,* 15, 251, 1986.
200. Kleinow, K.M., Duffin, K., Freel, M.W., Toth, B.R., and Henk, W.G., Fate of ^{3}H-benzo(a)pyrene in the intact catfish intestine, Proc. Soc. Environ. Toxicol. Chem., 1992.
201. James, M.O., Cytochrome P450 monooxygenases in crustacea, *Xenobiotica,* 19, 1063, 1989.

Chapter 2

Enzymology and Molecular Biology of Phase II Xenobiotic-Conjugating Enzymes in Fish

Stephen G. George

I. INTRODUCTION

Throughout their evolution, animals have been exposed to plant and microbial secondary metabolites, which they cannot completely metabolize and utilize for their own life processes. Thus, enzyme systems enabling detoxication and removal of multiple "foreign" or "xenobiotic" compounds from the body have evolved, which have relatively low activities and broad substrate specificities. Many of these xenobiotic compounds are biologically active and are used as drugs (e.g., digitalis, morphine, tetracycline, etc.), or as pesticides (e.g., pyrethrins), while others, including compounds such as benzo(a)pyrene (BaP) and heterocycles produced by pyrolysis of plant materials, are cytotoxic. Many thousands of xenobiotic compounds have been produced in the 20th century as a result of exploitation of petroleum, and the ultimate sink for many of them is the aquatic environment. Therefore, aquatic animals have become increasingly exposed to industrially derived xenobiotic pollutants. This chapter will focus upon enzyme systems involved in the detoxication of these xenobiotic compounds.

A. Pathways for Biotransformation of Pollutant Compounds

General descriptions of xenobiotic biotransformation systems and the metabolism of xenobiotic compounds in fishes have been presented in previous reviews by Chambers and Yarborough,[1] Bend et al.,[2] and James.[3] Tan and Melius[4] and Varanasi[5] have focused on metabolism of polyaromatic hydrocarbons (PAHs) and Khan et al.[6] have reviewed metabolism of pesticides by fish.

Some xenobiotic compounds possess the requisite functional groups for direct metabolism by the conjugative or "phase II" enzyme systems, while others are metabolized by an integrated process involving prior action of the so termed "phase I" enzymes (Figure 1). The first phase of metabolism, unmasking or adding reactive functional groups, involves oxidation, reduction, or hydrolysis. Quantitatively, the most important processes are oxidative. Hydroxylation of carbon atoms is catalyzed by the cytochrome P450-dependent monooxygenases (or mixed-function oxidases, MFOs), while hydroxylation of N- or S-heteroatoms is catalyzed by another family of oxidative enzymes, the flavine monooxygenases, which catalyze N-oxidation of amines (i.e., mixed function amine oxidases) or S-oxygenation of sulfides, thiols, and thiones. Alternatively, pre-existing functional groups can be exposed by a number of hydrolytic enzymes, including epoxide hydrolases, esterases, and amidases. A variety of redox enzymes (e.g., alcohol and aldehyde dehydrogenases) alter the oxidation state of a carbon atom, thus allowing more ready excretion or biotransformation of the parent compound, these can also be regarded as phase I enzymes.

Many compounds can be metabolized by alternative phase II pathways, the major route being dependent upon both the chemical nature of the compound and the characteristics of the enzymes (which, as will be discussed later, may determine substrate concentration effects). Certain generalizations of the relative importance, functionality, and substrate specificities of the phase II pathways in fish can be made. The major pathway for electrophilic compounds is conjugation with glutathione, while for nucleophilic compounds, conjugation with glucuronic acid is the major route. Other pathways play a minor role in fish and are the preferred route for only a few compounds. Thus, acetylation is largely confined to organic xenobiotics containing an amino group such as aminobenzoic acid,[3] while amino acid conjugation may be important for environmentally important carboxylic acids, including chlorinated phenoxyacetic acids (2,4-D and 2,4,5-T) and carboxylic acids such as the pyrethroid insecticides. Sulfation is a competing pathway to glucuronidation for metabolites of PAHs, but it is only effective at very low substrate concentrations.

Relatively little is known about the enzymology and molecular biology of piscine phase II systems, with our current knowledge being confined to those catalyzing glucuronidation and glutathione conjugation. When considering the role of these enzyme systems in metabolism and detoxication of xenobiotic pollutants, it is particularly important to recognize their functions in metabolism of endobiotic compounds, since any competition between xenobiotic pollutants and endogenous compounds may have profound consequences for the animal. In order to gain a meaningful overview of the possible functions of phase II systems in fish, I will first briefly summarize current knowledge of the much better

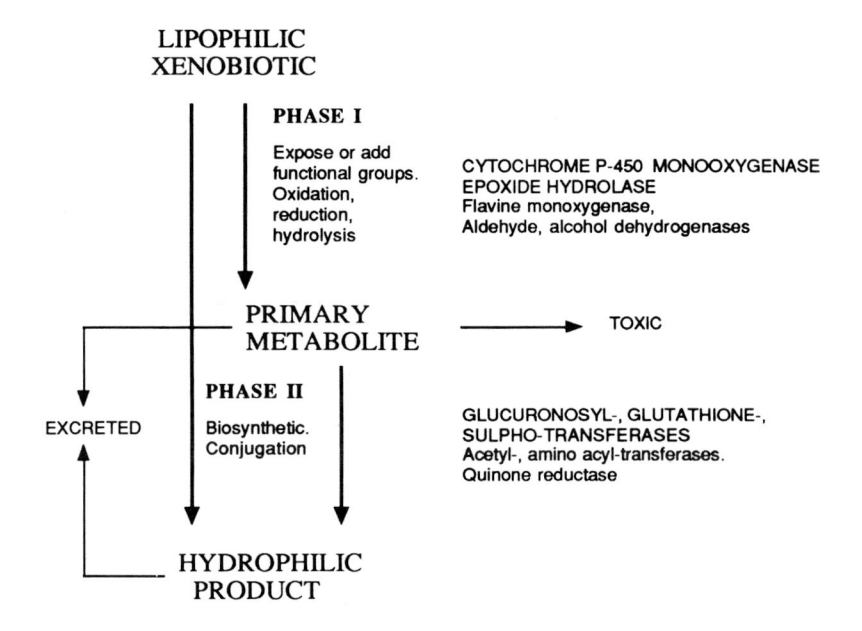

```
                    LIPOPHILIC
                    XENOBIOTIC
                         │
                         ▼
                    PHASE I
                    Expose or add         CYTOCHROME P-450 MONOOXYGENASE
                    functional groups.    EPOXIDE HYDROLASE
                    Oxidation,            Flavine monoxygenase,
                    reduction,            Aldehyde, alcohol dehydrogenases
                    hydrolysis
                         │
                         ▼
                    PRIMARY      ────────▶  TOXIC
                    METABOLITE
                         │
                    PHASE II
   EXCRETED         Biosynthetic.         GLUCURONOSYL-, GLUTATHIONE-,
                    Conjugation           SULPHO-TRANSFERASES
                                          Acetyl-, amino acyl-transferases.
                                          Quinone reductase
                         │
                         ▼
                    HYDROPHILIC
                    PRODUCT
```

Figure 1. Phase I and II pathways of xenobiotic metabolism.

understood mammalian systems to provide a conceptual framework for comparison and interpretation of data for piscine systems. Comparison of the enzymes and the scope of their reactions in fish and mammals will also provide some insights into the evolutionary changes that have taken place in the 350 million years since their divergence and the colonization of land.

II. UDP-GLUCURONOSYLTRANSFERASES

A. Mammalian UDPGTs

1. Reactions and Properties

The synthesis of glucuronides by microsomal UDP-glucuronosyltransferase (UDPGT, E.C. 2.4.1.17) is a major pathway for the inactivation and subsequent excretion of both endogenous and xenobiotic organic compounds. The enzyme catalyzes the transfer (conjugation) of glucuronic acid from the high energy nucleotide, UDP-glucuronic acid (UDPGA), to a wide variety of acceptor substrates (aglycones) to form O-glucuronides (conjugation of alcohols, phenols, and carboxylic acids), N-glucuronides (carbamates, amides, and amines), S-glucuronides (aryl mercaptans and thiocarbamates), and C-glucuronides (1-3, dicarbonyls) (Figure 2).[7,8] As with most enzymes that exhibit a broad specificity for structurally diverse compounds, multiple isoenzymes belonging to a number

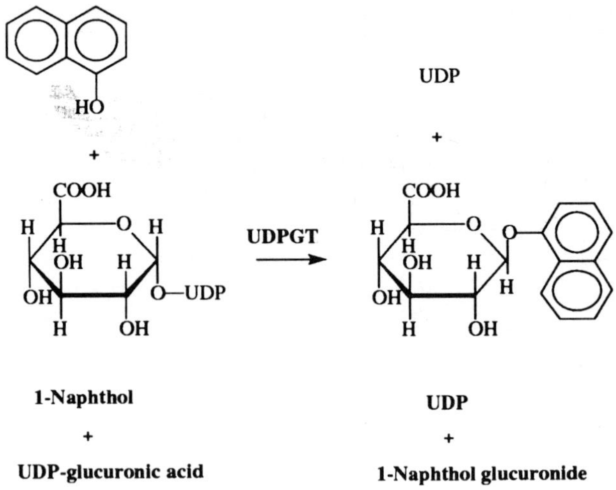

Figure 2. Glucuronidation of 1-naphthol by UDP-glucuronosyltransferase.

of multigene families are found. Several isoenzymes appear to show a fairly strict substrate specificity, while others display a degree of overlapping specificity. The heterogeneity and some characteristics of those rat hepatic UDPGTs, which have been identified by developmental, induction, purification, and cloning studies, are shown in Table 1.

Three UDPGTs are involved in the glucuronidation of steroids, displaying both positional and stereochemical specificity, and the enzyme conjugating 17β-hydroxysteroids is also active with planar aromatic phenols such as 1-naphthol and 4-nitrophenol. Two other UDPGTs are responsible for glucuronidation of bilirubin and 4-hydroxytryptamine, the latter also conjugating 4-methylumbelliferone. The endobiotic substrates for the other UDPGTs are so far unknown, although it is likely that the enzyme conjugating digitoxigenin monodigitoxide is also concerned with steroid metabolism. The UDPGTs display differential induction by xenobiotic compounds, and they also fall into four developmental clusters, which suggests differential regulation during development.[9]

2. Gene Structure and Regulation

The cDNAs of a number of rat and human UDPGTs, including those with specificities for planar phenols, bulky phenols, bilirubin (two isoforms), and steroids, have been cloned and sequenced; these cDNAs encode proteins larger than the purified enzymes (52 to 58 kDa), due to the presence of a 25 amino acid signal peptide, which is cleaved post-translationally. The COOH-terminus of the enzymes also contains a characteristic transmembrane domain, having a halt transfer signal of a hydrophobic segment, which is flanked by highly charged

Table 1. Rat Hepatic UDP-Glucuronosyltransferase Isoforms

Isoenzyme	Substrate(s)	MW (kDa)	Inducer(s)	Ontogeny
Phenol UDPGT	Hydroxy BaP 4-nitrophenol, 1-naphthol 4-methylumbelliferone, etc.	55	3-MC, BNF	Late fetal
Phenol UDPGT	1-naphthol, alkyl phenols	56	3-MC, BNF?	? Postnatal
Bilirubin UDPGT	Bilirubin	54	Clofibrate	Neonatal
17β-OH Steroid UDPGT	Testosterone (1-naphthol, 4-nitrophenol)	50	PB	Neonatal
3α-OH Steroid UDPGT	Androsterone lithocholic acid	52	PB	Postweaning
3β-OH Steroid UDPGT	Estrone, β-estradiol	?	?	Neonatal
Morphine UDPGT	Morphine	56	PB, PCN	Neonatal
5-OH Tryptamine UDPGT	5-OH tryptamine (4-methylumbelliferone)	55	?	Late fetal
4-OH Biphenyl UDPGT	4-OH biphenyl (4-methylumbelliferone)	52	PB	?
Digitoxigenin mono-digitoxide UDPGT	Digitoxigenin mono-digitoxide	?	PCN	Postweaning

Note: Identified by purification, cloning and induction studies. Activities toward other substrates are shown in parentheses. Inducers: 3-MC, 3-methylcholanthrene; BNF, β-naphthoflavone; PB, phenobarbital; PCN, pregnenolone 16α-carbonitrile.

amino acid residues. From both the sequence data and experiments with proteolytic enzymes, the accepted topology of the UDPGTs is as shown in Figure 3.[10] The asymmetrical arrangement of the phase I and II xenobiotic metabolizing enzymes is likely to account for the vectorial transport of metabolites from the cytosol to the lumen of the endoplasmic reticulum and then to the blood or bile.

cDNA sequencing studies in both rats and humans have shown that the 3-methylcholanthrene (3-MC)-inducible phenol UDPGT, another phenol UDPGT that preferentially catalyzes conjugation of halogenated and bulky aryl phenols, and two bilirubin UDPGTs all share a common highly homologous region (~95%) coding for 246 amino acids at the carboxyl terminal end of the enzyme and an amino terminal region that displays only 40 to 50% homology.[10-15] Present evidence indicates that there is only one copy of the conserved 3' region in the genome and that mRNAs for the bilirubin and the phenol UDPGTs (and probably also the digoxigenin monodigitoxoside UDPGT) are produced from a single primary transcript after an alternative splicing (Figure 4). The Gunn rat, a mutant strain of Wistar rat, which has unconjugated bilirubinemia due to the absence of hepatic bilirubin UDPGT activity, is also deficient in 3-MC-inducible phenol UDPGT and digoxigenin monodigitoxoside UDPGT activities. Molecular cloning studies have shown that the defect is due to a −1 frameshift deletion in the common 3' exon, which removes 115 amino acids from the -COOH terminus, resulting in inactive proteins.[10,16,140] The regulation of expression of this gene family appears to be quite complex, as the different isoforms are expressed at different times during development and also appear to be

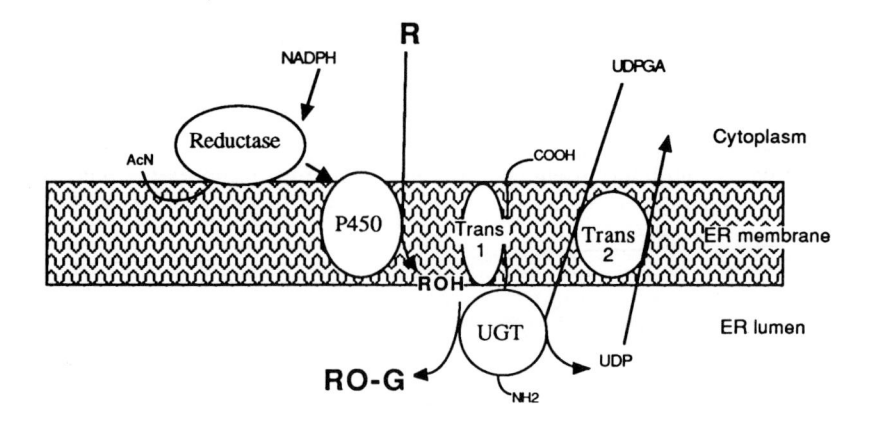

Figure 3. Topology of phase I and II biotransformation systems in the endoplasmic reticulum. The active site of UDP-glucuronosyltransferase (UGT) is in lumen of endoplasmic reticulum. UDPGA and the aglycone may be transported from cytoplasm by membrane translocases, or the substrate may be produced by cytochrome P450-dependent monooxygenase (P450) and associated reductase. Glucuronides produced are excreted via the bile or blood.[10]

differentially regulated by xenobiotics. Phenol UDPGT is induced by PAHs, inferring the presence of the Ah receptor.[57] Bilirubin UDPGT is refractive to PAH treatment; one isoform is specifically induced by hypolipidemic drugs such as clofibrate implying the presence of elements responsive to peroxisomal proliferators in the control region of the gene. At least one bilirubin UDPGT is

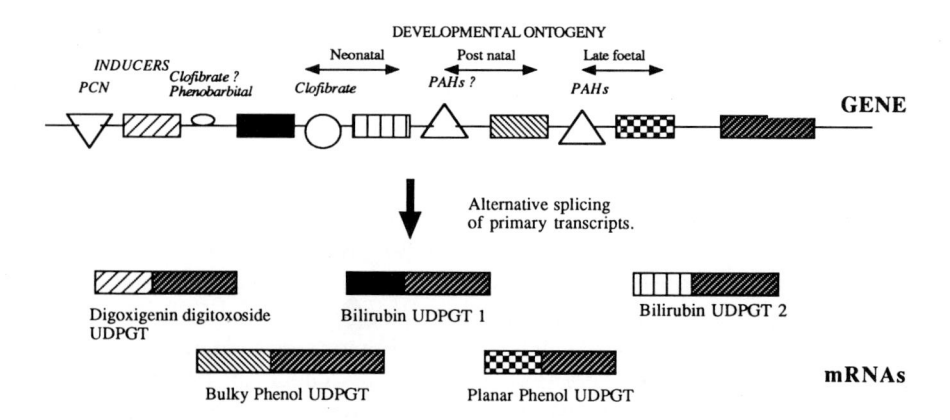

Figure 4. Possible organization of the rat phenol/bilirubin gene family. The UDPGTs of this gene family share a common 3' region of 738 nucleotides. Alternative splicing of the 5' end of the gene transcripts could produce five different mRNA molecules, coding for the different isoforms. Differential regulation by xenobiotics and expression at different developmental stages infer the presence of distinct regulatory elements for each enzyme.

induced by phenobarbital, while digitoxigenin monodigitoxide UDPGT activity is induced by the synthetic steroid, pregnenolone16α-carbonitrile, PCN (see Table 1 and Wooster et al.[15]). Current studies indicate that the bilirubin/phenol UDPGT gene may be very large, so that elucidation of its structure and the function of its regulatory elements presents an interesting and important challenge. Other UDPGT isoforms are also inducible by xenobiotics; PCN induces the glucuronidation of both morphine and digitoxigenin monodigitoxide, and phenobarbital-type inducers stimulate glucuronidation of morphine, 3α-hydroxy and 17β-hydroxy steroids, as well as 4-hydroxy biphenyl.

B. Piscine UDPGTs

1. Scope of Reactions

Most investigations of fish have been concerned with the identification of the glucuronides of endobiotic and xenobiotic compounds in vivo (reviewed by Clarke et al.).[17] Mono- and diglucuronide conjugates of bilirubin have been identified in the bile of several fish species, including two elasmobranchs, the dogfish *(Squalus acanthias)*[18] and the skate *(Raja erinacea)*,[18,19] and various teleosts, including the eel *(Anguilla japonica)*,[20] Atlantic salmon *(Salmo salar)*,[21,22] winter flounder *(Pseudopleuronectes americanus)*,[19] and plaice (*Pleuronectes platessa*).[23,24] Steroid hormone glucuronides formed in the testes and seminal vesicles of fish appear to be important pheromones, acting as stimulants of ovulation and sperm production and also enhancing vitellogenesis in many species (see syntheses by Schoonen et al.[25] and Lambert and Resinck).[26] Extensive studies of hormone biosynthesis in these organs have been carried out, especially by the group at the University of Utrecht, and glucuronides of 3α-, 3β-, and 17β-hydroxy steroids and their metabolites have been identified.[27–29] 17β-Estradiol glucuronide has been identified in trout,[30] and following injection of 17β-estradiol, biliary elimination of its glucuronide conjugate was demonstrated in English sole.[31] Formation of glucuronides may also be important in the termination of thyroid hormone action; biliary T4 glucuronide has been identified in some ten freshwater teleosts, and following intraperitoneal injection of thyroid hormones, glucuronides of thyroxine (T4) and triodothyronine (T3) were identified in the bile of brook trout.[32]

As found in mammals, glucuronidation is clearly an extremely important pathway for detoxication and excretion of xenobiotic compounds in fish. Glucuronide conjugates of a structurally diverse range of xenobiotic compounds have been detected in bile, urine, and various tissues (see Clarke et al.).[17] These include an extensive range of phenols, including phenol and pentachlorophenol, the lampreycide trifluoromethylnitrophenol, chlorophenolics present in paper pulp effluents, and phenolic metabolites of BaP, naphthalene, phenanthrene, di-2-ethyl hexylphthalate, and aflatoxin B_1. The majority of these conjugates are O-glucuronides, although N-glucuronides of a carbamate insecticide and a phenylthiazole fish anesthetic have also been identified in rainbow trout. To my knowledge C- and S-glucuronide formation has not been studied in fish.

Compared with the large number of compounds that are known to be glucuronidated in vivo, the glucuronidation of relatively few compounds has been studied in vitro. Due to the ease of assay, most studies have utilized 4-nitrophenol as acceptor substrate, and hepatic UDPGT activity with this compound has been established in a number of teleost species, including bream *(Abramis brama)*, carp *(Cyprinus carpio)*, channel catfish *(Ictalurus punctatus)*, perch *(Perca fluviatilis)*, roach *(Rutilus rutilus)*, and pike *(Esox lucius)*,[33-35] the salmonids, brown trout *(S. trutta)*, rainbow trout *(Onchorynchus mykiss)*, vendace *(Coregonus albula)*, and Atlantic salmon *(S. salar)*,[33,36-38] and two marine species, cod *(Gadus morhua)* and plaice *(Pleuronectes platessa)*[23,39] (Table 2).

Activities toward several other phenolic substrates have also been measured in hepatic microsomes from fish. Thus, the capacity to glucuronidate the lampreycide tetrafluoromethylnitrophenol,[40] 1-naphthol,[36,41,42] and phenolphthalein[41] has been detected in rainbow trout liver microsomes, while 4-methylumbelliferone UDPGT activity has been reported in channel catfish and bluegill[43] and glucuronidation of 1-naphthol, 2-aminophenol, and phenolphthalein has been reported in plaice.[23,24] Glucuronidation of bilirubin has been detected in plaice,[23,24] salmon,[21,22] and trout[37] microsomes and that of testosterone in catfish,[30] goldfish,[29] plaice,[23,24] trout,[36,37,49] and zebrafish.[38]

2. Properties and Relative Activities of Fish UDPGT

In vitro studies have shown that UDPGT in fish liver possesses many of the same properties as the mammalian enzymes. The activity is microsomal,[44] and maximal expression of activity is obtained at 37°C at neutral pH and in the presence of Mg^{2+}.[24] In vitro, activity in native fish microsomes displays the latency characteristic of mammalian microsomes, being removed by treating microsomes with detergents or by physical treatment, which destroys vesicle structure.[23,24,36,42,44,45] Compared with mammals, the degree of latency of fish hepatic UDPGT activity is somewhat lower,[24,36,42,44,45] as shown in Figure 5A, where hepatic phenol UDPGT activity was measured in rat and plaice microsomes under identical assay conditions. Curiously, the latency of plaice phenol UDPGT activity appears to vary with the tissue tested (Figure 5B).[24,46] The physiological significance of these results, if any, is not known. It is possible that these apparent differences are purely artifactual, since in mammals it has been found that latency is partially dependent upon substrate concentrations and thus may be a feature of assay conditions (which therefore makes interlaboratory comparisons of activities difficult). However, it has also been postulated that the decreased latency observed in fetal microsomes may be attributed to increased membrane fluidity[47] and kinetic data now indicate that there may be membrane translocases for both the aglycone and UDPGA. Thus,

Table 2. Hepatic Microsomal 4-Nitrophenol UDPGT Activity in Fish Species

Species		Activity, nmol. min^{-1}mg^{-1}	Ref.
Bream	*Abramis brama*	0.06	136
Brown trout	*Salmo trutta*	0.55	136
Carp	*Cyprinus carpio*	0.06[a]	137
Channel catfish	*Ictalurus punctatus*	0.8[a]	35
Cod	*Gadus morhua*	0.95[a,b]	134
Perch	*Perca*	0.09	33
Pike	*Esox lucius*	0.03	136
Plaice	*Pleuronectes platessa*	14[a,b]	23
Rainbow trout	*Onchorynchus mykiss*	0.18–0.66	33, 36, 37, 157
Roach	*Rutilus rutilus*	0.07–0.11	33, 157
Vendace	*Corygonus alba*	0.07	33, 157

[a] Measured in detergent disrupted microsomes.
[b] Activities expressed as nmol min^{-1} mg microsomal protein^{-1} corrected to 18°C for different assay temperatures using a Q^{10} of 2.

membrane structure and kinetic properties of the three protein components will all be factors controlling apparent reaction characteristics.

3. Substrate Specificity of Piscine Hepatic UDPGTs In Vitro

By using a variety of aglycone substrates, which are conjugated by different mammalian UDPGT isoforms (Table 1), studies in Klaasen's laboratory on rainbow trout[41] and in my own laboratory on plaice[23,41,48] have demonstrated a similar broad substrate specificity to mammals, a finding that infers the presence of multiple isoforms in fish (Table 3). Comparison of the relative activities toward these substrates shows that for nearly all substrates tested, the activities are considerably lower in the trout, while those in the plaice are generally as high or higher than in

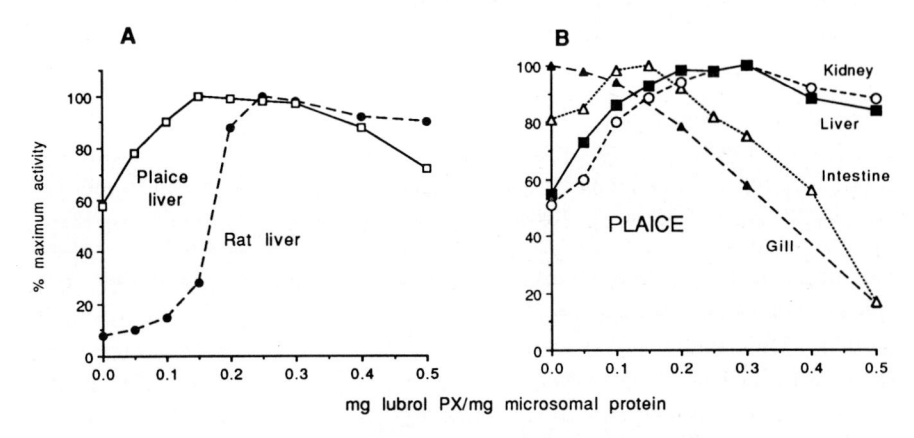

Figure 5. Latency of phenol UDPGT activity in different plaice tissues. (A) Detergent/activity curves for rat and plaice liver microsomes. (B) Detergent/activity curves for microsomes from different tissues of plaice.

Table 3. Aglycone Specificity of Glucuronidation in Trout, Rat, and Plaice Hepatic Microsomes

Substrate	Activity, nmol min⁻¹ mg microsomal protein⁻¹ at 37°C		
	Plaice[a]	Rainbow trout[b]	Rat[b]
1-Naphthol	14.3 ± 5.4	0.43 ± 0.02	7.76 ± 0.53
4-Nitrophenol	56.3 ± 18.5	0.36 ± 0.04	5.13 ± 0.76
Phenolphthalein	3.8 ± 1.2	0.06 ± 0.01	3.8 ± 1.2
2-Aminophenol	0.15 ± 0.05	—	—
(−)-Morphine	0.10 ± 0.04	0.09 ± 0.01	0.47 ± 0.04
Androsterone	0.03 ± 0.01	—	—
Estradiol	0.17 ± 0.05	—	—
Estrone	0.16 ± 0.06	0.02 ± 0.003	0.008 ± 0.001
Testosterone	0.18 ± 0.08	0.06 ± 0.01	0.013 ± 0.01
Bilirubin	0.11 ± 0.03	0.03 ± 0.01	0.06 ± 0.01
Chloramphenicol	—	0.025 ± 0.003	0.21 ± 0.02
Diethylstilbesterol	—	0.1 ± 0.01	0.39 ± 0.05
Digitoxigenin-mono-digitoxide	—	0.001 ± 0.0003	0.007 ± 0.001

[a] From Clarke et al.[24]
[b] From Gregus et al.[37] (raw data kindly supplied by J.B. Watkins, III).

the rat. Additional evidence for the multiplicity of piscine UDPGTs is derived from tissue expression purification, induction, and immunological studies.

4. Tissue Expression

While liver is quantitatively the most important site for glucuronidation of xenobiotics in fish, significant activity has also been detected in many other extrahepatic tissues, including kidney, gills, and intestine of several species[23,33,46] (Table 4). Interestingly, the relative tissue distribution appears to vary between species, with liver displaying the highest specific activity toward phenols in plaice[46] and trout.[33] Gills appear to be the most active in vendace and roach,[3] while intestinal activity is relatively low compared with the liver and kidney in plaice.[46] There are also tissue-specific differences in the expression of UDPGT activities involved in endobiotic metabolism. Testosterone UDPGT activity is present in liver,[25,36,46] and it has also been found in the testes of rainbow trout[49] and the intestine of plaice,[25,46] while synthesis of glucuronide metabolites of pregnenolone and androstenedione has been demonstrated in testicular tissues in vitro.[50] Unlike rodents, where bilirubin UDPGT is present in both liver and kidney,[7,9] glucuronidation of bilirubin was only detected in the livers of both plaice[46] and salmon.[22] Thus, both mammals and fish there exhibit a tissue-specific expression of different UDPGT activities, indicating a multiplicity of UDPGT isoforms.

5. Purification and Characterization of Plaice UDPGTs

At least six immunoreactive peptides can be visualized in plaice microsomes blotted with mammalian UDPGT antisera,[23] indicating a multiplicity of UDPGTs

Table 4. Tissue Distribution of Piscine UDPGTS

	Species					
	Rainbow trout[a]	Vendace[a]	Roach[a]	Plaice[b]	Plaice[b]	Plaice[b]
	Activities, pmol/min/mg microsomes					
Tissue	4-Nitrophenol			1-Naphthol	Testosterone	Bilirubin
Liver	380	70	110	17,000	240	80
Kidney	60	90	160	6,000	nd	nd
Gill	70	130	240	1,000	nd	nd
Intestine	250	90	75	1,000	50	nd

Note: nd = not detected.

[a] From reference 33, determined at 18°C with 4-nitrophenol as substrate.
[b] From reference 46, determined at 37°C.

with common structural motifs (Figure 6a). By ion exchange chromatography of solubilized plaice microsomes, we have partially resolved multiple isoforms with differing substrate specificities (bilirubin, phenol, 3α-, 3β-, and 17β-hydroxy steroids).[48] An antiserum raised against this copurified preparation recognizes multiple polypeptides of 52 to 57 kDa in plaice hepatic microsomes.[46] Recently, we have obtained highly purified preparations of bilirubin UDPGT and 17β-hydroxy steroid (testosterone) UDPGTs from plaice liver and have purified a phenol UDPGT to apparent homogeneity.[51] The purified 55-kDa phenol UDPGT, reconstituted with exogenous phosphatidyl choline, has a high activity toward phenols and no measurable activity with bilirubin or steroids. Immunoblot studies have shown that it is also expressed in other tissues, including intestine, kidney, and gills. In rats, both hepatic and renal phenol UDPGT are induced by β-naphthoflavone, whereas only the hepatic enzyme is induced by 3-MC. Similarly in fish, while 3-MC only induces the hepatic enzyme in plaice,[46] Pesonen et al.[52] reported induction of both hepatic and renal 4-nitrophenol UDPGT activity in β-naphthoflavone-treated trout. Possible explanations for the lack of effectiveness of 3-MC as an inducer in kidney are the lack of transport to this organ or that a metabolite is the active inducer and that this is only produced in the liver. Neither postulate is entirely satisfactory unless different critical threshold concentrations are required for upregulation of phenol UDPGTs compared with cytochrome P4501A1, since 3-MC treatment induces this enzyme in both liver and kidney of both rats and fish.[53,54] Immunochemical analysis of liver microsomes from xenobiotic-treated plaice using an anti-plaice UDPGT antibody demonstrated that increased amounts of the 55-kDa polypeptide correlated with the induction of phenol UDPGT activity in fish treated with a polyaromatic hydrocarbon (3-MC) and the polychlorinated biphenyl mixture Aroclor 1254.[46] Aroclor also induces renal phenol UDPGT expression (Figure 6b), further complicating a mechanistic explanation for the lack of 3-MC induction in this tissue. These studies have therefore shown that plaice possess an orthologous 3-MC inducible phenol UDPGT

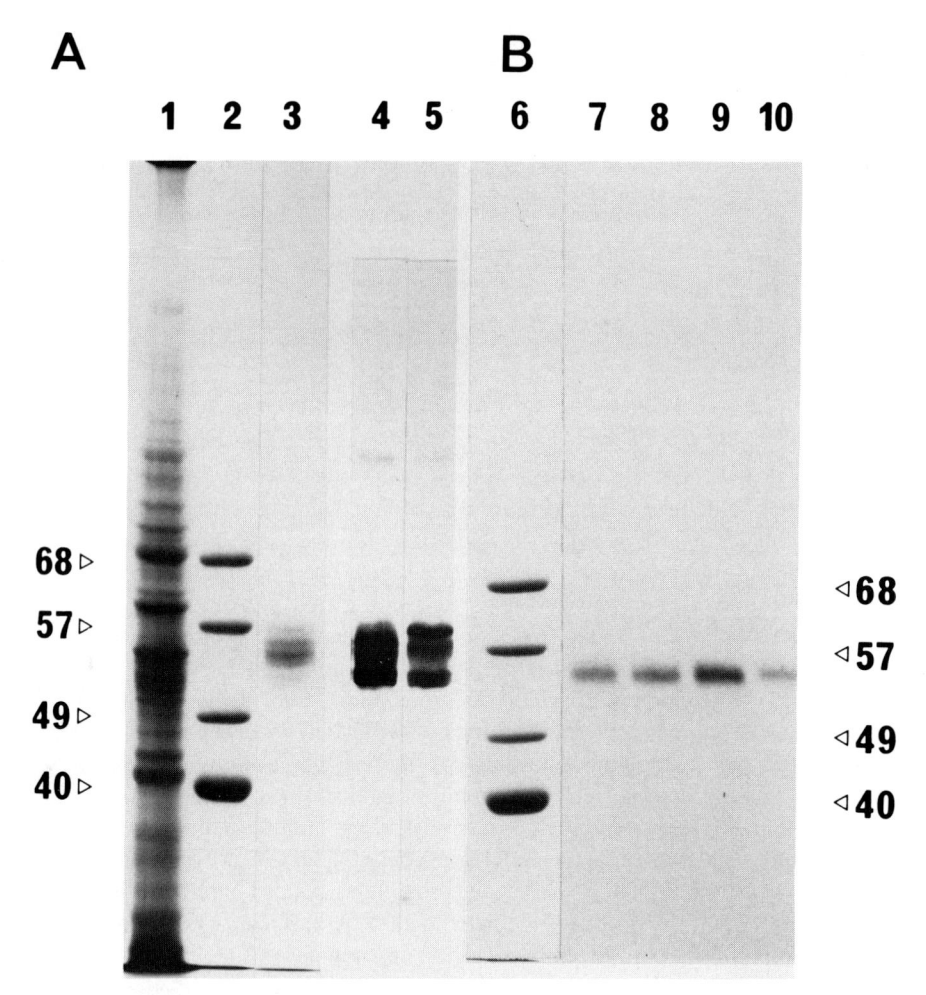

Figure 6. (A) Immunoblot analysis of copurified plaice liver UDPGTs with mammalian anti-UDPGT sera. Samples were fractionated by SDS-PAGE using 7% acrylamide gels. Lanes 1 to 3 are stained for protein. Lane 1 shows plaice microsomes, lane 2 molecular weight standards, lanes 3 to 5 co-purified plaice bilirubin, phenol and steroid UDPGTs. Lanes 4 and 5 are transferred to nitrocellulose and immunoblotted with anti-rat liver testosterone/4-nitrophenol UDPGT (lane 4) or anti-rat kidney bilirubin/1-naphthol UDPGT (lane 5) antibodies using alkaline phosphatase-labeled secondary antibodies for detection. (B) Immunoblot analysis of plaice renal microsomes from fish treated with various inducers using anti-plaice UDPGT antiserum. Lane 6 shows molecular weight standards, lanes 7 to 10 are immunoblotted, lane 7 shows control fish microsomes. Lane 8 shows 3-MC-treated fish, lane 9 shows Arochlor 1254™-treated fish, lane 10 shows clofibrate-treated fish kidney microsomes.

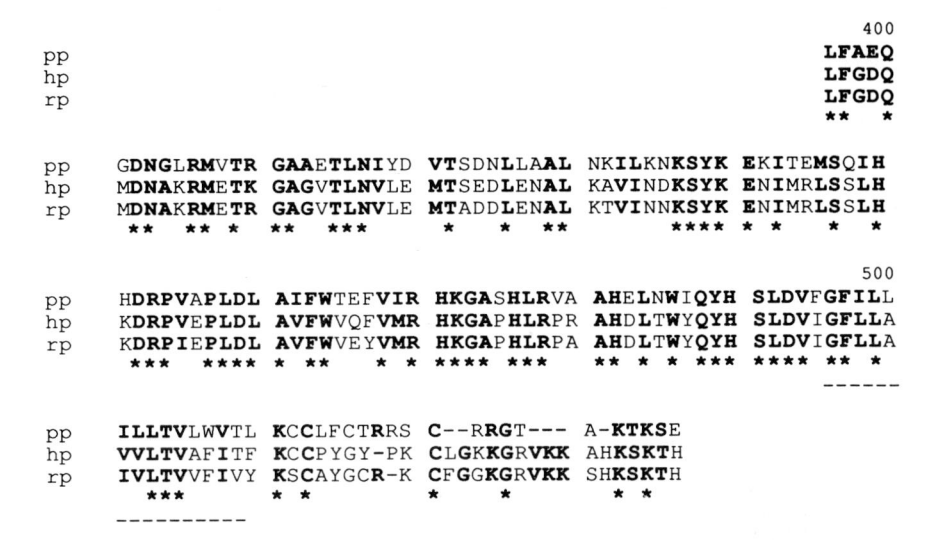

```
                                                                400
pp                                                              LFAEQ
hp                                                              LFGDQ
rp                                                              LFGDQ
                                                                **  *


pp    GDNGLRMVTR  GAAETLNIYD  VTSDNLLAAL  NKILKNKSYK  EKITEMSQIH
hp    MDNAKRMETK  GAGVTLNVLE  MTSEDLENAL  KAVINDKSYK  ENIMRLSSLH
rp    MDNAKRMETR  GAGVTLNVLE  MTADDLENAL  KTVINNKSYK  ENIMRLSSLH
      **   ** *   **   ***     *   *  **  ****  * *    *    *

                                                                500
pp    HDRPVAPLDL  AIFWTEFVIR  HKGASHLRVA  AHELNWIQYH  SLDVFGFILL
hp    KDRPVEPLDL  AVFWVQFVMR  HKGAPHLRPR  AHDLTWYQYH  SLDVIGFLLA
rp    KDRPIEPLDL  AVFWVEYVMR  HKGAPHLRPA  AHDLTWYQYH  SLDVIGFLLA
      ***   ****    *  **     *  *  ****  ***   ** *  *  ***   **** ** *
                                                                ------

pp    ILLTVLWVTL  KCCLFCTRRS  C--RRGT---  A-KTKSE
hp    VVLTVAFITF  KCCPYGY-PK  CLGKKGRVKK  AHKSKTH
rp    IVLTVVFIVY  KSCAYGCR-K  CFGGKGRVKK  SHKSKTH
      ***          *  *         *    *      *  *
      ----------
```

Figure 7. Carboxyl terminal amino acid sequence homologies of plaice and mammalian phenol UDPGTs. The deduced amino acid sequences are derived from the respective cDNAs. pp, Plaice (this lab, EMBL #); hp, Human;[11] rp, Rat.[10] The conserved exon 5 of the human *UGT1* gene is shown. Homologous amino acids shown in bold; conserved amino acids asterisked; putative transmembrane region underlined.

with comparable activity and which exhibits common structural motifs to the mammalian enzyme.

Very recently we have isolated a 2-kB UDPGT cDNA clone from a 3-MC-induced plaice liver cDNA library (EMBL data access in # X74116), which displays a high degree of homology with mammalian phenol UDPGT isoforms. Comparison of the deduced amino acid sequence of the carboxyl-terminal region of the plaice UDPGT (i.e., corresponding to the conserved exon 5 of the human UGT1 gene) with mammalian phenol UDPGTs shows over 60% sequence homology (Figure 7). It is interesting that the "cytoplasmic tails" (see Figure 3) are markedly different; there are about 60% homology in the transmembrane region and some 70% homology of the remainder of this region, indicating a very high degree of structural conservation throughout the vertebrates. While definitive assignment of this clone as the 56 kDa 3-MC-inducible phenol UDPGT we purified previously will require expression of a full length clone, Northern blot analysis with the isolated cDNA as probe has shown that there is a three-fold increase in a 2.5-kB mRNA in liver of a 3-MC-induced plaice compared with control fish, indicating that this is the isoform we have cloned.

We have purified the plaice bilirubin and steroid UDPGTs some 300-fold; full characterization has not been possible, as these preparations were not

Table 5. Inferred Characteristics of Plaice UDPGTS

Isoform	M_r, kDa	Substrates	Induction	Tissues
Phenol UDPGT	56	1-Naphthol, 4-nitrophenol	3-MC, ARO	L, K, G, I
Bilirubin UDPGT	57	Bilirubin, bilirubin glucuronide	Not Clofibrate	L
Steroid UDPGTs	52, 53	Testosterone (1-naphthol, 4-NP)	—	L, I
	52, 53	Androsterone	—	L
	52, 53	Estradiol	—	L

Source: From Clarke et al.[23,24,46,48,51]

Note: 3-MC = 3-methylcholanthrene; ARO = Aroclor 1254; G = gills; K = kidney; I = intestine L = liver.

homogeneous. However, immunoblotting with anti-plaice UDPGT and anti-rat UDPGT antisera, tissue distribution and induction studies indicate that the 57-kDa polypeptide is probably a bilirubin UDPGT. Interestingly, unlike the mammalian enzyme, we have not been able to demonstrate induction of plaice bilirubin UDPGT activity by treatment with the hypolipidemic drug clofibrate. Multiple hydroxysteroid steroid UDPGTs of 52 to 53 kDa appear to be expressed in plaice; one also appears to display phenol UDPGT activity like its mammalian orthologue, the 17β-hydroxy steroid (testosterone) UDPGT.[24,46,51] The postulated characteristics of the plaice UDPGTs are summarized in Table 5.

III. SULFOTRANSFERASES

A. Reactions

The sulfotransferases (ST, E.C. 2.8.2.1.) are cytosolic enzymes that conjugate hydroxyl groups of polyaromatic compounds, aliphatic alcohols, aromatic amines, and hydroxylamines with activated 3′-phosphoadenosine 5′ phosphosulfate (PAPS) to form a sulfate monoester (sulfonate or ethereal sulfate) (Figure 8).[55] In mammals, four classes of ST have been identified, displaying specificities for (1) phenolic steroids (estrone ST), (2) hydroxy steroids, (3) bile salts, and (4) phenols, catechols, and hydroxylamines (aryl ST).

Both glucuronide and sulfate conjugates can be formed with many xenobiotic compounds, (e.g., 1-naphthol), so that there are two potential pathways for detoxication and excretion of these compounds. In studies with isolated rat hepatocytes using prototypical substrates, Koster et al.[56] demonstrated that, at low doses, sulfation is the preferred pathway, while glucuronidation predominates at higher dose levels. While it has been suggested that this may be due to differing K_m values for the two enzymes, determination of K_m for isolated enzymes and apparent K_m in intact cells yields differing values.[57] Sulfates and glucuronides of MW <250 are generally excreted in urine, while the biliary route is favored for glucuronides of MW >350.

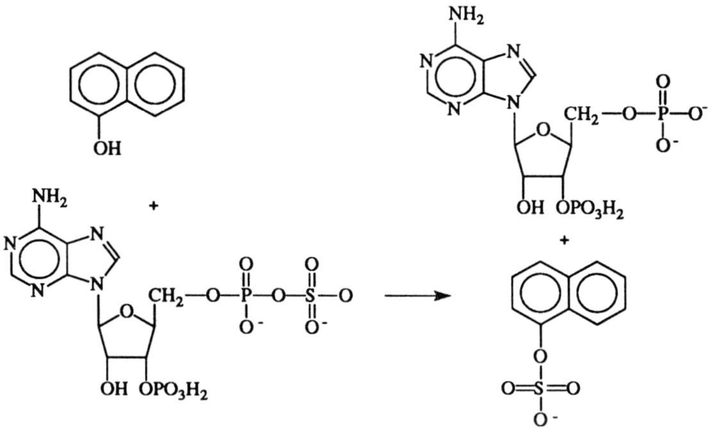

1-Naphthol
+
5'-Phosphoadenosine-5'-phosphosulphate

Phosphoadenosinephosphate
+
1-Naphthol sulphate

Figure 8. Sulfation of 1-naphthol by arylsulfotransferase.

B. Piscine Sulphotransferases

The relative importance of piscine ST in xenobiotic detoxication is difficult to assess from the comparatively few studies reported in the literature and the apparently conflicting results obtained. Clearly, marked species differences are present. Layiwola and Linnecar[58] reported high ST activity toward waterborne phenol in a number of species including bream *(A. brama),* goldfish *(Carassius auratus),* guppy *(Poecilia reticulata),* minnow *(Phoxinus phoxinus),* roach *(R. rutilis),* and tench *(Tinca tinca).* These findings were confirmed by Kobayashi et al.[34] using liver slices of carp *(Cyprinus carpio)* and goldfish. In southern flounder *(Paralicthys lethostigma),* Pritchard and Bend[59] reported greater secretion rates for sulfate conjugates of the ultimate carcinogen 9,10-epoxy-7,8-dihydroxybenzo(a)pyrene than for glucuronidated metabolites; however, several other studies with flatfish, including English sole *(Parophys vetulus),*[5,60] starry flounder *(Platicthys stellatus),*[5] and plaice *(Pleuronectes platessa),*[61,62] carried out both in vivo and with isolated hepatocytes, have shown that quantitatively, sulfates are very minor metabolites of PAHs such as naphthalene (1-naphthol) and BaP in these species. Other studies have also reported very little or no measurable ST activity toward other typical substrates of the mammalian enzymes, including compounds such as acetaminophen, 7-ethoxycoumarin, and pentachlorophenol in rainbow trout *(O. mykiss)*[34,37,63,64] and phenol in perch

(Perca fluviatilis).[58] Indeed, in a comparison of trout, a bird, and mammals, Gregus et al.[37] reported comparable ST activity toward the bile acid taurolithochloate and two- to sevenfold lower activities toward steroids and 2-naphthol in trout compared with the rat. In the study of Morrison et al.[61] using isolated hepatocytes of plaice, a species where activity of aryl ST is comparatively low,[38] sulfation became saturated at very low concentrations of 1-naphthol, and glucuronidation predominated at all substrate concentrations tested.

The present data are thus consistent with the conclusion that glucuronidation is the more important excretory pathway in salmonids and pleuronectid flatfish, where aryl ST activity is low. However, further metabolic studies using isolated hepatocytes and characterization of piscine aryl ST are required before the functional significance of sulfation in fish can be truly assessed.

IV. GLUTATHIONE S-TRANSFERASES

A. Reactions and Occurrence

The glutathione S-transferases (GSTs, E.C. 2.5.1.18) are a multigene superfamily of dimeric, multifunctional, primarily soluble enzymes, which occur ubiquitously, having been identified in prokaryotes, yeasts, higher plants, molluscs, crustaceans, insects, fish, amphibians, and mammals.[65-67] The GSTs are believed to play major roles in:

1. Biosynthesis of vasoactive metabolites of arachidonic acid (leukotrienes and prostaglandins) and isomerization of steroids (e.g., Δ5-androsterone 1,4 dione)
2. Se-independent GSH peroxidase activity toward organic peroxides, including lipid, steroid, and DNA hydroperoxides
3. Intercellular transport of endogenous compounds, such as heme, bilirubin, and steroid hormones
4. Irreversible binding of reactive electrophilic xenobiotics, such as azo dye carcinogens and PAHs
5. Catalysis of glutathione (GSH) conjugation with electrophilic centers in a variety of compounds (e.g., epoxides) as the first step in the formation of the excretory metabolites, the mercapturic acids (Figure 8)

With the exception of the θ-class enzymes, all other GSTs conjugate the artificial substrate 1-chloro-2,4-dinitrobenzene (CDNB).

The GSTs, therefore, play an essential role in protecting the organism from peroxidative damage and in the cellular transport and detoxication of both endogenous compounds[66,68] and reactive xenobiotics.[69] They have also been implicated in acquired drug and pesticide resistance,[70,71] while ethoxyquin-induced resistance to the carcinogenic mycotoxin Aflatoxin B_1 has been related to induction of a GST with high activity toward Aflatoxin B_1 8,9-epoxide.[141,142] Thus, toxicity of many exogenous compounds can be modulated by induction of GSTs.

B. Properties of Mammalian GSTs

The GSTs have been most extensively studied in terrestrial mammals, especially rats, and before current knowledge of the enzymology and molecular biology of piscine glutathione S-transferases is discussed, some important characteristics of the mammalian enzymes and their genes will be reviewed.

With the exception of a single microsomal enzyme, characterized by Morgenstern, et al.[144-147] the GSTs are soluble, dimeric proteins comprising identical or different subunits of 25 to 28 kDa.[65-67,71] They are particularly abundant in liver, constituting as much as 10% of the cytosolic protein in the rat. Multiple cytosolic GST isoenzymes have been purified and several of their genes cloned from the rat, mouse, and human (see reviews by Board[73] and Pickett).[74,75] On the basis of substrate specificity, immunological cross-reactivity and protein sequence data, the soluble GSTs have been grouped into four classes: α, μ, π, and θ.[67,76] Subunits from the same class form heterodimers, share 75 to 95% sequence identity, and have cDNAs that cross-hybridize at high stringency. Subunits from different classes do not form heterodimers. There is no cross-hybridization of cDNAs, and only 25 to 30% sequence identity is observed.[77] At least 13 subunits of cytosolic GST have been described in the rat, and 19 dimeric forms formed by cross-dimerization of subunits have so far been identified. Each subunit contains an active site, and the dimeric enzyme displays the combined catalytic properties of both subunits.[66] The classification, nomenclature, subunit structure, molecular weights, and characteristic activities of the rat GSTs, for which there are most data, are summarized in Table 6. A more detailed discussion follows.

1. α-Class

This consists in the rat of subunits 1, 2, 8, and 10, which characteristically have neutral/basic isoelectric points.

a. Substrate Specificity. They are the most active class in conjugating B(a)P quinones. Subunit 1 also conjugates cholesterol α-epoxides and binds nonsubstrate ligands, including bile acids (e.g., lithocholic acid), heme derivatives (hematin and bilirubin), and very reactive electrophilic xenobiotics such as bromosulphalein and azo dye carcinogens. From the latter characteristic, the 1-1 and 1-2 dimers were originally termed "ligandin."[65,68] Subunits 1 and 2 are also involved in prostaglandin (PGH_2) metabolism, catalyzing the reduction of PGH_2 to $PGF_{2\alpha}$ and the isomerization of PGH_2 to PGE_2 and PGD_2.[78] Subunit 2 is a highly active, selenium-independent GSH peroxidase, which acts on organic hydroperoxides,[79] while subunit 8 is the most active GST in conjugating 4-hydroxyalken-2-enals, products of lipid peroxidation.[80] Subunit 10 also has high GSH peroxidase activity with cumene hydroperoxide.[81,82] It has been proposed that a molecular variant of subunit 2 (Yc_2), with high catalytic activity toward Aflatoxin B_1 8,9-epoxide and inducible by ethoxyquin (via an antioxidant

Table 6. Summarized Characteristics of Rat Cytosolic GST Subunits

Class	Subunit	MW, kDa	Inducers	Activities in addition to CDNB conjugation	Tissues
Alpha	1	25.5	MC, tSOX, PB, tBHQ, H$_2$O$_2$	Cholesterol epoxide; aflatoxin B$_1$ 8,9 oxide; androsterone 3,17-dione and prostaglandin H2 isomerase; *ligandin*	L, K, sl
	2	28.5	MC, PB	Aflatoxin B$_1$ 8,9 oxide; *peroxidase* (Cumene hydroperoxide)	L, K, Lu, S, T
	8	25		*ETHA;* NBC; 4-hydroxynonenal	Lu, K, S, sl, C (T, L)
	10	25.7		*Peroxidase* (cumene hydroperoxide)	K, nL
Mu	3	27	PB, MC, tSOX	*BSP;* DCNB; NBC; B(a)P 4,5 oxide	L, C, H, T, (Lu, S, sl)
	4	27	PB, MC, tSOX	NBC, 4-hydroxy dodecenal, (4-OH nonenal); *BPDE;* leukotriene A4	L, C, K, Lu
	6	26.5		(Very high activity with CDNB)	
	9	26			
	11	26.5		4-OH nonenal, NBC, DCNB	T, B
Pi	7	24.8	AAF, DEN	DNA hydroperoxides; *acrolein;* BPDE (no ENPP, ETHA, DCNB, CuOOH)	K, B, Lu, C, S
Theta	5	26.3		(No CDNB) *ENPP;* B(a)P 4,5 oxide	L, T
	12	?		? (Very labile)	
	Yrs	27.3		(No CDNB) *ETHA*, tPBO, *menaphthyl sulfate,* cumene hydroperoxide	L, T, K, Lu, B

Inducers: AAF, 2-acetylaminofluorene; DEN, diethylnitrosamine; MC, 3-methycholanthrene; PB, phenobarbital; tBHQ, tert-butylhydroquinone; tSOX, trans-stilbene oxide.

Substrates: B(a)P, benzo(a)pyrene; BPDE, benzo(a)pyrene diol epoxide; BSP, bromosulfophthalein; CDNB, 1-chloro-2,4-dinitrobenzene; DCNB, 1,2-dichloro-4-nitrobenzene; ENPP, 1,2-epoxy-3-(p-phenoxy)propane; ETHA, ethacrynic acid; NBC, p-nitrobenzyl chloride; tPBO, trans-4-phenyl-3-buten-2one. Italicised activities may be classed as diagnostic for these isoforms.

Tissues: B, brain; C, colon; H, heart; K, kidney; L, liver; Lu, lung; nL, neonatal liver; S, spleen; sl, small intestine. Tissues with low concentrations in parentheses.

responsive element, ARE), is responsible for resistance to Aflatoxin B$_1$-induced carcinogenicity in rats.[141] In mice that are resistant to aflatoxin, an isoform (Ya$_3$) with this activity is expressed constitutively.

b. Gene Structure and Regulation. The α-class GSTs are a multigene family of enzymes. Three different subunit 1 cDNA clones have been sequenced (coding for 222 amino acids), exhibiting eight amino acid coding differences (which may merely represent allelic variations), and on Southern blot analysis there appear to be five subunit 1 genes. Pickett's group has sequenced one of these genes, which spans 11 kB and comprises 7 exons. It has multiple regulatory elements (Figure 9), a basal level element, a xenobiotic regulatory element, homologous to the Ah receptor in the cytochrome P4501A1 gene,[57] which is responsive to PAHs such as 3-MC and β-naphthoflavone, and also an ARE, responsive to radical-generating

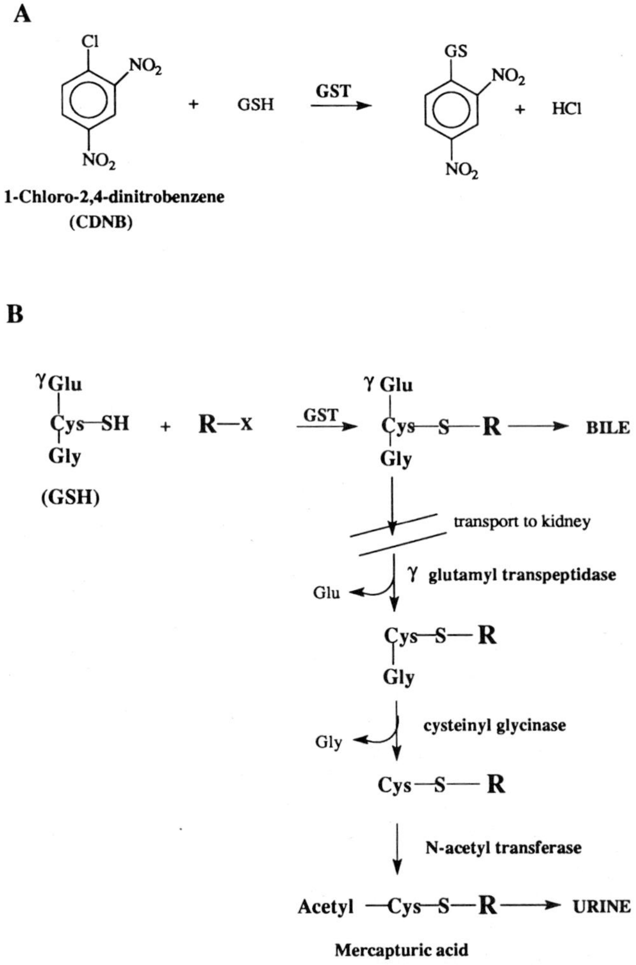

Figure 9. Glutathione S-transferase-catalyzed conjugation of 1-chloro-2,4-dinitrobenzene with glutathione (A), and (B) the mercapturic acid pathway for urinary excretion of GSH conjugates.

compounds such as *tert*-butylhydroquinone, *trans*-stilbene oxide, and hydrogen peroxide.[83,84] Phenobarbital also increases subunit 1 mRNA transcription some four- to sevenfold.[85] The structural genes of the other α-class GSTs have not been characterized. The cDNA for subunit 2 codes for a 221 amino acid protein (with 68% homology to the subunit 1 protein), and Southern blot analysis indicates three genes in the rat genome. It is probable that this finding is due to cross-hybridization of highly homologous sequences, such as the ethoxyquin-inducible isoform (Yc$_2$), which shows over 90% amino acid sequence homology for a number of

CNBr peptides with that of subunit 2.[141] Phenobarbital and 3-MC treatment only produce a slight increase (~1.5-fold) in subunit 2 (Yc$_1$) mRNA transcription,[74] while as noted above, antioxidants induce the Yc$_2$ isoform.[141] In a recent study, McLellan et al.[86] showed that, after treatment of mice with butylated hydroxyanisole or β-naphthoflavone, two α-class subunits (Ya$_1$ and Ya$_2$) were expressed, and the constitutive isoform (Ya$_3$), with high aflatoxin-conjugating activity, was not induced.

2. µ-Class

This consists in the rat of subunits 3, 4, 6, 9, and 11, having neutral/acidic isolectric points.

a. Substrate Specificity. Subunits 3 and 4 are the most active toward the mutagens BaP diol epoxide, styrene oxide, *trans*-stilbene oxide, nitropyrene oxide, and the DNA methylating carcinogen, 1-methyl-2-nitro-1-nitrosoguanidine.[87,88] Subunit 3 is a highly active aryltransferase and is the most active toward 1,2-dichloro-4-nitrobenzene (DCNB) but is inactive toward *trans*-4-phenylbut-3-en-2-one (tPBO), while subunit 4 has high activity toward tPBO, but does not conjugate DCNB. Subunit 3 is most noted for its biosynthetic function in conjugating leukotriene A$_4$ to form leukotriene C$_4$. Subunits 3 and 4 are both able to conjugate DNA hydroperoxides. Subunits 6, 9, and 11 are major GSTs of testicular tissues,[82,89,90] displaying extremely high activities toward chlorodinitrobenzene and measurable activities with 4-hydroxynonenals, p-nitrobenzylchloride, and DCNB.[82,90] They are also major components of brain GSTs[91] and are only very weakly expressed in liver, where only heterodimers with subunits 3 or 4 have been identified.[89]

b. Gene Structure and Regulation. The µ-class GSTs are also multigene families, with the two cDNA clones of subunit 3 that have been sequenced differing in only two amino acids.[92,93] The cDNA for subunit 4 shares 84% nucleotide identity (80% amino acid identity) with subunit 3 and its structural gene, comprising eight exons, spans 5 kB.[94] Subunit 6 displays 80% homology with subunit 4.[91] Subunits 3 and 4 are transcriptionally induced (~five- to eight-fold) by phenobarbital, 3-MC, and *trans*-stilbene oxide.[85] In Balb/c mice, only one µ-class GST is expressed constitutively (Yb$_1$). After dietary treatment with the anticarcinogenic chemoprotectant, butylated hydroxyanisole, two additional µ-class GSTs (Yb$_2$ and Yb$_5$) are also expressed.[95] In humans, at least four structurally related µ-class GST genes have been identified, which appear to be dispersed on different chromosomes.[96] Two of these genes, which are expressed in liver, differ by a single amino acid substitution and are polymorphic with a >50% frequency of the null phenotype in Caucasian, Chinese, and Indian populations. A third, which has an almost identical 3′ noncoding sequence, is muscle-specific, and the fourth, which shows 75% identity in the coding region and little homology in the 3' noncoding region, is brain specific.

3. π-Class

This consists of rat subunit 7. The single rat enzyme that has been identified in this class, a heterodimer of subunit 7 (commonly termed GST-P or π) is normally absent from liver but is strongly expressed during early stages of chemical carcinogenesis in preneoplastic foci. In mouse liver, this enzyme is expressed constitutively,[86] and in humans two isoenzymes may be present,[73] one a constitutive isoform and the other overexpressed in preneoplastic nodules. There have been detailed investigations of the regulation of this gene and its use as a diagnostic marker for carcinogenesis (reviewed by Sato).[97,98]

a. Substrate Specificity. Subunit 7 is highly active toward lipid hydroperoxides[81] and, unlike other GSTs, it conjugates acrolein, a product of lipid peroxidation.[99] This isoform is also highly efficient in conjugating 9,10-epoxy-7,8-dihydroxybenzo(a)pyrene, the metabolite that is the ultimate carcinogen of BaP.[100]

b. Gene Structure and Regulation. Independent sequences for two cDNA clones that code for a 210 amino acid protein have been reported.[100–102] The structural gene was isolated by Okuda et al.[103] and is only about 3 kB long, yet like the subunit 1 gene, contains seven exons. Muramatsu's group has characterized its regulatory elements (Figure 10) and shown that the major enhancer element (GPEI) contains two phorbol 12-O-tetradecanoate 13-acetate-responsive elements (TRE) in a 120 bp promoter at –2.5 kB in the 5' flanking region of the GST-P gene,[104–106] with which c-jun and c-fos oncogene products are supposed to transact. In common with the human metallothionein IIA gene, the GST-P gene also contains a TRE and a GC box in the promoter region (–60 to –40), which is believed to act as a basal level promoter.[106] During the early stages of chemical hepatocarcinogenesis induced by the Solt-Farber model with 2-acetylaminofluorene or diethylnitrosamine, mRNA levels of c-jun, c-fos, and GST-P are all increased.[98,107] Whether these oncogenes are expressed in immunologically GST-P positive foci and are naturally associated with the mechanism of GST-P gene expression has yet to be confirmed. Recently, a repressor region, which acts as a silencer in normal liver, has been defined in detail, and a silencer-binding protein (SF-A) of 50 to 55 kD molecular weight, which interacts with at least eight regions of this suppressor region, has been partially purified.[105] Sato et al.[98] reported that, uniquely among GSTs, GST-P contains a sensitive -SH group (the 47th cysteine residue), and in common with many other enzymes (reviewed by Ziegler),[108] its activity can be regulated by S-thiolation or by mixed disulfide formation with proteins; i.e., it is sensitive to reactive oxygen species (O_2^-, H_2O_2, and $OH·$) and GSH/GSSG levels. Thus, regulation of GST-P

GST-1 gene regulatory elements

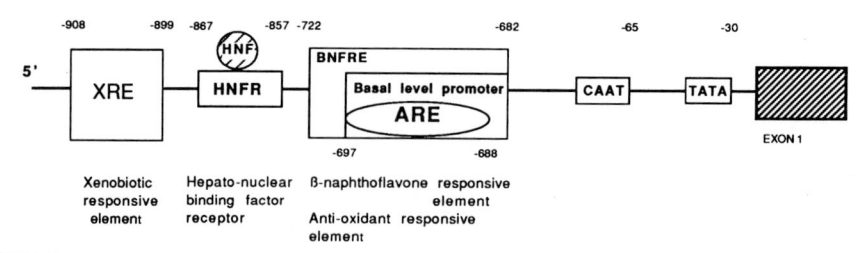

```
-908    -899 -867   -857 -722                -682              -65      -30
                   (HNF)  BNFRE
5'                        ┌─────────────────────┐
─────┬─────┬─────┬────────┤ Basal level promoter├──────┬────┬──────┐
     │ XRE │     │ HNFR   │         ARE         │      │CAAT│ TATA │////
─────┴─────┴─────┴────────┤                     ├──────┴────┴──────┘
                          └─────────────────────┘                EXON 1
                              -697        -688

  Xenobiotic    Hepato-nuclear   ß-naphthoflavone responsive
  responsive    binding          element
  element       factor           Anti-oxidant responsive
                receptor         element
```

GST-7 gene regulatory elements

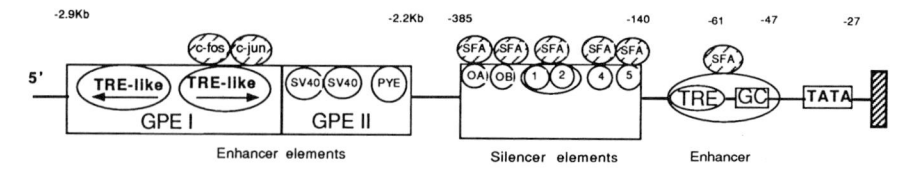

```
-2.9Kb                    -2.2Kb  -385              -140      -61  -47    -27
         (c-fos)(c-jun)          (SFA)(SFA)(SFA)(SFA)(SFA)
5'    ┌──────┬──────┐ ┌────┬────┬───┐ ┌───────────────────┐  (SFA)
──────┤TRE-like TRE-like├─┤SV40 SV40 PYE├─┤OA OB 1  2   4 5├──┤TRE GC├─┤TATA├─▨
      │  GPE I         │ │  GPE II       │ │                │
      └────────────────┘ └───────────────┘ └────────────────┘
          Enhancer elements                 Silencer elements   Enhancer
```

Figure 10. Schematic representation of the rat GST-1 and GST-P gene regulatory regions. GST-1 is expressed by basal level promoter, induced by hydrogen peroxide and antioxidants via the antioxidative responsive element (homologous with the ARE of quinone reductase), which shares common areas with the basal level promoter and an element responsible for β-naphthoflavone induction. The major xenobiotic responsive element is homologous to the Ah responsive element of CYP4501A1. The enhancer element of GST-P responsible for basal level of expression is normally subject to action of a soluble silencer protein (silencer factor A, SFA) and also contains a phorbol ester responsive site (TRE). The strongest upregulation is mediated by a TRE in site GPEI, which may involve oncogene interactions. Consensus sequences of SV40 and polyoma enhancer core-like are also present in enhancer GPEII.

activity is extremely complex. Apart from multiple regulatory elements in the gene, including those that might be controlled by oncogene expression, the activity of the enzyme can also be chemically modulated in a reversible manner. For such a complex regulatory system to have evolved, it is probable that the enzyme has an important and essential function. Since it is highly active toward lipid hydroperoxides and other products of lipid peroxidation such as acrolein, its natural function may be detoxication of these compounds, and its overexpression in preneoplastic lesions may also be a response to the increased lipid peroxidation found in early tumorogenesis.

4. θ-Class

This consists of highly labile neutral homodimeric enzymes, subunits Yrs, 5, and 12 of rat.

a. Substrate Specificity. The θ-class GSTs differ from GSTs of other classes in that, unlike all other isoforms, they do not conjugate the model substrate 1-chloro-2,4-dinitrobenzene, and they do not bind to GSH-affinity matrixes. Yrs is notable for its activity toward ethacrynic acid, peroxidase activity toward cumene hydroperoxide and especially for conjugation of reactive carcinogenic arylmethanols such as menaphthyl sulfate.[72] Subunit 5 displays a high activity toward some aromatic substrates (1,2-epoxy-3-(p-nitrophenoxy)propane and p-nitrobenzyl chloride) and epoxyeicosatrienoic acids.[109-111] A very high activity with DNA hydroperoxides was originally attributed to this enzyme, but recent purification studies have assigned this activity to another, as yet uncharacterized enzyme.[112] The partial sequence data of subunit 5 and subunit 12 show that they are very closely homologous to the Yrs subunit whose cDNA was recently sequenced.[162] The θ-class enzymes exhibit closer structural relationships with GSTs sequenced from the plants, maize[113] and carnation,[114] the insect *Drosophila*,[115] and a fish, plaice[164] than are found with other rat enzymes.

5. Microsomal

The microsomal GST has been studied in great detail; it is predominantly found in liver and is structurally distinct from the cytosolic enzymes.[144-146] The holomer appears to be a trimer of 154 amino acid (17 kDa) subunits. The active site is on the cytosolic side of the endoplasmic reticulum membrane, and the enzyme is activated by -SH reagents such as N-ethylmaleimide. While the microsomal and cytosolic GSTs conjugate a similar spectrum of substrates such as CDNB, the microsomal GST is most notable for its peroxidase activity with 4-hydroxynon-2-enal and phospholipid hydroperoxides, which is a strong indication for a functional role in inhibition of lipid peroxidation.[147]

6. Conclusions

The above summary clearly shows the great multiplicity of GSTs in rats and the diversity of activities they display with both endogenous and xenobiotic substrates (summarized in Table 6). The marked tissue, species, and even strain differences in expression that have been found in mammals are also potentially important determinants of the toxicity of xenobiotic compounds. While orthologous GSTs have been found in rodents and humans, there appear to be differences in their expression.[155,161,203] The polymorphism in human μ-class enzymes and the differential induction of isoforms in the mouse were noted earlier. Indeed the mouse, which is widely used for studies of carcinogenesis and effects of chemoprotectants, such as butylated hydroxyanisole, provides the best example of exogenously mediated changes in gene expression of biotransformation enzymes. These are highlighted in Table 7.

Table 7. Some Characteristics of Mouse Cytosolic GSTs

Class	Constitutive			Inducible by dietary butylated hydroxyanisole treatment		
	Subunit	MW, kDa	Activities in addition to CDNB conjugation	Subunit	MW, kDa	Activities in addition to CDNB conjugation
Alpha	Ya$_3$	25.8	Aflatoxin B$_1$ 8,9 oxide	Ya$_1$	25.6	
				Ya$_2$	25.6	
Mu	Yb$_1$	26.4	Dichloronitrobenzene	Yb$_2$	26.2	*trans*-phenylbutenone
				Yb$_5$	26.5	
Pi	Yf	24.8	Ethacrynic acid			

Note: CDNB, 1-chloro-2,4-dinitrobenzene.

Apart from their essential functions in intracellular transport (heme, bilirubin, and bile acids) and biosynthesis of leukotrienes and prostaglandins, a critical and important role for GSTs is obviously in defense against oxidative damage and peroxidation products of DNA and lipids. This peroxidase activity is a coupled reaction involving GSH reductase and glucose 6-phosphate dehydrogenase in regeneration of reduced GSH (Figure 11). At least seven GSTs are highly active against such compounds: subunits 2, 10, and Yrs are Se-independent glutathione peroxidases highly active toward organic hydroperoxides such as cumene hydroperoxide; subunits 3 and 4 and another so far uncharacterized isoenzyme are active with DNA hydroperoxides; subunit 8 is highly active toward 4-hydroxy alkenals, and subunit 7 is highly active with lipid hydroperoxides and compounds such as acrolein. The microsomal enzyme also displays peroxidase activity with 4-hydroxy alkenals, lipid, and phospholipid hydroperoxides.

C. Piscine GSTs

1. Occurrence and Comparative Activities

As noted above, GSTs are ubiquitous. Thus, activity toward the model substrate CDNB has been demonstrated in all piscine species so far examined,[3,18,116,117] and data derived from comparative studies where standardized assay conditions were used are given in Table 8. The finding that GST activities in farmed Atlantic salmon were induced by an order of magnitude compared with wild fish by dietary components, particularly antioxidants, which in mammals have been shown to be chemoprotectants,[118] highlights the problems in interpreting such comparative data. Although the specific activities vary quite markedly between species, the activity in fish is comparable with that in mammals. A number of studies with prototypical substrates other than CDNB have also highlighted interspecies differences (Table 9). A notable finding is the higher activity toward p-nitrobenzyl chloride than CDNB in shark cytosol.

Bend et al.[2] studied the metabolism of several arene oxides in various aquatic animals, and hepatic cytosolic GST activities toward styrene 7,8-oxide and

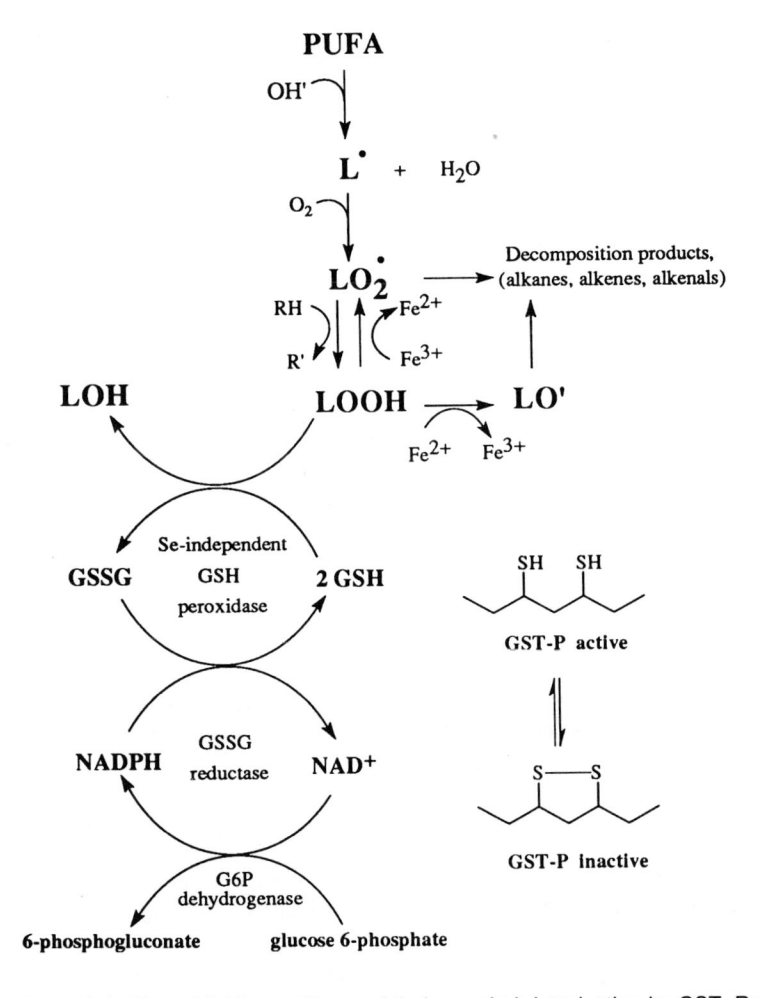

Figure 11. Formation of lipid peroxides and their coupled detoxication by GST. Radicals produced chemically (e.g., by redox cycling of quinonoid drugs such as adriamycin or by oxidative stress) attack double bonds of polyunsaturated fatty acids (PUFA) to produce lipid hydroperoxides (LOOH), which can be reduced by nucleophilic attack of GSH on the electrophilic oxygen. Reduced GSH can be regenerated. The activity of GST-π is also sensitive to GSH/GSSG levels.

benzo(a)pyrene-4,5-oxide are compared in Table 10. The sixfold higher activity of the skate toward benzo(a)pyrene-4,5-oxide as compared to the rat is striking.

Hepatic microsomal GST activity, constituting about 2% of the total CDNB-conjugating activity, has been reported in rainbow trout,[143,163] pike,[144] and plaice;[38] however, unlike the mammalian enzyme, it is not activated by thiol reagents and, therefore, is not readily distinguishable from adsorbed cytosolic activity. In a preliminary study, we found that this microsomal activity could be readily solu-

Interspecies Comparisons of CDNB-Conjugating Activities in Hepatic Cytosols

Species	Activity, nmol/min/mg protein	Ref.
Mammals		
Mouse, *Mus* sp.	6370	82
Rat, *Rattus* sp.	1430	82
Elasmobranchs		
Shark, *Platyrhinoides triseriata*	270	122
Skate, *Raja erinacea*	220–440	123
Teleosts		
Atlantic salmon, *Salmo salar*	100–200 (400–2000[a])	118
Bass, *Centropristis striata*	120–410	119
Bluegill, *Lepomis macrochirus*	350	117
Brown trout, *S. trutta*	300–400	118
Carp, *Cyrinus carpio*	210	117
Cod, *Gadus morhua*	1350–1550	118
English sole, *Parophys vetulus*	750–1050	138
Fathead minnow, *Pimephales promelas*	1000–2000	156
Flounder, *Platichthys flesus*	400–500	118
Guppy, *Poecilia reticulata*	1000	117
Killifish, *Fundulus grandis*	1000–1200	156
Mummichog, *F. heteroclitus*	800–3500	155
Pike, *Esox lucius*	450	135
Plaice, *Pleuronectes platessa*	650–850	118
Rainbow trout, *Onchorynchus mykiss*	500–2500	118
Sheepshead minnow, *Cyprinodon variegatus*	2000–4000	156
Sole, *Solea solea*	250–350	118
Starry flounder, *Platichthys stellatus*	1750–3250	138
Turbot, *Scophthalmus maximus*	100–250	118
Zebra fish, *Brachidanio reria*	360	117

[a] Farmed.

bilized and was activated some fivefold by treatment of plaice microsomes with Triton X100 or Lubrol PX. Unfortunately, it then became extremely labile, and further characterization was not achieved. Clearly, further work on the possible identity and properties of a microsomal GST in fish is required.

2. Tissue Distribution

Extrahepatic GST activity has been demonstrated in a number of species[2,3,116,119] and analyzed in greater detail in bass, rainbow trout, plaice, and skate.[2,119–121,163] In the bass and trout, CDNB-conjugating activity in tissues in contact with the environment, i.e., intestine and gills, is comparable with hepatic activity, while in plaice, the relative activity in the liver is an order of magnitude higher (Table 11). In the skate, activity toward three arene oxides showed clear tissue-specific expression (Table 12), with liver being most active toward styrene oxide, kidney most active toward benzo(a)pyrene-4,5-oxide, and gills most active toward octene-1,2-oxide.[2] In trout, the substrate specificity of the enzymes in extrahepatic tissues appears to be more limited than the hepatic enzymes, and in the gill, there was no

Table 9. Multisubstrate Comparisons of GST Activities in Hepatic Cytosols

Species	Bromo-sulfophthalein	Dichloro-nitrobenzene	Epoxy(phen-oxy) propane	Ethacrynic acid	p-Nitrobenzyl-chloride
Rat[a]	2	6	11	4	63
Shark[b]	<0.05	0.7	4	6	117
Skate[c]	—	2	—	—	—
Bluegill[d]	0.06	—	<0.05	3	13
Carp[d]	—	—	nd	3	—
Guppy[d]	0.03	—	0.7	2	2
Zebra fish[d]	0.02	—	nd	3	6
Trout[e]	0.05	0.8	1.2	2	1
Cod[e]	<0.01	0.3	—	0.6	1
Flounder[e]	<0.05	2	—	0.3	0.3
Plaice[e]	0.01	—	<0.01	0.1	0.1
Sole[e]	<0.05	—	3.6	1	0.6
Turbot[e]	0.6	4	7	0.6	0.3

Source: Calculated from data of [a]Mannervik, B.;[66] [b]Sugiyama, Y. et al.;[122] [c]Bend, J.R. et al.;[2] [d]Donnarumma, L. et al.;[117] [e]George, S. et al.[118]

Note: Activities relative to CDNB = 100.

measurable activity toward 1,2-epoxy-3-(p-phenoxy)propane, ethacrynic acid, or p-nitrobenzylchloride.[120] The expression of a varied spectrum of relative activities in the different tissues suggests that, as in mammals, different isoenzymes are expressed in a tissue-specific manner.

In plaice, CDNB-conjugating activity (i.e., an integration of all GST isoforms) has been found in all tissues that have been examined. However, when the tissue distribution of GST-A, the major hepatic isoform, was studied using a nucleic acid probe for determination of its mRNA levels,[121] immunoblotting,[142] and immunocytochemistry with a specific antiserum to detect the protein, it was found that

Table 10. GST Activity toward Arene Oxides in Different Species

		Activity, nmol/min/mg cytosol protein		
Species		Styrene-7,8-oxide	BaP-4,5-oxide	Octene-1,2-oxide
Mammals				
Rat	Rattus norvegicus	190	24	—
Rabbit	Oryctolagus cunnilingus	31	9	—
Elasmobranchs				
Atlantic stingray	Dasyatis sabina	5	2	5
Dogfish shark	Squalas acanthas	15	8	1
Little skate	Raja erinacea	5	131	1
Teleosts				
Black drum	Pogonius cromi	16	9	9
Eel	Anguilla rostrata	15	3	15
Mangrove snapper	Lutjanus griseus	4	2	—
Sheepshead	Archosargus probatocephalus	26	53	21
Winter flounder	Pseudopleuronectes americanus	5	5	5

Source: Bend, J.R. et al., Ann. N.Y. Acad. Sci., 298, 505, 1977.

Table 11. Tissue Expression of GST Activity toward Chlorodinitrobenzene in Different Fish

Species	Liver specific activity nmol/min/mg	Activities relative to chlorodinitrobenzene = 100							Ref.
		Liver	Kidney	Intestine	Gills	Brain	Muscle	Blood	
Rainbow trout	510	100	65	45	50	—	—	1	120
	350	100	34	—	46	—	—	—	163
Bass	406	100	60	—	—	50	16	17	119
Plaice	970	100	6	11	2	1	—	—	121

Note: —, not determined.

while this isoform is expressed in most tissues, there is a cell-specific expression. In liver, it is confined to the hepatocytes and is not found in blood vessels, bile ducts, endocrine, or exocrine pancreatic tissue (Figure 12A). In the kidney, it is strongly expressed in the P2 tubules and weakly in collecting ducts, but it is not found in the P1 tubules, glomeruli, or hemopoietic tissue (Figure 12B).

3. Purification Studies

Hepatic glutathione S-transferase activity has been purified and partially characterized from two elasmobranchs, the thorny-back shark *(Platyrhinoides triserata)*[122] and the little skate *(Raja erinacea)*[123] and from four teleost species, the rainbow trout *(Oncorynchus mykiss)*,[124,125] the carp *(Cyprinus carpio)*,[126] the plaice *(Pleuronectes platessa)*,[127] and the salmon *(Salmo salar)*.[128,129]

a. Shark (Platyrhinoides triseratia). Sugiyama et al.[122] resolved two isoforms of GST from the thorny back shark, a species where the p-nitrobenzyl chloride-conjugating activity of cytosol is higher than that with CDNB. The major isoform (Y2), accounting for approximately 70% of the CDNB conjugating activity, was purified by gel permeation, ion exchange, and hydroxylapatite chromatography to high specific activity. This homodimeric enzyme (subunit 24 kDa), with a pI of 7.7, exhibited comparable activities toward CDNB and p-nitrobenzylchloride, while it had very low activity toward 1,2-dichloro-4-ni-

Table 12. Tissue Distribution of GST Activity toward Arene Oxides in Little Skate *(Raja erinacea)*

Tissue	Styrene 7,8 oxide	B(a)P 4,5 oxide	Octene 1,2 oxide
Liver, nmol/min/mg cytosol	19.3	16.5	0.2
Kidney[a]	35	190	nd[b]
Gill[a]	19	64	650
Heart[a]	nd[b]	28	nd[b]
Pancreas[a]	nd[b]	18	200

Source: Bend, J.R. et al., *Ann. N.Y. Acad. Sci.*, 298, 505, 1977.

[a] Activities relative to liver = 100.
[b] nd = not detected.

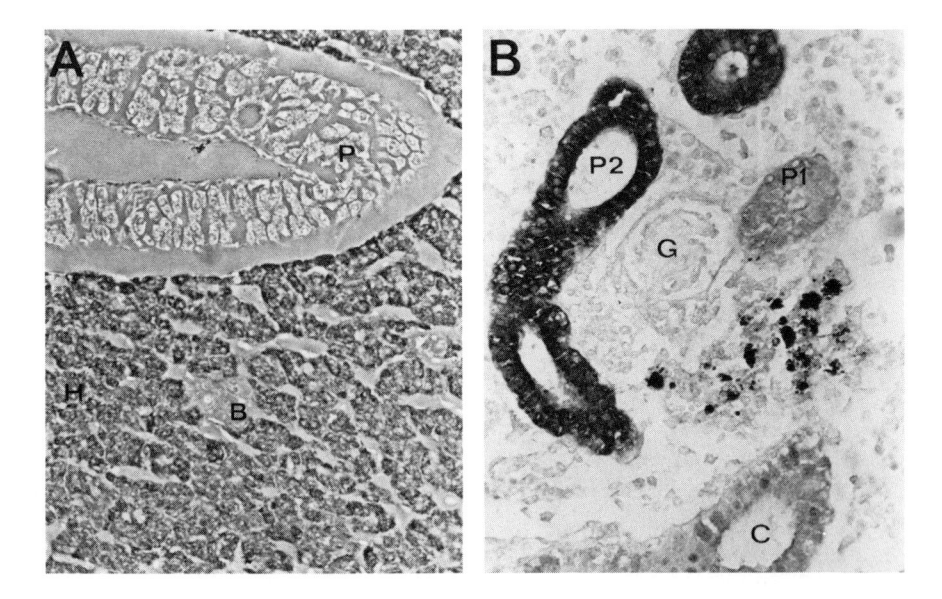

Figure 12. Immunocytochemical visualization of plaice GST-A. Formalin-fixed, wax-embedded sections stained with anti-plaice GST-A antiserum visualized by peroxidase-labeled secondary antibody. (A) Liver (phase contrast microscopy to highlight unstained tissues), showing positively reactive hepatocytes, unreactive exocrine and endocrine pancreas, blood vessels, and (B) kidney showing heavily stained P2 tubules, unstained glomerulus, collecting duct, and surrounding hemopoietic tissues. Abbreviations: B = blood vessel; C = collecting duct; G = glomerulus; H = hepatocyte; P = exocrine pancreatic tissue; P1 = proximal tubule 1; P2 = proximal tubule 2.

trobenzene and bromosulfothalein. Curiously, fractions were not monitored for activity toward p-nitrobenzylchloride; thus, characteristics of the isoform accounting for the majority of this activity were not identified. Organic anion binding and inhibition studies with partially purified GST fractions and the purified GST Y2 showed that binding affinities for bromosulfothalein, bilirubin, and Rose Bengal were an order of magnitude lower than those of rat ligandin (GST 1-1) although 1-anilino-8-naphthalene-sulfonate was bound with comparable affinity (Table 13). Thus, by the common definition of ligandin (as a protein that binds bromosulfothalein with a high affinity), the shark does not contain ligandin, although its GSTs may bind highly reactive electrophiles irreversibly. From the substrate specificity of the shark GSTY2, it may be predicted that the enzyme may be related to either the α-class subunit 8 or the μ-class subunit 4 of rat (Table 7).

b. Skate (R. erinacea). Foureman and Bend[123] resolved five cytosolic GSTs from skate by ion exchange chromatography and in a later study investigated their stereochemical specificity toward several K-region arene oxides.[130] The major isoform (E-

Table 13. Characteristics of Purified Elasmobranch GSTs

		Thorny back shark GST Y2	Little skate GST E-4
Subunit MWs, kDa		24 + 24	26 + 26
Isoelectric point, pI		7.7	5.4–5.6
Activities (mmol/min/mg)	Chlorodinitrobenzene	21.7	46.4
	p-Nitrobenzylchloride	18[a]	—
	Dichloronitrobenzene	—	<0.01
	B(a)P 4,5-oxide	—	12.0
K_ms (mM)	Chlorodinitrobenzene	0.3	0.42
	Glutathione	2.1	1.1
Binding constants, K_d (mM)	Hematin	—	1.7
	Bilirubin	9.1[b]	5.1
	Bromosulfophthalein	2.5[b]	—
	8-Anilonaphthalene sulfonate	25	—

Source: Data for the major skate GST (E-4) from Foureman, G.L. and Bend, J.R.[123] and from Foureman, G.L. et al.[130] Data for thorny back shark GST Y2 from Sugiyama, Y. et al.[122]

[a] Estimated from the elution profile on gel permeation chromatography.
[b] Determined on impure fraction.

4) was apparently homogeneous after purification by affinity chromatography. It consisted of a homodimer of 26 kDa subunits, had an acidic pI of 5.2 to 5.4, and displayed high activities toward CDNB and the arene oxides benzo(a)pyrene-4,5-oxide and styrene-oxide. It did not, however, conjugate dichloronitrobenzene. It was highly stereoselective for the conjugation of the R-oxirane carbon of the K-region arene oxides. The rate of conjugation of benzo(a)pyrene-4,5-oxide catalyzed by the skate GST E-4 is some 150 times greater than that reported for rat GST C (subunit 3-4 heterodimer), the most active rat GST toward this substrate.[131] The skate GST E-4 also bound hematin and bilirubin with moderate affinity (Table 13); however, the binding of these substrates was again an order of magnitude lower than for rat ligandin. On the basis of this evidence, it can be concluded that the skate does not possess a functional ligandin-like activity. Activities of two other acidic heterodimeric enzymes containing subunits of 26 and 27 kDa could be attributed to the presence of the same 26 kDa subunit as in E-4 unfortunately, no characteristic activities for the 27-kDa subunit were reported. Another acidic GST with an isoelectric point of 5.2, designated E-5, was not purified to homogeneity and conjugated styrene oxide at a comparable rate to E-4. However, its activity with benzo(a)pyrene-4,5-oxide was very low, and it lacked a homologous 26-kDa subunit. Another basic GST (pI >9.5) was identified that was a homodimer of 28 kDa subunits and conjugated CDNB but, unlike the other skate GSTs, exhibited a significant activity with dichloronitrobenzene. This isoform did not conjugate arene oxides. Unfortunately, activities toward p-nitrobenzylchloride and other diagnostic synthetic substrates have not been reported, so class assignment and comparison with the other elasmobranch species is not yet possible. The very high activity of skate GST E-4 toward benzo(a)pyrene-4,5-oxide warrants further molecular characterization of the enzyme.

c. *Trout (Onchorynchus mykiss)*. Hepatic GSTs have been purified from rainbow trout by Ramage and Nimmo[125] and by Dierickx[124] using chromatofocusing after affinity chromatography on S-hexyl GSH- or GSH-agarose, respectively. The former affinity matrix retained only 50% of the cytoplasmic CDNB-conjugating activity, while 5% of the activity (including the anionic isoform estimated to comprise 3% of the activity) did not bind to the GSH-affinity column. Therefore, direct comparability of results from the two studies is difficult. Different subunit molecular weight estimates by SDS PAGE were also made, although on the basis of comigration with rat subunit 7,[129] the original estimates of Nimmo and co-workers should be increased by 2.4 kDa. The results of these two studies have shown that trout liver appears to contain at least 6 or 7 GST isoforms separable by charge. The characteristics of the proteins purified in the two studies are summarized in Table 14.

The major isoforms in trout liver appear to be neutral and, in addition to CDNB, also conjugate a number of other substrates. Immunological studies have demonstrated that trout liver cytosol purified by GSH-agarose affinity chromatography contains a 25-kDa polypeptide recognized by antiserum raised against rat subunit 7.[118] Microheterogeneity of this subunit has been found, and four isoforms that exhibit differences in activity toward ethacrynic acid (a characteristic substrate for P class GST) have been resolved (see reference 129 and Hayes, personal communication). Neither of the apparently homogeneous preparations of 25 kDa homodimers conjugated 1,2-epoxy-3-(p-phenoxy)propane, and activities toward this substrate and p-nitrobenzylchloride have been found in preparations containing subunits with apparent Mrs of 24 and 26.6 kDa. The anionic form(s) that did not bind to S-hexyl GSH agarose catalyzed Δ^5 andosterone dione isomerization (an activity characteristic of the basic rat subunit 1) and conjugated p-nitrobenzylchloride at a greater rate than chlorodinitrobenzene, thus resembling the shark GSTs.[122] Bell et al.[132] reported that affinity-purified trout cytosol did not possess selenium-independent GSH peroxidase activity with cumene hydroperoxide, thus indicating the absence of a subunit 2 orthologue. Antisera raised against rat subunits 1, 2, and 3 did not recognize trout orthologues, while there was a weakly positive reaction with subunit 8 antiserum.[118] Thus, the only definitive homology identified between trout and mammalian GSTs appears to be the presence of a subunit 7 orthologue.

d. *Salmon (Salmo salar)*. Ramage et al.[128] also reported separation of three classes of GSTs from salmon liver, which exhibited different characteristics in their avidities for affinity matrices, their isoelectric points, subunit compositions, and catalytic properties (Table 14). The picture was complicated by the presence of different patterns of expression according to season and whether the fish were adapted to saltwater or freshwater. The predominant subunit had an apparent MW of 24.8 kDa (previously reported as 22.4), and others of 25.7 and 26.6 kDa were detected. The major isoforms were presumably closely related charge isomers,

Table 14. Some Characteristics of Purified Salmonid Liver GSTs

Enzyme	Subunit MW, kDa	pI	Other substrates apart from chorodinitrobenzene
Rainbow trout (*Onchorynchus mykiss*)			
From Dierickx.[124]			
Bound to GSH-agarose (40% total CDNB-conjugating activity)			
K1–3	25 + 26.5	9.3–9.5	—
K4–6	25 + 25 (tr. 26, 26.3)	8.6 –9.0	—
From Ramage and Nimmo.[125 a]			
Bound by S-hexyl GSH-agarose (97% total CDNB-conjugating activity)			
C1, 2	25 + 25	8.6–8.3	Bromosulfophthalein, ethacrynic acid
C4	25 + 26.6	7.7–7.8	1,2-epoxy-3-(p-phenoxy)propane, ethacrynic acid
C5	25 + 25 (tr. 24)	7.7	1,2-epoxy-3-(p-phenoxy)propane
Not bound by S-hexyl GSH-agarose (3% total CDNB-conjugating activity)			
Anionic[b]	?	<5.0	p-nitrobenzyl chloride, Δ5 androsterone dione
Atlantic salmon (*Salmo salar*)			
From Ramage et al..[128 c]			
Bound to S-hexyl GSH agarose			
Aug 2,4,5,	24.8	5.8–7.5	Dichloronitrobenzene, *trans*-2-phenyl-3-butene-2-one, p-nitrobenzyl chloride, (ethacrynic acid)
Not bound to S-hexyl GSH-agarose			
Aug basic	26.5	Basic	Dichloronitrobenzene, (p-nitrobenzyl chloride, ethacrynic acid)
Aug n	25, 26.5, 28	Neutral	1,2-epoxy-3-(p-phenoxy)propane, (ethacrynic acid)
Aug a	26.5	Acidic	p-nitrobenzyl chloride, 1,2-epoxy-3-(p-phenoxy)propane, (ethacrynic acid)

[a] Molecular weights adjusted upwards by 2.5 kDa following immunochemical and sequence data.[118,129]
[b] An impure preparation obtained from flow-through of affinity column.
[c] Data for enzymes purified from fish caught in August.[128] Marked interanimal and seasonal variations in salmon GST contents were reported.

since they were all homodimers of 24.8 kDa subunits and displayed moderately high activity toward p-nitrobenzylchloride, dichloronitrobenzene, and *trans*-4-phenyl-3-butene-2-one (1 to 7% of CDNB activity). However, in this study no activity was detected with 1,2-epoxy-3-(p-phenoxy)propane and very little with ethacrynic acid as substrates. The acidic enzyme(s) exhibited high activities toward p-nitrobenzylchloride and 1,2-epoxy-3-(p-phenoxy)propane, which were 53 and 37% of their CDNB-conjugating activities, respectively.

In a subsequent study, Dominey et al.[129] purified the major isoforms containing the 24.8 kDa subunits by GSH-agarose affinity chromatography and hydroxylapatite chromatography and confirmed that they displayed cross-immunoreactivity with antiserum to rat GST 7-7. These isoforms displayed varying levels of

```
Salmon    M P P Y T I T Y F G V R G R - G A M R I M M A D Q
Human π   M P P Y T V V Y F P V R G R C A A L R M L L A D Q
Mouse π     P P Y T V V Y F P V R G G C A A M R M L L A D Q
Rat 7     M P P Y T I V Y F P V R G R C E A T R M L L A D Q
```

Figure 13. Amino terminal amino acid sequence alignments of salmon and mammalian GSTs.

ethacrynic acid-conjugating activity (characteristic of subunit 7), and they also cross-dimerized, indicating microheterogeneity of a single subunit type.[132a] The amino acid sequences of a number of tryptic peptides have been determined, and over the regions sequenced, it was found that the salmon protein displayed 70% homology with the rat π-class enzyme (Figure 13).[129] This substantial conservation of structure in the π-class GST is not surprising, since the enzyme has previously been shown to be very highly conserved and has even been identified in prokaryotes. Purification and immunoblot studies have shown that this subunit is also expressed in a number of salmonid species and in the cod, but not in all flatfish.[118,129]

e. Plaice (Pleuronectes platessa). GST activity in plaice liver appears to be very labile, necessitating that ligand binding and inhibitor studies be carried out within 4 hr of sacrifice using GSH-affinity purified cytosols.[133] A functional role for plaice GSTs in ligand binding is highly questionable. While hematin binds to the plaice GSTs with an affinity that is similar to its binding to ligandin, it does not inhibit enzyme activity. Moreover, bilirubin is bound to plaice GST extremely weakly, and it is some 30 times less effective in inhibiting CDNB conjugation in plaice than it is with rat ligandin. Additional evidence for the absence of a subunit 1 orthologue is the apparent absence of a basic isoenzyme[127] and a lack of cross-reactivity with antiserum to rat subunit 1.[118]

Three neutral/acidic GSTs have been purified and characterized from plaice liver.[133] The major isoforms, GST-A, a homodimer of 25 kDa subunits (26 kDa by SDS PAGE), and GST-B, a homodimer of 23.5 kDa subunits, were resolved by ion exchange chromatography on DEAE cellulose. They were differentially retained on GSH-affinity matrices, GST-A binding preferentially to S-hexyl GSH-agarose and GST-B to GSH-agarose. The apparently homogeneous proteins had remarkably similar substrate specificities, only exhibiting significant activity toward CDNB and no other common synthetic substrates (Table 15). Neither possessed measurable peroxidase activity with cumene hydroperoxide. Apart from differences in apparent molecular weights the isoforms are clearly structurally distinct, since antisera raised to each of the enzymes were isoform specific. These antisera have been used to investigate phyllogenic relationships between GSTs.[118,133] Investigation of the immunological properties of the rat and plaice enzymes showed that GST-A was immunochemically distinct from GST-B and rat subunits 1, 2, 3, 7, or 8, while GST-B showed related structural motifs to rat

Table 15. Characteristics of Purified Hepatic GSTs from Plaice (Pleuronectes platessa)

Properties	Substrates	GST-A	GST-B[a]
Subunit MW, kDa		25 + 25	23.5 + 23.5
Activity, mmol/min/mg	Chlorodinitrobenzene	35	37
	Dichloronitrobenzene	0.13	0.15
	p-Nitrobenzylchloride	0.01	0.11
	Ethacrynic acid	nd[b]	0.03
	Cumene hydroperoxide	nd[b]	nd[b]
K_m	Chlorodinitrobenzene (mM)	—	0.83 ± 0.1
	Glutathione (mM)	—	2.73 ± 0.48
V_{max}, mmol/min/mg	Chlorodinitrobenzene	—	39.5 ± 6

		GST A + B	
Ligand-binding, K_d	Bilirubin (mM)	20	(I$_{50}$ 320 mM)
	Hematin (mM)	0.03	(I$_{50}$ 10 mM)

Source: George, S.G. and Buchanan, G.[133]

[a] Immunologically cross-reactive with rat subunit 8.
[b] nd = not detected.

GST subunits 2 and 8.[133] Therefore, plaice GST-B may be an α-class enzyme. The minor plaice GST isoform that was identified appears to be a heterodimer of 23.5 and 25 kDa subunits and displayed higher relative activities toward dichloronitrobenzene (13% of CDNB) and p-nitrobenzylchloride (3%) than the homodimeric isoforms, a substrate specificity indicative of a μ-class enzyme.

Recently, by using antibody screening of a plaice λ-GT11 expression library, we have isolated a cDNA clone of GST-A and determined its nucleotide sequence.[164] The cDNA codes for a protein of M_r 25.2 kDa and its deduced amino acid sequence bear little homology with mammalian α–, μ-, or π-class GSTs. Interestingly, it exhibits significant homologies with GSTs sequenced from maize *(Zea mais)*,[113] carnation *(Dianthus)*,[114] fruit fly *(Drosophila magna)*,[115] and a similar homology to the recently reported partial sequences of the rat θ-class GSTs.[112,162] The plant, insect, and plaice GST-AS all share very similar characteristics and substrate specificities, although unlike the rat θ-class GSTs, they do conjugate CDNB and bind to GSH affinity columns.[112] The overall homologies between the mammalian θ and nonmammalian forms are not sufficient to identify them as homologous θ-class enzymes, and further studies are required to characterize these genes.

4. Immunological Investigations of Species Homologies

It is difficult from our present knowledge to define relationships between various subunits, either within or between species. However, from the data quoted above for purified enzymes and a preliminary study using activity measurements and Western immunoblotting of GSH-agarose affinity-purified fish cytosols (Table 16), some general conclusions can be made. The major hepatic GST isoforms of

Table 16. Interspecies Distribution of Immunologically Orthologous GST Subunits in Fish (Intensity of Immunostaining on "Western" Blot)

	Plaice-GST		Antiserum, anti-rat-GST subunits				
			α-class			μ-class	π-class
Species	A	B	1	3	8	2	7
Cod	++++	±	+	ns	±	++	++++
Flounder	++++	+	—	—	±	—	—
Plaice	++++	++++	+++	—	++++	—	—
Turbot	++++	+++	±	±	+	±	—
Brook trout	+	—	ns	ns	±	ns	++++
Brown trout	+	±	ns	ns	+	ns	++++
Rainbow trout	+	±	—	—	±	—	++++
Sea trout	+	±	+++	+	+++	+++	++++
Salmon	—		+	—	++	—	++++

Source: Modified from George, S. et al. Mar. Environ. Res., 28, 1, 1989.

salmonids and the cod are related to the π-class enzymes, and turbot liver also appears to contain a π-class GST.[118] However, other flatfish appear to lack such an isoenzyme and instead contain a homodimeric enzyme (GST-B) of subunits that are immunologically related to rat subunit 8[118,133] and a major isoform (GST-A of plaice), which displays no immunological cross-reactivity with antisera to rat α-, μ-, π-, or θ-class subunits (see reference 118 and unpublished results). None of these isoforms exhibits significant GSH peroxidase activity toward cumene hydroperoxide. The lack of immunoreactivity with rat subunit 2 antisera[118] and lack of Se-independent GSH peroxidase activity toward cumene hydroperoxide of affinity-purified trout GST[132] or the major GSTs purified from plaice[133] indicate that a subunit 2 orthologue is absent from these piscine species. Indeed, another activity of the α-class GSTs, notably "ligandin-like" activity, also appears to be absent from both elasmobranchs and teleosts, such as the trout and plaice.[116,133] The possible role of piscine GSTs in protection from lipid peroxidation is of particular interest, since fish are particularly rich in polyunsaturated fatty acids. In some species this role may be fulfilled by the GST-P homologue, while in others, such as the plaice, the presence of GST-B, which is structurally similar to rat subunit 8, an isoform active toward lipid peroxidation products, and GST-A, which is induced by radical-generating compounds, may confer protection.

5. Inducibility of Piscine GSTs

Unlike CYP1A levels, which are induced by one to two orders of magnitude by PAHs such as 3-MC or β-naphthoflavone (BNF), levels of mammalian phase II enzymes UDPGT and GST are only induced by two- to fivefold at the most. The effects of inducing agents on total GST activity in fish livers, measured by CDNB-conjugation, have been reported in several studies, mostly in conjunction with studies of CYP1A induction (Table 17). Following treatment of several fish

Table 17. Effect of Inducer Treatments on Hepatic Glutathione S-Transferase Activity in Different Fish

Species	Agent	Effect	References
Cod	β-Naphthoflavone	No effect	134
Fathead minnow	3-Methylcholanthrene	No effect	156
Flounder	3-Methylcholanthrene	Repressed	149
	Arochlor 1254	No effect	149
	trans-Stilbene oxide	Induced	149
Killifish	3-Methylcholanthrene	No effect	156
Plaice	3-Methylcholanthrene	Induced	62
	Arochlor 1254	Induced	62
	trans-Stilbene oxide	Induced	62
	Butylated hydroxyanisole	Induced	62
Sheepshead minnow	3-Methylcholanthrene	No effect	156
Trout	β-Naphthoflavone	No effect	134
	β-Naphthoflavone	Slightly induced	36
	Clophen A50	Slightly induced	36
Anabas	Phenol	Induced	160

species by intraperitoneal injection with PAHs, a modest twofold induction of GST activity is observed.[36,62,127,134,148,155–158,161] However, this can sometimes go unnoticed with the wide interanimal variations observed in wild fish populations and the complex dose/response relationships that seem to occur. A single dose of BNF induced trout GST activity twofold, although a second treatment 3 days later repressed activity,[158] while treatment of plaice *(P. platessa)* with 3-MC-induced CDNB-conjugating activity one to two days after injection and later led to a decrease in activity.[148] This may be due to metabolite inhibition (e.g., a reactive arene oxide), as was suggested for the decrease in GST activity in stingray treated with a fairly high dose of 3-MC.[161] Another compounding factor is the possibility of differential effects on various GST isoforms, since when CDNB-conjugating activity is measured, it is an integration of the activities of multiple isoforms. Thus, isoform-specific probes are required to address the question. By use of GST-A antiserum and a GST-A cRNA probe in flounder *(Platichthys flesus)*, we have shown that expression of GST-A is in fact repressed by 3-MC and a commercial PCB mixture, Arochlor 1254 and that it is strongly induced by *trans*-stilbene oxide.[149] This induction of flatfish GST-A activity by *trans*-stilbene oxide and butylated hydroxyanisole,[62,149] agents which induce specific mammalian isoforms via an antioxidant responsive element, indicates that the GST-A gene may contain an antioxidant responsive element. This possibility is currently being studied in my laboratory.

Overexpression of GST-π in mammals is characteristic of preneoplastic nodules, and expression of GST has been examined in similar lesions in fish livers. An immunocytochemical study was conducted, using a polyclonal antiserum raised to the two major trout GST subunits (25 and 26.5 kDa) in fish with tumors experimentally induced by aflatoxin B_1 or the PAH, 1,2-dimethylbenzanthracene. The study demonstrated induced GST expression in approximately 20 to 30% of small altered foci, while GST was deficient in the more advanced stages of malignancy, in adenomas and carcinomas.[159] Analysis of CDNB-conjugating activity in visible hepatic lesions (altered foci and hepatocellular carcinomas) of mummichog from a creosote-contaminated environment showed no significant differences in activity between abnormal and adjacent normal tissue.[155]

V. FUTURE DIRECTIONS

Fish systems are important as models for investigating chemical interactions of xenobiotics in man. Additionally, the study of xenobiotic metabolism in fish is important in its own right in order to advance understanding of chemical toxicity at the individual, population, or community levels in aquatic ecosystems. A major objective in such studies is to understand how the toxicity of xenobiotics is influenced by the way in which they are biotransformed; that is, are they cleared from the body or metabolized to toxic derivatives? Potential interactions might also occur in biotransformation of xenobiotics and natural endobiotic compounds. For example, do xenobiotics affect conjugation of bile acids and pigments, hormones, or steroids? Indeed, to my knowledge, the latter is an area which has thus far not yet been studied, even in mammals. These objectives require an understanding at the molecular level of the mechanism of action of the enzymes involved and also of the mechanisms for regulating their expression. This, in turn, requires characterizing purified enzymes, their isolated genes, and regulatory elements, and defining the impact of environmental pollutants and signals on these systems. Essential tools for such studies are probes for measuring transcription (i.e., cDNA or cRNA probes to measure mRNA) and probes for measuring translation (i.e., antibody assays to measure levels of enzymes), in conjunction with conventional assays to measure their catalytic activity.

Such enzymological and molecular biological studies of piscine xenobiotic biotransformation systems are still in their infancy, although a very positive start has been made. These studies are currently most advanced for piscine cytochrome P4501A, which was first purified from a fish species in 1982,[150] to which monoclonal antibodies were first raised in 1986,[151] and the cDNA of which was first sequenced in 1988. These probes are now enabling studies of the control of cytochrome P4501A to be pursued at the RNA, protein, and active enzyme levels (see review by Stegeman, Chapter 3). The very high conservation of both structure and function of cytochrome P4501A has meant that its probes can be applied readily to many species without the need to replicate the original work on probe development. As shown in this review, efforts to purify, characterize, and develop probes for the much more complex and diverse phase II enzymes are lagging by nearly a decade for both fish and mammals. Contributing factors to this delay include the multiplicity of isoforms involved in xenobiotic metabolism and the variation in their expression with development, between species, as well as with other factors such as diet and the presence of foreign inducing agents. There are also severe difficulties in purifying active enzymes, especially of less abundant isoforms. Indeed, in most laboratories, the strategy for studies of rat and human UDPGTs has been to study their enzymology and substrate specificity by expression of their cloned cDNAs in cell lines. With our recent cloning of plaice CYP1A, UDPGT, and GST-A cDNAs, we hope to follow the same strategy for the fish homologues. The gene structures of mammalian phase II enzymes also appear to be much more complex than those of the phase I enzymes, the 100 kB human UGT1 gene with its alternative splicing and differential expression being a prime example. The structure of the corresponding plaice UDPGT gene is of great interest to us, since at first inspection, the bulky phenol and clofibrate-inducible bilirubin isoforms do not appear to be expressed, and therefore the piscine gene may be an ancestral form. Similarly, the apparent lack of some GST isoforms in fish poses extremely interesting questions. GST-A appears to show greater homologies with nonvertebrate enzymes and may also represent an ancestral form with a critical function.

Phase II systems are inherently more complex than their phase I counterparts, and to avoid superficiality, it is important that future enzymological and molecular biological studies of phase II systems focus on relatively few species, facilitating *inter alia* the deployment of relatively few probes in field situations where specific problems can be addressed. Obvious examples of the latter are areas where there is a high environmental impact of particular pollutants, such as specific episodic pollution by petroleum hydrocarbons (e.g., following shipwrecks of the Exxon Valdez in Alaska, the Braer in Shetland, and the Gulf War), or areas where epidemiological evidence firmly links longer term contamination with specific pathologies including hepatic neoplasia (e.g., in the Hudson River, New York,[153] and Puget Sound, Washington).[154] In areas such as these, particular fish species appear to be highly susceptible to pathological damage so that comparative studies of the expression and function of their biotransforming systems are likely to be particularly fruitful for advancing understanding of toxicological mechanisms.

The area is advancing rapidly as with all reviews, they are only current at the time of writing, and by publication, several other genes will have been cloned, in particular plaice GST and UDPGTs. However, these directions are likely to hold for some time to come.

VI. ACKNOWLEDGMENTS

I would like to thank my colleagues, past and present, Gordon Buchanan, Dougie Clarke, Mike Leaver, Tracy Gallagher, Helen Grant, Katherine Holmes, Karen Scott, Andras Strom, Joy Wright, and Peter Young, for their contributions to our work on these enzymes. Special thanks to my collaborators, especially Brian Burchell, John Hayes, and Ian Nimmo. I would also like to thank John

Sargent for his encouragement and support of the program and his help in assembling this chapter. Last but not least, I thank my long-suffering wife for putting up with the incessant clack of the keyboard in the evenings.

REFERENCES

1. Chambers, J.E. and Yarborough, J.D., Xenobiotic biotransformation systems in fishes, *Comp. Biochem. Physiol.,* 55C, 77, 1976.
2. Bend, J.R., James, M.O., and Dansette, P.M., *In vitro* metabolism of xenobiotics in some marine animals, *Ann. N.Y. Acad. Sci.,* 298, 505, 1977.
3. James, M.O., Conjugation of organic pollutants in aquatic species, *Environ. Health Perspect.,* 71, 97, 1987.
4. Tan, B. and Melius, P., Polynuclear aromatic hydrocarbon metabolism in fishes, *Comp. Biochem. Physiol.,* 83C, 217, 1986.
5. Varanasi, U., *Metabolism of Polyaromatic Hydrocarbons in the Aquatic Environment,* CRC Press, Boca Raton, FL, 1989.
6. Khan, M.A.Q., Forte, F., and Payne, J.F., Metabolism of pesticides by aquatic animals, in *Pesticides in Aquatic Environments,* Khan, M.A.Q., Ed., Plenum Press, New York, 1976, 191.
7. Dutton, G.J., *Glucuronidation of Drugs and Other Compounds,* CRC Press, Boca Raton, FL, 1980.
8. Kasper, C.B. and Henton, D., Glucuronidation, in *Enzymatic Basis of Detoxication,* Vol. II, Jakoby, W.B., Ed., Academic Press, New York, 1980, 3–36.
9. Burchell, B. and Coughtrie, M.W.H., UDP-glucuronosyltransferases, *Pharmac. Ther.,* 43, 261, 1989.
10. Iyanagi, T., Watanabe, T., and Uchiyama, Y., The 3-methylcholanthrene-inducible UDP-glucuronosyltransferase deficiency in the hyperbilirubinemic rat (Gunn rat) is caused by a −1 frameshift mutation, *J. Biol. Chem.,* 264, 21302, 1989.
11. Harding, D., Fournel-Gigleux, S., Jackson, M.R., and Burchell, B., Cloning and substrate specificity of a human phenol UDP-glucuronosyltransferase expressed in COS-7 cells, *Proc. Natl. Acad. Sci. U.S.A.,* 85, 8381, 1988.
12. MacKenzie, P.I., Rat liver UDP-glucuronosyltransferase. Identification of cDNAs encoding two enzymes which glucuronidate testosterone, dihydrotestosterone and β-estradiol., *J. Biol. Chem.,* 262, 9744, 1987.
13. Ritter, J.K., Crawford, J.M., and Owens, I.S., Cloning of two human liver bilirubin UDP-glucuronosyltransferase cDNAs with expression in COS-1 cells, *J. Biol. Chem.,* 266, 1043, 1991.
14. MacKenzie, P.I., Roy-Chowdhury, N., and Roy-Chowdhury, J., Characterization and regulation of rat liver UDP-glucuronosyltransferases, *Clin. Exp. Pharmacol. Physiol.,* 16, 501, 1989.
15. Wooster, R., Sutherland, L., Ebner, T., Clarke, D., and Da Cruz Silva, O., Cloning and stable expression of a new member of the human liver phenol/bilirubin:UDP-glucuronosyltransferase cDNA family, *Biochem. J.,* 278, 465, 1991.

16. Sato, H., Aono, S., Kashiwamata, S., and Koiwai, O., Genetic defect of bilirubin UDP-glucuronosyltransferase in the hyperbilirubinaemic Gunn rat, *Biochem. Biophys. Res. Commun.*, 177, 1161, 1991.
17. Clarke, D.J., George, S.G., and Burchell, B., Glucuronidation in fish, *Aquat. Toxicol.*, 20, 35, 1991.
18. Jansen, P.L.M. and Arias, I.M., Bilirubin metabolism in the spiny dogfish, *Squalus acanthias*, and the small skate, *Raja erinacea*, *Comp. Biochem. Physiol.*, 56B, 255, 1977.
19. Roy Chowdhury, J., Roy Chowdhury, N., and Arias, I., Bilirubin conjugation in the spiny dogfish, *Squalus acanthias*, the small skate, *Raja erinacea*, and the winter flounder, *Pseudopleuronectes americanus*, *Comp. Biochem. Physiol.*, 66B, 523, 1980.
20. Fang, L.S. and Huang, M.S., Purification and characterization of bilirubin UDP-glucuronosyltransferase from liver of the eel, *Anguilla japonica*, *Comp. Biochem. Physiol.*, 95C, 219, 1990.
21. Holmes, K., Characterization of haem metabolism in normal and hyperbilirubinaemic salmonids. M. Phil. Thesis, Dept. of Biochemistry, University of Leeds, 1989.
22. George, S.G., Brown, S., Groman, D., Holmes, K., and Pirie, J., Biochemical toxicology of a pollutant-induced hyperbilirubinaemia in Atlantic salmon, *Salmo salar*, *Mar. Environ. Res.*, 34, 81, 1992.
23. Clarke, D.J., Burchell, B., and George, S.G., Characterization and molecular analysis of hepatic microsomal UDP-glucuronosyltransferases in the marine fish plaice, *Pleuronectes platessa*, *Mar. Environ. Res.*, 24, 105, 1988.
24. Clarke, D.J., Burchell, B., and George, S.G., Functional and immunological comparison of hepatic UDP-glucuronosyltransferases in a piscine and a mammalian species, *Comp. Biochem. Physiol.*, 102B, 425, 1992.
25. Schoonen, W.G.E.J., Granneman, J.C.M., Lambert, J.G.D., and Van Oordt, P.G.W.J., Steroidogenesis in the testes and seminal vesicles of spawning and non-spawning African catfish, *Clarias gariepinus*, *Aquaculture*, 63, 77, 1987.
26. Lambert, J.G.D. and Resinck, J.W., Steroid glucuronides as male pheromones in the reproduction of African Catfish *Clarias gariepinus* — a brief review, *J. Steroid Biochem. Mol. Biol.*, 40, 4, 1991.
27. Kasper, C.B. and Henton, D., *Glucuronidation*, in *Enzymatic Basis of Detoxication*, Vol. II, Jakoby, W.B., Ed., Academic Press, New York, 1980, 3–36.
28. Van Den Hurk, R., Schoonen, W.G.E.J., Zoele, G.A., and Lambert, J.G.D., The biosynthesis of steroid glucuronides in the testis of the zebrafish, *Brachydanio rerio*, and their pheromonal function as ovulation inducers, *Gen. Comp. Endocrinol.*, 68, 179, 1987.
29. Van der Kraak, G., Sorensen, P.W.N., Stacey, N.E., and Duka, J.G., Periovulatory female goldfish release three potential pheromones: 17α,20ß-dihydroxyprogesterone, 17α,20β-dihydroxyprogesterone glucuronide and 17α-hydroxyprogesterone, *Gen. Comp. Endocrinol.*, 73, 452, 1989.
30. Forlin, L. and Haux, C., Increased excretion in the bile of 17β-(^{3}H) oestradiol-derived radioactivity in rainbow trout bile treated with β-naphthoflavone, *Aquat. Toxicol.*, 6, 197, 1985.
31. Stein, J.E., Hom, T., Sanborn, H., and Varanasi, U., The metabolism and disposition of 17β-oestradiol in English sole exposed to a contaminated sediment extract, *Mar. Environ. Res.*, 24, 252, 1988.

32. Sinclair, D.A.R. and Eales, J.G., Iodothyronine glucuronide conjugates in the bile of brook trout, *Salvelinus fontinalis* (Mitchill) and other freshwater teleosts, *Gen. Comp. Endocrinol.*, 552, 1972.
33. Lindstrom-Seppa, P., Koivusaari, U., and Hanninen, O., Extrahepatic metabolism in North European freshwater fish., *Comp. Biochem. Physiol.*, 69C, 259, 1981.
34. Kobayashi, K., Kimura, S., and Akitake, H., Studies on the metabolism of chlorophenols in fish. VII. Sulfate conjugation of phenol and PCP by fish livers, *Bull. Jpn. Soc. Sci. Fish.*, 42, 171, 1976.
35. Short, C.R., Flory, W., and Flynn, M., Hepatic drug metabolising enzyme activity in the channel catfish, *Ictalurus punctatus*, *Comp. Biochem. Physiol.*, 89C, 153, 1988.
36. Andersson, T., Pesonen, M., and Johansson, C., Differential induction of cytochrome P-450 dependent monooxygenase, epoxide hydrolase, glutathione transferase and UDP-glucuronosyltransferase activities in the liver of the rainbow trout by β-naphthoflavone or Clophen A50, *Biochem. Pharmacol.*, 34, 3309, 1985.
37. Gregus, Z., Watkins, J.B., Thompson,T.N., Harvey, M.J., Rozman, K., and Klaassen, C.D., Hepatic phase I and phase II biotransformations in quail and trout: comparison to other species commonly used in toxicity testing, *Toxicol. Appl. Pharmacol.*, 67, 430, 1983.
38. George, S., Young, P., Leaver, M., and Clarke, D., Activities of pollutant metabolising and detoxication systems in the liver of the plaice, *Pleuronectes platessa:* sex and seasonal variations in non-induced fish, *Comp. Biochem. Physiol.*, 96C, 185, 1990.
39. Goksoyr, A., Stenersen, J., Snowberger, E.A., Woodin, B.R., and Stegeman, J., Xenobiotic and steroid metabolism in adult and foetal piked (minke) whales, *Balaenoptera acutorostrata*, *Mar. Environ. Res.*, 24, 1, 1988.
40. Lech, J.J. and Statham, C.N., Role of glucuronide formation in the selective toxicity of 3-trifluoromethyl-4-nitrophenol (TFM) for the sea lamprey: comparative aspects of TFM uptake and conjugation in sea lamprey and rainbow trout, *Toxicol. Appl. Pharmacol.*, 31, 150, 1975.
41. Gregus, Z. and Klaasen, C.D., Effect of butylated hydroxyanisole on hepatic glucuronidation and biliary excretion of drugs in mice, *J. Pharm. Pharmacol.*, 40, 237, 1988.
42. Hanninen, O., Lindstrom-Seppa, P.U., Koivusaari, U., Vaisanen, M., Julkunen, A., and Juvonen, R., Glucuronidation and glucosidation reactions in aquatic species from boreal regions., *Biochem. Soc. Trans.*, 12, 13, 1984.
43. Ankley, G.T. and Agosin, M., Comparative aspects of hepatic UDP-glucuronosyltransferases and glutathione-S-transferases in Bluegill and Channel catfish, *Comp. Biochem. Physiol.*, 87B, 671, 1987.
44. Castren, M. and Oikari, A., Optimal assay conditions for liver UDP-glucuronosyltransferase from the rainbow trout, *Salmo gairdneri.*, *Comp. Biochem. Physiol.*, 76C, 365, 1983.
45. Koivusaari, U., Lindstrom, S.P., and Hanninen, O., Xenobiotic metabolism in rainbow trout intestine, *Adv. Physiol. Sci.*, 29, 433, 1980.
46. Clarke, D.J., Burchell, B., and George, S.G., Differential expression and induction of UDP-glucuronosyltransferase isoforms in hepatic and extrahepatic tissues of a fish: Immunochemical and functional characterization, *Toxicol. Appl. Pharmacol.*, 115, 130, 1992.

47. Kapitulnik, J., Weil, E., Rabinowitz, R., and Krausz, M.M., Fetal and adult human liver differ markedly in the fluidity and lipid composition of their microsomal membranes, *Hepatology,* 7, 55, 1987.
48. Clarke, D.J., Functional and molecular characterization of piscine UDP-glucuronosyltransferase, Ph.D.Thesis, Dept. of Biochemistry, University of Dundee, Scotland, 1990.
49. Hews, E.A. and Kime, D.E., Formation of testosterone glucuronide by testes of rainbow trout, *Salmo gairdneri, Gen. Comp. Endocrinol.,* 34, 116, 1978.
50. Resinck, J.W., Schoonen, W.G.E.J., Van Den Hurk, R., Viveen, W.J.A.R., and Lambert, J.G.D., Seasonal changes in steroid metabolism in the male reproductive organ system of the African catfish, *Clarias gariepinus, Aquaculture,* 63, 59, 1987.
51. Clarke, D.J., George, S.G., and Burchell, B., Multiplicity of UDP-glucuronosyltransferases in fish: purification and characterization of a phenol UDP-glucuronosyltransferase from liver of the plaice, *Biochem. J.,* 284, 417, 1992.
52. Pesonen, M., Celander, M., Forlin, L., and Andersson, T., Comparison of xenobiotic biotransformation enzymes in kidney and liver of rainbow trout *(Salmo gairdneri), Toxicol. Appl. Pharmacol.,* 91, 75, 1987.
53. Whitlock, J.P.J., The regulation of cytochrome P-450 gene expression, *Annu. Rev. Pharmacol. Toxicol.,* 26, 33, 1986.
54. Kleinow, K.M., Melancon, M.J., and Lech, J.J., Biotransformation and induction: implications for toxicity, bioaccumulation and monitoring of environmental xenobiotics in fish, *Environ. Health Perspect.,* 71, 1987.
55. Jakoby, W.B., *The Sulfotransferases,* Francis and Taylor, London, 1989.
56. Koster, H., Halsema, I., Scholtens, E., Knippers, M., and Mulder, G., Dose dependent shifts in the sulfation and glucuronidation of phenolic compounds in the rat *in vivo* and in isolated hepatocytes. The role of saturation of phenolsulfotransferase, *Biochem. Pharmacol.,* 30, 2569, 1981.
57. Landers, J.P. and Bunce, N.J., The *Ah* receptor and the mechanism of dioxin toxicity, *Biochem. J.,* 276, 273–287, 1991.
58. Layiwola, P.J. and Linnecar, D.F.C., The biotransformation of [^{14}C] phenol in some freshwater fish, *Xenobiotica,* 11, 167, 1981.
59. Pritchard, J. and Bend, J., Mechanisms controlling the renal excretion of xenobiotics in fish: effects of chemical structure, *Drug Metab. Rev.,* 15, 655, 1984.
60. Varanasi, U. and Gmur, D., Hydrocarbons and metabolites in English sole *(Parophys vetulus)* exposed simultaneously to [^{3}H] benzo(a)pyrene and [^{14}C]naphthalene in oil-contaminated sediment, *Aquat. Toxicol.,* 1, 49, 1981.
61. Morrison, H., Young, P., and George, S., Conjugation of organic compounds in isolated hepatocytes from a marine fish, the plaice, *Pleuronectes platessa, Biochem. Pharmacol.,* 34, 3933, 1985.
62. Leaver, M.J., Clarke, D.J., and George, S.G., Molecular studies of the phase II xenobiotic-conjugating enzymes of marine Pleuronectid flatfish, *Aquat. Toxicol.,* 22, 425, 1992.
63. Parker, R.S., Morrissey, M.T., Moldeus, P., and Selivonchick, D.P., The use of isolated hepatocytes from rainbow trout *(Salmo gairdneri)* in the metabolism of acetaminophen, *Comp. Biochem. Physiol.,* 70B, 631, 1981.
64. Andersson, T., Forlin, L., and Hansson, T., Biotransformation of 7-ethoxycoumarin in isolated perfused rainbow trout liver, *Drug. Metab. Disp.,* 11, 494, 1983.

65. Jakoby, W.B., The glutathione S-transferases, *Adv. Enzymol.*, 46, 383, 1978.
66. Mannervik, B., The isoenzymes of glutathione transferase, *Adv. Enzymol.*, 57, 357, 1985.
67. Hayes, J.D., Pickett, C.B., and Mantle, T.J., *Glutathione S-Transferases and Drug Resistance*, Hayes, J.D. Mantle, T.J., and Pickett, C.B., Eds., Taylor and Francis, London, 1990.
68. Listowsky, I., Abramowitz, M., Homa, H., and Niitsu, Y., Intracellular binding and transport of hormones and xenobiotics by glutathione S-transferases, *Drug Metab. Rev.*, 19, 305, 1988.
69. Smith, M.T., Evans, C.G., Doane, S.P., Castro, V.M., Tahir, M.K., and Mannervik, B., Denitrosation of 1,3-bis(2-chloroethyl)-1-nitrosourea by class μ-glutathione transferases and its role in cellular resistance in rat brain tumor cells, *Cancer Res.*, 49, 2621, 1989.
70. Hayes, J.D. and Wolf, C.R., Molecular mechanisms of drug resistance, *Biochem. J.*, 272, 281, 1990.
71. Clark, A.G., The glutathione S-transferases and resistance to insecticides, in *Glutathione S-Transferases and Drug Resistance*, Hayes, J.D., Mantle, T.J., and Pickett, C.B., Eds., Taylor and Francis, London, 1990, 369.
72. Hiratsuka, A., Sebata, N., Kawashima, K., Okuda, H., Ogura, K., Watabe, T., Satoh, K., Hatayama, I., Tsuchida, S., Ishikawa, T., and Sato, K., A new class of rat glutathione S-transferase Yrs-Yrs inactivating reactive sulfate esters as metabolites of carcinogenic arylmethanols, *J. Biol. Chem.*, 265, 11973, 1990.
73. Board, P., Genetic polymorphisms of glutathione S-transferase in man, in *Glutathione S-Transferases and Drug Resistance*, Hayes, J.D., Pickett, C.B., and Mantle, T.J., Eds., Taylor and Francis, London, 1990, 232.
74. Pickett, C.B., Structure and regulation of glutathione S-transferase genes, *Essays Biochem.*, 23, 116, 1987.
75. Pickett, C.B. and Lu, A.Y.H., Glutathione S-transferases: gene structure, regulation, and biological function, *Annu. Rev. Biochem.*, 58, 743, 1989.
76. Mannervik, B. and Danielson, U.H., Glutathione transferases — structure and catalytic activity, *Crit. Rev. Biochem.*, 23, 283, 1988.
77. Christ-Hazelhof, E. and Nutgeren, D.H., Purification and characterization of prostaglandin endoperoxide D-isomerase, a cytoplasmic glutathione-requiring enzyme, *Biochim. Biophys. Acta*, 572, 43, 1979.
78. Ketterer, B., Meyer, D.J., and Clark, A.G., Soluble glutathione transferase isoenzymes, in *Glutathione Conjugation. Mechanisms and Biological Significance*, Sies, H. and Ketterer, B., Eds., Academic Press, London, 1988, 73.
79. Ketterer, B., Meyer, D.J., Taylor, J.B., Pemble, S., Coles, B., and Fraser, G., GSTs and protection against oxidative stress, in *Glutathione S-Transferases and Drug Resistance*, Hayes, J.D., Pickett, C.B., and Mantle, T.J., Eds., Taylor and Francis, London, 1990, 97.
80. Alin, P., Danielson, U.H., and Mannervik, B., 4-Hydroxyalkl-2-enals are substrates for glutathione transferase, *FEBS Lett.*, 179, 267, 1985.
81. Meyer, D.J., Beale, D., Tan, K.H., Coles, B., and Ketterer, B., Glutathione transferases in primary rat hepatomas: the isolation of a form with GSH peroxidase activity, *FEBS Lett.*, 184, 139, 1985.
82. Hayes, J.D., Selective elution of rodent glutathione S-transferases and glyoxalase I from the S-hexylglutathione-sepharose affinity matrix, *Biochem. J.*, 255, 913, 1988.

83. Rushmore, T.H., King, R.G., Paulson, K.E., and Pickett, C.B., Regulation of glutathione S-transferase Ya subunit gene expression: identification of a unique xenobiotic-responsive element controlling inducible expression by planar aromatic compounds, *Proc. Natl. Acad. Sci. U.S.A.*, 87, 3826, 1990.
84. Rushmore, T.H., Morton, M.R., and Pickett, C.B., The antioxidant responsive element. Activation by oxidative stress and identification of the DNA consensus sequence required for functional activity, *J. Biol.Chem.*, 266, 11632, 1991.
85. Ding, V.D.-H. and Pickett, C.B., Transcriptional regulation of rat liver glutathione S-transferase genes by phenobarbital and 3-methylcholanthrene, *Arch. Biochem. Biophys.*, 215, 539, 1985.
86. McLellan, L.I., Kerr, L.A., Cronshaw, A.D., and Hayes, J.D., Regulation of mouse glutathione S-transferases by chemoprotectors — molecular evidence for the existence of 3 distinct α-class glutathione S-transferase subunits, Ya_1, Ya_2 and Ya_3, in mouse liver, *Biochem. J.*, 276, 461, 1991.
87. Jensen, D.E. and MacKay, R.L., Rat, mouse and hamster isoenzyme specificity in the glutathione transferase-mediated denitrosation of nitrosoguanidinium compounds, *Cancer Res.*, 50, 1440, 1990.
88. Stanley, J.S. and Benson, A.M., The conjugation of 4-nitroquinoline 1-oxide, a potent carcinogen, by mammalian glutathione transferases. 4-Nitroquinoline 1-oxide conjugation by human, rat, and mouse liver cytosols, extrahepatic organs of mice by purified mouse glutathione transferase isoenzymes, *Biochem. J.*, 256, 303, 1988.
89. Hayes, J.D., Purification and charterization of glutathione S-transferases P, S, and N. Isolation from rat liver of Yb_1Y_n protein, the existence of which was predicted by subunit hybridization *in vitro, Biochem. J.*, 224, 839, 1984.
90. Kispert, A., Meyer, D.J., Lalor, E., Coles, B., and Ketterer, B., Purification and characterization of a labile rat glutathione transferase of the μ-class, *Biochem. J.*, 260, 789, 1989.
91. Abramovitz, M., Ishigaki, S., Felix, A.M., and Listowsky, I., Expression of an enzymatically active Y_{b3} glutathione S-transferase in *Escherichia coli* and identification of its natural form in rat brain, *J. Biol. Chem.*, 263, 17627, 1988.
92. Lai, H.C., Grove, G., and Tu, C.P., Cloning and sequence analysis of a cDNA for a rat liver glutathione S-transferase Y_b subunit, *Nucl. Acids Res.* 14, 6101, 1986.
93. Lai, H.C., Qian, B., and Tu, C.P., Characterization of a variant rat glutathione S-transferase by cDNA expression in *Escherichia coli, Arch. Biochem. Biophys.*, 273, 423, 1989.
94. Tu, C.P.-D., Lai, H.-C., and Reddy, C.C., The rat glutathione S-transferases supergene family: molecular basis of gene multiplicity, in *Glutathione S-Transferases and Carcinogenesis,* Mantle, T.J., Pickett, C.B., and Hayes, J.D., Eds., Taylor and Francis, London, 1987, 87.
95. Hayes, J.D., Kerr, L.A., Peacock, S.D., Cronshaw, A.D., and McLellan, L.I., Hepatic glutathione S-transferases in mice fed on a diet containing the anticarcinogenic antioxidant butylated hydroxyanisole, *Biochem. J.*, 277, 501, 1991.
96. DeJong, J.L., Mohandas, T., and Tu, C.-P. D., The human H_b μ-class glutathione S-transferases are encoded by a dispersed gene family, *Biochem. Biophys. Res. Commun.*, 180, 15, 1991.

97. Sato, K., Glutathione S-transferase and hepatocarcinogenesis, *Jpn. J. Cancer Res.(Gann),* 79, 556, 1988.
98. Sato, K., Satoh, K., Tsuchida, S., Hatayama, I., Tamai, K., and Shen, H., Glutathione S-transferases and (pre)neoplasia, in *Glutathione S-Transferases and Drug Resistance,* Hayes, J.D., Pickett, C.B., and Mantle, T.J., Eds., Taylor and Francis, London, 1990, 389.
99. Berhane, K. and Mannervik, B., Inactivation of the genotoxic aldehyde acrolein by human glutathione transferases of classes α, μ, and π, *Mol. Pharmacol.,* 37, 251, 1990.
100. Robertson, I., Jensson, H., Mannervik, B., and Jernstrom, B., Glutathione transferases in rat lung: the presence of transferase 7-7, highly efficient in the conjugation of glutathione with the carcinogenic (+)-7β, 8 α-dihydroxy-9 α, 10 α-oxy-7,8,9,10-tetrahydrobenzo(a) pyrene, *Carcinogenesis,* 7, 295, 1986.
101. Sugioka, Y., Kano, T., Okuda, A., Sakai, M., Kitagawa, T., and Muramatsu, M., Cloning and the nucleotide sequence of rat glutathione S-transferase P cDNA, *Nucl. Acids Res.,* 13, 6049, 1985.
102. Pemble, S.E., Taylor, J.B., and Ketterer, B., Tissue distribution of rat glutathione transferase subunit 7, a hepatoma marker, *Biochem. J.,* 240, 885, 1986.
103. Okuda, A., Sakai, M., and Muramatsu, M., The structure of the rat glutathione S-transferase P gene and related pseudogenes, *J. Biol. Chem.,* 262, 3858, 1987.
104. Sakai, M., Okuda, A., and Muramatsu, M., Multiple regulatory elements and phorbol 12-O-tetradecanoate 13-acetate responsiveness of the rat placental glutathione transferase gene, *Proc. Natl. Acad. Sci. U.S.A.,* 85, 9456, 1988.
105. Imagawa, M., Osada, S., Okuda, A., and Muramatsu, M., Silencer binding proteins function on multiple *cis*-elements in the glutathione transferase-P gene, *Nucl. Acids Res.,* 19, 5, 1991.
106. Okuda, A., Imagawa, M., Maeda, Y., Sakai, M., and Muramatsu, M., Structural and functional analysis of an enhancer GPEI having a phorbol 12-O-tetradecanoate 13-acetate responsive element-like sequence found in the rat glutathione transferase P gene, *J. Biol. Chem.,* 264, 16919, 1989.
107. Sakai, M., Okuda, A., Hatayama, I., Sato, K., Nishi, S., and Muramatsu, M., Structure and expression of the rat c-jun messenger RNA: tissue distribution and increase during chemical carcinogenesis, *Cancer Res.,* 49, 5633, 1989.
108. Ziegler, D.M., Role of reversible oxidation-reduction of enzyme thiols-disulphides in metabolic regulation. *Annu. Rev. Biochem.,* 54, 305, 1985.
109. Fjellstedt, T.A., Allen, R.H., Duncan, B.K., and Jakoby, W.B., Enzymatic conjugation of epoxides with glutathione, *J. Biol. Chem.,* 248, 3702, 1973.
110. Meyer, D.J., Christodoulides, L.G., Tan, K.B., and Ketterer, B., Isolation, properties and tissue distribution of rat glutathione transferase E, *FEBS Lett.,* 173, 327, 1984.
111. Spearman, M., Prough, R., Estabrook, R., Falck, J., Manna, S., Leibman, K., Murphy, R., and Capdevila, J., Novel glutathione conjugates formed from epoxyeicosatrienoic acids (EETS), *Arch. Biochem. Biophys.,* 242, 225, 1985.
112. Meyer, D.J., Coles, B., Pemble, S.E., Gilmore, K.S., Fraser, G.M., and Ketterer, B., θ, a new class of glutathione transferases purified from rat and man, *Biochem. J.,* 274, 409, 1991.
113. Grove, G., Zarlengo, R.P., Timmerman, K.P., Li, N.Q., Tam, M.F., and Tu, C.P., Characterization and heterospecific expression of cDNA clones of genes in the maize GSH S-transferase multigene family, *Nucl. Acids Res.,* 16, 425, 1988.

114. Meyer, R.C.J., Goldsborough, P.B., and Woodson, W.R., An ethylene responsive flower senescence gene from carnation encodes a protein homologous to glutathione S-transferases, *Plant Mol. Biol.*, 17, 277, 1991.
115. Toung, Y.P., Hsieh, T.S., and Tu, C.-P.D., *Drosophila* glutathione S-transferase 1-1 shares a region of sequence homology with the maize glutathione S-transferase III, *Proc. Natl. Acad. Sci. U.S.A.*, 87, 31, 1990.
116. Nimmo, I.A., The glutathione S-transferases of fish, *Fish Physiol. Biochem.*, 3, 163, 1987.
117. Donnarumma, L., De Angelis, G., Gramenzi, F., and Vittozzi, L., Xenobiotic metabolizing enzyme systems in test fish: III. Comparative studies of liver cytosolic glutathione S-transferases, *Ecotoxicol. Environ. Saf.*, 16, 180, 1988.
118. George, S., Buchanan, G., Nimmo, I., and Hayes, J.D., Fish and mammalian liver cytosolic glutathione S-transferases: substrate specificities and immunological comparison, *Mar. Environ. Res.*, 28, 1, 1989.
119. Braddon, S.A., McIlvaine, C.M., and Balthrop, J.A., Distribution of GSH and GSH cycle enzymes in black sea bass *(Centropristis striata)*, *Comp. Biochem. Physiol.*, 80B, 213, 1985.
120. Bauermeister, A., Lewendon, A., Ramage, P., and Nimmo, I.A., Distribution and some properties of the glutathione S-transferase and γ-glutamyl transpeptidase activities of rainbow trout, *Comp. Biochem. Physiol.*, 74C, 89, 1983.
121. Leaver, M.J., Scott, K., and George, S.G., Expression and tissue distribution of plaice glutathione S-transferase A, *Mar. Environ. Res.*, in press, 1992.
122. Sugiyama, Y., Yamada, T., and Kaplowitz, N., Glutathione S-transferases in elasmobranch liver: molecular heterogeneity, catalytic and binding properties, and purification, *Biochem. J.*, 199, 749, 1981.
123. Foureman, G.L. and Bend, J.R., The hepatic glutathione transferases of the male little skate, *Raja erinacea*, *Chem. Biol. Interactions*, 49, 89, 1984.
124. Dierickx, P.J., Hepatic glutathione S-transferases in rainbow trout and their interaction with 2,4-dichlorophenoxyacetic acid and 1,4-benzoquinone, *Comp. Biochem. Physiol.*, 82C, 495, 1985.
125. Ramage, P. and Nimmo, I.A., The substrate specificities and subunit compositions of the hepatic glutathione S-transferases of rainbow trout *(Salmo gairdneri)*, *Comp. Biochem. Physiol.*, 78B, 1984.
126. Dierckx, P., Purification and partial characterization of the glutathione S-transferases in carp liver, and their interaction with 2,4-dichlorophenoxyacetic acid and 1,4-benzoquinone., *Biochem. Int.*, 11, 755, 1985.
127. George, S.G. and Young, P., Purification and properties of plaice liver cytosolic glutathione-S-transferases, *Mar. Environ. Res.*, 24, 1, 1986.
128. Ramage, P.I.N., Rae, G.H., and Nimmo, I.A., Purification and some properties of the hepatic glutathione S-transferases of the Atlantic salmon *(Salmo salar)*, *Comp. Biochem. Physiol.*, 83B, 23, 1986.
129. Dominey, R.J., Nimmo, I.A., Cronshaw, A.D., and Hayes, J.D., The major glutathione S-transferase in salmonid fish livers is homologous to the mammalian π-class GST, *Comp. Biochem. Physiol.*, 100B, 93, 1991.
130. Foureman, G.L., Hernandez, O., Bhatia, A., and Bend, J.R., The stereoselectivity of four hepatic glutathione S-transferases purified from a marine elasmobranch *(Raja erinacea)* with several K-region polycyclic arene oxide substrates, *Biochim. Biophys. Acta*, 914, 127, 1987.

131. Nemoto, N., Gelboin, H.V., Habig, W.H., and Ketley, J.N., K region benzo(a)pyrene 4,5-oxide is conjugated by homogeneous glutathione S-transferases, *Nature (Lond.)*, 255, 512, 1974.
132. Bell, J.G., Cowey, C.B., and Youngson, A., Rainbow trout liver microsomal lipid peroxidation. The effect of purfied glutathione peroxidase, glutathione S-transferase and other factors, *Aquaculture*, 65, 43, 1984.
132a. Dominey, R.J., Unpublished data.
133. George, S.G. and Buchanan, G., Isolation, properties and induction of plaice liver cytosolic glutathione S-transferases, *Fish Physiol. Biochem.*, 8, 437, 1990.
134. Goksoyr, A., Andersson, T., Hansson, T., Klungsoyr, J., Zhang, Y., and Forlin, L., Species characteristics of the hepatic xenobiotic and steroid biotransformation systems of two teleost fish, Atlantic cod *(Gadus morhua)* and rainbow trout *(Salmo gairdneri), Toxicol. Appl. Pharmacol.*, 89, 347, 1987.
135. Koss, M., Losekam, M., Schuler, E., and Schreiber, I., The hepatic glutathione content and glutathione S-transferase activity in the pike *(Esox lucius* L.) and rat, *Comp. Biochem. Physiol.*, 99B, 257, 1991.
136. Lindstrom-Seppa, P., Biotransformation in fish: monitoring inland water pollution caused by pulp and paper mill effluents, Original reports 8, Dept. of Physiology, University of Kuopio, Finland, 1990.
137. Sivarajah, K., Franklin, C.S., and Williams, W.P., The effects of polychlorinated biphenyls on plasma steroid levels and hepatic microsomal enzymes in fish, *J. Fish Biol.*, 13, 401, 1978.
138. Collier, T.K., Singh, S.V., Awasthi, Y.C., and Varanasi, U., Hepatic xenobiotic metabolising enzymes in 2 species of benthic fish showing different prevalences of contaminant-associated liver neoplasms, *Toxicol. Appl. Pharmacol.*, 113, 319, 1992.
139. Leaver, M.J., Scott, K., and George, S.G., Cloning and characterization of the major hepatic glutathione S-transferase from a marine teleost flatfish, the plaice, *Pleuronectes platessa*, with structural similarities to plant, insect and mammalian θ-class isoenzymes, *Biochem. J.*, 292, 189–195, 1993.
140. Roy-Chowdhury, J.R., Huang, T., Kesari, K., Lederstein, M., Arias, I.M., and Roy-Chowdhury, N., Molecular basis for the lack of bilirubin-specific and 3-methylcholanthrene-inducible UDP-glucuronosyltransferase activities in Gunn rats, *J. Biol. Chem.*, 266, 18924, 1991.
141. Hayes, J.D., Judah, D.J., McLellan, L.I., Kerr, L.A., Peacock, S.D., and Neal, G.E., Ethoxyquin-induced resistance to aflatoxin B_1 in rat is associated with the expression of a novel α-class glutathione S-transferase subunit,Yc_2, which possesses high catalytic activity toward aflatoxin B_1-8,9-epoxide, *Biochem, J.*, 279, 385, 1991.
142. Mandel, H.G., Manson, M.M., Judah, D.J., Simpson, J.L., Green, J.A., Forrester, L.M., Wolf, C.R., and Neal, G.E., Metabolic basis for the protective effect of the antioxidant ethoxyquin in Aflatoxin B_1 hepatocarcinogenesis in the rat, *Cancer Res.*, 47, 5218, 1987.
143. Nimmo, I.A., Coghill, D.R., Hayes, J.D., and Strange, R.C., A comparison of the subcellular distribution, subunit composition and bile-acid binding activity of glutathione S-transferases from trout and rat liver, *Comp. Biochem. Physiol.*, 68B, 579, 1981.

144. Morgenstern, R., Lundqvist, G., Balk, L., and De Pierre, J.W., The distribution of microsomal glutathione transferase among different organelles, different organs and different organisms, *Biochem. Pharmacol.*, 33, 3609, 1984.
145. Morgenstern, R. and De Pierre, J.W., Microsomal glutathione transferase, in *Reviews in Biochemical Toxicology,* Hodgson, E., Bend, J.R., and Philpot, R.M., Eds., Elsevier, New York, 1985, 67.
146. Morgenstern, R. and De Pierre, J.W., Membrane-bound glutathione transferases, in *Glutathione Conjugation: Mechanisms and Biological Significance,* Ketterer, B. and Sies, H., Eds., Academic Press, London, 1988, 157.
147. Morgenstern, R., Lundqvist, G., Andersson, C., and Mosialou, E., Membrane bound glutathione S-transferase: function and properties, in *Glutathione S-Transferases and Drug Resistance,* Hayes, J.D., Pickett, C., and Mantle, T.J., Eds., Taylor and Francis, London, 1990, 57.
148. George, S.G. and Young, P., The time course of effects of cadmium and 3-methylcholanthrene on activities of enzymes of xenobiotic metabolism and metallothionein levels in the plaice, *Pleuronectes platessa, Comp. Biochem. Physiol.*, 83C, 37, 1986.
149. Scott, K., Leaver, M., and George, S., Regulation of hepatic glutathione S-transferase expression in flounders, *Mar. Environ. Res.*, 34, 233, 1992.
150. Williams, D. and Buhler, D., Purification of cytochromes P-448 from β-naphthoflavone-treated rainbow trout, *Biochim. Biophys. Acta,* 717, 398, 1982.
151. Park, S., Miller, H., Klotz, A., Kloepper-Sams, P., Stegeman, J., and Gelboin, H., Monoclonal antibodies to liver microsomal cytochrome P-450-E of the marine fish *Stenomotus chrysops* (Scup), *Arch. Biochem. Biophys.*, 249, 339, 1986.
152. Heilmann, L.J., Sheen, Y.-Y., Bigelow, S.W., and Nebert, D.W., Trout P4501A1: cDNA and deduced protein sequence, expression in liver and evolutionary significance, *DNA,* 7, 379, 1988.
153. Wirgin, I., Kreamer, G.-L., and Garte, S.J., Genetic polymorphism of cytochrome P-4501A in cancer-prone Hudson River tomcod, *Aquat. Toxicol.*, 19, 205, 1992.
154. Myers, M.S., Landahl, J.T., Krahn, M.M., and McCain, B.B., Relationships between hepatic neoplasms and related lesions and exposure to toxic chemicals in marine fish from the U.S. west coast, *Environ. Health Perspect.*, 90, 7, 1991.
155. Van Veld, P.A., Ko, U.K., Vogelbein, W.K., and Westbrook, D.J., Glutathione S-transferase in intestine, liver and hepatic lesions of mummichog *(Fundulus heteroclitus)* from a creosote-contaminated environment, *Fish Physiol. Biochem.*, 9, 369, 1991.
156. James, M.O., Heard, C.S., and Hawkins, W.E., Effect of 3-methylcholanthrene on monooxygenase, epoxide hydrolase, and glutathione S-transferase activities in small estuarine and freshwater fish, *Aquat. Toxicol.*, 12, 1, 1988.
157. Pesonen, M. and Andersson, T., Subcellular localization and properties of cytochrome P-450 and UDP-glucuronosyltransferase in the rainbow trout kidney, *Biochem. Pharmacol.*, 36, 823, 1987.
158. Zhang, Y.S., Andersson, T., and Forlin, L., Induction of hepatic xenobiotic biotransformation enzymes in rainbow trout by β-naphthoflavone. Time course studies, *Comp. Biochem. Physiol.*, 95B, 247, 1990.
159. Kirby, G.M., Stalker, M., Metcalfe, C., Kocal, T., Ferguson, H., and Hayes, M.A., Expression of immunoreactive glutathione S-transferases in hepatic neoplasms induced by aflatoxin B1 or 1,2-dimethylbenzanthracene in rainbow trout *(Onchorynchus mykiss), Carcinogenesis,* 11, 2255, 1990.

160. Chattergee, S. and Bhattacharya, S., Detoxication of industrial pollutants by the glutathione S-transferase system in the liver of *Anabas testudineus* (Bloch), *Toxicol. Lett.*, 22, 187, 1984.
161. James, M.O. and Bend, J.R., Polycyclic aromatic hydrocarbon induction of cytochrome P-450 dependent mixed function oxidases in marine fish, *Toxicol. Appl. Pharmacol.*, 54, 117, 1980.
162. Ogura, K., Nishiyama, T., Okada, T., Kajita, J., Narihata, H., Watabe, T., Hiratsuka, A., and Watabe, T., Molecular cloning and amino acid sequencing of rat liver class θ-glutathione S-transferase Yrs-Yrs inactivating reactive sulfate esters of carcinogenic arylmethanols, *Biochem. Biophys. Res. Commun.*, 181, 1294, 1991.
163. Lauren, D.J., Halarnkar, P.P., Hammock, B.D., and Hinton, D.E., Microsomal and cytosolic epoxide hydrolase and glutathione transferase activities in the gills, liver and kidney of the rainbow trout, *Salmo gairdneri*, *Biochem. Pharmacol.*, 38, 881, 1989.

Chapter 3

Biochemistry and Molecular Biology of Monooxygenases: Current Perspectives on Forms, Functions, and Regulation of Cytochrome P450 in Aquatic Species

John J. Stegeman* and Mark E. Hahn*

I. PERSPECTIVE

By its nature, aquatic toxicology may be the major part of environmental toxicology. Seventy percent of the earth's surface and more than ninety percent of the biosphere is water. This environment is perceived to be at risk from chemical contaminants. Thousands of synthetic chemical compounds are currently registered for use in industry and agriculture, and thousands of tons of these are produced annually. Regardless of the source or original intended use, portions of these chemicals are released either deliberately or unintentionally into the environment. Thousands of tons more of unintended byproducts accompany the synthetic chemicals. If not placed deliberately into the aquatic environment, hydrologic and atmospheric processes distribute these chemicals, eventually depositing them in aquatic systems. These anthropogenic chemicals join a multitude of natural products, chemicals that are synthesized by plants and animals or, like hydrocarbons, are formed through geochemical or pyrocatalytic transformations. These anthropogenic and natural toxic compounds are referred to as xenobiotics; how they interact with aquatic life is the concern of aquatic toxicology.

* To Betsy and Rachel and our parents.

Enormous effort is devoted to dissecting the mechanisms of action of foreign chemicals (drugs, pollutants, carcinogens, etc.) in mammalian models such as rats and mice. The ultimate aim of that comparative research is to detect, control, and possibly intervene in chemical exposure and effects in humans. In aquatic toxicology, we are broadly concerned with the health and safety of aquatic species, for their own sake and as resources for human needs. This concern transcends a focus on the health or reproduction of a few species within an ecosystem. Properly functioning aquatic ecosystems, not only provide critical food and material resources, but also may be indicators of the "health" and well-being of the planet. We have a vital interest in evaluating and understanding the effects of chemicals in the aquatic environment, to judge the reality of the perceived risk. There is need to address these issues in a timely fashion, as some current perceptions are that the risk is great.[1] There is a further need for knowledge of how drugs or other therapeutic agents act in aquatic species, to develop a pharmacology in support of aquaculture, an increasingly important source of protein.

Many and diverse specific research objectives are pursued in aquatic toxicology. These are generally subsidiary to the broader aims, which include the use of aquatic species (1) as model systems for investigating fundamental biochemical and cellular processes involved in the action of xenobiotics, (2) in evaluating new chemicals for their environmental or human health hazard, (3) in evaluating and monitoring effluents and other waste streams, and (4) in delineating and evaluating the biological and ecological significance of local and global contamination. Progress in aquatic toxicology might be judged not only by how successfully the more focused objectives are met, but by whether those objectives, once met, have a lasting influence on how we use and protect the aquatic environment.

Understanding the principles that determine chemical or biological specificity of xenobiotics is particularly important to meeting the larger objectives in aquatic toxicology, as it is impossible to establish empirically the sensitivity, susceptibility, or resistance of every species to each type of compound. Knowledge of the comparative biochemistry and molecular biology of chemical effects is essential for judging when extrapolation between species is valid and when it is not. Determining the validity of species extrapolation is the foundation on which the use of all animal models should be built, whether mammalian or nonmammalian, terrestrial or aquatic. The comparative approach ultimately will define the more general principles involved and their evolution. It is thus imperative that we define the mechanisms of chemical action and the similarities and differences in how these mechanisms operate in different species.

II. BIOTRANSFORMATION

There is a common paradigm in toxicology (Table 1), regardless of the species or the system of concern. This paradigm encompasses the sources,

Table 1. Elements of the Common Paradigm in Toxicology

Chemical sources
Environmental fate
  Transport
  Partitioning
  Degradation
    (chemical, photochemical, and microbial)
Bioavailability
Uptake (passive or active; routes)
Distribution (intraorganismal; intracellular)
Toxicity (mechanisms)
Biotransformation
  Activation
  Inactivation
Redistribution
Elimination

environmental transport and transformations, biological uptake, distribution, action, detoxication, and elimination of xenobiotics. The structure of foreign compounds dictating their chemical and biological reactivity and effects is at the heart of that paradigm. Metabolic transformation (biotransformation) of those structures is able to dramatically alter their biological activity and, consequently, the outcome of interaction between chemical and cell. The metabolic machinery called into play in biotransformation includes a large number of different enzymes, which act on diverse types of substrates. Some of these enzymes are listed in Table 2. Many of these enzymes have in common a function of converting toxic structures to less toxic structures, and converting lipophilic or fat-soluble chemicals into structures that are more water soluble and thus more readily excreted. In earlier times, this aggregate function was known as the drug metabolizing system.

It has long been known that some metabolic products are themselves more reactive and toxic than the unmetabolized parent compound. Many carcinogens and toxic chemicals are activated by metabolism. Often such compounds go through a series of reactions, with several enzymes participating in sequential steps leading to activation, detoxification, and ultimately the excretion of the xenobiotic. R. T. Williams[2] proposed the concept of phase I reactions (oxidation and functionalization) and phase II reactions (conjugation and detoxication) to distinguish the phases in this sequential process. Though oxidative reactions predominate in activation, conjugation reactions (e.g., sulfation) can also participate directly in bioactivation.

Current directions in research on the biochemistry of oxidative metabolism of xenobiotics and its regulation in aquatic species are discussed here. In keeping with the objective of this volume, we try to look forward as much as to look back. We regret not being able to review in full the literature on biotransformation enzymes in aquatic species (and apologize for omissions). There are many excellent earlier

Table 2. Enzymes Involved in Xenobiotic Biotransformation

Acetyltransacetylase
Alcohol dehydrogenases
Aldehyde dehydrogenases
Aldehyde oxidase
D-Amino acid oxidase
Carbonyl reductase
Catalase
Cysteine-conjugate N-acetyltransferase
Cytochrome P450-containing monooxygenases
Dihydrodiol dehydrogenase
Epoxide hydrolase
Esterases/amidases
Flavin-containing monooxygenases
Glutathione peroxidase
Glutathione reductase
Glutathione transferases
Methyltransferases
Monoamine oxidase
Prostaglandin synthetase-hydroperoxidase
Quinone reductase
Rhodanese
Sulfotransferases
Superoxide dismutase
Thiol transferase
UDP-glucuronosyl transferases
Xanthine oxidase

and more recent contributions that provide such review;[3–13] we draw on and refer to these throughout this chapter. We summarize and illustrate some aspects of xenobiotic metabolism, but provide greater detail on other aspects, particularly regarding the diversity of P450 forms and their regulation. In each case, we point to research directions that we feel might yield rewards, in basic biochemical knowledge and in addressing the needs in aquatic toxicology. One objective here is to pose questions and offer generalizations, however tentative. The questions stated or implied are perhaps the most important feature of this chapter. But we hope the summaries and larger picture make it suitable also as an introduction to the field, illustrating the exciting possibilities for future research on the diversity, significance, and evolution of mechanisms by which organisms defend themselves against toxic chemical challenges.

What we discuss here is not fundamentally new. From the older literature, one appreciates that the same concerns were held by researchers 25 to 30 years ago. Indeed, Adamson,[3] in his review of 1967, offered similar reasons for pursuing metabolism of foreign chemicals in aquatic species. He stated, for example "Studies of the disposition of foreign compounds in various marine species will shed further light on the evolution of enzymes that metabolize drugs, on drug metabolic pathways, and drug excretion, and on factors affecting the biological half life of foreign compounds." He also pointed out that many factors involving, for example, organ and hormonal differences, complicate the issues. Early, major figures in drug metabolism (notably, R. Tecwyn Williams, who coined the term xenobiotic, and Bernard B. Brodie) raised questions concerning xenobiotic metabolism in aquatic species that influence the research today. (Williams' influence continues through his students, e.g., M. O. James.) Brodie and Maickel[14] suggested (unfortunately incorrectly) that aquatic species did not metabolize xenobiotics, and that elimination of unmetabolized xenobiotics across epithelial surfaces might be the dominant process. Potter and O'Brien,[15] Adamson,[3] and Buhler and Rasmussen,[16] described xenobiotic metabolism in numerous fish, avian, invertebrate, and amphibian species. Creaven et al.[17,18] described the dealkylation and hydroxylation of a series of model biphenyl and alkoxybiphenyl compounds by fish liver microsomes. Carcinogenesis by aflatoxin and by diethylnitrosamine was demonstrated in fish in the mid 1960s.[19,20] Based on such studies, Adamson[3] concluded that "The mechanisms for disposition of drugs and other foreign compounds are surprisingly similar in the evolutionarily more primitive water dwelling species and in mammals." How then have we progressed? How should we progress?

Since 1967, advances have come in discovery of an unanticipated and indeed extraordinary number and diversity of enzymes that transform xenobiotics, receptor mechanisms for their control, and linkages between these enzymes and endogenous functions. The biochemistry and molecular biology of these enzymes have attracted the attention of literally thousands of investigators, most working with mammals, who seek to understand and control the processes of xenobiotic biotransformation and its effects. The application of biochemical and molecular biological methods and probes is changing our understanding of mechanisms in aquatic toxicology. Long-standing and critical questions about the nature of biotransformation enzymes and their regulation now can be addressed in ways not possible a decade ago. These critical questions concern:

1. the identity and number of the enzymes of a given type,
2. the catalytic capability or function of specific enzymes, and relationships between structure and function,
3. the mechanisms by which genes for specific enzymes are regulated,
4. the modification of functions by both internal and external variables,
5. the cellular sites and timing of the gene expression, and the coordinate expression or function of different enzymes involved in biotransformation, and
6. linkages, either temporal or physiological, between the function or regulation of biotransformation enzymes and the function or regulation of other molecular events or processes in particular cell systems.

Each of these is important, and each is related to the other. Yet, it is in the last, item 6, where the full role of biotransformation enzymes in toxic mechanisms will be found. In each there is much we do not know. Indeed, as discussed below, the regulatory systems for these enzymes in some cases might be as important as the function(s) of the enzymes themselves in mediating the effects of toxic chemicals, and the cellular targets may determine the nature of systemic effects, but neither the regulatory mechanism nor the cellular targets are known.

Addressing these issues at the molecular level in different species obviously will deliver information for those individual species. But the phylogenetic viewpoint ultimately will provide the foundation for generalizations regarding toxic mechanisms. Having achieved those generalizations, it will be possible to use the resulting information and accompanying probes, cells, or transgenic systems to reliably address the objectives of chemical and environmental assessment. Phylogenetic studies also will, as Adamson suggested, reveal the evolution of the biotransformation systems. But such studies will reveal more than that, disclosing the role of toxic mechanisms and chemicals in the interplay of ecology, natural products, and evolution. That is where we should go. What follows is where we are.

III. OXIDATIVE BIOTRANSFORMATION AND MONOOXYGENASES

Oxidative metabolism involving molecular oxygen is the initial enzymatic process in the biotransformation of a majority of lipophilic organic foreign compounds. When molecular oxygen began to accumulate in the ancient atmosphere, organisms were confronted with the toxic effects of O_2 and the need to defend against that toxicity. However, the possibility of using O_2 as an electron acceptor allowed organisms to gain greater amounts of energy from reduced carbon. Organisms also gained the opportunity to use O_2 to generate novel molecular structures, incorporating oxygen into those structures in a process of controlled oxygen fixation by enzymes termed oxygenases. In eukaryotic organisms, these are monooxygenases or mixed-function oxidases. The general scheme for monooxygenase reactions is

$$RH + NADPH + O_2 + H^+ \rightarrow ROH + NADP^+ + H_2O$$

where RH = the substrate
ROH = the hydroxylated product

This scheme, or one similar in which R is converted to RO, is involved in the transformation of endogenous substrates and literally thousands of foreign chemicals, including natural products or allelochemicals, and organic chemical pollutants.

Two major groups of monooxygenases are involved in the transformation of xenobiotics, the flavoprotein monooxygenases (FMO) and the heme protein (cytochrome P450) monooxygenases. There are certain similarities between these enzymes. Both groups of enzymes cleave the O-O bond of molecular oxygen by a mechanism of heterolytic cleavage. In eukaryotic species, both types of enzymes are membrane-bound, occurring in various membrane structures in the cell, but most abundantly in the endoplasmic reticulum (ER). (Microsomes are fragments of the ER.) Both require electron transfer involving interaction between the catalyst and other proteins. Both exist in multiple forms, and many of the forms can act on multiple substrates. The metabolism of xenobiotics by both FMO and P450 enzymes can result in both detoxication and toxication of a substrate. In the one process, toxic substrates are converted to products that are less toxic than the parent compound. In the other, they are converted to intermediate or final products that are more toxic than the parent. This is, for example, a major process involved in carcinogenesis. Indeed, many of the chemicals that cause cancer are actually *pro*carcinogens and are converted in the body to the active carcinogen form of the molecule. Genes for FMO and P450 are subject to regulation by a variety of endogenous variables. However, only the P450s appear to be induced by xenobiotic chemical substrates, involving transcriptional and/or translational activation. The general properties of FMO and P450 enzymes in prokaryotes and eukaryotes are listed in Tables 3 and 4.

Both groups of enzymes are represented in aquatic vertebrates and invertebrates. However, the cytochromes P450 dominate the literature on monooxygenases, in mammalian as well as in aquatic species. This chapter will emphasize the cytochromes P450, reflecting that dominance. Yet, the FMO act preferentially on some important xenobiotics less readily attacked by P450. They also offer an intriguing introduction into the potential role that both endogenous compounds and natural products may have played in the evolution and diversification of monooxygenase enzymes and their regulatory systems.

A. Flavoprotein Monooxygenases

Flavoprotein monooxygenases were described first as the amine oxidase system in pig liver.[21] Studies on this enzyme by Ziegler and colleagues have played a major role in establishing the chemistry and the biochemistry of the FMO.[22,23]

1. Reaction Mechanism

The mechanism involves the reduction of the prosthetic group FAD and binding of molecular oxygen resulting in a hydroperoxyflavin that is stabilized by the protein structure. Substrate (R) is then introduced into the structure, and it attacks the terminal oxygen of the peroxyflavin, rupturing the O-O bond of the peroxide. The product, usually an RO, is released, and then water is released from the enzyme complex. It is the release of water and the regeneration of the enzyme-FAD that appears to be the rate-limiting step.

Any nucleophilic structure that can be oxidized by such a peroxide appears to be a potential substrate for FMO. Stabilization of the peroxide structure by the

Table 3. Flavoprotein Monooxygenases

	Bacterial hydroxylases	Eukaryotic monooxygenases
Prosthetic group:	FAD	FAD
Electron donor:	NADH	NADPH
Localization:	Soluble	Membrane bound
Substrates:	Varied, but often including aromatic rings, e.g., p-hydroxybenzoic acid	Secondary amines, tertiary amines, thiols, sulfides, etc.
Specificity:	Generally narrow	Broad
Regulation:	Inducible by substrate	Generally not inducible by xenobiotics
Phylogeny:	Aerobic culture	Mammals, fish (TMA oxidase), invertebrates (TMA oxidase)

protein thus seems to account for the wide range of substrates that can be metabolized. The enzymes carry out N-oxidations or S-oxidations with a wide range of aliphatic amines, secondary and tertiary aromatic amines, and sulfides, thiols, phosphines, and others. The substrate specificity (or latitude) of these enzymes is known mostly from mammalian systems. Selected reactions catalyzed by mammalian FMO are illustrated in Figure 1, with substrates of toxicological and/or ecological significance.

2. FMO in Aquatic Species

N-oxidation reactions typical of FMO have long been known in fish, from the metabolism of trimethylamine (TMA) to trimethylamine oxide (TMAO). TMAO is involved in osmoregulation in some species and is synthesized in some of those, but the nature of the enzyme involved was established only recently.[24] FMO-like activities detected in striped bass or rainbow trout liver microsomes and studied in some detail include trimethylamine N-oxidase, N,N-dimethylaniline-N-oxidase, methimazole N-oxidase, and p-chlorobenzyl N,N-diethylthiocarbamate S-oxidase.[24-30] The studies by Schlenk and Buhler in rainbow trout[26] also showed immunochemical cross-reactivity between antibodies to either pig or rabbit FMO and proteins in

Table 4. Cytochrome P450 Monooxygenases

	Bacterial P450	Eukaryotic P450
Prosthetic group:	Heme	Heme
Electron donor:	NAD(P)H	NADPH
Localization:	Soluble	Membrane bound
Electron transfer	Flavoprotein reductase; iron sulfur protein (redoxin)	Flavoprotein reductase (iron sulfur protein) (cytochrome b_5)
Specificity:	Narrow	Narrow to broad
Regulation:	Inducible by substrate	Diverse, including xenobiotic induction
Phylogeny:	Broad	All major phyla, but forms will differ

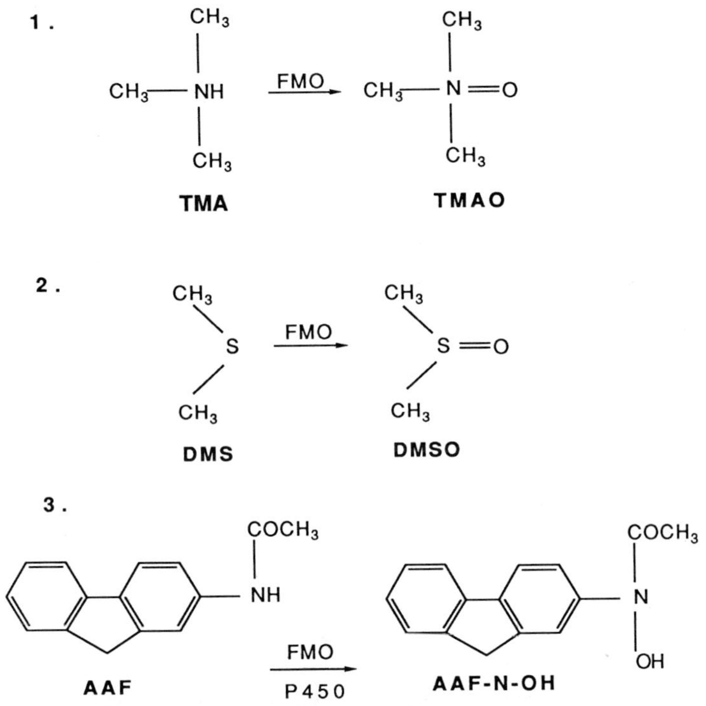

Figure 1. Representative reactions catalyzed by flavin monooxygenases (FMO). 1. Trimethylamine → trimethylamine oxide. 2. Dimethylsulfide → dimethylsulfoxide. 3. N-Acetylaminofluorene → N-hydroxyacetylaminofluorene. The P450 form catalyzing N-hydroxylation of AAF in some mammals is CYP1A2.

rainbow trout liver or gill. The detection of multiple bands in Western blot analysis suggests the presence of multiple forms of FMO in fish liver.

The presence of FMO in invertebrates was first suggested in studies of procarcinogen activation. A number of aromatic amine carcinogens, typified by 2-acetylaminofluorene (AAF), are metabolized by FMO. AAF is a procarcinogen that is activated by N-hydroxylation, as shown in Figure 1.[31] As with any enzyme reaction, the metabolism of AAF can be detected by direct analysis of substrate disappearance or the formation of product. However, the metabolism of procarcinogens can also be detected by in vitro mutation assays, in which tissue preparations convert a compound to a mutagen, detected as increased mutation frequency in bacterial or eukaryotic target cells incubated with the reaction mixture.[32] Such assays indicate that the tissue preparation is capable of the metabolic activation of the compound.

Molluscan (bivalve) hepatopancreas preparations were found to activate AAF and other aromatic amines to bacterial mutagens.[33] AAF can be N-hydroxylated and activated by P450 as well. Unlike P450, FMO lack heme and are not inhibited by carbon monoxide. The absence of carbon monoxide inhibition of AAF activation by the mulluscan preparations first indicated that there might be an active FMO system in these invertebrates. Subsequent studies by Kurelec and colleagues[34,35] and by Schlenk and Buhler[36] further indicated the presence of FMO in marine invertebrates. Additional information on the properties of these enzymes can be found in a recent review by Schlenk.[30]

3. Questions Regarding FMO

It is clear from catalytic studies that FMO-like enzymes occur in aquatic species. Immunological cross-reactivities suggest the presence of proteins structurally related to mammalian FMO. However, there is substantial uncertainty regarding the identity, number, functions, relationships and regulation of the enzymes. Studies in trout suggest that there are sex-linked and developmental changes in the expression of FMO in fish,[26] but the similarities in regulation of teleost and invertebrate FMO enzymes to one another and to those in mammals are still quite unclear. Uncovering the details of FMO activity with carcinogens in aquatic species is important for establishing the role of these enzymes in carcinogenesis in different taxa and is essential to using fish as models in carcinogenesis. Such details are also important to questions regarding the uncertain nature of chemical carcinogenesis in some invertebrates. Is, for example, their substantial capacity for activation of aromatic amines to mutagens likely to make mollusks susceptible to carcinogenesis by such compounds? If not, why not?

There are equally interesting questions regarding the evolution and ecological role of FMO. Some have speculated that endogenous substrates, perhaps cysteamine or TMA are the "true" substrates for the FMO.[30,37] However, reducing environments, such as those commonly found in sediments of marshes and other marine systems, generate an abundance of sulfides that could be FMO substrates and that could have provided an ecological pressure for development of these enzymes in aquatic species. A particularly intriguing possibility we suggest involves the role that dimethylsulfide, an excellent substrate for FMO, may have played in the evolution of molluscan FMO. When ruptured, plant cells including phytoplankton release dimethylsulfoniopropionate (DMSP), which appears to rapidly hydrolyze to acrylic acid and DMS.[38,39] Suspension feeders such as bivalve mollusks, pelagic zooplankton, and filter feeders that consume phytoplankton might experience frequent exposure to DMS, possibly released by hydrolysis of DMSP internally. Could this constitute a demand sufficient to require an efficient FMO? Interestingly, the possible metabolism of DMS may have a link to global change, as DMS is thought to play an important role in climate.[40]

Whether a need to metabolize natural products derived externally or endogenous substrate(s) involved in physiological pathways provided the greater pressure for evolution of the FMO is not known. This same question looms large in the consideration of P450 enzyme evolution.

B. Cytochromes P450

The cytochromes P450 comprise a large and expanding superfamily of heme proteins that catalyze oxygenase reactions. The unusual CO-binding pigment previously found in rat liver[41,42] was identified by Omura and Sato[43] as a heme protein, a b-type cytochrome, and termed "P450" after the characteristic CO-bound, reduced absorption peak at 450 nm (hence Pigment 450). That such a CO-binding pigment was catalytically active was first demonstrated by showing its involvement in steroid hydroxylation in adrenal cortex microsomes.[44] Cytochrome P450 enzymes have since been linked to the metabolism of a vast array of substrates, with more described regularly. Research on P450 gained momentum when the membrane-bound eukaryotic enzymes were finally purified.[45] P450s occur in prokaryotes and eukaryotes, where they are soluble and membrane bound, respectively. Each year since 1980 there have been more than 1000 publications concerning the chemistry, biochemistry, molecular biology, and functions of P450 forms in microbial, plant, and animal species. Unraveling the still bewildering complexities of multiple substrates, multiple catalysts, multiple regulatory mechanisms, and multiple biological functions is a continuing and stimulating challenge. There is a further challenge involving the topology of P450 enzymes in the membrane and its influence on substrate access to the active site. We do not discuss this aspect further, but the reader might bear it in mind while going through the chapter.

1. Reaction Mechanism

The catalytic cycle of P450 activity has been worked out over years of investigation largely on the bacterial camphor-metabolizing P450cam.[46] However, the presence of strongly conserved sequences in functional regions of bacterial and eukaryotic P450s suggests similarities in the catalytic mechanism. Recent reviews are recommended.[47,48] The sequence of events in catalysis is depicted in Figure 2. In its resting, unoccupied state, the prosthetic heme iron is a low spin Fe^{3+}. The first step is binding of substrate, which usually results in conversion to high-spin iron and a shift in the redox potential that favors reduction of the iron. (In P450cam, the binding of camphor causes a shift from near −300 mV to near −180 mV.) Following substrate binding, the iron is reduced by electron transfer from the flavoprotein NADPH-cytochrome P450 reductase, in some cases with involvement of iron-sulfur proteins (redoxins). Subsequently, O_2 is bound, a critical point at which catalysis may proceed or be interrupted, resulting in release of active oxygen (superoxide). The next steps involve addition of a second electron, in some cases via cytochrome b_5, and formation of a peroxide, followed by cleavage of the O-O bond, the formation of a substrate radical, and the hydroxylation of that radical and release of the product. The exact chemistry of the intermediates involved in the steps between

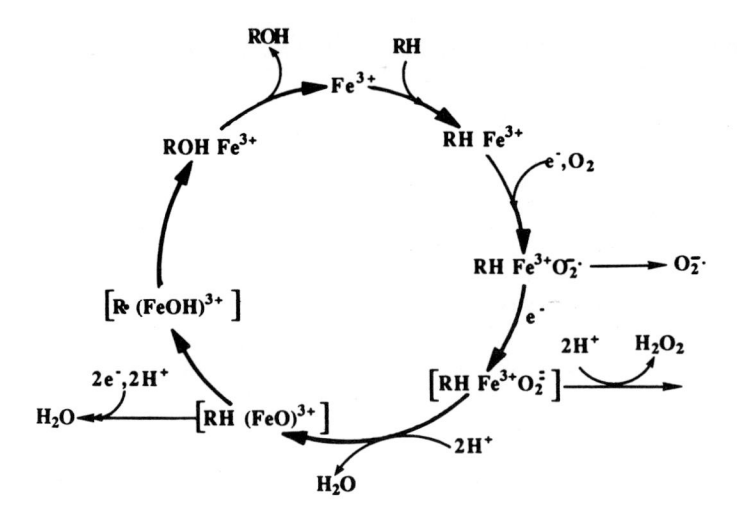

Figure 2. The P450 catalytic cycle. Reproduced with permission from Poulos, T.L. and Ragg, R., *FASEB J.*, 6, 674, 1992.

the addition of the second electron and the hydroxylation of the substrate is not known. However, addition of organic hydroperoxides such as cumene hydroperoxide to P450 in vitro can circumvent the need for electrons and oxygen and accomplish substrate hydroxylations via the so-called peroxide shunt. Continuing studies on mechanistic parallels between P450 and various peroxidases[48,49] may reveal the nature of the intermediates.

2. Diversity of Functions

The P450 enzymes catalyze many types of seemingly disparate, primarily oxidative reactions. Hydroxylation, epoxidation, dealkylation, and other oxidative reactions (Table 5) all share at some point the fixation of oxygen, though it need not appear in each product. Thus, in dealkylation reactions, O appears in the aldehyde that is removed.

The biological roles of these P450 reactions can be grouped according to the type of substrate and can be divided largely into (1) the synthesis and degradation of endogenous substrates and (2) the metabolism of foreign chemical (xenobiotic) substrates. In eukaryotes the endogenous compounds synthesized or degraded by P450 include steroids and steroid-derived compounds (cholesterol, androgens, estrogens, corticosteroids, ecdysteroids, bile acids, vitamin D), fatty acids and fatty acid derivatives (arachidonic acid, lauric acid), and plant secondary metabolites (cinnamic acid, tannins, flavonoids), to name but a few. The breadth and significance of endogenous P450 functions are illustrated by the activities and catalysts shown in Table 6. Many P450s involved in endogenous functions are highly specific.

Table 5. Types of P450 Reactions and Representative Substrates

Aromatic hydroxylation (e.g., benzo[a]pyrene)
Aliphatic hydroxylation (e.g., n-hexane)
Epoxidation (e.g., aflatoxin B_1; arachidonic acid)
N-dealkylation (e.g., aminopyrine, ethylmorphine)
O-dealkylation (e.g., 7-ethoxyresorufin)
S-dealkylation (e.g., methylmercaptan)
N-dealkylation (e.g., carbaryl)
N-oxidation (e.g., aniline, amphetamine)
S-oxidation (e.g., aldicarb)
P-oxidation (e.g., parathion)
Desulfuration (e.g., parathion)
Oxidative deamination (e.g., arginine)
Oxidative dehalogenation (e.g., halothane)
Reductive dehalogenation (e.g., hexachlorobenzene, carbon tetrachloride)

Xenobiotic substrates of P450 in animals include drugs, industrial chemicals, pesticides, synthetic intermediates and byproducts, chemical carcinogens, fungal and plant-derived compounds (phytoallexins), and other natural products. The diversity and number of xenobiotic chemicals that can be substrates of one or another P450 may well be limited only by the number of those chemicals possibly synthesized by man and nature. For example, of the hundreds to thousands of compounds that occur in petroleum, most can be expected to be substrates for one or another P450 form. Examples of the diversity of chemicals transformed by P450 are shown in Figure 3. There are numerous excellent reviews of the biochemistry and toxicological implications of cytochrome P450 activities in the metabolism of such compounds.[50–54]

Table 6. Diversity of Cytochrome P450 Functions

P450 designation	Catalytic role	Processes involved	Ref.
1A1/1A2 (CYP1A1/2) (vertebrates)	Epoxidation and hydroxylation of PAH and others	Detoxication; chemical carcinogenesis (inadvertent)	52, 54
3A (CYP3A)	6β-hydroxylation of corticosterone	Salt balance	161, 163
4A1 to 4A7 (CYP4A1-7)	Hydroxylation of fatty acids, oxidation of prostaglandins	Lipid metabolism	432, 433
Aromastase (CYP19) (animals)	Conversion of androgens to estrogens	Sex determination and reproductive processes	113, 262
26/25-hydroxylase (CYP27) (animals)	25-hydroxylation of vitamin D to active form	Calcium metabolism	434
20-hydroxylase (invertebrates)	Conversion of ecdysone to 20-hydroxyecdysone	Insect and crustacean molting	183
P450olfl (CYP2G1) (mammals)	Undescribed substrates, in olfactory mucosa	Olfaction?	326–328
CYP71 (plants)	Metabolism of plant hormone	Ripening of fruit (avocado)	435
Cinnamate 4-hydroxylase, others	Conversion of cinnamic acid to a variety of secondary metabolites	Phytoallexin synthesis	436, 437

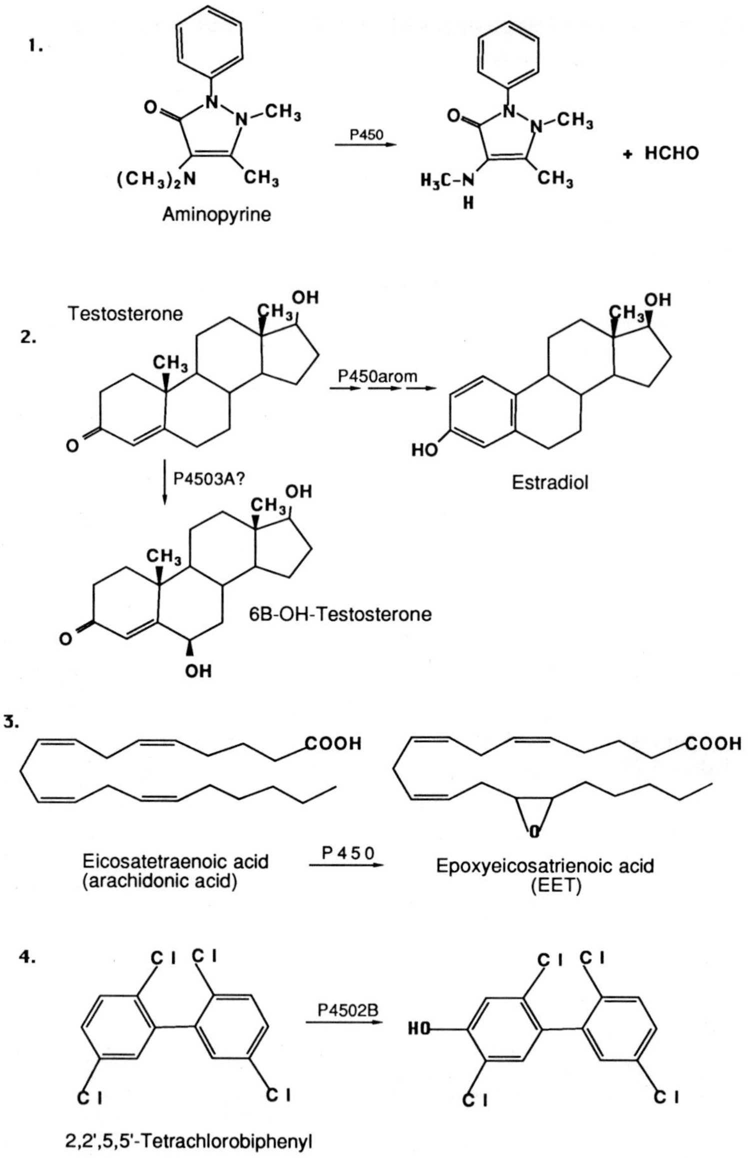

Figure 3. Representative reactions catalyzed by cytochromes P450. The reactions shown are in order: 1. Aminopyrine N-demethylation. 2. Different fates of testosterone; aromatization, which is a three step process, and 6β-hydroxylation, the dominant hepatic hydroxylation in fish. 3. Epoxidation of arachidonic acid. 4. Hydroxylation of a chlorobiphenyl congener. In mammals there is a 3,4-epoxide formed and rearrangement to 3-OH and 4-OH-TCB. The 4-OH-TCB is formed by fish.[473] The catalysts indicated are known from mammalian studies, and are indicated in studies with fish.

Thus, cytochrome P450 enzymes have essential roles in much of endocrinology, toxicology, pharmacology, and carcinogenesis, and in ecology. The catalytic mechanism is likely similar in most if not all of P450 reactions. The distinctions between the catalysts thus depend on the latitude for substrate binding, necessarily a feature of the structure of the proteins.

3. The P450 Gene Superfamily

Prior to 1987 there were as many as 14 different names in use for what now is known as the same P450 from rat liver, and more than 20 names for the homologous proteins from different species. Sequencing of P450 genes and proteins offered a way out of this confusion, making possible a classification system based on the measured or inferred P450 amino acid sequences.[55-58] P450s are now known by their systematic and common names. As of late 1992 there were 221 P450 cDNAs or proteins sequenced and classified in 36 P450 gene families in the superfamily, many with multiple subfamilies. The classification is still evolving. At present, it includes the designation of CYP, followed by symbols denoting gene family (Arabic numeral), subfamily (capital letter), and specific gene (Arabic numeral).[57] The number of P450 genes occurring in any one species is not known. However, there is direct evidence for at least 20 and presumptive evidence for 40 distinct P450 genes in rat, and estimates generally agree that there are likely to be more than 100 distinct P450 genes in a given mammalian species.

In mammals the cytochromes P450 in gene families 1 to 4 are the most prominent in the metabolism of xenobiotics. Many genes in these families are also induced by xenobiotics. Table 7 gives examples of xenobiotic induction in different gene families, illustrating the spectrum of inducers and known or suspected mechanism of induction, and prominent catalytic activities of the induced proteins. P450s in families 1 to 4 each appear to catalyze many activities, with much overlap, and there are few substrates known that are exclusive to one protein (diagnostic substrates). Moreover, orthologous P450s in different species may have different substrate specificities. As discussed in Section VII. C, the regulation of P450 genes is being increasingly recognized as a critical factor in toxicity of chemicals, not only by altering rates of biotransformation, but also through other avenues that may involve the induction mechanism itself.[59,60]

a. Note on Nomenclature. The current P450 classification system[58] is based on proposed evolutionary relationships between genes, as inferred from the degree of amino acid sequence identity between pairs of enzymes. For example, P450 proteins >40% identical are defined as being within a single family, and >55% identical are defined as in the same subfamily.[58] This system is derived from studies in many species but in few nonmammalian ones. This presents some difficulties in classification and nomenclature for the forms in aquatic species. Without the benefit of sequences from diverse taxonomic groups, the current

Table 7. Induction in Selected P450 Gene Subfamilies

Gene family and subfamily[a]	Selected protein members	Prominent substrates[b]	Common inducers[c]	Mechanisms
CYP1A	1A1	PAH, planar PCB, 7-ethoxyresorufin	PAH, planar PCB, BNF, chlorinated dioxins (e.g., TCDD), and furans	Mostly transcriptional
	1A2	Acetanilide, estradiol, caffeine	PAH, planar PCB, BNF, chlorinated dioxins (e.g., TCDD), and furans, ISF	Transcriptional, post-transcriptional (protein stabilization)
CYP2B	2B1	Barbiturates, steroids	Barbiturates, non-planar PCBs, DDT	Transcriptional
CYP2E	2E1	Ethanol, alkylnitrosamines	Ethanol, ketones, starvation, diabetes	Transcriptional, post-transcriptional (protein stabilization)
CYP3A	3A1	Steroids (6β-hydroxylase)	PCN	Transcriptional
CYP4A	4A1	Lauric acid, arachidonic acid	Clofibrate, phthalates, PCBs	Transcriptional

[a] According to Nebert and colleagues.[55-58]
[b] The substrates listed are common ones for the forms indicated.
[c] Abbreviations used: PAH — polynuclear aromatic hydrocarbons; BNF — β-naphthoflavone; DDT — dichloro-diphenyltrichloroethane (1,1,1-trichloro-2,2-bis(p-chlorophenyl)ethane); ISF — isosafrole; PB — phenobarbital; PCB — polychlorinated biphenyls; PCN — pregnenolone-α-carbonitrile; TCDD — 2,3,7,8-tetrachlorodibenzo-p-dioxin.

guidelines may require that orthologous (linearly related) genes be assigned different classification. This is also a problem in mammalian systems, where for example the rat CYP2B and rabbit CYP2B forms are classified differently, and probable orthologues are not identified. Thus, a deficiency of this system is its failure to take into account the evolutionary distance between two species when comparing the sequences of their potentially orthologous P450 forms. For example, CYP1A1 proteins in rats and mice are 93% identical, but CYP1A1 proteins in rat and trout are 59% identical. The explanation, of course, is that the most recent common ancestor of rats and mice lived 10 MYA, whereas mammals and fish diverged 360 MYA. The trout protein, which shares 51 to 59% amino acid identity with mammalian CYP1A1 and CYP1A2 forms, is considered a CYP1A only as the result of an exception to the 55% rule.[61] Put another way, the current system is a one-factor (% identity) system, while a two-factor system (% identity + species distances) is needed.

The sections below refer to several P450 forms by classification, as presented in detail in Section V. These should be viewed in light of the difficulties described above. With regard to putative members of the CYP1A subfamily in aquatic species, it has been proposed[62] that the terms CYP1A1 and P4501A1 be reserved for those forms and species where sequence information confirms this identity, consistent with the classification recommendations.[58] For forms exclusively detected by specific antibodies or nucleic acid probes, only the family and subfamily designations (i.e., CYP1A) are used. We also refer to CYP1A in some older literature, where the studies involved catalytic assays that now are very strongly linked to this subfamily. Additional sequences will be important in refining the incorporation of an evolutionary perspective in the classification and nomenclature of P450 genes and proteins.

IV. CYTOCHROME P450 IN AQUATIC SPECIES

There is an astounding diversity of known mammalian P450 forms and functions. Considering the great variety of habitats, life histories, and ecological relationships represented by fish, amphibians, other vertebrates, and aquatic invertebrates, P450s in aquatic species are likely to be as diverse as known mammalian forms, if not more so. Nonmammalian species almost certainly have evolved forms of P450 whose functions are related to distinctive aspects of their physiology. For example, steroid or other hormones that control molting in arthropods are likely synthesized by P450 forms distinct to arthropods. Perhaps more important in generating P450 diversity are the unique chemical environments that aquatic species inhabit. The number and types of marine natural products and their demonstrated roles in ecological interactions (see Section XIII) and the need for aquatic organisms to deal with such compounds in various ways must have been an effective driving force for the evolution of new P450 forms and regulatory mechanisms. There is a great opportunity to enhance our fundamental understanding of this group of enzymes through research in aquatic species and, conversely, to better understand the physiology of these species through study of their P450s.

Research on aquatic species monooxygenase systems grew rapidly in the mid 1960s, so that by the late 1970s and early 1980s, several major reviews of microsomal P450 systems in aquatic species appeared.[5,10,63] Research in these areas has continued to grow; from 1980 to 1992 there were more than 800 papers published on P450 or monooxygenase activity in aquatic species.

A. Microsomal P450 Systems

Cytochromes P450 and attendant enzymes in aquatic species are known mostly from studies in microsomal preparations. Cytochrome P450, the flavoprotein NADPH-P450 reductase, cytochrome b_5, and NADH cytochrome b_5 reductase are integral proteins of various membranes, primarily the endoplasmic reticulum. General features of these enzymes in the microsomal preparations from fish, including spectral and electron paramagnetic resonance properties of P450, have been described in earlier reviews.[5,10,63] Similarly, properties of microsomal enzymes and rates and patterns of xenobiotic metabolism in marine invertebrates have been detailed in recent reviews, particularly the excellent compilations by Livingstone and James.[8,9,12] Specific contents of total P450, representing the sum of all forms present, have been measured in tissue microsomes from numerous groups of aquatic vertebrates, including fish, amphibians, reptiles, birds, and mammals. These values range from <0.05 to >2.0 nmol/mg microsomal protein. Without other information, the content of total P450 itself does not reveal the identity or function of any specific form. However, the molar ratios of P450 to P450 reductase are important. In liver of fish and mammals, these ratios typically are 20 moles of P450 to 1 mole of reductase.[64] Lesser relative amounts of reductase are seen in some extrahepatic organs, for example in heart, where the P450 to reductase molar ratio is 200 to 1. A lesser capacity for electron transfer to P450 can result in a less efficient catalytic function of some or all P450 forms present.

Some features of microsomal electron transport components are commonly seen. The levels of total P450 in mollusks, crustaceans, echinoderms, polychaetes, and other invertebrates are generally less than the levels typical of vertebrates. Differences in P450 content between individuals of any one species may be associated with differences in strain, sex, developmental status, chemical exposure, or unknown variables and presumably indicate variable expression of one or more P450 forms. Large differences between species could suggest differences in the levels of expression of homologous enzymes or the expression of different P450 forms. Evaluating these possibilities requires additional information on catalytic or structural properties of the proteins.

B. Catalytic Functions and Rates

Certain generalizations might be made regarding endogenous as opposed to exogenous functions. The roles demonstrated for P450 in synthesis of regulatory molecules in mammals are likely to occur in other groups where those products occur. Thus, we can predict that similar steroids in species even where the synthetic pathways have not been described, such as many invertebrates,[65,66] will involve synthesis by P450 forms similar to the catalysts known in other species. Pathways involved in degradation of xenobiotics might not be so conserved. Tissue microsomes of aquatic vertebrates and invertebrates catalyze a spectrum of reactions broadly similar to those in mammalian systems. But similarities in the catalysts are less certain. The same activities in different species could be catalyzed by P450 proteins that are not closely related. Conversely, P450 forms that are structural homologues in different species, whether fish, birds, mammals, or others, could have different functional and/or regulatory properties. Even closely related proteins can have distinct functional properties; the clearest examples involve allelic variants (Section IX.C).

1. Vertebrates

Rates of in vitro and/or in vivo metabolism of numerous compounds by vertebrates have been compiled by several authors.[67,68] Many studies deal with substrates or issues of specific toxicological interest, such as the role of metabolism in toxicity of an insecticide in a particular nontarget species experiencing exposure. Others deal with substrates of more general interest, such as the polynuclear aromatic hydrocarbons (PAH). In vitro rates of PAH metabolism by fish liver preparations vary greatly depending on the sex, reproductive status, chemical treatment, and other variables.[5,10,63] Rates of PAH metabolism can be strongly influenced by the degree of CYP1A expression, such that species comparisons are difficult for animals of undefined exposure history. The turnover number (mole of substrate transformed per minute per mole of the specific catalyst) would be a valid basis on which to compare the efficiency of the enzymes in different species, but this value has been directly determined for few species.

Regardless of such uncertainties, some features appear to be widely applicable:

1. PAH metabolism in fish is strongly induced by polynuclear and planar halogenated aromatic hydrocarbons. The rates of in vitro metabolism of PAH are rapid in most fish exposed to inducers (often exceeding 2 nmol/min/mg for BaP).
2. Halogenated aromatic hydrocarbons, including PCBs, dioxins, and others, are metabolized more slowly than PAH in fish, even in animals that are induced.
3. PCB congeners with unsubstituted meta, para-positions may be metabolized more rapidly than others, but rates are still slight (i.e., <10 pmol/min/mg for 2,2′,4,4′-tetrachlorobiphenyl).[69]

Metabolic rates with numerous substrates have been summarized for some other vertebrate groups (e.g., birds),[68,70] but too little is known for many groups, for example cetaceans, to warrant strong generalizations. Yet, even the limited descriptions of microsomal metabolism of xenobiotics and/or steroids suggest that many of the same enzymes will be detected in marine mammals as in other vertebrate species. This is described further in Section V.C.

2. Invertebrates

Microsomal preparations from all groups studied have been found to transform a range of xenobiotics, including aromatic hydrocarbons, that are common substrates for P450 in vertebrates. Despite substantial effort, there is still uncertainty regarding the role of P450 forms in the metabolism of xenobiotics in many aquatic invertebrates. Yet, we can offer generalizations additional to those above:

1. The rates of xenobiotic metabolism by P450 in many invertebrate groups are substantially less than in vertebrates. The rates of PAH metabolism detected in vitro in some molluscan tissues, for example, are two to three orders of magnitude lower than those seen in most teleost fish liver preparations.
2. Difficulties involved in preparation of catalytically competent microsomes from some crustacean tissues[71] complicate interpretations of their relative rates of in vitro hydrocarbon metabolism, yet data[8,71] indicate that the potential rates of PAH metabolism in crustaceans fall between those in molluscan and fish groups.
3. The relative rates of metabolism of compounds in vitro are reflected in relative rates of elimination in vivo, seen in the amounts of metabolites excreted or of parent compound residues in tissues. Invertebrates such as mollusks, with low capacities for biotransformation of many substrates, accumulate higher steady state levels of many xenobiotics.

It remains to be seen how many exceptions to these generalizations appear as additional species are examined. There are many uncertainties regarding interactions of different chemicals with the catalyst, the inhibition of P450 by some substrates, and the identity of the catalysts, in all aquatic groups. Establishing the identity and diversity of P450 forms and their regulation, and then linking those to specific substrates is essential to understanding the role of a given P450 form in biotransformation and effects of xenobiotics.

C. Substrate Structure/Activity Relationships and Diagnostic Substrates

Determining the catalytic properties of a particular P450 form is fundamental to evaluating the significance of that form in xenobiotic effects. Ultimately, the objective is to define the structure/function relationship or the *substrate structure/activity relationships* (substrate SAR) for different P450s. There are several complementary ways that the catalytic functions of a P450 form can be determined, or inferred. These include:

1. analysis of the activity of purified proteins in reconstitution;
2. correlation of the rates of microsomal metabolism of different substrates;
3. correlation of rates of metabolism of a given substrate with the content of a particular P450;
4. use of specific chemical inhibitors;
5. use of specific and inhibitory antibodies.

Additional approaches employing site-directed mutagenesis and expression of recombinant proteins in cells in culture are discussed in subsequent sections. Of the approaches listed above, the last is the surest way to determine the contribution of a P450 to microsomal activities. Antibodies to purified proteins or to synthetic oligopeptides representing conserved amino acid sequences can be used. Assignment of catalytic functions to P450s from aquatic species described in the next section, reflects studies with several of the above methods, including antibody inhibition. Here we consider briefly the use of specific substrates. In Section IV.D we consider specific inhibitors.

Catalytic rates with specific substrates have been used extensively to indicate differential regulation of distinct P450 forms. However, inferring the identity of a P450 solely on the basis of catalytic activity is risky. Homologous proteins, even in closely related species, could catalyze different reactions. Nonetheless, catalytic changes after treatment with inducers in mammals have identified several potential "diagnostic" substrates or substrate probes for certain P450 forms that may be useful in characterizing microsomal systems and the regulation of P450 forms, particularly when more direct structural probes are not available. For example, BaP metabolism or AHH activity,[72] catalyzed primarily by CYP1A forms, was one of the first activities strongly linked to an induction mechanism. Catalytic activity of fish liver microsomes has supported the utility of such substrates, and indeed first indicated some of the important similarities and dissimilarities between fish and mammals. Thus, induction of AHH activity in fish by hydrocarbons[73] offered the first indication that a similar catalyst occurs in fish and mammals and also first indicated the use of induction as a marker for environmental exposure to aromatic hydrocarbons. The lack in fish of a phenobarbital (PB) effect on the metabolic rates with drug substrates commonly induced by PB in mammals indicated differences between fish and mammals either in the catalyst(s) or their regulation.[74]

Alkoxy-derivatives of the dye resorufin, described by Burke and Mayer,[75,76] have seen the greatest use as diagnostic substrates. Resorufin is a multi-ring structure whose shape (Figure 4) is similar to those of several CYP1A inducers. Substituted resorufins are dealkylated by P450, and the identity of the substituent apparently can determine the access to active sites of different P450 forms, as indicated in the responses of different activities to different inducers (Table 8).

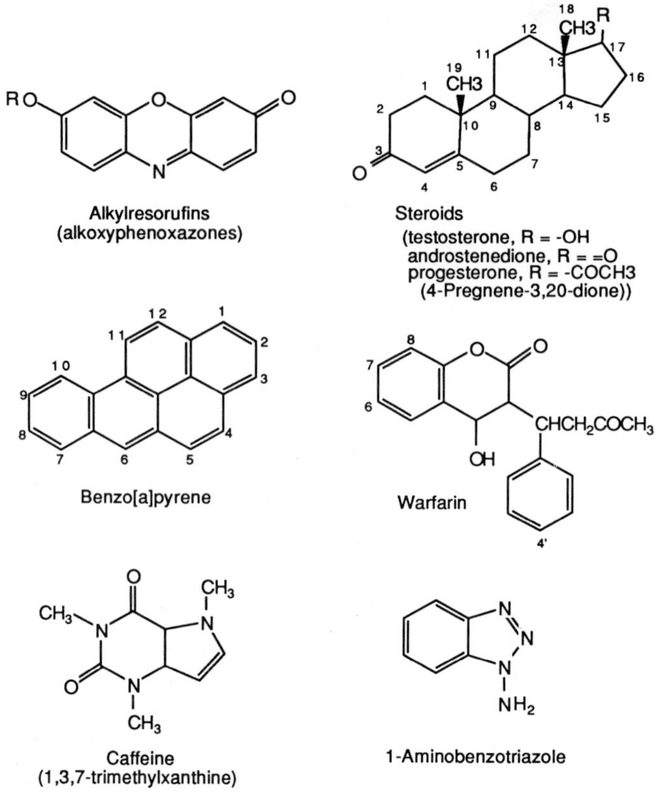

Figure 4. Candidate diagnostic substrates for different P450s. Included is a prominent inhibitor, aminobenzotriazole. Most of the substrates are acted on by a diversity of P450 forms, although caffeine N-demethylation is an activity apparently restricted to CYP1A forms in some mammals, including humans.

Table 8. Specificity of Alkylresorufins (a.k.a. Alkoxyphenoxazones) as Probes of Induced P450 in Mammals

Alkyl substituent	Induction by 3-MC-type inducers	Induction by PB-type inducers
Methyl	++	—
Ethyl	+++	+/–
Propyl	+++	+/–
Pentyl	+	+++
Benzyl	+	+++

Note: Pluses indicate approximate, relative degree of induction in rats or mice, as estimated from Burke, M.D. and Mayer, R.T.[75,76] Highly induced (+++), moderately induced (++), somewhat induced (+), not induced (–). See text for discussion.

Induction of ethoxyresorufin O-deethylase (EROD) activity in vertebrates is generally accepted as an indicator of CYP1A induction, while pentoxyresorufin O-deethylase (PROD) activity generally marks CYP2B forms. Yet, there are uncertainties in extrapolating from mammalian results to newly studied species, even with such promising substrates. Thus, PROD activity is slightly induced by CYP1A inducers,[75-79] and the presence of this activity cannot be interpreted strictly as a marker of CYP2B. In mammals, EROD is largely specific to CYP1A1 and acetanilide hydroxylation to CYP1A2. Acetanilide hydroxylation is catalyzed by purified scup P450E (CYP1A1) in reconstitution, but the catalyst in microsomes has not been established. In rats the inducible CYP1A2 is the dominant estradiol hydroxylase.[80] But that is not the case in all mammals,[80a] and hepatic CYP1A is not a microsomal E_2-2-hydroxylase in some fish.[81]

Many substrates can be metabolized at different sites on the molecule, and the patterns of metabolism with such compounds can be used to evaluate the function or expression of different P450 forms. Such substrates usually have multiple alkyl-substituent groups, which can be hydroxylated or removed by dealkylation, and/or multiple sites where epoxidation or hydroxylation can occur. Prominent examples of the latter include steroids and polynuclear aromatic hydrocarbons (PAH). In mammals, steroids including testosterone are hydroxylated at 10 or more sites on the molecule, and different hepatic P450s show a strong selectivity not only for specific sites (e.g., C6 vs. C16), but for attack from a different face at a given site (e.g., 16β vs. 16α). Fish liver microsomes hydroxylate testosterone at many of the same sites (e.g., 6β, 16β, 16α) as do those of mammals. BaP also is metabolized to 10 or more products, As summarized more than 10 years ago,[63] numerous teleost fish species form a suite of BaP metabolites similar to that formed by mammalian species. Teleosts in general show a stronger preference than some mammals for oxidation at the 7,8- and 9,10-positions,[63] associated with activation of BaP to a carcinogen. CYP1A1 is the catalyst for this benzo-ring metabolism in some fish.[82,83] Metabolite profiles for BaP also have been obtained for several invertebrate groups. In some species, these metabolite profiles differ substantially from those in vertebrates. Mollusks, for example, form predominantly quinone derivatives of BaP in vitro.[84,85] The different patterns apparently reflect different catalysts acting in PAH transformation in invertebrates and vertebrates (discussed further in Section VIII.A).

Determining common structural features of preferred sites of attack on different substrates will (if the catalyst has been identified) help to establish the topology of P450 active sites and to predict how novel compounds may fit. Thus, a regio-specificity of scup CYP1A for structures similar to the benzo-ring of BaP has been seen with other compounds. For example, Figure 5 illustrates the structural similarity between metabolism at the 7,8-positions of α-naphthoflavone (ANF) and BaP.

Fully defining the latitude of the active site of P450 forms is still a daunting task. The full substrate specificity is not known for *any* xenobiotic metabolizing

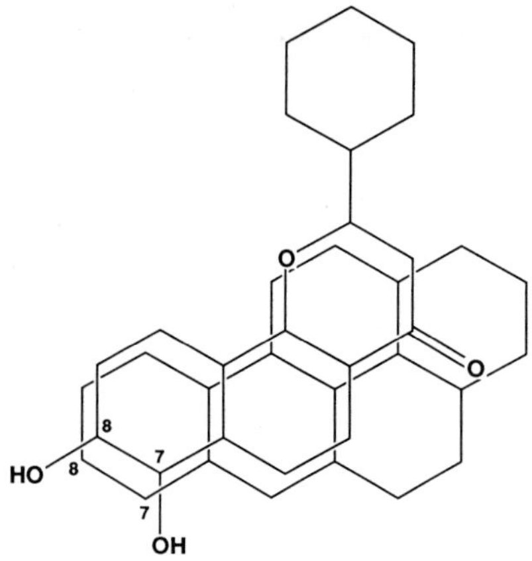

Figure 5. 7,8-Dihydrodiols of BaP and ANF formed by scup liver CYP1A1. The structures of BaP and ANF have been overlain to show the major sites of hydroxylation by cytochrome P4501A1. Formation of the dihydrodiols involves both CYP1A1 and epoxide hydrolase. For the sake of clarity, the double bonds are not shown, and the hydroxyl groups are shown only for ANF. The 7 and 8 carbons indicated are correct for both BaP and ANF.

P450 in any species. Structure/function relationships based on similarities of known substrates have long been sought.[86] Expanding libraries of substrates linked to specific forms of P450 have begun to show the common structural features of substrates for those forms. More than 45 substrates have been identified for mammalian CYP2E1;[87] all are small lipophilic molecules, including alcohols, ketones, aromatic structures (e.g., benzene), halogenated alkanes (trichloroethane), nitrosamines, and ethers. Similarly, more than 20 substrates are known for mammalian CYP2D6 and for CYP3A. The CYP 2D6 substrates are basic nitrogenous compounds that are hydroxylated in a lipophilic domain near the N.[88] Whether the substrate SAR developed for these mammalian proteins will hold for homologous P450 forms in aquatic species is unknown. Establishing substrate SAR for P450s from aquatic species is essential to interpreting the consequences of increases or decreases in the amounts and activities of these proteins, and an area badly in need of research.

D. Inhibition/Inactivation and Inhibitor Structure/Activity Relationships

An increasing diversity of chemicals have been found to inhibit P450 activity.[89] Inhibitors of P450 also have been used extensively to characterize microsomal functions and to suggest P450 identity. Inhibitors specific for different forms can help to define the substrate specificity of those forms. As with substrate SAR, *inhibitor structure/activity relationships* (inhibitor SAR) can help to define the latitude of the active sites. Inhibitors also have significance in vivo. In clinical settings, a reduced rate of metabolism could either potentiate or reduce the efficacy of drugs, depending on whether a compound is inactivated or activated by P450. In agriculture, P450 inhibitors, particularly methylenedioxyphenyl (MDP) compounds (a group including the inhibitor piperonyl butoxide; PBO), have been added to pesticide formulations to inhibit the detoxication and thus potentiate the toxic action in insect target species. There is growing recognition that environmental chemicals also can inhibit P450 activity. A selection of inhibitors of P450 known to act in aquatic species is listed in Table 9. The majority of these act either by competitive inhibition or by mechanism-based (or "suicide") inactivation, in which the inhibitor is metabolized by the P450 into a product that covalently modifies the active site and thereby inactivates the enzyme. A few examples will be considered.

1. Metyrapone, SKF-525-A (β-diethylaminoethyl-diphenylpropylacetate), and α-naphthoflavone (7,8-benzoflavone; ANF) have been used extensively as in vitro inhibitors to characterize microsomal systems in mammals and fish. Metyrapone and SKF-525-A inhibit a variety of MO activities in fish[90] but seem not to be selective for particular P450 forms. ANF, on the other hand, is highly specific for CYP1A forms and acts as a competitive inhibitor, consistent with its metabolism by this form.[91] It is not clear whether in some species ANF acts on other forms and by other mechanisms. Thus, ANF inhibition of MO activity in systems where there is uncertainty regarding the identity of forms, in mollusks for example, cannot itself show that CYP1A is involved.

2. Common mechanism-based inhibitors include structures with allyl groups and nitrogen heterocycles such as aminobenzotriazole (ABT) or naphthalene aminotriazole (NAT).[92] The latter compounds were used in studies to establish that mechanism-based inhibitors (also called "suicide" substrates) commonly are metabolized to products that alkylate the heme, thereby inactivating P450. ABT and NAT both inhibit P450 activities in fish[64] but have an unknown but apparently broad selectivity. Thiono-sulfur compounds also inhibit P450, apparently in mechanism-based processes.[93] These have been little studied in aquatic species. Decreases in microsomal P450 or catalytic function in fish given the anesthetic tricaine methyl-methanesulfonate (MS-222), shown by several groups,[94,95] might involve similar mechanisms.

3. Several inducers of CYP1A synthesis can also inhibit the activity of this enzyme. Such inhibition is evident as decreased CYP1A catalytic activity (EROD or AHH) in liver microsomes from fish or fish cells receiving high doses of BaP,[96] BNF,[97] or planar chlorobiphenyls,[98–100] even though CYP1A protein and/or mRNA levels may remain elevated at those same high doses. These inducers are also substrates and appear to be retained in microsomes and to competitively inhibit the enzyme.[97,98] Thus, incubation of the microsomes with NADPH prior to EROD assay is able to alleviate the inhibition by BNF, presumably by clearing the BNF substrate.[97] However, decreases in immunodetected protein at the highest doses indicate other processes are also involved (see Section VI.A). There is evidence that such inhibition of P450 does occur in fish from sites highly contaminated by PCBs, or by some pulp mill effluents.[101,102,381]

Table 9. Compounds that Inhibit or Inactivate Monooxygenase Activity in Fish

Compound	Activity inhibited	In vivo or in vitro	P450 form involved	Probable mechanism	Ref.
Heterocyclic compounds					
α-naphthoflavone	AHH, EROD	In vitro	1A	Competitive	82, 134, 143, 314, 371, 379, 438–442
β-naphthoflavone	EROD	Both	1A	Competitive, other	97
Naphthalene aminotriazole	EROD, ECOD, AHH	In vitro	1A	Mechanism-based inact.	64
Aminobenzotriazole	EROD, ECOD	In vitro	1A	Mechanism-based inact.	64
Metyrapone	AnH, BeND, ECOD	In vitro	Unknown	Unknown	90
Piperonyl butoxide	PCDD metabolism, rotenone metabolism	In vivo	Unknown	Unknown	443–445
Aromatic Hydrocarbons					
Benzo(a)pyrene	AHH, EROD	In vivo	1A	Competitive, other	96
Naphthalene	AHH	Both	1A?	Unknown	446
Benzene	EROD	In vivo	1A?	Unknown	447
Chlorinated aromatics					
3,3',4,4'-TCB	EROD, AHH	Both	1A	Competitive, other	98, 99, 448
3,3',4,4',5,5'-HCB	AHH, AFB$_1$ activation, P-6β-OHase	In vivo	1A, 2K, 3A?	Unknown	100
Metals and alkyl metals					
Cadmium	EROD	In vivo	1A	Unknown	107, 108
Tributyltin	EROD	Both	1A	Unknown	103, 104
Other compounds					
MS-222 (tricaine methanesulfonate)	EROD, AHH	Both	1A	Unknown	94, 95
SKF-525A	AnH, BeND, ECOD, LAH	In vitro	Unknown	Unknown	90, 258
Phenobarbital	Various	In vivo	Unknown	Unknown	90

Note: Abbreviations used: AHH, aryl hydrocarbon (benzo[a]pyrene) hydroxylase; AFB$_1$, Aflatoxin B$_1$; AnH, aniline hydroxylase; BeND, benzphetamine N-demethylase; ECOD, ethoxycoumarin O-deethylase; EROD, ethoxyresorufin O-deethylase; HCB, hexachlorobiphenyl; LAH, lauric acid (ω-1)-hydroxylase; PCDD, polychlorinated dibenzo-p-dioxin; P-6β-OHase, progesterone 6β-hydroxylase; TCB, tetrachlorobiphenyl.

Recently, piperonyl butoxide (PBO), an inhibitor of P450, has been found to induce CYP1A in fish liver (see section VI.C). The mechanism by which methylene dioxyphenyl compounds such as isosafrole or PBO induce EROD activity in fish is not known.

4. Some organometallics such as the alkyltins (e.g., tributyltin; TBT) are potent inhibitors of P450 in vitro and in vivo in mammals and fish.[103,104] There appears to be some selectivity of TBT for CYP1A, but the mechanism(s) of this effect are not known. Alkyltins and heavy metals such as cadmium can also affect the activity of P450 by induction of heme oxygenase,[105,106] resulting in a decrease in heme available for insertion into the P450 apoenzyme to form the active holoenzyme. Based on this mechanism alone, it is predictable that metals will affect P450 activities in aquatic species, borne out by studies of Forlin et al.[106a] showing decreased ethoxycoumarin O-deethylase (ECOD) activity in renal and hepatic microsomes of Cd-treated trout, and George[107,108] showing decreased EROD activity in Cd-treated plaice. Den Besten,[109] on the other hand, found no effect of Cd on AHH in the sea star *Asterias rubens*, although the total P450 content was decreased.

There are many questions concerning the mechanisms but also significance of P450 inhibition in aquatic species. That inhibitors can affect the substrate residue dynamics in fish was shown nearly 20 years ago. The addition of piperonyl butoxide to an aquatic microcosm also dosed with PAH resulted in the accumulation of PAH residues in fish tissues, not seen in the microcosm without the piperonyl butoxide.[110] Would such an inhibitor alter the toxicity or carcinogenesis of PAH in aquatic systems? Can less specific inhibitors affect P450s, such as P450 aromatase, involved in reproduction or other critical endogenous functions?[111,112] Evidence in birds and reptiles suggests such a possibility.[113,114] Do some inhibitors of CYP1A act as antagonists of the Ah receptor?[115] Finally, certain physiological molecules can inhibit some P450s. Compounds such as carbon monoxide (CO) and nitric oxide (NO) bind avidly to heme[116-118] and can be expected to inhibit heme proteins indiscriminately. As both NO and CO are products of enzyme reactions (of NO synthase and heme oxygenase, respectively),[119,120] these molecules could be produced in the same cells where microsomal P450 occurs and could bind and inhibit such P450 in vivo. This represents an important and poorly understood area for investigating linkages between P450 and endogenous functions.

V. CYTOCHROME P450 DIVERSITY IN AQUATIC SPECIES

The anticipated diversity of P450 forms in aquatic species has yet to be established, but the pace is increasing. Several approaches have been used, separately and in concert, to study P450 forms in vertebrates and invertebrates and the relationship of those forms to P450s in other groups, chiefly terrestrial mammals. The approaches can be grouped into three types:

1. protein purification and characterization/amino acid sequencing
2. immunological relationships
3. cloning and nucleic acid hybridizations

None of these approaches applied independently allows a complete picture of P450 structure, function, and regulation; used together, however, they provide complementary information. As discussed above, purification and reconstitution of a P450 form can permit an assessment of some of its intrinsic functional capability. Purification is still commonly used to provide immunogens for generating specific antibodies. Purification alone does not enable one to determine how expression of the gene is regulated. On the other hand, cloning of a P450 cDNA or gene will provide nucleic acid and deduced amino acid sequences, often revealing evolutionary relationships, but will not necessarily provide information concerning function. Functional characterization following expression of cloned P450 cDNA sequences in heterologous systems, such as bacteria, yeast, or mammalian cells, has been carried out with several mammalian P450s,[121] but similar studies have been conducted for only three aquatic P450s, trout CYP11A (P450scc),[122] CYP19 (aromatase)[123] and CYP17 (17α-hydroxylase).[124] A comprehensive characterization — purification, raising of antibodies, determination of catalytic function, and cloning — has been achieved with two "aquatic" P450s, CYP1A1 and CYP2K (see below). Nevertheless, each of the approaches listed above has on its own provided interesting and useful information.

A. P450 Diversity in Fish

Cytochrome P450 forms have been studied in one or more species in over 10 families of fish. Several fish P450 forms have been purified to various degrees of homogeneity, including several forms each from rainbow trout, scup, cod, and perch (Table 10). Some of these P450s show catalytic activity with multiple substrates. Reconstituted trout LMC2, for example, metabolizes estradiol, progesterone, testosterone, lauric acid, benzphetamine, and aflatoxin B_1. Monoclonal and/or polyclonal antibodies have been prepared against several of the more pure proteins and used to define more accurately the contributions of a given P450 to activities in native membranes (microsomes), which otherwise show the contribution of all forms present. Catalytic profiles, immunological cross-reactivities, and actual or deduced amino acid sequences have indicated relationships between several of the fish forms and between fish and mammalian forms. Fish P450s that are proven or likely members of families (subfamilies) 1(A), 2(B, E, K), 3(A), 4(A), 11(A), 17, and 19 have been described. The familial relationships of other fish forms have not yet been determined. Table 10 and the following discussion summarize current knowledge of P450 forms in fish and aquatic invertebrates and their relationships to mammalian enzymes. Further details can be found in several recent reviews.[7,11,125,126]

Table 10. Purified or Cloned P450 Forms in Aquatic Species

Species/P450 form	Protein or cloned	P450 Family/ subfamily	Regulation	Tissue source; other reported sites of expression	Prominent microsomal catalytic activities[a]	Ref.
Rainbow trout (BNF-treated)						
P450LM1	Prot.		Liver		138, 441	138, 152, 238, 258, 441
P450LM2	Prot. cloned	2K	M>F	Liver; kidney	LA-(ω-1)-OHase, AFB$_1$ activation	
P450LM3	Prot.		Liver		138, 441	61, 83, 138, 305, 441, 442
P450LM4b	Prot. cloned	1A1	Induced by PAH, HAH	Liver; kidney	EROD, AHH	
P450DS-1	Prot.		Liver		130	
P450DS-2	Prot.		Liver		130	
P450DS-3	Prot.		Liver		130	
Rainbow trout (untreated)						
P450LMC1	Prot. 2B		Liver	LA-(ω)-OHase	127, 449	127, 152, 449
P450LMC2 (=LM2)	Prot. cloned	2K		Liver	T-OHase, AFB$_1$ activation, LA-(ω-1)-OHase, BeND	
P450LMC3	Prot.		Liver		127	
P450LMC4	Prot.		Liver		127	
P450LMC5	Prot. 3A?	M>F	Liver; kidney	P-6β-OHase, E$_2$-2-OHase, BeND	127, 147	
P450con	Prot. 3A?		M>F, induced by cortisol. PCN	Liver; kidney, gut		129
P450KM1	Prot.		Kidney			
P450KM2	Prot.		Male-specific, induced by androgens	Kidney	135	135
P450scc	Cloned	11		Ovary	Side-chain cleavage	122
P450arom	Cloned	19		Ovary	Aromatase	123
P450c17	Cloned	17		Ovary	17α-OHase 17,20-lyase	124

Table 10. (Continued)

Species/P450 form	Protein or cloned	P450 Family/ subfamily	Regulation	Tissue source; other reported sites of expression	Prominent microsomal catalytic activities[a]	Ref.
Scup (feral)						
P450A	Prot.	3A?	Suppressed by E_2	Liver	T-6β-OHase	82, 131, 275, 276
P450B	Prot.	2B		Liver		82, 131
P450C	Prot.			Liver		82, 131
P450D	Prot.			Liver		82, 131
P450E	Prot.	1A1	Induced by PAH, HAH; suppressed by E_2	Liver; kidney, gut, nasal epithelium, all endothelium	EROD, AHH, ANF and BNF-OHase	82, 131, 278, 316, 320, 321
Cod (BNF-treated)						
P450a	Prot.			Liver		132
P450b	Prot.			Liver		132
P450c	Prot.	1A	Induced by PAH	Liver; kidney	EROD, AHH, biphenyl 4-OHase, phenanthrene OHase	132, 317, 450
P450d	Prot.			Liver		132
Perch (BNF-treated)						
P450I	Prot.			Liver		133
P450II	Prot.			Liver		133
P450III	Prot.			Liver		133
P450IV	Prot.			Liver		133
P450V	Prot.	1A?		Liver		133
Plaice	Cloned	1A1		Liver		137
Little skate (DBA-treated)						
P450 DBA-I	Prot.	1A?		Liver		134
P450 DBA-II	Prot.			Liver		134

Spiny lobster[b]				
P450 D1	Prot. cloned	?	P-6α-OHase	71, 183, 196
P450 D2	Prot.		N-demethylase (benzphetamine; aminopyrine)	71, 183
Mussel	Prot.		Digestive gland	190
Pond snail	Cloned	10	Dorsal body	195

Note: Abbreviations: E_2, estradiol; HAH, halogenated aromatic hydrocarbon; LA, lauric acid; PAH, polynuclear aromatic hydrocarbon; PCN, pregnenolone 16α-carbonitrile; T, testosterone; others as in Table 9.

[a] Contribution to microsomal activities as determined with inhibitory antibodies and by correlation of P450 content with activity. Activities measured in reconstituted systems are not listed, as they are not always in agreement with microsomal activities (for example, see Miranda et al.).[127] However, the most prominent reconstituted activities of the spiny lobster proteins are listed, to illustrate the prominent functional capacity of those proteins.

[b] The partial purification of P450 forms from several other crustaceans has been reviewed by James, M. O.[8] and Livingstone, D.R.[12]

1. Protein Purification

Multiple cytochrome P450 forms have been isolated from a number of fish species in recent years; the number of forms and species of origin are given in Table 10. The most extensive work to date on isolation of fish P450s has been with the rainbow trout *(Oncorhynchus mykiss)*. Buhler and colleagues[83,127] resolved several fractions from BNF-treated and control trout, designated LM1 through LM4, and LMC1 through LMC5, respectively. Form LM_{4b} has since been assigned to the CYP1A subfamily,[61] and form LM2 to a new subfamily, CYP2K. Relationships of the other forms have not been determined with certainty (but see *Immunological Relationships,* below). Other investigators also have purified or partially purified one or more forms from rainbow trout. Arinc and Adali[128] prepared two forms (not fully characterized), Celander et al.[129] purified a form, P450con, from control trout and Celander and Forlin[130] isolated 3 forms, DS-1, 2 and 3, from BNF-treated trout; DS-3 is probably equivalent to Buhler's $P450LM_{4b}$.

Klotz et al.[64,82,131] prepared five P450 fractions (A to E) from scup *(Stenotomus chrysops)* hepatic microsomes, using animals that had been induced by PAH-type compounds in the environment. Demonstrated or suggested microsomal activities catalyzed by these forms are shown in Table 10. One of these forms, P450E, is a scup CYP1A1. Four P450 fractions were prepared from BNF-treated cod *(Gadus morhua)* by Goksøyr.[132] Of these, P450c — a presumptive CYP1A form — has received the greatest attention. Førlin and co-workers[133] resolved five P450 fractions (I to V) with different spectral, electrophoretic, or immunological properties from perch *(Perca fluviatilis)*.

As indicated in Table 10 and the above discussion, studies of fish P450s have focused on teleosts. The only attempts at purification of P450 forms from other groups of fish have been those of Bend and co-workers,[134] who isolated two forms, DBA-I and DBA-II, from the livers of dibenzanthracene-treated little skate *(Raja erinacea)*. These forms showed BaP hydroxylase activity upon reconstitution. DBA-I was suggested to be the 1A form in skates, but relationships to other fish or mammalian forms have not been determined with certainty. Further attention to elasmobranchs and other "primitive" groups of fish (e.g., holocephalians, agnathans) may provide new insights into the evolutionary relationships of P450 forms.

All of the P450 forms or fractions described above were isolated from liver, thought to be the richest source of this group of enzymes because of its role in detoxification. Recently, however, two P450 forms were partially purified from rainbow trout kidney by Andersson.[135] One of these forms, P450 KM2, was preferentially expressed in mature males but could be induced by androgen treatment of juvenile fish.

2. Immunological Relationships

a. Subfamily 1A. The immunochemical relatedness of putative CYP1A forms in different species of fish, and of these forms to mammalian CYP1A representatives, has now been shown by a number of investigators. These studies have utilized monoclonal and polyclonal antibodies to the fish and mammalian forms, in conjunction with microsomes or purified CYP1A proteins, to demonstrate reciprocal recognition in Western blots or ELISA. Overall, the findings support a close immunochemical relationship between "CYP1A" forms in mammals and fish, and among vertebrates generally.[7,136] This relationship has been confirmed at the level of deduced or actual amino acid sequence following the cloning of CYP1A forms from rainbow trout[61] and plaice[137] and the N-terminal sequencing of CYP1A (P450E) from scup.

Antibodies have been raised against purified CYP1A proteins from trout (polyclonal α-P450LM$_{4b}$), cod (polyclonal α-P450c), and scup (polyclonal and monoclonal α-P450E). The antibodies show a range of specificities; some only recognize CYP1A proteins from species closely related to the source of the immunogen, while others have a much broader specificity, recognizing epitopes common to CYP1A forms in widely divergent groups. Monoclonal antibody 1-12-3 to scup CYP1A1 is of the latter type; it recognizes presumptive CYP1A forms in animals from many vertebrate taxa, including mammals, birds, amphibians, reptiles, and fish (both bony and cartilaginous) (Table 11). Reagent antibodies such as monoclonal 1-12-3 have proven useful for examining CYP1A regulation (induction, localization, etc.) in a wide variety of species. The epitope recognized by MAb 1-12-3 is not known, but is obviously highly conserved.

b. Subfamily 2B. The greatest number of mammalian CYP genes is in family 2.[58] There could be a large number of homologous genes in fish, but there is little information except for three subfamilies, 2B, 2E, and 2K.

An enigma in studies of monooxygenase systems in fish for 15 years has been the lack of any clear-cut induction by PB or PB-type inducers. Two fish forms now have been found to bear close immunological relationships to mammalian PB-inducible 2B proteins. Antibodies to trout LMC1 and LMC2 strongly cross-react with 2B1 from rats,[138] and antibodies to scup P450B cross-react strongly with rat 2B1, 2B2, and 2B3.[139,140] Anti-rat 2B1 recognizes each of the fish proteins, but LMC1 more so than LMC2. Furthermore, the N-terminal amino acid sequence of the scup P450B is about 50% identical to the same region of rat 2B1/2 (Figure 6). This indicates that scup B and trout LMC1 may be homologous to one another and that they are likely members of the 2B subfamily or a closely related subfamily. We have seen that proteins detected by antibodies to both scup P450B and to rat 2B1 occur in numerous aquatic species (Table 12).

Where studied, these proteins are not inducible by common 2B inducers. For example, trout P450LM2 is not induced by phenobarbital and 2,2′,4,4′,5,5′-hexachlorobiphenyl, which induce 2B1 in rats. However, Southern blots of trout DNA show the presence of sequences recognized by a rat 2B1 cDNA probe.[141,142] Thus, the lack of PB-type induction in fish appears not to involve the lack of structural gene(s) coding for 2B homologues, but rather a different regulatory mechanism from that controlling expression of these genes in mammals. Fish have apparently experienced little pressure either to evolve or to maintain responsiveness

Table 11. A Selection of Species Showing Recognition of CYP1A Forms by Monoclonal Antibody 1-12-3

Group	Species	Common name
Cartilaginous fish		
	Mustelus canis	Smooth dogfish
	Raja erinacea	Little skate
	Squalus acanthias	Spiny dogfish
Bony fish		
	Abudefduf saxatilis	Sergeant major
	Anguilla anguilla	American eel
	Chaetodon capistratus	Foureye butterflyfish
	Cyprinus carpio	Carp
	Dissostichus mawsoni	
	Fundulus heteroclitus	Killifish (mummichog)
	Gadus morhua	Cod
	Haemulon sciurus	Blue-striped grunt
	Holocentrus rufus	Squirrelfish
	Ictalurus nebulosus	Brown bullhead catfish
	Lagodon rhomboides	Pinfish
	Latimeria chalumnae	Coelacanth
	Limanda limanda	Dab; grey (lemon) sole
	Microgadus tomcod	Tomcod
	Micropterus salmoides	Largemouth bass
	Morone saxitilis	Striped bass
	Notothenia coriiceps	
	Oncorhynchus mykiss	Rainbow trout
	Oreochromis mossambicus	Tilapia
	Pagothenia borchgrevinki	
	Perca flavescens	Yellow perch
	Pimephales promelas	Fathead minnow
	Platichthys flesus	Flounder
	Platichthys stellatus	Starry flounder
	Pleuronectes americanus	Winter flounder
	Poeciliopsis lucida	Topminnow
	Salmo trutta	Brown trout
	Salvelinus fontinalis	Brook trout
	Scarus sordidus	Bullethead parrotfish
	Sebastes ruberrimus	Yelloweye red rockfish
	Sparisoma viride	Stoplight parrotfish
	Stenotomus chrysops	Scup (porgy)
	Yidisha manischewitzi	Gefilte fish
	Zebrasoma flavescens	Yellow tang
Amphibians		
	Rana catesbeiana	Bullfrog
Reptiles		
	Alligator mississippiensis	Alligator
	Chrysemys picta picta	Eastern painted turtle
	Chrysemys scripta elegans	Southern painted turtle
Birds		
	Anas platyrhynchos	Mallard duck
	Columba fasciata	Pigeon
	Gallus domesticus	Chicken
	Larus argentatus	Herring gull
	Nycticorax nycticorax	Black-crowned night heron
	Phalocrocorax aristotelis	Shag
	Phalocrocorax carbo	Cormorant
	Sterna hirunda	Common tern
	Tyto alba	Barn owl

Table 11. A Selection of Species Showing Recognition of CYP1A Forms by Monoclonal Antibody 1-12-3 (Continued)

Group	Species	Common name
Mammals		
	Balaenoptera acutorostrata	Minke whale
	Callorhinus ursinus	Northern fur seal
	Delphinapterus leucas	Beluga whale
	Globicephala malaena	Pilot whale
	Homo sapiens	Human
	Meriones unguiculatus	Gerbil
	Mus musculus	Mouse
	Mustela vison	Mink
	Oryctolagus cuniculus	Rabbit
	Phoca vitulina	Harbor seal
	Rattus rattus	Rat
	Sus scrofa	Pig
	Ursus maritimus	Polar bear

Note: All samples were analyzed by traditional immunoblot (Western blot) techniques. Nearly 100 species of fish have been analyzed and found to have a protein cross-reacting with MAb 1-12-3; not all are listed. All of the species listed were experimentally treated by various methods with known CYP1A inducers and/or were analyzed directly after being taken from the environment. The latter species showing positive reaction are assumed to have been induced by environmental chemicals.

to PB-type inducers. A common biological feature in groups lacking PB-induction (but still expressing 2B-related forms; see below) is a general absence of terrestrial plant materials in the diet,[63] but the link between fish P450s and natural products has not been directly studied. Nebert and Gonzalez[144] speculated that radiation in the mammalian gene family 2 reflects evolutionary adaptation to the presence of plant products in the diet. Similarly, novel 2B forms may have evolved in selected species of fish exposed to secondary metabolites of marine origin. Further analysis of these genes and their regulation will be important to considerations of the fundamental biological significance of 2B proteins and the environmental forces contributing to radiation within this group.

c. Subfamily 2E. A teleost P450 related to 2E proteins was indicated in studies showing that diethylnitrosamine (a fish carcinogen) was dealkylated by fish liver microsomes.[145] Moreover, antibodies to rat 2E1 recognized a single band in liver of *Poeciliopsis* spp., and an oligonucleotide (49 mer) probe specific for rat 2E1 detected a 3.3 kb mRNA in samples from the same animals. The mRNA was also

Comparison of N-terminal amino acid sequences of scup P450B and rat CYP2B1/2B2

Scup P450B: M E L S T T L I L E G L I L A L L X L V
Rat CYP2B1: M E P S I L L L L A L L V G F L L L L V

Figure 6. Comparison of N-terminal amino acid sequences of scup P450B and rat P4502B1/2.

Table 12. Aquatic Species Examined for Recognition of CYP Forms by Antibodies to Rat CYP2B1 and Scup P450B

Group	Species	Common name	Anti-rat CYP2B1	Anti-scup P450B
Cartilaginous fish				
	Raja erinacea	Little skate	+	+
Bony fish				
	Abudefduf saxatilis	Sergeant major	+	+
	Chaetodon spp.	Butterflyfish		+
	Fundulus heteroclitus	Killifish (mummichog)	+	+
	Gadus morhua	Cod	+	−
	Haemulon airolineum	Tomtate	+	
	Haemulon flavolineatum	French grunt		+
	Haemulon sciurus	Blue-striped grunt	+	+
	Ictalurus punctatus	Channel catfish		+
	Kyphosus sectatrix	Bermuda sea chub	+	+
	Lagodon rhomboides	Pinfish		+
	Lutjanus griseus	Grey snapper	+	
	Oncorhynchus mykiss	Rainbow trout		+
	Pleuronectes americanus	Winter flounder		+
	Salvelinus fontinalis	Brook trout		+
	Sparisoma viride	Stoplight parrotfish	+	
	Stenotomus chrysops	Scup	+	+
Reptiles				
	Alligator mississippiensis	Alligator	+	+
Birds				
	Columba fasciata	Pigeon	+	+
	Nycticorax nycticorax	Black-crowned night heron		+
Mammals				
	Balaenoptera acutorostrata	Minke whale	−	
	Callorhinus ursinus	Northern fur seal	−	
	Delphinapterus leucas	Beluga whale	−	−
	Globicephala malaena	Pilot whale	+	+
	Mustela vison	Mink		+
	Phoca vitulina	Harbor seal	−	
	Rattus rattus	Rat	+	+

Notes: +, positive cross-reactivity.
−, no cross-reactivity detected.

induced by ethanol.[145] Proteins cross-reacting with antibodies to mammalian 2E1 occur in other fish species (unpublished), but the degree of relationship is not known.

d. Subfamily 3A. Several lines of evidence indicate a relationship of teleost forms to members of P450 subfamily 3A. Microsomal activities attributed to trout LMC5 include steroid 6β-hydroxylase (Table 10), catalyzed in mammals largely by 3A proteins.[146] Immunological cross-reactivity has been detected between LMC5 and both rat 3A1 and human 3A4.[147] Antibodies to LMC5 recognized both rat 3A1 and human 3A4 in immunoblots of hepatic microsomal protein from these species, and purified LMC5 was recognized by anti-human 3A4 (but not anti-rat 3A1). The structural similarity between the trout and human forms was supported by inhibition of progesterone 6β-hydroxylase in trout liver microsomes by anti-human 3A4. The evidence that scup P450A (Table 10) is a steroid 6β-hydroxylase suggests that the scup and trout proteins may be related. A relationship between P450con and rat subfamily 3 proteins was suggested by Celander et al.,[129] who found the amount of this protein was induced about 40% by high-dose PCN. Immunochemical relatedness of trout P450con, scup P450A, and human 3A4 forms has recently been demonstrated (unpublished). However, these suggested relationships to CYP3A forms require further study.

3. Cloning/Nucleic Acid Hybridizations

a. Gene Family 1. In mammals there are two known members of the CYP1A subfamily, CYP1A1 and CYP1A2. Sequencing of cDNA for the trout P450 induced by 3-MC[61] confirmed the identity of this trout gene as a member of the 1A subfamily. The authors inferred a 57 to 59% sequence identity of the trout 1A and mammalian 1A1 and a 51 to 53% identity with mammalian 1A2 forms, concluding that the trout gene was a 1A1. The inferred sequence in specific regions of the trout 1A1, such as that corresponding to the heme binding region, was greater than 60% identical to rat 1A1. Divergence of the mammalian 1A1 and 1A2 genes is suggested to have occurred subsequent to divergence of the teleost and mammalian lines.[148] If so, then fish would not have diverged CYP1A genes that would be paralogous to each other and orthologous to mammalian 1A1 and 1A2. Some regions of the inferred trout 1A1 sequence are identical to all mammalian 1A1 but not 1A2, and others are identical to all 1A2 but not 1A1.[7,61] This "hybrid" condition could be expected if fish have but single 1A genes, representing a type ancestral to both 1A1 and 1A2 in mammals. Whether the structural relationship of teleost CYP1A to mammalian 1A1 and 1A2 proteins is reflected also in catalytic and other properties is uncertain. Catalytic activities with 7-ER and BaP indicate a closer similarity to 1A1 than to 1A2, but the hydroxylation of acetanilide, a mammalian 1A2 substrate, by scup CYP1A1 and by trout CYP1A1[125] is consistent with a 1A1/1A2 hybrid character. Further examination of this question is warranted.

Whether multiple 1A genes occur in fish is still uncertain. In mammals, 1A1 and 1A2 show a number of regulatory, structural, and functional distinctions.[149] As discussed earlier,[7,130,150] distinctions in temporal patterns of AHH and EROD induction and disparate results regarding multiple protein or mRNA products induced in fish by preferential 1A1 or 1A2 inducers (BNF or isosafrole) could support either position. Wirgin et al.[151] found evidence for induction of allelic variants of CYP1A in tomcod *(Microgadus tomcod)* but not two distinct genes. Berghard and Chen (personal communication) recently cloned two closely related CYP1A genes from rainbow trout liver. These may have diverged quite recently.

Comparison of inferred trout 1A1 sequence with its homologue from scup (P450E) shows that about 85% of 25 residues in the N-terminal region are identical (Figure 7). This degree of similarity identifies the scup protein as a likely 1A1. The

Comparison of N-terminal amino acid sequences of fish CYP1A1 forms

Trout	M V L M I L P I I G S V S A S E G L V A M V T L C L
Scup	(M) V L M I L P V I G S V S V S E G L V A M I T M C L
Plaice	M M L M M L P F I G S V S V S E S L V A M T T M C L

Figure 7. Comparison of N-terminal amino acid sequences of fish CYP1A forms. Sequences are from Heilmann et al.[61] (trout), Klotz et al.[82] and Stegeman[126] (scup), and George[137] (plaice).

recent sequencing of a cDNA for a hydrocarbon-inducible P450 from plaice[137] showed a close structural similarity between plaice, trout, and scup 1A forms. At the N-terminus of the three forms (trout, scup, plaice) there is an 80 to 84% amino acid identity, and the full-length identity between plaice and trout is still about 78%. This is higher than might be expected based on the phylogenetic relationship of the families containing these species, which diverged as long ago as 200 m.y. before present. By comparison, more recently diverged mammalian groups show greater difference in 1A1 sequence in the N-terminal region, and overall.

The greater than expected degree of sequence similarity in the N-terminus of CYP1A from distant teleost families supports the idea that the rate of evolution could be slower in the P450 genes in fish than in mammals.[7,126] Whether the suggested slower rate of change in fish CYP1A merely reflects a slower rate of sequence change common to many fish proteins or is unique to fish P450s (or even CYP1A alone, perhaps related to some physiological or ecological variable peculiar to fish) is not yet known. This issue will be resolved as additional teleost P450 full-length sequences are obtained. Many fish species show single proteins cross-reacting with MAb 1-12-3 to scup P450E (1A1),[7] implying the presence of 1A1 in elasmobranchs and all teleost groups examined. Sequencing is necessary to establish the extent of similarity and identity and to address the evolutionary relationships.

b. Gene Family 2. As mentioned earlier, most of the evidence for family 2 P450s in fish comes from immunochemical studies. Two groups have obtained additional evidence through the use of nucleic acid probes for mammalian family 2 forms. Haasch and co-workers[141,142] showed that a rat CYP2B1 cDNA probe hybridized with trout DNA on Southern blots. Kaplan et al.[145] used an oligonucleotide (49-mer) specific for rat CYP2E1 to probe Northern blots of hepatic mRNA from *Poeciliopsis* spp.; a 3.3 kb band was detected. The first family 2 cDNA to be cloned from fish was recently reported by Buhler.[152] This form, which corresponds to P450LM2 purified by his group, has been classified as a CYP2K.

c. Gene Family 11. Most recently, a P450 cholesterol side-chain cleavage enzyme (P450scc) has been cloned from trout.[122] The sequence places it in the CYP11A subfamily.

d. Gene Family 17. A cDNA clone encoding a P450c17 (17α-hydroxylase/ 17,20-lyase) has been isolated from a rainbow trout ovarian follicle cDNA library.[124] The amino acid sequence of the 522-residue protein encoded by this cDNA shows a high degree of sequence identity (64%) to chicken CYP17 and lesser identity (46 to 48%) to CYP17 sequences from human, cow, and rat. When expressed in COS (monkey kidney) cells, the fish cDNA encoded a protein with both 17α-hydroxylase and 17,20-lyase activities.

e. Gene Family 19. Recently, a cDNA encoding a P450 aromatase (CYP19) was isolated from a rainbow trout ovarian cDNA library.[123] The sequence of this cDNA appears to encode a protein of 522 amino acids that is slightly more than 50% identical with the previously determined aromatase coding sequences from human, rat, mouse, and chicken. The fish CYP19 cDNA, when expressed in COS cells, encoded a protein that catalyzed the conversion of testosterone to 17β-estradiol (i.e., aromatase activity). Interestingly, the trout CYP19 cDNA encodes a longer leader sequence or membrane anchor at the N-terminal region than occurs in the other aromatases. The N-terminus of the trout gene also is more hydrophobic than the mammalian or chicken genes, suggesting that there may be some feature of its interaction with poikilotherm membranes that is distinct from that in homeotherms. As with the CYP1A from trout, there are regions of the trout aromatase that are highly similar to the other aromatases. Thus, the heme-binding region, an aromatic region, and a region of undefined function (amino acids 133 to 154) were each more than 70% identical among the various species.

Identification of P450scc, aromatase, and 17α-hydroxylase/17, 20-lyase genes in fish supports speculations that other steroidogenic P450s known in terrestrial mammals (e.g., the P450 11β-hydroxylase) will have homologues in fish and other aquatic vertebrates.

B. P450 Diversity in Amphibians and Reptiles

There have been few reports describing the P450-dependent monooxygenase systems in amphibians and reptiles, and no P450 forms have been purified or cloned from members of these classes. Nevertheless, there is some information regarding similarities and differences between amphibian and reptile P450s and those in other vertebrate groups.

Not unexpectedly, most of what is known concerns putative CYP1A forms. Several studies in amphibians and reptiles have shown induction of AHH or other catalytic activities associated with 1A forms in other vertebrates. Schwen and Mannering[153] found increased AHH in hepatic microsomes from frogs *(Rana pipiens)* and snakes *(Thamnophis* spp.) treated with 3-MC. Likewise, increased

AHH activity has been observed in 3-MC-treated salamanders *(Ambystoma tigrinum)*,[154] frogs *(Xenopus laevis)*,[155] and newts *(Pleurodeles waltl)*.[156,157] In the latter study, increased metabolism of alkyl phenoxazones (resorufins) and alkoxycoumarins was also measured in the treated animals. Jewell et al.[158] obtained induction of AHH, EROD, and other activities by 3-MC in the American alligator *(Alligator mississippiensis)*. In some studies, CYP1A-specific antibodies have been used as probes for likely CYP1A forms and their induction. Although Marty et al.[157] obtained negative results with an anti-rat 1A1/1A2 monoclonal antibody in newts, another amphibian *(Rana catesbaeiana)* contained at least one and possibly two hepatic microsomal proteins recognized by MAb 1-12-3 to scup CYP1A1 (Figure 8). In immunoblots of alligator microsomal proteins, Jewell et al.[158] detected two bands that were increased by 3-MC treatment; one was recognized preferentially by anti-rat 1A1 and the other by anti-rat 1A2. Immunoreactive 1A protein was also detected in the spectacle caiman *(Caiman crocodylus)*.[159] Yawetz et al.[160] found elevated levels of a MAb 1-12-3-reactive protein in hepatic microsomes of turtles *(Chrysemys picta* and *Chrysemys scripta)* treated with a PCB mixture or the coplanar PCB congener 3,3',4,4'-TCB. As further evidence for a 1A form in amphibians, a mammalian CYP1A1 cDNA hybridized to a single band on Northern blots of RNA isolated from a toad kidney cell line.[161]

Much less is known about the possible existence in amphibians and reptiles of forms related to mammalian CYP2Bs. Marty et al.[157] used a monoclonal antibody to rat 2B1/2B2 to probe Western blots of newt microsomal protein; a single cross-reactive band of 52 kDa molecular mass was found. Treatment of the newts with phenobarbital did not affect the intensity of this band and did not alter any of several monooxygenase activities, including PROD, catalyzed largely by 2B1 in some mammals. An earlier report by Schwen and Mannering[153] had shown a lack of PB effect on other 2B-associated activities in frog or snake. The apparent refractoriness of amphibian and reptilian P450s to PB induction resembles that seen in fish. However, Winston[162] has recently obtained evidence in alligator for PB induction of a protein immunochemically related to mammalian 2B forms.

The only other amphibian P450 for which evidence currently exists is the possible 3A form described by Schuetz et al.[161] in the A6 toad kidney cell line. A6 cells contain steroid-inducible corticosterone 6β-hydroxylase activity that is inhibited by polyclonal antibodies to rat P4503A1 and rabbit P4503A6. These antibodies also recognize single polypeptides on Western blots of A6 microsomal protein. In addition, human 3A cDNA probes hybridize with A6 RNA and genomic DNA on Northern and Southern blots, respectively. These data provide strong evidence for the existence of a 3A form in amphibians. The importance of this P450 is related to its postulated role in the regulation of Na$^+$ transport in A6 cells (and possibly in cells of other vertebrates) via synthesis of 6β-hydroxycorticosterone, an agonist for the type IV glucocorticoid/mineralocorticoid receptor involved in salt transport.[161,163] This work illustrates the potential

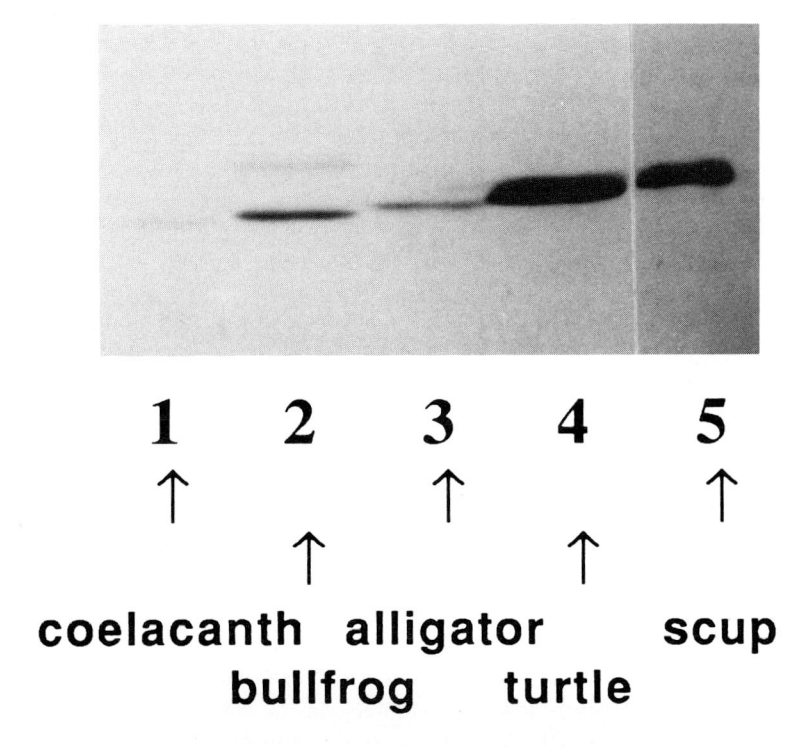

BIOCHEMISTRY AND MOLECULAR BIOLOGY OF MONOOXYGENASES

Figure 8. Immunoblot showing recognition of CYP1A forms by MAb 1-12-3 in multiple species of vertebrates. Hepatic microsomal protein (100 µg) from coelacanth *(Latimeria chalumnae)*, bullfrog *(Rana catesbaeiana)*, alligator *(Alligator mississippiensis)*, and turtle *(Chrysemys picta)* was analyzed by denaturing gel electrophoresis and immunoblotting with monoclonal antibody 1-12-3. Scup: 2 pmol purified scup CYP1A1 (P450E).

advances in fundamental biochemical and physiological understanding that are possible through studies in animals other than mammals.

C. P450 Diversity in Marine Mammals

Reports describing the existence and characteristics of P450s in marine mammals are few. Studies have identified P450 or P450-associated catalytic activities in pinnipeds[164–168] and a few species of cetaceans (minke, pilot, killer, and beluga whales and dolphins).[169–173] Other investigators have inferred P450 characteristics based on patterns of PCB congeners detected in cetacean tissues.[174–176] The quality of marine mammal tissues obtained for study is often less than optimal, as reflected by the substantial quantities of P420 (interpreted as degraded P450) that are commonly present.[169] Nevertheless, several catalytic activities typical of vertebrate P450s have been detected in preparations of liver or kidney of marine mammals (reviewed by Boon et al.)[177] Thus, marine mammal tissue fractions (microsomes or postmitochondrial supernates) have been shown to transform benzo(a)pyrene, 7-ethoxyresorufin, 7-pentoxyresorufin, 7-ethoxycoumarin, 7-methoxycoumarin, aminopyrine, phenanthrene, biphenyl, estradiol, androstenedione, aldrin, and aniline. Comparison of the rates of these reactions between terrestrial and aquatic mammals reveals that some (e.g., AHH, EROD) in marine species are similar to those in induced rats, whereas others (e.g., PROD, aldrin epoxidase) appear to be somewhat lower in marine species. In certain species (killer whale,[170] beluga[169]), the presence of appreciable EROD activity in some individuals is suggestive of induction by environmental chemicals.

Purification of P450 has not been reported for any marine mammal, but several investigators have used antibodies to probe for the presence of P450 forms related to forms in other vertebrates. Goksøyr et al.[172,173] used polyclonal anti-cod P450c and monoclonal anti-scup CYP1A1 (MAb 1-12-3) to identify immunochemically related forms in minke whale. White et al.[169] identified a MAb 1-12-3-reactive form in beluga. Inhibition of EROD and AHH by antibodies to rat 1A1 and 1A2 was shown by Watanabe et al.[170] Within the pinnipeds, putative 1A forms have been demonstrated in harbor,[167] fur,[178] harp,[168] and hooded seals.[168]

Questions regarding the presence and activity of CYP2B forms in marine mammals are of interest in light of inferences, from patterns of PCB congener residues in cetacean tissues, that these animals may be deficient in some functions typically associated with P4502B in terrestrial mammals. For example, Tanabe et al.[174] and Norstrom et al.[176] hypothesized, based on the relative persistence of *meta-para* unsubstituted PCB congeners, that monooxygenase activities characteristic of CYP2B forms were reduced in cetaceans compared to terrestrial mammals. Results of initial studies using anti-2B antibodies have provided a measure of support for this hypothesis, suggesting that cetacean livers do not express a P450 form closely related to CYP2B1 or 2B2. Using polyclonal antibodies to rat CYP2B1/2, Goksøyr et al.[172,173] and Stegeman[179] could detect no cross-reactive bands in hepatic microsomes from minke whales. Likewise, White et al.[169] found no such protein in beluga hepatic microsomes. Interestingly, a polyclonal antibody raised against rabbit CYP2B4 did recognize a band in beluga, suggesting the presence in beluga of a 2B-type P450 more closely related to rabbit 2B4 than to rat 2B1 or 2B2.

Results of immunochemical studies in pinnipeds have been similar to those in cetaceans. Anti-rat 2B1 failed to recognize any bands in hepatic microsomes of harbor seals[167] or fur seals,[178] but a single band was detected in the latter species with anti-rabbit 2B4, as with beluga. Antibodies to a PCB-metabolizing P450 from dog (CYP2B11) were found by Goksøyr et al.[168] to recognize multiple bands in harp and hooded seals. Thus, there is some evidence for the presence and expression of 2B forms in pinnipeds and cetaceans, but the metabolic capabilities of these forms and their exact relationship to 2B forms in other mammalian species remain to be determined.

There is also some evidence for a cetacean P450 in family 3. Polyclonal antibodies against trout P450con, a putative 3A-related protein, cross-reacted with two bands in minke whale liver.[173] One of these bands was also recognized by anti-rabbit P4503C2.[173]

Based on the studies cited above, some generalizations can be made regarding P450 forms in marine mammals:

1. Pinnipeds and cetaceans possess proteins recognized by antibodies to mammalian and fish CYP1A1 forms.
2. Good correlations between the amount of such proteins and the degree of contamination by PCBs suggest that these putative CYP1A forms are inducible, as in other vertebrates.[169]
3. Patterns of PCB congeners in tissues of cetaceans suggest a reduced capacity for metabolism of certain types of structures, in particular those that are transformed by P450s in subfamily 2B of terrestrial mammals.[174–176] In vitro rates of 3,3′,4,4′-TCB and 2,2′,5,5′-TCB metabolism[69] support this.
4. Catalytic and immunochemical data obtained so far are consistent with reduced expression and/or altered structural specificity of 2B forms in aquatic mammals.

Needless to say, there is a great deal more to learn regarding the regulation and catalytic function of P450 forms in aquatic mammals. Given the social/political barriers to experimental work with marine mammals, a better understanding of P450s in this group of animals will require creative approaches. Marine mammal tissues are sometimes available as a result of strandings or other types of accidental mortality. Such tissues can be used for biochemical or immunochemical studies of P450 characteristics. Existing cell lines derived from marine mammals[180] may prove useful if P450 expression in them can be demonstrated. However, the greatest progress will most likely be made through molecular studies, including cloning and sequencing of marine mammal P450 forms and their expression and subsequent study in heterologous systems such as COS cells.

D. P450 Diversity in Invertebrates

1. Protein Purification

Numerous studies have measured monooxygenases in aquatic invertebrates,[84,85,181–186] but there have been few successful attempts at purification and characterization of the P450 forms involved. This is due in part to difficulties resulting from the presence of P450 inhibitors and high proteolytic activity in some invertebrate species and tissues.

James and Shiverick[183] isolated one major and two minor P450 forms from hepatopancreas of untreated spiny lobster *(Panulirus argus)*. Two of these forms, when reconstituted with purified NADPH-cytochrome P450 reductase from mammalian sources, acted on a variety of substrates, including benzphetamine, progesterone (16α-hydroxylation), aminopyrine, testosterone,

BaP, and 7-ethoxycoumarin. P450-containing fractions have also been isolated from several species of crabs.[187–189] A detailed review of crustacean P450s has been published.[8]

The other group of aquatic invertebrates whose P450s have received significant attention are mollusks. Cytochrome P450 and associated catalytic activities have been measured in a number of molluscan species, where they appear to be concentrated in the endoplasmic reticulum of the digestive gland.[9,12] Molluscan digestive gland microsomes show the unusual property of NADPH-independent catalytic activities and variable effects of NADPH on the rate of these reactions. The reasons for this are not completely understood,[9,12,85] but this phenomenon serves to illustrate some of the novelties and complexities of working in invertebrate systems. A goal of future work should be to exploit these differences to increase our fundamental understanding of biotransformation mechanisms.

Attempts to purify P450 forms from molluscan sources have been limited. Kirchin et al.[190] obtained an eightfold purified, P450-containing fraction from digestive gland microsomes of *Mytilus edulis*. The catalytic function of this partially purified form was not reported.

Monooxygenase function has been examined in animals from several other invertebrate groups, including sponges, coelenterates, annelids, and echinoderms (reviewed in Livingstone),[12] but purification of P450s has not been reported for any of these groups.

2. Immunological Relationships

Knowledge of immunological relationships, or lack thereof, between invertebrate and vertebrate P450s is scant. Most likely, several groups have sought evidence for such relationships; the dearth of published reports on this topic might thus reflect a lack of positive results. For example, we have examined microsomal fractions from a number of invertebrate species for cross-reactivity to MAb 1-12-3, with little success. This could merely mean that the epitope recognized by MAb 1-12-3 appeared more recently than the vertebrate-invertebrate divergence. Schlenk and Buhler[191] used polyclonal antibodies to rainbow trout P450 forms LM2 (P4502K) and LM4b (CYP1A1) in Western blots of digestive gland microsomes from untreated and BNF-treated chitons *(Cryptochiton stelleri)*. Anti-LM4b IgG recognized a single band that appeared to increase in intensity upon BNF treatment, although the number of animals in which this occurred is not clear. Anti-LM2 IgG recognized two bands at 54 and 60 kDa, suggesting the presence in chitons of multiple P450 forms related to vertebrate family 2 P450s. Den Besten et al.[191a] reported proteins in pyloric caecae from *Asterias rubens* (Seastar) that showed weak cross-reactivity with antibodies to three forms of scup P450 (putative members of families 1, 2, and 3).

3. Cloning and Nucleic Acid Hybridizations

The application of molecular cloning techniques to research on invertebrate P450s is likely to yield substantial rewards in such areas as P450 gene evolution, invertebrate reproductive physiology, and chemical ecology. Until now, however, there has been little effort in this area. Spry et al.[192] reported hybridization of a mammalian P4504A probe to total RNA from *Mytilus edulis*. Interestingly, a member of P450 family 4 (P4504C1) has recently been cloned from a terrestrial invertebrate, the cockroach *Blaberus discoidalis*,[193] further indicating that the P450 4 family is an ancient one. Livingstone and co-workers have also detected cross-hybridization between *Mytilus* RNA and a CYP1A cDNA probe.[194]

The only P450 thus far cloned from an aquatic invertebrate (the mollusk *Lymnaea stagnalis*) has been assigned to family 10.[195] This form was isolated by differential hybridization of a *L. stagnalis* dorsal body cDNA library. It encodes a protein of 545 amino acids with highest sequence identity (28 to 31%) to mammalian mitochrondrial proteins in families 11 and 27. Preferential expression of CYP10 mRNA in the dorsal bodies led the authors to postulate its involvement in hormone synthesis.

Recently, James et al.[196] have succeeded in cloning a cDNA for a P450 from spiny lobster hepatopancreas. Based on partial sequence (300 amino acids from the N-terminus), this P450 could not be assigned to any known gene family.

It is clear that there is abundant opportunity for progress in the field of invertebrate P450 research. With the advances in biochemical, immunological, and molecular techniques that have occurred in the past 10 years, questions of long-standing interest can now be addressed. For example, over 15 years ago, Lee et al.[181,197] demonstrated high AHH activity in the antennal gland of female blue crab *(Callinectes sapidus)*. This activity fluctuated dramatically during the molt cycle, suggesting reciprocal regulation of a P450 form and molting hormones (ecdysteroids). Other workers have studied ecdysone metabolism in microsomal systems of crustaceans.[183] Our understanding of the activity and regulation of P450 forms involved in the synthesis and degradation of ecdysteroids might be greatly enhanced through the approaches mentioned above.

VI. CYTOCHROME P450 GENE REGULATION

The capacity of a cell for xenobiotic metabolism may depend largely on the content of P450 forms. In mammals, there are transcriptional, translational, and post-translational mechanisms involved in regulation of different P450 forms. Rates of synthesis and degradation (turnover) of both heme and apoprotein will determine the amount of active enzyme. There is little direct information on the turnover of P450 in aquatic species, a serious deficiency in our knowledge. But even though the library of known P450 genes in aquatic systems is small, there is evidence that similar regulatory motifs are involved in controlling the expression of similar P450s. Most results pertain to CYP1A, for which studies are yielding important information on the identity of inducers and on the temporal aspects of transcription and translation during induction. Most of the results have been obtained with liver, although cells in culture and extrahepatic tissues are being used increasingly.

A. Temporal Aspects of CYP1A Induction

Hydrocarbon induction of hepatic monooxygenase activity in some fish species is rapid, and rates above control values can persist for a long period (e.g., Kurelec).[198] Only recently has the temporal pattern of CYP1A1 transcription and translation been described in induced fish. In the first studies of freshwater and marine fish, single doses of the inducer BNF elicited detectable increases in mRNA content by 6 hr, a peak in mRNA at 30 to 40 hr, and decline to near control by 4 to 5 days.[199,200] Functional protein was evident by 18 hr and maximal at near 3 days. Unlike the mRNA, functional protein can persist at elevated levels for 2 to 3 weeks.[201] The long persistence of protein while mRNA declines is somewhat paradoxical; the half-lives for P450 in mammals are generally less than 48 hr.[202] Direct studies on turnover indicate similar half-lives for fish CYP1A (Kloepper-Sams and Stegeman, unpublished).

The induction of mRNA is strongly influenced by the nature of the inducing compound. In the studies just cited, the inducer, BNF, is a readily metabolized compound. Similar studies with less readily metabolized, halogenated inducing compounds produce different results. With those compounds, a single treatment induces high levels of mRNA that are sustained for as long as 2 weeks.[203,204] A comparison of mRNA content in samples treated with BNF or TCDF (Figure 9) illustrates this result in vivo. Studies in cell cultures have provided similar results,[205] indicating that we might validly generalize concerning this response to single doses of inducers.

There are uncertainties concerning the responses to very high dose and/or prolonged exposure. Treatment with high doses of 3,3′,4,4′-TCB resulted in a decline not only in activity but also in the amount of CYP1A protein.[98] Studies in cell cultures are providing further evidence that such declines occur at high doses.[99] In an important recent study by Haasch and colleagues,[97] prolonged exposure to the readily metabolized inducer BNF resulted in eventual declines in the CYP1A protein but not in the amount of message. The mechanism(s) involved in the decline of protein, even while levels of message are sustained, are not known. Moreover, the pattern of responses over long time periods and continuous exposure may hold surprises.[97] These are important questions to be pursued regarding CYP1A, but also regarding other xenobiotic-metabolizing forms, for which regulatory mechanisms and events are unknown.

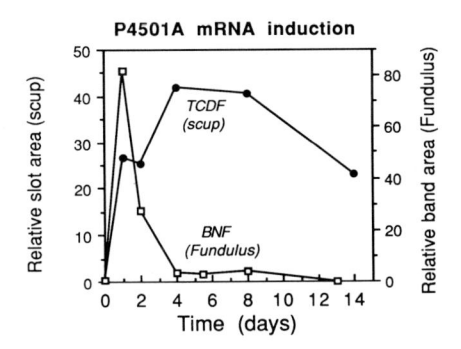

Figure 9. Time course of hepatic CYP1A mRNA content after single doses of BNF or 2,3,7,8-TCDF. Fish were given a single intraperitoneal injection of inducer and sampled at various times following treatment. CYP1A mRNA was determined on Northern blots or slot blots of total hepatic RNA, using the trout CYP1A1 cDNA probe pfP$_1$450-3'.[61] Data from Kloepper-Sams and Stegeman[199] *(Fundulus)* and Hahn et al.[203,204] (scup).

B. Steroidogenic P450 Gene Regulation

Regulation of the P450 aromatase gene in trout ovary appears to be closely linked to the production of estradiol by the ovary and consequently to the stimulation of vitellogenesis by estradiol.[123] The pattern is consistent with an interpretation of transcriptional activation. Transcriptional activation is recognized as a mechanism of gene activation in mammalian aromatase genes in steroidogenic organs.[206] It will be interesting to determine whether P450 aromatase in other teleost organs where the activity occurs is regulated in a similar fashion to that in ovary. It is conceivable that there would be a different demand and possibly different sets of *trans*-activating factors (receptors?) and response elements brought into play, for example in the brain as compared to ovary. Whether the other steroidogenic P450s described above will have regu-latory mechanisms similar to those in mammals is likely but as yet unknown.

There are other steroid hydroxylations that occur in fish that have no known counterpart in mammalian systems. Among these are hydroxylations that produce novel steroid hormones (di- and tri-hydroxy-progesterones, e.g., 17α, 20β, 21-trihydroxyprogesterone) that function in the maturation of oocytes.[207] We suggest that novel P450 forms will catalyze those hydroxylations in fish. The transitory appearance of the final maturation hormone[207] suggests that there may be interesting regulatory aspects of such P450s as well, in brain and in the major steroidogenic organs. The involvement of receptors for these P450s is an important area for research.

C. CYP1A Inducer Structure/Activity Relationships in Fish

Over the past 25 years, quite a large number of compounds have been evaluated as inducers of CYP1A in fish. Initially, studies focused on measurements of catalytic activity; more recent efforts have looked also at CYP1A protein and mRNA levels as measures of induction. Table 13 lists many of these compounds and their effects on CYP1A gene expression at these three levels. In general, the structural features associated with CYP1A induction in fish are similar to those seen in mammals. Thus, certain large PAH such as BaP, dimethylbenzanthracene, and 3-MC are able to induce 1A, but aliphatic (isooctane) and small aromatic hydrocarbons (benzene, naphthalene, phenanthrene) are not. Among the chlorinated aromatics, lateral substitution and a planar configuration, or the ability to attain a planar configuration, are important. For dioxins and dibenzofurans, chlorination in the 2,3,7, and 8 positions appears to be essential (Figure 10), though only a very limited number of congeners has been examined so far. The number of chlorinated biphenyl structures able to induce CYP1A in fish may even be more restricted than in mammals. For example, mono-ortho substituted derivatives of potent nonortho substituted congeners are inactive[98] or very weakly active in fish.[208] However, there may be species differences, suggested by differences in the ability of the mono-ortho 2,3′,4,4′,5-pentachlorobiphenyl to induce CYP1A in scup and trout.[98,208] There are also reports that the methylene dioxyphenyl compounds isosafrole and piperonyl butoxide can induce EROD activity in fish (see Table 13). The mechanism by which this induction occurs is not known.

Important questions remain concerning structure/activity relationships for CYP1A induction in fish:

1. Even for the best-known groups of compounds (PAHs and HAHs), our understanding is not sufficient to identify the most important contributors to environmental induction. There is a need for a systematic effort to define the structural features necessary for induction and to assess their constancy across species and higher taxonomic lines.

2. The possibility, long discussed,[63] that natural products found in marine and freshwater environments might include potent CYP1A inducers is as yet unexplored (see Section XIII).

3. The mechanistic basis for structure/induction relationships is also not known. Presumably, induction SARs are at least partly a reflection of receptor-binding SARs, but these have not been directly examined in any aquatic species. Such studies might provide an explanation for some of the species differences mentioned above. Moreover, studies in mammals have shown that receptor binding alone is not sufficient for induction to occur; some compounds bind but do not produce functional ligand-receptor complexes.[115] There is a need to define the full range of molecular events linking the uptake of various structures to the resulting induction response.

Table 13. Aromatic and Chlorinated Hydrocarbons Evaluated as Inducers of CYP1A in Fish

Compound	Activity	Protein	mRNA	Reference
β-naphthoflavone	+	+	+	74, 97, 199, 200, 439, 451–454
3-Methylcholanthrene	+	+	+	61, 143, 300, 451, 452, 455, 456
Benzo(a)pyrene	+	+	+	96, 143, 321, 457, 458
Benzanthracene	+			74, 459
Dibenzanthracene	+			379, 451, 460
Dimethylbenzanthracene	+			451
Naphthalene	–			446, 461
Phenanthrene	–	–		450, 461
Pyrene	–			457, 461
Benzene	–			461
Xylene	–			461
Dibenzothiophene	–			461
Isooctane	–			461
Chrysene	–			457
Fluoranthene	–			457
1,2,4-Trimethylnaphthalene	–			457
PCB mixtures	+			74, 291, 448, 462
3,3′,4,4′-TCB (#77)	+	+	+	98, 99, 448, 463, 464
3,3′,4,4′,5-PCB (#126)	+	+		221, 448, 463
3,3′,4,4′,5,5′-HCB (#169)	+	+		291, 452
2,3,3′,4,4′-PCB (#105)	–	–	–	98
2,3′,4,4′,5-PCB (#118)	–/+	–/+	–	98, 208
2,2′,3,3′,4,4′-HCB (#128)	–			98
2,2′,3,4,4′,5′-HCB (#138)	–	–	–	98
2,2′,4,4′,5,5′-HCB (#153)	–		–	200, 291, 464
2,2′,4,4′-TCB (#47)	–			439
2,3,7,8-TCDD	+			79, 205, 221, 300, 456, 460, 463–466
2,3,7,8-TCDF	+	+	+	203, 204, 320, 467
2,3,6,8-TCDF	–	–	–	204
2,3,4,7,8-PCDF	+			468
p,p′-DDT	–			469
p,p′-DDE	–			469
Pregnenolone-16α-carbonitrile	–	–		450, 455
Isosafrole	+	+		130, 150, 464, 470
Piperonyl butoxide	+			444, 445
Mirex	–			464
Kepone	–			464
Acrylamide	–	–		471
Phenylbutazone	–			179
Phenobarbital	–	–		74, 291, 450, 453, 455
Indole-3-carbinol	–	–		414

Note: Abbreviations: DDT, dichlorodiphenyltrichloroethane; HCB, hexachlorobiphenyl; PCB, pentachlorobiphenyl; TCB, tetrachlorobiphenyl; TCDD, tetrachlorodibenzo-p-dioxin; TCDF, tetrachlorodibenzofuran.

VII. Ah RECEPTOR IN AQUATIC SPECIES

The fundamental mechanism by which mammalian CYP1A forms are regulated involves a ligand-activated transcription factor known as the Ah (Aromatic

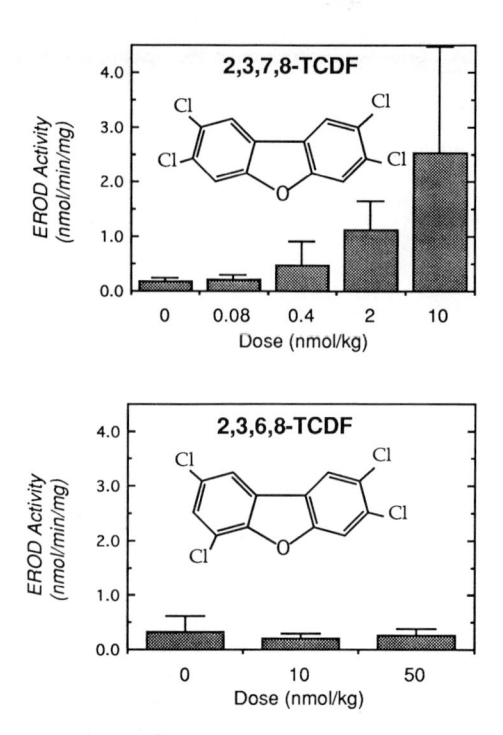

Figure 10. Comparison of 2,3,7,8-TCDF and 2,3,6,8-TCDF as inducers of CYP1A in scup. Fish were given a single intraperitoneal injection of tetrachlorodibenzofuran at the indicated doses. Four days following treatment, hepatic microsomal EROD activity was measured. Further details can be found in Hahn et al.[203,204]

hydrocarbon) receptor. Details of this mechanism have been described in earlier reviews,[60,209] and will not be repeated here. The Ah receptor (AhR) has been shown to exist in numerous species of terrestrial mammals,[210-212] and studies elucidating its role in the regulation of CYP1A gene expression have dealt almost exclusively with this group of animals. Knowledge of Ah receptors in aquatic species is comparatively scant (Table 14), but the presence and properties of such receptors could determine the susceptibility of these animals to CYP1A induction and other biochemical and toxic alterations resulting from exposure to PAH and HAH.

A. Species Distribution

The first extensive search for the AhR in nonmammalian species was carried out by Denison et al.[213,214] Using [1,6-³H]2,3,7,8-tetrachlorodibenzo-p-dioxin ([³H]TCDD) and a velocity sedimentation assay, they found no evidence for this receptor in several species of fish, a frog *(Rana),* or two invertebrates. Subsequently, Heilmann et al.[61] reported the presence of an AhR in rainbow trout, and

Table 14. Aquatic Species in Which Ah Receptors Have Been Detected

Group/species	Ref.
Mammals	
Beluga *(Delphinapterus leucas)*	217, 218
Reptiles	
Painted turtle *(Chrysemys picta)*	218
Amphibians	
Newt *(Pleurodeles waltl)*	156
Fish	
Rainbow trout *(Oncorhynchus mykiss)*	61, 217, 218
Brown trout *(Salmo trutta)*	472
Killifish *(Fundulus heteroclitus)*	217, 218
Winter flounder *(Pleuronectes americanus)*	217, 218
Scup *(Stenotomus chrysops)*	217, 218
Smooth dogfish *(Mustelus canis)*	217, 218
Spiny dogfish *(Squalus acanthias)*	217, 218
Fish cell lines	
RTH-149 (Rainbow trout hepatoma)	215
RTG-2 (Rainbow trout gonad)	216
PLHC-1 *(Poeciliopsis lucida* hepatoma)	99

Lorenzen and Okey[215] characterized the specific binding of [^{3}H]TCDD in cytosol from a rainbow trout hepatoma cell line (RTH-149). The latter authors described a number of technical modifications that facilitated the detection of the receptor in these cells. The AhR in RTH-149 cell cytosol had a sedimentation coefficient (~9 S) and apparent binding affinity (<1 nM) similar to those in rat or mouse, but the concentration of receptor (~20 fmol/mg protein) was less than that seen in rodents. These authors also showed that the AhR could be isolated as a 6 S complex from the nuclei of cells treated with [^{3}H]TCDD in culture, suggesting that the fish AhR undergoes "nuclear translocation" in the presence of ligand, a characteristic of mammalian AhR proteins.

More recently, two groups have used 2-azido-3-[^{125}I]iodo-7,8-dibromodibenzo-p-dioxin ([^{125}I]N$_3$Br$_2$DD) and a photoaffinity labeling technique to identify the AhR in aquatic species. Swanson and Perdew[216] detected the AhR in the RTG-2 rainbow trout embryonic gonad cell line following the addition of [^{125}I]N$_3$Br$_2$DD to cells in culture. Both nuclear and cytosolic forms were described. Hahn et al.[217,218] used in vitro incubation with [^{125}I]N$_3$Br$_2$DD to determine the phylogenetic distribution of the AhR in 20 species of aquatic vertebrates and invertebrates. Specific labeling of cytosolic proteins was observed in seven species of teleost and elasmobranch fish, in PLHC-1 fish hepatoma cells[99] (derived from the tropical topminnow *Poeciliopsis lucida),* and in beluga whales. In contrast, the AhR was not detected in jawless fish (hagfish and lamprey) nor in any of eight species representing four invertebrate phyla (see below).

One interesting aspect of the proteins detected by photoaffinity labeling with [^{125}I]N$_3$Br$_2$DD is the size heterogeneity that is evident both within and between terrestrial and aquatic groups of vertebrates (Figure 11). The apparent molecular mass of the AhR in terrestrial mammals ranges from 95 kDa (C57BL/6J mouse)

Ah Receptor Size

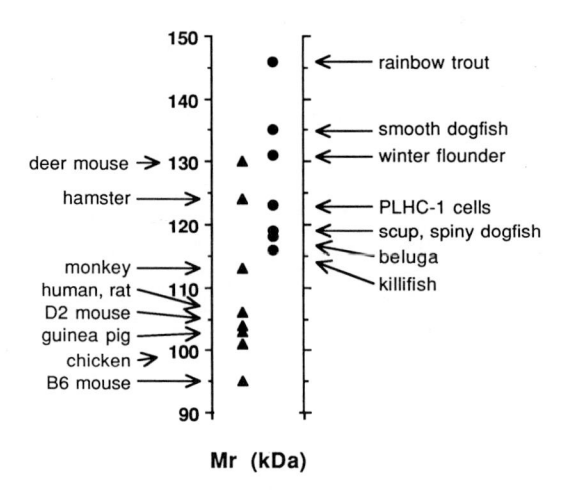

[Chart showing molecular weights in kDa:
- rainbow trout: ~145
- smooth dogfish: ~135
- winter flounder: ~130
- deer mouse: 130
- hamster: ~125
- PLHC-1 cells: ~125
- scup, spiny dogfish: ~120
- beluga: 118
- killifish: ~116
- monkey: ~113
- human, rat: 110
- D2 mouse: ~108
- guinea pig: ~105
- chicken: 100
- B6 mouse: ~98

Mr (kDa)]

Figure 11. Heterogeneity in molecular sizes of Ah receptor ligand-binding subunits in terrestrial and aquatic vertebrates. Ah receptors were identified by photoaffinity labeling with [^{125}I]N$_3$Br$_2$DD. Data from Poland and Glover[220] (terrestrial species) and Hahn et al.[217,218] (aquatic species).

to 130 kDa (deer mouse); in beluga whales, the only aquatic mammals in which the receptor has been identified, the apparent molecular mass is 118 kDa. Among the fish examined so far, Ah receptor sizes range from the 116 kDa protein seen in *Fundulus heteroclitus* to 145 kDa in rainbow trout.[216,218] This range of sizes suggests the existence of considerable heterogeneity in the structure of this protein. However, as discussed earlier,[219,220] at least some of its functional characteristics, such as the ability to control CYP1A1 induction, appear to have been conserved in all of these species. Future studies aimed at identifying and exploring conserved and variable regions of Ah receptors in diverse vertebrate animals could reveal structure-function relationships.

Knowledge of the AhR in amphibians and reptiles is minimal. Specific binding of [^{3}H]TCDD has been reported in the newt *Pleurodeles waltl*,[156] but examination of other members of this class (*Rana* spp.) have yielded negative results.[214,221] A single band of specific photoaffinity labeling was seen in the turtle *Chrysemys picta* but not in the alligator.[221] Examination of the presence and properties of Ah receptors in these and other groups of aquatic animals remains an important area for future research.

Studies to date have found no direct evidence for the presence of Ah receptors in any aquatic invertebrate.[213,214,217,218] [A possible exception is the report of a photoaffinity labeled peptide in bivalve *(Mercenaria)* gonad.][221a] These mostly negative results suggest one or more of the following possibilities: (1) the Ah receptor is absent from these groups; (2) it is present at very low concentrations; (3) it is expressed in tissues other than hepatopancreas/digestive gland; or (4) it is expressed at specific developmental stages not yet examined. It is also possible that invertebrate animals possess an Ah receptor homologue that is different from the vertebrate Ah receptor in structure and/or function. Two of the three proteins most closely related to the recently cloned mouse Ah receptor[222,223] are the *Drosophila* proteins Sim and Per, and the region of highest amino acid identity includes the potential Ah receptor ligand-binding domain.[223] There are reports of a [^{3}H]TCDD-binding protein in *Drosophila* and in the bollworm *Heliothis zea*.[224,225] However, the bollworm protein has ligand-binding characteristics that are significantly different from those of vertebrate Ah receptors.[225] Moreover, in different strains of *Drosophila,* there was no correlation between aryl hydrocarbon hydroxylase inducibility and the presence of a detectable [^{3}H]TCDD-binding protein.[224] Thus, the functional significance of the [^{3}H]TCDD-binding proteins detected in insects and their relationship, if any, to vertebrate Ah receptors remain to be demonstrated.

As noted above, the lack of evidence for an Ah receptor in aquatic invertebrates does not preclude its existence. Nevertheless, the information currently available suggests that this receptor, if present in invertebrates, has properties or patterns of expression that differ significantly from those of vertebrate Ah receptors. Based on this, one can predict that the invertebrates examined so far, and perhaps all invertebrates, may be less sensitive than mammals or teleosts to the effects of TCDD and related compounds. Similarly, the HAH structure/activity relationships determined in mammalian systems may have less relevance for aquatic invertebrates. Studies on the ability of individual PCB congeners to affect survival, growth, and reproduction in some crustaceans support this hypothesis.[226,227] Further evidence for invertebrate insensitivity to TCDD is emerging.[228] Clearly, additional research will be needed to define the phylogenetic distribution of the *AhR* gene and related genes in various aquatic groups and the roles of these genes and their protein products in regulation of CYP1A induction and dioxin toxicity. Indications are that this will be an exciting area of inquiry in aquatic toxicology in the coming years.

B. Relationship to Induction of CYP1A

There is an extensive and growing body of literature demonstrating conclusively that in mammals the AhR is necessary, though not sufficient, for transcriptional activation of *CYP1A* genes.[60] One would expect that a similar process governs CYP1A induction in those aquatic species where AhR occurs. Although there is as yet no direct evidence for this, everything we know currently about CYP1A regulation in aquatic animals is entirely consistent with an Ah receptor-dependent mechanism. It is thus interesting to compare the phylogenetic profiles for CYP1A inducibility and AhR presence.

1. Vertebrates

In fish, the correspondence between AhR presence and CYP1A inducibility is nearly perfect. CYP1A is inducible, and by the same suite of PAH and HAH compounds, in all bony and cartilaginous fish possessing an AhR. In contrast, lamprey and hagfish, in which the AhR is undetectable, do not respond to BNF[136,159] or 3,3′,4,4′-tetrachlorobiphenyl.[221] The situation is not as clear in other aquatic vertebrates. Induction of CYP1A-associated catalytic activities[153,157] and immunodetectable CYP1A protein[160,184] has been seen in certain amphibians and reptiles, while AhR measurements have sometimes been negative. It will be important to determine whether this apparent discrepancy represents a fundamental difference in induction mechanisms in these groups or merely the result of imperfect receptor assay techniques.

2. Invertebrates

The relationship between AhR presence and CYP1A inducibility in aquatic invertebrates is less clear. Although results of photoaffinity labeling studies suggest that aquatic invertebrates do not possess vertebrate-type Ah receptors,[217,218] there is uncertainty regarding the presence and inducibility of a CYP1A orthologue in invertebrates.[8,9,12] Increases in total P450 (detected spectrally) or monooxygenase activities (typically AHH) have been seen in invertebrate tissues following exposure to compounds or mixtures known to induce CYP1A in vertebrates.[8,9,12,109,181,191a,229–234] Although AHH activity is characteristic of CYP1A1 in vertebrates, it is not *exclusively* so, and the invertebrate catalyst(s) of AHH have yet to be isolated. Whether invertebrate AHH is catalyzed by an orthologue of vertebrate CYP1A and whether that form is inducible remain unknown. A more serious reservation regarding the data that have been cited as evidence for CYP1A induction in invertebrates concerns the conditions under which "induction" has been seen, and its magnitude. Typically, experiments showing increased AHH or other activities in invertebrates have involved prolonged treatment (for weeks) with high levels of potential inducers, conditions that differ substantially from those effective in vertebrates. Moreover, the changes in catalytic activities that have been reported are much lower in magnitude than those achieved in fish and mammals. Thus, there seem to be fundamental differences between vertebrates and invertebrates in the nature of the "induction" response.

Resolution of the questions surrounding CYP1A in invertebrates will likely be achieved through studies performed at the molecular level. Recognition of invertebrate P450 genes and their protein products by vertebrate CYP1A nucleic acid or antibody probes is one avenue of research. Such studies have yielded mixed results thus far. Three reports suggest, however, that some molluscan and echinoderm species may possess a P450 with some degree of similarity to vertebrate CYP1A forms. Schlenk and Buhler,[191] examining microsomes from untreated and BNF-treated gumboot chitons *(Cryptochiton stelleri),* found a protein that was recognized by polyclonal antibodies to trout CYP1A1, and Den Besten et al[191a] found in sea stars a band cross-reacting with polyclonal anti-scup CYP1A1. More recently, bands in total RNA from mussel have been detected by a CYP1A cDNA probe,[194] but the nature of those bands requires confirmation. In fact, such results should not be surprising, given that a gene ancestral to modern vertebrate CYP1A genes must have been present in the most recent common ancestor of mollusks (for example) and chordates. A descendent of this gene, possessing regions of sequence similarity to vertebrate CYP1A, might be present in modern invertebrate species, though its functional and regulatory characteristics could be quite different from those of vertebrate 1A forms. Further research, especially the isolation and characterization of invertebrate P450s and their genes, is clearly warranted.

C. Role of the Ah Receptor in Toxicity of CYP1A Inducers

More than 10 years ago, Poland and co-workers[235] proposed a model to describe the mechanism by which TCDD and related compounds regulate gene expression through the Ah receptor. They suggested that there were two types of responses to TCDD: (1) a "limited pleiotropic response", the induction of enzymes involved in foreign compound metabolism, which is produced in most tissues, and (2) a response involving altered expression of genes controlling cell proliferation and differentiation, occurring in a tissue-specific manner. An updated rendition of their model is shown in Figure 12. These authors further suggested[59] that "...the present, almost Ptolemaic preoccupation with the regulation of cytochrome P-450 may be unwarranted, and may preclude our understanding of the larger regulatory response controlled by the [Ah] receptor." Since then, the Poland model has continued as the dominant paradigm within which research on the toxicity of Ah receptor ligands is conducted, and the study of CYP1A regulation has provided much of our current understanding of the details by which the Ah receptor participates in altered gene expression and, presumably, toxicity. Yet we have not come much closer to understanding the "larger regulatory response" and its role in toxicity.

Consistent with the Poland model and the available evidence, compounds that induce CYP1A expression may produce toxicity through both CYP1A-dependent and CYP1A-independent pathways. The former are described in the following section. A comprehensive discussion of the latter is beyond the scope of this chapter. Indeed, knowledge of CYP1A-independent mechanisms of toxicity for such compounds in aquatic species[236] is almost nonexistent, and those of us working in this field might do well to once again consider the advice that Poland provided more than 10 years ago.

VIII. TOXICITY ASSOCIATED WITH P450 FUNCTION

Though primarily a pathway of detoxication, there are multiple avenues whereby cytochrome P450 activity or regulation can elicit toxic effects:

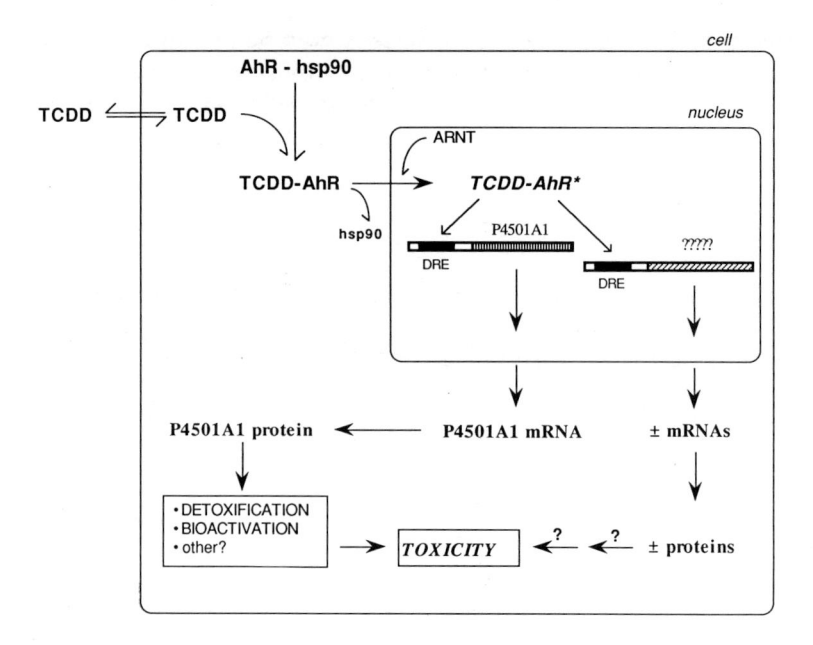

Figure 12. Model for Ah receptor-mediated changes in gene expression. Ligands such as TCDD bind to the Ah receptor (AhR), which is subsequently transformed to its transcriptionally active form *(TCDD-AhR*)*. Transformation involves the dissociation of a 90-kDa heat shock protein (hsp90) and formation of a heterodimer between the AhR and the Ah receptor nuclear translocator (ARNT) protein. Interaction of *TCDD-AhR** with dioxin responsive enhancer (DRE) sequences near the promoter region of specific genes initiates the transcription of those genes. Changes in gene expression can lead to toxicity through altered CYP1A1 expression or other effector pathways yet to be described.

1. Many protoxicants, promutagens, and procarcinogens are converted to reactive, toxic products by P450. Electrophilic metabolites can bind to nucleophilic centers in DNA, leading to mutations and, in proteins, leading to cellular toxicity.
2. Induction of cytochromes P450 can alter the rates of endogenous substrate metabolism, either directly as catalysts or possibly indirectly by competition for reducing equivalents from P450 reductase.
3. Direct or indirect inhibition or inactivation of P450 by substrates or nonsubstrates could affect the capacity for xenobiotic metabolism or for endogenous functions, depending on the P450 form inhibited.
4. Failure to complete a catalytic cycle following substrate binding, electron transfer, and O_2 binding can result in the formation and release of oxygen radicals, themselves toxic and mutagenic. Quinone metabolites could also participate in redox cycling, also contributing to the formation of radicals.
5. Receptor mechanisms involved in the regulation of some P450 genes could act in toxic mechanisms independent of the catalytic role of the P450.

Evidence exists for each of these processes; some have been discussed in detail elsewhere.[237] The involvement of receptor mechanisms was discussed in the preceding section. Here we consider briefly the roles of P450 in carcinogenesis, altered metabolism of endogenous compounds, and an integrative view of toxic mechanisms.

A. Carcinogen Bioactivation

There are compelling pictures derived from many studies showing catalyst identity, regulation, and function of different enzymes that activate or detoxify carcinogens. In fish (trout) the activation of aflatoxin B_1 to the 2,3-epoxide is catalyzed by P450LM2 (CYP2K1).[238] Detoxication of aflatoxin B_1 is a prominent function of CYP1A forms, recently indicated in fish[238a] and long indicated in mammals.[239,240] A similar story is being developed concerning the role of putative CYP2E forms in the activation of alkyl nitrosamines in fish.[145] However, benzo(a)pyrene (BaP) offers the most complete paradigm for the role of P450 in metabolic activation of a carcinogen.

The elegant and aggressive attack by several groups in the 1970s established the identity of carcinogenic metabolites of BaP and showed the essential oxygenation and hydration steps, catalyzed by P450 and by epoxide hydrolase, leading to formation of several isomeric dihydrodiol-epoxide structures.[241–243] One of these, the 7R,8S-dihydroxy-9S,10R-epoxy-7,8,9,10-tetrahydrobenzo(a)pyrene ((+)-anti-benzo(a)pyrene-7,8-diol-9,10-epoxide or (+)-anti-BaPDE), is the most potent carcinogenic derivative of BaP[241] (Figure 13). Activation via similar diol-epoxides was confirmed for other angular PAHs, leading to the "bay region" theory of carcinogenesis involving these compounds,[241] a bay region being the structural "inlet" at the angle. In mammals, CYP1A1 is involved in formation of many bay region diol epoxides.

Several lines of evidence indicate that CYP1A in fish is similarly responsible for metabolic activation of BaP:

1. Treatment with inducers of CYP1A greatly increases the rates of BaP metabolism in most fish.[63]
2. Antibodies to fish CYP1A almost completely inhibit the metabolism of BaP by liver microsomes of various fish species.[241a]
3. Similar preference for benzo-ring metabolism is exhibited by CYP1A1 forms purified from fish[64,83] and by CYP1A1 proteins from mammalian species.[242,243]
4. Regardless of total activity, microsomal preparations from fish liver and other organs (e.g., gill, kidney, heart) where CYP1A is induced produce similar relative abundances of BaP metabolites, with a high proportion (+)-7,8-dihydrodiols.[244]
5. Metabolism of the (−) BaP-7,8-dihydrodiol by fish yields the (+)-anti-diolepoxide.[245]

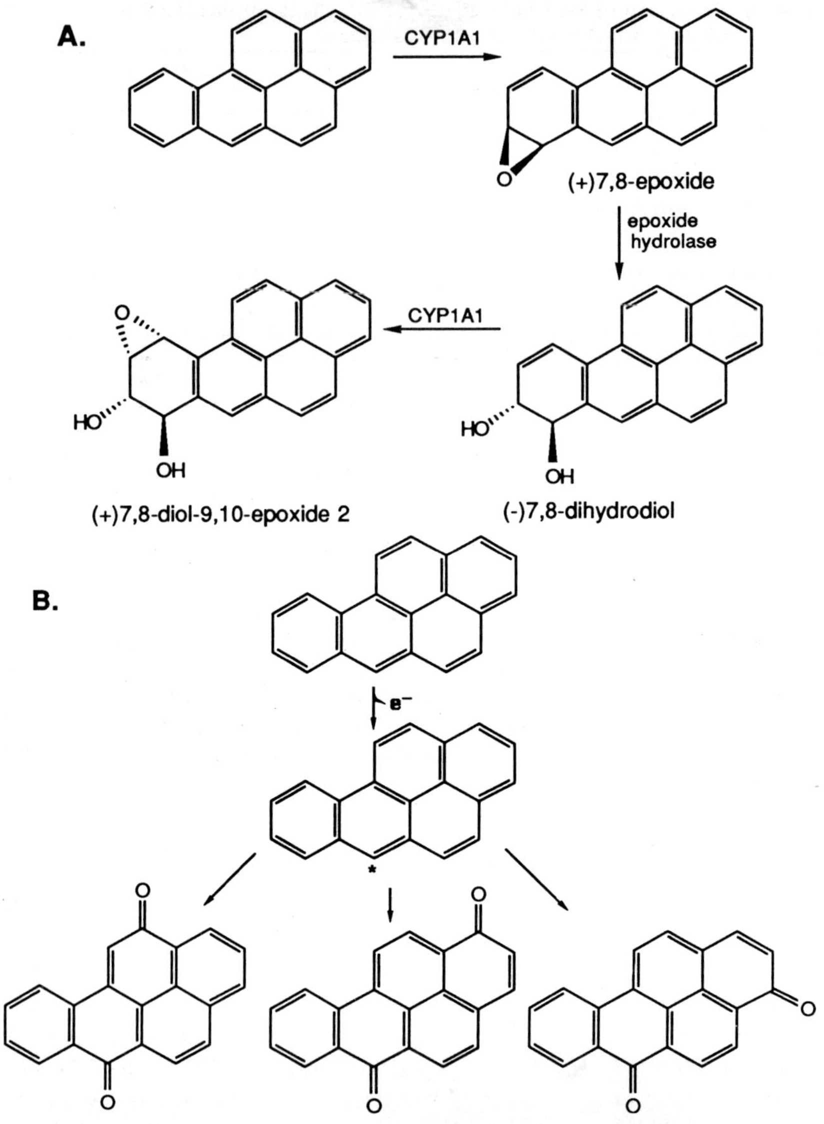

Figure 13. Dominant pathways of BaP metabolism in fish and mollusks. (A) Metabolic activation to dihydrodiol-epoxide metabolites via CYP1A1 and epoxide hydrolase. The structures shown are major products at each step, typical of mammals and fish. (B) Presumed pathway leading to the dominant products formed by mollusks, the 1,6-, 3,6-, and 6,12-quinones. Many fish form little or no 6,12-Q, indicating that such a one-electron pathway is not prominent in teleosts. The 1,6-Q and 3,6-Q that are in vitro metabolites formed by fish are apparently oxidation products of the 1-OH and 3-OH derivatives.[63]

Table 15. Activation of PAH to Mutagenic Derivatives by Hepatic Post Mitochondrial Supernatants from Scup

Compound	Concentration	Relative survival × 100	8-AG resistant fraction × 10^5
Benzo[a]pyrene	10 µg/ml	42	85.5
1,2,3,4-Dibenzanthracene	10 µg/ml	70	32.6
7,12-Dimethylbenzanthracene	5 µg/ml	85	14.2
Phenanthrene	100 µg/ml	70	5.0
Naphthalene	100 µg/ml	38	0
Aflatoxin B_1	200 µg/ml	100	0
Dimethylnitrosamine	68 mM	93	0
2-Acetylaminofluorene	5 µg/ml	100	7.0
Aroclor 1254	5 µg/ml	46	1.0

Note: Results were obtained using an assay for forward mutation to 8-azaguanine resistance in *S. typhimurium*. Relative survival refers to bacterial survival. 8-AG resistance was determined relative to the survival. Details of procedures can be found elsewhere.[250] The concentrations shown were selected from dose-response curves obtained using empirically-established optimal conditions. Some of these results have appeared elsewhere.[63] The lack of activation of AFB_1 or DEN and slight activation of AAF could reflect either the absence of the requisite catalyst(s), or the result of a more efficient competing reaction. Thus, ring hydroxylation of AAF could proceed more efficiently than the N-hydroxylase reaction involved in activation.

6. Consistent with the patterns of PAH metabolism, preparations of fish liver also activate BaP to products that bind covalently to DNA.[246] Furthermore, the structure of those adducts indicates that they are derived from anti-BaPDE.[247] Fish from sites where liver tumors are highly prevalent show similar DNA adducts.[248]
7. Preparations of fish liver can efficiently activate BaP to mutagenic products,[249,250,250a,b] consistent with the scheme in Figure 13, and with results in mammals.

Although the catalysts activating BaP in fish generally and that activating aflatoxin B_1 in trout are reasonably well established, the catalysts that activate other carcinogens are uncertain. Fish liver preparations activate PAHs other than BaP (Table 15). CYP1A could be responsible for activation of many of these PAHs, but this requires demonstration. In mammals, some alkylated PAHs (e.g., DMBA) can be activated via conjugation reactions; this, too, is yet to be shown in fish. Are alkylnitrosamines (DEN and DMN) activated via dealkylation catalyzed by CYP2E homologues? Is AAF activated via N-hydroxylation by P450 forms or by FMO? Which P450s catalyze competing and detoxication reactions, such as ring hydroxylation of AAF? Do these various functions operate similarly in different cell types?

Activation of PAHs to mutagens has been detected in other vertebrates. Thus, liver microsomes from MC-induced reptiles also activate BaP to a mutagen.[251] Given the similarities in various properties of CYP1A forms, we conclude that tissues of vertebrates with induced CYP1A will activate the procarcinogenic BaP, as a general rule. Conformity to this rule should not be surprising. Exceptions are likely to be rare but will be interesting for mechanistic reasons and still should be sought.

1. Stereoselectivity of P450

Metabolism of BaP by mammalian CYP1A and epoxide hydrolase occurs with an enantiomeric selectivity favoring formation of the enantiomer that is the most active carcinogen, the (+)-diolepoxide-2, or (+)-anti-BaPDE.[241] As suggested above, metabolite profiles, the structure of DNA adducts, mutation assays, and boronate column chromatography all indicate that the metabolism on the benzo-ring of BaP by fish involves a similar enantiomeric selectivity and that this stereochemistry is a feature of CYP1A. Despite the clear example of BaP, the isomeric selectivity of metabolism and possible differences in the biological reactivity associated with different chiral structures are often overlooked in the analysis of xenobiotic metabolism and effects. Knowing the enantiomeric and regioselectivity of a reaction can be important also in distinguishing the involvement of different catalysts. For example, both P450 and FMO enzymes catalyze S-oxidation of p-methoxyphenyl-1,3-dithiolane but appear do so with different enantioselectivity.[25] Further studies of the functional similarity of the enzymes in different species, addressing questions of regioselectivity and chiral specificity, are highly important. Accurate assessment of the latitude for substrate fit into the active site would aid in the predictions about which substrates would be acted on by CYP1A or another catalyst.

2. Invertebrate Carcinogenesis

There has been no concrete demonstration of carcinogenesis in aquatic invertebrates resulting from activation of a procarcinogen. Carcinogen metabolism has been most studied in mollusks. Mollusks only minimally metabolize BaP to benzo-ring derivatives; thus, activation of such PAHs via diol-epoxide pathways involving P450 is probably insignificant in these organisms. The quinone formation prominent in mollusks appears to involve predominantly one-electron or radical oxidation (e.g., Figure 13B). BaP quinones could exert some mutagenic activity, possibly involving active oxygen produced via redox cycling.[252] It is not yet known which pathways of PAH metabolism operate in mollusks in vivo. As stated above, several reports have shown that molluskan tissue preparations can activate aromatic amines to mutagenic derivatives.[33,35] DNA adducts of some of these compounds have been detected in mollusks.[253] Involvement of procarcinogen activation in invertebrate diseases is an important area to pursue, one which will be facilitated by description of the enzymes responsible for bioactivation in these animals.

B. Altered Metabolism of Endogenous Substrates

Numerous P450 forms in mammals and in aquatic species metabolize important regulatory molecules, including steroids, arachidonic acid, prostaglandins, and others. This, and the observation of reproductive effects in fish also showing CYP1A induction,[254-256] suggest that induction of some P450s could adversely affect the metabolism or action of these regulatory molecules. Inhibition of endogenous P450 forms by xenobiotics (see Section IV.D) could also affect biological processes.

In fish, the CYP1A forms are not prominent in metabolism of either estradiol or testosterone. The dominant activity in testosterone metabolism in livers of many fish, 6β-hydroxylase, is apparently catalyzed by CYP3A counterparts. Microsomal catalysts for the other prominent steroid hydroxylations in fish (e.g., 16α-,16β-, 7α-, 15α-hydroxylase activities) are not established. Arachidonic acid and prostanoids are metabolized by a variety of P450 forms in mammals (e.g., CYP4A, which is prominent in omega-hydroxylation of fatty acids including arachidonic acid).[257] Other mammalian forms, including CYP1A and CYP2B enzymes also metabolize arachidonic acid, forming hydroxyeicosatetraenoic acids (HETEs) and epoxyeicosatrienoic acids (EETs). Lauric acid hydroxylation by trout P450s has been described,[258] but otherwise this picture developing in mammals is still largely blank in aquatic species. Establishing the influence of xenobiotics on endogenous substrate metabolism requires a knowledge of the catalysts for specific reactions.

Many studies of endogenous substrate metabolism have concerned hepatic systems. It is crucial to bear in mind that induction of P450 forms and the alteration of endogenous functions in extrahepatic sites could be far more important to organismal effects than induction in the liver. Xenobiotic-induced alterations in endogenous substrate metabolism in extrahepatic organs are still poorly known, even in mammals. There could be unknown xenobiotic-inducible P450s that might also metabolize endobiotics. Recent studies[259] suggesting that planar HAH induce in chickens a novel P450 forming EETs and HETEs from arachidonic acid support this possibility. The metabolism of arachidonic acid and other endogenous substrates in aquatic species deserves critical attention.

C. Linkages in Toxic Mechanisms

Establishing the network of interactions between P450 enzyme function and/or regulation and other molecular processes in different cell types ultimately will reveal the full significance of these enzymes in toxic mechanisms. Interactions among and between a variety of molecular systems were described recently.[237] A thorough description of such interactions would be overwhelming, but a listing of selected potential sites of integration can convey the breadth of interactions. Thus, as shown in Figure 14, linkages exist between the P450 system and biochemical systems, including those involving heme synthesis and degradation, heat shock proteins, steroid receptors, antioxidant enzymes, metallothioneins, oncogene and tumor suppressor genes, synthesis and binding of gaseous biological messengers (NO and CO), etc. One promising area is the potential for linkages between P450 gene expression and the action of growth factors; for example, epidermal growth factor receptors are suppressed in animals treated with TCDD.[236,260] As additional linkages are discovered, then will mechanisms of toxicity be found.

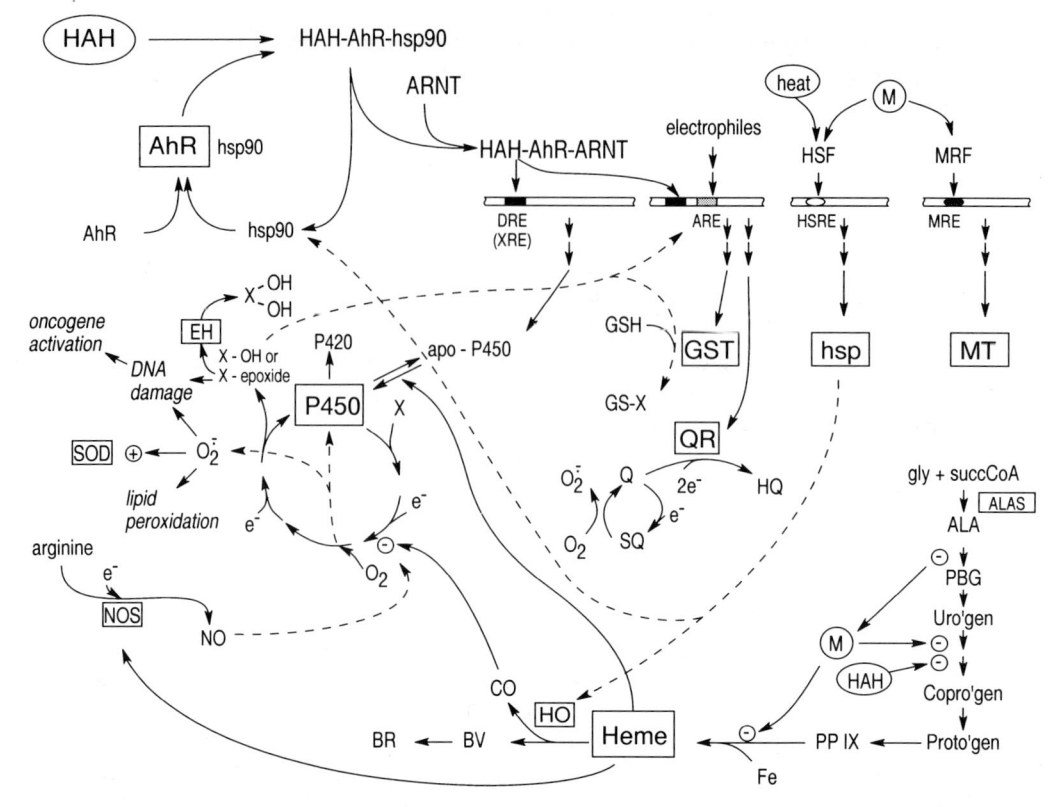

Figure 14. Linkages between P450 and other biochemical systems. This figure illustrates the complex interactions that are known to occur between biochemical systems involved in responses to pollutant exposure. Further linkages remain to be discovered. Abbreviations used: AhR, Ah receptor; ALAS, δ-amino levulinic acid synthase; ARE, antioxidant response element (electophile response element); ARNT, Ah receptor nuclear translocator; BR, bilirubin; BV, biliverdin; CO, carbon monoxide; DRE, dioxin response element; EH, epoxide hydrolase; GSH, glutathione; GST, glutathione S-transferase; HAH, halogenated aromatic hydrocarbon; HO, heme oxygenase; HQ, hydroquinone; HSF, heat shock factor; hsp, heat shock protein; hsp90, 90-kDa hsp; HSRE, heat shock response element; M, metal; MRE, metal responsive element; MRF, metal response factor; MT, metallothionein; NO, nitric oxide; NOS, nitric oxide synthase; P450, cytochrome P450; PP, protoporphyrin; Q, quinone; QR, quinone reductase (a.k.a. DT-diaphorase); SOD, superoxide dismutase; SQ, semiquinone radical; XRE, xenobiotic response element.

IX. VARIABLES AFFECTING P450 EXPRESSION

A number of variables have long been known to influence monooxygenase systems in mammals. These include the same variables of health, condition, nutritional status, and reproductive and developmental status that influence many metabolic systems. As one can readily predict, these variables more or less strongly influence the expression or function of P450 forms in aquatic species (e.g., Williams et al.).[261] That hormonal status related to sex and reproduction, diet, development, or disease status affect MO systems should no longer be questioned. The mechanisms and degree of effect in different species are less certain. Thus, there are strong sex differences in P450 content in mammals and fish, but the forms involved and mechanisms by which these differences are achieved may be different in these groups. Likewise, dietary differences can affect P450 function in fish,[261a] but the mechanisms are not known. In poikilothermic aquatic species, external variables, most prominently temperature, also can affect P450 systems. Endogenous substrate metabolism as well as xenobiotic metabolism might be affected by these variables. For example, the temperature effects on sex determination in some groups (e.g., turtles)[262] could involve temperature effects on regulation or function of steroidogenic P450s. Studies indicate that several of the above variables operate through post-transcriptional or pretranslational events, but the precise mechanisms are unknown and offer interesting areas for study.

A. Temperature Acclimation

The influence of temperature on the disposition of foreign compounds and on microsomal activities of untreated fish has been studied since the early 1970s.[263] Such studies have indicated that cold acclimation and the assay temperature can variably influence MO activity. Temperature appears to affect the function of P450 directly and possibly through effects on the saturation of membrane lipids.[264] Temperature also strongly influences the rate and magnitude of CYP1A induction in fish. Induction of MO activities (e.g., AHH activity) associated with CYP1A is suppressed in cold-acclimated animals, at least for several days after treatment.[265,266]

Recent studies[267] addressed whether this suppression in *Fundulus heteroclitus* resulted from temperature effects on transcriptional or translational processes by measuring CYP1A mRNA and protein content and the activity of the enzyme. In those studies, liver microsomal EROD activity and CYP1A were strongly induced by BNF in 16°C-acclimated fish but only slightly in 6°C-acclimated fish, even by 25 days after BNF-treatment. However, CYP1A mRNA content was similarly elevated in fish at both temperatures but had a longer apparent half-life in the cold-adapted fish (Figure 15). Repeated studies gave similar results, with cold suppression of CYP1A protein induction but not mRNA content. The indicated effect of temperature at a post-transcriptional site suggests interesting possibilities for study of the mechanisms by which temperature influences the translation and/or stability of mRNA. Whether the persistence of mRNA in cold-acclimated fish is the result of a slower turnover of mRNA or the result of slower metabolism of inducer and hence retention of active levels is not clear. It is possible that the mRNA accumulating in the cold-adapted fish is not translatable. Whether temperature effects might differ with different types of inducers (i.e., slowly vs. rapidly metabolized Ah receptor ligands) is not yet known.

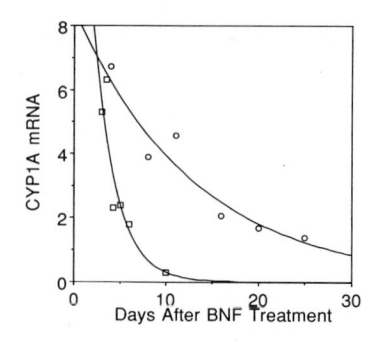

Figure 15. Induction and decay of CYP1A mRNA content following BNF treatment of *Fundulus heteroclitus* acclimated to 6°C and 16°C. CYP1A mRNA was determined on slot blots of total hepatic RNA, using the trout CYP1A1 cDNA probe pfP₁450-3'.[61] Squares represent 16°C BNF-treated fish, and circles indicate 6°C BNF-treated fish. Reproduced with permission from Kloepper-Sams, P.J. and Stegeman, J.J., *Arch. Biochem. Biophys.*, 299, 38, 1992.

B. Sex Differences and Hormonal Interactions

Since the first demonstrations of a sex difference in liver MO activity in trout,[268,269] there have been numerous reports that reproductively active mature female fish have a lower total P450 content and lesser rates of several MO activities than do males. Among the activities that are strongly differentiated are AHH and EROD, usually paralleling differences in CYP1A content.[199] This is not a uniform circumstance, as some species show different patterns of sex difference in some activities.[270] In those species exhibiting a strong sex difference, the similarities between reproductively active males, reproductively inactive males, and reproductively inactive females suggest that the sex differences are due to a suppression of CYP1A expression in reproductively active females.

Treatment of juvenile fish or of hypophysectomized fish with hormones, both androgens and estrogens, has shown that estradiol (E_2) is a regulator of monooxygenase activity, presumably altering the content and/or activity of various P450s.[271–274] Later studies confirmed a suppression of CYP1A.[275,276] The mechanism by which E_2 exerts this effect on CYP1A is not known. It could reflect an indirect effect, resulting from the protein translation machinery being subsumed by the translation of estradiol-induced vitellogenin.[277] However, observations that mRNA for CYP1A is suppressed in mature, PCB-exposed female fish[278] indicate that there could be a specific pretranslational effect. Moreover, the effect of BNF recently was also found to be reduced in extrahepatic organs of winter flounder,[279] indicating that whatever hormonal mechanism is acting, it is not restricted to the vitellogenic liver.

CYP1A is not the only form that shows sex differentiation. In liver of several species, the testosterone 6β-hydroxylase enzyme shows a sex difference that is not linked to the differences in CYP1A.[270] This difference is possibly in a CYP3A-type enzyme. Proteins seen by antibodies to trout P450con or scup P450A are sexually differentiated in several species, with opposite patterns in some, e.g., salmonids and winter flounder.[129,275,276] In trout, the content of CYP2K1 (LM2 or LMC2) also shows a sex difference in mature fish in liver and kidney.[280]

Studies with hypophysectomized fish[281,282] indicate that the actions of estradiol are not mediated by pituitary hormones, though the possibility for action of pituitary hormones on the liver still exists. In particular, the role of growth hormone as a regulator of hepatic P450 in fish is intriguing, as GH is a strong effector of P450 gene expression in mammals. Experiments with growth hormone (unpublished) have been and are likely to be inconclusive until the pattern of GH release from pituitary in fish is known. The pattern of GH release in mammals is critical to the effect on P450 expression in liver.[283]

Whether estradiol receptors mediate estradiol action(s) on P450 expression in fish is not clear. An involvement of other steroid receptors in the regulation of CYP1A in fish has been suggested. Devaux et al.[284] showed enhancement of CYP1A induction in trout hepatocytes by dexamethasone, suggesting that glucocorticoid response elements (GRE) occur in fish CYP1A genes. In contrast, Lee et al.[284a] reported a negative effect of dexamethasone on levels of CYP1A1, but not CYP3A, in trout. In mammals, GRE occur in the first intron of the CYP1A1 gene.[285] Do other regulatory elements occur in P450 gene introns? Whether estradiol response elements (ERE) occur in fish CYP1A genes is an intriguing possibility. A counterpoint is the recent finding of a dioxin-response element in the mammalian estrogen receptor gene,[286] suggesting an avenue for the dioxin effect on E_2 receptor levels. There is thus an interplay between steroids, steroid receptors, P450 genes, and the receptors controlling P450 gene expression. The biological significance of such interplay is not clear.

An intriguing possibility for further investigation involves the control of neurosecretory hormone release. In mammals there is an imprinting of the hypothalamo-pituitary axis shortly after birth that determines the pattern of hormone release later in the life of the animal. The imprinting apparently depends on the conversion of testosterone to estradiol in the brain, a P450 reaction; it determines the sex-linked pattern of P450 gene expression and activity in the liver at the time of sexual maturation.[287,288] The observation that many fish can undergo spontaneous sex-reversal or be induced to undergo sex reversal by hormone treatment[289] suggests that there may be fundamental differences between mammals and fish in the role and regulation of some steroid transformations in the brain.

C. Individual Variation in Biotransformation

Differences in xenobiotic metabolism between individuals of the same species could render some individuals much more susceptible or resistant than others to xenobiotic effects. Differences in the response to inducers of CYP1A are well known in mammals. Indeed, the differences in AHH induction between D2 and

B6 mice strains provided information crucial to determining the role of the Ah receptor.[59,60] Two examples illustrate the possible sources of individual variation in biotransformation enzymes in fish.

In one study the pattern of in vitro microsomal metabolites of BaP formed by the marine fish scup differed substantially between individuals. Liver microsomes of some individuals formed primarily benzo-ring dihydrodiols while others formed primarily the 7- and 9-hydroxy derivatives, yet the total amounts of benzo-ring metabolites were similar.[290] The differences suggest individual variation in the fate of the 7,8-epoxide and 9,10-epoxide intermediates, and indeed, there was a positive correlation between the amounts of the dihydrodiols formed and the rates of epoxide hydrolase activity. The implication is that some factor was affecting EH activity. There are chemicals known to affect EH activity in fish.[291] If the in vitro results reflect an in vivo condition, it would imply that EH activity could influence susceptibility to carcinogenesis by PAH activated via diolepoxide formation.

Detection of different mRNA species in tomcod by Wirgin and colleagues provided evidence for allelic variation in CYP1A in fish.[151] Allelic variation in P450 in fish should be expected; genes for many enzymes have multiple alleles. A significant question, however, is whether the different mRNA species detected in tomcod are translated, and if so, whether the proteins are functionally different. Profound functional differences are seen in products of some alleles, for example human hemoglobins or human red cell glucose-6-phosphate dehydrogenase variants differing by a single amino acid. More to the point, at least two allelic variants of CYP2D6 occur in humans, and the proteins have catalytic differences associated with clinically different responses to drug substrates for CYP2D6.[292] There are several presumptive alleles of CYP2B1 in rats.[293] We anticipate that additional allelic variation will be detected in P450s of aquatic species.

D. CYP1A in Developmental Stages of Fish

There is a general consensus that early developmental stages of fish are more sensitive to toxic effects of xenobiotics than are juveniles or adults.[294] Xenobiotic metabolism could contribute to that sensitivity. The first studies of the ontogeny and induction of P450 in fish embryos were carried out with salmonids and with *F. heteroclitus*. Exposure of embryos to oil or to PCBs resulted in the induction of AHH activity in embryos and larvae, indicating induction of CYP1A.[295,296] However, the studies in *F. heteroclitus* also disclosed a strong difference in the sensitivity of response to inducers before and after hatching. Induction prior to hatching occurred to the same extent, but with less sensitivity (Figure 16).[297] The mechanisms underlying that difference could involve differences in the nature, function, or abundance of Ah receptors, or some other factor.

Subsequent studies with salmonids disclosed strong induction of CYP1A in many cells of embryos exposed to inducer (BNF) by injection into the egg. The cellular sites of induction in the embryos of rainbow trout were much like those in the adults.[298] Studies in lake trout also showed a strong induction in multiple cells of embryos, including endothelial cells. Studies of embryos injected with TCDD as eggs revealed a difference between pre- and posthatch sensitivity to induction,[299] not unlike that seen in *Fundulus*. From these studies, the following generalizations can be tendered:

1. Xenobiotic metabolism does occur in fish embryos.
2. CYP1A is inducible before and after hatching in fish.
3. Dose/response relationships indicate that there is an increase in sensitivity after hatching.
4. Induction in pre- or posthatch embryos occurs commonly in endothelial cells (see also below).

X. EXTRAHEPATIC P450

Description of P450 systems in fish has focused primarily on the liver. This is reasonable, given that the liver is the primary site of xenobiotic metabolism and elimination. In some feral fish, the liver also shows cellular pathologies, including neoplasms, associated with chemical effects. We note that the relative roles of different organs in xenobiotic metabolism are poorly known and may be addressed in part by studies with perfused organs. Xenobiotic metabolism in extrahepatic sites is likely to be involved in systemic effects such as those on reproduction, immune dysfunction, and others.

A. Target Organs

Earlier studies disclosed that some organs, such as kidney, gut, and heart, had substantial specific content of P450, but that nearly all organs of feral fish had measurable rates of AHH or other activities normally catalyzed by P450.[300,301,301a] While various P450-dependent activities have been seen in most organs of fish, many recent studies have focused on CYP1A and its activities. CYP1A can be strongly induced in fish in organs proximal to the environment (gill, gut) and in excretory organs (kidney), and CYP1A substrates can also be metabolized by isolated extrahepatic tissue preparations.[302] However, differences in profiles with diagnostic substrates, appreciable content of total P450 in microsomes that have little CYP1A,[303,304] and immunodetection of proteins by antibodies to P450 forms other than CYP1A[280,305] all show that multiple P450 forms occur in extrahepatic organs of fish. Three organs are discussed, to illustrate some important points.

1. Gut

The induction of CYP1A in gut has been studied extensively (reviewed by Van Veld).[306] Interestingly, low concentrations of BaP in the diet of fish can produce a

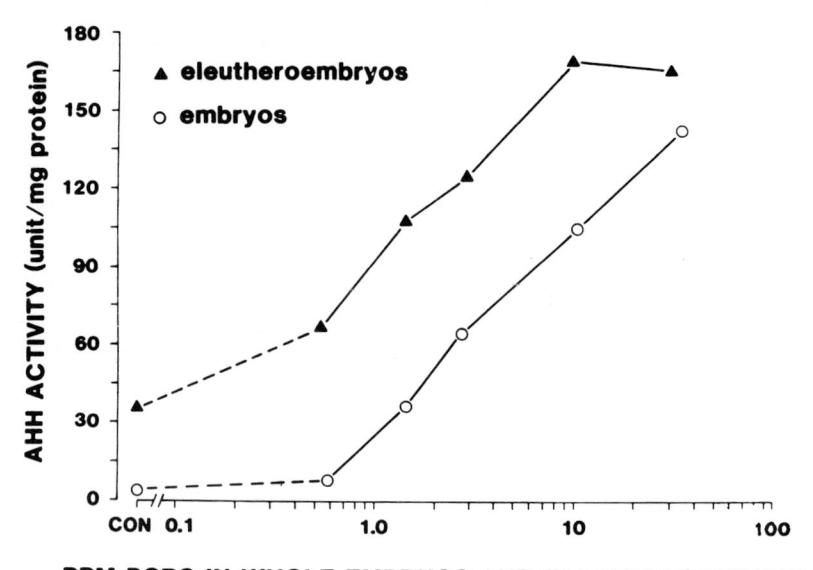

Figure 16. Developmental expression of CYP1A activity and its inducibility in PCB-treated *Fundulus heteroclitus*. Embryos were exposed to varying levels of a PCB (Aroclor 1254) mixture. The exposure was initiated at 10 days after fertilization; at 20 days after fertilization, microsomes were prepared separately from the embryos (not yet hatched) and eleutheroembryos (hatched) at each dose level. Reproduced with permission from Binder et al., *Chem.-Biol. Interact.*, 55, 185, 1985.

very strong induction of intestinal CYP1A, without induction occurring in the liver, while at higher dietary concentrations, there is induction in the liver also. The low dose effect presumably results from an elevated rate of metabolism of BaP in the gut sufficient to lower the concentrations of unmetabolized inducer passing into the hepatic portal system and therefore reaching the liver.[307] However, at higher doses, the capacity for metabolism at the site of uptake, the gut, appears not to meet the demand, and active (unmetabolized) inducer reaches the liver. Such first pass metabolism as seen with BaP would not be expected with HAH that are not readily metabolized. It should be interesting to compare the pharmacokinetics and the responses in gut and liver of fish given equipotent doses of rapidly or slowly metabolized inducers. At low doses, this first pass metabolism would appear to offer a mechanism of protection. However, the metabolism of BaP by P450 and EH in the gut could also export proximate carcinogenic BaP-7,8-dihydriols to the liver or other sites (see James and Kleinow, this volume).

2. Kidney

Induction of renal AHH activity in fish has been known for nearly 20 years.[300,301] Since that time, there has been substantial attention devoted to the kidney, in trout and other species.[280,308–313a] The kidney attracts attention partly because of its role in excretion and partly because of the function of the interrenal (= adrenal) as a steroidogenic organ. The kidney is the only fish extrahepatic organ from which P450 forms other than CYP1A have been purified. Andersson[135] purified two forms from trout kidney, KM1 and KM2. KM2 is highly expressed in kidney of mature male trout but not in mature females or in juveniles. However, KM2 can be induced in juvenile fish by treatment with 11-ketotestosterone (a potent androgen in fish). A similar response was detected in progesterone 16α-hydroxylase activity. The sex difference is similar to that seen by Williams et al.[280] and Miranda et al.,[127] who found that LMC2 (= LM2) and LMC5 both are present at high content in the kidney of mature male fish, but not in females or juveniles. LMC2 (LM2) also metabolizes progesterone at the 16α-position[127] (in reconstitution), suggesting that LM2 and KM2 could be similar proteins. As discussed above, LM2 has been sequenced and classified as CYP2K1. A CYP3A is expressed in amphibian kidney.[161]

3. Heart

In an early organ survey in scup from the environment, cardiac microsomes displayed a chromophore resembling P450, with a specific content second only to that in liver microsomes.[301,314] The properties of cardiac microsomes, both atrial and ventricular, were unusual compared to the other organs.[315] Questions as to the identity of the cardiac P450 remained until immunoassay and induction studies showed that it was predominantly CYP1A1.[316] Questions on the significance of the strongly induced CYP1A in cardiac tissues are becoming clearer with finding its cellular localization. Indeed, adequate interpretation concerning the significance of CYP1A or other P450 forms in any organ depends on knowing the cell type(s) where those enzymes are expressed or induced. We illustrate this with CYP1A.

B. Cellular Localization of CYP1A Expression

All organs are heterogeneous in cellular composition. Determining the regulation and role of P450 forms in specific cell types can reveal target cells for the action of inducers or substrates for those enzymes. The significance of CYP1A induction, for example, will depend on:

1. the catalytic function(s) of the protein(s),
2. the relative rates of activation and detoxification of the inducer and other compounds,
3. inducer avidity for Ah receptors and, hence, efficacy in eliciting CYP1A induction or other gene regulatory changes,
4. the cellular sites where these events occur.

Cellular expression of P450 genes can be identified by immunohistochemical or immunocytochemical detection of the protein or by in situ hybridization of the mRNA. Since 1987 there has been an increasing effort employing immunohistochemistry to identify organ- and cell-specific expression of CYP1A in fish. The first studies were carried out in adults of rainbow trout and cod.[317] In those studies, the expression in hepatocytes was demonstrated. CYP1A was also seen in kidney tubular cells.[305] Subsequent studies in trout and scup treated with BNF[316,318,319] and scup treated with 3,3',4,4'-TCB or TCDF[320] began to disclose the extent of CYP1A induction. Profound induction was detected in some cells of each organ examined.

Table 16 summarizes two studies[304,320] showing the cell types in various organs where CYP1A is expressed. These and other results[321] both answer and pose questions regarding CYP1A induction. First, tissue or whole animal sections stained with CYP1A-specific monoclonal antibody 1-12-3 show that CYP1A is commonly induced in epithelial cells of many organs, as indicated in Table 16. Second, some studies have shown that preferential induction in some of these sites is related to dose and route of exposure. Several general conclusions are offered:

1. There commonly is strong induction in the liver and kidney, consistent with the role of these organs in xenobiotic metabolism or excretion. However, there may be species or developmental differences in the responsiveness of hepatocytes. The basis for such differences will be interesting.
2. Exposure in the diet is associated with a strong induction in the mucosa of the gut.[307,322] Cellular patterns of induction in the mucosa can be predicted to reflect transit of inducers and turnover of cells in the gut.
3. Inducer delivered by external routes in the water or via sediment elicits a strong induction in the epithelial layers of the gill but induction in these cells can also occur when inducer arrives via the circulation.[321]

External routes of exposure appear also to be associated with induction in some specialized structures. Topminnows exposed to BaP in the water respond with a strong induction in the olfactory epithelium in the nose (see Figure 17A) and dermal chemosensory organs.[321] This epithelium almost certainly experiences a direct exposure to inducer in the water but would get no exposure from dietary or other internal routes. In a second example, direct exposure of pink salmon alevins to sediments highly contaminated by crude oil elicited an induction in the single epithelial cell layer of the lens.[179]

The detection of CYP1A in a cell type is strong evidence that induction mechanisms function in those cells. It is possible that responses in some cell types reflect induction of a protein similar but not identical to that in other cells. It is also possible that the induction mechanisms in different cell types involve different sets of regulatory elements additional to or in place of the AhR. Such elements might include repressors binding to the 1A genes or even different receptors. There is a continuing controversy over the possible role of a 4S "carcinogen binding protein" in 1A induction.[323,324] There is abundant 4S protein in some fish

Table 16. Occurrence and Intensity of CYP1A1 Protein in Various Tissues and Cell Types of Scup as Determined by Immunohistochemical Staining with MAb 1-12-3

Organ	Cell type	Control scup Occurrence[a]	Control scup Intensity[b]	Control scup Stain index[c]	TCB-treated scup Occurrence[a]	TCB-treated scup Intensity[b]	TCB-treated scup Stain index[c]	New Bedford scup Occurrence[a]	New Bedford scup Intensity[b]	New Bedford scup Stain index[c]
Liver	Hepatocyte	++	−/+	2	+++	+++	9	+++	+++	9
	Bile duct epithelium	−	−	−	+	+/++	2	−	−	−
	Pancreatic duct epithelium	−	−	−	+	+/++	2	−	−	−
	Pancreatic acinar cells	−	−	−	++	+	2	+	+	1
	Fibrous connective tissue	−	−	−	−	−	−	−	−	−
	Endothelium of portal veins and arteries	−	−	−	+	++/+++	3	++	++	4
Heart	Atrial endothelium	−	−	−	+++	+++	9	+++	+++	9
	Ventricular endothelium	−	−	−	+++	++	6	+++	++	6
	Bulbous arteriosus endothelium	−	−	−	+++	++	6	+++	++	6
	Endothelium of great vessels	−	−	−	+++	++	6	+++	++	6
	Coronary vessel endothelium	−	−	−	+++	++	6	+++	++	6
	Cardiac muscle cells	−	−	−	−	−	−	−	−	−
Gill	Pillar cells	++	+	2	++	++	4	++	+++	6
	Epithelial cells	−	−	−	+++	+++	9	−	−	−
	Chloride cells	−	−	−	−	−	−	−	−	−
Kidney	Nephronic duct epithelium	+	−/+	1	+++	++/+++	9	++	+++	6
	Collecting duct epithelium	−	−	−	++	++	4	−	−	−
	Opistonephric duct epithelium	−	−	−	++	++	4	−	−	−
	Glomerular endothelium	−	−	−	++	++	4	++	++	4
	Sinusoidal endothelium	−	−	−	+	+	1	+	+	1
	Vascular endothelium	−	−	−	+	+	1	++	++	4
	Hematopoietic tissue	−	−	−	−	−	−	−	−	−
	Endocrine tissue	−	−	−	−	−	−	−	−	−

Table 16. (Continued)

Organ	Cell type	Control scup Occurrence[a]	Control scup Intensity[b]	Control scup Stain index[c]	TCB-treated scup Occurrence[a]	TCB-treated scup Intensity[b]	TCB-treated scup Stain index[c]	New Bedford scup Occurrence[a]	New Bedford scup Intensity[b]	New Bedford scup Stain index[c]
Gastro-intestinal tract	Gastric mucosal epithelium	–	–	–	++	+/++	4	+	+	1
	Cecal mucosal epithelium	–	–	–	+++	+++	9	++	++	4
	Intestinal mucosal epithelium	–	–	–	+++	+++	9			
	Submucosal vascular endothelium	–	–	–	++	++	4	++	++	4
	Submucosal connective tissue	–	–	–	–	–	–	–	–	–
	Myocytes	–	–	–	–	–	–	–	–	–
Gall bladder[d]	Fibrous connective tissue	–	–	–	–	–	–			
	Epithelium	–	–	–	–	–	–			
Gonad	Vascular endothelium	–	–	–	++	++	4	+	++	2
	Ovigerous lamellar epithelium	–	–	–	+	+	1	–	–	–
	Spermatic duct epithelium	–	–	–	+	+	1	–	–	–
	Eggs	–	–	–	–	–	–	–	–	–
	Sperm	–	–	–	–	–	–	–	–	–
	Endocrine cells	–	–	–	–	–	–	–	–	–
Spleen	Vascular endothelium	–	–	–	++	++	4	++	++	4
	Hematopoietic tissue	–	–	–	–	–	–	–	–	–
Red muscle	Vascular endothelium	–	–	–	+	+/++	2			
	Myocytes	–	–	–	–	–	–			
Brain[d]	Vascular endothelium	–	–	–	+	+/++	2	+	+	1
	Neuroglia	–	–	–	–	–	–	–	–	–
	Neurons	–	–	–	–	–	–	–	–	–

Nasal tissues	Vascular endothelium	—	—	—	—	+	—	++	2
	Sustentacular cells	—	—	—	—	—	—	—	—
	Sensory cells	—	—	—	—	—	—	—	—
Other	Fibrocartilage[e]					++	—	+++	6

Source: Data from Smolowitz, R.M. et al.,[320] Stegeman, J.J. et al.,[304] and unpublished results.

[a] Occurrence is the relative frequency of stained cells (+ = seldom, ++ = frequent; +++ = always).
[b] Intensity is the density of the peroxidase-stained cells (+ = mild; ++ = moderate; +++ = heavy).
[c] Stain index = occurrence × intensity.
[d] Brain and gall bladder were collected from TCDF-induced fish only.
[e] Cells appearing to be fibrocartilage were heavily stained in some head regions. These were not examined in control or TCB-treated fish.

(e.g., scup).[325] In part, answering such questions will require that equipotent doses of different inducers be tested for effect. Such an approach, at various doses, could reveal differential sensitivity of different cell types, including specialized cells such as those of the immune system. The use of small fish that afford analysis of sections of whole animals (done first with Poeciliopsis [321]) could help to distinguish cellular differences in timin or sensitivity of induction.

Having detected the induction of CYP1A in various cell types, questions then focus on the significance of the expression in those cells. The significance of CYP1A induction in specific cells will depend on the function of those cells. For example, does induction of olfactory CYP1A affect olfaction? Other P450 forms, especially the CYP2G proteins,[326–328] may have a role in removing odorants from the olfactory system and possibly even in signal transduction.[328] Is induction of CYP1A in lens epithelium associated with the occurrence of cataracts reported in fish from sites heavily contaminated by PAH?[329] Conceivably, the activation of PAH or the generation of active O_2 could be involved. Does induction in the gut mucosa proceed in waves, reflecting enterohepatic circulation of inducers? Induction has recently been detected in neurosecretory cells of the pituitary of BNF-treated trout.[330] That induction could be involved in reproductive effects of Ah-receptor agonists. Do cells showing strong induction of CYP1A have sufficient P450 reductase activity to fully support the potential catalytic rate? Data from heart suggest that some do not.[316] Is there in each cell type a full complement of enzymes to detoxify the products of CYP1A activity, whether activated xenobiotics (conjugating enzymes) or active oxygen (superoxide dismutase)?

C. Endothelium as a Primary Target

Studies such as those above using MAb 1-12-3 have shown that CYP1A can be induced in epithelia of many organs. However, the common site of induction in extrahepatic organs is in endothelial cells, regardless of inducer or route of exposure. Seen first in scup and rainbow trout gill and heart,[316,319] this localization has now been seen in many species. Situated between the blood and all underlying tissues, endothelium is a potentially crucial target site where xenobiotics could be intercepted and exert toxicity. Figure 18 illustrates the strength of endothelial cell induction in the heart, where we calculate that the content of CYP1A in the endothelial cell microsomes would be near 5.0 nmol/mg (25% of the ER protein!).[316]

The prevalence of CYP1A induction in endothelium indicates that this is a fundamental response in this cell type in fish and implies that Ah receptors are operating in endothelium. The presence of this coupled and functional system could have quite different consequences for systemic effects depending on the nature of the interaction of P450 with xenobiotics being delivered via the blood:

1. CYP1A induction in endothelium throughout the organism could serve a protective role, eliminating some CYP1A substrates from the blood in a "first pass"

BIOCHEMISTRY AND MOLECULAR BIOLOGY OF MONOOXYGENASES 161

A

B

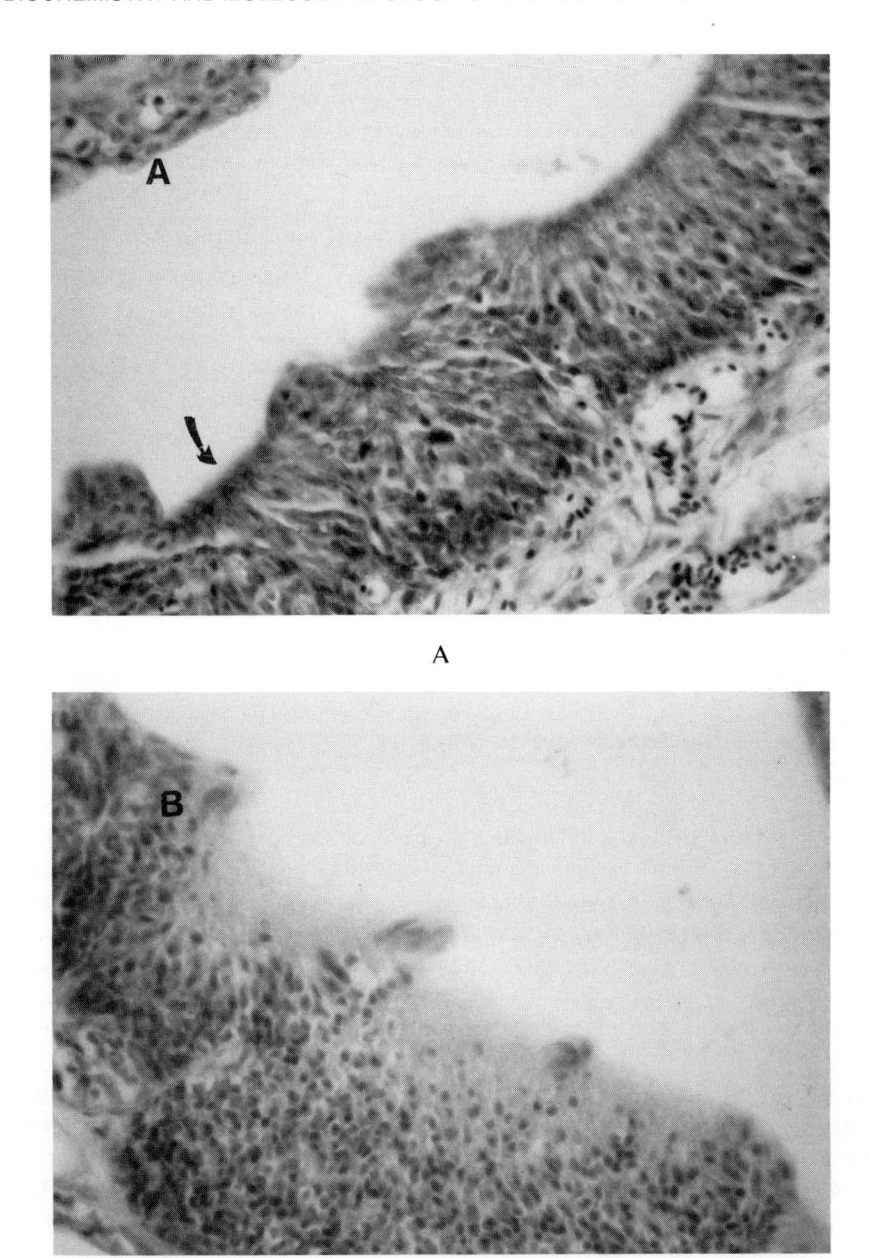

Figure 17. Immunohistochemical localization of CYP1A in olfactory mucosa of *Poeciliopsis*. (A) Animal exposed to BaP in the water, as described elsewhere.[321] (B) Animal in water only. Both sections were stained with MAb 1-12-3, as in Figure 18. Stained structure in A is the olfactory sensory epithelium.

effect. Thus, endothelial CYP1A could detoxify or sequester pollutants, influencing the structure and/or amount of toxicant passing to the underlying cells. Induction in the endothelium of the brain could constitute a component of the blood-brain barrier, which structurally is the brain microvasculature (capillary endothelium).
2. CYP1A activity could lead to enhanced activation of foreign compounds in the endothelium. This could produce toxicity in the endothelium directly or could result in the export or passage of potentially toxic metabolites to the underlying cells.
3. Induced CYP1A could alter the normal metabolism or active content of endothelial regulatory factors, including arachidonic acid and/or nitric oxide (NO), that control vascular integrity and function. CYP1A inducers can alter the metabolism of arachidonic acid,[331] and P450 does bind NO.
4. Occupation of Ah receptors by external ligands and subsequent regulation of non-P450 genes could mediate toxic effects in the endothelium.

As in adults, the endothelium is also a prominent site of induction in the embryos and larvae of fish. Evidence indicating a role in toxicity includes the hemorrhage and edema linked to TCDD toxicity in salmonid larvae,[332–334] conditions signaling endothelial cell dysfunction. The TCDD LD_{50} and EC_{50} for endothelial CYP1A induction are similar in those fish.[299]

D. CYP1A Expression and Neoplasms

Hepatic neoplasms in feral fish show a diversity of phenotypes, including hepatocellular carcinoma, cholangiocellular carcinoma, and hemangiosarcomas.[335,336] It is possible that within a species, the relative capacities for carcinogen bioactivation, detoxication, DNA repair, and other processes may render some cell types more vulnerable to the production of relevant mutations. For example, biliary cells in winter flounder or rainbow trout liver show induction of CYP1A but apparently less strongly than in hepatocytes.[337,338] The predominant tumor phenotype in winter flounder is cholangiocellular. There are suggestions that biliary cells may be stem cells in fish liver.[339,340] It is intriguing to ask whether there is some stage in the growth and differentiation of liver at which these putative stem cells may be not only more disposed to mutagenesis but also more inclined toward the cholangial phenotype than a hepatocytic phenotype. This could involve P450 in the activation of procarcinogens in the stem cells. However, the appearance of cellular abnormalities can be expected to occur well after bioactivation of a promutagen by induced P450.

The capacity for CYP1A expression is altered in neoplastic cells and in abnormal cells that are possibly preneoplastic cells. A decreased capacity for xenobiotic metabolism in mammalian tumors has been known for some time.[341] Not unexpectedly then, there is also a decrease in the content of CYP1A in hepatic tumors in fish. Studies in *F. heteroclitus,* rainbow trout, and winter flounder have shown that nodular tissue, including neoplastic but also nonneoplastic lesions, shows lower rates of microsomal EROD activity and lesser amounts of CYP1A

BIOCHEMISTRY AND MOLECULAR BIOLOGY OF MONOOXYGENASES 163

A

B

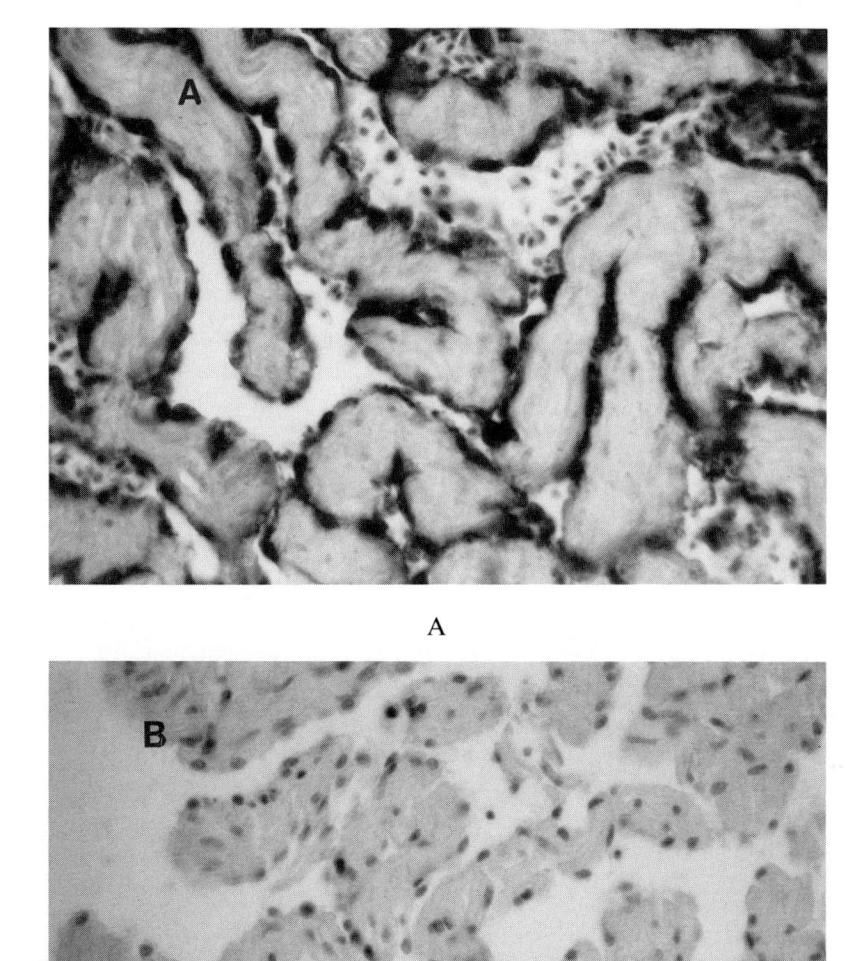

Figure 18. Immunohistochemical localization of CYP1A in scup heart. (A) Animal treated i.p. with 2,3,7,8-tetrachlorodibenzofuran, as described elsewhere.[320] (B) Control animal. Both sections were stained with MAb 1-12-3 to scup CYP1A1 and developed with a peroxidase-labeled second antibody. Staining visible is in the endothelial cells.

than are seen in the surrounding tissue or in normal tissue.[337,342–344] Analysis by IHC shows that tumor cells are those impoverished in P450.[337,342,343] The full significance of a decreased CYP1A content or function in neoplasia is not clear. It could render tumor cells resistant to effects from bioactivation of protoxicants, a process thought to be one component of the drug resistance of tumors. However, it is also possible that the decreased P450 capacity is related to the status of these cells in proliferation and differentiation, with an incidental reduction in the capacity to respond to inducers. The relationships between differentiation and/or cell proliferation and P450 gene regulation are important areas for study.

XI. CYTOCHROME P450 IN ISOLATED CELLS

Isolated cells have been used in studies of mammalian P450 systems for over 20 years. Much of the work has been conducted in cells of hepatic origin, including primary cell cultures (e.g., see references 345–347) and cell lines (e.g., see references 60, 348), and has focused on aspects of P450 regulation. More recently, expression of cloned P450 forms in heterologous cells has been increasingly used to obtain information on catalytic function.[121]

In vitro approaches using cells in culture offer distinct advantages for study of a number of aspects of P450 systems. For example, they allow the investigator to closely control the environment of the cells and to quickly obtain results from many replicates by use of multiwell plates. Several topics are particularly amenable to such research approaches.

Unmodified cells can be used for:

1. regulation of P450 gene expression, including hormonal, temporal, thermal, and mechanistic aspects;
2. determination of structure/activity relationships for induction potency and efficacy;
3. the use of primary cultures for comparing the response of different species to the same compounds, to identify more sensitive or resistant species;

Transgenic systems, in which cell lines devoid of a particular function might be invested with that function by gene transfer, can be used for:

4. the study of mechanisms of P450-dependent cell toxicities;
5. studies of xenobiotic substrate specificity.

A. Primary Isolates

Primary cultures of fish hepatocytes have been used by several groups in the study of CYP1A regulation and function. Miller et al.[349,350] and Pesonen and co-workers[205,351] studied the induction of CYP1A1 in primary cultures of rainbow trout hepatocytes after exposure to BNF and TCDD. In the latter study, induction was observed at the levels of mRNA, protein, and catalytic activity, comparable to results seen in vivo. Temporal differences in the response to the two compounds were also equivalent to those observed in fish in vivo[203] and in mammalian cells.[352] Devaux et al.[284] found that glucocorticoids could potentiate BNF induction of CYP1A activity in trout hepatocytes (see Section IX.B). Sikka and colleagues[245,353] used freshly isolated hepatocytes from brown bullheads to examine the metabolism of BaP and BaP-7,8-dihydrodiol. The use of isolated fish hepatocytes in toxicology has been reviewed elsewhere.[354]

B. Cell Lines

While fresh cultures of cells may more accurately reflect the in vivo condition of the liver, such cells are difficult to establish in continuous culture. Extensive use of primary cultures requires repeated isolation from individual organisms that may differ in their physiology and responsiveness. For this and other reasons, transformed cell lines derived from fish have begun to be used in studies of P450 regulation. Considering the wealth of information regarding CYP1A1 regulation that has been provided by studies in mammalian hepatoma cells (e.g., mouse Hepa-1, rat H4IIE, human HepG2), it is only natural that those working with aquatic animals would seek similar systems. RTH-149, a rainbow trout hepatoma cell line, has been shown by Lorenzen and Okey[215] to possess Ah receptors and exhibit induction of AHH activity by TCDD and benzanthracene. Hightower and Renfro[355] described the isolation of a hepatoma cell line (PLHC-1) from the tropical topminnow *Poeciliopsis lucida*. This cell line was subsequently shown to contain immunochemically detectable CYP1A protein and catalytic (EROD) activity that are inducible by 3,3',4,4'-TCB.[99] Dose-response experiments revealed both induction and, at higher doses, inhibition or inactivation of CYP1A activity like that seen in vivo. Recently, a nontransformed liver cell line from zebrafish *(Brachydanio rerio)* has been established in continuous culture.[356] Treatment of these cells with TCDD increased the amounts of two proteins recognized by antibodies to trout CYP1A1. Do these represent two CYP1A gene products?

Nonhepatic cell lines have also been used to study P450 function. Smolarek et al.[357] examined several such lines for the ability to activate two PAHs to DNA-binding species. Although the fish cells were quite active in PAH metabolism, amounts of DNA adducts were lower than those observed in rodent systems.

Mammalian cell lines — especially the H4IIE rat hepatoma cell line — have been used extensively to define the structural specificity for induction of CYP1A by halogenated aromatic compounds.[358-360] Strong correlations between in vitro CYP1A inducing potency and in vivo toxic potency have contributed to the *Toxic Equivalency Factor* (TEF) concept, in which the potency of individual congeners is expressed relative to that of 2,3,7,8-TCDD.[361] Evidence for differences between mammals and fish in the relative potencies of various halogenated compounds[98,208,362] has encouraged the development of fish cell lines that could be used to quantitate such differences. PLHC-1 cells[99,355] represent one such line; fish-specific TEFs have begun to be determined using this and other cell lines.[363,364] There are, however, some uncertainties in this approach. A strong correlation between inducing potency and toxic potency like that seen in mammals has not yet been demonstrated conclusively in fish. Another unresolved question clouding the accurate determination of TEFs, in any species, is how to accommodate partial agonists — compounds whose efficacy or intrinsic activity is less than that of full agonists.[365] Determining the inducing potency of compounds that differ in the level of maximal induction (efficacy) is problematic. Current approaches ignore this difficulty, which will only be resolved through an improved mechanistic understanding of the processes of CYP1A induction and toxicity.

An area in which cell culture studies could provide information simply and quickly is in studies of the interactive effects of chemical mixtures. A method originally developed by Kennedy et al.[366] for growing avian hepatocytes in multiwell plates is now being adapted for fish cell lines, facilitating the rapid analysis of large numbers of chemical mixtures.[363,367] Such studies will elucidate synergistic, antagonistic, and other interactive effects that occur following exposure to complex mixtures and will allow a more realistic appraisal of the compounds responsible for P450 induction in a "real world" setting.

We have dealt exclusively with fish cells in this section, for that is where the majority of emphasis has been placed. However, there are isolated cell systems from other aquatic groups that are, or could be, exploited for P450 research. Studies in an amphibian cell line[161] were discussed earlier (Section V.B). Culture of invertebrate cells has been much more difficult. To our knowledge, such cultures have not been used in P450 research. However, as culture methods improve (e.g., see Baksi and Nelson[368]), we foresee significant promise in the use of invertebrate cell systems.

Increased use of nonhepatic cells (e.g., gill epithelia) and of transgenic cells will enhance our understanding of the regulation and function of a variety of P450 forms in aquatic species. For example, heterologous expression of cDNAs coding for P450s from aquatic species will provide much needed information on the catalytic specificity of such proteins, as it has in mammals.[121] In vitro mutagenesis of P450 cDNAs using PCR or other techniques, followed by expression in cultured cells and catalytic assay, will allow the identification of amino acid residues that are essential for catalytic function. Ultimately, it will become possible to infer catalytic function, and thus chemical effect, from sequence identities or similarities between species.

XII. P450 INDUCTION AS A BIOMARKER IN THE ENVIRONMENT

There is great and growing interest in using CYP1A induction as a biomarker to indicate the exposure of organisms in the wild to inducing compounds, to evaluate the degree and possible risk of environmental contamination.[237] The induction of monooxygenase (AHH) activity in fish liver was first suggested as an indicator of exposure to petroleum in the early 1970s.[369] The first environmental induction in fish larvae was found in the progeny spawned from salmonids taken from contaminated sites in the Great Lakes.[370] Analysis of environmental induction by immunoassay with antibodies to fish CYP1A was demonstrated in 1986,[371] and data suggesting that mRNA detection could be used appeared in 1989[372] and again more recently.[373] The first immunohistochemical analysis showing the multiple organ and cellular localization of environmentally induced CYP1A in fish appeared in 1991.[304] The results of that study (see Table 16) establish that immunohistochemistry can be applied usefully to detect the cellular targets where induction occurs in environmentally-exposed fish. In situ hybridization is also being developed.[374] These approaches complement one another, and all have value in detecting induction as a marker of exposure.

The burgeoning literature continues to indicate the value of the approach. Induction is being repeatedly tested and proven as a marker for contamination in coastal waters, rivers, and lakes around the world, not only in fish, but in many vertebrate groups, including reptiles, amphibians, birds, and mammals.[169,179,375-377] Unfortunately, based on our present knowledge, there is little potential for soon using monooxygenase activity or P450 levels in many invertebrates, such as mollusks or crustaceans, to indicate their exposure to compounds such as the aromatic and chlorinated hydrocarbons. This is due to the current lack of convincing evidence for induction of known P450 forms or of monooxygenase activity by any known mechanism (however, see Section VII.B.2). Recent suggestion of a CYP1A-like gene product in mussels[194] suggests that this circumstance could change.

A. Distribution: Coastal and Global

Many studies have shown that fish from coastal regions near population centers have induced levels of CYP1A in their livers.[101,255,303,304,314,378-387] A great weight of evidence by now indicates that we can reliably predict this condition to occur in fish from such areas. At greater distances from urban centers, there is usually a decrease in the content of CYP1A. However, disseminated sources and local sources both have been associated with strong induction signals in fish at remote sites. The apparent induction in the liver of fish in the deep ocean is a prime example.[371] Slight induction seen in fish at sites distant from major sources of contaminants could reflect a global background of contamination. In the case of marine mammals showing high levels of apparent CYP1A,[168-170,172] it is likely that the signal reflects their position at the top of the food chain and an integration of signal over large regions. Information concerning the global condition will require additional geographic coverage, linkage to the inducer identity, and analysis of different trophic levels. Studies of animals from polar, (e.g., see reference 388) deep ocean, (e.g., see reference 371) and open ocean regions might be particularly revealing in this regard.

B. Causes of Induction

Studies have shown that the degree of CYP1A induction in fish from the wild can be closely linked to the degree of contamination by chemicals such as PCBs or PAH in the animals themselves or in other biota or sediments at the same site.[255,303,371,380,381,383,389] The studies of Collier et al.[390] should soon provide abundant information of this type. Studies in marine mammals are also revealing close correlations between tissue residues of candidate inducers and the amount of CYP1A in liver.[169] Indeed, the detection of putative CYP1A proteins in cetaceans most likely reflects an environmental (and not experimental) effect. The results generally are so convincing that we can safely predict that such use of CYP1A induction will become commonplace, not only in aquatic species but in many terrestrial species. The development of biopsy or other nondestructive sampling methods, coupled with immunohistochemical or other technically simple analytical approaches, will ensure a widespread use.

There is a continuing need to establish the identity of the most active agents contributing to induction in specific sites and environments. Given growing knowledge of induction mechanisms and the potency of compounds capable of eliciting induction, it is increasingly possible, with knowledge of chemistry in a given area, to name suspect active agents in that area, a possibility suggested long ago. The transition from suspected to confirmed cause of environmental induction is much more difficult, as illustrated by studies of induced CYP1A in fish or fish cells exposed to bleached and unbleached pulp mill effluent,[391,392,392a] in which the unbleached effluents also appear to elicit induction.

C. Significance of Induction

Challenging questions concern the significance of environmental induction. Indeed, the significance of this or any biochemical effect in wild species is at the core of aquatic toxicology; the questions cannot be answered lightly or briefly. At minimum, we can say that the induction will aid in clearing compounds that are substrates from the body. How far we can interpret the induction to indicate toxic consequences, as it does indicate exposure, is less certain. Adequate answers must consider the full range of possible toxic mechanisms involving P450 function or Ah receptor function discussed above (Section VIII). We can point to many signs that inducers or substrates of P450 are toxic and tentatively generalize about some effects. The most obvious involves carcinogenesis.

As detailed above, activation of many procarcinogens requires the function of P450. A cell or organ devoid of the requisite catalyst(s) will not transform such a compound into a carcinogen. As CYP1A is a prominent catalyst through which some PAHs are activated in fish, some degree of CYP1A induction is a likely prerequisite for the initial steps in environmental carcinogenesis involving those compounds. Carcinogenic PAHs that are active inducers are at high concentrations in some regions.[380,393,394] In such environments, greater P450 induction could contribute to a higher steady-state level of activated carcinogens and, consequently, to a greater formation of DNA adducts or to enhanced oxidative DNA damage; there is a correlation between induction and the degree of DNA binding by carcinogens in mammalian systems.[395] But highly induced P450 is not necessarily associated with a greater risk of carcinogenesis. Formation and persistence of critical genetic lesions may be influenced as much by detoxication or repair processes as by the oxidative metabolism creating the activated carcinogenic derivative.

Many CYP1A inducers are acutely toxic. In many environments, though, there is likely to be long-term, low-level exposure to compounds producing sustained induction of CYP1A.[97] There is substantial uncertainty regarding the effects at such low doses, and dose extrapolation is currently being contested.[396] Determining the significance of binding of the Ah receptor by external ligands at low environmental doses that are yet capable of inducing CYP1A is an important concern in environmental toxicology.

XIII. THE NATURAL PRODUCTS CONNECTION

The interactions between plants or sessile invertebrates and their respective predators often involve the elaboration of natural products by the prey, acting as chemical defenses to protect against predation. There is a wealth of information describing the metabolism of natural products, in many cases by hydroxylation, in animals.[397] Natural products, e.g., aflatoxins and PAHs, are substrates for P450 in aquatic systems. Chemical/biological interactions between natural products and cytochrome P450 could govern chemical ecology, contribute to chemical effects in aquatic systems, and force the evolution of xenobiotic-metabolizing enzymes.

In the terrestrial environment, interactions between insects and their plant hosts include many examples of the coevolution of the plant's production of phytoalexins or secondary metabolites and the predator's metabolic capability to detoxify the otherwise toxic compounds.[398] Some ecological interactions involve P450 induction, allowing the animal to utilize a food resource that might otherwise be unavailable because of its toxicity.[399,400] It has long been known that insect larvae possess or develop a generally greater capacity for monooxygenase activity as a result of feeding on their plant host.[401,402] A recent study explicitly showed that the ability of an insect (swallowtail butterfly larvae) to feed on an otherwise toxic plant (milkweed) involves the induction of a novel cytochrome P450 form.[403]

Marine invertebrates and plants synthesize many natural products, including toxins of environmental and public health significance.[404–406] Not surprisingly, there are examples of selective predation in marine systems not unlike selective predation seen in terrestrial plant-insect pairs.[407] Cytochrome P450 could have a dual function involving many natural products, first in their synthesis in prey organisms, and second in their metabolism in the predator. Synthesis of some marine invertebrate and plant compounds, particularly those that are steroid derivatives or hydroxylated structures will likely involve P450, as was suggested earlier. Two examples illustrate the importance of these considerations to questions of chemical ecology and evolution and also to practical concerns involving environmental assessment in aquatic systems.

First there have been speculations for years that natural products might induce AHH activity (CYP1A) in fish,[63] among other things interfering with the use of such induction as a biomarker. The prospect for natural products to act as inducers of CYP1A is clear; many marine natural products have structures similar to known Ah receptor ligands and/or substrates for CYP1A. Some, such as terpenoids synthesized by gorgonian corals, are active inducers and substrates of P450 forms in mammals,[408] and similar responses are likely in fish. Halogenated diphenyl ethers and diphenyl methanes have been isolated from sponges and green algae, respectively.[409] Brominated indoles found in many hemichordates are similar in structure to indole derivatives such as indole-3-carbinol (I3C) derived from terrestrial plants;[410] condensation products of the latter have been shown to bind the Ah receptor with high affinity.[411] I3C is an inducer of CYP1A in mammals[412,413] but apparently not in trout.[414] Flavonoid compounds synthesized in certain terrestrial and marine plants are related to synthetic Ah receptor ligands such as BNF.[409] In freshwater systems, the induction of CYP1A is used increasingly to detect effects of effluents from pulp mills, in which bleaching can produce a variety of chlorinated compounds, including dioxins and furans.[415,416] Recent data indicate that processing of the wood without bleaching also results in formation or release of some unidentified CYP1A inducers.[391]

Second, the diversity of mammalian genes in some subfamilies, including CYP2B, and possibly in others, has been suggested to have been driven by the consumption of terrestrial plants with their evolving chemical defenses.[144,417] Earlier Stegeman[63] suggested that a general lack of terrestrial plant material in the diet was related to the lack of PB responsiveness in fishes and some other groups. Groups that do consume terrestrial plants or groups that consume an abundance of marine plants could have a spectrum of P450 genes or regulation different from those common in the temperate carnivorous fish most often studied.

There has been little direct attention to the role of biosynthesized natural products as inducers or substrates for aquatic systems until recently. Indications that there may be some regulation of P450 gene expression linked to the diet have come from studies of reef fishes. Surveys of hepatic microsomal P450 content and monooxygenase activities in reef fishes from Bermuda revealed some species with total P450 content far exceeding the content typically seen in many temperate and freshwater species.[384] Recently, Vrolijk et al.[418,418a] also observed a high content of hepatic P450 in butterfly fish that feed on gorgonians containing high concentrations of prostaglandins and several terpenoid structures, whereas congeneric species that do not feed on gorgonians contain lesser content of total P450. These fish also displayed different amounts of proteins cross-reacting with antibodies to various fish P450 forms.

Empirical demonstration that marine natural products can induce one or another of the known P450 genes in fish is yet to come. Likewise, the P450 complement in any of these species is not yet known, but the possibility for novel P450 genes and proteins is clear. Elucidating biochemical interactions involving natural products is important to questions regarding the roles that external chemicals or internal physiological processes have played in the evolution and diversification of P450 systems. This is a fruitful area for molecular biological investigation. Natural products as inducers or substrates of P450 would also have practical importance in the debate regarding the relative significance of natural products or anthropogenic chemicals as toxicants, mutagens, or carcinogens.[419,420]

XIV. EVOLUTIONARY ASPECTS

The evolution of genes for the cytochromes P450 and for receptors that may govern their expression is most intriguing. One might address these issues by considering:

1. the origin and the original function of P450 and regulatory factor genes, and their ancestors;
2. the forces contributing to diversification of the genes in evolution, and in different phylogenetic groups;
3. factors contributing to maintenance or to loss of either ancient or more recent functions.

How old are the genes that we recognize as P450 or receptor? What were the original functions? Are the original functions still represented in contemporary species? Did the ancestors to the receptors and the P450 enzymes that they regulate have unrelated functions that were only later combined? If so, which came first, and how long have they evolved in tandem? Or did the two systems evolve in tandem from the earliest appearance?

A. Evolution of Cytochrome P450

Estimates based on all P450 sequences suggest that these genes originated more than 1.5 and possibly as long as 3 billion years ago.[58] Cladistic analysis of P450 sequences has revealed certain relationships and the possible times of divergence of some genes.[421] Consideration of structure and extant function of various P450s, how ancient those functions might be, and the time at which different animal groups may have diverged compared to when different P450 lines diverged can suggest whether homologous genes might occur in different groups. Thus, some mitochondrial P450s involved in steroidogenesis are related to bacterial P450s, inferred from sequence data but also from localization and a similar involvement of iron-sulfur redoxins (adrenodoxin in adrenal cortex; putidaredoxin in *Pseudomonas*) in electron transfer to the P450. Yet, the original functions for mitochondrial and microsomal P450s remain clouded.

Of two major extant functions served by liver microsomal P450s in animals, the metabolism of steroid compounds and the metabolism of xenobiotics, steroid hydroxylation is thought to be the more ancient. However, xenobiotic or natural product substrates for P450 also could have been present in the ancient environment. Further, BaP, other PAH, and possibly even dioxins are natural products resulting from incomplete combustion and/or diagenesis.[422,423,423a] Such compounds would have been present in the environment shortly after the appearance of organic matter that could be burned. Diversification of the microsomal P450 genes seems to have accelerated more than 800 million years ago, possibly, as some have suggested,[144,417] associated with defense against terrestrial plant products being elaborated at that time. The presence of toxic secondary metabolites has been suggested as a major ecological factor driving the evolution and diversity of cytochromes P450 in terrestrial vertebrate groups that have recently or still include herbivores.[417] Yet, acquisition by animals of xenobiotic metabolism functions and pressures associated with xenobiotics could have preceded the suggested diversification involving adaptation to dietary plant toxins. Most likely, biogenic, pyrogenic, and diagenic compounds each have contributed to the diversification of P450 genes, and continue to do so.

B. Evolution of the Ah Receptor

The original function of the P450 genes must be considered in concert with the evolution and function of the regulatory mechanisms. Chlorinated dibenzo-p-dioxins, the most potent and well-known ligands for the Ah receptor, are primarily anthropogenic in origin and have been present in appreciable amounts only since about 1940.[424] The phylogenetic data on the Ah receptor[217-219] and CYP1A, however, suggest that these two genes have existed, both structurally and functionally, for over 450 million years, far longer than the reign of human influence on the chemical environment. Elucidation of the origin of the receptor, then, must involve the identity of the "natural" Ah receptor ligand and the normal physiological function of this signal transduction system.

Two attractive hypotheses have been put forth to explain the origin and evolution of the Ah receptor/CYP1A system. Nebert[425] proposed that drug metabolizing enzymes, including CYP1A, are responsible for controlling the steady-state levels of small endogenous ligands that modulate growth and differentiation.

Foreign chemicals that induce CYP1A, then, might mimic an endogenous ligand, whose identity is not yet known.

This hypothesis was especially attractive when the Ah receptor was thought to be a member of the superfamily of nuclear receptors for steroid, thyroid, and retinoid "hormones." A number of ligands for these receptors (e.g., glucocorticoids, fatty acids) are involved in regulating the expression of P450 genes, which in turn catalyze the synthesis or degradation of the ligand. However, the recent cloning of an Ah receptor cDNA[222,223] has shown that, despite its physicochemical and other similarities to steroid hormone receptors, the AhR is completely unrelated to these receptors at the level of nucleic acid or deduced amino acid sequence. Rather, it appears to be similar to a group of transcription factors bearing a basic region/helix-loop-helix (bHLH) dimerization and DNA-binding motif.[222,223] The Ah receptor nuclear translocator (ARNT), a component of the transformed Ah receptor complex, is also a member of this group.[426] Other than the Ah receptor, none of the known members of this family of proteins appears to require ligand binding for transcriptional activity. Despite the discovery that the Ah receptor is not structurally related to the steroid receptors, the existence of a regulatory pathway involving an endogenous ligand for the Ah receptor remains a viable possibility for the origin of this system. The recent identification of biological systems in which CYP1A proteins are "constitutively" expressed[426a-c] may contribute to identifying such a pathway.

A second hypothesis, discussed above, is that natural products have been a driving force behind the evolution of the Ah receptor/CYP1A system. Thus, the Ah receptor may have evolved through a need for certain animals to detect the presence of specific compounds in their diet or environment and to respond in an appropriate fashion, for example by increasing the synthesis of enzymes able to degrade the compounds or to initiate some new phase of their life cycle. In support of this hypothesis are some striking similarities between certain marine natural products and the known Ah receptor ligands,[427] discussed above. A goal of future work should be to identify natural products that are able to bind the Ah receptor and activate CYP1A transcription and to determine their ecological significance.

Whether the Ah receptor evolved together with CYP1A function in a detoxification context or alternatively, as part of a separate signal transduction system unrelated to xenobiotic metabolism (which may have been later appropriated for regulation of CYP1A), remains an intriguing and as yet untested area of inquiry. Similar questions concern the evolution of other P450/regulatory system pairs. For example, there could be a receptor involved in induction of CYP2B genes.[428-430] If so, is the induction of CYP2B genes by phenobarbital-type compounds related to ecological pressures of natural products during evolution? Did this type of responsiveness originate in a role in cell growth/differentiation involving endogenous ligands? Many of the inducers of CYP2B also are tumor promoters and may affect cell proliferation. If evidence suggests that mammalian-like regulatory mechanisms for the 2B genes arose in groups ancestral to fish, then would similar mechanisms be expected in fish, or have they been lost?

If there are fewer genes for xenobiotic metabolizing P450s, or "missing" regulatory mechanisms in some groups, it could reflect evolution in environments with a less abundant or diverse natural chemical challenge. The relative lack of terrestrial plant material in many, though not all, aquatic environments suggests that such possibilities may be tested. Determining whether the P450s involved in xenobiotic metabolism diversified from (presumably ancient) catalysts that functioned primarily in metabolism of ligands for receptors that have critical roles in growth and differentiation[425] is an aim for the future. Indeed, biotransformation systems in aquatic species could serve as important subjects in addressing this issue; they could have fewer extant forms of P450 that would yet include representatives of those more ancient catalysts. Direct consideration of these questions in a diversity of species and a diversity of P450 families is essential for a rational study of the evolution of the regulatory systems as well as the P450 genes themselves.

XV. CONCLUSION

We know a great deal more about xenobiotic metabolism and its regulation in aquatic species than was known 25 years ago, and we know it in greater detail. Yet, much of this knowledge simply underscores Adamson's[3] conclusion about the basic similarity of these processes in different groups. Despite long effort,[12] our understanding of these systems in invertebrates is much less. We can, however, see novel features emerging and the directions where substantial answers may lie in vertebrates and invertebrates. As the catalysts themselves become known, their roles in the protection against and mediation of toxic effects will become known. As the regulatory mechanisms become clear, P450 genes that are inducible by known external chemicals will be increasingly useful for dissecting the mechanism by which other variables, such as temperature, influence receptor mechanisms, transcription, and translation. As linkages to other processes involved in signal transduction, growth, and /or differentiation become known, so will we know the significance of P450 function and regulation in the physiology and health of animals. As phylogenetic and/or ecological differences in the expression of different P450 forms are established, the ability to interpret the effects of anthropogenic chemicals in ecosystems will progress.

Research on monooxygenase systems must continue to be an amalgam of molecular, biochemical, in vitro, and in vivo studies. Molecular biology of these enzymes is of obvious importance if we are to understand the origin of P450 gene diversity and regulatory mechanisms and their significance in aquatic species. Much of this research in aquatic systems will follow largely predictable pathways. We cannot help but tread paths laid out by studies with mammals. Yet that is not only acceptable but desirable, if considered within an evolutionary context and if researchers try to probe the species similarities. However, studies in nonmammalian species have disclosed and will continue to disclose new aspects of P450 systems! (Getting mammalian researchers to attend to such findings is sometimes more difficult.)

At times, agencies that conduct or support studies of chemical effects in the aquatic environment have discounted both mechanistic and comparative research, on the one hand because direct links between studies of biotransformation systems and the endpoints the agencies were interested in could not already be shown and on the other because comparative research was thought (mistakenly) not to be fundamental. We should not treat our ignorance of the precise pathways by which biotransformation enzymes and their regulatory systems function in toxicopathic mechanisms as evidence that no such functions exist. Rather, we should rely on data that point to the possibility of such linkages as reason enough to expect them and indeed to expect to be surprised. The questions posed and promise suggested by Adamson[3] still provide impetus to understand xenobiotic metabolism in aquatic species. A recent review by Greenberg[431] extolling the value of comparative biochemistry was brightly entitled, "Ex bouillabaisse lux." We concur and state again that the comparative biochemistry and molecular biology of biotransformation systems will continue to illuminate mechanistic linkages and generalizations essential to aquatic toxicology.

ACKNOWLEDGMENTS

Support was provided in part by U.S. E.P.A. grant R817988, U.S.P.H.S. (N.I.H.) grants ES-04220 and ES-06272, and the S. W. Watson Chair. We thank Bruce R. Woodin for valuable assistance through the years and assistance in preparation of this manuscript. We acknowledge the collaboration of many colleagues in generating data discussed in this chapter, but in particular Pamela J. Kloepper-Sams, Roxanna Smolowitz, Harry V. Gelboin, and Alan Poland. We also wish to thank Margaret James, Dave Hinton, Pam Kloepper-Sams, and Karl Fent for comments on earlier drafts of the manuscript. This is contribution number 8366 from the Woods Hole Oceanographic Institution.

REFERENCES

1. Stone, R., Swimming against the PCB tide, *Science,* 255, 798, 1992.
2. Williams, R.T., *Detoxication Mechanisms,* 2nd ed., Chapman & Hall, London, 1959.
3. Adamson, R.H., Drug metabolism in marine vertebrates, *Fed. Proc.,* 26, 1047, 1967.
4. Chambers, J.E. and Yarbrough, J.D., Xenobiotic transformation systems in fishes, *Comp. Biochem. Physiol.,* 55C, 77, 1976.

5. Bend, J.R. and James, M.O., Xenobiotic metabolism in marine and freshwater species, in *Biochemical and Biophysical Perspectives in Marine Biology,* Malins, D.C. and Sargent, J.R., Eds., Academic Press, New York, 1978, 4, 125.
6. Stegeman, J.J., Monooxygenase systems in marine fish, in *Pollutant Studies in Marine Animals,* Giam, C.S. and Ray, L.E., Eds., CRC Press, Boca Raton, FL, 1987, 65.
7. Stegeman, J.J., Cytochrome P450 forms in fish: catalytic, immunological and sequence similarities, *Xenobiotica,* 19, 1093, 1989.
8. James, M.O., Cytochrome P450 monooxygenases in crustaceans, *Xenobiotica,* 19, 1063, 1989.
9. Livingstone, D.R., Kirchin, M.A., and Wiseman, A., Cytochrome P450 and oxidative metabolism in mollusks, *Xenobiotica,* 19, 1041, 1989.
10. Lech, J.J., Vodicnik, M.J., and Elcombe, C.R., Induction of monooxygenase activity in fish, in *Aquatic Toxicology,* Raven Press, 1982, 107.
11. Goksøyr, A. and Forlin, L., The cytochrome P-450 system in fish, aquatic toxicology and environmental monitoring, *Aquat. Toxicol.,* 22, 287, 1992.
12. Livingstone, D.R., Organic xenobiotic metabolism in marine invertebrates, in *Advances in Comparative and Environmental Physiology, Vol. 7,* Gilles, R., Ed., Springer-Verlag, Berlin, 1991, 45.
13. Andersson, T. and Forlin, L., Regulation of the cytochrome P450 enzyme system in fish, *Aquat. Toxicol.,* 24, 1, 1992.
14. Brodie, B.B. and Maickel, R.P., Comparative biochemistry of drug metabolism, in *First International Pharmacology Meeting,* Macmillan, New York, 1962.
15. Potter, J.L. and O'Brien, R.D., Parathion activation by livers of aquatic and terrestrial vertebrates, *Science,* 144, 55, 1964.
16. Buhler, D.R. and Rasmussen, M.E., The oxidation of drugs by fishes, *Comp. Biochem. Physiol.,* 25, 223, 1968.
17. Creaven, P.J., Parke, D.V., and Williams, R.T., A fluorimetric study of the 7-hydroxylation of coumarin by liver microsomes, *Biochem. J.,* 96, 879, 1965.
18. Creaven, P.J., Davies, W.H., and Williams, R.T., Dealkylations of alkoxybiphenyls by trout and frog liver preparations, *Life Sci.,* 6, 105, 1967.
19. Ashley, L.M., Halver, J.E., and Wogan, G.N., Hepatoma and aflatoxicosis in trout, *Fed. Proc.,* 23, 105, 1964.
20. Stanton, M.F., Diethylnitrosamine-induced hepatic degeneration and neoplasia in the aquarium fish, *Brachydanio rerio, J. Natl. Cancer Inst.,* 34, 117, 1965.
21. Ziegler, D.M. and Mitchell, C.H., Microsomal oxidase IV: properties of a mixed-function amine oxidase isolated from pig liver microsomes, *Arch. Biochem. Biophys.,* 150, 116, 1972.
22. Ziegler, D.M., Flavin-containing monooxygenases: catalytic mechanism and substrate specificities, *Drug Metab. Rev.,* 6, 1, 1988.
23. Ziegler, D.M., Flavin-containing monooxygenases: enzymes adapted for multisubstrate specificity. *Trends Pharmacol. Sci.,* 11, 321, 1990.
24. Augustsson, I. and Strom, A.R., Biosynthesis and turnover of trimethylamine oxide in the teleost cod, *Gadus morhua, J. Biol. Chem.,* 256, 8045, 1981.
25. Cashman, J.R., Olsen, L.D., Nishioka, R.S., Gray, E.S., and Bern, H.A., S-oxygenation of thiobencarb (Bolera) in hepatic preparations from striped bass *(Morone saxitalis)* and mammalian systems, *Chem. Res. Toxicol.,* 3, 433, 1990.

26. Schlenk, D. and Buhler, D.R., Immunological characterization of flavin-containing monooxygenases from the liver of rainbow trout *(Oncorhynchus mykiss)* — sexual-dependent and age-dependent differences and the effect of trimethylamine on enzyme regulation, *Biochim. Biophys. Acta*, 1156, 103, 1993.
27. Schlenk, D. and Buhler, D.R., Flavin-containing monooxygenase activity in liver microsomes of rainbow trout *(Oncorhynchus mykiss)*, *Aquat. Toxicol.*, 20, 13, 1991.
28. Schlenk, D. and Buhler, D.R., Role of flavin-containing monooxygenase in the in vitro biotransformation of aldicarb in rainbow trout, *Oncorhynchus mykiss*, *Xenobiotica*, 21, 1583, 1991.
29. Schlenk, D., Ronis, M.J.J., Miranda, C.L., and Buhler, D.R., Channel catfish liver monooxygenases. Immunological characterization of constitutive cytochromes P450 and the absence of active flavin-containing monooxygenases, *Biochem. Pharmacol.*, 45, 217, 1993.
30. Schlenk, D., A comparison of endogenous and exogenous substrates of the flavin-containing monooxygenases in aquatic organisms, *Aquat. Toxicol.*, 26, 157, 1993.
31. Miller, E.C. and Miller, J.A., Mechanisms of chemical carcinogenesis, *Cancer*, 47, 1055, 1981.
32. Ames, B.N., Durston, W.E., Yamasaki, E., and Lee, F.D., Carcinogens are mutagens: a single test system combining liver homogenate for activation and bacteria for detection, *Proc. Natl. Acad. Sci. U.S.A.*, 70, 2281, 1973.
33. Anderson, R.S. and Doos, J.E., Activation of chemical carcinogens to bacterial mutagens by microsomal enzymes from a pelecypod mollusc, *Merceneria merceneria*, *Mutat. Res.*, 116, 247, 1983.
34. Kurelec, B., Exclusive activation of aromatic amines in the marine mussel *Mytilus edulis* by FAD-containing monooxygenase, *Biochem. Biophys. Res. Commun.*, 126, 773, 1985.
35. Kurelec, B. and Krca, S., Metabolic activation of 2-aminofluorene, 2-acetyl-aminofluorene and N-hydroxyacetylaminofluorene to bacterial mutagens with mussel *(Mytilus galloprovincialis)* and carp *(Cyprinus carpio)* subcellular preparations, *Comp. Biochem. Physiol.*, 88C, 171, 1987.
36. Schlenk, D. and Buhler, D.R., Flavin-containing monooxygenase activity in the gumboot chiton *Cryptochiton stelleri*, *Mar. Biol.*, 104, 47, 1990.
37. Sipes, I.G. and Gandolfi, A.J., Biotransformation of toxicants, in *Casarett and Doull's Toxicology, The Basic Science of Poisons*, Amdur, M.O., Doull, J., and Klaassen, C.D., Eds., Pergamon Press, New York, 1991, 88.
38. Wakeham, S.G., Howes, B.L., Dacey, J.W.H., Schwarzenbach, R.P., and Zeyer, J., Biogeochemistry of dimethylsulfide in a seasonally stratified coastal salt pond, *Geochim. Cosmochim. Acta*, 51, 1675, 1987.
39. Dacey, J.W.H. and Wakeham, S.G., Oceanic dimethylsulfide: production during zooplankton grazing on phytoplankton, *Science*, 233, 1314, 1986.
40. Lovelock, J.E., Maggs, R.J., and Rasmussen, R.A., Atmospheric dimethylsulfide and the natural sulfur cycle, *Nature*, 237, 452, 1972.
41. Klingenberg, M., Pigments of rat liver microsomes, *Arch. Biochem. Biophys.*, 75, 376, 1958.
42. Garfinkel, D., Studies on pig liver microsomes. I. Enzymic and pigment composition of different microsomal fractions, *Arch. Biochem. Biophys.*, 77, 493, 1958.

43. Omura, T. and Sato, R., The carbon monoxide-binding pigment of liver microsomes, *J. Biol. Chem.*, 239, 2370, 1964.
44. Estabrook, R.W., Cooper, D.Y., and Rosenthal, O., The light-reversible carbon monoxide inhibition of the steroid C-21 hydroxylase of the adrenal cortex, *Biochem. J.*, 338, 741, 1963.
45. Lu, A.Y.H., Strobel, H.W., and Coon, M.J., Properties of a solubilized form of cytochrome P-450-containing mixed-function oxidase of liver microsomes, *Mol. Pharmacol.*, 6, 213, 1970.
46. Sligar, S.G. and Murray, R.I., Cytochrome P450$_{cam}$ and other bacterial P-450 enzymes, in *Cytochrome P-450: Structure, Mechanism, and Biochemistry*, Montellano, P.R.O.d., Ed., Plenum Press, New York, 1986, 429.
47. Poulos, T.L. and Ragg, R., Cytochrome P450cam:crystallography, oxygen activation and electron transfer, *FASEB J.*, 6, 674, 1992.
48. Guengerich, F.P. and MacDonald, T.L., Mechanisms of cytochrome P450 catalysis, *FASEB J.*, 4, 2453, 1990.
49. Hollenberg, P.F., Mechanisms of cytochrome P450 and peroxidase-catalyzed xenobiotic metabolism, *FASEB J.*, 6, 686, 1992.
50. Conney, A., Pharmacological implications of microsomal enzyme induction, *Pharmacol. Rev.*, 19, 317, 1967.
51. Conney, A.H. and Burns, J.J., Metabolic interactions among environmental chemicals and drugs, *Science*, 178, 1972.
52. Conney, A.H., Induction of microsomal enzymes by foreign chemicals and carcinogenesis by polycyclic aromatic hydrocarbons: G.H.A. Clowes memorial lecture, *Cancer Res.*, 42, 4875, 1982.
53. Gonzalez, F.J., The molecular biology of cytochrome P-450s, *Pharmacol. Rev.*, 40, 243, 1989.
54. Ioannides, C. and Parke, D.V., The cytochrome P450 I gene family of microsomal hemoproteins and their role in the metabolic activation of chemicals, *Drug Metab. Rev.*, 22, 1, 1990.
55. Nebert, D.W., Adesnik, M., Coon, M.J., Estabrook, R.W., Gonzalez, F.J., Guengerich, F.P., Gunsalus, I.C., Johnson, E.F., Kemper, B., Levin, W., Phillips, I.R., Sato, R., and Waterman, M.R., The P450 gene superfamily: recommended nomenclature, *DNA*, 6, 1, 1987.
56. Nebert, D.W., Nelson, D.R., Adesnik, M., Coon, M.J., Estabrook, R.W., Gonzalez, F.J., Guengerich, F.P., Gunsalus, I.C., Johnson, E.F., Kemper, B., Levin, W., Phillips, I.R., Sato, R., and Waterman, M.R., The P450 superfamily: updated listing of all genes and recommended nomenclature for the chromosomal loci, *DNA*, 8, 1, 1989.
57. Nebert, D.W., Nelson, D., Coon, M., Estabrook, R., Feyereisen, R., Fujii-Kuriyama, Y., Gonzalez, F., Guengerich, F., Gunsalus, I., Johnson, E., Loper, J., Sato, R., Waterman, M., and Waxman, D., The P450 superfamily: update on new sequences, gene mapping, and recommended nomenclature, *DNA Cell Biol.*, 10, 1, 1991.
58. Nelson, D.R., Kamataki, T., Waxman, D.J., Guengerich, F.P., Estabrook, R.W., Feyereisen, R., Gonzalez, F.J., Coon, M.J., Gunsalus, I.C., Gotoh, O., Okuda, K., and Nebert, D.W., The P450 superfamily — update on new sequences, gene mapping, accession numbers, early trivial names of enzymes, and nomenclature, *DNA Cell Biol.*, 12, 1, 1993.

59. Poland, A. and Knutson, J.C., 2,3,7,8-Tetrachlorodibenzo-p-dioxin and related halogenated aromatic hydrocarbons: examination of the mechanism of toxicity, *Annu. Rev. Pharmacol. Toxicol.*, 22, 517, 1982.
60. Whitlock, J.P., Genetic and molecular aspects of 2,3,7,8-tetrachlorodibenzo-p-dioxin action, *Annu. Rev. Pharmacol. Toxicol.*, 30, 251, 1990.
61. Heilmann, L.J., Sheen, Y.-Y., Bigelow, S.W., and Nebert, D.W., Trout P450IA1: cDNA and deduced protein sequence, expression in liver, and evolutionary significance, *DNA*, 7, 379, 1988.
62. Stegeman, J.J., Nomenclature for hydrocarbon inducible cytochrome P450 in fish, *Mar. Environ. Res.*, 34, 133, 1992.
63. Stegeman, J.J., Polynuclear aromatic hydrocarbons and their metabolism in the marine environment, in *Polycyclic Hydrocarbons and Cancer*, Gelboin, H.V. and Ts'o, P.O.P., Eds., Academic Press, New York, 1981, 3, 1.
64. Klotz, A.V., Purification and characterization of the hepatic microsomal monooxygenase system from the coastal marine fish *Stenotomus chrysops*, Ph.D. Thesis, Massachusetts Institute of Technology/Woods Hole Oceanographic Institution, Woods Hole, MA, 1983.
65. Goad, L.J., The steroids of marine algae and invertebrate animals, in *Biochemical and Biophysical Perspectives in Marine Biology*, Malins, D.C. and Sargent, J.R., Eds., Academic Press, New York, 1976, 3, 213.
66. Sandor, T. and Mehdi, A.Z., Steroids and evolution, in *Hormones and Evolution*, Barrington, E.J.W., Ed., Academic Press, New York, 1979, 1.
67. Sijm, D.T.H.M. and Opperhuizen, A., Biotransformation of organic chemicals by fish: a review of enzyme activities and reactions, in *Handbook of Environmental Chemistry, Vol. 2, Part E: Reactions and Processes*, Hutzinger, O., Ed., Springer-Verlag, Heidelberg, 1989, 2, 163.
68. Walker, C.H. and Ronis, M.J.J., The monooxygenases of birds, reptiles, and amphibians, *Xenobiotica*, 19, 1111, 1989.
69. White, R.D. and Stegeman, J.J., unpublished results.
70. Ronis, M.J.J. and Walker, C.H., The microsomal monooxygenases of birds, *Rev. Biochem. Toxicol.*, 10, 1989.
71. James, M.O., Isolation of cytochrome P450 from hepatopancreas microsomes of the spiny lobster, *Panulirus argus*, and determination of catalytic activity with NADPH cytochrome P450 reductase from vertebrate liver, *Arch. Biochem. Biophys.*, 282, 8, 1990.
72. Nebert, D.W. and Gelboin, H.V., Substrate-inducible microsomal aryl hydrocarbon hydroxylase in mammalian cell culture. I. Assay and properties of induced enzyme, *J. Biol. Chem.*, 243, 6242, 1968.
73. Payne, J.F. and Penrose, W.P., Induction of aryl hydrocarbon hydroxylase in fish by petroleum, *Bull. Environ. Contam. Toxicol.*, 14, 112, 1975.
74. Elcombe, C.R. and Lech, J.J., Induction and characterization of hemoproteins P-450 and monooxygenation in rainbow trout, *Toxicol. Appl. Pharmacol.*, 49, 437, 1979.
75. Burke, M.D. and Mayer, R.T., Differential effects of phenobarbitone and 3-methylcholanthrene induction on the hepatic microsomal metabolism and cytochrome P450-binding of phenoxazone and a homologous series of its n-alkyl ethers (alkoxyresorufins), *Chem.-Biol. Interact.*, 45, 243, 1983.

76. Burke, M.D., Thompson, S., Elcombe, C.R., Halpert, J., Haaparanta, T., and Mayer, R.T., Ethoxy-, pentoxy- and benzyloxyphenones and homologues: a series of substrates to distinguish between different induced cytochromes P-450, *Biochem. Pharmacol.*, 34, 3337, 1985.
77. Addison, R.F., Sadler, M.C., and Lubet, R.A., Absence of hepatic microsomal pentyl- or benzyl-resorufin O-dealkylase induction in rainbow trout (Salmo gairdneri) treated with phenobarbitone, *Biochem. Pharmacol.*, 36, 1183, 1987.
78. Ankley, G.T., Blazer, V.S., Reinert, R.E., and Agosin, M., Effects of Aroclor 1254 on cytochrome P-450-dependent monooxygenase, glutathione-S-transferase, and UDP-glucuronosyltransferase activities in channel catfish liver, *Aquat. Toxicol.*, 9, 91, 1986.
79. Lindstrom-Seppa, P. and Oikari, A., Bleached kraft pulpmill effluents and TCDD: effects on biotransformation and testerone status of mature male rainbow trout *(Salmo gairdneri), Mar. Environ. Res.*, 28, 99, 1989.
80. Dannan, G.A., Porubek, D.J., Nelson, S.D., Waxman, D.J., and Guengerich, F.P., 17β-estradiol 2- and 4-hydroxylation catalyzed by rat hepatic cytochrome P450: roles of individual forms, inductive effects, developmental patterns, and alterations by gonadectomy and hormone replacement, *Endocrinology*, 118, 1952, 1986.
80a. Spink, D.C., Eugster, H.-P., Lincoln, D.W., Schuetz, J.D., Schuetz, E.G., Johnson, J.A., Kaminsky, L.S., and Gierthy, J.F., 17β-estradiol hydroxylation catalyzed by human cytochrome P4501A1: a comparison of the activities induced by 2,3,7,8-tetrachlorodibenzo-p-dioxin in MCF-7 cells with those from heterologous expression of the cDNA, *Arch. Biochem. Biophys.*, 293, 342, 1992.
81. Snowberger, E.A. and Stegeman, J.J., Patterns and regulation of estradiol metabolism by hepatic microsomes from two species of marine teleosts, *Gen. Comp. Endocrinol.*, 66, 256, 1987.
82. Klotz, A.V., Stegeman, J.J., and Walsh, C., An aryl hydrocarbon hydroxylating hepatic cytochrome P-450 from the marine fish *Stenotomus chrysops, Arch. Biochem. Biophys.*, 226, 578, 1983.
83. Williams, D.E. and Buhler, D.R., Benzo[a]pyrene hydroxylase catalyzed by purified isozymes of cytochrome P-450 from β-naphthoflavone-fed rainbow trout, *Biochem. Pharmacol.*, 33, 3743, 1984.
84. Livingstone, D.R. and Farrar, S.V., Tissue and subcellular distribution of enzyme activities of mixed-function oxygenase and Benzo[a]pyrene metabolism in the common mussel *Mytilus edulis* L., *Sci. Total Environ.*, 39, 209, 1984.
85. Stegeman, J.J., Benzo[a]pyrene oxidation and microsomal enzyme activity in the mussel *(Mytilus edulis)* and other bivalve mollusc species from the western N. Atlantic, *Mar. Biol.*, 89, 21, 1985.
86. Dus, K.M., Insights into the active site of the cytochrome P450 haemoprotein family — a unifying concept based on structural considerations, *Xenobiotica*, 12, 745, 1982.
87. Koop, D.R., Oxidative and reductive metabolism by cytochrome P450 2E1, *FASEB J.*, 6, 724, 1992.
88. Smith, D.A. and Jones, B.C., Speculations on the substrate structure-activity relationship (SSAR) of cytochrome P450 enzymes, *Biochem. Pharmacol.*, 44, 2080, 1992.
89. Murray, M. and Reidy, G.F., Selectivity in the inhibition of mammalian cytochromes P-450 by chemical agents, *Pharmacol. Rev.*, 42, 85, 1990.

90. Bend, J.R., Pohl, R.J., and Fouts, J.R., Further studies of the microsomal mixed-function oxidase system of the little skate, *Raja erinacea*, including its response to some xenobiotics, *Bull. Mt. Desert Isl. Biol. Lab.*, 13, 9, 1973.
91. Stegeman, J.J. and Woodin, B.R., The metabolism of α-naphthoflavone (7,8 benzoflavone) by hepatic microsomes from the marine fish scup (*Stenotomus chrysops*), *Biochem. Biophys. Res. Commun.*, 95, 328, 1980.
92. Ortiz de Montellano, P.R., Ed., *Cytochrome P-450: Structure, Mechanism and Biochemistry*, Plenum Press, New York, 1986, 556.
93. Neal, R.A., Metabolism and mechanisms of toxicity of compounds containing thiono-sulfur, in *The Scientific Basis of Toxicity Assessment*, Witschi, H., Ed., Elsevier/North-Holland, Amsterdam, 241, 1980.
94. Chevion, M., Stegeman, J.J., Peisach, J., and Blumberg, W.E., Electron paramagnetic resonance studies on the hepatic microsomal cytochrome P-450 from a marine teleost, *Life Sci.*, 20, 895, 1977.
95. Kleinow, K.M., Haasch, M.L., and Lech, J.J., The effect of tricaine anesthesia upon induction of select P-450 dependent monooxygenase activities in rainbow trout (Salmo gairdneri), *Aquat. Toxicol.*, 8, 231, 1986.
96. Goddard, K.A., Schultz, R.J., and Stegeman, J.J., Uptake, toxicity, and distribution of benzo[a]pyrene and monooxygenase induction in the topminnows *Poeciliopsis monacha* and *Poeciliopsis lucida*, *Drug Metab. Dispos.*, 15, 449, 1987.
97. Haasch, M.L., Quardokus, E.M., Sutherland, L.A., Goodrich, M.S., and Lech, J.J., Hepatic CYP1A1 induction in rainbow trout by continuous flowthrough exposure to β-naphthoflavone, *Fundam. Appl. Toxicol.*, 20, 72, 1993.
98. Gooch, J.W., Elskus, A.A., Kloepper-Sams, P.J., Hahn, M.E., and Stegeman, J.J., Effects of *ortho* and non-*ortho* substituted polychlorinated biphenyl congeners on the hepatic monooxygenase system in scup *(Stenotomus chrysops)*, *Toxicol. Appl. Pharmacol.*, 98, 422, 1989.
99. Hahn, M.E., Lamb, T.M., Schultz, M.E., Smolowitz, R.M., and Stegeman, J.J., Cytochrome P4501A induction and inhibition by 3,3',4,4'-tetrachlorobiphenyl in an Ah receptor-containing fish hepatoma cell line (PLHC-1), *Aquat. Toxicol.*, 26, 185, 1993.
100. Miranda, C.L., Wang, J.-L., Chang, H.-S., and Buhler, D.R., Multiple effects of 3,4,5,3',4',5'-hexachlorobiphenyl administration on hepatic cytochrome P450 isozymes and associated mixed-function oxidase activities in rainbow trout, *Biochem. Pharmacol.*, 40, 387, 1990.
101. Monosson, E. and Stegeman, J.J., Cytochrome P450E (P450IA) induction and inhibition in winter flounder by 3,3',4,4'-tetrachlorobiphenyl: comparison of response in fish from Georges Bank and Narragansett Bay, *Environ. Toxicol. Chem.*, 10, 765, 1991.
102. Ahokas, J.T., Karki, N.T., Oikari, A., and Soivio, A., Mixed function monooxygenase of fish as an indicator of pollution of aquatic environment by industrial effluent, *Bull. Environ. Contam. Toxicol.*, 16, 270, 1976.
103. Fent, K. and Stegeman, J.J., Effects of tributyltin chloride on the hepatic microsomal monooxygenase system in the fish *Stenotomus chrysops*, *Aquat. Toxicol.*, 20, 159, 1991.
104. Fent, K. and Stegeman, J.J., Effects of tributyltin in vivo on hepatic cytochrome P450 forms in marine fish, *Aquat. Toxicol.*, 24, 219, 1993.

105. Maines, M.D., New developments in the regulation of heme metabolism and their implications, *CRC Crit. Rev. Toxicol.*, 12, 241, 1982.
106. Maines, M.D. and Kappas, A., Metals as regulators of heme metabolism, *Science*, 198, 1215, 1977.
106a. Forlin, L.C., Haux, C., Karlsson-Norrgren, L., Runn, P., and Larsson, A., Biotransformation enzyme activities and histopathology in rainbow trout *(Salmo gairdneri)* treated with cadmium, *Aquat. Toxicol.*, 8, 51, 1986.
107. George, S.G. and Young, P., The time course of effects of cadmium and 3-methylcholanthrene on activities of enzymes of xenobiotic metabolism and metallothionein levels in the plaice, *Pleuronectes platessa, Comp. Biochem. Physiol.*, 83C, 37, 1986.
108. George, S.G., Cadmium effects on plaice liver xenobiotic and metal detoxication systems: dose-response, *Aquat. Toxicol.*, 15, 303, 1989.
109. den Besten, P.J., Elenbaas, J.M.L., Maas, J.R., Dieleman, S.J., Herwig, H.J., and Voogt, P.A., Effects of cadmium and polychlorinated biphenyls (Clophen A50) on steroid metabolism and cytochrome P-450 monooxygenase system in the sea star Asterias rubens L., *Aquat. Toxicol.*, 20, 95, 1991.
110. Lu, P.-Y., Metcalf, R.L., Plummer, N., and Mandel, D., The environmental fate of three carcinogens: benzo(a)pyrene, benzidene, and vinyl chloride evaluated in laboratory model ecosystems, *Arch. Environ. Contam. Toxicol.*, 6, 129, 1977.
111. Lee, R.F., Metabolism of tributyltin by marine animals and possible linkages to effects, *Mar. Environ. Res.*, 32, 29, 1991.
112. Gibbs, P.E., Pascoe, P.L., and Bryan, G.W., Tributyltin-induced imposex in stenoglossan gastropods: pathological effects on the female reproductive system, *Comp. Biochem. Physiol.*, 100C, 231, 1991.
113. Elbrecht, A. and Smith, R.G., Aromatase enzyme activity and sex determination in chickens, *Science*, 255, 467, 1992.
114. Lance, V.A. and Bogart, M.H., Disruption of ovarian development in alligator embryos treated with an aromatase inhibitor, *Gen. Comp. Endocrinol.*, 86, 59, 1992.
115. Gasiewicz, T.A. and Rucci, G., α-Naphthoflavone acts as an antagonist of 2,3,7,8-tetrachlorodibenzo-*p*-dioxin by forming an inactive complex with the Ah receptor, *Mol. Pharmacol.*, 40, 607, 1991.
116. Chevion, M., Peisach, J., and Blumberg, W.E., Imidazole, the ligand *trans* to mercaptide in ferric cytochrome P450: an EPR study of model compounds, *J. Biol. Chem.*, 252, 3637, 1977.
117. Stern, J.O. and Peisach, J., A model compound study of the CO-adduct of cytochrome P450, *J. Biol. Chem.*, 249, 7495, 1974.
118. Wink, D.A., Osawa, Y., Darbyshire, J.F., Jones, C.R., Eshenaur, S.C., and Nims, R.W., Inhibition of cytochromes P450 by nitric oxide and a nitric oxide-releasing agent, *Arch. Biochem. Biophys.*, 300, 115, 1993.
119. Maines, M.D., Heme oxygenase: function, multiplicity, regulatory mechanisms, and clinical applications, *FASEB J.*, 2, 2557, 1988.
120. Bredt, D.S. and Snyder, S.H., Isolation of nitric oxide synthetase, a calmodulin-requiring enzyme, *Proc. Natl. Acad. Sci. U.S.A.*, 87, 682, 1990.
121. Langenbach, R., Smith, P.B., and Crespi, C., Recombinant DNA approaches for the development of metabolic systems used in in vitro toxicology, *Mutat. Res.*, 277, 251, 1992.

122. Takahashi, M., Tanaka, M., Sakai, N., Adachi, S., Miller, W.L., and Nagahama, Y., Rainbow trout ovarian cholesterol side-chain cleavage cytochrome-P450 (P450scc) — cDNA cloning and messenger RNA expression during oogenesis, *FEBS Lett.*, 319, 45, 1993.
123. Tanaka, M., Telecky, T.M., Fukuda, S., Adachi, S., and Chen, S., Cloning and sequence analysis of the cDNA encoding P-450 aromatase (P450arom) from a rainbow trout *(Oncorhynchus mykiss)* ovary: relationship between the amount of P450arom mRNA and the production of oestradiol-17β in the ovary, *J. Mol. Endocrinol.*, 8, 53, 1992.
124. Sakai, N., Tanaka, M., Adachi, S., Miller, W.L., and Nagahama, Y., Rainbow trout cytochrome — P-450c17 (17-alpha-Hydroxylase/17,20-Lyase) — cDNA cloning, enzymatic properties and temporal pattern of ovarian P-450c17 messenger RNA expression during oogenesis, *FEBS Lett.*, 301, 60, 1992.
125. Stegeman, J.J. and Kloepper-Sams, P.J., Cytochrome P-450 isozymes and monooxygenase activity in aquatic animals, *Environ. Health Perspect.*, 71, 87, 1987.
126. Stegeman, J.J., Cytochrome P450 forms in fish, in *Cytochrome P450,* Handbook of Experimental Pharmacology, Vol. 105, Schenkman, J.B. and Greim, H., Eds., Springer-Verlag, Berlin, 1993, 105.
127. Miranda, C.L., Wang, J.-L., Henderson, M.C., and Buhler, D.R., Purification and characterization of hepatic steroid hydroxylases from untreated rainbow trout, *Arch. Biochem. Biophys.*, 268, 227, 1989.
128. Arinc, E. and Adali, O., Solubilization and partial purification of two forms of cytochrome P-450 from trout liver microsomes, *Comp. Biochem. Physiol. B*, 76B, 653, 1983.
129. Celander, M., Ronis, M., and Forlin, L., Initial characterization of a constitutive cytochrome P-450 isoenzyme in rainbow trout liver, *Mar. Environ. Res.*, 28, 9, 1989.
130. Celander, M. and Forlin, L., Catalytic activity and immunochemical quantification of hepatic cytochrome P-450 in beta-naphthoflavone and isosafrol treated rainbow trout *(Oncorhynchus mykiss), Fish Physiol. Biochem.*, 9, 189, 1991.
131. Klotz, A.V., Stegeman, J.J., Woodin, B.R., Snowberger, E.A., Thomas, P.E., and Walsh, C., Cytochrome P-450 isozymes from the marine teleost *Stenotomus chrysops*: their roles in steroid hydroxylation and the influence of cytochrome b_5, *Arch. Biochem. Biophys.*, 249, 326, 1986.
132. Goksøyr, A., Purification of hepatic microsomal cytochromes P-450 from β-naphthoflavone-treated Atlantic cod *(Gadus morhua)*, a marine teleost fish, *Biochim. Biophys. Acta*, 840, 409, 1985.
133. Zhang, Y.S., Goksøyr, A., Andersson, T., and Førlin, L., Initial purification and characterization of hepatic microsomal cytochrome P-450 from BNF-treated perch (Perca fluviatilis), *Comp. Biochem. Physiol.*, 98B, 97, 1991.
134. Ball, L.M., Elmamlouk, T.H., and Bend, J.R., Metabolism of benzo[a]pyrene in little skate mixed-function oxidase systems, in *Microsomes, Drug Oxidations, and Chemical Carcinogenesis, Vol. II,* Coon, M.J., Conney, A.H., Estabrook, R.W., Gelboin, H.V., Gillette, J.R., and O'Brien, P.J., Eds., Academic Press, New York, 1203, 1980.
135. Andersson, T., Purification, characterization and regulation of a male-specific cytochrome P450 in the rainbow trout kidney, *Mar. Environ. Res.*, 34, 109, 1992.

136. Gøksoyr, A., Andersson, T., Buhler, D.R., Stegeman, J.J., Williams, D.E., and Forlin, L., Immunochemical cross-reactivity of β-naphthoflavone-inducible cytochrome P450 (P450IA) in liver microsomes from different fish species and rat, *Fish Physiol. Biochem.*, 9, 1, 1991.
137. George, S., Personal communication, 1992.
138. Miranda, C.L., Wang, J.-L., Henderson, M.C., and Buhler, D.R., Immunological characterization of constitutive isozymes of cytochrome P-450 from rainbow trout. Evidence for homology with phenobarbital-induced rat P-450s, *Biochim. Biophys. Acta*, 1037, 155, 1990.
139. Stegeman, J.J., Woodin, B.R., and Waxman, D.J., Structural relatedness of mammalian cytochromes P450 IIB and cytochrome P450B from the marine fish scup *(Stenotomus chrysops), FASEB J.*, 4, A739, 1990.
140. Stegeman, J.J., Woodin, B.R., and Smolowitz, R.M., Structure, function and regulation of cytochrome P-450 forms in fish, *Biochem. Soc. Trans.*, 18, 19, 1990.
141. Haasch, M.L., Wejksnora, P.J., and Lech, J.J., Molecular aspects of cytochrome P450 induction in rainbow trout, *Toxicologist*, 10, 322, 1990.
142. Kleinow, K.M., Haasch, M.L., Williams, D.E., and Lech, J.J., A comparison of hepatic P450 induction in rat and trout *(Oncorhynchus mykiss)*: delineation of the site of resistance of fish to phenobarbital-type inducers, *Comp. Biochem. Physiol. C*, 96C, 259, 1990.
143. Stegeman, J.J., Klotz, A.V., Woodin, B.R., and Pajor, A.M., Induction of hepatic cytochrome P-450 in fish and the indication of environmental induction in scup, *Aquat. Toxicol.*, 1, 197, 1981.
144. Nebert, D.W. and Gonzalez, F.J., P450 genes: structure, evolution, and regulation, *Annu. Rev. Biochem.*, 56, 945, 1987.
145. Kaplan, L.A.E., Schultz, M.E., Schultz, R.J., and Crivello, J.F., Nitrosodiethylamine metabolism in the viviparous fish Poeciliopsis: evidence for the existence of liver P450pj activity and expression, *Carcinogenesis*, 12, 647, 1991.
146. Halvorson, M., Greenway, D., Eberhart, D., and Fitzgerald, K., Reconstitution of testosterone oxidation by purified rat cytochrome P450p (IIIA1), *Arch. Biochem. Biophys.*, 277, 166, 1990.
147. Miranda, C.L., Wang, J.-L., Henderson, M.C., Zhao, X., Guengerich, F.P., and Buhler, D.R., Comparison of rainbow trout and mammalian cytochrome P450 enzymes: evidence for structural similarity between trout P450 LMC5 and human P450IIIA4, *Biochem. Biophys. Res. Commun.*, 176, 558, 1991.
148. Jaiswal, A.K., Gonzalez, F., and Nebert, D.W., Human dioxin-inducible cytochrome P_1-450: complementary DNA and amino acid sequence, *Science*, 228, 80, 1985.
149. Fagan, J.B., Pastewka, J.V., Chalberg, S.C., Gozukara, E., Guengerich, F.P., and Gelboin, H.V., Noncoordinate regulation of the mRNAs encoding cytochromes P-450 BNF/MC-3 and ISF/BNF-G, *Arch. Biochem. Biophys.*, 244, 261, 1986.
150. Celander, M. and Forlin, L., Quantification of cytochrome P4501A1 and catalytic activities in liver microsomes of isosafrole- and β-naphthoflavone-treated rainbow trout *(Oncorhynchus mykiss), Mar. Environ. Res.*, 34, 123, 1992.
151. Wirgin, I., Kreamer, G.-L., and Garte, S.J., Genetic polymorphism of cytochrome P-450IA in cancer-prone Hudson River tomcod, *Aquat. Toxicol.*, 19, 205, 1991.
152. Buhler, D., Personal communication, 1992.

153. Schwen, R.J. and Mannering, G.J., Hepatic cytochrome P-450-dependent monooxygenase system of the trout, frog and snake. III. Induction, *Comp. Biochem. Physiol.*, 71B, 445, 1982.
154. Busbee, D., Colvin, D., Muijsson, I., Rose, F., and Cantrell, E., Induction of aryl hydrocarbon hydroxylase in *Ambystoma tigrinum*, *Comp. Biochem. Physiol.*, 50C, 33, 1975.
155. Noshiro, M. and Omura, T., Microsomal monoxygenase system in frog livers, *Comp. Biochem. Physiol.*, 77B, 761, 1984.
156. Marty, J., Lesca, P., Jaylet, A., Ardourel, C., and Riviere, J.L., In vivo and in vitro metabolism of benzo(a)pyrene by the larva of the newt, Pleurodeles waltl, *Comp. Biochem. Physiol.*, 93C, 213, 1989.
157. Marty, J., Riviere, J.L., Guinaudy, M.J., Kremers, P., and Lesca, P., Induction and characterization of cytochromes P450IA and -IIB in the newt, *Pleurodeles waltl, Ecotoxicol. Environ. Saf.*, 24, 144, 1992.
158. Jewell, C.S.E., Cummings, L.E., Ronis, M.J.J., and Winston, G.W., The hepatic microsomal mixed-function oxygenase (MFO) system of *Alligator mississippiensis*: induction by 3-methylcholanthrene (3-MC), *Xenobiotica*, 19, 1181, 1989.
159. Ronis, M.J.J., Andersson, T., Hansson, T., and Walker, C.H., Differential expression of multiple forms of cytochrome P-450 in vertebrates: antibodies to purified rat cytochrome P-450 as molecular probes for the evolution of P-450 gene families I and II, *Mar. Environ. Res.*, 28, 131, 1989.
160. Yawetz, A., Woodin, B.R., Smolowitz, R.M., and Stegeman, J.J., Induction, fractionation and localization of cytochrome P450 isozymes in the liver of the fresh water turtle, *Chrysemes picta picta, Mar. Environ. Res.*, 35, 205, 1993.
161. Schuetz, E.G., Schuetz, J.D., Grogan, W.M., Narayfejestoth, A., Fejestoth, G., Raucy, J., Guzelian, P., Gionela, K., and Watlington, C.O., Expression of cytochrome-P450 3A in amphibian, rat, and human kidney, *Arch. Biochem. Biophys.*, 294, 206, 1992.
162. Winston, G., Personal communication, 1992.
163. Duncan, R., Grogan, W., Kramer, L., and Watlington, C.O., Corticosterone's metabolite is an agonist for Na+ transport stimulation in A6 cells, *Am. J. Physiol.*, 255, F736, 1988.
164. Engelhardt, F., Hydrocarbon metabolism and cortisol balance in oil-exposed ringed seals, Phoca hispida, *Comp. Biochem. Physiol.*, 72C, 133, 1982.
165. Addison, R. and Brodie, P., Characterization of ethoxyresorufin O-deethylase in grey seal *Halichoerus grypus, Comp. Biochem. Physiol.*, 79C, 261, 1984.
166. Addison, R., Brodie, P., Edwards, A., and Sadler, M., Mixed function oxidase activity in the harbour seal *(Phoca vitulina)* from Sable Is., N.S., *Comp. Biochem. Physiol.*, 121, 1986.
167. Hahn, M.E., Steiger, G.H., Calambokidas, J., Shaw, S.D., and Stegeman, J.J., Immunochemical characterization of cytochrome P450 in harbor seals *(Phoca vitulina)*, Ninth Biennial Conference on the Biology of Marine Mammals, (Abstract), 1991, 30.
168. Goksøyr, A., Beyer, J., Larsen, H.E., Andersson, T., and Forlin, L., Cytochrome P450 in seals: monooxygenase activities, immunochemical cross-reactions and response to phenobarbital treatment, *Mar. Environ. Res.*, 34, 113, 1992.
169. White, R.D., Hahn, M.E., Lockhart, W.L., and Stegeman, J.J., Catalytic and immunochemical characterization of hepatic microsomal cytochromes P450 in beluga whales *(Delphinapterus leucas), Toxicol. Appl. Pharmacol.*, in press, 1993.

170. Watanabe, S., Shimada, T., Nakamura, S., Nishiyama, N., Yamashita, N., Tanabe, S., and Tatsukawa, R., Specific profile of liver microsomal cytochrome P-450 in dolphin and whales, *Mar. Environ. Res.*, 27, 51, 1989.
171. Goksøyr, A., Solbakken, J., Tarlebø, J., and Klungsøyr, J., Initial characterization of the hepatic microsomal cytochrome P-450-system of the piked whale (minke) *Balaenoptera acutorostrata, Mar. Environ. Res.*, 19, 185, 1986.
172. Goksøyr, A., Andersson, T., Forlin, L., Stenersen, J., Snowberger, E.A., Woodin, B.R., and Stegeman, J.J., Xenobiotic and steroid metabolism in adult and foetal piked (minke) whales, *Balaenoptera acutorostrata, Mar. Environ. Res.*, 24, 9, 1988.
173. Goksøyr, A., Andersson, T., Forlin, L., Snowberger, E.A., Woodin, B.R., and Stegeman, J.J., Cytochrome P-450 monooxygenase activity and immunochemical properties of adult and foetal piked whales, *Balaenoptera acutorostrata*, in *Cytochrome P-450, Biochemistry and Biophysics,* Schuster, I., Ed., Taylor & Francis, London, 1989, 698.
174. Tanabe, S., Watanabe, S., Kan, H., and Tatsukawa, R., Capacity and mode of PCB metabolism in small cetaceans, *Mar. Mamm. Sci.*, 4, 103, 1988.
175. Duinker, J.C., Hillebrand, M.T.J., Zeinstra, T., and Boon, J.P., Individual chlorinated biphenyls and pesticides in tissues of some cetacean species from the North Sea and the Atlantic Ocean; tissue distribution and biotransformation, *Aquat. Mammals*, 15, 95, 1989.
176. Norstrom, R.J., Muir, D.C.G., Ford, C.A., Simon, M., Macdonald, C.R., and Beland, P., Indications of P450 monooxygenase activities in beluga *(Delphinapterus leucas)* and narwhal *(Monodon monoceros)* from patterns of PCB, PCDD and PCDF accumulation, *Mar. Environ. Res.*, 34, 267, 1992.
177. Boon, J.P., Van Arnhem, E., Jansen, S., Kannan, N., Petrick, G., Schulz, D., Duinker, J.C., Reijnders, P.J.H., and Goksøyr, A., The toxicolkinetics of PCBs in marine mammals with special reference to possible interactions of individual congeners with the cytochrome P450-dependent monooxygenase system: an overview, in *Persistent Pollutants in Marine Ecosystems,* Walker, C.H. and Livingstone, D.R., Eds., Pergamon Press, New York, 1992, 119.
178. Hart, C.A. and Hahn, M.E., Unpublished results, 1993.
179. Stegeman, J.J., Unpublished results, 1993.
180. Nielsen, O., Kelly, R.K., Lillie, W.R., Clayton, J.W., Fujioka, R.S., and Yoneyama, B.S., Some properties of a finite cell line from beluga whale *(Delphinapterus leucas), Can. J. Fish. Aquat. Sci.*, 46, 1472, 1989.
181. Lee, R.F., Mixed function oxygenase (MFO) in marine invertebrates, *Mar. Biol. Lett.*, 2, 87, 1981.
182. den Besten, P.J., Herwig, H.J., van Donselaar, E.G., and Livingstone, D.R., Cytochrome P450 monooxygenase system and benzo(a)pyrene metabolism in echinoderms, *Mar. Biol.*, 107, 171, 1990.
183. James, M.O. and Shiverick, K.T., Cytochrome P-450-dependent oxidation of progesterone, testosterone, and ecdysone in the spiny lobster, *Panulirus argus, Arch. Biochem. Biophys.*, 233, 1, 1984.
184. Jewell, C.S.E. and Winston, G.W., Characterization of the microsomal mixed-function oxygenase system of the hepatopancreas and green gland of the red swamp crayfish, *Procambrius clarkii, Comp. Biochem. Physiol.*, 92B, 329, 1989.
185. Schlenk, D. and Buhler, D.R., Cytochrome P-450 and phase II activities in the gumboot chiton Cryptochiton stelleri, *Aquat. Toxicol.*, 13, 167, 1988.

186. Lindstrom-Seppa, P., Koivusaari, U., and Hanninen, O., Metabolism of foreign compounds in freshwater crayfish *(Astacus astacus* L.) tissues, *Aquat. Toxicol.*, 3, 35, 1983.
187. Quattrochi, L.C. and Lee, R.F., Microsomal cytochromes P-450 from marine crabs, *Comp. Biochem. Physiol.*, 79C, 171, 1984.
188. Quattrochi, L.C. and Lee, R.F., Purification and characterization of microsomal cytochrome P-450 from spider crabs, *Libinia* sp., *Mar. Environ. Res.*, 14, 399, 1984.
189. Batel, R., Bihari, N., and Zahn, R.K., Purification and characterization of a single form of cytochrome P450 from the spiny crab *Maja crispata, Comp. Biochem. Physiol. C*, 83C, 165, 1986.
190. Kirchin, M.A., Wiseman, A., and Livingstone, D.R., The partial purification of cytochrome P-450 from the digestive gland of the common mussel *Mytilus edulis, Biochem. Soc. Trans.*, 15, 1100, 1987.
191. Schlenk, D. and Buhler, D.R., Determination of multiple forms of cytochrome P-450 in microsomes from the digestive gland of *Cryptochiton stelleri, Biochem. Biophys. Res. Commun.*, 163, 476, 1989.
191a. den Besten, P.J., Lemaire, P., Livingstone, D.R., Woodin, B.R., Stegeman, J.J., Herwig, H.J., and Seinen, W., Time-course and dose-response of the apparent induction of the cytochrome P450 monooxygenase system of pyloric caeca microsomes of the female sea star *Asterias rubens* L. by benzo[a]pyrene and polychlorinated biphenyls, *Aquat. Toxicol.*, 26, 23, 1993.
192. Spry, J.A., Livingstone, D.R., Wiseman, A., Gibson, G.G., and Goldfarb, P.S., Cytochrome P-450 gene expression in the common mussel Mytilus edulis, *Biochem. Soc. Trans.*, 17, 1013, 1989.
193. Bradfield, J.Y., Lee, Y.-H., and Keeley, L.L., Cytochrome P450 family 4 in a cockroach: molecular cloning and regulation by hypertrehalosemic hormone, *Proc. Natl. Acad. Sci. U.S.A.*, 88, 4558, 1991.
194. Wootton, A.N., Herring, C., Spry, J.A., Wiseman, A., Livingstone, D.R., and Goldfarb, P.S., Evidence for the existence of cytochrome P450 gene families, CYP1A1, 3A, 4A1, 11A1 in the common mussel, *Mytilus edulis* L., in Seventh International Symposium on Responses of Marine Organisms to Pollutants. 1993.
195. Teunissen, Y., Geraerts, W.P.M., Heerikhuizen, H.V., Plants, R.J., and Joosse, J., Molecular cloning of a cDNA encoding a member of a novel cytochrome P450 family in the mollusc *Lymnaea stagnalis, J. Biochem.*, 112, 249, 1992.
196. James, M.O., Boyle, S.M., Trapido-Rosenthal, H.G., Carr, W.E.S., and Shiverick, K.T., Identification of a cytochrome P450 sequence in cDNA from the hepatopancreas of the Florida spiny lobster, *FASEB J.*, 7, A1201, 1993.
197. Singer, S.C. and Lee, R.F., Mixed function oxygenase activity in blue crab, *Callinectes sapidus:* tissue distribution and correlation with changes during molting and development, *Biol. Bull.*, 153, 377, 1977.
198. Kurelec, B., Britvic, S., Rijavec, M., Muller, W.E.G., and Zahn, R.K., Benzo(a)pyrene monooxygenase induction in marine fish — molecular response to oil pollution, *Mar. Biol.*, 44, 211, 1977.
199. Kloepper-Sams, P.J. and Stegeman, J.J., The temporal relationships between cytochrome P-450E protein content, catalytic activity and mRNA levels in the teleost *Fundulus heteroclitus* following treatment with β-naphthoflavone, *Arch. Biochem. Biophys.*, 268, 525, 1989.

200. Haasch, M.L., Kleinow, K.M., and Lech, J.J., Induction of cytochrome P-450 mRNA in rainbow trout: in vitro translation and immunodetection, *Toxicol. Appl. Pharmacol.*, 94, 246, 1988.
201. Kloepper-Sams, P.J., Molecular regulation of the induction of cytochrome P-450E in the estuarine fish *Fundulus heteroclitus*, Ph.D. thesis, Woods Hole Oceanographic Institution/Massachusetts Institute of Technology, Woods Hole, MA, 1989.
202. Parkinson, A., Thomas, P.E., Ryan, D.E., and Levin, W., The in vivo turnover of rat liver epoxide hydrolase and both the apoprotein and heme moieties of specific cytochrome P-450 isozymes, *Arch. Biochem. Biophys.*, 225, 216, 1983.
203. Hahn, M.E., Woodin, B.R., and Stegeman, J.J., Induction of cytochrome P450E (P450IA1) by 2,3,7,8-tetrachlorodibenzofuran (2,3,7,8-TCDF) in the marine fish scup *(Stenotomus chrysops)*, *Mar. Environ. Res.*, 28, 61, 1989.
204. Hahn, M.E. and Stegeman, J.J., Sustained induction of cytochrome P450IA1 mRNA, protein, and catalytic activity by 2,3,7,8-tetrachlorodibenzofuran (2,3,7,8-TCDF) in the marine teleost *Stenotomus chrysops*, *Toxicologist*, 10, 28, 1990.
204a. Hahn, M.E. and Stageman, J.J., Regulation of cytochrome P4501A1 in teleosts: sustained induction of CYP1A1 mRNA, protein and catalytic activity by 2,3,7,8-tetrachlorodibenz furan in the marine fish *Stenotomus chrysops*, *Tox. Appl. Pharm.*, submitted.
205. Pesonen, M., Goksøyr, A., and Andersson, T., Expression of P4501A1 in a primary culture of rainbow trout hepatocytes exposed to β-naphthoflavone or 2,3,7,8-tetrachlorodibenzo-p-dioxin, *Arch. Biochem. Biophys.*, 292, 228, 1992.
206. Evans, C.T., Corbin, C.J., Saunders, C.T., Merril, J.C., Simpson, E.R., and Mendelson, C.R., Regulation of estrogen biosynthesis by human adipose stromal cells: effects of dibutyryl cyclic AMP, epidermal growth factor, and phorbol ester on the synthesis of aromatase cytochrome P450, *J. Biol. Chem.*, 262, 6914, 1987.
207. Thomas, P. and Trant, J.M., Evidence that 17alpha-20beta-21-trihydroxy-4-pregnen-3-one is a maturation-inducing steroid in spotted seatrout, *Fish Physiol. Biochem.*, 7, 185, 1989.
208. Skaare, J.U., Jensen, E.G., Goksøyr, A., and Egaas, E., Response of xenobiotic metabolizing enzymes of rainbow trout (Oncorhynchus mykiss) to the mono-ortho substituted polychlorinated PCB congener 2,3',4,4',5-pentachlorobiphenyl, PCB-118, detected by enzyme activities and immunochemical methods, *Arch. Environ. Contam. Toxicol.*, 20, 349, 1991.
209. Landers, J.P. and Bunce, N.J., The *Ah* receptor and the mechanism of dioxin toxicity, *Biochem. J.*, 276, 273, 1991.
210. Poland, A., Glover, E., and Kende, A.S., Stereospecific, high-affinity binding of 2,3,7,8-tetrachlorodibenzo-p-dioxin by hepatic cytosol, *J. Biol. Chem.*, 251, 4936, 1976.
211. Gasiewicz, T.A. and Rucci, G., Cytosolic receptor for 2,3,7,8-tetrachlorodibenzo-p-dioxin. Evidence for a homologous nature among various mammalian species, *Mol. Pharmacol.*, 26, 90, 1984.
212. Denison, M.S. and Wilkinson, C.F., Identification of the Ah receptor in selected mammalian species and induction of aryl hydrocarbon hydroxylase, *Eur. J. Biochem.*, 147, 429, 1985.
213. Denison, M.S., Hamilton, J.W., and Wilkinson, C.F., Comparative studies of aryl hydrocarbon hydroxylase and the Ah receptor in nonmammalian species, *Comp. Biochem. Physiol.*, 80C, 319, 1985.

214. Denison, M.S., Wilkinson, C.F., and Okey, A.B., Ah receptor for 2,3,7,8-tetrachlorodibenzo-p-dioxin: comparative studies in mammalian and nonmammalian species, *Chemosphere*, 15, 1665, 1986.
215. Lorenzen, A. and Okey, A.B., Detection and characterization of [^{3}H]2,3,7,8-tetrachlorodibenzo-*p*-dioxin binding to Ah receptor in a rainbow trout hepatoma cell line, *Toxicol. Appl. Pharmacol.*, 106, 53, 1990.
216. Swanson, H.I. and Perdew, G.H., Detection of the Ah receptor in rainbow trout. Use of 2-azido-3-[^{125}I]iodo-7,8-dibromodibenzo-p-dioxin in cell culture, *Toxicol. Lett.*, 58, 85, 1991.
217. Hahn, M.E., Poland, A., Glover, E., and Stegeman, J.J., The Ah receptor in marine animals: phylogenetic distribution and relationship to P4501A inducibility, *Mar. Environ. Res.*, 34, 87, 1992.
218. Hahn, M.E., Poland, A., Glover, E., and Stegeman, J.J., Photoaffinity labeling of the Ah receptor: phylogenetic survey of diverse vertebrate and invertebrate species, submitted, 1993.
219. Hahn, M.E. and Stegeman, J.J., Phylogenetic distribution of the Ah receptor in non-mammalian species: implications for dioxin toxicity and Ah receptor evolution, *Chemosphere*, 25, 931, 1992.
220. Poland, A. and Glover, E., Variation in the molecular mass of the Ah receptor among vertebrate species and strains of rats, *Biochem. Biophys. Res. Commun.*, 146, 1439, 1987.
221. Hahn, M.E., Unpublished results, 1992.
221a. Brown, D. and Van Beneden, R.J., Investigations of molecular mechanisms of gonadal tumors in herbicide-exposed bivalves, *Proc. Amer. Assoc. Cancer Res.*, 34, 168, 1993.
222. Ema, M., Sogawa, K., Watanabe, N., Chujoh, Y., Matsushita, N., Gotoh, O., Funae, Y., and Fuji-Kuriyama, Y., cDNA cloning and structure of mouse putative Ah receptor, *Biochem. Biophys. Res. Commun.*, 184, 246, 1992.
223. Burbach, K.M., Poland, A., and Bradfield, C.A., Cloning of the Ah receptor cDNA reveals a distinctive ligand-activated transcription factor, *Proc. Natl. Acad. Sci. U.S.A.*, 89, 8185, 1992.
224. Bigelow, S.W., Zijlstra, J.A., Vogel, E.W., and Nebert, D.W., Measurements of the cytosolic Ah receptor among four strains of *Drosophila melanogaster*, *Arch. Toxicol.*, 56, 219, 1985.
225. Muehleisen, D.P., Plapp, F.W., Benedict, J.H., and Carino, F.A., High affinity TCDD binding to fat body cytosolic protein of the bollworm, *Heliothis zea*, *Pest. Biochem. Physiol.*, 35, 50, 1989.
226. Borgmann, U., Norwood, W.P., and Ralph, K.M., Chronic toxicity and bioaccumulation of 2,5,2',5' and 3,4,3',4' tetrachlorobiphenyl and Aroclor 1242 in the amphipod *Hyallela azteca*, *Arch. Environ. Contam. Toxicol.*, 19, 558, 1990.
227. Dillon, T.M., Benson, W.H., Stackhouse, R.A., and Crider, A.M., Effects of selected PCB congeners on survival, growth, and reproduction in *Daphnia magna*, *Environ. Toxicol. Chem.*, 9, 1317, 1990.
228. Cook, P., Personal communication, 1992.
229. Lee, R.F., Singer, S.C., and Page, D.S., Response of cytochrome P-450 systems in marine crab and polychaetes to organic pollutants, *Aquat. Toxicol.*, 1, 355, 1981.
230. Reily, L.A., Means, J.C., Yan, Z.M., and Winston, G.W., Induction of ethoxyresorufin O-deethylase (EROD) in the sandworm *(Nereis virens)* exposed to petroleum-contaminated sediment, in Society of Environmantal Toxicology and Chemistry, 13th Annual Meeting, 1992.

231. Batel, R., Bihari, N., and Zahn, R.K., 3-Methylcholanthrene does induce mixed function oxidase activity in hepatopancreas of spiny crab *Maja crispata, Comp. Biochem. Physiol.*, 90C, 435, 1988.
232. Fries, C.R. and Lee, R.F., Pollutant effects on the mixed function oxygenase (MFO) and reproductive systems of the marine polychaete *Nereis virens, Mar. Biol.*, 79, 187, 1984.
233. Anderson, R.S., Benzo(a)pyrene metabolism in the American oyster *Crassostrea virginica*, U.S. Environmental Protection Agency, Washington, D.C., 1978, EPA-600/3-78-009.
234. James, M.O. and Little, P.J., 3-Methylcholanthrene does not induce *in vitro* xenobiotic metabolism in spiny lobster hepatopancreas, or affect *in vivo* disposition of benzo(a)pyrene, *Comp. Biochem. Physiol.*, 78C, 241, 1984.
235. Knutson, J.C. and Poland, A., Response of murine epidermis to 2,3,7,8-tetrachlorodibenzo-*p*-dioxin: interaction of the *Ah* and *hr* loci, *Cell*, 30, 225, 1982.
236. Newsted, J.L. and Giesey, J.P., The effects of 2,3,7,8-tetrachlorodibenzo-p-dioxin on epidermal growth factor binding and protein kinase activity in the RTH-149 rainbow trout hepatoma cell line, *Aquat. Toxicol.*, 23, 119, 1992.
237. Stegeman, J.J., Brouwer, M., DiGiulio, R.T., Forlin, L., Fowler, B.M., Sanders, B.M., and Van Veld, P., Molecular responses to environmental contamination: enzyme and protein systems as indicators of contaminant exposure and effect, in *Biomarkers, Biochemical, Physiological, and Histological Markers of Anthropogenic Stress*, Huggett, R.J., Kimerle, R.A., Mehrle, P.M., and Bergman, H.L., Eds., Lewis Publishers, Boca Raton, FL, 1992, 235.
238. Williams, D.E. and Buhler, D.R., Purified form of cytochrome P-450 from rainbow trout with high activity toward conversion of aflatoxin B1 to aflatoxin B1-2,3 epoxide, *Cancer Res.*, 43, 4752, 1983.
238a. Goeger, D.E., Shelton, D.W., Hendricks, J.D., Pereira, C., and Bailey, G.S., Comparative effect of dietary butylated hydroxyanisole and β-naphthoflavone on aflatoxin B1 metabolism, DNA adduct formation, and carcinogenesis in rainbow trout, *Carcinogenesis*, 9, 1793, 1988.
239. Gurtoo, H.L., Dahms, R.P., Kanter, P., and Vaught, J.B., Association and dissociation of the Ah locus with the metabolism of aflatoxin B_1 by mouse liver, *J. Biol. Chem.*, 253, 3952, 1978.
240. Koser, P., Faletto, M.B., Maccubbin, A.E., and Gurtoo, H.L., The genetics of aflatoxin B_1 metabolism. Association of the induction of aflatoxin B_1-4-hydroxylase with the transcriptional activation of cytochrome P_3-450 gene, *J. Biol. Chem.*, 263, 12584, 1988.
241. Jerina, D.M., Yagi, H., Thakker, D.R., Sayer, J.M., van Bladeren, P.J., Lehr, R.E., Whalen, D.L., Levin, W., Chang, R.L., Wood, A.W., and Conney, A.H., Identification of the ultimate carcinogenic metabolites of the polycyclic aromatic hydrocarbons: bay-region (R,S)-diol-(S,R)-epoxides, in *Foreign Compound Metabolism*, Caldwell, J. and Paulson, G., Eds., Taylor & Francis, London, 1984, 1, 257.
241a. Stegeman, J.J., Woodin, B.R., Park, S.S., Kloepper-Sams, P.J., and Gelboin, H.V., Microsomal cytochrome P-450 function in fish evaluated with polyclonal and monoclonal antibodies to cytochrome P-450E from scup (Stenotomus chrysops), *Mar. Environ. Res.*, 17, 83, 1985.
242. Gelboin, H.V., Benzo(a)pyrene metabolism, activation, and carcinogenesis: role and regulation of mixed-function oxidases and related enzymes, *Physiol. Rev.*, 60, 1107, 1980.

243. Levin, W., Wood, A., Lu, A., Ryan, D., West, S., Conney, A.H., Yhakker, D., Yagi, H., and Jerina, D., Role of purified cytochrome P448 and epoxide hydrase in the activation and detoxification of benzo(a)pyrene, in *American Chemical Society Symposium Series 44,* American Chemical Society, Washington, D.C., 1976.
244. Stegeman, J.J., Woodin, B.R., and Binder, R.L., Patterns of benzo[a]pyrene metabolism by varied species, organs, and developmental stages of fish, *J. Natl. Cancer Inst.,* 65, 371, 1984.
245. Zaleski, J., Steward, A.R., and Sikka, H., Comparative metabolism of (–)benzo(a)pyrene-7,8-dihydrodiol by hepatocytes isolated from two species of bottom dwelling fish, *Mar. Environ. Res.,* 28, 153, 1989.
246. Varanasi, U. and Gmur, D.J., Metabolic activation and covalent binding of benzo(a)pyrene to deoxyribonucleic acid catalyzed by liver enzymes of marine fish, *Biochem. Pharmacol.,* 29, 753, 1980.
247. Varanasi, U., Stein, J.E., Nishimoto, M., Reichert, W.L., and Collier, T.K., Chemical carcinogenesis in feral fish: uptake, activation, and detoxification of organic xenobiotics, *Environ. Health Perspect.,* 71, 155, 1987.
248. Varanasi, U., Reichert, W., and Stein, J., ^{32}P Postlabelling analysis of DNA adducts in liver of wild english sole *(Parophrys vetulus)* and winter flounder *(Pseudopleuronectes americanus), Cancer Res.,* 49, 1171, 1989.
249. Kurelec, B., Matijasevic, Z., Rijavec, M., Alecevic, M., Britvic, S., Rijavec, M., Muller, W.E.G., and Zahn, R.K., Induction of benzo(a)pyrene monooxygenase in fish and the Salmonella test as a tool for detecting mutagenic/carcinogenic xenobiotics in the aquatic environment, *Bull. Environ. Contam. Toxicol.,* 21, 799, 1979.
250. Stegeman, J.J., Skopek, T.R., and Thilly, W.G., Bioactivation of polynuclear aromatic hydrocarbons to cytotoxic and mutagenic products by marine fish, in Carcinogenic Polynuclear Aromatic Hydrocarbons in the Marine Environment, U.S. Environmental Protection Agency, Gulf Breeze, FL, 1982.
250a. Balk, L., DePierre, J.W., Sundvall, A., and Rannug, U., Formation of mutagenic metabolites from benzo(a)pyrene and 2-aminoanthracene by the S-9 fraction from the liver of the Northern pike *(Esox lucius):* inducibility with 3-methylcholanthrene and correlation with benzo(a)pyrene monooxygenase activity, *Chem.-Biol. Interact.,* 41, 1, 1982.
250b. Michel, X.R., Cassand, P.M., Ribera, D.G., and Narbonne, J.-F., Metabolism and mutagenic activation of benzo(a)pyrene by subcellular fractions from mussel *(Mytilus galloprovincialis)* digestive gland and sea bass *(Discenthrarcus labrax)* liver, *Comp. Biochem. Physiol.,* 103C, 43, 1992.
251. Kirchin, M.A. and Winston, G.W., Microsomal activation of benzo[a]pyrene to mutagens by *Alligator mississippiensis in vitro*: induction by 3-methylcholanthrene, *Mar. Environ. Res.,* 34, 273, 1992.
252. Marnett, L.J., Reed, J.A., and Dannison, D.J., Prostaglandin synthase dependent activation of 7,8-dihydro-7,8-dihydroxy-benzo(a)pyrene to mutagenic derivatives, *Biochem. Biophys. Res. Commun.,* 82, 210, 1978.
253. Farley, C.A., Plutschak, D.L., and Scott, R.F., Epizootiology and distribution of transmissible sarcoma in Maryland softshell clams, *Mya arenaria,* 1984–1988, *Environ. Health Perspect.,* 90, 35, 1991.
254. Spies, R.B. and Rice, D.W., Effects of organic contaminants on reproduction of the starry flounder Platichthys stellatus in San Francisco Bay. II. Reproductive success of fish captured in San Francisco Bay and spawned in the laboratory, *Mar. Biol.,* 98, 191, 1988.

255. Spies, R.B., D.W. Rice, J., and Felton, J., Effects of organic contaminants on reproduction of the starry flounder Platichthys stellatus in San Francisco Bay. I. Hepatic contamination and mixed-function oxidase (MFO) activity during the reproductive season, *Mar. Biol.*, 98, 181, 1988.
256. Johnson, L.L., Casillas, E., Collier, T.K., McCain, B.B., and Varanasi, U., Contaminant effects on ovarian development in English sole *(Parophyrs vetulus)* from Puget Sound, Washington, *Can. J. Fish. Aquat. Sci.*, 45, 2133, 1988.
257. Gibson, G.G., Comparative aspects of the mammalian P450 IV gene family, *Xenobiotica*, 19, 123, 1989.
258. Williams, D.E., Okita, R.T., Buhler, D.R., and Masters, B.S., Regiospecific hydroxylation of lauric acid at the (omega-1) position by hepatic and kidney microsomal cytochromes P-450 from rainbow trout, *Arch. Biochem. Biophys.*, 231, 503, 1984.
259. Nakai, K., Ward, A.M., Gannon, M., and Rifkind, A.B., β-Naphthoflavone induction of a cytochrome P-450 arachidonic acid epoxygenase in chick embryo liver distinct from the aryl hydrocarbon hydroxylase and from phenobarbital-induced arachidonate epoxygenase, *J. Biol. Chem.*, 267, 19503, 1992.
260. Hudson, L.G., Toscano, W.A., and Greenlee, W.F., 2,3,7,8-Tetrachlorodibenzo-p-dioxin (TCDD) modulates epidermal growth factor (EGF) binding to basal cells from a human keratinocyte cell line, *Toxicol. Appl. Pharmacol.*, 82, 481, 1986.
261. Williams, D.E., Carpenter, H.M., Buhler, D.R., Kelly, J.D., and Dutchuk, M., Alterations in lipid peroxidation, antioxidant enzymes, and carcinogen metabolism in liver microsomes of vitamin E-deficient trout and rat, *Toxicol. Appl. Pharmacol.*, 116, 78, 1992.
261a. Ankley, G.T., Blazer, V.S., Plakas, S.M., and Reinert, R.E., Dietary lipid as a factor modulating xenobiotic metabolism in channel catfish *(Ictalurus punctatus), Can. J. Fish. Aquat. Sci.*, 46, 1141, 1989.
262. Jeyasuria, P. and Place, A.R., The role of P450 aromatase in sex determination of the diamond-back terrapin, *Am. Zool.*, 32, 16, 1992.
263. Dewaide, J.H. and Henderson, P.T., Seasonal variation of hepatic drug metabolism in the roach, *Leuciscus rutilus* L., *Comp. Biochem. Physiol.*, 32, 489, 1970.
264. Williams, D.E., Becke, R.R., Potte, D.E., Guengerich, F.P., and Buhler, D.R., Purification and comparative properties of NADPH-cytochrome P450 reductase from rat and rainbow trout: differences in temperature optima between reconstituted and microsomal trout enzymes, *Arch, Biochem. Biophys.*, 225, 55, 1983.
265. Forlin, L., Andersson, T., Koivusaari, U., and Hansson, T., Influence of biological and environmental factors on hepatic steroid and xenobiotic metabolism in fish: interaction with PCB and β-naphthoflavone, *Mar. Environ. Res.*, 14, 47, 1984.
266. Stegeman, J.J., Temperature influence on basal activity and induction of mixed function oxygenase activity in *Fundulus heteroclitus, J. Fish. Res. Board Can.*, 36, 1400, 1979.
267. Kloepper-Sams, P.J. and Stegeman, J.J., The effect of temperature acclimation on the expression of cytochrome P4501A mRNA and protein in the fish *Fundulus heteroclitus, Arch. Biochem. Biophys.*, 299, 38, 1992.
268. Stegeman, J.J. and Chevion, M., Sex differences in cytochrome P-450 and mixed function oxidase activity in gonadally mature trout, *Biochem. Pharmacol.*, 28, 1686, 1980.

269. Hansson, T. and Gustafsson, J.-A., Sex differences in the hepatic metabolism of 4-androstene-3,17-dione in rainbow trout, Salmo gairdneri, *Gen. Comp. Endocrinol.*, 49, 490, 1981.
270. Stegeman, J.J. and Woodin, B.R., Differential regulation of hepatic xenobiotic and steroid metabolism in marine teleost species, *Mar. Environ. Res.*, 14, 422, 1984.
271. Stegeman, J.J., Pajor, A.M., and Thomas, P., Influence of estradiol and testosterone on cytochrome P-450 and monooxygenase activity in immature brook trout, *Salvelinus fontanalis, Biochem. Pharmacol.*, 31, 3979, 1982.
272. Hansson, T., Forlin, L., Rafter, J., and Gustafsson, J.-A., Regulation of hepatic steroid and xenobiotic metabolism in fish, in *Cytochrome P-450, Biochemistry, Biophysics and Environmental Implications,* Hietanen, E., Laitinen, M., and Hanninen, O., Eds., Elsevier Biomedical Press B.V., Amsterdam, 1982, 217.
273. Hansson, T., Androgenic regulation of hepatic metabolism of 4-androstene-3,17-dione in the rainbow trout, *Salmo gairdneri, J. Endocrinol.*, 92, 409, 1982.
274. Vodicnik, M.J. and Lech, J.J., The effect of sex steroids and pregnenolone-16a-carbonitrile on the microsomal monooxygenase system of rainbow trout *(Salmo gairdneri), J. Ster. Biochem.*, 18, 323, 1983.
275. Pajor, A.M., Stegeman, J.J., Thomas, P., and Woodin, B.R., Feminization of the hepatic microsomal cytochrome P-450 system in brook trout by estradiol, testosterone, and pituitary factors, *J. Exp. Zool.*, 253, 51, 1990.
276. Gray, E.S., Woodin, B.R., and Stegeman, J.J., Sex differences in hepatic monooxygenases in winter flounder *(Pseudopleuronectes americanus)* and scup *(Stenotomus chrysops)* and regulation of P450 forms by estradiol, *J. Exp. Zool.*, 259, 330, 1991.
277. Ho, S., Endocrinology of vitellogenesis, in *Hormones and Reproduction in Fishes, Amphibians, and Reptiles,* Norris, D.O. and Jones, R.E., Eds., Plenum Press, New York, 1987, 145.
278. Elskus, A.A., Pruell, R., and Stegeman, J.J., Endogenously-mediated, pretranslational suppression of cytochrome P4501A in PCB-contaminated flounder, *Mar. Environ. Res.*, 34, 97, 1992.
279. Lindstrom-Seppa, P. and Stegeman, J.J., Unpublished.
280. Williams, D., Master, B., Lech, J.J., and Buhler, D., Sex differences in cytochrome P-450 isozyme composition and activity in kidney microsomes of mature rainbow trout, *Biochem. Pharmacol.*, 35, 2017, 1986.
281. Hansson, T. and Gustafsson, J.-A., In vitro metabolism of 4-androstene-3,17-dione by hepatic microsomes from the rainbow trout *(Salmo gairdneri):* effects of hypophysectomy and oestradiol-17B, *J. Endocrinol.*, 90, 103, 1981.
282. Elskus, A.A., Polychlorinated biphenyl (PCB) effects, PCB congener distributions, and cytochrome P450 regulation in fish, Ph.D. dissertation, Boston University, MA, 1992.
283. Waxman, D.J., Pampori, N.A., Ram, P.A., Agrawal, A.K., and Shapiro, B.H., Interpulse interval in circulating growth hormone patterns regulates sexually dimorphic expression of hepatic cytochrome P450, *Proc. Natl. Acad. Sci. U.S.A.*, 88, 6868, 1991.
284. Devaux, A., Pesonen, M., Monod, G., and Andersson, T., Glucocorticoid-mediated potentiation of P450 induction in primary culture of rainbow trout hepatocytes, *Biochem. Pharmacol.*, 43, 898, 1992.

284a. Lee, P.C., Yoon, H.I., Haasch, M.L., and Lech, J.J., Negative control of cytochrome P450 1A1 (CYP1A1) by glucocorticoids in rainbow trout liver, *Compar. Biochem. Physiol.*, 104C, 457, 1993.
285. Mathis, J.M., Houser, W.H., Bresnick, E., Cidlowski, J.A., Hines, R.N., Prough, R.A., and Simpson, E.R., Glucocorticoid regulation of the rat cytochrome P450c (P450IA1) gene: receptor binding within intron 1, *Arch. Biochem. Biophys.*, 269, 93, 1989.
286. White, T.E.K. and Gasiewicz, T.A., The human estrogen receptor structural gene contains a DNA sequence that binds activated mouse and human Ah receptors: a possible mechanism of estrogen receptor regulation by 2,3,7,8-tetrachlorodibenzo-p-dioxin, *Biochem. Biophys. Res. Commun.*, 193, 956, 1993.
287. Gustafsson, J.-A. and Stenberg, A., Irreversible androgenic programming at birth of microsomal and soluble rat liver enzymes active on 4-androstene-dione-3,17-dione and 5a-androstane-3a,17b-diol, *J. Biol. Chem.*, 240, 711, 1974.
288. Gustafsson, J.-A., Mode, A., Norstedt, G., Hokfelt, T., Sonnenschein, C., Eneroth, P., and Skett, P., The hypothalamo-pituitary-liver axis: a new hormonal system in control of hepatic steroid and drug metabolism, *Biochem. Act. Horm.*, 7, 47, 1980.
289. Reinboth, R., Spontaneous and hormone induced sex-inversion in wrasses (Labridae), *Publ. Staz. Zool. Napoli, Suppl.*, 39, 550, 1975.
290. Stegeman, J.J. and James, M.O., Individual variations in patterns of benzo[a]pyrene metabolism in the marine fish scup (Stenotomus chrysops), *Mar. Environ. Res.*, 17, 122, 1985.
291. James, M.O. and Little, P.J., Polyhalogenated biphenyls and phenobarbital: evaluation as inducers of drug metabolizing enzymes in the sheepshead, *Archosargus probatocephalus, Chem.-Biol. Interact.*, 36, 229, 1981.
292. Kagimoto, M., Heim, M., Kagimoto, K., Zeugin, T., and Meyer, U.A., Multiple mutations of the human cytochrome P450IID6 gene (CYP2D6) in poor metabolizers of debrisoquine, *J. Biol. Chem.*, 265, 17209, 1990.
293. Kedzie, K.M., Balfour, C.A., Escobar, G.Y., Grimm, S.W., He, Y., Pepper, D.J., Regan, J.W., Stevens, J.C., and Halpert, J.R., Molecular basis for a functionally unique cytochrome IIB1 variant, *J. Biol. Chem.*, 266, 22515, 1991.
294. Weis, J.S. and Weis, P., Effects of environmental pollutants on early fish development, *Rev. Aquat. Sci.*, 1, 45, 1989.
295. Binder, R.L. and Stegeman, J.J., Basal levels and induction of hepatic aryl hydrocarbon hydroxylase activity during the embryonic period of development in brook trout, *Biochem. Pharmacol.*, 32, 1324, 1983.
296. Binder, R.L. and Stegeman, J.J., Induction of aryl hydrocarbon hydroxylase activity in embryos of an estuarine fish, *Biochem. Pharmacol.*, 29, 949, 1980.
297. Binder, R., Stegeman, J.J., and Lech, J., Induction of cytochrome P-450-dependent monooxygenase systems in embryos and eleutheroembryos of the killifish Fundulus heteroclitus, *Chem.-Biol. Interact.*, 55, 185, 1985.
298. Lauren, D.J., Okihiro, M.S., Hinton, D.E., and Stegeman, J.J., Localization of cytochrome P450 IA1 induced by β-naphthoflavone (BNF) in rainbow trout *(Oncorhynchus mykiss)* embryos, *FASEB J.*, 4, A739, 1990.
299. Guiney, P.D., Stegeman, J.J., Smolowitz, R.M., Walker, M.K., and Peterson, R.E., Localization of 2,3,7,8-tetrachlorodibenzo-p-dioxin (TCDD)-induced cytochrome P450 1A1 in vascular endothelium of early life stages (ELS) of lake trout, in Society of Environmantal Toxicology and Chemistry, 13th Annual Meeting, 1992.

300. Bend, J.R., Pohl, R.J., Davidson, N.P., and Fouts, J.R., Response of hepatic and renal microsomal mixed-function oxidases in the little skate, *Raja erinacea,* to pretreatment with 3-methylcholanthrene or TCDD (2,3,7,8-tetrachlorodibenzo-p-dioxin), *Bull. Mt. Des. Isl. Biol. Lab.,* 14, 7, 1974.
301. Stegeman, J.J., Binder, R.L., and Orren, A., Hepatic and extrahepatic microsomal electron transport components and mixed-function oxygenases in the marine fish *Stenotomus versicolor, Biochem. Pharmacol.,* 28, 3431, 1979.
301a. Lindström-Seppa, P., Koivusaari, U., and Hanninen, O., Extrahepatic xenobiotic metabolism in North-European freshwater fish, *Comp. Biochem. Physiol.,* 69C, 259, 1981.
302. Andersson, T. and Part, P., Benzo(a)pyrene metabolism in isolated perfused rainbow trout gills, *Mar. Environ. Res.,* 28, 3, 1989.
303. Van Veld, P.A., Westbrook, D.J., Woodin, B.R., Hale, R.C., Smith, C.L., Huggett, R.J., and Stegeman, J.J., Induced cytochrome P-450 in intestine and liver of spot *(Leiostomus xanthurus)* from a polycyclic aromatic hydrocarbon contaminated environment, *Aquat. Toxicol.,* 17, 119, 1990.
304. Stegeman, J.J., Smolowitz, R.M., and Hahn, M.E., Immunohistochemical localization of environmentally induced cytochrome P450IA1 in multiple organs of the marine teleost *Stenotomus chrysops* (scup), *Toxicol. Appl. Pharmacol.,* 110, 486, 1991.
305. Lorenzana, R.M., Hedstrom, O.R., and Buhler, D.R., Localization of cytochrome P-450 in the head and trunk kidney of rainbow trout *(Salmo gairdneri), Toxicol. Appl. Pharmacol.,* 96, 159, 1988.
306. Van Veld, P.A., Absorption and metabolism of dietary xenobiotics by the intestine of fish, *Rev. Aquat. Sci.,* 2, 185, 1990.
307. Van Veld, P.A., Patton, J.S., and Lee, R.F., Effect of preexposure to dietary benzo[a]pyrene (BaP) on the first-pass metabolism of BaP by the intestine of toadfish *(Opsamus tau): in vivo* studies using portal vein-catheterized fish, *Toxicol. Appl. Pharmacol.,* 92, 255, 1988.
308. Payne, J.F., Bauld, C., Dey, A.C., Kiceniuk, J.W., and Williams, U., Selectivity of mixed-function oxygenase enzyme induction in flounder *(Pseudopleuronectes americanus)* collected at the site of the Baie Verte, Newfoundland oil spill, *Comp. Biochem. Physiol.,* 79C, 15, 1984.
309. Pesonen, M. and Andersson, T., Subcellular localization and properties of cytochrome P-450 and UDP glucuronosyl transferase in the rainbow trout kidney, *Biochem. Pharmacol.,* 36, 823, 1987.
310. Pesonen, M., Celander, M., Førlin, L., and Andersson, T., Comparison of xenobiotic biotransformation enzymes in kidney and liver of rainbow trout *(Salmo gairdneri), Toxicol. Appl. Pharmacol.,* 91, 75, 1987.
311. Pesonen, M., Hansson, T., Førlin, L., and Andersson, T., Regional distribution of microsomal xenobiotic and steroid metabolism in kidney microsomes from rainbow trout, *Fish Physiol. Biochem.,* 8, 141, 1990.
312. Andersson, T. and Rafter, J., Progesterone metabolism in the microsomal fraction of the testis, head kidney, and trunk kidney from the rainbow trout, *Gen. Comp. Endocrinol.,* 79, 130, 1990.
313. Andersson, T., Sex differences in cytochrome P-450-dependent xenobiotic and steroid metabolism in the mature rainbow trout kidney, *J. Endocrinol.,* 126, 9, 1990.

313a. Balk, L., Maner, S., Bergstrand, A., Birberg, W., Pilotti, A., and DePierre, J.W., Preparation and characterization of subcellular fractions from the head kidney of the Northern pike *(Esox lucius)*, with particular emphasis on xenobiotic-metabolizing enzymes, *Biochem. Pharmacol.*, 34, 789, 1985.
314. Stegeman, J.J. and Binder, R.L., High benzo[a]pyrene hydroxylase activity in the marine fish *Stenotomus versicolor, Biochem. Pharmacol.*, 28, 1686, 1979.
315. Stegeman, J.J., Woodin, B.R., Klotz, A.V., Wolke, R.E., and Orme-Johnson, N.R., Cytochrome P-450 and monooxygenase activity in cardiac microsomes from the fish *Stenotomus chrysops, Mol. Pharmacol.*, 21, 517, 1982.
316. Stegeman, J.J., Miller, M.R., and Hinton, D.E., Cytochrome P450IA1 induction and localization in endothelium of vertebrate (teleost) heart, *Mol. Pharmacol.*, 36, 723, 1989.
317. Goksøyr, A., Andersson, T., Hansson, T., Klungsoyr, J., Zhang, Y., and Forlin, L., Species characteristics of the hepatic xenobiotic and steroid biotransformation systems of two teleost fish, Atlantic cod *(Gadus morhua)* and rainbow trout *(Salmo gairdneri), Toxicol. Appl. Pharmacol.*, 89, 347, 1987.
318. Miller, M.R., Hinton, D.E., Blair, J.J., and Stegeman, J.J., Immunohistochemical localization of cytochrome P450E in liver, gill and heart of scup *(Stenotomus chrysops)* and rainbow trout *(Salmo gairdneri), Mar. Environ. Res.*, 24, 37, 1988.
319. Miller, M.R., Hinton, D.E., and Stegeman, J.J., Cytochrome P-450E induction and localization in gill pillar (endothelial) cells of scup and rainbow trout, *Aquat. Toxicol.*, 14, 307, 1989.
320. Smolowitz, R.M., Hahn, M.E., and Stegeman, J.J., Immunohistochemical localization of cytochrome P450IA1 induced by 3,3′,4,4′-tetrachlorobiphenyl and by 2,3,7,8-tetrachlorodibenzofuran in liver and extrahepatic tissues of the teleost *Stenotomus chrysops* (scup), *Drug Metab. Dispos.*, 19, 113, 1991.
321. Smolowitz, R.M., Schultz, M.E., and Stegeman, J.J., Cytochrome P4501A induction in tissues, including olfactory epithelium, of topminnows (Poeciliopsis spp.) by waterborne benzo(a)pyrene, *Carcinogenesis*, 13, 2395, 1992.
322. Van Veld, P.A., Stegeman, J.J., Woodin, B.R., Patton, J.S., and Lee, R.F., Induction of monooxygenase activity in the intestine of spot *(Leiostomus xanthurus)*, a marine teleost, by dietary polycyclic aromatic hydrocarbons, *Drug Metab. Dispos.*, 16, 659, 1988.
323. Merchant, M., Wang, X., Kamps, C., Rosengren, R., Morrison, V., and Safe, S., Mechanism of benzo[a]pyrene-induced Cyp1a-1 gene expression in mouse hepa 1c1c7 cells: Role of the nuclear 6 s and 4 s proteins, *Arch. Biochem. Biophys.*, 292, 250, 1992.
324. Houser, W.H., Cunningham, C.K., Hines, R.N., Schaeffer, W.I., and Bresnick, E., Interaction of the 4S polycyclic aromatic hydrocarbon-binding protein with the cytochrome P-450c gene, *Arch. Biochem. Biophys.*, 259, 215, 1987.
325. Barton, H.A. and Marletta, M.A., Kinetic and immunochemical studies of a receptor-like protein that binds aromatic hydrocarbons, *J. Biol. Chem.*, 263, 5825, 1988.
326. Nef, P., Heldman, J., Lazard, D., Margalit, T., Jaye, M., Hanukoglu, I., and Lancet, D., Olfactory-specific cytochrome P-450: cDNA cloning of a novel neuroepithelial enzyme possibly involved in chemoreception, *J. Biol. Chem.*, 264, 6780, 1989.
327. Nef, P., Larabee, T.M., Kagimoto, K., and Meyer, U.A., Olfactory-specific cytochrome P-450 (P-450olf1; IIG1). Gene structure and developmental regulation, *J. Biol. Chem.*, 265, 2903, 1990.

328. Ding, X., Porter, T.D., Peng, H.-W., and Coon, M.J., cDNA and derived amino acid sequence of rabbit nasal cytochrome P450NMb (P450IIG1), a unique isozyme possibly involved in olfaction, *Arch. Biochem. Biophys.*, 285, 120, 1991.
329. Hargis, W.J. and Zwerner, D.E., Effects of certain contaminants on eyes of several estuarine fishes, *Mar. Environ. Res.*, 24, 265, 1988.
330. Andersson, T., Forlin, L., Olsen, S., Fostier, A., and Breton, B., Pituitary as a target organ for toxic effects of P4501A1 inducing chemicals, *Mol. Cell. Endocrinol.*, 91, 99, 1993.
331. Rifkind, A.B., Gannon, M., and Gross, S.S., Arachidonic acid metabolism by dioxin-induced cytochrome P-450: a new hypothesis on the role of P-450 in dioxin toxicity, *Biochem. Biophys. Res. Commun.*, 172, 1180, 1990.
332. Guiney, P.D., Walker, M.K., and Peterson, R.E., The edema in TCDD-exposed lake trout sac fry is an ultrafiltrate of blood, in Society of Environmental Toxicology and Chemistry, 11th Annual Meeting, 1990.
333. Walker, M.K., Spitsbergen, J.M., Olson, J.R., and Peterson, R.E., 2,3,7,8-Tetrachlorodibenzo-*p*-dioxin (TCDD) toxicity during early life stage development of lake trout *(Salvelinus namaycush), Can. J. Fish. Aquat. Sci.*, 48, 875, 1991.
334. Spitsbergen, J.M., Walker, M.K., Olson, J.R., and Peterson, R.E., Pathologic alterations in early life stages of lake trout, *Salvelinus namaycush,* exposed to 2,3,7,8-tetrachlorodibenzo-*p*-dioxin as fertilized eggs, *Aquat. Toxicol.*, 19, 41, 1991.
335. Harshbarger, J.C. and Clark, J.B., Epizootiology of neoplasms in bony fish of North America, *Sci. Total Environ.*, 94, 1990.
336. Vogelbein, W.K., Fournie, J.W., Veld, P.A.V., and Huggett, R.J., Hepatic neoplasms in the mummichog *Fundulus heteroclitus* from a creosote-contaminated site, *Cancer Res.*, 50, 5978, 1990.
337. Smolowitz, R.M., Moore, M.J., and Stegeman, J.J., Cellular distribution of cytochrome P-450E in winter flounder liver with degenerative and neoplastic disease, *Mar. Environ. Res.*, 28, 441, 1989.
338. Lester, S.M., Braunbeck, T.A., Teh, S.J., Stegeman, J.J., Miller, M.M., and Hinton, D.E., Hepatic cellular distribution of cytochrome P4501A1 in rainbow trout *Oncorhynchus mykiss*: An immunohisto- and cytochemical study, *Cancer Res.*, 53, 3700, 1993.
339. Hinton, D.E. and Pool, C.R., Ultrastructure of the liver in channel catfish *Ictalurus punctatus* (Rafinesque), *J. Fish. Biol.*, 8, 209, 1976.
340. Moore, M., Smolowitz, R., and Stegeman, J., Cellular alterations preceding neoplasia in *Pseudopleuronectes americanus* from Boston Harbor, *Mar. Environ. Res.*, 28, 425, 1989.
341. Roomi, M.W., Ho, R.K., Sarma, D.S.R., and Farber, E., A common biochemical pattern in preneoplastic hepatocyte nodules generated in four different models in the rat, *Cancer Res.*, 45, 564, 1985.
342. Van Veld, P.A., Vogelbein, W.K., Smolowitz, R., Woodin, B.R., and Stegeman, J.J., Cytochrome P4501A1 in hepatic lesions of a teleost fish *(Fundulus heteroclitus)* collected from a polycyclic aromatic hydrocarbon-contaminated site, *Carcinogenesis*, 13, 505, 1992.
343. Lorenzana, R.M., Hedstrom, O.R., Gallagher, J.A., and Buhler, D.R., Cytochrome P450 isozyme distribution in normal and tumor-bearing tissue from rainbow trout *(Salmo gairdneri), Exp. Mol. Pathol.*, 50, 348, 1989.

344. Parker, L.M., Lauren, D.J., Hammock, B.D., Winder, B., and Hinton, D.E., Biochemical and histochemical properties of hepatic tumors of rainbow trout, *Oncorhynchus mykiss, Carcinogenesis*, 14, 211, 1993.
345. Nebert, D.W. and Gelboin, H.V., The *in vivo* and *in vitro* induction of aryl hydrocarbon hydroxylase in mammalian cells of different species, tissues, strains, and developmental and hormonal states, *Arch. Biochem. Biophys.*, 134, 76, 1969.
346. Steward, A.R., Wrighton, S.A., Pasco, D.S., Fagan, J.B., Li, D., and Guzelian, P.S., Synthesis and degradation of 3-methylcholanthrene-inducible cytochromes P-450 and their mRNAs in primary monolayer cultures of adult rat hepatocytes, *Arch. Biochem. Biophys.*, 241, 494, 1985.
347. Schuetz, E.G., Li, D., Omiecinski, C.J., Muller-Eberhard, U., Kleinman, H.K., Elswick, B., and Guzelian, P.S., Regulation of gene expression in adult rat hepatocytes cultured on a basement membrane matrix, *J. Cell. Physiol.*, 134, 309, 1988.
348. Niwa, A., Kumaki, K., and Nebert, D.W., Induction of aryl hydrocarbon hydroxylase activity in various cell cultures by 2,3,7,8-tetrachlorodibenzo-p-dioxin, *Mol. Pharmacol.*, 11, 399, 1975.
349. Miller, M.R., Hinton, D.E., and Blair, J.B., Characterization of BNF-inducible cytochrome P450IA1 in cultures of rainbow trout liver cells, *Mar. Environ. Res.*, 28, 105, 1989.
350. Stegeman, J.J., Miller, M.J., Woodin, B.R., and Blair, J.D., Effect of β-naphthoflavone on multiple cytochrome P450 forms in primary cultures of rainbow trout hepatocytes, *Mar. Environ. Res.*, 35, 209, 1993.
351. Pesonen, M. and Andersson, T., Characterization and induction of xenobiotic metabolizing enzyme activities in a primary culture of rainbow trout hepatocytes, *Xenobiotica*, 21, 461, 1991.
352. Xu, L.-C. and Bresnick, E., Induction of cytochrome P450IA1 in rat hepatoma cell by polycyclic hydrocarbons and a dioxin, *Biochem. Pharmacol.*, 40, 1399, 1990.
353. Steward, A.R., Zaleski, J., and Sikka, H.C., Metabolism of benzo(a)pyrene and (–) trans-benzo(a)pyrene-7,8-dihydrodiol by freshly isolated hepatocytes of brown bullheads, *Chem.-Biol. Interact.*, 74, 119, 1990.
354. Baksi, S.M. and Frazier, J.M., Isolated fish hepatocytes — model systems for toxicology research, *Aquat. Toxicol.*, 16, 229, 1990.
355. Hightower, L.E. and Renfro, J.L., Recent applications of fish cell culture to biomedical research, *J. Exp. Zool.*, 248, 290, 1988.
356. Miranda, C.L., Collodi, P., Zhao, X., Barnes, D.W., and Buhler, D.R., Regulation of cytochrome P450 expression in a novel liver cell line from zebrafish, *Arch. Biochem. Biophys.*, 305, 320, 1993.
357. Smolarek, T.A., Morgan, S.L., Moynihan, C.G., Lee, H., Harvey, R.G., and Baird, W.M., Metabolism and DNA adduct formation of benzo[a]pyrene and 7,12-dimethylbenz[a]anthracene in fish cell lines in culture, *Carcinogenesis*, 8, 1501, 1987.
358. Bradlaw, J.A. and Casterline, J.L., Induction of enzyme activity in cell culture: a rapid screen for detection of planar polychlorinated organic compounds, *J. Assoc. Off. Anal. Chem.*, 62, 904, 1979.
359. Sawyer, T. and Safe, S., PCB isomers and congeners: induction of aryl hydrocarbon hydroxylase and ethoxyresorufin O-deethylase enzyme activities in rat hepatoma cells, *Toxicol. Lett.*, 13, 87, 1982.

360. Tillitt, D.E., Giesy, J.P., and Ankley, G.T., Characterization of the H4IIE rat hepatoma cell bioassay as a tool for assessing toxic potency of planar halogenated hydrocarbons in environmental samples, *Environ. Sci. Technol.*, 25, 87, 1991.
361. Safe, S., Polychlorinated biphenyls (PCBs), dibenzo-p-dioxins (PCDDs), dibenzofurans (PCDFs), and related compounds: environmental and mechanistic considerations which support the development of toxic equivalency factors (TEFs), *CRC Crit. Rev. Toxicol.*, 21, 51, 1990.
362. Walker, M.K. and Peterson, R.E., Potencies of polychlorinated dibenzo-*p*-dioxin, dibenzofuran, and biphenyl congeners, relative to 2,3,7,8-tetrachlorodibenzo-*p*-dioxin, for producing early life stage mortality in rainbow trout *(Oncorhynchus mykiss), Aquat. Toxicol.*, 21, 219, 1991.
363. Tillitt, D.E. and Cantrell, S.M., Planar halogenated hydrocarbon (PHH) structure activity relationship in a teleost (PLHC) cell line, in Society of Environmental Toxicology and Chemistry, 13th Annual Meeting, 1992.
364. Clemons, J.H., van den Heuvel, M.R., Bols, N.C., and Dixon, D.G., Toxic equivalent factors for selected congeners of PCDDs, PCDFs and PCBs using a rainbow trout liver cell line (RTL-W1), in Society of Environmental Toxicology and Chemistry, 13th Annual Meeting, 1992.
365. Goldstein, A., Aronow, L., and Kalman, S.M., *Principles of Drug Action: The Basis of Pharmacology,* 2nd ed., John Wiley & Sons, New York, 1974.
366. Kennedy, S.W., Lorenzen, A., James, C.A., and Collins, B.T., Ethoxyresorufin-O-dccthylasc and porphyrin analysis in chicken embryo hepatocyte cultures with a fluorescence multi-well plate reader, *Anal. Biochem.*, 211, 102, 1993.
367. Hahn, M.E., Patel, A.B., and Stegeman, J.J., Rapid assessment of cytochrome P4501A induction in fish hepatoma cells grown in multi-well plates, in Seventh International Symposium on Responses of Marine Organisms to Pollutants, 1993.
368. Baksi, S. and Nelson, S., Primary cell cultures from the oyster *(Crassostrea virginica)* and their applications in biomarkers research, in Society of Environmental Toxicology and Chemistry, 13th Annual Meeting, 1992.
369. Payne, J.F., Field evaluation of benzopyrene hydroxylase induction as a monitor for marine pollution, *Science*, 191, 945, 1976.
370. Binder, R.L. and Lech, J.J., Xenobiotics in gametes of Lake Michigan lake trout (Salvelinus namaycush) induce hepatic monooxygenase activity in their offspring, *Fundam. Appl. Toxicol.*, 4, 1042, 1984.
371. Stegeman, J.J., Kloepper-Sams, P.J., and Farrington, J.W., Monooxygenase induction and chlorobiphenyls in the deep-sea fish *Coryphaenoides armatus, Science*, 231, 1287, 1986.
372. Haasch, M.L., Wejksnora, P.J., Stegeman, J.J., and Lech, J.J., Cloned rainbow trout liver P1450 complementary DNA as a potential environmental monitor, *Toxicol. Appl. Pharmacol.*, 98, 362, 1989.
373. Haasch, M.L., Prince, R., Wejksnora, P.J., Cooper, K.R., and Lech, J.J., Caged and wild fish: induction of hepatic cytochrome P-450 (CYP1A1) as an environmental monitor, *Environ. Toxicol. Chem.*, 12, 885, 1993.
374. Lech, J.J., Personal communication.
375. Bellward, G.D., Norstrom, R.J., Whitehead, P.E., Elliot, J.E., Bandiera, S.M., Dworschak, C., Chang, T., Forbes, S., Cadario, B., Hart, L.E., and Cheng, K.M., Comparison of polychlorinated dibenzodioxin levels with hepatic mixed-function oxidase induction in great blue herons, *J. Toxicol. Environ. Health*, 30, 33, 1990.

376. Ellenton, J., Brownlee, L., and Hollebone, B., Aryl hydrocarbon hydroxylase levels in herring gull embryos from different locations on the Great Lakes, *Environ. Toxicol. Chem.*, 4, 615, 1985.
377. Hoffman, D.J., Rattner, B.A., Sileo, L., Doucherty, D., and Kubiak, T.J., Embryotoxicity, teratogenicity, and aryl hydrocarbon hydroxylase activity in Forster's terns on Green Bay, Lake Michigan, *Environ. Res.*, 42, 176, 1987.
378. Varanasi, U., Collier, T.K., Williams, D.E., and Buhler, D.R., Hepatic cytochrome P-450 isozymes and aryl hydrocarbon hydroxylase in English sole *(Parophyrs vetulus)*, *Biochem. Pharmacol.*, 35, 2967, 1986.
379. Foureman, G., White, N., and Bend, J., Biochemical evidence that winter flounder (Pseudopleuronectes americanus) have induced hepatic cytochrome P-450-dependent monooxygenase activities, *Can. J. Fish. Aquat. Sci.*, 40, 854, 1983.
380. Elskus, A.A. and Stegeman, J.J., Induced cytochrome P-450 in *Fundulus heteroclitus* associated with environmental contamination by polychlorinated biphenyls and polynuclear aromatic hydrocarbons, *Mar. Environ. Res.*, 27, 31, 1989.
381. Elskus, A.A., Stegeman, J.J., Susani, L.C., Black, D., Pruell, R.J., and Fluck, S.J., Polychlorinated biphenyls concentration and cytochrome P-450E expression in winter flounder from contaminated environments, *Mar. Environ. Res.*, 28, 25, 1989.
382. Stegeman, J.J., Teng, F.Y., and Snowberger, E.A., Induced cytochrome P-450 in winter flounder *(Pseudopleuronectes americanus)* from coastal Massachusetts evaluated by catalytic assay and monoclonal antibody probes, *Can. J. Fish. Aquat. Sci.*, 44, 1270, 1987.
383. Stegeman, J.J., Woodin, B.R., and Goksøyr, A., Apparent cytochrome P-450 induction as an indication of exposure to environmental chemicals in the flounder Platichthys flesus, *Mar. Ecol. Prog. Ser.*, 46, 55, 1988.
384. Stegeman, J.J., Renton, K.W., Woodin, B.R., Zhang, Y.-S., and Addison, R.F., Experimental and environmental induction of cytochrome P450E in fish from Bermuda waters, *J. Exp. Mar. Biol. Ecol.*, 138, 49, 1990.
385. Spies, R., Felton, J., and Dillard, L., Hepatic mixed function oxidases in California flatfishes are increased in contaminated environments and by oil and PCB ingestion, *Mar. Biol.*, 70, 117, 1982.
386. Kreamer, G.L., Squibb, K., Gioeli, D., Garte, S.J., and Wirgin, I., Cytochrome P450IA mRNA expression in feral Hudson River tomcod, *Environ. Res.*, 55, 64, 1991.
387. Luxon, P.L., Hodson, P.V., and Borgmann, U., Hepatic aryl hydrocarbon hydroxylase activity of lake trout (Salvelinus namaycush) as an indicator of organic pollution, *Environ. Toxicol. Chem.*, 6, 649, 1987.
388. Focardi, S., Fossi, C., Leonzio, C., and Di Simplicio, P., Mixed-function oxidase and conjugating enzymes in two species of Antarctic fish, *Mar. Environ. Res.*, 28, 31, 1989.
389. Payne, J., Fancey, L., Rahimtula, A., and Porter, E., Review and perspective on the use of mixed-function oxygenase enzymes in biological monitoring, *Comp. Biochem. Physiol.*, 86C, 233, 1987.
390. Collier, T.K., Connor, S.D., Eberhart, B.-T.L., Anulacion, B.F., Goksøyr, A., and Varanasi, U., Using cytochrome P450 to monitor the aquatic environment: initial results from regional and national surveys, *Mar. Environ. Res.*, 34, 195, 1992.
391. Lindstrom-Seppa, P., Huuskonen, S., Pesonen, M., Muona, P., and Hanninen, O., Unbleached pulp mill effluents affect cytochrome P450 monooxygenase enzyme activities, *Mar. Environ. Res.*, 34, 157, 1992.

392. Munkittrick, K.R., Kraak, G.J.V.D., McMaster, M.E., and Portt, C.B., Response of hepatic MFO activity and plasma sex steroids to secondary treatment of bleached kraft pulp mill effluent and mill shutdown, *Environ. Toxicol. Chem.*, 11, 1427, 1992.
392a. Pesonen, M. and Andersson, T., Toxic effects of bleached and unbleached paper mill effluents in primary cultures of rainbow trout hepatocytes, *Ecotox. Environ. Safety*, 24, 63, 1992.
393. Malins, D.C., McCain, B.B., Brown, D.W., Chan, S.L., Myers, M.S., Landahl, J.T., Prohaska, P.G., Friedman, A.J., Rhodes, L.D., Burrows, D.G., Gronlund, W.D., and Hodgins, H.O., Chemical pollutants in sediments and diseases of bottom-dwelling fish in Puget Sound, Washington, 18, 705, 1984.
394. Malins, D.C., Krahn, M.M., Brown, D.W., Rhodes, L.D., Myers, M.S., McCain, B.B., and Chan, S.-L., Toxic chemicals in marine sediment and biota from Mukilteo, Washington: relationships with hepatic neoplasms and other hepatic lesions in English sole (Parophrys vetulus), *J. Natl. Cancer Inst.*, 74, 487, 1985.
395. Eberhart, J., Baird, W.M., Park, S.S., Gelboin, H.V., and Stegeman, J.J., Role of cytochrome P4501A1 in the binding of benzo(a)pyrene to DNA in a human hepatoma cell line Hep G2, *Chem. Res. Toxicol.*, in press.
396. Roberts, L., Dioxin risks revisited, *Science*, 251, 624, 1991.
397. Scheline, R.R., Handbook of Mammalian Metabolism of Plant Compounds, CRC Press, Boca Raton, FL, 1991.
398. Sondheimer, E. and Simeone, J.B., *Chemical Ecology,* Academic Press, New York, 1970.
399. Brattsten, L.B., Ecological significance of mixed-function oxidations, *Drug Metab. Rev.*, 10, 35, 1979.
400. Frank, M.R. and Fogleman, J.C., Involvement of cytochrome-P450 in host-plant utilization by sonoran desert drosophila, *Proc. Natl. Acad. Sci. U.S.A.*, 89, 11998, 1992.
401. Ivie, G.W., Bull, D.L., Beier, R.C., Pryor, N.W., and Oertli, E.H., Metabolic detoxification: mechanism of insect resistance to plant psoralens, *Science*, 221, 374, 1983.
402. Cohen, M.B., Berenbaum, M.R., and Schuler, M.A., Induction of cytochrome P-450-mediated detoxification of xanthotoxin in the black swallowtail, *J. Chem. Ecol.*, 15, 2347, 1989.
403. Cohen, M.B., Schuler, M.A., and Berenbaum, M.R., A host-inducible cytochrome P-450 from a host-specific caterpillar: molecular cloning and evolution, *Proc. Natl. Acad. Sci. U.S.A.*, 89, 10920, 1992.
404. Fenical, W., Natural products chemistry in the marine environment, *Science*, 215, 923, 1982.
405. Scheuer, P.J., Ed., *Marine Natural Products: Chemical and Biological Perspectives,* Vol. 1 to 4, Academic Press, New York, 1978–1981.
406. Scheuer, P.J., Some marine ecological phenomena: chemical basis and biomedical potential, *Science*, 248, 173, 1990.
407. Paul, V.J., Ed., *Ecological Roles of Marine Natural Products,* Cornell University Press, Ithaca, NY, 1992.
408. Austin, C.A., Shephard, E.A., Pike, S.F., Rabin, B.R., and Phillips, I.R., The effect of terpenoid compounds on cytochrome P-450 levels in rat liver, *Biochem. Pharmacol.*, 37, 2223, 1988.
409. Bakus, G.J., Targett, N.M., and Schulte, B., Chemical ecology of marine organisms: an overview, *J. Chem. Ecol.*, 12, 951, 1986.

410. Higa, T., Fujiyama, T., and Scheuer, P.J., Halogenated phenol and indole constituents of acorn worms, *Comp. Biochem. Physiol.*, 65B, 525, 1980.
411. Bjeldanes, L.F., Kim, J.-Y., Grose, K.R., Bartholomew, J.C., and Bradfield, C.A., Aromatic hydrocarbon responsiveness-receptor agonists generated from indole-3-carbinol *in vitro* and *in vivo*: comparisons with 2,3,7,8-tetrachlorodibenzo-*p*-dioxin, *Proc. Natl. Acad. Sci. U.S.A.*, 88, 9543, 1991.
412. Loub, W.D., Wattenberg, L.W., and Davis, D.W., Aryl hydrocarbon hydroxylase induction in rat tissues by naturally occurring indoles of cruciferous plants, *J. Natl. Cancer Inst.*, 54, 985, 1975.
413. Vang, O., Jensen, M.B., and Autrup, H., Induction of cytochrome P450IA1 in rat colon and liver by indole-3-carbinol and 5,6-benzoflavone, *Carcinogenesis*, 11, 1259, 1990.
414. Fong, A.T., Swanson, H.I., Dashwood, R.H., Williams, D.E., Hendricks, J.D., and Bailey, G.S., Mechanisms of anti-carcinogenesis by indole-3-carbinol. Studies of enzyme induction, electrophile-scavenging, and inhibition of aflatoxin B_1 activation, *Biochem. Pharmacol.*, 39, 19, 1990.
415. Swanson, S.E., Rappe, C., Malmstrom, J., and Kringstad, K.P., Emissions of PCDDs and PCDFs from the pulp industry, *Chemosphere*, 17, 681, 1988.
416. Kuehl, D.W., Butterworth, B.C., DeVita, W.M., and Sauer, C.P., Environmental contamination by polychlorinated dibenzo-*p*-dioxins and dibenzofurans associated with pulp and paper mill discharge, *Biomed. Environ. Mass. Spectrom.*, 14, 443, 1987.
417. Gonzalez, F.J. and Nebert, D.W., Evolution of the P450 gene superfamily, *Trends Genet.*, 6, 182, 1990.
418. Vrolijk, N.H. and Targett, N.M., Comparison of in vivo detoxification enzyme activities among three members of the genus Chaetodon (butterflyfish), in Second International Marine Biotechnology Conference (IMBC '91), 1991.
418a. Vrolijk, N.H., Targett, N.M., Woodin, B.R., and Stegeman, J.J., Toxicological and ecological implications of biotransformation enzymes in the tropical teleost, *Chaetodon capistratus*, *Mar. Biol.*, in press, 1993.
419. Ames, B.N., Profet, M., and Gold, L.S., Dietary pesticides (99.99% all natural), *Proc. Natl. Acad. Sci. U.S.A.*, 87, 7777, 1990.
420. Ames, B.N., Profet, M., and Gold, L.S., Nature's chemicals and synthetic chemicals: comparative toxicology, *Proc. Natl. Acad. Sci. U.S.A.*, 87, 7782, 1990.
421. Nelson, D. and Strobel, H., Evolution of cytochrome P450 proteins, *Mol. Biol. Evol.*, 4, 572, 1987.
422. Hunt, J.M., *Petroleum Geochemistry and Geology*, W.H. Freeman, San Fransisco, CA, 1979.
423. Guerin, M.R., Energy sources of polycyclic aromatic hydrocarbons, in *Polycyclic Hydrocarbons and Cancer*, Gelboin, H.V. and Ts'o, P.O.P., Eds., Academic Press, New York, 1978, 1.
423a. Bumb, R.R., Crummett, W.B., Cutie, S.S., et al., Trace chemistries of fire: a source of chlorinated dioxins, *Science*, 210, 385, 1980.
424. Czuczwa, J.M., McVeety, B.D., and Hites, R.A., Polychlorinated dibenzo-p-dioxins and dibenzofurans in sediments from Sikiwit Lake, Isle Royale, *Science*, 226, 568, 1984.
425. Nebert, D.W., Proposed role of drug-metabolizing enzymes: regulation of steady state levels of the ligands that effect growth, homeostasis, differentiation, and neuroendocrine functions, *Mol. Endocrinol.*, 5, 1202, 1991.

426. Hoffman, E.C., Reyes, H., Chu, F.-F., Sander, F., Conley, L.H., Brooks, B.A., and Hankinson, O., Cloning of a factor required for activity of the Ah (dioxin) receptor, *Science*, 252, 954, 1991.
426a. Kimura, S., Donovan, J.C., and Nebert, D.W., Expression of the mouse $P_1 450$ gene during differentiation without foreign chemical stimulation, *J. Exper. Pathol.*, 3, 61, 1987.
426b. Kapitulnik, J. and Gonzalez, F.J., Marked endogenous activation of the *CYP1A1* and *CYP1A2* genes in the congenitally jaundiced Gunn rat, *Mol. Pharmacol.*, 43, 722, 1993.
426c. Lindstrom-Seppa, P., Korytko, P.J., Hahn, M.E., and Stegeman, J.J., Uptake of waterborne 3,3',4,4'-tetrachlorobiphenyl and organ and cell-specific induction of cytochrome P4501A in the fathead minnow *Pimephales promelas*, *Aquat. Toxicol.*, in press, 1993.
427. Gribble, G.W., Naturally occurring organohalogen compounds — a survey, *J. Nat. Prod.*, 55, 1353, 1992.
428. He, J.-S. and Fulco, A.J., A barbiturate-regulated protein binding to a common sequence in the cytochrome P450 genes of rodents and bacteria, *J. Biol. Chem.*, 266, 7864, 1991.
429. Rangarajan, P.N. and Padmanaban, G., Regulation of cytochrome P450b/e gene expression by a heme and phenobarbitone-modulated transcription factor, *J. Biol. Chem.*, 86, 3963, 1989.
430. Shaw, G.-C. and Fulco, A.J., Inhibition by barbiturates of the binding of Bm3R1 repressor to its operator site on the barbiturate-inducible cytochrome $P450_{BM-3}$ gene of *Bacillus megaterium*, *J. Biol. Chem.*, 268, 2997, 1993.
431. Greenberg, M., Ex bouillabaisse lux: the charm of comparative physiology and biochemistry, *Am. Zool.*, 25, 737, 1985.
432. Hardwick, J.P., Song, B.-J., Huberman, E., and Gonzalez, F.J., Isolation, complementary DNA sequence, and regulation of rat hepatic lauric acid ω-hydroxylase (cytochrome $P-450_{LA\omega}$). Identification of a new P-450 gene family, *J. Biol. Chem.*, 262, 801, 1987.
433. Palmer, C.N.A., Griffin, K.J., and Johnson, E.F., Rabbit prostaglandin ω-hydroxylase (CYP4A4): gene structure and expression, *Arch. Biochem. Biophys.*, 300, 670, 1993.
434. Jefcoate, C.R., Cytochrome P-450 enzymes in sterol biosynthesis and metabolism, in *Cytochrome P-450. Structure, Mechanism, and Biochemistry*, Montellano, P.R.O.D., Ed., Plenum Press, New York, 1986, 387.
435. Bozak, K.R., Yu, H., Sirevag, R., and Christoffersen, R.E., Sequence analysis of ripening-related cytochrome P-450 cDNAs from avocado fruit, *Proc. Natl. Acad. Sci. U.S.A.*, 87, 3904, 1990.
436. Kochs, G. and Grisebach, H., Phytoalexin synthesis in soybean: purification and reconstitution of cytochrome P450 3,9-dihydroxypterocarpan 6a-hydroxylase and separation from cytochrome P450 cinnamate 4-hydroxylase, *Arch. Biochem. Biophys.*, 273, 543, 1989.
437. Kochs, G., Werck-Reichhart, D., and Grisebach, H., Further characterization of cytochrome P450 involved in phytoalexin synthesis in soybean: cytochrome P450 cinnamate 4-hydroxylase and 3,9-dihydroxypterocarpan 6a-hydroxylase, *Arch. Biochem. Biophys.*, 293, 187, 1992.
438. Ahokas, J.T., Pelkonen, O., and Karki, N.T., Metabolism of polycyclic hydrocarbons by a highly active aryl hydrocarbon hydroxylase in the liver of a trout species, *Biochem. Biophys. Res. Commun.*, 63, 635, 1975.

439. Melancon, M.J., Elcombe, C.R., Vodicnik, M.J., and Lech, J.J., Induction of cytochromes P-450 and mixed-function oxidase activity by polychlorinated biphenyls and β-naphthoflavone in carp *(Cyprinus carpio), Comp. Biochem. Physiol.*, 69C, 219, 1981.
440. Stegeman, J.J., Hepatic microsomal monooxygenase activity and the biotransformation of hydrocarbons in deep benthic fish from the western North Atlantic, *Can. J. Fish. Aquat. Sci.*, 40 (Suppl. 2), 78, 1983.
441. Williams, D.E. and Buhler, D.R., Purification of cytochromes P-448 from β-naphthoflavone-treated rainbow trout, *Biochim. Biophys. Acta*, 717, 398, 1982.
442. Williams, D.E. and Buhler, D.R., Comparative properties of purified cytochrome P-448 from b-naphthoflavone treated rats and rainbow trout, *Comp. Biochem. Physiol.*, 75C, 25, 1983.
443. Sijm, D.T.H.M., Yarechewski, A.L., Muir, D.C.G., Webster, G.R.B., Seinen, W., and Opperhuizen, A., Biotransformation and tissue distribution of 1,2,3,7-tetrachlorodibenzo-p-dioxin, 1,2,3,4,7-pentachlorodibenzo-p-dioxin, and 2,3,4,7,8-pentachlorodibenzofuran in rainbow trout, *Chemosphere*, 21, 845, 1990.
444. Erickson, D.A., Goodrich, M.S., and Lech, J.J., The effect of piperonyl butoxide on hepatic cytochrome P450-dependent monooxygenase activities in rainbow trout, *Toxicol. Appl. Pharmacol.*, 94, 1, 1988.
445. Erickson, D.A., Laib, F.E., and Lech, J.J., Biotransformation of rotenone by hepatic microsomes following pretreatment of rainbow trout with inducers of cytochrome P450, *Pest. Biochem. Physiol.*, 42, 140, 1992.
446. McKee, M.J., Hendricks, A.C., and Ebel, R.E., Effects of naphthalene on benzo[a]pyrene hydroxylase and cytochrome P-450 in *Fundulus heteroclitus, Aquat. Toxicol.*, 3, 103, 1983.
447. Arinc, E. and Alaattin, S., Effects of *in vivo* benzene treatment on cytochrome P450 and mixed-function oxidase activities of gilthead seabream *(Sparus auratus)* liver microsomes, *Comp. Biochem. Physiol.*, 104C, 61, 1993.
448. Melancon, M.J. and Lech, J.J., Dose-effect relationship for induction of hepatic monooxygenase activity in rainbow trout and carp by Aroclor 1254, *Aquat. Toxicol.*, 4, 51, 1983.
449. Miranda, C.L., Wang, J.-L., Henderson, M.C., Williams, D.E., and Buhler, D.R., Regiospecificity in the hydroxylation of lauric acid by rainbow trout hepatic cytochrome P450 isozymes, *Biochem. Biophys. Res. Commun.*, 171, 537, 1990.
450. Goksøyr, A., Solbakken, J.E., and Klungsøyr, J., Regioselective metabolism of phenanthrene in Atlantic cod *(Gadus morhua):* studies on the effects of monooxygenase inducers and role of cytochromes P-450, *Chem.-Biol. Interact.*, 60, 247, 1986.
451. James, M.O. and Bend, J.R., Polycyclic aromatic hydrocarbon induction of cytochrome P-450-dependent mixed-function oxidases in marine fish, *Toxicol. Appl. Pharmacol.*, 54, 117, 1980.
452. Kloepper-Sams, P.J., Park, S.S., Gelboin, H.V., and Stegeman, J.J., Specificity and cross-reactivity of monoclonal and polyclonal antibodies against cytochrome P450E of the marine fish scup, *Arch. Biochem. Biophys.*, 253, 268, 1987.
453. Elskus, A.A. and Stegeman, J.J., Further consideration of phenobarbital effects on cytochrome P-450 activity in the killifish, *Fundulus heteroclitus, Comp. Biochem. Physiol. C*, 92C, 223, 1989.

454. Gooch, J.W. and Matsumura, F., Characteristics of the hepatic monooxygenase system of the goldfish (Carassius auratus) and its induction with β-naphthoflavone, *Toxicol. Appl. Pharmacol.*, 68, 380, 1983.

455. Förlin, L., Effects of Clophen A50, 3-methylcholanthrene, pregnenolone 16-alpha carbonitrile and phenobarbital on the hepatic microsomal cytochrome P-450-dependent monooxygenase system in rainbow trout, *Salmo gairdneri*, of different sex and age, *Toxicol. Appl. Pharmacol.*, 54, 420, 1980.

456. Bend, J.R., Pohl, R.J., Arinc, E., and Philpot, R.M., Hepatic microsomal and solubilized mixed-function oxidase systems from the little skate, *Raja erinacea*, a marine elasmobranch, in *Microsomes and Drug Oxidations*, Ullrich, V., Roots, I., Hildebrandt, A., Estabrook, R.W., and Conney, A.H., Eds., Pergamon Press, Oxford, 1977, 160.

457. Gerhart, E.H. and Carlson, R.M., Hepatic mixed-function oxidase activity in rainbow trout exposed to several polycyclic aromatic hydrocarbons, *Environ. Res.*, 17, 284, 1978.

458. Jiminez, B.D. and Burtis, L.S., Influence of environmental variables on the hepatic mixed-function oxidase system in bluegill sunfish, *Lepomis macrochirus*, *Comp. Biochem. Physiol.*, 93C, 11, 1989.

459. Statham, C.N., Elcombe, C.R., Szyjka, S.P., and Lech, J.J., Effect of polycyclic aromatic hydrocarbons on hepatic microsomal enzymes and disposition of methylnaphthalene in rainbow trout in vivo, *Xenobiotica*, 8, 65, 1978.

460. Pohl, R.J., Fouts, J.R., and Bend, J.R., Response of hepatic microsomal mixed-function oxidases in the little skate, *Raja erinacea*, and the winter flounder, *Pseudopleuronectes americanus* to pretreatment with TCDD (2,3,7,8-tetrachlorodibenzo-p-dioxin) or DBA (1,2,3,4-dibenzanthracene), *Bull. Mt. Des. Isl. Biol. Lab.*, 15, 64, 1975.

461. Payne, J.F. and May, N., Further studies on the effect of petroleum hydrocarbons on mixed-function oxidases in marine organisms, in *Pesticide and Xenobiotic Metabolism in Aquatic Organisms*, Khan, M.A.Q., Lech, J.J., and Mann, J.J., Eds., American Chemical Society, Washington, D.C., 1979, 339.

462. Addison, R.F., Zinc, M.E., and Willis, D.E., Induction of hepatic mixed-function oxidase (MFO) enzymes in trout *(Salvelinus fontinalis)* by feeding Aroclor 1254 or 3-methylcholanthrene, *Comp. Biochem. Physiol.*, 61, 323, 1978.

463. Janz, D.M. and Metcalfe, C.D., Relative induction of aryl hydrocarbon hydroxylase by 2,3,7,8-TCDD and two coplanar PCBs in rainbow trout *(Oncorhynchus mykiss)*, *Environ. Toxicol. Chem.*, 10, 917, 1991.

464. Vodicnik, M.J., Elcombe, C.R., and Lech, J.J., The effect of various types of inducing agents on hepatic microsomal monooxygenase activity in rainbow trout, *Toxicol. Appl. Pharmacol.*, 59, 364, 1981.

465. Wisk, J.D. and Cooper, K.R., Effect of 2,3,7,8-tetrachlorodibenzo-p-dioxin on benzo(a)pyrene hydroxylase activity in embryos of the Japanese medaka *(Oryzias latipes)*, *Arch. Toxicol.*, 66, 245, 1992.

466. van der Weiden, M.E.J., Kolk, J.v.d., Bleumink, R., Seinen, W., and van den Berg, M., Concurrence of P4501A1 induction and toxic effects after administration of a low dose of 2,3,7,8-tetrachlorodibenzo-p-dioxin (TCDD) in the rainbow trout *(Oncorhynchus mykiss)*, *Aquat. Toxicol.*, 24, 1992.

467. Muir, D.C.G., Yarechewski, A.L., Metner, D.A., and Lockhart, W.L., Dietary 2,3,7,8-tetrachlorodibenzofuran in rainbow trout: accumulation, disposition, and hepatic mixed-function oxidase enzyme induction, *Toxicol. Appl. Pharmacol.*, 117, 65, 1992.
468. Muir, D.C.G., Yarechewski, A.L., Metner, D.A., Lockhart, W.L., Webster, G.R.B., and Friesen, K.J., Dietary accumulation and sustained hepatic mixed function oxidase induction by 2,3,4,7,8-pentachlorodibenzofuran in rainbow trout, *Environ. Toxicol. Chem.*, 9, 1463, 1990.
469. Addison, R.F., Zinc, M.E., and Willis, D.E., Mixed-function oxidase in trout liver: absence of induction following feeding of p,p'-DDT or p,p'-DDE, *Comp. Biochem. Physiol.*, 57C, 39, 1977.
470. Leaver, M.J., Burke, M.D., George, S.G., Davies, J.M., and Raffaelli, D., Induction of cytochrome P450 monooxygenase activities in plaice by 'model' inducers and drilling muds, *Mar. Environ. Res.*, 24, 27, 1988.
471. Petersen, D.W. and Lech, J.J., Hepatic effects of acrylamide in rainbow trout, *Toxicol. Appl. Pharmacol.*, 89, 249, 1987.
472. Poland, A., Personal communication.
473. Melancon, M.J. and Lech, J.J., Isolation and identification of a polar metabolite of tetrachlorobiphenyl from bile of rainbow trout exposed to ^{14}C-tetrachlorobiphenyl, *Bull. Environ. Contam. Toxicol.*, 15, 181, 1976.

Chapter 4

Cells, Cellular Responses, and Their Markers in Chronic Toxicity of Fishes

David E. Hinton

"Disease is not associated with new, different structure and function; rather quantitative alterations (increase or decrease) of existing structure and function occur." W.D. Forbus, 1952.

"No matter how we twist and turn we shall always come back to the cell." Rudolf Virchow, *"Cellular-Pathologie," Virchows Arch.,* 8, 3, 1855.

I. INTRODUCTION

As aquatic toxicologists, we are introduced to a large number of invertebrate and vertebrate species. For example, when Roesijadi[1] reviewed the metallothioneins, including their role in metal regulation and toxicity of aquatic animals, he presented information on 34 bony and cartilaginous fishes, 4 echinoderms, 2 insects, 11 crustaceans, 18 mollusks, and 5 annelids. Similarly, Stegeman[2] reviewed cytochromes P450 in fishes and included findings from more than 20 species. A recent listing of the bony and cartilaginous fishes in which tumors of the liver have been reported included 74 species.[3]

I have become increasingly aware of the magnitude of our tasks to design research approaches that enable us to bridge from one species to another or across families or classes. With such diverse and numerous organisms, are there common features on which we may focus? As a comparative morphologist, I have observed greater similarity between diverse organisms at the levels of tissues, cells, and their organelles. In fact, only four basic tissues, epithelium, connective tissue, muscle, and nerve tissue are described in histology texts.[4,5] Each is a collection of cells of similar type, which, together with their associated extracellular substances, are specialized to perform a common function or functions. Each basic tissue is present, in variable amounts, in the organs that collectively make-up the entire organism. Recent texts on comparative anatomy and histology that include ample fish material and the major histology references from fish literature[6-18] indicate that, like other vertebrates, fish contain a limited number of specific cell types.

Cells are the component units of organization in each tissue, and they show a variety of structural and functional specializations, based on quantitative differences in individual organelles. Cellular organelles may be considered highly conserved structures, since they are recognizable in vertebrates as well as invertebrates. It is at this ultrastructural level that perhaps, despite their location within different portions of the bodies of various organisms, the most common features can be determined. In toxicity, the perturbation of complex molecules, followed by organelle and cellular responses, leads to alterations that we ascribe to a toxicant or complex mixture. Since genes control cell products and therefore appearance of specific cells, it is apparent that a focus on cells would enhance our detection of common mechanisms of toxicity.[19-21]

The levels of biological organization are shown in Table 1. From Table 1, it is apparent that those levels of biological organization common to individuals, whether invertebrates or vertebrates, include molecules through cells and, for all but unicellular organisms, tissues. However, some reviewers correctly point to the fact that it is at the higher levels of biological organization (e.g., populations, communities, and ecosystems) that aquatic toxicological research is being applied to monitor environmental effects, conduct hazard assessments, and make decisions of a regulatory nature.[22,23] These reviewers are also quick to point out that the predictive utility of research findings at the higher levels may be questioned because ecologically important effects have already occurred at lower levels by the time they are detected. We all seek a timely and cost-effective manner to verify exposure and to establish an adverse health endpoint.[24,25] Some of the problems in determination of exposure, adverse health or ecological effect, and linkage of exposure to effect are presented in Table 2. Given the uncertainties discussed above, there is growing interest in a different approach, one that examines exposure and effect in aquatic toxicology. This approach uses biological markers of exposure and effect that can be defined as "measurements of body fluids, cells, or tissues that indicate in biochemical or cellular terms the presence of contaminants or the magnitude of the host response."[24] Other workers are equally enthusiastic about a biomarker approach:[21] "Precise diagnostic tools with a predictive capabil-

Table 1. Levels of Biological Organization

Molecules
  Pathways
    Organelles
      Cells
        Tissues
          Organs
            Organ systems
              Organisms
                Populations
                  Communities
                    Ecosystem

ity in contaminant impact assessment, biomarkers have begun to be used with marine animals and the value of this multi-disciplinary mechanistic and molecular cell biological method is emerging as a powerful tool in environmental pathology." The cellular emphasis in aquatic toxicology is entirely consistent with a biomarker approach. In addition, cellular studies are likely to lead to detection of alteration prior to changes at the organism and higher levels of organization and to provide an opportunity for intervention. Since cellular changes are early and sensitive, they may serve as markers of both exposure and effect. Cellular changes may shed light on important mechanistic considerations and lead to improved understanding of the toxic agent and the host's response. Cells may be cultured *in vitro* and provide rapid ways to sort out effects of individual components of complex mixtures.

This chapter focuses on the liver of teleost fishes, paying particular attention to normal and altered cell and tissue morphology, including histo- and cytochemistry.

Table 2. Difficulties in Determination of Exposure, Adverse Health or Ecological Effect, and Linking Exposure to Effect

Estimate desired	Source of variation
Extent of exposure	Differences in exposure routes.
	Differences in biological availability of contaminant in various environmental media.
	Differences in pharmacodynamics of toxicant and its disposition in animal.
	Dynamic seasonal changes in environmental media and host altered metabolism.
	Differences in bioaccumulation.
	Diversity and possible interactions due to presence of complex mixture.
Adverse health or ecological effect	Differences in magnitude and duration of exposure.
	Mode(s) of action of toxicants differ.
	Latency periods for disease state(s) differ.
	Susceptibility of organism(s) differ(s).
	Changes in host age, nutrition, and prior history.
	Time between exposure and effect (if long duration, linkage less likely).

Source: Adapted from review by McCarthy, J.F. and Shugart, L.R., Eds. (1990a).[24]

Markers used to differentiate specific liver cell types are presented. Tissue alterations inherent with chronic toxicity (hepatocarcinogenicity) are reviewed, and markers (used in rodent studies) that may be brought to aquatic toxicological investigations are presented and discussed. Although these are morphological tools, many of their bases are firmly established in protein biochemistry. The information presented illustrates both the type of cellular biomarkers currently used and also those markers that are needed for future toxicological investigations.

II. TARGETS OF TOXICANTS AND LEVELS OF BIOLOGICAL ORGANIZATION

Organ systems of individuals are the highest levels of organization that are commonly studied in laboratory exposures to various toxicants, and the concept of target organ toxicology is firmly established in mammals[26,27] and in aquatic organisms.[28,29] While this concept is useful in drawing attention to the methods and results determined from analysis of a single organ, it is becoming increasingly apparent that the actual targets of toxicity are molecules and that these hits are closely followed by alterations in pathways, organelles, and cell structure/function.[21,30,31] Figure 1 illustrates the importance of considering even small volumes of specific tissues and cells within a given organ. Peroxidase localization of cytochrome P4501A within the brain of a β-naphthoflavone (βNF)-treated rainbow trout embryo is shown. Although an appreciable portion of this longitudinal section through the developing brain is included, only the cells lining blood vessels appear to contain this isozyme. Homogenate values of brain cytochrome P4501A (e.g., ethoxyresorufin-O-deethylase) would tend to indicate that this protein was of negligible importance. However, the restricted localization of this enzyme system to cells that line blood vessels may result in bioactivation of certain toxicants with possible resultant damage to endothelium. Exudation and even hemorrhage into brain tissue might follow. By the use of modern tools to localize specific proteins and antigens within cells, a better appreciation of the precise location of toxic insults might result. It is just this focus on specific cells within tissues of individual organs that is required to provide an understanding of the cellular and subcellular mechanisms of toxicity.

Molecular responses within aquatic organisms exposed to environmental contaminants have been reviewed recently.[31] These responses occur early and are usually the first detectable quantifiable response to environmental change. Linkage of such molecular events to damage at the cell and higher levels of organization is only beginning to emerge, and more focus is being brought in an effort to determine the significance of these molecular events to subsequent forms of injury and response. Molecular markers which have been recommended due to their potential for application in monitoring programs include

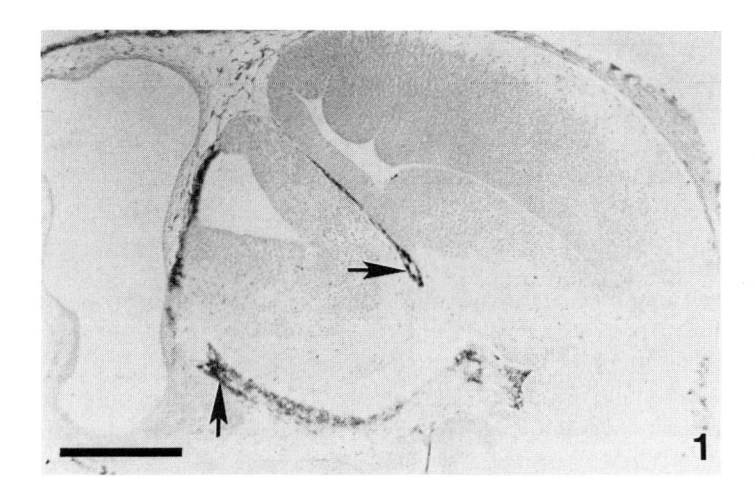

Figure 1. Section of β-naphthoflavone-injected trout embryo reacted with anti-scup P4501A. Note the strong reaction in the endothelial cells (horizontal arrow) and in the choroid plexus of ventricle (vertical arrow). Frozen section, Streptavidin-peroxidase complex technique (Vector Laboratories, Inc., Burlingame, CA) counterstained with hematoxylin. Bar = 299 μm.

cytochrome P450 monooxygenases, metallothioneins, stress proteins or heat shock proteins, phase II-conjugating enzymes, oxidant-mediated responses, and heme-porphyrin systems.

Organelles and their alteration following exposure to environmental contaminants have been determined by electron microscopy, which has the advantage of precise localization and characterization of the structural alterations but lacks the quantifiable response, unless newer technology employing computer-assisted morphometric devices is used.[32] The electron microscope coupled with cytochemical techniques can be used to localize specific proteins or antigens within organelles of individual cell types. Such an application has recently been made with the localization of cytochrome P4501A in specific cells and organelles of rainbow trout liver.[33] The major glutathione S-transferase isoform of flounder liver was shown to be antigenically related to the structural homologue of plaice GST-A, and its synthesis was induced following in vivo exposure of flounder to an epoxide, *trans*-stilbene oxide.[34] In addition, Northern blot analysis of various plaice organs demonstrated the presence of glutathione S-transferase in the spleen, liver, kidney, intestine, heart, gill, gonad, and brain.[34] Since the cytosolic glutathione S-transferases conjugate reactive electrophiles, their presence and relative amounts may signal exposure and host response to various xenobiotics. Immunohisto- and cytochemistry of reactive GSTs may form additional markers for monitoring responses of feral organisms to environmental stressors.

The tissue, organ, organ system, and organism levels of responses have been addressed by routine and, more recently, quantitative light microscopy.[35]

Histopathology has been a standard tool in aquatic toxicology investigations, although its application in integrated field studies has, until recently, lagged behind other approaches. Recent reviews on cellular, tissue, and organ level responses determined by histopathology have focused on the history of the usage of this technique, advantages, possible pitfalls, and description of biomarkers recommended for use in current field investigations.[19,20,36-40] Many of the markers developed through laboratory investigations have been shown to have field relevance and are currently recommended for use in field studies.[36] These new field studies will require integration and communication between toxicologists, hydrologists, soil scientists, ecologists, fisheries biologists, modelers, and analytical chemists as efforts are made to build the bridges that will connect laboratory investigations at molecular through organism levels of organization with those at the population, community, and possibly ecosystem levels. Epidemiological approaches have shown that one specific defect in late stage, lake trout embryos is strongly associated with polychlorinated biphenyls (PCBs), while another is linked to organochlorines, and that each defect is correlated with reproductive failure and population decline.[41]

III. THE LIVER AS A SITE OF BIOMARKER RESPONSES

Although results of studies with any of several major organ systems could illustrate cellular responses and markers, it is the liver with which I am most familiar. This organ will be used to demonstrate cells, cellular responses, and their markers. However, the approach used here is no different from similar approaches in other organs. The liver is a primary organ site for biomarker responses at the molecular through tissue levels of organization. Reasons for this include the following:

1. This organ is the major site of the cytochrome P450-mediated, mixed-function oxidase system,[42] and, while this system inactivates or detoxifies some xenobiotics, it activates others to their toxic forms.
2. Nutrients derived from intestinal absorption are stored in hepatocytes and released for further catabolism by other tissues.[43,44]
3. Bile synthesized by hepatocytes[45,46] aids in the digestion of fatty acids and carries conjugated metabolites of toxicants[47] into the intestine for excretion or intrahepatic recirculation.
4. The yolk protein, vitellogenin, destined for incorporation into the oocyte, is synthesized entirely within the liver.[48].

Given the role of this organ in various key functions and its metabolic capacity, the hepatotoxic effects of various toxicants is not surprising. Liver toxicological pathology has been the subject of recent reviews including findings in mammalian[49] and teleost[19,20,30,50] species.

IV. HEPATIC EMBRYOLOGY AND ORGANOGENESIS — ORIGIN OF SPECIFIC CELL TYPES

Cells of common embryological origin often reveal the presence of specific macromolecules that may form appropriate markers for identification and differentiation from other cell types. Resident hepatic cells are of two distinct embryonic sources, forming part of the rationale for marker development. In addition, recent aquatic toxicity exposures have shifted to early life stages, including embryonated eggs and hatchlings (see reference 51 for review). For these reasons, it is important to gain an appreciation of aspects of hepatic, exocrine pancreatic, and caudal foregut embryology and organogenesis.

Most embryology texts regard the liver as an entodermal derivative that arises by way of an evagination of the floor of the foregut.[52] Cells within this evagination form tubules or cords that undergo further development reaching the septum transversum, where they establish contact with cells derived from the mesoderm. Another mesodermal derivative, the yolk sac, is in close proximity to the septum transversum. These three elements give rise to the adult liver. Hepatic mesodermal derivatives include endothelial cells lining blood vessels (sinusoids, venules, veins, arterioles, and arteries) and the fat-storing, perisinusoidal, stellate cells of Ito et al.[53,54] Scanning electron microscopic observations on embryogenesis of chick liver clearly substantiate work with sections of fish liver.[55] Development in each uses tubules of entodermally derived epithelial cells. Elias[56] reviewed the origin and early development of the liver in various vertebrates, including 30 species from 16 orders of 8 classes. He found that teleost entoderm was the sole contributor to liver parenchymal development and that internal rearrangements of liver cells and yolk-associated blood islands occurred, giving rise to cellular arrangements consistent with future adult relationships. The manner by which the teleost bile duct system made contact with the gut lumen was not defined in the investigations of Elias.[56] Hinton[3] studied liver formation in embryonic medaka *(Oryzias latipes)*. The foregut epithelium at 9 days after fertilization was moderately well differentiated and revealed mucosal cells with a prominent brush border. Serosal cells of the foregut and cells of the future muscularis propria were flattened and undifferentiated. Even at this early stage of liver bud development, hepatocyte polarity was evident; basal portions differed in their organelle components from apical portions. Regular spacing occurred between adjacent hepatocytes. Endothelial cells revealed pinocytotic vesicles, and fat-storing cells of Ito were present. Processes of the fat-storing cells were in close proximity to collagen. Further evidence for differentiation of the liver mass included the presence of bile canaliculi and bile preductules. Hepatic tubular architecture was apparent. Based on the studies of Elias[56] and on the findings in the medaka, hepatocytes, biliary epithelial cells, and exocrine pancreatic cells (when present), as well as the cells lining their drainage duct systems, are entodermal derivatives arising from the foregut floor.

V. CONTROL ADULT MORPHOLOGY

The histology of the liver in control or normal teleosts has been described in a variety of species and recently reviewed.[3] The epithelial cells of the teleost liver include hepatocytes; biliary epithelial cells lining bile passageways, including preductules, ductules, and intrahepatic ducts; and, in certain fishes, the acini of exocrine pancreatic cells (the so-called hepatopancreas). In addition, endothelial cells (specialized epithelial cells lining blood vessels) line all intrahepatic branches of the portal vein and the hepatic artery and the tributaries of the hepatic veins. Finally, mesothelial cells are incorporated as the covering of the outside surface of the liver, the hepatic capsule.

In the rainbow trout and selected other species, hepatic architecture, cellular composition, and quantitative relationships are well characterized.[32,57-60] The liver is, in three dimensions, a tunneled continuum.[3,61,62] The solid portion is arranged as tubules of hepatocytes, with canaliculi, bile preductules, or ductules as the lumen. Individual tubules are incompletely separated by sinusoids that, with their perisinusoidal space, fill the tunnels or labyrinth. This arrangement, similar to that of birds and reptiles, has some distinct features from mammals. Biliary epithelial cells are numerous within hepatic tubules. Hampton et al.[32] estimated that for every seven (male trout) or nine (female trout) hepatocytes there was a single biliary epithelial cell. Canaliculi, the principal parenchymal biliary passageway of mammalian livers, are short in teleosts and connect with transitional structures, bile preductules. Ductules (completely lined by biliary epithelial cells) course within the hepatic tubules and, upon leaving the continuum, pick up a basal lamina and course within the hepatic labyrinth in the space of Disse. Intrahepatic bile ducts are within the hepatic labyrinth, make-up approximately 8% of the stroma, and are lined by cuboidal through columnar epithelial cells. Sinusoids and larger vessels are lined by endothelial cells. The lumens of sinusoids are generally free of fixed macrophages (Küpffer cells). In the perisinusoidal space of Disse, thin extensions of stellate, perisinusoidal cells of Ito are seen. These apparently extend into hepatic tubules, making a form of support.[63] Occasional macrophage centers lie in the stromal space occupying about 1% of this volume.[32] Since bile ductules and hepatic arterioles do not lie in close proximity to portal (afferent) venules, it is usually not possible to distinguish hepatic venules (efferent) from venules distributing blood to the sinusoids. Sinusoids contain both arterial and venous blood. Acini of exocrine pancreatic cells and their centroacinar and ductular epithelial cells are found at the hilus of the liver in all species of teleosts studied to date. In some, for example, garfish, English sole, starry flounder, rock sole, and catfish, these structures course through the stromal compartment in close proximity to large veins of the hepatic portal system.

VI. LIVER CELL TYPES

A. Hepatocytes

Being the most numerous, hepatocytes account for up to 80% of the total liver volume.[32] Sexual differences in control hepatocyte ultrastructure have been demonstrated in several species.[64-66] Hepatocytes of actively spawning female rainbow trout are smaller and more numerous than those of their male counterparts.[32] Although nuclear volume is equivalent in each sex, cytoplasmic volume of trout female hepatocytes is significantly less. Apparently, these differences are also seen when female and male hepatocytes of similarly aged rats are compared.[67] The female vitellogenin-producing hepatocyte shows extensive and parallel arrays of cisternae of granular endoplasmic reticulum (GER) in two zones, immediately adjacent to the nucleus and at the cell periphery. Lakes of glycogen are apparently reduced in size at this time, and lipid droplets of medium electron density are seen in most hepatocytes. Given the glycogen reduction and relatively enhanced GER, these cells will likely show more basophilia than during periods of less synthetic activity. When compared to the fine structure of the synthetically active female hepatocyte, hepatocytes from male medaka of similar age appear larger and have more glycogen and less GER.

1. Differentiating Features of Hepatocytes

These cells contain lysosomes,[30,68] and they are readily marked by enzyme histochemical procedures for acid phosphatase. The lysosomes have a preferential pericanalicular localization within the apices of hepatocytes. Hepatocytes are lightly positive with the histochemical procedure for the enzyme, aldehyde dehydrogenase. Similarly, hepatocyte cytoplasm in medaka and rainbow trout is lightly positive for glucose-6-phosphate dehydrogenase, uridine diphosphate glucuronosyl dehydrogenase, and DT diaphorase.[69,70] In addition, hepatocytes typically contain fairly abundant amounts of glycogen, and this region of the cells and their plasma membranes are usually positive by the periodic acid Schiff's reagent stain. During periods of vitellogenin production, female hepatocytes might stain much more positively with lipid stains, due to the known association of abundant lipid in the liver during this process.[71] Polyclonal antibodies against the LM_2 and the LM_4 isoforms of cytochrome P450 also mark hepatocytes, as does the monoclonal antibody (anti-CYP1A1 MAb 1,12,3) directed against cytochrome P4501A.[72-76] When present, abundant glycogen is a strong differentiating feature of hepatocytes vs. the other resident liver cell types. Unfortunately, an early and general response of teleost hepatocytes to toxicity is the loss of glycogen.[20] Perhaps the second most recognizable feature of hepatocytes is their abundant GER. As will be presented later, alterations following chronic toxicity may induce appreciable changes in hepatocyte shape and staining, making unequivocal identification problematic in routine histopathologic preparation. Cytochemical markers for control or normal teleost hepatocytes are summarized in Table 3.

B. Biliary Epithelial Cells

Recently, the hierarchy of biliary passageways in the liver of the rainbow trout *(Oncorhynchus mykiss)* was described.[58] Relationships of canaliculi, intratubular preductules, and ductules and subsequent passageways coursing in the hepatic labyrinth were presented[58] and detailed by Hinton.[3]

The bile preductular epithelial cell appears elongated when the hepatic tubule is sectioned longitudinally and small, with only a narrow rim of cytoplasm, if the tubular orientation is transverse. These cells share cell junctions with hepatocytes to make a portion of the wall of this transitional passageway. Bile preductular cells contribute few if any microvilli to the wall of the bile passageway. This is in contradistinction to hepatocytes, whose contribution is great. The cytoplasm immediately beneath the lumenal plasma membranes of biliary epithelial cells is the site of numerous cytofilaments. Cell junctions are of the desmosomal type, and associated cytofilaments are numerous.

The bile ductule is completed when the entire mural elements are of ductular epithelial cells. Ductular epithelial cell ultrastructure is similar to that of the preductule in cell size, and nuclei retain the elongation of their preductular counterparts. Broad but few microvilli are present. Apical and lateral cytoplasm contains abundant cytofilaments.

Hampton et al.[58] defined the transition from ductule to small bile duct in trout at the point where basal lamina is present. I have found this to occur when the ductule leaves the intratubular site to enter the perisinusoidal space of the hepatic labyrinth. Here, stellate fat-storing cells of Ito lie immediately outside the basal lamina. Epithelial cell height increases as the lumenal diameter of the ducts enlarges. The tallest columnar epithelial cells show a prominent brush border, tight junctions at the apical-most edge of the lateral plasma membrane, large accumulations of mucus in the upper half of the cell, and a prominent nucleus in the lower third of the cell. Nuclei are frequently indented. Macrophages are often seen between bases of the adjacent ductal epithelial cells. A prominent basal lamina separates epithelium and macrophages from smooth muscle cells, nerve endings, fibroblasts, and a capillary plexus.[58]

1. Differentiating Features of Biliary Epithelial Cells

The circumferential arrangement of ductular and ductal epithelial cells may be visible in sections. Chemical markers for biliary epithelial cells are provided in Table 3. The enzyme histochemical reaction for alkaline phosphatase[69] is

Table 3. Cells of Control Teleost Liver and Associated Markers

Cell type	Marker
Hepatocytes	Acid phosphatase
	Aldehyde dehydrogenase
	Glucose-6-phosphate dehydrogenase
	Uridine diphosphate glucuronosyl dehydrogenase
	DT diaphorase
	Periodic acid Schiff's reagent
	Lipid stains
	Anti-P450 LM$_2$ IgG
	Anti-P450 LM$_4$ IgG
	Anti-CYP1A
Bile ductular and ductal epithelium	Alkaline phosphatase
	γ-glutamyl transpeptidase
	Glucose-6-phosphate dehydrogenase
	Mg^{2+}-Adenosine triphosphatase
	Uridine diphosphate glucuronosyl dehydrogenase
	DT diaphorase
	Periodic acid Schiff's reagent
	Anti-P450 LM$_2$ IgG
	Anti-P450 LM$_4$ IgG
	Anti-CYP1A
Endothelial cell	Alkaline phosphatase
	Anti-P450 LM$_2$ IgG
	Anti-P450 LM$_4$ IgG
	Anti-CYP1A
Perisinusoidal cell of Ito	Lipid stains
	Apple green autofluorescence
	Alcian blue pH 2.5 positive products
Macrophages of macrophage aggregates	Acid phosphatase
	Periodic acid Schiff's reagent

particularly strong in the connective tissue sheath of medium-sized and larger intrahepatic bile ducts. Biliary epithelial cells are the single resident hepatic cell type in which γ-glutamyl transpeptidase is normally found. Mg^{2+}-dependent adenosine triphosphatase is particularly strongly reactive over ductular and ductal epithelium.[77] The other enzymes (glucose-6-phosphate dehydrogenase, uridine diphosphate glucuronosyl dehydrogenase, and DT diaphorase) all mark biliary epithelial cells.[69] Plasma membrane of biliary epithelial cells is usually positive with the periodic acid Schiff's reagent, and the mucus granules of tall columnar biliary epithelial cells stain positively. Biliary epithelial cells are positive for cytochrome P450 when immunohistochemical procedures using Anti-P450 LM$_2$ IgG, Anti-P450 LM$_4$ IgG, and Anti-CYP1A are used. Electron microscopical localization of the latter antibody using immunogold procedures has confirmed the light microscopic observations.[33]

C. Endothelial Cells

Little of the comparative microscopic anatomy literature has been devoted to endothelium of fishes in general and to venous and arterial endothelium specifically. The few accounts have dealt almost exclusively with sinusoidal endothelial cells.

Freeze-etch replicas[78] and scanning electron microscopy of liver following vascular perfusion fixation[79] were used to study fenestrae. In goldfish,[80] approximately 15 to 20 fenestrae were grouped in forms resembling sieve plates described in other vertebrates.[81] In trout,[79] fenestrae were not as tightly grouped. When their diameters were estimated, ranges from 75 to 150 nm were obtained. In both species, cytofilaments surrounded fenestrae. Ultrastructure of fenestrae and endocytotic vesicles in endothelium from trout were illustrated and described.[36] Sinusoidal endothelium has proven to be particularly reactive with alkaline phosphatase in histochemical preparations.[69] In addition, the antibodies directed against cytochrome P450 isozymes, LM_2, LM_4, and CYP1A, are positive[74] over endothelial cells.

1. Differentiating Features of Endothelial Cells

Location immediately adjacent to blood cells and attenuated cytoplasm help differentiate these cells. The strong response with alkaline phosphatase will assist in differentiating endothelial cells from fat-storing cells of Ito.

D. Stellate, Fat-Storing Cells of Ito

Morphology and functional properties of this cell in various vertebrates were reviewed.[82] The connective tissue cells of the teleost liver include stellate, fat-storing, perisinusoidal cells of Ito;[53] fibroblasts associated with the hilus of the liver; and cells incorporated as a sheath around major blood vessels and bile ducts and located immediately beneath the liver capsule and the blood cells. The cells of Ito and the fibroblasts provide a supportive framework for the liver. Perisinusoidal, fat-storing cells contain an elongated and frequently indented (by lipid droplets, the solvent of vitamin A) nucleus with peripheral heterochromatin. Well-developed GER with dilated cisternae also characterize these cells. Ito et al.[83] studied livers of 48 species of fishes and reported these cells in all but three. Since then, Hinton et al.[13] and Laurén et al.[84] have demonstrated the presence of fat-storing cells in one of the three originally reported to lack them, medaka *(Oryzias latipes)*.

"Empty" fat-storing cells have been distinguished from those containing lipid droplets.[78] Empty fat-storing cells are fibroblast-like cells with well-developed GER.[54] First described in human adults and embryos,[85] these cells have subsequently been described in various fishes and reptiles. In teleosts, injection of vitamin A causes the number of fibroblast-like cells to decrease while the number of lipid-laden cells increases.[86,87] Perhaps the heaviest accumulation of lipid is found in the Antarctic fish *Dissostichus mawsoni*.[88] Numerous desmosomal junctions between perisinusoidal, fat-storing cells and other liver cells[78,86,87,89,90] suggest a mechanical, supporting role for these cells in teleosts. Numerous cytofilaments are another distinguishing feature of these cells. If the cytofilaments are contractile, as has been suggested, the orientation of the fat-storing cells would suggest they have a role in the regulation of blood flow within the sinusoid.[90] The contribution of fat-storing cells to the volume of the parenchyma was estimated for trout.[32] Actual volume may be underestimated, since their processes are elongated and the terminal portions may be difficult to differentiate from portions of the other cell types.[32]

Vitamin A shows apple green autofluorescence, which fades upon continued excitation. Alcian blue pH 2.5 is positive for acidic glycosaminoglycans of fat-storing cells. When present, lipid vacuoles stain positively with any of a variety of lipid stains. Semi-thin sections of epon-embedded material stained with toluidine blue reveal lipid of fat-storing cells as aquamarine droplets.

1. Differentiating Features of Ito Cells

Location, shape, and contents (or lack thereof) have found usage in the nomenclature of this cell type. To differentiate fat-storing cells from endothelial cells appears to be the greatest challenge. Their common mesodermal origin and the presence of cytoplasmic vesicles and cytofilaments illustrate the degree of "shared features" of these cell types. Alkaline phosphatase, positive in endothelial but negative in fat-storing cells, may be the strongest differentiator at present.[36]

E. Macrophages of Macrophage Aggregates

Descriptions of hepatic macrophages outside the sinusoids of teleost liver are far more common than in their mammalian counterparts. Two sites are involved: first within the hepatic tubule and second in aggregates or collections sometimes referred to as centers. Yamamoto[91] and Tanuma and Ito[89] analyzed livers of crucian carp *(Carassius carassius)* and described macrophages in the interhepatocytic space (i.e., intratubular), especially adjacent to bile preductules and ductules. These have also been reported in rainbow trout *(Oncorhynchus mykiss)*.[58,79] Perisinusoidal, interhepatocytic macrophages were more numerous and their cytoplasm was of greater volume in female vs. male trout at spawning,[32] suggesting a role for the cells in the remodeling of trout liver after vitellogenin synthesis. Perisinusoidal, interhepatocytic macrophages are particularly well demonstrated by their staining for glucose-6-phosphate dehydrogenase.[77,92]

Hepatic macrophages of macrophage aggregates are considered stromal components.[32] These cells contain large heterophagic vacuoles, residual bodies, and variable amounts of pigment, including melanin and lipofuscin.[68] Macrophages of aggregates react positively for periodic acid Schiff's reagent and for acid phosphatase and presumably other lysosomal enzymes.

1. Differentiating Features of Macrophages

For those of aggregates, the location and proximity to other macrophages are strong differentiating features. The strongly positive glucose-6-phosphate dehydrogenase reaction in perisinusoidal, interhepatocytic macrophages distinguishes them from those of aggregates. The former may be newly arriving cells, possibly having entered the interhepatocytic space after traversing the sinusoidal wall.[50] These macrophages usually occur singly or as a pair and may have processes in close approximation to bases of bile ductular epithelial cells at the center of hepatic tubules.[58]

F. Concluding Perspective — Resident Hepatic Cells

From the above, it is apparent that we have markers for most of the resident cell populations within the teleost liver. Perhaps a single cell type that needs additional work is the Ito cell and, more specifically, the differentiation of this cell from endothelial cells. Since both are of mesodermal origin, it may be difficult to differentiate these cell types in toxicant induced lesions without use of additional markers. Teh and Hinton[69] have developed a technique of freeze-drying and embedment in complete monomer of glycolmethacrylate (GMA). This procedure employs no fixative and no chemical dehydration and thereby retains enzyme activity for extended lengths of time when stored under dry conditions at room temperature. Higher resolution than cryostat sections is obtained when sectioning is at 4 μm or thinner. With this technique, individual organs and their microscopic entities can be analyzed in serial sections, using immunohistochemical, enzyme and determinative histochemical, and routine staining preparations.

VII. INTERCELLULAR MATRIX OF TELEOST LIVER — A RESEARCH NEED

Components of the intercellular matrix of various organs play important roles in both sending signals to and receiving them from epithelial cells. The end result of these signals helps to regulate such important cellular processes as replication, biosynthesis, differentiation, spacing, and shaping.[93] Important components of the matrix include elastin, collagen, glycoproteins (fibronectin, laminin, chondronectin, and osteonectin), and proteoglycans, containing various different glycosaminoglycan (GAG) chains. Of the GAGs, hyaluronic acid is important in migration of cells through spaces within matrix. This activity is important in embryogenesis and neoplasia, specifically tumor invasion.[94] In general, GAG chains are found covalently attached to protein (as proteoglycans). Hyaluronic acid, however, occurs as a free GAG chain of very large molecular weight. Other GAGs in connective tissue proteoglycans include heparin sulfate, heparin, chondroitin 4-sulfate, chondroitin 6-sulfate, dermatin sulfate, and keratin sulfate.[93] In studies with teleosts, it is important to know that while the protein core of the many different proteoglycans found in connective tissues may differ, the structure of the sugar units in each GAG is usually relatively constant in all animal species. Studies with various tumor cells have shown changes in sulfation of the cell surface with transformation.[95] For example, mucosal cells of normal colon contained little chondroitin sulfate, while the other proteoglycans (hyaluronate, heparin sulfate, and dermatin sulfate) were high. By contrast, in neoplasms, chondroitin sulfate was by far the most abundant.[96] Chondroitin sulfate is increased in nephroblastoma and, following removal of the tumor from patients, serum level for this GAG decreases.[93,95] All of the above implicate various GAGs in tumor biology. Despite the importance of the matrix in neoplasia, literature search revealed no information on presence and relative role of matrix components in teleost liver. If the matrix of the teleost liver affects such important events as cell replication, cell shape, and metastasis of tumor cells, characterization is most important.

Another important aspect of this type of analysis is the potential for detection, by sensitive immunohistochemical methods, of component portions of the matrix and the cells that produce them. Such information could aid in determining cellular components of the liver involved in various ill-defined neoplastic lesions. For example, in rat liver made cirrhotic by repeated carbon tetrachloride (CCl_4) injection,[97,98] certain matrix components increased, and electron microscopic histochemistry traced laminin, a basement membrane component, to endothelial, smooth muscle, and Ito cells. Since neoplastic transformation affects synthesis and secretion of proteoglycans, it is reasonable to conclude that tumor cells surround themselves and grow within an extracellular matrix that is quite different from normal cells. The importance of the intercellular matrix in embryogenesis and neoplasia, coupled with the timing of exposures in fish carcinogenesis, underscores the need for further characterization of this teleost hepatic component.

VIII. REQUIREMENTS FOR CELL MARKERS IN CHRONIC TOXICITY OF FISHES

Chronic toxicity places certain demands on the toxicologist that are not normally encountered in acute toxicity. Cell alterations in acute lethal toxicity are generally restricted to changes within target cells. The time frame is such that host responses are few, and sequelae lesions have not developed. With chronic toxicity, sublethal adaptation of target cells occurs, leading to a variety of alterations in organelles of toxicant altered, but surviving cells. The spectrum of these alterations has been described in detail.[99]

Endpoints of chronic toxicity include various lesions that differ appreciably from pre-exposure morphology. Studies on the histogenesis and progression of these lesions require the identification of component cell types. Even cells whose control or normal morphology is well understood may assume shapes in chronically altered states that make them difficult to resolve without electron microscopy (see Reference 20 for examples of altered hepatocytes or special techniques).

The flounder nephron and its changes in cell injury have been examined in structure/function studies. Such studies in other organs in fish are apparently lacking (compare teleost renal toxicity review by Pritchard and Renfro[100] to the hepatic toxicology review of Gingerich)[47]. Now, with isolation and maintenance of fish hepatocytes[101] and other resident liver cells[102] in defined media, the opportunity exists for precise evaluation of coupled structure/function studies. The liver of teleost fishes is a common target organ in aquatic toxicology,[16] but exposure to proven mammalian hepatotoxicants such as CCl_4,[103–105] bromobenzene,[106] acetaminophen,[107] or acrylamide[108] has failed to induce appreciable histopathology. Only allyl formate[109,110] induced severe hepatocellular necrosis.

Agents that produce hepatic neoplasia in fish may cause appreciable cytotoxicity early after exposure. However, because hepatic neoplasms and not cytotoxicity were the endpoints in most hepatocarcinogenesis studies in fish, early changes have not been investigated in detail. Recent studies of progression of hepatic neoplasia following exposure to diethylnitrosamine included serial analysis of alterations, and the toxicity was appreciable.[84,111] Braunbeck et al.[30] focused on these alterations in an ultrastructural study employing vascular perfusion fixation of medaka. Nuclear and cytoplasmic alterations were observed. Swollen, condensed, and variously shaped mitochondria with alterations of cristae were seen. Lysosomes and peroxisomes were verified using cytochemical procedures, and the former showed abundant increases in number and ultrastructural alterations. Smooth endoplasmic reticulum was abundant in some hepatocytes, and the cisternae of endoplasmic reticulum showed steatosis. This work is important, since it illustrates the nature and extent of organelle alteration possible in teleost hepatotoxicity.

Serial progression studies of fish liver carcinogenesis[50,84,112,113] not only provide important information on histogenesis of eventual neoplasms, but they illustrate changes at the cell and tissue levels that eventually result in grossly distorted livers. The next sections will focus on some specific lesions, currently unresolved questions, and the types of markers that may facilitate our understanding of the cellular composition and histogenesis of these lesions.

A. Hepatic Histopathologic Biomarkers

Liver lesions that signal effects resulting from prior or ongoing exposure to one or more toxic agent(s) are candidates for hepatic histopathologic biomarkers. When field investigations detect a lesion in higher than anticipated prevalence at a highly localized site those toxicants consistently shown to be associated with that lesion in laboratory exposures emerge as suspects. The relationship of liver histopathologic biomarkers to other types of biomarkers has been described.[36] Briefly, histopathologic biomarkers are higher level responses often signifying prior metabolism and macromolecular binding; i.e., they occur after chemical and cellular interaction. Exposure to a xenobiotic might induce the formation of a specific enzyme. If subsequent exposure led to increased metabolism by the induced enzyme, levels of toxic intermediates that exceed cellular protective mechanisms could arise, resulting in cellular toxicity and death. In this case, the histopathologic biomarker is necrosis.

The hepatic biomarkers recently recommended as presently applicable[36] are presented in Table 3. Each is defined in brief, and more extensive definition is available.[19,20,36] Readers should also consult Chapter 8.

B. Relationship of Specific Liver Cells to Histopathologic Biomarkers

Liver cellular distribution for cytochrome P4501A includes hepatocytes, biliary epithelial cells, and endothelial cells of arterioles and sinusoids.[33,121] It is not surprising that under certain conditions, each of these cell types reflects alterations consistent with injury and cytotoxicity. In the medaka, where studies have included detailed light and electron microscopic analysis of diethylnitrosamine toxicity, it appears that hepatocytes are the major target. However, biliary epithelial cells also show alteration. Perhaps the lack of application of electron microscopy in the majority of histopathologic investigations of teleost liver toxicity explains in part why few studies have demonstrated endothelial toxicity and necrosis. With the compound allyl formate, Droy et al.[110] demonstrated primary endothelial damage with resultant hemorrhage and necrosis of hepatocytes.

C. Foci of Cellular Alteration and Role of Cytochemistry

Liver carcinogenesis induced in laboratory rodents is preceded by a variety of cellular changes, some of which have been related to the development of defined neoplastic endstages and are regarded as preneoplastic lesions.[122] Characteristic alterations in the biochemical and morphological phenotype of hepatocytes emerge in various species, including primates, long before hepatocellular adenomas and carcinomas become manifest.[123] To the extent that it is understood, histogenesis of liver neoplasia following exposure of spheepshead minnow (*Cyprinodon variegatus*) or of medaka to diethylnitrosamine involves a similar series of cellular changes.[113,124] The abnormal hepatocytes that precede hepatic neoplasms usually form foci that are perfectly integrated into the normal liver parenchymal architecture. These foci of cellular alteration (FCA) were described in Table 4. Different types of FCA have been distinguished in rats[125] and fish (see review[3] and Chapter 8). The abnormal morphology of FCA is regularly associated with a variety of biochemical aberrations, as demonstrated by cytochemical methods in rats[67,126] and in medaka.[69,113] In rats and in medaka, changes in glycogen content and a decrease or increase in the activity or content of certain enzymes have been used as "negative" or "positive" markers for FCA.

Table 4. Hepatic Histopathologic Biomarkers

Biomarker	Definition	Tool needed for detection
Hepatocellular necrosis and sequelae	Coagulative necrosis in which the shape of cells and their tissue arrangement is maintained, facilitating recognition of organ and tissue. Different from liquefactive necrosis associated with bacterial infection. May be associated with inflammatory response. May be focal or multifocal in extent.	Conventional light microscopy/histopathology
Hyperplasia of regeneration	Follows necrosis. Surviving cells undergo hyperplasia to regenerate needed epithelial cells lost due to necrosis.	Light microscopy focusing on cell size and basophilic nature, as well as the formation of islands of regenerating cells. May be quantified by light microscopic autoradiography using tritiated thymidine or immunohistochemistry using monoclonal antibodies directed against the synthetic thymidine analog, bromodioxyuridine, or other markers such as proliferating cell nuclear antigen (PCNA) and K_i67.
Bile ductular/ductal hyperplasia	Profiles of these biliary passageways are numerous and contiguous with abundant branching and coiling. Cytologic features of hyperplastic epithelial cells are normal. Of chronic duration, this condition has been a consistent finding in wild fish from chemically contaminated sites.[114,115]	Light microscopic histopathology with enhanced mitotic activity as determined above
Hepatocytomegaly	Hepatocellular hypertrophy characterized by organelle enlargement or hyperplasia, with enlarged cellular diameter but without nuclear changes.	Light microscopy with confirmatory, selective electron microscopy
Variant — megalocytosis	Marked cellular and nuclear enlargement. Enlarged nuclei often contain false and real inclusions, and multinucleated megalocytes may be seen.	Light microscopy with selective transmission electron microscopy for confirmation

Table 4. (Continued)

Biomarker	Definition	Tool needed for detection
Hydropic vacuolation of hepatic epithelial cells[116-118]	Seen in vacuolated cells of liver. Affected cells have clear cytoplasm, small compound nuclei, and are markedly vacuolated.	Histopathology with selective confirmatory electron microscopy
Foci of cellular alternation — staining or tinctorial change	Foci of hepatocytes detection by conventional hematoxylin and eosin preparations as being basophilic, eosinophilic, clear cell, or fatty. Foci are in early stage in the stepwise histogenesis of hepatic neoplasia.	Conventional light microscopy
Hepatocellular adenoma	Perhaps an enlargement of the basophilic focus previously described. Apparently, clear cell and eosinophilic variants also occur in English sole[119] and in brown bullheads.[120] May be quite large, forming a bulge at the surface of the organ. Mitoses are rare, but proliferation is evident through compression of surrounding cells, especially with larger lesions.	Conventional histopathology[a]
Hepatocellular carcinoma	Varies from microscopic to very large; often occupies a major portion of the organ. Mitotic figures are numerous and can be bizarre. The masses of cells are usually solid in pattern, resembling engorged tubules. Tumor cells may be quite pleomorphic, with some assuming spindle shapes, while others are small, polyhedral in shape, and arranged as tight collections with distinct intercellular spaces. The margin of the tumor is less well-defined, and invasion into otherwise normal parenchyma is common.	Conventional light microscopy
Cholangioma	Tumors of bile duct origin characterized by retention of their ductular/ductal architecture. Possess distinct margins and have a nodular, well-defined	Conventional light microscopy

Table 4. (Continued)

Biomarker	Definition	Tool needed for detection
Cholangioma (continued)	mass. Focal ductal elements may be present, some of which may be cystic with papillary projections. Lining epithelium is cuboidal to low columnar, simple, and well differentiated.	
Cholangiocarcinoma	A larger lesion; component cells are pleomorphic, and mitotic figures are common. Invasion of surrounding tissues is common. At the margin, columns of invading cells interdigitate with nontumorous liver tissue. Biliary epithelial cells of this neoplasm may form sheets or may be in the form of undifferentiated ductules.	Conventional light microscopy
Mixed hepato-cholangiocellular carcinoma	Of hepatic neoplasms, these carcinomas are seen in some species at equal or even greater frequency than those composed purely of hepatocytes. Evidence has been presented suggesting that both cell types of the tumor are indeed neoplastic.	Conventional light microscopy

[a] Baumann et al. (1990)[120] present a different classification. From hepatocellular focal alteration, they describe a larger, more clearly defined subpopulation of hepatocytes, which they term the "hepatocellular nodule." They do not accept hepatocellular adenoma as a substitute for hepatocellular nodule. Whatever the term, the bridging lesion between focal cellular alteration and the more advanced lesions described are all considered presently applicable biomarkers.

Only a few studies in the rat have been conducted in such a manner as to provide a comparison of both morphological and biochemical phenotypes in individual foci systematically.[125]

To perform enzyme cytochemistry, it is necessary to use procedures that retain the activity of the enzymes. Teh and Hinton developed methodology that employs freeze-drying of nonfixed specimens and embedment in cold-polymerized glycolmethacrylate.[69] The enzyme histochemical markers developed to date mark foci and portions of subsequent neoplastic lesions. Enzymes that have been proven useful in medaka include acid phosphatase (ACP), alkaline phosphatase (ALKP), γ-glutamyl transpeptidase (GGT), uridine diphosphate glucose dehydrogenase (UDPGDH), quinone oxidoreductase or DT diaphorase (DTD), and glucose-6-phosphate dehydrogenase (G6PDH). GGT and G6PDH commonly mark toxic and inflammatory phases, respectively, while GGT, DTD, ALKP, ACP, G6PDH, and UDPGDH have proven useful in differentiating phenotypically altered hepatocytes during the later stages of neoplastic development.[69] Biochemical and histochemical properties of hepatic tumors of rainbow trout were determined[70] using procedures similar to those of the medaka studies reported above. Specific activities of phase I enzymes, ethoxyresorufin-O-deethylase (EROD), microsomal and cytosolic epoxide hydrolase (mEH and cEH), and aldehyde dehydrogenase (ALDH) and DT-diaphorase, and the phase II enzymes, γ-glutamyltranspeptidase (GGT), glutathione transferase (GST), and uridine diphosphoglucuronyl transferase (UDPGT), were measured. Cryostat sections of tumor and surrounding liver were analyzed immunohistochemically for cytochrome P4501A1 and histochemically for ALDH (benzaldehyde and hexanal), DT-diaphorase, GGT, and uridine diphosphoglucuronosyl dehydrogenase (UDPGDH). In tumor tissues, the largest biochemical changes were found with benzaldehyde dehydrogenase, where activity increased from undetectable levels to 7.4 nmol/min/mg protein, and GGT, where activity increased 12-fold over controls. Increases in other enzymes ranged from 1.26 to 2.84 times that of control liver. EROD decreased, and cEH and mEH were unchanged. Histochemistry showed tumors were enriched in ALDH, GGT, DT-diaphorase, and UDPGDH and depressed in cytochrome P4501A1.

IX. LECTINS AS MARKERS OF SPECIFIC CELL TYPES IN TROUT LIVER

Using a panel of 14 different lectins, we screened normal liver and hepatic neoplasms in trout.[126a] The lectin soybean agglutinin was a positive marker for neoplastic bile ducts and allowed detection of very small malignancies. The lectin succinylated wheat germ agglutinin was a positive marker for neoplastic bile ductules and ducts. Examples of this lectin histochemistry are shown in Figure 2. Binding to cytosol, succinylated wheat germ agglutinin may be used in sections where the bile duct and/or ductular lumens are obscured. This work assumes added significance when one considers the common finding of mixed hepato-cholangiolar tumors of trout and other fishes, including medaka.

X. FATE OF PHENOTYPICALLY ALTERED HEPATOCYTE POPULATIONS

As stated above, phenotypically altered hepatocyte populations immediately follow cytotoxicity and are an early change encountered in medaka serially sampled following exposure to diethylnitrosamine.[50] To show that phenotypically altered populations within foci can become a component of the eventual tumor, it is necessary to demonstrate their resistance to toxicity, their selective growth with respect to the remaining elements of the liver, and their inclusion in cells of the resultant neoplasm. Two major factors appear to affect growth. These are (1)

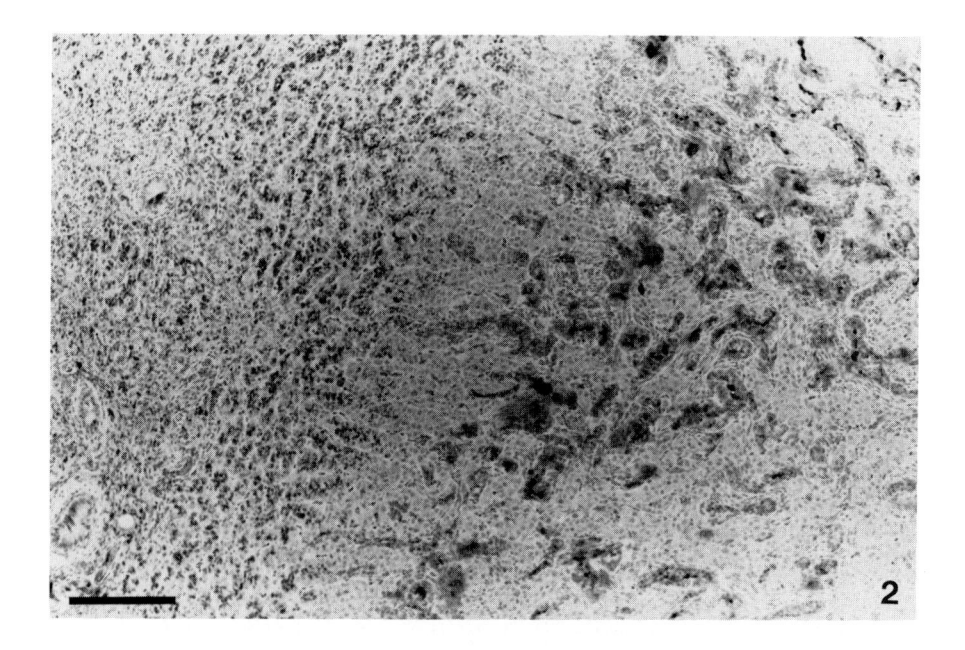

Figure 2. Right-hand side of the field is a mixed hepatocholangiocellular carcinoma from rainbow trout exposed to tumorigenic concentration of aflatoxin B_1. Intermediate to dark gray cells in this area are biliary ductular and ductal epithelial cells. Left-hand side of the field shows normal liver adjacent to the tumor. Darkly reacting biliary epithelial cells were marked by succinyl wheat germ agglutinin (WGA). This lectin illustrates cell-specific binding. Frozen section, Streptavidin-peroxidase complex technique counterstained with hematoxylin. Bar = 227 µm.

the capacity of the cells to undergo mitotic division and (2) the life span of the phenotypically altered cells. If the latter is longer than those of the surrounding, noninvolved liver, net growth of the small lesion may occur. The markers for mitotic division in tissue sections include autoradiography with tritiated thymidine, use of a monoclonal antibody directed against the thymidine analogue bromodeoxyuridine, or recently developed antibodies against nuclear antigens. The latter include proliferating cell nuclear antigen (PCNA) and Ki-67.[127-129] Figures 3 and 4 illustrate the PCNA test applied to sections from a medaka hepatocellular carcinoma. The proliferating cell nuclear antigen test has several advantages over other methods. First, no use of radioactive isotope is necessary and, second, the intact animals do not have to be exposed to isotope or bromodeoxyuridine prior to their sampling.

In addition, markers of apoptosis may be used to quantify the number of cells in a given lesion that are undergoing controlled cell death.[130] If apoptosis is diminished in a given lesion, this lesion would be expected to grow with respect to the remainder of the liver.

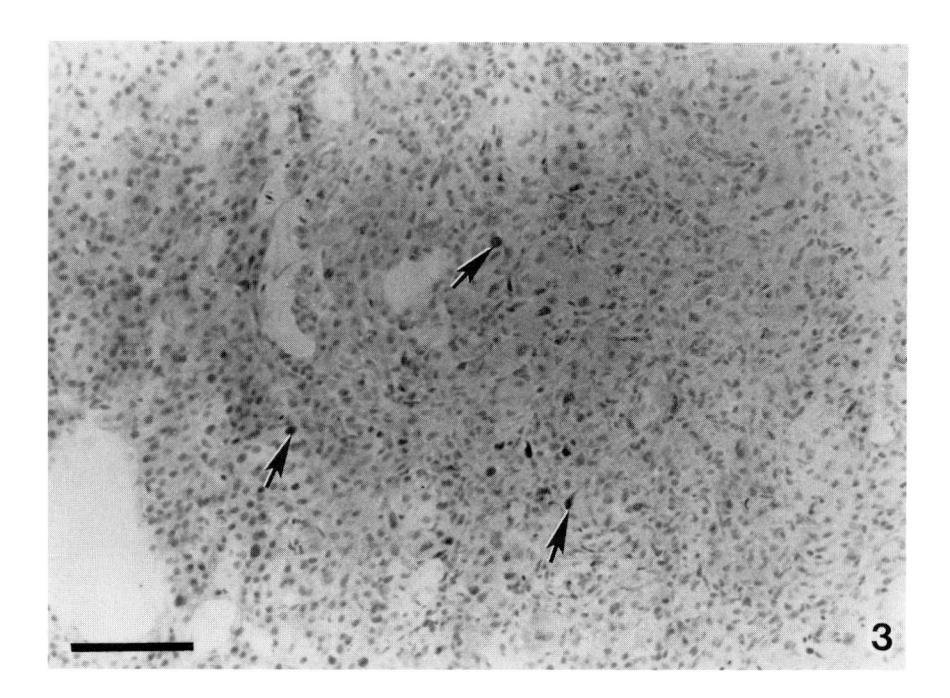

Figure 3. Proliferating cell nuclear antigen (PCNA, PC10) (DAKO, Santa Barbara, CA) is localized over nuclei of cell in hepatocellular carcinoma of medaka *(Oryzias latipes)* 27 weeks after exposed to 350 ppm DENA for 48 hr (arrows). Note the decrease in number of positive staining nuclei seen in the nontumor part of the liver (left side of the field). Section was treated with saponin for 30 min, incubated in 1:100 dilution of anti-PCNA overnight at 4°C, Streptavidin-peroxidase complex followed by counterstaining with hematoxylin. Freeze-dried glycolmethacrylate-embedded fresh liver (FDGE). Bar = 114 μm.

XI. GENETIC MARKERS AND HEPATIC LESIONS

Markers specific for individual gene expression are now becoming available and provide additional means of detecting important molecular events at the cell level and characterizing specific lesions. In situ hybridization may indicate those lesions in which component cells are expressing mutated *ras* or other oncogenes. Moore and Evans[131] employed an immunohistochemical reaction in which the N-terminal peptide of N-*ras* oncoprotein was detected. Small groups of hepatocytes and circular foci of altered cells in all livers of dab *(Limanda limanda)* from a contaminated site in the German Bight proved positive. The finding of low or zero incidence when the same test was applied to liver sections from dab at less-contaminated sites lends strength to the positive results. Important genes such as the p53 family, which control differentiation and growth, are highly conserved,[132] and their mutation during carcinogenesis may be important in teleost neoplasia.

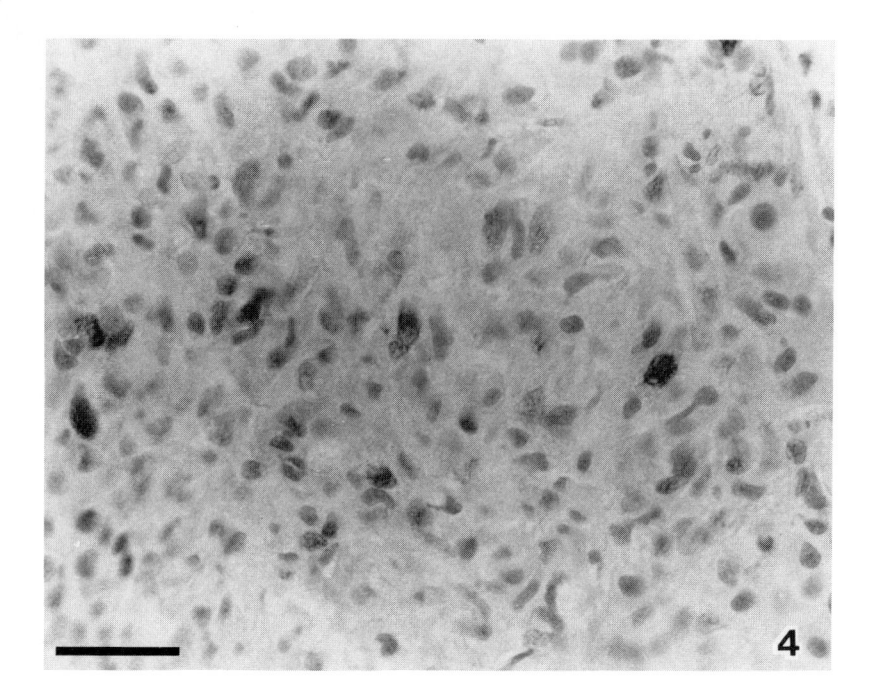

Figure 4. Higher magnification of portion of Figure 3, showing the positive staining nuclei. Note the absence of staining in G_0 nuclei. Approximately 11 nuclei in stages late G_1 through G_2 are labeled. FDGE. Bar = 38 µm.

XII. CONCLUSIONS

1. Cells are the component units of all living matter and permit a focus that enables comparison of effects in various organisms exposed to contaminants.
2. Although certain cells and tissues may comprise a small portion of the overall volume of an organ, their location and contents may render them of extreme importance in states of toxicity.
3. Methods that overlook the role of specific cell types and their localization in chronic toxicity will likely miss key events that may ultimately lead to an understanding of the mechanism(s) of injury associated with a given toxicant.
4. Sufficient information is at hand to assemble cellular biomarker approaches and to apply them in integrated field studies. Approaches presented herein using examples from the livers are equally useful in other organs.
5. Molecular alterations are manifest at the cell and higher levels of biological organization. Organisms currently employed to determine lethality, acute toxicity, bioaccumulation, and primary productivity are amenable to cellular bioassay.

6. Gene products can be localized within discrete cells and tissues during normal growth and development, and their alterations and resultant products in toxicity should assist us in determining exposure and effect.
7. The development of cellular markers will likely need the continued interaction of molecular biologists, protein chemists, and morphologists. These products of their traditional laboratory disciplines are needed by ecologists and field biologists to demonstrate exposure to and effects of xenobiotics.
8. The more field-oriented disciplines have accepted "body burden" concepts and will likely be even more accepting of methodology linking exposure to effect.

ACKNOWLEDGMENTS

This work was partially funded by U.S. P.H.S. Grant CA 45131 from the National Cancer Institute, ES O4699 from the National Institute of Environmental Health Sciences (Superfund Basic Research Program), and by Contract No. DAMD17-90-C-0003 from the U.S. Army Medical Research and Development Command.

REFERENCES

1. Roesijadi, G., Metallothioneins in metal regulation and toxicity in aquatic animals, *Aquat. Toxicol.*, 22, 81, 1992.
2. Stegeman, J.J., Cytochrome P-450 forms in fish: catalytic, immunological and sequence similarities, *Xenobiotica,* 19, 1093, 1989.
3. Hinton, D.E., Structural considerations in teleost hepatocarcinogenesis: gross and microscopic features including architecture, specific cell types and focal lesions, in *Atlas of Neoplasms and Related Disorders*, Dawe, C.J., Ed., Academic Press, New York, in press.
4. Elias, H. and Pauly, J., *Human Microanatomy,* 3rd ed., F.A. Davis Co., Philadelphia, PA, 1966.
5. Ham, A.W., *Histology,* 7th ed., J.B. Lippincott Co., Philadelphia, PA, 1974.
6. Ashley, L.M., Comparative fish histology, in *The Pathology of Fishes,* Ribelin, W.E. and Migaki, G., Eds., The University of Wisconsin Press, Madison, WI, 1975, 3.
7. Bucke, D., Some histological techniques applicable to fish tissues, *Symp. Zool. Soc. Lond.,* 30, 153, 1972.
8. Ellis, A.E., Roberts, R.J., and Tytler, P., The anatomy and physiology of teleosts, in *Fish Pathology,* Roberts, R.J., Ed., Balliere Tindall, London, 1978, 13.
9. Grizzle, J.M. and Rogers, W.A., *Anatomy and Histology of the Channel Catfish,* Auburn Printing, Auburn, AL, 1976.
10. Groman, D.B., Histology of the striped bass, in *Am. Fish Monogr.,* Vol. 3, American Fisheries Society, Bethesda, MD, 1982, 116.

11. Harder, W., *Anatomy of Fishes. E. Schweizer-bar'sche Verlogsbuchhandlung (Nagele u. Obermiller)*, Stuttgart, West Germany, 1975.
12. Hinton, D.E., Lantz, R.C., and Hampton, J.A., Effect of age and exposure to a carcinogen on the structure of the medaka liver: a morphometric study, *Natl. Cancer Inst. Monogr.*, 65, 239, 1984a.
13. Hinton, D.E., Walker, E.R., Pinkstaff, C.A., and Zuchelkowski, E.M., Morphological survey of teleost organs important in carcinogenesis with attention to fixation, *Natl. Cancer Inst. Monogr.*, 65, 291, 1984b.
14. Kent, G.C., *Comparative Anatomy of the Vertebrates*, 7th ed., Mosby-YearBook, Inc., St. Louis, MO, 1992.
15. Kubota, S.S., Miyazaki, T., and Egusa, S., *Color Atlas of Fish Histopathology*, Shin-Suisan Shingun-sha, Tokyo, 1982.
16. Meyers, T.R. and Hendricks, J.D., Histopathology, in *Fundamentals of Aquatic Toxicology*, Rand, G.M. and Petrocelli, S.R., Eds., Hemisphere Publishing, Washington, D.C., 1985, 283.
17. Walker, J., W.F., *Functional Anatomy of the Vertebrates, An Evolutionary Perspective*, CBS College Publishing, Philadelphia, PA, 1987, 781.
18. Yasutake, W.T. and Wales, J.H., Microscopic Anatomy of Salmonids: An Atlas, U.S. Dept. of the Interior, Fish and Wildlife Service, Washington, D.C., 1983, 1–89.
19. Hinton, D.E. and Laurén, D.J., Liver structural alterations accompanying chronic toxicity in fishes: potential biomarkers of exposure, *Biomarkers of Environmental Contamination*, McCarthy, J.F. and Shugart, L.R., Eds., Lewis Publishers, Chelsea, MI, 1990a, 17.
20. Hinton, D.E. and Laurén, D.J., Integrative histopathological approaches to detecting effects of environmental stressors on fishes, *Am. Fish. Soc. Symp.*, 8, 51, 1990b.
21. Moore, M.N. and Simpson, M.G., Molecular and cellular pathology in environmental impact assessment, *Aquat. Toxicol.*, 22, 313, 1992.
22. Mayer, F.L., Versteeg, D.J., McKee, M.J., Folmar, L.C., Graney, R.L., McCume, D.C., and Rattner, B.A., Physiological and nonspecific biomarkers, *Biomarkers: Biochemical, Physiological, and Histological Markers of Anthropogenic Stress*, Huggett, R.J., Kimerle, R.A., Mehrle, P.M., Jr., and Bergman, H.J., Eds., Lewis Publishers, Chelsea, MI, 1992, 5.
23. Varanasi, U., Stein, J.E., Johnson, L.L., Collier, T.K., Casillas, E., and Myers, M.S., Evaluation of bioindicators of contaminant exposure and effects in coastal ecosystems, *Ecological Indicators, Vol. 1, Proceedings of the International Symposium on Ecological Indicators*, McKenzie, D.H., Hyatt, D.E., and McDonald, V.J., Eds., Elsevier Applied Science, London, 1992, 461.
24. McCarthy, J.F. and Shugart, L.R., Eds., Biomarkers of environmental contamination, in *Biomarkers of Environmental Contamination*, Lewis Publishers, Chelsea, MI, 1990a.
25. McCarthy, J.F. and Shugart, L.R., Biological markers of environmental contamination, *Biomarkers of Environmental Contamination*, Lewis Publishers, Chelsea, MI, 1990b, 3.
26. Plaa, G.L. and Hewitt, W.R., *Target Organ Toxicology Series*, Raven Press, New York, 1982.
27. Hook, J.B., *Toxicology of the Kidney*, Raven Press, New York, 1993.
28. Weber, L.J., *Aquatic Toxicology*, Vol. 1, Raven Press, New York, 1982.

29. Weber, L.J., *Aquatic Toxicology*, Vol. 2, Raven Press, New York, 1984.
30. Braunbeck, T.A., Teh, S., Lester, S.M., and Hinton, D.E., Ultrastructural alterations in hepatocytes of medaka *(Oryzias latipes)* exposed to diethylnistrosamine, *Toxicol. Pathol.*, 20, 179, 1992.
31. Stegeman, J.J., Brouwer, M., Di Giulio, R.T., Förlin, L., Fowler, B.A., Sanders, B.M., and Van Veld, P.S., Molecular responses to environmental contamination: enzyme and protein systems as indicators of chemical exposure and effect, *Biomarkers: Biochemical, Physiological, and Histological Markers of Anthropogenic Stress*, Huggett, R.J., Kimerle, R.A., Mehrle, P.M., Jr., and Bergman, H.L., Eds., Lewis Publishers, Chelsea, MI, 1992, 235.
32. Hampton, J.A., Lantz, R.C., and Hinton, D.E., Functional units in rainbow trout *(Salmo gairdneri,* Richardson) liver: III. Morphometric analysis of parenchyma, stroma, and component cell types, *Am. J. Anat.*, 185, 58, 1989.
33. Lester, S.M., Braunbeck, T.A., Teh, S.J., Stegeman, J.J., Miller, M.R., and Hinton, D.E., Immunocytochemical localization of cytochrome P450IA1 in liver of rainbow trout *(Oncorhynchus mykiss), Mar. Environ. Res.*, 34, 117, 1992.
34. Scott, K., Leaver, M.J., and George, S.G., Regulation of hepatic glutathione S-transferase expression in flounder, *Mar. Environ. Res.*, 34, 233, 1992.
35. Adams, S.M., Shepard, K.L., Greeley, J.M.S., Jimenez, B.D., Ryon, M.G., Shugart, L.R., and McCarthy, J.F., The use of bioindicators for assessing the effects of pollutant stress on fish, *Mar. Environ. Res.*, 28, 459, 1989.
36. Hinton, D.E., Baumann, P.C., Gardner, G.R., Hawkins, W.E., Hendricks, J.D., Murchelano, R.A., and Okihiro, M.S., Histopathological biomarkers, in *Biomarkers: Biochemical, Physiological, and Histological Markers of Anthropogenic Stress*, Huggett, R.J., Kimerle, R.A., Mehrle, P.M., Jr., and Bergman, H.L., Eds., Lewis Publishers, Chelsea, MI, 1992a, 155.
37. Hinton, D.E., Toxicologic histopathology of fishes: a systemic approach and overview, in *Pathobiology of Marine and Estuarine Organisms*, Couch, J.A. and Fournie, J.W., Eds., CRC Press, Boca Raton, FL, 1993, 177.
38. Kranz, H. and Dethlefsen, V., Liver anomalies in dab *(Limanda limanda)* from the southern North Sea with special consideration given to neoplastic lesions, *Dis. Aquat. Org.*, 9, 171, 1990.
39. Malins, D.C., Krahn, M.M., Myers, M.S., Rhodes, L.D., Brown, D.W., Krone, C.A., McCain, B.B., and Chan, S.-L., Toxic chemicals in sediments and biota from creosote-polluted harbor: relationships with hepatic neoplasms and other hepatic lesions in English sole *(Parophrys vetulus), Carcinogenesis*, 6, 1463, 1985b.
40. Sindermann, C.J., *Principal Diseases of Marine Fish and Shellfish*, Vol. 1, 2nd ed., Academic Press, New York, 1990.
41. Mac, M.J. and Edsall, C.C., Environmental contaminants and the reproductive success of lake trout in the Great Lakes: an epidemiological approach, *J. Toxicol. Environ. Health*, 33, 375, 1991.
42. Stegeman, J.J., Binder, R.L., and Orren, A., Hepatic and extrahepatic microsomal electron transport components and mixed-function oxygenases in the marine fish *Stenotomus versicolor, Biochem. Pharmacol.*, 28, 3431, 1979.
43. Moon, T.W., Walsh, P.J., and Mommsen, T.P., Fish hepatocytes: a model metabolic system, *Can. J. Fish. Aquat. Sci.*, 42, 1772, 1985.

44. Walton, M.J. and Cowey, C.B., Aspects of intermediary metabolism in salmonid fish, *Comp. Biochem. Physiol.,* 73B, 59, 1982.
45. Schmidt, D.C. and Weber, L.J., Metabolism and biliary excretion of sulfobromophthalein by rainbow trout *(Salmo gairdneri), J. Fish. Res. Board Can.,* 30, 1301, 1973.
46. Boyer, J.L., Swartz, J., and Smith, N., Biliary secretion in elasmobranchs. II. Hepatic uptake and biliary excretion of organic anions, *Am. J. Physiol.,* 230, 974, 1976.
47. Gingerich, W.H., Hepatic toxicology of fishes, in *Aquatic Toxicology,* Vol. 1, Weber, L.J., Ed., Raven Press, New York, 1982, 55.
48. Vaillant, C., Le Guellec, C., and Padkel, F., Vitellogenin gene expression in primary culture of male rainbow trout hepatocytes, *Gen. Comp. Endocrinol.,* 70, 284, 1988.
49. Arias, I.M., Jakoby, W.B., and Poller, H., *The Liver: Biology and Pathology,* 2nd ed., Raven Press, New York, 1988.
50. Hinton, D.E., Teh, S.J., Okihiro, M.S., Cooke, J.B., and Parker, L.M., Phenotypically altered hepatocyte populations in diethylnitrosamine-induced medaka liver carcinogenesis: resistance, growth and fate, *Mar. Environ. Res.,* 34, 1, 1992b.
51. Metcalfe, C.D., Tests for predicting carcinogenicity in fish, *Rev. Aquat. Sci.,* 1, 111, 1989.
52. Arey, L.B., *Developmental Anatomy,* 7th ed., W.B. Saunders, Philadelphia, PA, 1974, 255.
53. Ito, T. and Nemoto, M., Uber die küpfferechen sternzellen und die "fettspeicherungszellen" (fat storing cells) in der blutsapillarenwand der menschlichen leber, *Okajimas Foba Anat., (Jpn),* 24, 243, 1952.
54. Ito, T. and Shibasaki, S., Electron microscopic study on the hepatic sinusoidal wall and the fat-storing cells in the normal human liver, *Arch. Histol. Jpn.,* 29, 137, 1968.
55. Overton, J. and Meyer, R., Aspects of liver and gut development in the chick, *Scan. Electron Microsc. II,* 293, 1984.
56. Elias, H., Origin and early development of the liver in various vertebrates, *Acta. Hepatol.,* 3, 1, 1955.
57. Hampton, J.A., McCuskey, P.A., McCuskey, R.S., and Hinton, D.E., Functional units in rainbow trout *(Salmo gairdneri,* Richardson) liver. I. Histochemical properties and arrangement of hepatocytes, *Anat. Rec.,* 213, 166, 1985.
58. Hampton, J.A., Lantz, R.C., Goldblatt, P.J., Laurén, D.J., and Hinton, D.E., Functional units in rainbow trout *(Salmo gairdneri,* Richardson) liver: II. The biliary system, *Anat. Rec.,* 221, 619, 1988.
59. Braunbeck, T., Storch, V., and Bresch, H., Species-specific reaction of liver ultrastructure in zebra fish *(Brachydanio rerio)* and trout *(Salmo gairdneri)* after prolonged exposure to 4-chloroaniline, *Arch. Environ. Contam. Toxicol.,* 19, 405, 1990a.
60. Segner, H. and Braunbeck, T., Hepatocellular adaptation to extreme nutritional conditions in ide, *Leuciscus idus melanotus* L. (Cyprinidae). A morphofunctional analysis, *Fish Physiol. Biochem.,* 5, 79, 1988.
61. Elias, H. and Sherrick, J.C., *Morphology of the Liver,* Academic Press, New York, 1969.
62. Simon, R.C., Dollar, A.M., and Smuckler, E.A., Descriptive classification of normal and altered histology of trout livers, in Trout Hepatoma Research Conference Papers, Halver, J.E. and Mitchell, I.A., Eds., Research Report #70, U.S. Fish and Wildlife Service, Washington, D.C., 1967, 18.

63. Fujita, H., Tatsumi, H., and Ban, T., Fine-structural characteristics of the liver of the cod *(Gadus macrocephalus)* with special regard to the concept of a hepatoskeletal system formed by its cells, *Cell Tissue Res.*, 244, 63, 1986.
64. Braunbeck, T., Storch, V., and Nagel, R., Sex-specific reaction of liver ultrastructure in zebra fish *(Brachydanio rerio)* after prolonged sublethal exposure to 4-nitrophenol, *Aquat. Toxicol.*, 14, 185, 1989.
65. Braunbeck, T., Görge, G., Storch, V., and Nagel, R., Hepatic steatosis in zebra fish *(Brachydanio rerio)* induced by long-term exposure to γ-hexachlorocyclohexane, *Ecotoxicol. Environ. Safety*, 19, 355, 1990b.
66. Ishii, K. and Yamamoto, K., Sexual differences of the liver cells in the goldfish, *Carassius auratus, Bull. Fac. Fish Hokkaido Univ.*, 21, 161, 1970.
67. Pitot, H.C., Campbell, H.A., Maronpot, R., Bawa, N., Rizvi, T.A., Xu, Y., Sargent, L., Dragan, Y., and Pyron, M., Critical parameters in the quantitation of the states of initiation, promotion, and progression in one model of hepatocarcinogenesis in the rat, *Toxicol. Pathol.*, 17, 594, 1989.
68. Hinton, D.E. and Pool, C.R., Ultrastructure of the liver in channel catfish *Ictalurus punctatus* (Rafinesque), *J. Fish Biol.*, 8, 209, 1976.
69. Teh, S.J. and Hinton, D.E., Detection of enzyme histochemical markers of hepatic preneoplasia and neoplasia in medaka *(Oryzias latipes), Aquat. Toxicol.*, 24, 163, 1993.
70. Parker, L.M., Laurén, D.J., Hammock, B.D., Winder, B., and Hinton, D.E., Biochemical and histochemical properties of hepatic tumors of rainbow trout, *Oncorhynchus mykiss, Carcinogenesis,* 14, 211, 1993.
71. van Bohemen, C.G., Lambert, J.G.D., and Peute, J., Annual changes in plasma and liver in relation to vitellogenesis in the female rainbow trout, *Salmo gairdneri, Gen. Comp. Endocrinol.*, 44, 94, 1981b.
72. Kloepper-Sams, P.J., Park, S.S., Gelboin, H.V., and Stegeman, J.J., Immunochemical analysis of cytochrome P-450 in teleosts with polyclonal and monoclonal antibodies to cytochrome P-450E, *Fed. Proc.*, 45, 320, 1986.
73. Kloepper-Sams, P.J., Park, S.S., Gelboin, H.V., and Stegeman, J.J., Specificity and cross-reactivity of monoclonal and polyclonal antibodies against cytochrome P-450E of the marine fish scup, *Arch. Biochem. Biophys.*, 253, 268, 1987.
74. Lorenzana, R.M., Hedstrom, O.R., Gallagher, J.A., and Buhler, D.R., Cytochrome P450 isozyme distribution in normal and tumor-bearing hepatic tissue from rainbow trout *(Salmo gairdneri), Exp. Mol. Pathol.*, 50, 348, 1989.
75. Park, S.S., Miller, H., Klotz, A.V., Kloepper-Sams, P.J., Stegeman, J.J., and Gelboin, H.V., Monoclonal antibodies to liver cytochrome P450 E of the marine fish scup, *Arch. Biochem. Biophys.*, 249, 339, 1986.
76. Stegeman, J.J., Woodin, B.R., Park, S.S., Kloepper-Sams, P.J., and Gelboin, H.V., Microsomal cytochrome P-450 function in fish evaluated with polyclonal and monoclonal antibodies to cytochrome P450E from scup *(Stenotomus chrysops), Mar. Environ. Res.*, 17, 83, 1985.
77. Hampton, J.A., Klaunig, J.E., and Goldblatt, P.J., Resident sinusoidal macrophages in the liver of the brown bullhead *(Ictalurus nebulosus):* an ultrastructural, functional and cytochemical study, *Anat. Rec.*, 219, 338, 1987.
78. Nopanitaya, W., Carson, J.L., Grisham, J.W., and Aghajanian, J.G., New observations on the fine structure of the liver in goldfish *(Carassius auratus), Cell Tissue Res.*, 196, 249, 1979b.

79. McCuskey, P.A., McCuskey, R.S., and Hinton, D.E., Electron microscopy of cells of the hepatic sinusoids in rainbow trout *(Salmo gairdneri)*, in *Cells of the Hepatic Sinusoid*, Vol. 1, Kirn, A., Knook, D.L., and Wisse, E., Eds., Kupffer Cell Foundation, Leiden, The Netherlands, 1986, 489.
80. Nopanitaya, W., Aghajanian, J., Grisham, J.W., and Carson, J.L., An ultrastructural study on a new type of hepatic perisinusoidal cell in fish, *Cell Tissue Res.*, 198, 35, 1979a.
81. Wisse, E., Observations on the fine structure and peroxidase cytochemistry of normal rat liver Kupffer cells, *J. Ultrastruct. Res.*, 38, 528, 1974.
82. Wake, K., Perisinusoidal stellate cells (fat-storing cells, interstitial cells, lipocytes), their related structure in and around the liver sinusoids, and vitamin A-storing cells in extrahepatic organs, *Int. Rev. Cytol.*, 66, 303, 1980.
83. Ito, T., Watanabe, A., and Takahashi, Y., Histologische und cytologische untersuchungen der leber bei fisch und cyclostomata, nebst bemerkungen uber die fettspeicherungszellen, *Arch. Histol. Jpn.*, 22, 429, 1962.
84. Laurén, D.J., Teh, S.J., and Hinton, D.E., Cytotoxicity phase of diethylnitrosamine-induced hepatic neoplasia in medaka, *Cancer Res.*, 50, 5504, 1990b.
85. Yamamoto, M. and Enzan, H., Morphology and function of Ito cell (fat-storing cell) in the liver, *Rec. Adv. RES Res.*, 15, 54, 1975.
86. Takahashi, Y., Tsubouchi, H., and Kobayashi, K., Effects of vitamin A administration upon Ito's fat-storing cells of the liver in the carp, *Arch. Histol. Jpn.*, 41, 339, 1978.
87. Fujita, S., Watanabe, T., and Kitajima, C., Nutritional quality of *Artemia* from different localities as a living feed for marine fish from the viewpoint of essential fatty acids, in *The Brine Shrimp Artemia, Vol. 3, Ecology, Culturing, Use in Aquaculture*, Persoone, G., Sorgeloos, P., and Roels, O., Eds., Universa Press, Wetteren, Belgium, 1980, 277.
88. Eastman, J.T. and DeVries, A.L., Hepatic ultrastructural specialization in antarctic fishes, *Cell Tissue Res.*, 219, 489, 1981.
89. Tanuma, Y. and Ito, T., Electron microscopic study on the sinusoidal wall of the liver of the crucian, *Carassius carassius,* with special remarks on the fat-storing cell (FSC), *Arch. Histol. Jpn.*, 43, 241, 1980.
90. Tanuma, Y., Ohata, M., and Ito, T., Electron microscopic study on the sinusoidal wall of the liver in the flatfish, *Kareius bicoloratus:* demonstration of numerous desmosomes along the sinusoidal wall, *Arch. Histol. Jpn.*, 45, 453, 1982.
91. Yamamoto, T., Some observations of the fine structure of the intrahepatic biliary passages in the goldfish *(Carassius auratus)*, *Z. Zellforsch.*, 65, 319, 1965.
92. Laurén, D.J., Okihiro, M.S., Hinton, D.E., and Stegeman, J.J., Localization of cytochrome P4501A1 induced by β-naphthoflavone (BNF) in rainbow trout *(Oncorhynchus mykiss)* embryos, *FASEB J.*, 4, A739, 1990a.
93. Trelstad, R.L., Glycosaminoglycans: mortar, matrix, mentor (Editorial), *Lab. Invest.*, 53, 1, 1985.
94. Liotta, L.A., Rao, C.N., and Barsky, S.H., Tumor invasion and the extracellular matrix, *Lab. Invest.*, 49, 636, 1983.
95. Iozzo, R.V., Biology of disease, Proteoglycans: structure, function, and role in neoplasia, *Lab Invest.*, 53, 373, 1985.
96. Iozzo, R.V. and Wight, T.N., Isolation and characterization of proteoglycans synthesized by human colon and colon carcinoma, *J. Biol. Chem.*, 257, 11135, 1982.

97. Martinez-Hernandez, A., The hepatic extracellular matrix: I. Electron immunohistochemical studies in normal rat liver, *Lab. Invest.*, 51, 57, 1984.
98. Martinez-Hernandez, A., The hepatic extracellular matrix: II. Electron immunohistochemical studies in rats with CCl_4-induced cirrhosis, *Lab. Invest.*, 53, 166, 1985.
99. Trump, B.F., McDowell, E.M., and Arstila, A.U., Cellular reaction to injury, in *Principles of Pathobiology*, 3rd ed., Hill, R.B. and LaVia, M.F., Eds., Oxford University Press, New York, 1980, 20.
100. Pritchard, J.B. and Renfro, J.L., Interactions of xenobiotics with teleost renal function, in *Aquatic Toxicology*, Vol. 2, Weber, L.J., Ed., Raven Press, New York, 1982, 51.
101. Baksi, S.M. and Frazier, J.M., Isolated fish hepatocytes — model systems for toxicology research, *Aquat. Toxicol.*, 16, 229, 1990.
102. Blair, J.B., Miller, M.R., Pack, D., Barnes, R., Teh, S.J., and Hinton, D.E., Isolated trout liver cells: establishing short-term primary cultures exhibiting cell-cell interactions *in vitro, Cell Dev. Biol.*, 26, 237, 1990.
103. Gingerich, W.H., Weber, L.J., and Larson, R.E., Carbon tetrachloride-induced retention of sulfobromophthalein in the plasma of rainbow trout, *Toxicol. Appl. Pharmacol.*, 43, 147, 1978.
104. Pfeifer, K.F., Weber, L.J., and Larson, R.E., Carbon tetrachloride-induced hepatotoxic response in rainbow trout, *Salmo gairdneri*, as influenced by two commercial fish diets, *Comp. Biochem. Physiol.*, 67C, 91, 1980.
105. Racicot, J.G., Gaudet, M., and Leray, C., Blood and liver enzymes in rainbow trout *(Salmo gairdneri* Rich.) with emphasis on their diagnostic use: study of CCl_4 toxicity and a case of *Aeromonas* infection, *J. Fish Biol.*, 7, 825, 1975.
106. Weber, L.J., Gingerich, W.H., and Pfeifer, K.F., Alterations in rainbow trout liver function and body fluids following treatment with carbon tetrachloride or monochlorobenzene, in *Pesticide and Xenobiotic Metabolism in Aquatic Organisms*, Khan, M.A.Q., Lech, J.J., and Menn, J.J., Eds., ACS Symposium Series 99, Washington, D.C., 1979, 401.
107. Droy, B.F., Effect of reference hepatotoxicants on rainbow trout liver, Ph.D. Thesis, University of Morgantown, West Virginia, 1988.
108. Petersen, D.W. and Lech, J.J., Hepatic effects of acrylamide in rainbow trout, *Toxicol. Appl. Pharmacol.*, 89, 249, 1987.
109. Droy, B.F. and Hinton, D.E., Allyl formate-induced hepatotoxicity in rainbow trout, *Mar. Environ. Res.*, 24, 259, 1988.
110. Droy, B.F., Davis, M.E., and Hinton, D.E., Mechanism of allyl formate-induced hepatotoxicity in rainbow trout, *Toxicol. Appl. Pharmacol.*, 98, 313, 1989.
111. Bunton, T.E., Hepatopathology of diethylnitrosamine in the medaka *(Oryzias latipes)* following short-term exposure, *Toxicol. Pathol.*, 18, 313, 1990.
112. Couch, J.A. and Courtney, L.A., N-Nitrosodiethylamine-induced hepatocarcinogenesis in estuarine sheepshead minnow *(Cyprinodon variegatus)*: neoplasms and related lesions compared with mammalian lesions, *J. Natl. Cancer Inst.*, 79, 297, 1987.
113. Hinton, D.E., Couch, J.A., Teh, S.J., and Courtney, L.A., Cytological changes during progression of neoplasia in selected fish species, *Aquat. Toxicol.*, 11, 77, 1988b.
114. Murchelano, R.A. and Wolke, R.E., Epizootic carcinoma in the winter flounder *Pseudopleuronectes americanus, Science,* 228, 587, 1985.

115. Hayes, M.A., Smith, I.R., Rushmore, T.H., Crane, T.L., Thorn, C., Kocal, T.E., and Ferguson, H.W., Pathogenesis of skin and liver neoplasms in white suckers from industrially polluted areas in Lake Ontario, *Sci. Total Environ.*, 94, 105, 1990.
116. Moore, M.J., Vacuolation, proliferation and neoplasia in the liver of Boston Harbor winter flounder *(Pseudopleuronectes americanus)*, Ph.D. Dissertation, Woods Hole Oceanographic Institution and The Massachusetts Institute of Technology (WHOI-91-28), Woods Hole, MA, 1991.
117. Moore, M.J. and Stegeman, J.J., Bromodeoxyuridine uptake in hydropic vacuolation and neoplasms in winter flounder liver, *Mar. Environ. Res.*, 34, 13, 1992.
118. Myers, M.S., Olson, O.P., Johnson, L.L., Stehr, C.S., Hom, T., and Varanasi, U., Hepatic lesions other than neoplasms in subadult flatfish from Puget Sound, Washington: Relationships with indices of contaminant exposure, *Mar. Environ. Res.*, 34, 45, 1992.
119. Myers, M.S., Rhodes, L.D., and McCain, B.B., Pathologic anatomy and patterns of occurrence of hepatic neoplasms, putative preneoplastic lesions, and other idiopathic hepatic conditions in English sole *(Parophrys vetulus)* from Puget Sound, Washington, *J. Natl. Cancer Inst.*, 78, 333, 1987.
120. Baumann, P.C., Harshbarger, J.C., and Hartman, K.J., Relationship between liver tumors and age in brown bullhead populations from two Lake Erie tributaries, *Sci. Total Environ.*, 94, 71, 1990.
121. Lester, S.M., Braunbeck, T.A., Teh, S.J., Stegeman, J.J., Miller, M.R., and Hinton, D.E., Hepatic cellular distribution of cytochrome CYP1A in rainbow trout *(Oncorhynchus mykiss)*: an immunohisto- and cytochemical study, *Cancer Res.*, 3701, 1993.
122. Bannasch, P. and Zerban, H., Tumours of the liver, in *Pathology of Tumours in Laboratory Animals, Vol. 1, Tumours of the Rat*, 2nd ed., Turosov, V.S. and Mohr, U., Eds., IARC Scientific Publication No. 99, International Agency for Research on Cancer (IARC), Lyon, France, 1990, 199.
123. Bannasch, P., Pathobiology of chemical hepatocarcinogenesis: recent progress and perspectives. Part I. Cytomorphological changes and cell proliferation. Part II. Metabolic and molecular changes, *J. Gastroenterol. Hepatol.*, 5, 149, 1990.
124. Courtney, L.A. and Couch, J.A., Usefulness of *Cyprinodon variegatus* and *Fundulus grandis* in carcinogenesis testing: advantages and special problems, *Natl. Cancer Inst. Monogr.*, 65, 83, 1984.
125. Bannasch, P. and Zerban, H., Predictive value of hepatic preneoplastic lesions as indicators of carcinogenic response, in *Mechanisms of Carcinogenesis in Risk Identification*, Vainio, H., Magee, P.N., McGregor, D.B., and McMichael, A.J., Eds., International Agency for Research on Cancer (IARC), Lyon, France, 1992, 389.
126. Popp, J.A. and Goldsworthy, T.L., Defining foci of cellular alteration in short-term and medium-term rat liver tumor models, *Toxicol. Pathol.*, 17, 561, 1989.
126a. Okihiro, M.S., et al., Unpublished observations.
127. Apte, S.S., Ki-67 monoclonal antibody (MAb) reacts with a proliferation associated nuclear antigen in the rabbit *Oryctolagus cuniculus*, *Histochemistry*, 94, 201, 1990.
128. Yonemura, Y., Ohoyama, S., Kimura, H., Kamata, T., Yamaguchi, A., and Miyazaki, I., Assessment of tumor cell kinetics by monoclonal antibody Ki-67, *Eur. Surg. Res.*, 22, 365, 1990.

129. Zeymer, U., Fishbein, M.C., Forrester, J.S., and Cercek, B., Proliferating cell nuclear antigen immunohistochemistry in rat aorta after balloon denudation. Comparison with thymidine and bromodeoxyuridine labeling, *Am. J. Pathol.*, 141, 685, 1992.
130. Wyllie, A.H., Kerr, J.F.G., and Cumi, A.R., Cell death: the significance of apoptosis, *Int. Rev. Cytol.*, 68, 251, 1980.
131. Moore, M.N. and Evans, B., Detection of *ras* oncoprotein in liver cells of flatfish (dab) from a contaminated site in the North Sea, *Mar. Environ. Res.*, 34, 33, 1992.
132. Hollstein, M., Sidransky, D., Vogelstein, B., and Harris, C.C., p53 mutations in human cancers, *Science,* 253, 49, 1991.

Chapter 5

Modulation of Blood Cell-Mediated Oxyradical Production in Aquatic Species: Implications and Applications

Robert S. Anderson

I. BACKGROUND

A. Reactive Oxygen Intermediates of Mammalian Blood Cells

It has been known for some time that the process of phagocytosis by mammalian polymorphonuclear leukocytes (PMNs) and macrophages is accompanied by an abrupt increase in oxygen uptake, followed by the production of cytotoxic reactive oxygen intermediates (ROIs). The biochemical events associated with this respiratory burst have also been elucidated and will be mentioned in more detail later in this chapter. It is accepted that the first reaction in the respiratory burst is the one electron reduction of oxygen to superoxide (O_2^-), catalyzed by NADPH oxidase associated with the phagocyte membrane. Superoxide anions are converted to hydrogen peroxide (H_2O_2) by the cytoplasmic enzyme superoxide dismutase (SOD). Superoxide and, to a great extent, H_2O_2 are highly reactive and toxic ROIs; H_2O_2, in conjunction with myeloperoxidase (MPO) and a halide, forms the basis of a potent antibacterial system.[1] Other toxic ROIs can also be generated, such as hydroxyl radicals (.OH) and singlet oxygen ($^1O_2$). Singly or collectively, these ROIs can participate in the cell-mediated destruction of bacteria, fungi, and protozoa. The phagocytes have detoxification enzymes to protect themselves against autooxidative damage; these include SOD, glutathione peroxidase, catalase, certain vitamins that are radical scavengers, etc. Nevertheless, overproduction of ROIs can saturate these protective mechanisms and lead to damage to cells and tissues in the vicinity of the phagocytes. The main purpose of this chapter is to cite evidence for similar respiratory activity in the phagocytes of aquatic invertebrates and fish. The beneficial and the potentially less desirable properties of ROIs will be discussed, as will speculation on the physiological consequences of the modulation of ROI production after exposure of aquatic species to environmental chemicals or during the course of certain infectious diseases.

Several phagocyte functions are linked to recognition phenomena taking place at the plasma membrane via surface receptors. The binding of immunocomplexes to the Fc receptor can result in their internalization, presentation of antigens to lymphocytes, secretion of arachidonic acid metabolites, and formation of hydrolases and ROIs. At least three distinct Fc receptors have been described on macrophages for different immunoglobulin G (IgG) subclasses,[2-4] as well as IgE[5] and IgM.[6] The C3b receptors, or complement receptors type 1 (CR1) and type 3 (CR3), are present on phagocytes; they are involved in clearance and processing of immunocomplexes[7] and in the adherence of C3b-opsonized microorganisms to the cells.[8] Other receptors involved in phagocytosis include carbohydrate receptors, such as those specific for mannose[9] and glucan,[10] and the fibronectin receptor.[11] Many other receptors for regulatory factors, interleukins, colony stimulating factor, etc., have been characterized on mammalian phagocytes, but their description falls outside the scope of this chapter.

In addition to particle ingestion, the respiratory burst of phagocytes can be triggered by interaction with various soluble ligands, such as phorbol myristate acetate and lectins like concanavalin A. The details of signal transduction linking membrane stimulation and activation of the NADPH oxidase system have yet to be fully elucidated. During phagocyte stimulation, there is increased phosphoinositide hydrolysis, increased cytosolic free Ca^{2+}, activation of protein kinase C, and arachidonic acid release; however, recently the role of these reactions in signal transduction has been questioned.[12]

B. Common ROI Methods: A General Overview

Stimulation of phagocytes by membrane receptor perturbations or ingestion of foreign particles triggers increased oxygen consumption, NADPH-oxidase activity, and hexose monophosphate shunt activity. The ROIs produced during the respiratory burst can be measured by numerous methods, and many of these assays have been carried out with hemocytes from mollusks and fish. This short section introduces some of the more common of these procedures. It is not intended to be complete or exhaustively referenced.

One of the earliest ROI methods developed was that of nitroblue tetrazolium (NBT) reduction to detect O_2^-. Like most O_2^- assays, NBT reduction is not totally specific for superoxide; the specificity of the test is controlled by the use of superoxide dismutase, and the results are reported as "SOD-inhibitable activity." NBT reduction was first used to demonstrate O_2^- production cytochemically by microscopy.[21] In the presence of O_2^-, pale yellow NBT is converted to a dark blue, water-insoluble formazan product that can easily be visualized in the cytoplasm of phagocytes. In this way, one can get an idea of the percent of leukocytes producing O_2^- and the intensity of the activity in the cells. More quantitative methods for measuring NBT reduction are also available. These are simple colorimetric assays of formazan concentration in pyridine or dimethyl-sulfoxide extracts of leukocytes.[20,120] Probably the most often used O_2^- assay measures the reduction of ferricytochrome c taking place extracellularly, after stimulation of the cells.[121] The optical density of the cytochrome c-containing medium is read at 550 nm, and the reaction product is quantified using the extinction coefficient ($\Delta E_{550} = 2.1 \times 10^4 \, M^{-1} \, cm^{-1}$) for the reduced cytochrome. Both of these assays can be run in conventional 96-well microtiter plates and read with an ELISA plate-scanning spectrophotometer, using as few as 10^5 cells per well.[122]

In vivo, O_2^- rapidly reacts with other molecules or is dismutated to H_2O_2 by SOD. Hydrogen peroxide is thought to be an ROI of major toxic significance and is commonly quantified by the peroxidase-catalyzed oxidation of a substrate; the specificity of the assays are tested by the inclusion of catalase in the medium. Diaminobenzidine (DAB), in the presence of H_2O_2 and peroxidase, is oxidized to a brown product conspicuous in the phagolysosomes of cells that are involved in phagocytosis. The DAB reaction is used in microscopic evaluation of the oxidative burst. More quantitative data are obtained by measuring the oxidation of homovanillic acid by fluorescence[33] or the oxidation of phenol red to a product with an absorbance maximum at 610 nm.[123] The phenol red assay for H_2O_2 is the preferred method and has been successfully adopted for use with 96-well microtiter plates.[122]

ROIs produced during the respiratory burst will react with several chemiluminogenic probes to produce photons that can be quantified in a liquid scintillation counter set in the out-of-coincidence mode or in a standard luminometer. Thus, chemiluminescence (CL) has become a popular method to measure the ROI production and microbial killing potential of blood cells.[36] The most frequently used probe is luminol (5'-amino-2,3-dihydro-1,4-phtalazinedione), which will react with O_2^-, H_2O_2, singlet oxygen, etc., to give a strong CL signal, but it is generally thought that luminol-dependent CL measures mainly myeloperoxidase-dependent activity.[124] Lucigenin (9,9'-*bis*-N-methylacridinium nitrate) is also used as a CL probe that is specific for the superoxide anion.[125]

A number of fluorescent probes are now available to measure the respiratory burst in leukocytes, using a flow cytometer or fluorescence concentration analyzer.

For example, 2′,7′-dichlorofluorfluorescin-diacetate (DCFH-DA) will enter the leukocytes, where it is metabolized to nonfluorescent, membrane-impermeable 2′,7′-dichlorofluorescin (DCFH). Intracellular DCFH is oxidized to fluorescent 2′,7′-dichlorofluorescein (DCF) upon activation of the respiratory burst and is mainly a measure of the phagosomal formation of ROIs. In leukocytes, DCFH is oxidized by H_2O_2 and phagosomal peroxidases, whereas hydroethidine (HE) is probably oxidized at an earlier step in ROI metabolism by O_2^-.[127] Similar to DCFH-DA, HE permeates the cell membrane and is oxidized to ethidium bromide (EB), which is trapped in the nucleus by intercalation into DNA. Another useful fluorescent probe for respiratory burst activity is dihydrorhodamine 123 (DHR), an uncharged, nontoxic, nonfluorescent derivative of the laser dye rhodamine 123 (R123). During oxidative activity of phagocytes, DHR is intracellularly oxidized to the highly fluorescent R123, to provide an assay about threefold more sensitive than DCFH-DA.[128] The use of fluorescent methods for measuring the respiratory burst of mammalian leukocytes is increasing rapidly as more probes are developed. These methods have not often been used to study comparable phenomena in molluscan or fish leukocytes, but should find more application in the future.

C. ROI Assays in the Clinical Setting

One of the most interesting and instructive series of studies of phagocyte ROI production has resulted from attempts to understand the basis of chronic granulomatous disease (CGD). This clinical syndrome affects predominantly male children, who present with recurrent bacterial and fungal infections, granulomatous infiltration of many organs, and early deaths.[13,14] The neutrophils and macrophages of these patients phagocytize the microorganisms normally but fail to kill them. The continued presence of the microbes in intracellular phagosomal vacuoles induces a chronic inflammatory state that triggers massive granuloma formation in an attempt to isolate the affected areas. Neutrophils from CGD patients were shown to have no major defects in their ability to migrate, phagocytize, and degranulate and had a typical suite of lysosomal enzymes.[15] However, subsequent work showed that CGD neutrophils did not undergo the respiratory burst upon ingestion of bacteria and did not produce O_2^- during the process.[16,17] Indeed, it is now accepted that CGD is actually caused by defects in phagocyte NADPH oxidase. The use of cells from CGD patients has contributed much of our understanding of the composition and functions of this oxidase and the role that ROI formation plays in the antimicrobial and cytotoxic activities of phagocytes (see Reference 18 for a current review). Recent evidence suggests that the oxidase in the unstimulated cell has four components compartmentalized within both the cytosol and the membrane. The membrane component includes two oligomeric subunits of cytochrome b_{558} and a *ras*-like G protein; the cytosol contains two additional protein components of the phagocyte oxidase, a NADPH-binding protein, and an as yet unidentified component. Stimulation of the cell causes assembly of all these components into a functional enzyme on the cell membrane.[19] About two thirds of CGD patients have inherited abnormalities in membrane-bound cytochrome b_{558} subunits, and the rest have defects in the cytosolic protein components of the oxidase. Most defects in membrane-located components are inherited in an X-linked fashion; apparently all defects in the cytosolic components are transmitted via autosomal recessive inheritance.

As indicated above, CGD has impelled considerable study of the respiratory burst oxidase using the techniques of molecular biology and genetics; however, initial detection of its presence was frequently based on the results of basic ROI measurements. For example, the analysis of superoxide anion by quantifying SOD-inhibitable ferricytochrome c reduction by intact leukocytes was used to detect low oxidase levels typical of CGD. However, the NBT test was widely used because of its simplicity and availability.[20,21] As will become apparent, ROI analyses such as these can also be performed with blood cells from fish and aquatic invertebrates.

II. PRODUCTION OF ROIs BY AQUATIC SPECIES

A. Oxidant-Antioxidant Systems in Mollusks

Mussels and other bivalves are frequently used to monitor the concentrations of certain xenobiotics in aquatic environments.[22] While they provide chemical data, it is difficult to predict the significance of tissue concentrations of chemicals on animal health. One of the techniques currently finding wide acceptance as an indicator of biological effects involves measuring increased MFO activity as a function of exposure to pollutants.[23,24] In addition to metabolically activating/ detoxifying polycyclic aromatic hydrocarbons (PAHs) and other xenobiotics, the MFO system is also involved in oxyradical generation,[25] and this activity is enhanced by chemicals that are capable of redox cycling. A number of inducible enzymes that protect the host organism against ROIs have been described in mollusks.[26-28] In fact, increases in the activities of antioxidant enzymes, such as SOD, glutathione peroxidase, and catalase, exposed to organic pollutants have been measured.[29,30] Therefore, it is possible that quantitation of the antioxidase system may prove useful in detecting biological impacts of pollution, in addition to assays of MFO activity and oxyradical production.

The above studies of molluscan MFO activity and the antioxidant enzyme system were performed primarily on digestive gland preparations. The digestive gland is the primary site of uptake of organic pollutants, as well as the principal location of both the MFO system and antioxidase enzymes. It is likely that overproduction of oxyradicals via redox cycling and general MFO activity could exhaust the inducible protective antioxidant defense system, contributing to pollutant-mediated toxicological responses. It is also possible that abnormally elevated production of oxyradicals by blood cells could further exacerbate oxidative stress. Examples of such increased production of ROIs by blood cells from several aquatic species, after exposure to xenobiotics and during parasitic infection, will be cited later in this chapter.

B. ROI Production by Aquatic Invertebrates

Hemocytes from several molluscan species seem to undergo a typical respiratory burst with concomitant ROI production. Early attempts to measure increased oxygen uptake and/or MPO-halide-H_2O_2 system activity by phagocytozing clam hemocytes were unsuccessful.[31] However, in 1985, O_2^- production in stimulated snail hemocytes was reported by Dikkeboom et al.,[32] and H_2O_2 production by resting and stimulated scallop hemocytes was shown by Nakamura et al.[33] Papers on ROI production by molluscan hemocytes have proliferated since that time. In a 1991 review of this subject, Adema et al.[34] cited 14 references on ROI activity in 13 species of gastropods and bivalves. In these papers, O_2^- was measured by reduction of NBT or cytochrome c, and H_2O_2 was measured by colorimetric assays (diaminobenzidine or phenol red oxidation) or in fluorescence assays (homovanillic acid). CL in the presence of signal-amplifying probes was also used as an indicator of ROI production. The CL method does not yield quantitative information on particular ROI species; the contribution of O_2^- or H_2O_2 to the CL signal can be estimated by the use of appropriate enzymatic inhibitors. It is widely accepted that the CL response of phagocytes is correlated with bactericidal activity associated with the respiratory burst,[35,36] and the method has been specifically recommended for use with fish and shellfish.[37,38] The detection of luminol-augmented CL using stimulated hemocytes has been used to estimate ROI production in several bivalves such as *Crassostrea virginica*,[39,40] *C. gigas*,[41] *Ostrea edulis*,[41] and *Pectin maximus*.[42] Likewise, CL assays indicate ROI induction after stimulation of hemocytes from gastropods including *Lymnaea stagnalis*,[43] *Helix aspersa*,[44] and *Achatina achatina*.[45] Hemocytes from other species *(Biomphalaria glabrata* and *Planorbarius corneus)* failed to show phagocytosis-enhanced CL, although other more specific O_2^- tests were positive.[44] The CL assay has been used in several studies of effects of the interactions of xenobiotics and parasites with molluscan hemocytes; these will be described in subsequent portions of this chapter.

The zymosan-induced CL activity of *Mytilus edulis* hemocytes has been described recently.[46] Luminol-dependent CL predominantly measures H_2O_2/MPO activity, and sodium azide (a MPO inhibitor) was shown to inhibit CL by *M. edulis* cells. However, a less pronounced inhibition was seen in the presence of SOD. The CL response of mussels was reported to be lower than oysters and scallops and showed considerable variation from animal to animal, as well as indications of seasonality. Chemical inhibitors of NADPH-oxidase activity also produced CL inhibition in *M. edulis;* these catechol-like phenols also inhibit oxidative activities of gastropod hemocytes when stimulated with zymosan.[129] Discontinuous Percoll density gradient centrifugation was used to obtain several mussel hemocyte cell subpopulations, and cell types classified by morphology and via monocolonal antibodies. Stimulated CL was most strongly associated with the subpopulation enriched for eosinophilic granulocytes, as compared to the basophilic granulocytes or hyalinocytes.

Recently, Pipe[47] has quantified NBT reduction by *M. edulis* hemocytes. The activation of the nonfluorescent probe dihydrorhodamine 123 was used to validate the production of O_2^- by hemocytes. Although it is often stated that O_2^- production requires activation of the respiratory burst oxidase, Pipe showed that NBT reduction by these cells could proceed to some extent without phagocytic (or other) stimuli. A similar result was reported with *C. virginica* hemocytes by Anderson et al.,[48] who likened the response to the formation of so-called "formazan cells" in NBT preparations of human leukocytes.[49] Pipe[47] also demonstrated H_2O_2 generation by zymosan-stimulated hemocytes via the oxidation of phenol red; however, no spontaneous H_2O_2 production was noted. Aside from direct cytotoxic properties, H_2O_2 could likely be involved in antibacterial HOCl production, because myeloperoxidase has been shown to be present in mussel hemocytes.[50]

Reduction of NBT by molluscan hemocytes was typically evaluated by microscopic examination of cell monolayers. Anderson et al.[48] also used a quantitative NBT reduction technique that spectrophotometrically measured the production of blue formazan extracted from the cells; this method was adapted from that of Baehner and Nathan.[20] The NBT reduction was partially inhibited by SOD and occurred in both resting and phagocytically stimulated hemocytes. This activity was recorded over the course of 1 year in cells from field-collected oysters, as well as oysters held in flow-through tanks in the laboratory. Both stimulated and unstimulated levels of NBT reduction tended to increase with elevation of the ambient water temperature (2 to 29°C) but seemed to be independent of salinity in the limited range studied (~10 to 14%). Cells from oysters collected at 2 to 13°C responded to phagocytosis with a small, statistically insignificant increment in NBT reduction. However, hemocytes from oysters at 21 to 29°C showed significant induction of this activity. These data suggest a temperature-dependent seasonal variation of the NBT reduction reaction, which should probably be considered during the planning and interpretation of ROI induction studies.

Recently, Ito et al.[51] demonstrated, via homovanillic acid oxidation, basal and phagocytically induced H_2O_2 generation by phagocytes of the sea urchin *Strongylocentrotus nudus*. The first demonstration of the respiratory burst by crustacean hemocytes has just been reported for crab *(Carcinus maenas)* cells that produce O_2^- following stimulation.[130] There are few other reports of ROI production by hemocytes of aquatic invertebrates other than mollusks; however, this is probably more a result of lack of investigation rather than lack of the phenomenon.

C. ROI Production by Fish

Aerobic killing by mammalian phagocytes is mediated by ROIs produced subsequent to the reduction of molecular oxygen to O_2^- via a membrane-associated NADPH oxidase. Its components form an electron transport chain with a flavoprotein component accepting electrons from NADPH, transferring them to O_2 via cytochrome b. Secombes et al.[52] presented evidence that fish macrophages also have a cytochrome b component of the NADPH oxidase similar to that typical of mammals. Furthermore, specific inhibitors of the flavoprotein and cytochrome b-$_{245}$ components of NADPH oxidase were both shown to inhibit the respiratory burst activity of trout macrophages.

As previously noted, luminol-enhanced blood cell CL is one of the most frequently used techniques to quantify ROI production. This method has been applied to blood cells from various species of fish. The CL response of pronephros cells from *Morone saxatilis* was measured after exposure to bacterial pathogens.[53] Various phagocyte to bacteria ratios were employed, each producing somewhat different CL responses with regard to the magnitude of the activity and the individual kinetics. In addition to bacteria, both particulate (zymosan) and soluble (PMA) classical activators of CL by mammalian leukocytes were shown to be effective with fish cells.[54]

Prior incubation of test particulates with various serum preparations has been generally found to enhance, or opsonize, phagocytic activity of fish leukocytes based on measurements of CL activity.[37] For example, channel catfish peripheral blood leukocytes, when mixed with *Edwardsiella ictaluri* previously incubated with autologous immune serum, produced an elevated CL response, as compared to that elicited by untreated or nonimmune serum-treated *E. ictaluri*.[55] The opsonic activity of immune serum was attributed to the presence of both specific antibodies and complement. Treatment of immune serum with heat or absorption with antigen caused significant reduction in its opsonic activity, as measured by CL. The effect of incubation temperature on the CL response of rainbow trout peripheral neutrophils was compared to that of human neutrophils.[56] The peak CL response of both species could be stimulated by either *Staphylococcus aureus* or PMA. However, maximal CL occurred at 4 to 15°C in fish and 23 to 37°C in human cells. In both species, the time of peak CL was retarded at lower than optimal incubation temperatures, possibly due to decreased fluidity of the cell membrane at temperatures lower than the physiological norm. Elevated temperatures are known to enhance host defense systems in ectothermic animals; the results of Sohnle and Chusid[56] suggest that initiation of neutrophil ROI production occurs more rapidly at higher temperatures in trout, although the peak response was not increased. The effects of temperature on CL by peripheral blood phagocytes for channel catfish exposed to bacteria *(E. ictaluri)* were further investigated by Scott et al.[55] Peak light emission was seen at about 15:1, bacteria to hemocyte ratio, using bacteria opsonized with ~10 µg/mL immune serum. Increased temperature (20 to 30°C) was accompanied by increased cell-mediated bactericidal activity and peak CL response. Noting discrepancies between the findings of Sohnle and Chusid[56] and Scott et al.[55] with regard to the effect of temperature on peak CL, Angelidis et al.[57] carried out additional studies on sea bass head kidney phagocytes. Their results were similar to those of Sohnle and Chusid.[56] Presumably, stimulation by opsonized zymosan particles elicited greater CL from phagocytes at lower temperatures (5 to 20°C) than higher (25 to 40°C), but CL peak values occurred more rapidly at higher temperatures.

Clearly, the total Cl response and its kinetics should be thoroughly understood in any fish model under study before attempting to generalize on particular experimental findings. Existing evidence suggests that, in addition to individual variation, the CL response of a given sample of fish phagocytes may depend on a variety of factors, including the species of fish, the anatomical source of the cells, the type of stimulating agent, opsonization of particulate stimuli, cell to particle ratio, incubation temperature, and media components.

III. MODULATION OF HEMOCYTE-MEDIATED ROI PRODUCTION BY EXPOSURE TO ENVIRONMENTAL CHEMICALS

A. Examples in Molluscan Species

The CL response of *C. virginica* blood cells, either exposed in vitro or collected from oysters after various periods of in vivo exposure to selected heavy metals, pesticides or other organic compounds, was measured by Larson et al.[39] They found that, in both in vivo and in vitro studies, copper was the most effective agent tested with regard to ability to depress the CL response, although most of the compounds appeared to suppress CL, particularly at high exposure levels. Certain compounds, such as cadmium, aluminum, zinc, dieldrin, and naphthalene, apparently caused increased CL at low levels, but this effect was usually reversed at higher concentrations.

Fisher et al.[58,59] reported that in vitro tributyltin (TBT) treatment of hemocytes from *C. virginica* or *C. gigas* produced slight stimulation at low concentrations (0.4 ppb), followed by dose-dependent suppression of CL at higher concentrations. The suppressive TBT concentrations (40 to 400 ppb) exceeded those found in most environmental samples; nonetheless, hemocytes of field-exposed oysters are probably exposed to high TBT levels as a result of bioaccumulation.

Work in this laboratory has demonstrated the immunosuppressive effect of cadmium on oyster hemocyte CL.[40] When the cells were exposed in vitro to sublethal cadmium levels, a dose-dependent inhibition of ROI production was consistently recorded. Several CL parameters were measured, resting or basal CL (the activity of phagocytically unstimulated cells), peak CL (the maximal CL response induced by phagocytosis of yeast particles), and the total CL response (obtained by integrating the area under the curve of the induced CL response). Peak and total CL were significantly inhibited by 10 and 2 ppm cadmium, respectively; but the basal CL activity was not affected by any concentration tested (1 to 50 ppm). The availability of cadmium to the cells was influenced by the composition of the medium, especially the presence of oyster serum, which contained metal-binding proteins. Regardless of the nominal cadmium concentration in the medium, the actual CL inhibition experienced was shown to be a function of the intracellular cadmium concentration. An attempt was made to duplicate the in vitro immunosuppressive effects by in vivo exposure of oysters. The oysters were exposed for 2 weeks to 0 to 0.25 ppm cadmium; toxicity, as manifested by decreased condition index[60] and reduced shell growth,[61] starts to become evident at ~0.20 ppm. Effects on CL responses of oyster hemocytes collected after exposure to all cadmium concentrations tested (≤0.25 ppm) were variable and showed no significant dose dependency. This finding was probably explained by the fact that in no case did intrahemocytic cadmium levels recorded in the in vivo study approach those produced in the CL-suppressed cells in vitro (2.2 vs. ~15.0 μg Cd per milligram protein, respectively).

Similar chemically induced inhibition of luminol-augmented CL responses of oyster hemocytes have also been produced by exposure to particulate brass, copper, and pentachlorophenol.[62-64] Such evidence implicates these xenobiotics as potential immunotoxicants by extension of the criteria already developed for mammals.[65]

B. Examples in Fish

Preliminary studies suggest that fish phagocytes show significantly reduced CL activity following in vitro or in vivo exposure to xenobiotics. Elsasser et al.[66] studied the CL response of rainbow trout pronephros phagocytes to *S. aureus* after exposure to various metals. When metals were added immediately before the bacterial stimulus, the CL response was increased by Cd but reduced by Al or Cu. Metal concentrations tested were one tenth of the lowest concentration cytotoxic to the trout cells. One hour of metal exposure produced essentially the same effects as those following immediate exposure. After 24 hr of exposure, both Al, and particularly Cu, were immunosuppressive; Cd-exposed cells no longer showed increased CL, but the degree of reduction was minimal. If the CL response was initiated by phagocytosis of *S. aureus* prior to addition of metals, added Cu was strongly suppressive, and added Cd had a stimulatory effect. This direct inhibition of CL by Cu may partially explain the observations that Cu exposure can predispose fish to infection when challenged with viral[67] or bacterial[68] pathogens. Warinner et al.[69] examined the response of kidney phagocytes from normal and pollutant-exposed fish. In initial studies of the response to various stimulatory agents, typical interspecies differences were observed with regard to time of peak CL and the magnitude of the response. However, the zymosan-elicited CL response in spot *(Leiostomus xanthrus)* collected from the PAH-polluted Elizabeth River was negligible compared to the response from spot from the York River (a more pristine site). This same suppression of phagocyte CL could be induced under laboratory conditions by exposing spot to contaminated Elizabeth River sediments. Exposure of dab *(Limanda limanda)* to sewage sludge also inhibited the respiratory burst, as indicated by reduced O_2^- production.[70]

The effects of in vitro exposure to TBT on CL of phagocytes from kidneys of three species of estuarine fish were studied by Wishkovsky et al.[71] The cells were incubated in the presence of TBT for up to 18 hr prior to CL determination. If the CL reaction was initiated by the addition of a phagocytic stimulant (zymosan) immediately after TBT addition, 400 µg/L was generally required to produce significant inhibition. Lower effective concentrations were seen after 18 hr exposure, and some evidence of dose dependency appeared in these data. Whereas the effective TBT concentration used in these in vitro studies is at least two orders of magnitude greater than what might be considered an environmental concentration, it is possible that fish blood cells might be exposed to comparable TBT levels in vivo as a result of bioconcentration.

Other investigators[72] also found that similar concentrations of TBT (~500 µg/L) inhibited CL, but a lower dose (50 µg/L) was stimulatory when oyster toadfish macrophages were exposed in vitro. The opposing effects were attributed to changes in calcium flux across the hemocyte membrane induced by different concentrations of TBT. Peritoneal macrophages removed after in vivo exposure of the fish to TBT showed a dose-dependent suppression of CL.[73] Clearly, the effects of TBT on CL are complex; at various concentrations in vitro, it may stimulate or inhibit CL, at 50 µg/L it enhances PMA- and calcium ionophore (A23187)-stimulated CL but not zymosan-induced CL, and generally a dose-dependent inhibition of CL follows in vivo exposure of fish.

Exposure of *(Oryzias latipes or Fundulus heteroclitus)* pronephros phagocytic blood cells to a ubiquitous environmental contaminant pentachlorophenol (PCP) produced a dose-dependent inhibition of phagocytically induced CL.[74,132] Although it is likely that contaminants such as dioxins and furans present in technical grade PCP play significant roles in suppression of lymphocyte-dependent immune functions,[75,76] pure PCP was strongly immunotoxic to medaka phagocytes, as measured by reduced CL responsiveness. Analytical grade PCP also produced decreased resistance to bacterial infection in the clam, *Mercenaria mercenaria,* as a consequence of impaired hemocyte-mediated antibacterial capacity.[77,78] One could reasonably speculate that this was due to decreased ROI production. With regard to the effects of analytical grade PCP on phagocyte function in fish and bivalves, it is interesting to note that 2,4-dinitrophenol, another uncoupler of oxidative phosphorylation, will inhibit superoxide release and CL by rat and rabbit alveolar macrophages.[79,80]

IV. MODULATION OF ROI PRODUCTION BY MOLLUSCAN DISEASES

A. Neoplasia

Putative neoplastic diseases are reported in several species of bivalve mollusks. A hemic proliferative disease, sometimes likened to leukemia, occurs in the soft shell clam, *Mya arenaria,*[81–84] and the bay mussel, *Mytilus edulis.*[85–88] In both species, the disease can be progressive, invasive, and ultimately fatal. In contrast to normal hemocytes, transformed or neoplastic cells have enlarged nuclei, minimal cytoplasm, and reduced ability to spread on glass surfaces.[88] A comparable condition characterized by the proliferation of abnormal hemocytes has also been described in cockles and oysters of several species.[89-91]

Neoplasia in higher animals is sometimes associated with immunological defects that may impair normal ability to recognize and destroy transformed cells. A recent report describes impaired defense mechanisms in *M. edulis* with hemic neoplasia.[92] In mussels with advanced disease, the ability to clear injected bacteria *(Cytophaga* sp.) was significantly impaired. However, this defect seems to be an expression of the inability of transformed hemocytes to phagocytize foreign particles, rather than impairment of the phagocytic capacity of those morphologically normal hemocytes still present in the clams. Serum components apparently did not influence the phagocytic capabilities of normal or transformed hemocytes. It was unlikely that the development and/or progression of this disease resulted from reduced immunosurveillance, since the untransformed hemocytes continued to function normally, at least with regard to phagocytic activity. However, the phagocytic ability of the neoplastic cells was severely limited, and since these cells make up ~90% of the hemocytes in heavily infected mussels, their presence could lead to serious impairment of defense mechanisms. This may account for the observations of bacterial septicemia reported in mussels with terminal hemic neoplasia.[92]

The original observations of impaired phagocytic ability of transformed hemocytes have been confirmed and enlarged upon by a report of suppressed ROI production by these cells, as measured by reduced CL responses.[131] In this study, the CL responses of cells from nine neoplastic *Mytilus* were compared to those of cells from normal individuals. Generally (in 7/9 of the mussels), the neoplastic cells failed to respond or showed an untypically low response to the CL stimulators PMA or zymosan. The presence of CL responses in hemolymph samples from neoplastic animals was attributed to the presence of normal, phagocytically active hemocytes in the preparations. Attempts to directly correlate CL activity in hemolymph samples from neoplastic mussels to the stage of the disease (i.e., ratio of neoplastic to normal cells in the hemolymph) have not been reported.

B. Parasitism

Certain gastropod mollusks serve as intermediate hosts for metazoan parasites, which can evade the usual encapsulation and cytotoxic activities of the snail's hemocytes. The role of ROIs in the process of blood cell/parasite interaction is currently under study. The *Biomphalaria glabrata/Schistosoma mansoni* host/parasite relationship is one of the most thoroughly investigated from an immunological point of view.[93] Mammalian phagocytes use ROIs and peroxidase to kill schistosomes;[94] the possible participation of these agents in the destruction of an incompatible parasite *(S. mansoni* sporocysts) by hemocytes of the snail *Lymnaea stagnalis* was studied by Dikkeboom et al.[95] The sporocysts were encapsulated and killed within 24 hr following contact with the hemocytes; plasma had no effect on the parasites. Early in the process, O_2^- and H_2O_2 activities were evident around the encapsulated parasites, as evidenced by a positive diaminobenzidine reaction and NBT reduction. These responses were inhibited by catalase or SOD. Peroxidase was also demonstrated at the parasite/capsule interface. However, luminol-augmented CL was not recorded. Possibly, the ROI production occurred only within the hemocyte capsule at the contact points between the cells and the parasite, or perhaps the interior of the capsule was not accessible to the luminol in the medium. Nevertheless, the histochemical evidence implicates ROIs in hemocyte-mediated antiparasitic mechanisms in this in vitro model.

S. mansoni sporocysts produce a number of glycoproteins that have been isolated and characterized.[96,97] Some of these excretory-secretory (E-S) products can modulate the host's defense mechanisms, interfering with hemocyte killing mechanisms,[98] phagocytosis,[99,100] motility,[101] and biosynthetic activity.[102] Immunomodulation of the host by the presence of a parasite or its products may determine the degree of susceptibility/resistance characteristic of particular snail strains. An example of this phenomenon of direct relevance to this paper is the fact that E-S products have been shown to inhibit phagocytosis and O_2^- production by *B. glabrata* hemocytes.[103,104] Hemocytes from a parasite-resistant snail strain had greater ability to generate SOD-inhibitable O_2^- than did cells from a susceptible strain. *S. mansoni* E-S products inhibited phagocytosis and O_2^- production by hemocytes from both strains, but the cells from the resistant strain maintained higher levels of activity. This implies that parasite E-S products may further reduce the already low level of protection afforded by the hemocytes' ability to ingest and destroy the parasite present in the susceptible snails. It is possible that similar mechanisms may explain the ability of *Perkinsus marinus* and other parasites to escape destruction by molluscan hemocytes and to fail to elicit a CL response upon phagocytosis by the hemocytes. It will be interesting to see if *P. marinus* generates E-S products capable of scavenging or dismutating O_2^-.

Bonamia ostreae is an intracellular parasite of the oyster *Ostrea edulis*. Hervio et al.[105] and Boulo et al.[106] reported that this parasite could be phagocytized by hemocytes of both *O. edulis* and *Crassostrea gigas,* but infections were established only after uptake by *O. edulis*. The cells from both oyster species took up zymosan in vitro and produced CL responses;[41] however, the uptake of *B. ostreae* did not trigger CL activity by the cells of either species. The same apparent lack of CL activity is seen after uptake of *P. marinus* by *C. virginica* hemocytes.[107,108] A similar phenomenon was reported for Rickettsiales-like organisms (RLO) that infect the gill tissues and blood of the scallop *Pecten maximus*. Hemocytes of *P. maximus* were readily stimulated by phagocytic stimuli such as zymosan, but showed no CL response after ingestion of RLO.[42,109]

Phagocytosis of protozoan parasites by molluscan hemocytes in vitro may fail to elicit the immediate production of ROIs, as measured by luminol-augmented

CL, but the longer-term presence of intracellular parasites may actually increase the cells' responsiveness to oxidative burst stimuli. This appears to be the case for *C. virginica* hemocytes withdrawn from individuals with *Perkinsus marinus* infections.[110] In this study, the number of circulating hemocytes per milliliter of hemolymph, their CL response to a standardized phagocytic stimulus, and the level of *Perkinsus* infection were determined for individual oysters. The intensities of infection were scored as the number of *Perkinsus* cells per milliliter of hemolymph. Although there was little difference in the zymosan-induced CL responses of lightly and moderately infected groups, the peak and total CL responses of cells from heavily infected oysters were significantly higher than the lightly and moderately infected animals. In similar fashion, if the total hemocyte count data were grouped according to diagnostic scores of individual oysters, the correlation between heavy intensity of infection and elevated hemocyte numbers was significant. As with the CL data, significant differences in hemocyte numbers were encountered when comparing light or moderate vs. heavy, but not when comparing light vs. moderate groups. Therefore, advanced cases of this disease appear to be characterized by hemocyte recruitment into the circulating hemolymph and activation. These factors may work together to generate oxidant loads that could produce various forms of inflammatory cell damage and could contribute to the pathogenesis of *P. marinus* infection.

V. CONCLUSIONS: PROBLEMS AND PROMISES

The production of ROIs by blood cells of invertebrate and vertebrate representatives of aquatic animal species has been convincingly demonstrated by a variety of assays that have already gained wide acceptance in studies of mammalian leukocytes. However, much more work is needed to understand fully the mechanisms underlying the phenomenon and their control. The physiological significance of modulation of ROI production by hemocytes has yet to be established. This is complicated because ROIs can also arise from other sources and processes. An expanding appreciation of the significance of active oxygen species in health and disease will undoubtedly stimulate additional research on ROI generation and antioxidants in aquatic animals. The following paragraphs are meant to point out a few of the problems and promises of such studies.

Since this area of research is comparatively new, it is important to spend extra effort to optimize the experimental conditions. The physiological understanding gained in the process will pay dividends when the time comes to interpret the data. The medium used should not only permit the hemocytes to remain viable, but must also allow the cells to retain the ability to produce ROIs. If particulate stimuli are to be used, divalent cations (especially calcium) and other factors necessary for phagocytosis should be present. Phagocytosis should be verified by microscopic or other means. As in other phagocytic assays, the effects of varying the particle to cell ratio on the magnitude and kinetics of the response in question need to be established. Once determined, this ratio should be held constant in future work with that particular system; the ratio is likely to be different for hemocytes from different species and for different kinds of target particles. It will probably be useful to look at effects of varying the key components of the medium. For example, the presence of antibiotics in oyster hemocyte medium, at concentrations found to be appropriate for the primary culture of fish leukocytes, produced a marked suppression of CL activity.[111] In addition, since the activity of hemocytes is often very labile and a certain degree of activation or suppression may occur during short-term in vitro maintenance, specific controls should be run to quantify this "background" activity in order to accurately quantify the changes in activity produced by an experimental situation. The magnitude and kinetics of the CL response to a standard phagocytic stimulus are often different for cells maintained in culture for 20 vs. 1 hr, perhaps due to some kind of in vitro activation. It is also likely that the response of hemocytes to phagocytic stimulation may vary in suspension cultures vs. adhered cell monolayers;[49] the process of adhesion itself may trigger an oxidative burst response.[112] Knowledge of the adhesion properties of the cells under study can be essential from several points of view. Several frequently used ROI assays, such as superoxide anion determination via cytochrome c, require that the cells remain firmly attached to glass or plastic substrates during a number of washing steps and medium changes. This may be difficult with certain kinds of cells; oyster hemocytes, for example, initially adhere firmly but tend to loosen after about 1 to 2 hr. This process can lead to variable numbers of cells present at the time of ROI assay. None of the above-mentioned problems are insurmountable; with care, assays can be developed that permit accurate quantitation and comparison of ROI activities between individual animals.

After developing an acceptable experimental procedure to quantify ROI generation by leukocytes, it is desirable to learn something of the normal ranges of this activity. Interpretation of "immunosuppression" or "immunoactivation" requires statistical proof that the mean ROI values measured after some treatment or exposure lie outside the "normal" range. This analysis may be difficult or impossible unless an appreciation of natural variation in ROI responses is known and taken into account during the experimental design. Individual differences in activity may be expected to be greater than those usually seen when comparing data from within genetically defined strains of laboratory rodents because the majority of aquatic species studied will be from outbred, wild populations. Furthermore, blood samples often contain various blood cell types, which may have vastly different ROI-generating capabilities and may comprise different percentages of the total cell count in particular individuals. Therefore, it may be helpful to work with purified blood cell types. There seems to be little evidence for differences in ROI activity in hemocytes from different sexes within a species, but this should be a consideration. On the other hand, our data indicate that considerable seasonal variation in O_2^- responses in oyster hemocytes can be measured.

This could result from physiological changes accompanying the reproductive cycle and/or cyclic alterations in temperature. By keeping these and other sources of potential variation in mind, one can more precisely identify and measure experimental effects. When characterizing effects of in vitro exposure to xenobiotics, it is useful to express the activity of treated hemocytes as a percent of the activity of untreated hemocytes, treatments being made on aliquots of a common cell pool. These percentages can then be readily compared statistically to similar data from other pools.

In my view, there are many reasons to study ROI production by blood cells in order to increase understanding of the basic physiology, pathobiology, immunology, and toxicology of aquatic organisms. Phagocytic blood cells comprise an important line of defense against pathogenic microorganisms and parasites, destroying them after internalization or encapsulation. ROIs, in conjunction with lysosomal hydrolases, seem to be involved in these protective killing activities of hemocytes in most, if not all, animals. Clearly, agents that modify the cells' ROI responses probably will also alter the resistance/susceptibility of the host to infectious disease. This, in turn, will affect the general health and survival of individuals and populations. Cells comparable to monocytes or macrophages of higher animals can be found in the circulation of most invertebrates. Like their mammalian counterparts, these cells seem to be readily activated or suppressed by various physical and chemical means. The respiratory burst oxidase has been described in oysters[113] and fish.[52] In mononuclear phagocytes the K_m of the respiratory burst oxidase for NADPH varies with the degree of cell activation. Oxidase preparations from nonactivated macrophages and deactivated monocytes have a K_m of 150 to 300 μmol/L, whereas for activated cells the $K_m \simeq 30$ μmol/L.[114,115] It is interesting that the NADPH oxidase in neutrophils from certain patients with CGD also has a high K_m for NADPH.[116] If such observations hold for fish and shellfish leukocytes, they may help to explain the changes in ROI production that seem to accompany macrophage activation/suppression.

Not only are ROIs interesting because of their involvement in immunological responses, but also because of their role in oxidant-mediated pathology. Oxidants are known to be generated in tissues under many circumstances, including the stimulation of leukocytes during the inflammatory process. These oxidants are known to induce injury to cells in the vicinity of their release. Activated phagocytes produce ROIs in large quantities for use as microbicidal agents that can also flood neighboring tissues when the blood cells become activated.[117,118] Many targets of oxidants have been identified, including DNA, membrane-associated lipids and proteins, enzymes of the glycolytic pathway, cytoskeletal elements, and mitochondria. In mammals, oxidant damage has been implicated in the pathogenesis of respiratory distress syndrome, emphysema, myocardial infarction, and cancer. It is likely that oxidant damage will eventually become associated with specific pathologies in aquatic species. As mentioned in a previous section, phagocytically derived ROIs might play a role in the pathogenesis of a major parasitic disease of oysters.[110] At this time, the opportunities for additional work to define the roles of ROIs in health and disease in aquatic organisms seem virtually unlimited.

In conclusion, it is apparent that exposure of blood cells from aquatic animals to sublethal concentrations of environmental contaminants can reduce the cells' ability to produce ROIs. This form of immunosuppression may increase susceptibility to disease. In vitro studies of the effects of chemicals on hemocytic ROI generation can find application in screening for potential immunomodulators. It is possible that significantly reduced hemocytic ROI activity could serve as a biomarker of chemical stress; however, this requires much additional study. Extracellular release of ROIs contributes to various forms of oxidant damage, induces antioxidant enzymes (SOD, catalase, glutathione peroxidase), and depletes free radical scavengers (glutathione; vitamins A, C, and E; β-carotene). Levels of antioxidant enzymes, in conjunction with responses of the mixed-function oxygenase system, have been suggested in formulating environmental impact assessments.[119] The possible uses of ROIs and/or antioxidant levels in quantifying the biological impacts of environmental pollutants should add a practical impetus to this area of aquatic toxicology.

REFERENCES

1. Klebanoff, S.J., Myeloperoxidase-halide-hydrogen peroxide antibacterial system, *J. Bacteriol.*, 95, 2131, 1968.
2. Diamond, B. and Scharff, M.D., IgG1 and IgG2b share the Fc receptor on mouse macrophages, *J. Immunol.*, 125, 631, 1980.
3. Heusser, C.M., Anderson, C.L., and Grey, H.M., Receptors for IgG: subclass specificity of receptors on different mouse cell types and the definition of two distinct receptors on a macrophage cell line, *J. Exp. Med.*, 145, 1316, 1977.
4. Diamond, B. and Yelton, D.E., A new Fc receptor on mouse macrophages binding IgG$_3$, *J. Exp. Med.*, 153, 514, 1981.
5. Boltz-Nitulescu, G., Plummer, J.M., and Spiegelberg, H.L., Fc receptors for IgE on mouse macrophages and macrophage-like cell lines, *J. Immunol.*, 128, 2265, 1982.
6. Rhodes, J., Receptor for monomeric IgM on guinea-pig splenic macrophages, *Nature (London)*, 243, 527, 1973.
7. Cornacoff, J.B., Hebert, L.A., Smead, W.L., Vanaman, M.E., Birmingham, D.J., and Waxman, F.J., Primate erythrocyte-immune complex-clearing mechanism, *J. Clin. Invest.*, 71, 236, 1983.
8. Ehlenberger, A.G. and Nussenzweig, V., The role of membrane receptors for C3b and C3d in phagocytosis, *J. Exp. Med.*, 145, 357, 1977.
9. Stephenson, J.D. and Shepherd, V.L., Purification of the human alveolar macrophage mannose receptor, *Biochem. Biophys. Res. Commun.*, 148, 883, 1987.

10. Czop, J.K. and Austen, K.F., A β-glucan inhibitable receptor on human monocytes: its identity with the phagocytic receptor for particulate activators of the alternative complement pathway, *J. Immunol.*, 134, 2588, 1985.
11. Van De Water, L., III, Phagocytosis, in *Plasma Fibronectin: Structure and Function*, McDonagh, J., Ed., Marcel Dekker, New York, 1985, 175.
12. Della Bianca, V., Grzeskowiak, M., and Rossi, F., Studies on molecular regulation of phagocytosis and activation of the NADPH oxidase in neutrophils, *J. Immunol.*, 144, 1411, 1990.
13. Landing, B.H. and Shirkey, H.S., A syndrome of recurrent infection and infiltration of viscera by pigmented lipid histiocytes, *Pediatrics*, 20, 431, 1957.
14. Bridges, R.A., Berendes, H., and Good, R.A., A fatal granulomatous disease of childhood, *Am. J. Dis. Child.*, 97, 387, 1959.
15. Curnutte, J.T. and Babior, B.M., Chronic granulomatous disease, in *Advances in Human Genetics*, Vol. 16, Harris, H. and Hirschhorn, K., Eds., Plenum Press, New York, 1987, 229.
16. Holmes, B., Page, A.R., and Good, R.A., Studies of the metabolic activity of leukocytes from patients with a genetic abnormality of phagocytic function, *J. Clin. Invest.*, 46, 1422, 1967.
17. Curnutte, J.T., Whitten, D.M., and Babior, B.M., Defective superoxide production by granulocytes from patients with chronic granulomatous disease, *N. Engl. J. Med.*, 290, 593, 1974.
18. Smith, R.M. and Curnutte, J.T., Molecular basis of chronic granulomatous disease, *Blood*, 77, 673, 1991.
19. Clark, R.A., Volpp, B.D., Leidal, K.G., and Nauseef, W.M., Two cytosolic components of the human neutrophil respiratory burst oxidase translocate to the plasma membrane during cell activation, *J. Clin. Invest.*, 85, 714, 1990.
20. Baehner, R.L. and Nathan, D.G., Quantitative nitroblue tetrazolium test in chronic granulomatous disease, *N. Engl. J. Med.*, 278, 971, 1968.
21. Park, B.H., Holmes, B.M., Rodey, G.E., and Good, R.A., Nitroblue-tetrazolium test in children with fatal granulomatous disease and newborn infants, *Lancet*, 1, 157, 1969.
22. Goldberg, E.D., Bowen, V.T., Farrington, J.W., Harvey, G., Martin, J.H., Parker, P.L., Riseborough, R.W., Robertson, W., Schneider, W., and Gamble, E., The mussel watch, *Environ. Conserv.*, 5, 101, 1978.
23. Lee, R.F., Davies, J.M., Freeman, H.C., Ivanovici, A., Moore, M.N., Stegeman, J., and Uthe, J.F., Biochemical techniques for monitoring biological effects of pollution in the sea, *Rapp. P.-v. Reun. Cons. Int. Explor. Mer*, 179, 48, 1980.
24. Livingstone, D.R., Biochemical measurements, in *The Effects of Stress and Pollution on Marine Animals*, Bayne, B.L. et al., Eds., Praeger Publishers, New York, 1985, 81.
25. Kappus, H., Overview of enzyme systems involved in bioreduction of drugs and in redox cycling, *Biochem. Pharmacol.*, 35, 1, 1986.
26. Blum, J. and Fridovich, I., Enzymatic defenses against oxygen toxicity in hydrothermal vent animals *Rifta pachyptila* and *Calyptogena magnifica*, *Arch. Biochem. Biophys.*, 228, 617, 1984.
27. Wenning, R.J. and DiGiulio, R.T., Microsomal enzyme activities, superoxide production, and antioxidant defenses in ribbed mussels *(Geukensia demissa)* and wedge clams *(Rangia cuneata)*, *Comp. Biochem. Physiol.*, 90C, 21, 1988.

28. Livingstone, D.R., Garcia-Martinez, P., and Winston, G.W., Menadione-stimulated oxyradical formation in digestive gland microsomes of the common mussel, *Mytilus edulis* L., *Aquat. Toxicol.*, 15, 213, 1989.
29. Livingstone, D.R., Garcia-Martinez, P., Michel, X., Narbonne, J.F., O-Hara, S., Ribera, D., and Winston, G.W., Oxyradical generation as a pollution-mediated mechanism of toxicity, *Funct. Ecol.*, 4, 415, 1990.
30. Porte, C., Solé, M., Albaigés, J., and Livingstone, D.R., Responses of mixed-function oxygenase and antioxidase enzyme system of *Mytilus* sp. to organic pollution, *Comp. Biochem. Physiol.*, 100C, 183, 1991.
31. Cheng, T.C., Aspects of substrate utilization and energy requirement during molluscan phagocytosis, *J. Invertebr. Pathol.*, 27, 263, 1976.
32. Dikkeboom, R., Mulder, E.C., Tijnagel, J.M.G.H., and Sminia, T., Phagocytic blood cells of the pond snail *Lymnaea stagnalis* generate active forms of oxygen, *Eur. J. Cell. Biol. Suppl.*, 11, 12, 1985.
33. Nakamura, N., Mori, K., Inooka, S., and Nomura, T., *In vitro* production of hydrogen peroxide by the amoebocytes of the scallop, *Patinopecten yessoensis* (Jay), *Dev. Comp. Immunol.*, 9, 407, 1985.
34. Adema, C.M., van der Knaap, W.P.W., and Sminia, T., Molluscan hemocyte-mediated cytotoxicity: the role of reactive oxygen intermediates, *Rev. Aquat. Sci.*, 4, 201, 1991.
35. Welsh, W.D., Correlation between measurements of the luminol-dependent chemiluminescence response and bacterial susceptibility to phagocytosis, *Infect. Immun.*, 30, 370, 1980.
36. Horan, T.D., English, D., and McPherson, T.A., Association of neutrophil chemiluminescence with microbicidal activity, *Clin. Immunol. Immunopathol.*, 22, 259, 1982.
37. Scott, A.L. and Klesuis, P.H., Chemiluminescence: a novel analysis of phagocytosis in fish, in *Developments in Biological Standardization,* Anderson, D.P. and Hennessen, W., Eds., S. Krager, Basel, 1981, 49, 245.
38. Wishkovsky, A., Chemiluminescence: an advanced tool for measuring phagocytosis, in *Disease Processes in Marine Bivalve Molluscs,* Fisher, W.S., Ed., *Am. Fish. Soc. Spec. Publ.* 18, 1988, 292.
39. Larson, K.G., Roberson, B.S., and Hetrick, F.M., Effect of environmental pollutants on the chemiluminescence of hemocytes from the American oyster *Crassostrea virginica, Dis. Aquat. Org.*, 6, 131, 1989.
40. Anderson, R.S., Oliver, L.M., and Jacobs, D., Immunotoxicity of cadmium for the eastern oyster *(Crassostrea virginica* [Gmelin, 1791]): effects on hemocyte chemiluminescence, *J. Shellfish Res.*, 11, 31, 1992.
41. Bachère, E., Hervio, D., and Mailhe, E., Luminol-dependent chemiluminescence by hemocytes of two marine bivalves, *Ostrea edulis* and *Crassostrea gigas, Dis. Aquat. Org.*, 11, 173, 1991.
42. LeGall, G., Bachère, E., and Mailhe, E., Chemiluminescence analysis of the activity of *Pecten maximus* hemocytes stimulated with zymosan and host-specific Rickettsiales-like organisms, *Dis. Aquat. Org.*, 11, 181, 1991.
43. Adema, C.M., van Deutekom-Mulder, E.C., van der Knapp, W.P.W., Meuleman, E.A., and Sminia, T., Generation of oxygen radicals in hemocytes of the snail *Lymnaea stagnalis* in relation to the rate of phagocytosis, *Dev. Comp. Immunol.*, 15, 17, 1991.

44. Dikkeboom, R., van der Knapp, W.P.W., van den Bovenkamp, W., Tijnagel, J.M.G.H., and Bayne, C.J., The production of toxic oxygen metabolites by haemocytes of different snail species, *Dev. Comp. Immunol.*, 12, 509, 1988.
45. Adema, C.M., Harris, R.A., and van Deutekom-Mulder, E.C., A comparative study of hemocytes from 6 different snails: morphology and functional aspects, *J. Invertebr. Pathol.*, 59, 24, 1992.
46. Noel, D., Bachère, E., and Mialhe, E., Phagocytosis associated chemiluminescence of hemocytes in *Mytilus* (Bivalvia), *Dev. Comp. Immunol.*, 17, 483, 1993.
47. Pipe, R.K., Generation of reactive oxygen metabolites by the haemocytes of the mussel *Mytilus edulis, Dev. Comp. Immunol.*, 16, 111, 1992.
48. Anderson, R.S., Oliver, L.M., and Brubacher, L.L., Superoxide anion generation by *Crassostrea virginica* hemocytes as measured by nitroblue tetrazolium reduction, *J. Invertebr. Pathol.*, 59, 303, 1992.
49. Gifford, R.H. and Malawista, S.E., The nitroblue tetrazolium reaction in human granulocytes adherent to a surface, *Yale J. Biol. Med.*, 45, 119, 1972.
50. Schlenk, D., Garcia Martinez, P., and Livingstone, D.R., Studies on myeloperoxidase activity in the common mussel, *Mytilus edulis* L., *Comp. Biochem. Physiol.*, 99C, 63, 1991.
51. Ito, T., Matsutani, T., Mori, K., and Nomura, T., Phagocytosis and hydrogen peroxide production by phagocytes of the sea urchin *Stronglyocentrotus nudus, Dev. Comp. Immunol.*, 16, 287, 1992.
52. Secombes, C.J., Cross, A.R., Sharp, G.J.E., and Garcia, R., NADPH oxidase-like activity in rainbow trout *Oncorhynchus mykiss* (Walbaum) macrophages, *Dev. Comp. Immunol.*, 16, 405, 1992.
53. Stave, J.W., Roberson, B.S., and Hetrick, F.M., Chemiluminescence of phagocytic cells isolated from the pronephros of striped bass, *Dev. Comp. Immunol.*, 7, 269, 1983.
54. Stave, J.W., Roberson, B.S., and Hetrick, F., Factors affecting the chemiluminescent response of fish phagocytes, *J. Fish Biol.*, 25, 197, 1984.
55. Scott, A.L., Rogers, W.A., and Klesius, P.H., Chemiluminescence by peripheral blood phagocytes from channel catfish: function of opsonin and temperature, *Dev. Comp. Immunol.*, 9, 241, 1985.
56. Sohnle, P.G. and Chusid, M.J., The effect of temperature on the chemiluminescence response to neutrophils from rainbow trout and man, *J. Comp. Pathol.*, 93, 493, 1983.
57. Angelidis, P., Baudin-Laurencin, F., and Youinou, P., Effects of temperature on chemiluminescence of phagocytes from sea bass, *Dicentrarchus labrax* L., *J. Fish Dis.*, 11, 281, 1988.
58. Fisher, W.S., Chu, F.-L.E., and Wishkovsky, A., Immunosuppression of oysters by tributyltin, *J. Shellfish Res.*, 8, 437, 1989.
59. Fisher, W.S., Wishkovsky, A., and Chu, F.-L.E., Effects of tributyltin on defense-related activities of oyster hemocytes, *Arch. Environ. Contam. Toxicol.*, 19, 354, 1990.
60. Roesijadi, G. and Klerks, P.L., Kinetic analysis of cadmium binding to metallothionein and other intracellular ligands in oyster gills, *J. Exp. Zool.*, 251, 1, 1989.
61. Shuster, C.N., Jr. and Pringle, B.H., Trace metal accumulation by the American eastern oyster, *Creassostrea virginica, Proc. Natl. Shellfish Assoc.*, 59, 91, 1969.

62. Anderson, R.S., Mora, L.M., and Thomson, S.A., Exposure of oyster macrophages to particulate brass suppresses luminol-augmented chemiluminescence, *Toxicologist*, 12, 391, 1992.
63. Anderson, R.S., Mora, L.M., and Thomson, S.A., Copper inhibits oxyradical production by oyster macrophages, *Toxicologist*, 13, in press, 1993.
64. Roszell, L.E. and Anderson, R.S., Effects of pentachlorophenol on the chemiluminescent response of phagocytes from two estuarine species, *Toxicologist*, 12, 392, 1992.
65. Tam, P.E. and Hinsdill, R.D., Screening for immunomodulators: effects of xenobiotics on macrophage chemiluminescence in vitro, *Fund. Appl. Toxicol.*, 4, 542, 1990.
66. Elsasser, M.S., Roberson, B.S., and Hetrick, F.M., Effects of metals on the chemiluminescent response of rainbow trout *(Salmo gairdneri)* phagocytes, *Vet. Immunol. Immunopathol.*, 12, 243, 1986.
67. Hetrick, F.M., Knittel, M.D., and Fryer, J.L., Increased susceptibility of rainbow trout to infectious hematopoietic necrosis virus after exposure to copper, *Appl. Environ. Microbiol.*, 37, 198, 1979.
68. Knittel, M.D., Susceptibility of steelhead trout *Salmo gairdneri* Richardson to redmouth infection *Yersenia ruckeri* following exposure to copper, *J. Fish Dis.*, 4, 33, 1981.
69. Wariner, J.E., Mathews, E.S., and Weeks, B.A., Preliminary investigations of the chemiluminescent response in normal and pollutant-exposed fish, *Mar. Environ. Res.*, 24, 281, 1988.
70. Secombes, C.J., Fletcher, T.C., O'Flynn, J.A., Costello, M.J., Stagg, R., and Houlihan, D.F., Immunocompetence as a measure of the biological effects of sewage sludge pollution in fish, *Comp. Biochem. Physiol.*, 100C, 133, 1991.
71. Wishkovsky, A., Mathews, E.S., and Weeks, B.A., Effect of tributyltin on the chemiluminescent response of phagocytes from three species of estuarine fish, *Arch. Environ. Contam. Toxicol.*, 18, 826, 1989.
72. Rice, C.D. and Weeks, B.A., Influence of tributyltin on *in vitro* activation of oyster toadfish macrophages, *J. Aquat. Anim. Health*, 1, 62, 1989.
73. Rice, C.D. and Weeks, B.A., The influence of *in vitro* exposure to tributyltin on reactive oxygen formation in oyster toadfish macrophages, *Arch. Environ. Contam. Toxicol.*, 19, 854, 1990.
74. Anderson, R.S. and Brubacher, L.L., In vitro inhibition of medaka phagocyte chemiluminescence by pentachlorophenol, *Fish Shellfish Immunol.*, 2, 299, 1992.
75. Holsapple, M.P., McNerney, P.J., and McKay, J.A., Effects of pentachlorophenol on the in vitro and in vivo antibody response, *J. Toxicol. Environ. Health*, 20, 229, 1987.
76. Kerkvliet, N.I., Brauner, J.A., and Matlock, J.P., Humoral immunotoxicity of polychlorinated diphenyl ethers, phenoxyphenols, dioxins and furans present as contaminants of technical grade pentachlorophenol, *Toxicology*, 36, 307, 1985.
77. Anderson, R.S., Giam, C.S., Ray, L.E., and Tripp, M.R., Effects of environmental pollutants on immunological competency of the clam, *Mercenaria mercenaria*: impaired bacterial clearance, *Aquat. Toxicol.*, 1, 187, 1981.
78. Anderson, R.S., Effects of pollutant exposure on bactericidal activity of *Mercenaria mercenaria* hemolymph, Extended Abstracts, Division of Environmental Chemistry, *Am. Chem. Soc. Natl. Meet.*, 28, 248, 1988.

79. Miles, P.R., Lee, P., Trush, M.A., and Van Dyke, K., Chemiluminescence associated with phagocytosis of foreign particles in rabbit alveolar macrophages, *Life Sci.*, 20, 165, 1977.
80. Castranova, V., Lee, P., Ma, J.Y.C., Weber, K.C., Pailes, W.H., and Miles, P.R., Chemiluminescence from macrophages and monocytes, in *Cellular Chemiluminescence*, Van Dyke, K. and Castranova, V., Eds., CRC Press, Boca Raton, FL, 1987, 3.
81. Brown, G., Wolke, R.E., Saila, S.B., and Brown, C., Prevalence of neoplasia in ten New England populations of the soft-shelled clam *(Mya arenaria)*, *Ann. N.Y. Acad. Sci.*, 298, 522, 1977.
82. Cooper, K.R., Brown, R.S., and Chang, P.W., The course and mortality of a hematopoietic neoplasm in the soft-shell clam, *Mya arenaria*, *J. Invertebr. Pathol.*, 39, 149, 1982a.
83. Cooper, K.R., Brown, R.S., and Chang, P.W., Accuracy of blood cytological screening techniques for the diagnosis of a possible hematopoietic neoplasm in the bivalve mollusc, *Mya arenaria*, *J. Invertebr. Pathol.*, 39, 281, 1982b.
84. Farley, C.A., Otto, S.V., and Reinisch, C.L., New occurrence of epizootic sarcoma in Chesapeake Bay soft shell clams, *Mya arenaria*, *Fish. Bull.*, 84, 851, 1986.
85. Farley, C.A., Sarcomatoid proliferative disease in a wild population of blue mussels *(Mytilus edulis)*, *J. Natl. Cancer Inst.*, 43, 590, 1968.
86. Cosson-Mannevy, M.A., Wong, C.S., and Cretney, W.J., Putative neoplastic disorders in mussels *(Mytilus edulis)* from southern Vancouver Island waters, British Columbia, *J. Invertebr. Pathol.*, 44, 151, 1984.
87. Mix, M.C., Haemic neoplasms of bay mussels, *Mytilus edulis* L., from Oregon: occurrence, prevalence, seasonality and histopathological progression, *J. Fish. Dis.*, 6, 239, 1983.
88. Elston, R.A., Kent, M.L., and Drum, A.S., Progression, lethality and remission of hemic neoplasia in the bay mussel, *Mytilus edulis*, *Dis. Aquat. Org.*, 4, 135, 1988.
89. Twomey, E. and Mulcahy, M.F., A proliferative disorder of possible hemic origin in the common cockle, *Cerastoderma edule*, *J. Invertebr. Pathol.*, 44, 109, 1984.
90. Alderman, D.J., Van Banning, P., and Perez-Colomer, A., Two European oyster *(Ostrea edulis)* mortalities associated with an abnormal haemocytic condition, *Aquaculture*, 10, 335, 1977.
91. Farley, C.A., Probable neoplastic disease of the hematopoietic system in oysters, *Crassostrea virginica* and *Crassostrea gigas*, *Natl. Cancer Inst. Monogr.*, 31, 541, 1968.
92. Kent, M.L., Elston, R.A., Wilkinson, M.T., and Drum, A.S., Impaired defense mechanisms in bay mussels, *Mytilus edulis*, with hemic neoplasia, *J. Invertebr. Pathol.*, 53, 378, 1989.
93. Loker, E.S. and Bayne, C.J., Immunity to trematode larvae in the snail *Biomphalaria*, in *Immune Mechanisms in Invertebrate Vectors*, Lackie, A.M., Ed., Oxford University Press, Oxford, 1986, 199.
94. Kazura, J.W., Fanning, M., and Blumer, J.L., Role of cell-generated hydrogen peroxide in granulocyte-mediated killing of schistosomula of *Schistosoma mansoni* in vitro, *J. Clin. Invest.*, 67, 93, 1981.
95. Dikkeboom, R., Bayne, C.J., van der Knaap, W.P.W., and Tijnagel, T.M.G.H., Possible role of reactive forms of oxygen in in vitro killing of *Schistosoma mansoni* sporocysts by hemocytes of *Lymnaea stagnalis*, *Parasitol. Res.*, 75, 148, 1988.

96. Lodes, M.J. and Yoshino, T.P., Characterization of excretory-secretory proteins synthesized in vitro by *Schistosoma mansoni* primary sporocysts, *J. Parasitol.*, 75, 853, 1989.
97. Lodes, M.J., Connors, V.A., and Yoshino, T.P., Isolation and functional characterization of snail hemocyte-modulating polypeptide from primary sporocysts of *Schistosoma mansoni, Mol. Biochem. Parasitol.*, 49, 1, 1991.
98. Loker, E.S., Bayne, C.J., and Yui, M.A., *Echinostoma paraensei*: hemocytes of *Biomphalaria glabrata* as targets of echinostome mediated interference with host snail resistance to *Schistosoma mansoni, Exp. Parasitol.*, 62, 149, 1986.
99. van der Knaap, W.P.W., Meuleman, E.A., and Sminia, T., Alterations in the internal defense system of the pond snail *Lymnaea stagnalis* induced by infection with the schistosome *Trichobilharzia ocellata, Parasitol. Res.*, 73, 57, 1987.
100. Noda, S. and Loker, E.S., Effects of infection with *Echinostoma paraensei* on the circulating haemocyte population of the snail host *Biomphalaria glabrata, Parasitology,* 98, 35, 1989.
101. Lodes, M.J. and Yoshino, T.P., The effect of schistosome excretory-secretory products on *Biomphalaria glabrata* hemocyte motility, *J. Invertebr. Pathol.*, 56, 75, 1990.
102. Yoshino, T.P. and Lodes, M.J., Secretory protein biosynthesis in snail hemocytes: in vitro modulation by larval schistosome excretory-secretory products, *J. Parasitol.*, 74, 538, 1988.
103. Connors, V.A. and Yoshino, T.P., In vitro effect of larval *Schistosoma mansoni* excretory-secretory products on phagocytosis-stimulated superoxide production in hemocytes from *Biomphalaria glabrata, J. Parasitol.*, 76, 895, 1990.
104. Connors, V.A., Lodes, M.J., and Yoshino, T.P., Identification of a *Schistosoma mansoni* sporocyst excretory-secretory antioxidant molecule and its effect on superoxide production by *Biomphalaria glabrata* hemocytes, *J. Invertebr. Pathol.*, 58, 387, 1991.
105. Hervio, D., Bachère, E., Mialhe, E., and Grizel, H., Zymosan and specific-rickettsia activation of oxygen free radicals production in *Pecten maximus* hemocytes, *Dev. Comp. Immunol.*, 13, 448, 1989.
106. Boulo, V., Hervio, D., Morvan, A., Bachère, E., and Mialhe, E., In vitro culture of mollusc hemocytes. Functional study of burst respiratory activity and analysis of interactions with protozoan and procaryotic pathogens, *In Vitro,* 27, 42A, 1991.
107. Anderson, R.S., Unpublished data, 1992.
108. LaPeyre, J.F., Chu, F.-L. E., and Vogelbein, W.K., In vitro interaction of *Perkinsus marinus* with hemocytes from eastern and Pacific oysters, *Crassostrea virginica* and *Crassostrea gigas, J. Shellfish Res.*, 11, 200, 1992.
109. La Gall, G., Bachère, E., Mialhe, E., and Grizel, H., Zymosan and specific-rickettsia activation of oxygen free radicals production in *Pecten maximus* hemocytes, *Dev. Comp. Immunol.*, 13, 448, 1989.
110. Anderson, R.S., Paynter, K.T., and Burreson, E.M., Increased reactive oxygen intermediate production by hemocytes withdrawn from *Crassostrea virginica* infected with *Perkinsus marinus, Biol. Bull.*, 183, 476, 1992.
111. Anderson, R.S., Mora, L.M., and Brubacher, L.L., Sensitivity of oyster hemocyte chemiluminescence to antibiotic-antimycotic agents, *J. Invertebr. Pathol.*, 60, 317, 1992.

112. Helmke, R., Hidalgo, H., German, V., and Mangos, J., The macrophage oxidative burst *in vitro:* effects of adherence, *In Vitro,* 26, 39A, 1990.
113. Chagot, D.J., Characterisation morphologique et fonctionelle des hemocytes d'*Ostrea edulis* et de *Crassostrea gigas,* mollusques bivalves. Etude *in vitro* de leurs interactions avec le protozoaire *Bonamia ostrea* (Ascetospora), Ph.D. Thesis, Ministere de l'Education Nationale, Ecole Pratique des Hautes Etudes, France, 1989.
114. Berton, G., Cassatella, M., Cabrini, G., and Rossi, F., Activation of mouse macrophages causes no change in expression and function of phorbol diester receptors, but is accompanied by alterations in activity and kinetic parameters of NADPH oxidase, *Immunology,* 54, 371, 1985.
115. Tsunawaki, S. and Nathan, C.F., Enzymatic basis of macrophage activation. Kinetic analysis of superoxide production in lysates of resident and activated mouse peritoneal macrophages and granulocytes, *J. Biol. Chem.,* 259, 4305, 1984.
116. Shurin, S.B., Cohen, H.J., Whitin, J.C., and Newburger, P.E., Impaired granulocyte superoxide production and prolongation of the respiratory burst due to a low-affinity NADPH-dependent oxidase, *Blood,* 62, 564, 1983.
117. Babior, B.M., Microbial oxidant production by phagocytes, in *Oxy-Radicals in Molecular Biology and Pathology,* Cerutti, P.A., Fridovich, I., and McCord, J.M., Eds., Alan R. Liss, New York, 1988, 39.
118. Cochrane, C.G., Schraufstätter, I.U., Hyslop, P.A., and Jackson, J.H., Cellular and biochemical events in oxygen injury, in *Oxy-Radicals in Molecular Biology and Pathology,* Cerutti, P.A., Fridovich, I., and McCord, J.M., Eds., Alan R. Liss, New York, 1988, 125.
119. Livingstone, D.R., Toward a specific index of impact by organic pollution for marine invertebrates, *Comp. Biochem. Physiol.,* 100C, 151, 1991.
120. Secombes, C.J., Isolation of salmonid macrophages and analysis of their killing activity, in *Techniques in Fish Immunology,* Stolen, J.S., et al., Eds., SOS Publications, NJ, 1990, 137–154.
121. Secombes, C.J., Chung, S., and Jeffries, A.H., Superoxide anion production by rainbow trout macrophages detected by the reduction of ferricytochrome c, *Dev. Comp. Immunol.,* 12, 201, 1988.
122. Pick, E. and Mizel, D., Rapid microassays for the measurement of superoxide and hydrogen peroxide production by macrophages in culture using an automatic enzyme immunoassay reader, *J. Immunol. Meth.,* 46, 211, 1981.
123. Pick, E. and Keisari, Y.A., A simple colorimetric method for the measurement of hydrogen peroxide produced by cells in culture, *J. Immunol. Meth.,* 30, 161, 1980.
124. Allen, R.C. and Loose, L.D., Phagocytic activation of a luminol-dependent chemiluminescence in rabbit alveolar and peritoneal macrophages, *Biochem. Biophys. Res. Commun.,* 69, 245, 1976.
125. Allen, R.C., Lucigenin chemiluminescence: a new approach to the study of polymorphonuclear leukocyte redox activity, in *Bioluminescence and Chemiluminescence, Basic Chemistry and Analytical Applications,* DeLuca, M.A. and McElroy, W.D., Eds., Academic Press, New York, 1981, 63.
126. Burow, S. and Valet, G., Flow-cytometric characterization of stimulation, free radical formation, peroxidase activity and phagocytosis of human granulocytes with 2, 7-dichlorofluorescein (DCF), *Eur. J. Cell Biol.,* 43, 128, 1987.

127. Rothe, G. and Valet, G., Flow cytometric analysis of respiratory burst activity in phagocytes with hydroethedine and 2′,7′-dichlorofluorescein, *J. Leuk. Biol.*, 47, 440, 1990.
128. Rothe, G., Oser, A., and Valet, G., Dihydrorhodamine 123: a new flow cytometric indicator for respiratory burst activity in neutrophil granulocytes, *Naturwissenschaften*, 75, 354, 1988.
129. Adema, C.M., van Deute Kom-Mulder, E.C., van der Knaap, W.P.W., and Sminia, T., NADPH-oxidase activity: the probable source of reactive oxygen intermediate generation in hemocytes of the gastropod *Lymnaea stagnalis*, *J. Leuk. Biol.*, 54, 379, 1993.
130. Bell, K.L. and Smith, V.J., In vitro superoxide production by hyaline cells of the shore crab *Carcinus maenas* (L.), *Dev. Comp. Immunol.*, 17, 211, 1993.
131. Elston, R.A., personal communication, 1993.
132. Roszell, L.E. and Anderson, R.S., In vitro immunomodulation by pentachlorophenol in phagocytes from an estuerine teleost, *Fundulus heteroclitus*, as measured by chemiluminescence activity, *Arch. Environ. Contam. Toxilcol.*, 25, 493, 1993.

Chapter 6

DNA Adduct Analysis in Fish: Laboratory and Field Studies

Alexander E. Maccubbin

I. INTRODUCTION

In an overview of chemical carcinogenesis, Miller[1] described historical events that provided a foundation for modern research on chemical carcinogenesis. Among the important landmarks were the observations of scrotal skin cancer in chimney sweeps by Pott,[2] work by Japanese researchers demonstrating induction of skin cancer by coal tar,[3-5] isolation of benzo(a)pyrene (BaP) from coal tar,[6] and the description of initiation and promotion stages of skin cancer with tar and BaP.[7-10] In addition to these studies, the recognition and demonstration of covalent binding of chemicals to macromolecules, especially DNA, provided impetus for evaluating the role of such interactions in the process of carcinogenesis.[11-13] In the years that followed, numerous adducts formed by a variety of chemical agents were described.[14-16]

The study of carcinogenesis in fish has similar landmark studies. Although there were reports of tumors in the early literature,[17] the first epizootic of liver neoplasia in wild fish was reported in 1964 by Dawe et al.[18] In that report, the authors suggested the use of bottom-feeding fish to detect environmental carcinogens. Since then, the number of reports of epizootics of fish tumors in North America has been ever increasing, with many of them describing liver tumors in bottom-dwelling fish from areas with high levels of environmental pollutants.[17]

Around the same time as the first description of liver tumors in wild fish, worldwide occurrences of liver tumors in hatchery-raised rainbow trout were described.[19-22] Subsequent research pinpointed the cause of the cancer as a potent hepatocarcinogen, Aflatoxin B_1 (AFB_1), which had contaminated the vegetable protein component of the commercial diet being fed to the trout.[23-25] The discovery of the sensitivity of rainbow trout to chemically induced liver tumors thus led to the development of a rainbow trout laboratory model for studying chemical carcinogenesis and encouraged the development of other fish model systems.[26-33]

Within the discipline of genetic toxicology, research has taken place at three levels: (1) identifying the kind and determining the frequency of genetic diseases, (2) studying the mechanisms of how chemical and physical agents cause genetic disorders, and (3) evaluating agents for their potential to cause genetic damage.[34] The study of DNA adducts in humans and animal models has been an important part of the latter two levels of research. From such studies, several generalities have been made regarding the value of DNA adducts in the study of carcinogenesis. Many chemical carcinogens are metabolized to reactive species that bind covalently to DNA.[35] There is a good but not perfect correlation between the level of chemical binding to DNA and carcinogenic potency and organ specificity.[15,36-38] DNA adducts are important in chemical carcinogenesis during initiation and first-stage promotion and may be important in later stages.[35,39-41] Finally, DNA adduct formation can be used as a measure of exposure to genotoxins[39,42] and may provide information about the biological effect and potential risk of a chemical.[36,38,43] Indeed, it has been suggested that any chemical that forms DNA adducts even at very low levels should be considered to have carcinogenic and mutagenic potential.[44]

The study of genetic toxicology in aquatic systems has mainly focused on carcinogenesis in fish and shellfish. Numerous field studies have been conducted documenting the species affected, the frequency of occurrence, and type of tumors.[17] In the laboratory, fish carcinogenesis models have been used to evaluate chemicals for carcinogenicity and to study the factors influencing the carcinogenic process. DNA adducts analysis has been an important component of both laboratory and field studies. In laboratory studies, analysis of DNA adducts has been used to validate model systems using fish, so that comparisons can be made with other animal models and humans. In these studies, the ability of a variety of fish species to activate carcinogens to reactive products that bind to DNA has been examined. The major adducts have been characterized, rates of formation and persistence have been determined, and factors (including anticarcinogens) that modulate adduct formation have been evaluated. Finally, laboratory studies have evaluated adduct formation after exposure to complex mixtures, such as sediment extracts. These studies were designed to see if the adducts formed were comparable to those observed in wild fish. In field studies, fish have been used as environmental monitors for carcinogens, especially those found in contaminated sediments. These studies (see Section II) have focused on the use of fish to monitor exposure to carcinogens and the effects they elicit. They have been undertaken in the context of identifying environmental carcinogens that might present a risk to the fish population or to consumers of the fish (including humans).

In this chapter, we will examine some of the studies of DNA adducts in fish species that have been conducted in the laboratory and in the field. The objectives of such an examination are to determine what we have learned thus far, to point out areas that need more study, and to provide suggestions for future directions that should be taken. It is hoped that by critical evaluation of the literature, important advances and deficiencies will be recognized.

II. DNA ADDUCT FORMATION AND THE TOXICOLOGICAL PARADIGM

The toxicological paradigm suggests that there is a continuum from exposure to a toxicant to a possible toxic endpoint in an organism.[34] Throughout this continuum, many factors may increase or decrease the likelihood of the occurrence of a toxic endpoint. This is especially true when the toxic endpoint is cancer. In the multistage carcinogenesis model that has been derived from laboratory studies, the stages of initiation, promotion, conversion, and progression are well recognized.[40] These stages involve a series of genetic and epigenetic changes that ultimately result in neoplastic disease. In the first stage, initiation, exposure to a carcinogenic stimulus and interaction with DNA result in genetic damage in some cells. This damage may occur in a nondividing cell and be of no further consequence. However, if a round of cell division occurs and the damage is not repaired or repaired incorrectly, then initiated cells will exist.[45] In the promotion stage, initiated cells grow to a greater extent than normal cells and thus undergo clonal expansion. During such growth, cells have an increased chance to receive additional genetic damage from exogenous or endogenous sources. Conversion of benign lesions to malignant tumors may be enhanced by additional exposure to genotoxins.[46] Finally, during progression, spread of the malignancy may occur as cells acquire altered phenotypic characteristics and genetic instability.[47]

The formation of DNA adducts in fish has been mainly studied in the context of initiation by genotoxic chemicals. In this stage (as well as in others), the level of adducts that can be detected at any given time is the net result of three major factors: exposure, metabolism (activation and detoxification), and repair.[39] Before examining the studies of DNA adduct analysis in fish, a brief review of what is known about these factors in fish is in order.

A. Environmental Exposure and Uptake

The occurrence of chemical carcinogens in the environment has been well documented in many of the areas that have been found to have epizootics of fish tumors; it is beyond the scope of this chapter to review all these studies. However, the extensive studies in Puget Sound and the Great Lakes provide good examples of chemically contaminated waterways.[48-56] With the existence of contaminated environments, exposure and uptake of the contaminants by the biota is necessary for these contaminants to evoke a response.

Fish may be exposed to chemical contaminants contained in water, food sources, or sediment. Water serves as a route of exposure for those chemicals that are water soluble or are adsorbed to fine particles suspended in the water column. A continuous exposure is obtained by the passage of water through the gills and by contact with the water that always surrounds fish. A more concentrated exposure to chemical contaminants can be received via food sources, especially by bottom-feeding species that ingest sediment and organisms living in contaminated sediments. The bioavailability of chemical contaminants has been studied, and there are many studies that demonstrate the uptake of chemical contaminants found in waters and sediments. For example, analysis of fish tissues from Great Lakes species has documented the presence of a wide spectrum of organic xenobiotics.[57] Varanasi and co-workers[58-61] have demonstrated in laboratory studies that xenobiotic chemicals contained in sediments can be assimilated and metabolized by bottom-feeding fish. Moreover, several field studies in chemically contaminated areas have demonstrated that the stomach contents of fish have chemical profiles that match the chemical profiles of sediments in the areas from which the fish were collected.[49,50,62] Finally, evidence for exposure to chemicals comes from studies that demonstrate the existence of metabolites of a variety of chemicals in the bile of wild fish collected from polluted sites.[63-66]

B. Metabolism

As a determinant of carcinogen/DNA adduct formation, metabolism plays a key role. Many chemical carcinogens are inactive per se and require some form of metabolic activation to reactive species that can interact with target macromolecules such as DNA. In addition to activation reactions carried out by enzymes in fish, detoxification and deactivation reactions can take place. The balance between activation and detoxification/deactivation has a direct impact on the amount of chemical/DNA adduct that can be detected at any given time.[39] The metabolic competence of various fish species for both activating and deactivating chemicals has been studied extensively, and fish have been shown to possess essentially all of the enzyme systems found in mammals that are involved in metabolizing xenobiotics.[67-75]

The mixed function oxidase (MFO) enzyme system has probably been the most widely studied activating system. Among the fish species studied, there exists a wide variation in MFO activity, as measured by arylhydrocarbon hydroxylase activity (AHH), and the ability to activate a compound such as BaP does not seem to correlate with the sensitivity of the species to BaP-induced liver cancer. For example, laboratory studies with BaP have demonstrated hepatocarcinogenicity in rainbow trout and medaka,[76–79] and yet there is a 25- to 30-fold difference in AHH activity found in liver microsomes of these two species.[80,81] Similarly, both brown bullheads and carp are able to metabolize BaP,[82] and yet only brown bullheads have been found to have liver tumors when collected from BaP-contaminated waterways. Significant variability in AHH activity among strains within the same species has also been documented. For example, among six strains of rainbow trout, a 57-fold difference in AHH activity was measured.[83] Finally, individual fish of a given species may have vastly different levels of AHH activity. Bend et al.[84] found a 76- to 220-fold difference in AHH activity in individual winter flounder caught at different times of the year.

As mentioned previously, detoxification and deactivation reactions are important in determining the availability of reactive DNA-binding species. These reactions may detoxify a compound by metabolizing it to a less reactive species, as is the case for one pathway of AFB_1 metabolism that results in the formation of the less toxic Aflatoxin M_1 (AFM_1). Alternatively, so-called phase II enzymes may conjugate reactive metabolites, making them more water soluble, so they are readily excreted.[75] The role of some of these detoxification mechanisms has been studied relative to DNA adduct formation and will be discussed in Section III.

C. DNA Repair

The repair of DNA damage caused by carcinogens has not been studied widely in fish. In general, studies have suggested that repair, as measured by unscheduled DNA synthesis (UDS) in cultured or freshly isolated cells, is relatively low in fish.[85–88] In addition, studies of DNA repair measured in vivo by UDS, using medaka and other small aquarium fish, demonstrated that UV-type damage and alkylation damage could be repaired in brain ganglion cells.[89] However, no evidence for excision repair was observed in liver or intestinal cells, and the authors suggested that this might make these cells more susceptible to the effects of chemical carcinogens.[89] Recently, attempts have been made to assess the activity of specific repair enzymes such as O^6-methylguanine-DNA methyltransferase.[90,91] This enzyme activity can be detected in fish liver but appears to be inefficient in removing O^6-alkylguanine adducts.[91] Finally, lack repair of BaP and AFB_1 DNA adducts in fish has been inferred because of the persistence of these adducts.[92–94,138]

III. THE STUDY OF DNA ADDUCTS IN FISH

A. Methods

The most widely used method for examining the interaction of chemicals and DNA has been the use of radiolabeled chemicals.[95] This technique has proven to be valuable for the study of those chemicals that can be synthesized in radioactive forms with high specific activities. However, there are limitations to application of this method because of the lack of radiolabeled forms of chemicals or the inability to obtain high specific activity chemicals. Moreover, in studies that are designed to detect and measure DNA adducts in whole animals or humans, alternative methods that do not require administration of radiolabeled compounds are needed. Over the past several years, a number of sensitive methods that do not use radiolabeled chemicals have been developed to detect and quantitate DNA adducts (see reviews in References 16 and 96 to 99) Some of these methods have been used to study DNA adducts in fish and are summarized in Table 1. All of these sensitive methods have advantages and disadvantages, and the method chosen for adduct analysis depends on the types of adducts to be detected, the availability of reagents (such as antibodies), and the access to special instrumentation that might be required for a method.

B. Laboratory Studies

1. In Vitro Studies

Studies that have evaluated DNA adducts using in vitro systems have employed radiolabeled compounds and have concentrated on determining the formation of DNA adducts by BaP and AFB_1 (Table 2). In most of the studies, the overall binding of a particular carcinogen was determined. However, several studies also identified the major adduct that was formed.[114,115,117–124] In all cases, the major adduct was the same as one that had been previously described in rodent systems. For example, the major BaP/DNA adduct (anti-BPDE/dG) found in brown bullhead, English sole, starry flounder, and blue gill is the same adduct that has been observed in mouse skin treated with BaP.[127,128] Moreover, this BaP/DNA adduct was formed whether hepatic microsomes, cultured cell lines, or fresh hepatocytes were used for activating the carcinogen.[117–123] In contrast, however, studies with two trout species demonstrated that the anti-BPDE/dG adduct was formed in smaller amounts relative to unidentified polar adducts[123] and presumptive 9-hydroxy-benzo(a)pyrene-4,5-oxide-deoxyguanosine adduct.[115] This latter adduct is the major one formed by hepatic microsomes from rats.[105]

In addition to determining DNA binding of activated carcinogens and identification of the major adducts formed, in vitro studies have been used to compare the level of adduct formation of a specific carcinogen in various species,[94,114–116,123] to examine the relative DNA adduct formation of carcinogens and their metabolites,[119–122,126] and to evaluate the effect of antitumor agents on DNA adduct formation.[129–131] Results of studies comparing species confirm that although carcinogen binding to DNA may be an important event in initiation of carcinogenesis, it is by no means the sole determinant in tumor formation. BaP/DNA adduct formation in species of fish that inhabit chemically contaminated waterways and

Table 1. Sensitive Methods not Requiring Radiolabeled Carcinogens that have been Used to Detect DNA Adducts in Fish

Method[a]	Sensitivity[b]	DNA(μg)	Ref.
HPLC/FL	$1/10^7$	100	91, 100, 101
GC-MS/SIM	$1/10^5$	<1	102–104
^{32}P-postlabeling	$1/10^9$	1–10	92, 93, 105–112
Immunochemical	$1/10^7$	1	113

[a] Method: HPLC/FL = high performance liquid chromatography with flourescence detection; GC-MS/SIM = gas chromatography-mass spectrometry with single-ion monitoring; immunochemical = competitive enzyme-linked immunosorbant assay.

[b] Sensitivity: adduct/normal nucleotides.

Table 2. Selected In Vitro Studies of DNA Adducts in Fish

Species	System[a]	Binding	Adduct[e]	Ref.
Rainbow trout	^{14}C-DMN/liver slices	2.5[b]	N7-Gua	114
Goldfish	^{14}C-DMN/liver slices	2.5[b]	N7-Gua	
Trout	^{3}H-BaP/microsomes	680[c]	BPDE/dN, 9-OHBaP4,5/dN	115
Roach	^{3}H-BaP/microsomes	20[c]	nd	
English sole	^{3}H-BaP/10K super	600[c]	nd	116, 117
Starry flounder	^{3}H-BaP/10K super	150[c]	nd	
Coho salmon	^{3}H-BaP/10K super	20[c]	nd	
English sole	^{3}H-BaP/microsomes	nd	Anti-BPDE/dG	117–120
Starry flounder	^{3}H-BaP/microsomes	nd	Anti-BPDE/dG	
Brown bullhead	^{3}H-BaP/hepatocytes	37[b]	Anti-BPDE/dG	121, 122
Carp	^{3}H-BaP/hepatocytes	212[b]	Anti-BPDE/dG, one unknown	
Blue gill	^{3}H-BaP/cell line	5[b]	Anti-BPDE/dG	123
	^{3}H-DMBA/cell line	8[b]	Polar adducts	
Rainbow trout	^{3}H-BaP/cell line	44[b]	Polar adducts	123
	^{3}H-DMAB/cell line	4.7[b]	Polar adducts	
Brown bullhead	^{3}H-BaP/cell line	17[b]	Anti-BPDE/dG	123
	^{3}H-DMBA/cell line	8.6[b]	DMBADE/dN	
Rainbow trout	^{3}H-AFB$_1$/hepatocytes	164–208[d]	AFB$_1$-N7-G	124–126
	^{3}H-AFL/hepatocytes	87–95[d]	nd	125, 126
	^{3}H-AFM$_1$/hepatocytes	124[d]	nd	126
	^{3}H-AFL-M$_1$/hepatocytes	134[d]	nd	126
Coho salmon	^{3}H-AFB$_1$/hepatocytes	10.6[d]	AFB$_1$-N7-G	94
Rainbow trout	^{3}H-AFB$_1$/microsomes	132[b]	nd	130

Note: nd = not determined.

[a] System: ^{14}C-DMN = ^{14}C-dimethylnitrosamine; ^{3}H-BaP = ^{3}H-benzo(a)pyrene; 10K super = 10,000 × g supernatant from liver homogenate; ^{3}H-DMBA = ^{3}H-7,12-dimethylbenz(a)anthracene; cell line = cultured cell line derived from species indicated; ^{3}H-AFB$_1$ = ^{3}H-Aflatoxin B$_1$; ^{3}H-AFL = ^{3}H-Aflatoxicol; ^{3}H-AFM$_1$ = ^{3}H-Aflatoxin M$_1$; ^{3}H-AFL-M$_1$ = ^{3}H-Aflatoxicol-M$_1$; hepatocytes = freshly isolated hepatocytes from the indicated species.

[b] pmol substrate bound/mg DNA.

[c] pmol substrate bound/μg DNA/mg protein.

[d] pmol substrate bound/μg DNA/ μmol substrate added.

[e] Adduct: major adduct observed, N7-Gua = N7-methylguanine; anti-BPDE/dG = anti-benzo(a)pyrene-7,8-diol-9,10 epoxide/deoxyguanosine; BPDE/dN = benzo(a)pyrene-7,8-diol-9,10-epoxide/deoxynucleoside; 9-OHBaP4,5/dN = 9-hydroxy-benzo(a)pyrene-4,5-oxide/deoxynucleoside; DMBADE/dN = 7,12-dimethylbenz(a)anthracene-3,4-diol-1,2-epoxide/deoxyribonucleoside; AFB$_1$-N7-G = 8,9-dihydro-8-(N7-guanyl)-9-hydroxylaflatoxin B$_1$.

have been found to have liver tumors has been compared with that of species of fish from the same habitat that have lower or no incidence of liver tumors. For example, when BaP/DNA adduct formation in hepatocytes from brown bullheads (a species that has been found with liver tumors)[53,132,133] was compared to that in hepatocytes from carp (tumor-resistant species),[134-136] the level of BaP/DNA adduct formation was lower in the bullheads.[121,122] Thus, if BaP is among the causative agents of liver tumors in these fish species from chemically contaminated areas, processes that take place after the formation of the DNA adduct influence the endpoint of tumor formation. In another study comparing BaP/DNA binding in English sole and starry flounder using a liver homogenate system,[116,117] the resistant starry flounder had higher levels of adduct formation. However, in vivo studies demonstrated that English sole had higher BaP/DNA binding rates (see below). In contrast to BaP, studies on AFB_1/DNA adduct formation in sensitive and resistant species demonstrated that DNA adduct formation did reflect tumor formation. Total binding of AFB_1 in salmon hepatocytes (resistant species) was more than tenfold lower than that found in the sensitive rainbow trout.[94] In both species, however, activation of AFB_1 led to the formation of the same major adduct, namely AFB_1-N7-guanine.

Loveland et al.[125,126] have used primary hepatocytes to evaluate the DNA binding of AFB_1 and its major phase I metabolites, Aflatoxicol (AFL), AFM_1, and Aflatoxicol-M_1 (AFL-M_1). These studies demonstrated linear binding of all four compounds with time and a relative binding in the order of AFB_1>AFL-M_1>AFM_1>AFL. These studies also demonstrated that the binding of AFL was primarily via reconversion to AFB_1 and subsequent metabolism to AFB_1-8,9-oxide.[125] Finally, these results suggested that in vitro measurements of DNA binding may not reflect the potential of a compound to cause tumors in vivo, as the binding of AFM_1 and AFL-M_1 was higher than expected based on mutagenicity and carcinogenicity studies.

Studies on the effect of several anticancer compounds on DNA binding have been conducted using rainbow trout in vitro systems.[129-131] In a study using hepatic microsomes, Fong et al.[130] found that products of indole-3-carbinol (I3C) formed by acid treatment caused a significant decrease in binding AFB_1 to DNA. These products apparently acted as inhibitors of AFB_1 activation. A combined in vivo and in vitro approach was used to look at the effect of butylated hydroxyanisole (BHA) and β-naphthoflavone (β-NF) on AFB_1/DNA binding.[129,131] Rainbow trout were fed the anticarcinogens, and primary hepatocyte cultures from the trout were then tested in vitro for their ability to mediate AFB_1 binding to DNA. Hepatocytes from trout pretreated with β-NF had a 60% decrease in AFB_1/DNA binding, whereas the BHA treatment had no effect. These results paralleled the results for tumor formation, and it was suggested that β-NF inhibited DNA adduct formation through the induction of detoxifying metabolic pathways.[129]

2. In Vivo Studies

The results of studies using in vitro systems demonstrated carcinogen activation and binding to DNA by using radiolabeled chemical carcinogens. In studies that have examined DNA adduct formation in vivo, detection techniques other than radiometric methods have been employed (Table 1). As discussed above for in vitro methods, many of the studies of DNA adduct formation in vivo measured total adduct formation, and several identified the major adduct formed. In all cases, the major adduct found with a specific carcinogen was the same as one that had been described in rodent model systems, and the major adducts identified in vivo were the same ones that had been described in vitro. For example, AFB_1-N7-gua[137] and anti-BPDE/dG[92,93,101,138] were the major adducts formed in vivo by AFB_1 and BaP, respectively. In several studies measuring BaP DNA adducts, adducts other than anti-BPDE/dG were observed.[92,93,101,106,138] For example, English sole or brown bullheads that received BaP by injection[92,93,106,138] at doses of 20 mg/kg or greater had at least two prominent adducts (Figure 1) and as many as six minor adducts.[92] The adducts shown in Figure 1 were also observed in rainbow trout that had received 50 mg/kg BaP.[106]

Fish have been exposed to carcinogens in vivo via several routes, including i.p. injection, feeding, and aqueous immersion. In general, regardless of the route of exposure, carcinogen binding to DNA was found to be linear with the dose administered. However, the route of exposure did have an effect on overall adduct formation. In experiments with English sole, starry flounder, and brown bullheads,[106,138] BaP adduct formation in hepatic DNA was 10 to 28 times greater when BaP was given by i.p. injection compared to administration by feeding. Varanasi et al.[138] suggested that less reactive species of BaP may reach the liver in a feeding protocol when compared to i.p. injection. Results of work by James et al.[139] supported this idea by the demonstration of significant binding of BaP to DNA in the intestine of southern flounder after exposure by gavage. Thus, uptake and metabolism of BaP in the intestine of fish[140] may allow DNA adduct formation in these extrahepatic tissues.

Several studies have compared species differences with regard to DNA adduct formation in vivo. In contrast to results of in vitro adduct analysis, starry flounder were found to have a two- to fourfold lower level of BaP adducts when compared to English sole.[105] Although the overall adduct formation was different in these two species, the same major adduct was observed, and BaP was activated to reactive species equally well in both fish. Thus, it was suggested that the differences in DNA adduct formation were due to increased conjugation of reactive intermediates to glutathione in starry flounder.[120] Differences in AFB_1 adduct formation in vivo in coho salmon and rainbow trout were more dramatic and in agreement with in vitro studies. Depending on the route of exposure, the level AFB_1 adducts in hepatic DNA in rainbow trout was 7 to 56 times higher than that found in coho salmon.[94,137] Unlike

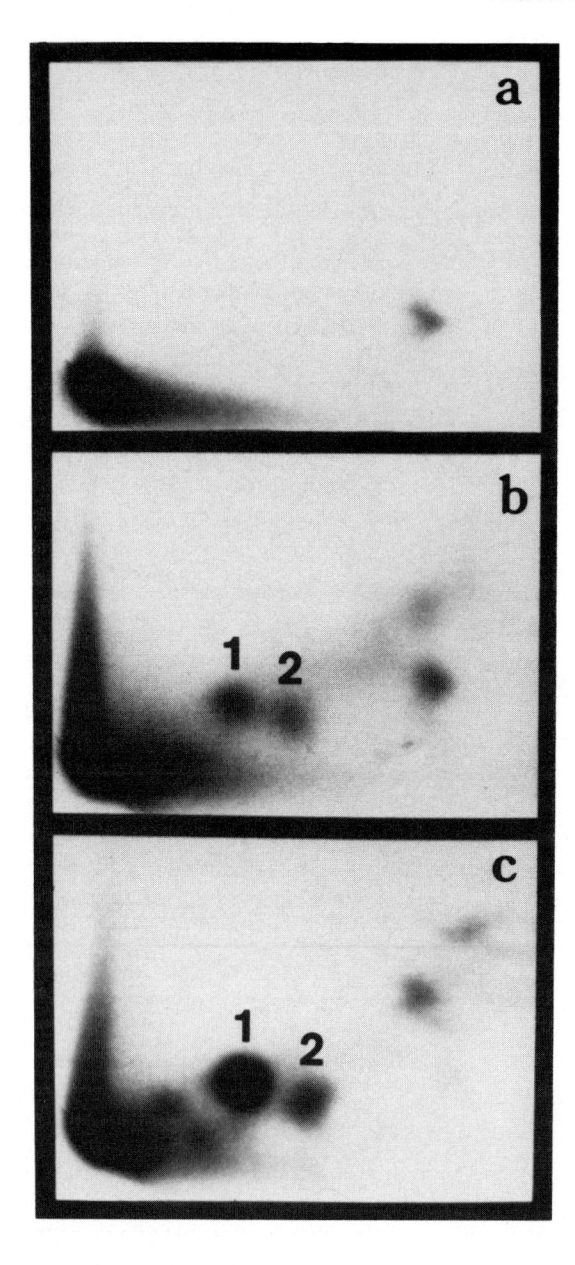

Figure 1. Autoradiograms of thin layer chromatography maps of ^{32}P-postlabeled DNA adducts in the liver of brown bullheads. The fish in panel a had received an injection of solvent only, whereas the fish in panels b and c had received 50 mg/kg BaP, 24 and 48 hr prior to sacrifice. The DNA adducts were enriched using the micrococcal nuclease/spleen phosphodiesterase/nuclease P1 procedure. The spots labeled 1 and 2 were the major adducts observed.

the case of BaP in English sole and starry flounder, conjugation of AFB_1 metabolites to glutathione in rainbow trout is a very minor pathway,[146] and the apparent reason for DNA adduct differences was the lower levels of the cytochrome P450 responsible for activation of AFB_1 in the coho salmon.[94]

The effect of anticarcinogens on DNA adduct formation has also been determined in vivo.[94,100,131,142,143] Feeding of I3C prior to AFB_1 or diethylnitrosamine (DEN) treatment in rainbow trout caused a decrease in the level of adducts.[100,142] Similar results were observed when β-NF or chlorophyllin were administered prior to AFB_1 treatment.[131,143,144] Decreases observed in DNA binding paralleled decreases in tumor formation when I3C or β-NF were administered prior to the carcinogen,[100,141,145] which suggested that these anticarcinogens were acting at the initiation stage of carcinogenesis, either by blocking activation, inducing detoxification pathways of metabolism, or scavenging reactive species. Although an anticarcinogenic effect of β-NF was observed when AFB_1 was the carcinogen, in the case of DEN, β-NF caused an increase in adduct formation and a subsequent increase in tumor formation.[100]

The persistence of DNA adducts in vivo has also been explored, and the results generally support the idea of poor DNA repair in fish. Although O^6-alkylguanine-DNA alkyltransferase activity was observed in rainbow trout,[91] the rate and extent of removal of O^6-ethylguanine was low. It was suggested that the enzyme was depleted rapidly and was not induced in rainbow trout exposed to 250 ppm DEN for 24 hr. The major AFB_1 DNA adduct in rainbow trout and salmon was shown to persist for up to 21 days after a single i.p. injection.[94] Finally, the major BaP adduct formed in English sole and brown bullhead liver was shown to persist for up to 60 and 70 days, respectively.[92,93,139]

In attempts to duplicate the pattern of DNA adducts observed in wild fish species from polluted waterways, fish were exposed to sediment extracts from these sites. In both English sole and brown bullheads treated with sediment extracts by i.p. injection, ^{32}P-postlabeling revealed adducts in hepatic DNA.[111,147,148] The pattern of adducts, a diagonal zone of overlapping adducts, was similar to that observed in wild fish collected from the chemically contaminated waterways that were the source of the sediments. These experiments support the contention that the sediments are a possible source for at least some of the genotoxic chemicals to which the fish are exposed.

A unique approach using fish to study the relationship of DNA adducts to tumor formation was developed by Hart et al.[149] Rather than expose fish to a carcinogen, they introduced a suspension of thyroid cells bearing a specific DNA adduct into a small aquarium fish species, the Amazon molly. This adduct, a pyrimidine dimer formed during exposure to UV radiation, caused a widespread and rapid growth of invasive thyroid tissue. Moreover, when the UV-irradiated suspension was treated to photoreactivate the dimer adduct, the tumor incidence was reduced to control levels. These experiments provided direct evidence that the pyrimidine dimer, if unrepaired, led to tumors.

C. Field Studies

It has been suggested that the measurement of DNA adducts in fish can provide an estimate of exposure to carcinogens and their effects.[150] As a measure of exposure, DNA adducts integrate the factors of uptake, metabolism, and repair and provide an assessment of the "dose" of carcinogen reaching a target tissue such as liver. The detection of DNA adducts in wild fish provides a measurable endpoint of genetic damage that does not have to be inferred.[150] The most widely used technique in studying DNA adducts in wild fish has been the ^{32}P-postlabeling method. Typically, when hepatic DNA from fish is analyzed by this technique, the adduct pattern is one of a diagonal radioactive zone, consisting of multiple overlapping spots (Figure 2). This pattern can vary from species to species, with more discrete spots being observed in some species.[109,112] Comparison of the level of DNA adducts in fish from polluted and unpolluted waterways has produced conflicting conclusions. Brown bullheads [107–110] and English sole[111] from chemically contaminated waterways had elevated levels of hepatic DNA adducts detected by ^{32}P-postlabeling when compared to fish from uncontaminated areas, and it was suggested that the adducts resulted from exposure to sediment-bound chemicals. In contrast, Kurelec et al.[112] found no differences in hepatic DNA adduct levels in five species of fish collected from polluted and unpolluted sites and suggested that natural causes of DNA adducts may be greater than pollution-related causes. These studies examined different fish species of unspecified age and used different criteria to define a polluted system, thus making it difficult, if not impossible, to resolve the conflicting conclusions with the available data. In fact, the results of these studies underline the need for caution in generalizing the results of a specific study to all situations.

In general, because of the prevalence of liver tumors in fish from many polluted areas, analyses of adducts found in hepatic DNA have been the focus of most studies. However, extrahepatic tissues from wild fish can contain detectable levels of DNA adducts. A study examining DNA adducts in the intestine of brown bullheads found elevated levels of adducts in fish from the Buffalo River when compared to those found in fish from a pristine lake.[151] The level of adducts in fish from the unpolluted site was below the limit of detection (<1 nmol/mol), whereas the fish from the Buffalo River had an average of 40 nmol/mol. Studies that have examined DNA adducts by ^{32}P-postlabeling have not determined exact chemical structures for any of the detected adducts. However, it has been suggested, because of the methods used for isolation of the adducts and their chromatographic characteristics, that these adducts were formed by covalent binding of bulky, hydrophobic, or aromatic compounds.[107,111,112]

Recently, the detection of specific adducts in liver DNA of wild fish has been reported.[102–104] In these studies, a gas chromatography-mass spectrometry (GC-MS) method with single-ion monitoring was used to detect DNA adducts that are the result of hydroxyl radical attack on DNA bases. When hepatic tumor tissue from

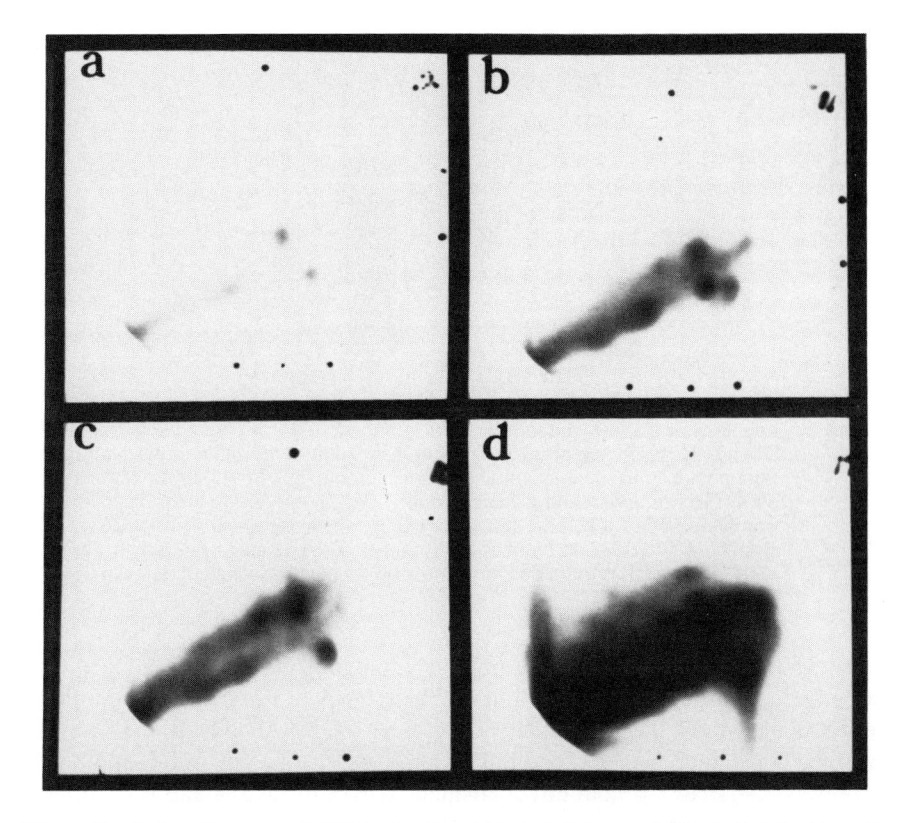

Figure 2. Autoradiograms of thin layer chromatography maps of ^{32}P-postlabeled hepatic DNA from brown bullheads collected from the Detroit River in Michigan. The level of adducts in the diagonal radioactive zone ranged from 100 nmol/mol (a) to 1661 nmol/mol (d). The nuclease P1 procedure was used to isolate the adducts.

English sole from a polluted site was compared either to nontumor tissue or to tissue from fish from control sites, elevated levels of four DNA lesions were found (Figure 3). These lesions, 4,6-diamino-5-formamidopyrimidine (FapyAde), 8-hydroxyguanine (8-OH-G), 8-hydroxyadenine (8-OH-A), and 2,6-diamino-4-hydroxy-5-formamidopyrimidine (FapyGua), were from 7 to 280 times higher in the tumor tissue.[102,103] These studies also demonstrated that nontumor tissue from fish collected from a site that had an intermediate level of contamination and tumor incidence had levels of the damaged bases that were higher than control sites but lower than tumor tissue.[104] Thus, it was suggested that these lesions represented "highly relevant biomarkers for cytogenetic change."[104] In fact, the level of DNA adducts resulting from oxidative damage, FapyAde, 8-OH-A, 8-OH-G, and FapyGua, were more than 2000 times greater than the levels of presumptive aromatic hydrocarbon adducts observed in English sole.[102,103,111] These are the only studies that we are aware of that have identified and measured specific DNA adducts in wild fish.

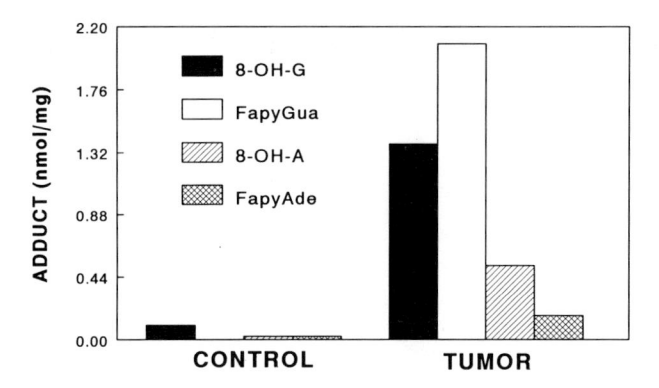

Figure 3. Levels of oxidatively damaged bases in hepatic DNA of English sole from Puget Sound. Control samples were from fish without hepatic tumors and from nonneoplastic tissue surrounding hepatic tumors. DNA from tumor samples was extracted from raised neoplastic nodules. Data are average values summarized from references 102 and 103. 8-OH-G = 8-hydroxyguanine, FapyGua = 4,6-diamino-4-hydroxy-5-formamidopyrimidine, 8-OH-A = 8-hydroxyadenine, FapyAde = 4,6-diamino-5-formamidopyrimidine, and * = below the limit of detection of 0.01 nmol/mg.

The observations of both presumptive aromatic hydrocarbon adducts and oxidatively damaged DNA bases in the hepatic DNA of English sole from Puget Sound demonstrate the multiplicity of sources of DNA adducts that may be involved in carcinogenesis. Using BaP as a model, metabolism of this compound can provide sources for both direct covalent adduct formation and indirect base damage caused by reactive oxygen species (Figure 4). One major pathway of BaP metabolism, mediated by cytochromes P450 and epoxide hydrolase, results in the formation of benzo(a)pyrene-7,8-diol-9,10-epoxide, which covalently binds to the N-2 position of guanine in DNA to form the major stable adduct.[152] In addition, it has been demonstrated that H_2O_2 can be generated during the redox cycling of BaP-diols and BaP-diones.[153] This H_2O_2, in turn, can be reduced by Fe^{2+} to yield highly reactive hydroxyl radicals. The hydroxyl radical can attack DNA bases, resulting in modified bases such as 8-OH-G and FapyGua. Finally, a recently described P450-mediated one electron oxidation of BaP that results in the formation of an unstable N-7 guanine/BaP adduct[154] is another potential source of DNA damage. This adduct easily undergoes depurination, which leads to the formation of potentially mutagenic apurinic sites within the DNA.[155] To date, the occurrence of this type of DNA damage has not been evaluated in fish.

IV. SUMMARY AND CONCLUSIONS

It is clear from the research using fish conducted to date that, at least for the strong carcinogens studied, fish can be suitable model systems for studying DNA

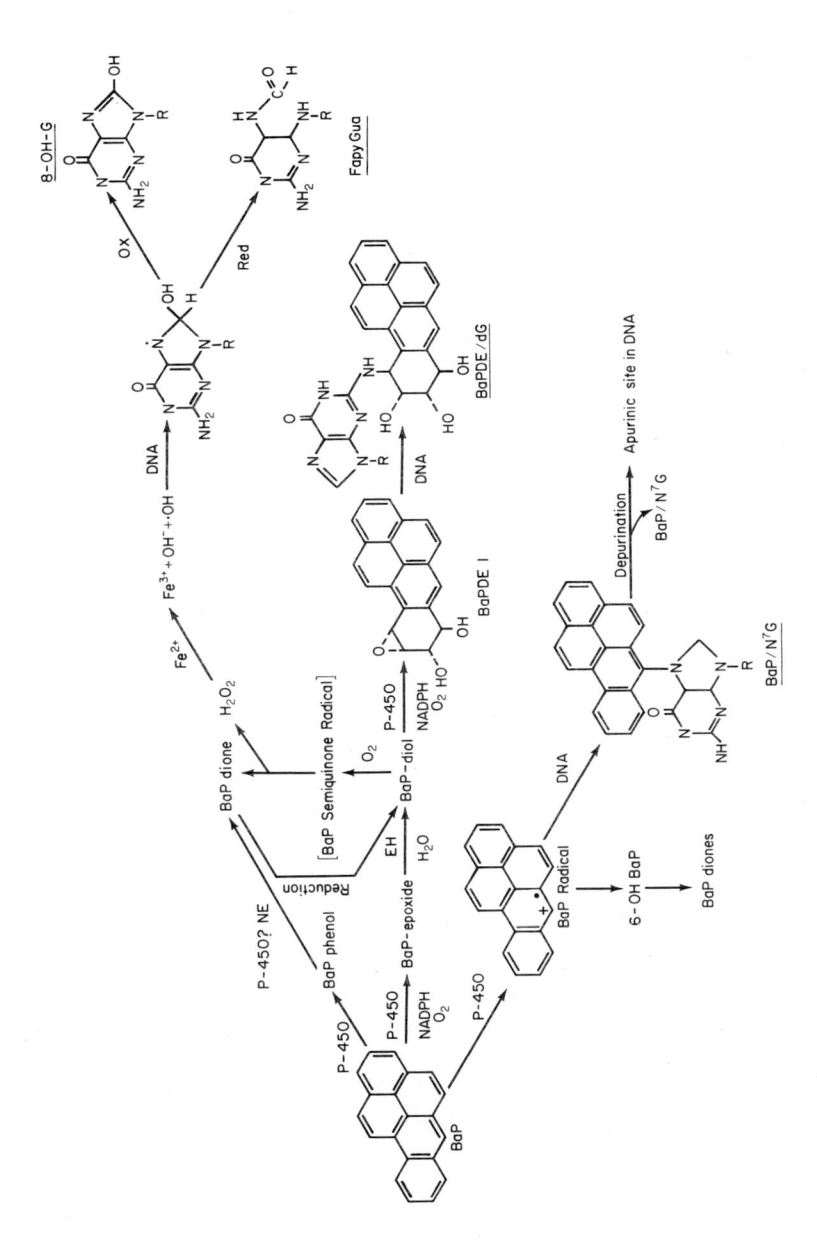

Figure 4. Summary of some of the possible pathways by which BaP may lead to the formation of DNA adducts. This scheme was derived from data presented in references 152–154 and 168–172.

adduct formation. The results of the experiments discussed in this chapter reinforce concepts that were originally developed from mammalian models and provide new insight on mechanisms that have not been described previously.

1. When studied in vitro and in vivo, fish have been shown to activate carcinogens to products that form DNA adducts, and these adducts are the same ones described in rodent systems.
2. In some fish species, the amount of an adduct formed in a target tissue is modulated by phase II enzymatic reactions, while in other species, the extent of activation of the carcinogen by phase I enzymes is the major determinant in adduct formation. Thus, when studies evaluating adduct formation relative to species sensitivity are conducted, a thorough characterization of the metabolic pathways for the carcinogen being examined is necessary.
3. The route and duration of exposure are important aspects to be considered when attempting to extrapolate from laboratory studies to real world situations in order to predict the ultimate biological effect of a chemical.
4. Anticarcinogens can modulate DNA adduct formation in different ways that, in turn, can affect tumorigenicity of a compound. When administered prior to carcinogen exposure, some compounds, such as I3C, are inhibitors of carcinogen binding to DNA, and others, β-NF for example, shunt the metabolism of a carcinogen toward detoxification. However, the effect of these anticarcinogens is timing and carcinogen dependent, and thus, these variables must be determined for each carcinogen.
5. The repair of DNA damage, at least in the total genome, appears to be at low levels in fish species and thus may contribute to the sensitivity of many species to carcinogens.
6. Wild fish from chemically contaminated waterways have elevated levels of at least two very different types of DNA adducts. Some of these adducts apparently are the result of direct binding of bulky, aromatic compounds to DNA, while others are due to oxidative damage that may be related to chemical exposure. Both types of lesions may have a role in different stages of carcinogenesis, although on a quantitative basis, oxidative lesions seem to be more important.

V. FUTURE CONSIDERATIONS

As stated earlier, the study of DNA adducts in fish is still a relatively new area of research. To date, most of the studies conducted in the laboratory have focused on strong carcinogens that have been widely studied in other systems. Now that several sensitive methods for detecting DNA adducts have been developed, the spectrum of compounds being examined should be expanded. This is especially important if fish models are to be used at some level in the screening/testing of chemicals for genotoxicity. There is the need for comparison of the various detection methods to determine if they are able to detect a given adduct with comparable sensitivity and specificity. This could be accomplished by interlaboratory comparisons of known adducts analyzed by the same technique in different laboratories or by analysis of known adducts by different techniques to assess variations in a given method and between different methods.

In general, the studies discussed have examined DNA adducts that are formed as a result of covalent binding of chemicals to DNA. However, in a larger sense, DNA adducts can be thought of as any modification that may occur in the fundamental structure of DNA resulting either directly or indirectly from exposure to chemical or physical agents. As was seen in fish from Puget Sound,[102–104] adducts that may result from oxygen radicals can be detected in liver DNA; the level of these adducts was increased in tumor tissue when compared to nontumor tissue. The role of active oxygen species in carcinogenesis is well documented,[156] and one source of active oxygen species is the metabolism of xenobiotics. In the metabolism of xenobiotics, reactive oxygen species can be generated along with xenobiotic radicals, such as those that have been described in English sole from polluted sites.[157,158] Thus, in the case of English sole from chemically contaminated waters, metabolism of xenobiotic chemicals may be a source for DNA-damaging active oxygen species. The existence of DNA adducts due to oxidative damage in other fish species from other sites needs to determined, as does the exact cause of the damage and the role it may play in tumor formation. In addition, other types of DNA damage, such as deamination of DNA bases[159] and their potential as a mechanism to cause genotoxic effects in wild fish, need to explored. Finally, the importance of secondary adducts, such as apurinic sites and ring-opened bases that have been shown to have important genotoxic effects,[155,160] remains unknown in fish species exposed to environmental carcinogens. This is an area that warrants expanded effort.

With respect to field studies, more work is needed in identifying the adducts observed in wild fish. This is especially true for the complex pattern observed by ^{32}P-postlabeling. Suitable standards need to be generated and characterized so that cochromatography with unknowns can be conducted. In addition to thin layer chromatography (TLC) separation, analysis of ^{32}P-postlabeled adducts by high performance liquid chromatography (HPLC) could be used for identification. For example, fluoranthene-DNA adducts have been analyzed directly by HPLC after ^{32}P-postlabeling.[161] Similarly, several small, nonbulky adducts have been first separated by TLC and then eluted from the plate and analyzed by HPLC, along with authentic markers, thus providing definitive identification of specific DNA adducts in maps generated by ^{32}P-postlabeling.[162,163] Finally, a new postlabeling strategy using ^{35}S may prove useful when HPLC separation of products is desired.[164] Alternatively, wider application of antibodies to specific adducts, GC-MS techniques, and other adduct-specific detection methods must be employed, so that the chemical nature of DNA adducts can be determined.

The process of DNA repair is poorly understood in relation to adduct formation and persistence in fish. As mentioned earlier, analysis of repair in the entire genome suggests that fish are relatively repair deficient. However, such analysis may miss gene-specific repair. The importance of this type of repair and methods to study it have been reviewed recently.[165] It seems feasible that these techniques could be applied to fish and could help to provide a better understanding of repair processes in these animals. It is also possible that persistence could be evaluated by using surrogate tissues such as blood cells. In this way, adduct persistence in individual fish could be followed after exposure to a carcinogen. However, before this approach is used, it must be determined if formation of a particular adduct in a surrogate tissue parallels adduct formation in the target tissue.

It has been proposed that DNA adducts are only relevant if they cause genetic changes in critical targets such as tumor suppressor genes or oncogenes.[40] The role of oncogenes and tumor suppressor genes in carcinogenesis in fish represents a very new field of research (see Chapter 7). In addition to oncogenes and tumor suppressor genes, several other types of genes may be potential targets for mutation and may be involved in the process of carcinogenesis.[166] Methods now exist for detecting mutations in vivo.[167] These methods have potential application to studies in fish and may be powerful tools to evaluate the role of specific adducts in genotoxic events. It may be possible to compare mutation rates and the level of specific adducts to determine correlations between the two.

Finally, the laboratory studies conducted with rainbow trout and aflatoxin and the field studies conducted on fish from the Puget Sound can serve as examples of how DNA adduct analysis can be part of an integrated study. The combining of the measurement of DNA adduct formation as a molecular dosimeter with analysis of carcinogen metabolism and determination of tumor formation can provide insights on the mechanisms involved in chemical carcinogenesis. This type of integrated study may also provide information about potential exposure and risk of environmental carcinogens that may be found in contaminated waterways.

REFERENCES

1. Miller, J.A., Carcinogenesis by chemicals: an overview — G.H.A. Clowes memorial lecture, *Cancer Res.*, 30, 559, 1970.
2. Pott, P., Chirurgical observations relative to the cancer of the scrotum, London 1775, reprinted in *Natl. Cancer Inst. Monogr.*, 10, 7, 1963.
3. Yamagiwa, K. and Ichikawa, K., Experimentelle studie uber die pathologenese der epitheliageschwulste, *Mitt. Med. Fak. Kaiserl. Univ.*, 15, 295, 1915.
4. Yamagiwa, K. and Ichikawa, K., Experimental study of the pathogenesis of carcinoma, *J. Cancer Res.*, 3, 1, 1918.
5. Tsutsui, H., Uber das kunstlich erzeugte cancroid bei der maus, *Gann*, 12, 17, 1918.
6. Cook, J.W., Hewett, C.L., and Hieger, I., The isolation of a cancer-producing hydrocarbon from coal tar, Parts I, II, and III, *J. Chem. Soc.*, 395, 1933.
7. Berenblum, I., The cocarcinogenic action of croton resin, *Cancer Res.*, 1, 44, 1941.

8. Berenblum, I., The mechanism of cocarcinogenesis. A study of the significance of cocarcinogenic action and related phenomena, *Cancer Res.,* 1, 807, 1941.
9. MacKenzie, I. and Rous, P., The experimental disclosure of latent neoplastic changes in tarred skin, *J. Exp. Med.,* 73, 391, 1941.
10. Rous, P. and Kidd, J.G., Conditional neoplasms and subthreshold neoplastic states. A study of the tar tumors in rabbits, *J. Exp. Med.,* 73, 365, 1941.
11. Wheeler, G.P. and Skipper, H.E., Studies with mustards. III. In vivo fixation of C^{14} from nitrogen mustard-$C^{14}H_3$ in nucleic acid fractions of animal tissues, *Arch. Biochem. Biophys.,* 72, 465, 1957.
12. Brookes, P. and Lawley, P.D., The reaction of mustard gas with nucleic acids in vitro and in vivo, *Biochem. J.,* 77, 478, 1960.
13. Brookes, B. and Lawley, P.D., Evidence for the binding of polynuclear aromatic hydrocarbons to the nucleic acids of mouse skin: relation between carcinogenic power of hydrocarbons and their binding to deoxyribonucleic acid, *Nature (London),* 218, 652, 1964.
14. Singer, B. and Grunberger, D., *Molecular Biology of Mutagens and Carcinogens,* Plenum Press, New York, 1983.
15. Hemminki, K., Nucleic acid adducts of carcinogens and mutagens, *Arch. Toxicol.,* 52, 249, 1983.
16. Jeffery, A.M., DNA modifications by chemical carcinogens, *Pharmac. Ther.,* 28, 237, 1985.
17. Harshbarger, J.C. and Clark, J.B., Epizootiology of neoplasms in bony fish of North America, *Sci. Total Environ.,* 94, 1, 1990.
18. Dawe, C.J., Stanton, M.F., and Schwartz, F.J., Hepatic neoplasms in native bottom-feeding fish of Deep Creek Lake, Maryland, *Cancer Res.,* 24, 1194, 1964.
19. Rucker, R.R., Yasutake, W.T., and Wolf, H., Trout hepatomas — a preliminary report, *Progr. Fish Cult.,* 23, 3, 1961.
20. Hueper, W.C. and Payne, W.W., Observations on the occurrence of hepatomas in rainbow trout, *J. Natl. Cancer Inst.,* 27, 1123, 1961.
21. Wood, E.M. and Larson, C.P., Hepatic carcinoma in rainbow trout, *Arch. Pathol.,* 71, 471, 1961.
22. Ghitttino, P. and Ceretto, F., Studio sulla eziopatogenesi dell'epatoma della trota iridea di allevamento, *Tumori,* 48, 393, 1962.
23. Halver, J.E., Crystalline aflatoxin and other vectors for trout hepatoma, in Trout Hepatoma Research Conference Papers, Halver, J.E. and Mitchell, I.A., Eds., Research Report #70, U.S. Fish and Wildlife Service, Washington, D.C., 1967, 78.
24. Sinnhuber, R.O., Aflatoxin in cottonseed meal and liver cancer in rainbow trout, in Trout Hepatoma Research Conference Papers, Halver, J.E. and Mitchell, I.A., Eds., Research Report #70, U.S. Fish and Wildlife Service, Washington, D.C., 1967, 48.
25. Sinnhuber, R.O., Wales, J.L., Ayres, J.L., Engebrecht, R.H., and Amend, D.L., Dietary factors and hepatomas in rainbow trout (Salmo gairdneri). I. Aflatoxins in vegetable protein feedstuffs, *J. Natl. Cancer Inst.,* 41, 711, 1968.
26. Hendricks, J.D., Wales, J.H., Sinnhuber, R.O., Nixon, J.E., Loveland, P.M., and Scanlan, R.A., Rainbow trout (Salmo gairdneri) embryos: a sensitive animal model for experimental carcinogenesis, *Fed. Proc.,* 39, 3222, 1980.
27. Hendricks, J.D., Chemical carcinogenesis in fish, in *Aquatic Toxicology,* Weber, L.J., Ed., Raven Press, New York, 1982, 149.

28. Meyers, T.R. and Hendricks, J.D., A summary of tissue lesions in aquatic animals induced by controlled exposures to environmental contaminants, chemotherapeutic agents, and potential carcinogens, *Mar. Fish. Rev.,* 44, 1, 1982.
29. Bailey, G.S., Hendricks, J.D., Nixon, J.E., and Pawlowski, N.E., The sensitivity of rainbow trout and other fish to carcinogens, *Drug Metab. Rev.,* 15, 725, 1984.
30. Hoover, K.L., Ed., Use of small fish species in carcinogenicity testing, *Natl. Cancer Inst. Monogr.,* 65, 1984.
31. Couch, J.A. and Harshbarger, J.C., Effects of carcinogenic agents on aquatic animals: an environmental and experimental overview, *Environ. Carcinogen. Rev.,* 3, 63, 1985.
32. Black, J.J., Carcinogenicity tests with rainbow trout embryos: a review, *Aquat. Toxicol.,* 11, 129, 1988.
33. Metcalfe, C.D., Tests for predicting carcinogenicity in fish, *CRC Crit. Rev. Aquat. Sci.,* 1, 111, 1989.
34. Thilley, W.G. and Call, K.M., Genetic toxicology, in *Casarett and Doull's Toxicology. The Basic Science of Poisons,* 3rd ed., Klaassen, C.D., Amour, M.O., and Doull J., Eds., Macmillan, New York, 1986, 6.
35. Miller, E.C. and Miller, J.A., Mechanisms of chemical carcinogenesis, *Cancer,* 47, 1055, 1981.
36. Lutz, W.K., *In vivo* covalent binding of organic chemicals to DNA as a quantitative indicator in the process of chemical carcinogenesis, *Mutat. Res.,* 65, 289, 1979.
37. Neumann, H.-G., Dose-response relationships in the primary lesions of strong electrophilic carcinogens, *Arch. Toxicol.,* 3, 69, 1980.
38. Lutz, W.K., Quantitative evaluation of DNA-binding data in vivo for low dose extrapolations, *Arch. Toxicol.,* 11 (Suppl.), 66, 1987.
39. Harris, C.C., Future directions in the use of DNA adducts as internal dosimeters for monitoring human exposure to environmental carcinogens, *Environ. Health Perspect.,* 62, 185, 1985.
40. Harris, C.C., Chemical and physical carcinogenesis: advances and perspectives for the 1990's, *Cancer Res.,* 51 (Suppl.), 5023s, 1991.
41. Lutz, W.K., Dose-response relationship and low dose extrapolation in chemical carcinogenesis, *Carcinogenesis,* 11, 1243, 1990.
42. Wogan, G.N., Detection of DNA damage in studies on cancer etiology and prevention, in *Methods for Detecting DNA Damaging Agents in Humans: Applications in Cancer Epidemiology and Prevention,* Bartsch, H., Hemminki, K., and O'Neill, I.K., Eds., IARC Scientific Publication #89, International Agency for Research on Cancer, Lyon, France, 1988, 32.
43. Perera, F.P. and Weinstein, I.B., Molecular epidemiology and carcinogen-DNA adduct detection, *J. Chron. Dis.,* 35, 581, 1982.
44. de Serres, F.J., Banbury Center DNA adducts workshop, meeting report, *Mutat. Res.,* 203, 55, 1988.
45. Farber, E., Cellular biochemistry of the stepwise development of cancer with chemicals: G.H.A. Clowes memorial lecture, *Cancer Res.,* 44, 5463, 1984.
46. Hennings, H., Shores, R., Wenk, M.L., Spangler, E.F., Tarone, R., and Yuspa, S.H., Malignant conversion of mouse skin tumours is increased by tumour initiators and unaffected by tumour promotors, *Nature (London),* 304, 67, 1983.

47. Nowell, P.C., The clonal evolution of tumor cell populations, *Science,* 194, 23, 1976.
48. Malins, D.C., McCain, B.B., Brown, D.W., Chan, S.-L., Myers, M.S., Landahl, J.T., Prohaska, P.G., Friedman, A.J., Rhodes, L.D., Burrows, D.G., Gronlund, W.D., and Hodgins, H.O., Chemical pollutants in sediments and diseases of bottom-dwelling fish in Puget Sound, Washington, *Environ. Sci. Technol.,* 18, 705, 1984.
49. Malins, D.C., Krahn M.M., Brown D.W., Rhodes L.D., Myers M.S., McCain, B.B., and Chan S.-L., Toxic chemicals in marine sediment and biota from Mukilteo, Washington: relationships with hepatic neoplasms and other hepatic lesions in English sole (Parophrys vetulus), *J. Natl. Cancer Inst.,* 74, 487, 1985.
50. Malins, D.C., Krahn, M.M., Myers, M.S., Rhodes, L.D., Brown, D.W., Krone, C.A., McCain, B.B., and Chan S.-L., Toxic chemicals in sediments and biota from a creosote-polluted harbor: relationships with hepatic neoplasms and other hepatic lesions in English sole (Parophrys vetulus), *Carcinogenesis,* 6, 1463, 1985.
51. Malins, D.C., McCain, B.B., Myers, M.S., Brown, D.W., Krahn, M.M., Roubal, W.T., Schiewe, M.H., Landahl, J.T., and Chan, S.-L., Field and laboratory studies of the etiology of liver neoplasms in marine fish from the Puget Sound, *Environ. Health Perspect.,* 71, 5, 1987.
52. Malins, D.C., McCain, B.B., Landahl, J.T., Myers, M.S., Krahn, M.M., Brown, D.W., Chan, S.-L., and Roubal, W.T., Neoplasms and other diseases in fish in relation to toxic chemicals: an overview, *Aquat. Toxicol.,* 11, 43, 1988.
53. Black, J.J., Field and laboratory studies of environmental carcinogenesis in Niagara River fish, *J. Great Lakes Res.,* 9, 326, 1983.
54. Baumann, P.C., Smith, W.D., and Ribbick, M., Hepatic tumor rates and polynuclear hydrocarbons in two populations of brown bullheads (Ictalurus nebulosus), in *Polynuclear Aromatic Hydrocarbons: Sixth International Symposium on Physical and Biological Chemistry,* Cooke, M.C., Dennis, A.J.,and Fisher, G.L., Eds., Battelle Press, Columbus, OH, 1982, 93.
55. Furlong, E.T., Carter, D.S., and Hites, R.A., Organic contaminants from the Trenton Channel of the Detroit River, Michigan, *J. Great Lakes Res.,* 14, 489, 1988.
56. Metcalfe, C.D., Balch, G.C., Cairns, V.W., Fitzsimons, J.D. and Dunn, B.P., Carcinogenic and genotoxic activity of extracts from contaminated sediments in western Lake Ontario, *Sci. Total Environ.,* 94, 125, 1990.
57. Baumann, P.C. and Whittle, D.M., The status of selected organics in the Laurentian Great Lakes: an overview of DDT, PCBs, dioxins, furans, and aromatic hydrocarbons, *Aquat. Toxicol.,* 11, 241, 1988.
58. Varanasi, U. and Gmur, D.J., Hydrocarbons and metabolites in English sole (Parophrys vetulus) exposed simultaneously to [^{3}H]benzo[a]pyrene and [^{14}C]naphthalene in oil-contaminated sediment, *Aquat. Toxicol.,* 1, 67, 1981.
59. Stein, J.E., Hom, T., and Varanasi U., Simultaneous exposure of English sole (Parophrys vetulus) to sediment-associated xenobiotics. Part I. Uptake and disposition of [^{14}C]-polychlorinated biphenyls and [^{3}H]-benzo[a]pyrene, *Mar. Environ. Res.,* 13, 97, 1984.
60. Varanasi, U., Richert, W.L., Stein, J.E., Brown, D.W., and Sanborn H.R., Bioavailability and biotransformation of aromatic hydrocarbons in benthic organisms exposed to sediment from an urban estuary, *Environ. Sci. Technol.,* 19, 836, 1985.

61. Stein, J.E., Hom, T., Casillas, E., Friedman, A., and Varanasi, U., Simultaneous exposure of English sole (Parophrys vetulus) to sediment-associated xenobiotics. Part II. Chronic exposure to an urban estuarine sediment with added [^{3}H]benzo(a)pyrene and [^{14}C]polychorinated biphenyls, *Mar. Environ. Res.*, 22, 123, 1987.
62. Maccubbin, A.E., Black, P., Trzeciak, L., and Black, J.J., Evidence for polynuclear aromatic hydrocarbons in the diet of bottom-feeding fish, *Bull. Environ. Contam. Toxicol.*, 34, 876, 1985.
63. Krahn, M.M., Myers M.S., Burrows D.G., and Malins D.C., Determination of metabolites of xenobiotics in the bile of fish from polluted waterways, *Xenobiotica*, 14, 633, 1984.
64. Krahn, M.M., Rhodes, L.D., Myers, M.S., Moore, L.K., MacLeod, W.D., Jr., and Malins, D.C., Associations between metabolites of aromatic compounds in bile and the occurrence of hepatic lesions in English sole (Parophrys vetulus) from Puget Sound, Washington, *Arch. Environ. Contam. Toxicol.*, 15, 61, 1986.
65. Krahn, M.M., Burrows, D.G., MacLeod, W.D., Jr., and Malins, D.C., Determination of individual metabolites of aromatic compounds in hydrolyzed bile of English sole (Parophrys vetulus) from polluted sites in Puget Sound, Washington, *Arch. Environ. Contam. Toxicol.*, 16, 511, 1987.
66. Maccubbin, A.E., Chidambaram, S., and Black, J.J., Metabolites of aromatic hydrocarbons in the bile of brown bullheads, Ictalurus nebulosus, *J. Great Lakes Res.*, 114, 101, 1988.
67. Chambers, J.E. and Yarbrough, J.D., Xenobiotic biotransformation systems in fish, *Comp. Biochem. Physiol.*, 55C, 77, 1976.
68. Bend, J.R. and James M.O., Xenobiotic metabolism in marine and freshwater species, in *Biochemical and Biophysical Perspectives in Marine Biology*, Malins, D.C. and Sargent, J.R., Eds, Academic Press, New York, 1978, 125.
69. Lech, J.J. and Bend J.R., Relationship between biotransformation and the toxicity and fate of xenobiotic chemicals in fish, *Environ. Health Perspect.*, 34, 115, 1980.
70. Stegeman, J.J., Polynuclear aromatic hydrocarbons and their metabolism in the marine environment, in *Polycyclic Hydrocarbons and Cancer,* Gelboin, H.V. and Ts'O, P.O., Eds. Academic Press, New York, 1981, 1.
71. Lech, J.J., Vodicnik, M.J., and Elcombe, C.R., Induction of monooxygenase activity in fish, *Aquatic Toxicology,* Weber, L.J., Ed, Raven Press, New York, 1982, 107.
72. Binder, R.L., Melancon, M.J., and Lech, J.J., Factors influencing the persistence and metabolism of chemicals in fish, *Drug Metab. Rev.*, 15, 697, 1984.
73. Lech, J.J. and Vodicnik, M.J., Biotransformation of chemicals in fish: an overview, *Natl. Cancer Inst. Monogr.*, 65, 355, 1984.
74. Tan, B. and Melius P., Polynuclear aromatic hydrocarbon metabolism in fishes, *Comp. Biochem. Physiol.*, 83C, 217, 1986.
75. James, M.O., Conjugation of organic pollutants in aquatic species, *Environ. Health Perspect.*, 71, 97, 1987.
76. Hendricks, J.D., Meyers, T.R., Shelton, D.W., Casteel, J.L., and Bailey, G.S., Hepatocarcinogenicity of benzo(a)pyrene to rainbow trout by dietary exposure and intraperitoneal injection, *J. Natl. Cancer Inst.*, 74, 839, 1985.
77. Black, J.J. Maccubbin, A.E., and Johnston, C.J., Carcinogenicity of benzo(a)pyrene in rainbow trout resulting from embryo injection, *Aquat.Toxicol.*, 13, 297, 1988.

78. Hawkins, W.E., Walker, W.W., Overstreet, R.M., Lytle, T.F., and Lytle, J.S., Dose-related carcinogenic effects of water-borne benzo(a)pyrene on livers of two small fish species, *Ecotoxicol. Environ. Safety,* 16, 219, 1988.
79. Hawkins, W.E., Walker, W.W., Overstreet, R.M., Lytle, T.F., and Lytle, J.S., Carcinogenic effects of some polycyclic aromatic hydrocarbons on the Japanese medaka and guppy in waterborne exposures, *Sci. Total Environ.,* 94, 155, 1990.
80. Franklin, R.B., Elcombe, C.R., Vodicnik, M.J., and Lech J.J., Comparative aspects of the disposition and metabolism of xenobiotics in fish and mammals, *Fed. Proc.,* 39, 3144, 1980.
81. Funari, E., Zoppini, A., Verdina, A., De Angelis, G., and Vittozzi L., Xenobiotic-metabolizing enzyme systems in test fish. I. Comparative studies of liver microsomal monooxygenases, *Ecotoxicol. Environ. Safety,* 13, 24, 1987.
82. Sikka, H.C., Rutkowski, J.P., and Kandaswami, C., Comparative metabolism of benzo(a)pyrene by liver microsomes from brown bullheads and carp, *Aquat. Toxicol.,* 16, 101, 1990.
83. Pederson, M.D., Hershberger, W.K., Zachariah, P.K., and Juchau, M.R., Hepatic biotransformation of xenobiotics in six strains of rainbow trout (Salmo gairdneri), *J. Fish. Res. Board Can.,* 33, 666, 1976.
84. Bend, J.R., Fourman, G.L., Ben-Zvi, Z., and Albro, P.W., Heterogeneity of hepatic arylhydrocarbon hydroxylase activity in feral winter flounder: relevance to carcinogenicity testing, *Natl. Cancer Inst. Monogr.,* 65, 359, 1984.
85. Woodhead, A.D., Setlow, R.B., and Grist, E., DNA repair and longevity in three species of cold-blooded vertebrates, *Exp. Gerontol.,* 15, 301, 1980.
86. Walton, D.G., Acton, A.B., and Stich, H.F., DNA repair synthesis in cultured mammalian and fish cells following exposure to chemical mutagens, *Mutat. Res.,* 124, 153, 1983.
87. Walton, D.G., Acton, A.B., and Stich, H.F., Comparison of DNA-repair synthesis, chromosome aberrations and induction of micronuclei in cultured human fibroblast, Chinese hamster ovary and central mudminnow (Umbra limi) cells exposed to chemical mutagens, *Mutat. Res.,* 129, 129, 1984.
88. Walton, D.G., Acton, A.B., and Stich, H.F., DNA repair synthesis following exposure to chemical mutagens in primary liver, stomach, and intestinal cells isolated from rainbow trout, *Cancer Res.,* 44, 1120, 1984.
89. Ishikawa, T., Masahito, P., and Takayama, S. Usefulness of the medaka, Oryzias latipes, as a test animal: DNA repair processes in medaka exposed to carcinogens, *Natl. Cancer Inst. Monogr.,* 65, 35, 1984.
90. Nakatsuru, Y., Nemoto, N., Nakagawa, K., Masahito, P., and Ishikawa, T., O^6-methylguanine-DNA methyltransferase activity in the liver from various fish species, *Carcinogenesis,* 8, 1123, 1987.
91. Fong, A.T., Hendricks, J.D., Dashwood, R.H., Van Winkle, S., and Bailey, G.S., Formation and persistence of ethylguanine in liver DNA of rainbow trout (Salmo gairdneri) treated with diethylnitrosamine by water exposure, *Food Chem. Toxicol.,* 26, 699, 1988.
92. Sikka, H.C., Rutkowski, J.P., Kandaswami, C., Kumar, S., Earley, K., and Gupta, R.C., Formation and persistence of DNA adducts in the liver of brown bullheads exposed to benzo(a)pyrene, *Cancer Lett.,* 49, 81, 1990.

93. Sikka, H.C., Steward, A.R., Kandaswami, C., Rutkowski, J.P., Kumar, S., Zaleski, J., Earley, K., and Gupta, R.C., Metabolism of benzo(a)pyrene and persistence of DNA adducts in the brown bullhead (Ictalurus nebulosus), *Comp. Biochem. Physiol.,* 100C, 25, 1991.
94. Bailey, G.S., Williams, D.E., Wilcox, J.S., Loveland, P.M., Coulombe, R.A., and Hendricks, J.D., Aflatoxin B1 carcinogenesis and its relation to DNA adduct formation and adduct persistence in sensitive and resistant salmonid fish, *Carcinogenesis,* 9, 1919, 1988.
95. Baird, W.M., The use of radioactive carcinogens to detect DNA modifications, in *Chemical Carcinogens and DNA,* Grover, P.L., Ed., CRC Press, Boca Raton, FL, 1979, 59.
96. Kriek, E., Den Englelse, L., Scherer, E., and Westra, J.G., Formation of DNA modifications by chemical carcinogens. Identification, localization, and quantification, *Biochim. Biophys. Acta,* 738, 181, 1984.
97. Farmer, P.B., Neumann, H.-G., and Henschler, D., Estimation of exposure of man to substances reacting covalently with macromolecules, *Arch. Toxicol.,* 60, 251, 1987.
98. Hemminki, K. and Randerath, K., Detection of genetic interaction of chemicals by biochemical methods: determination of DNA and protein adducts, in *Mechanisms of Cell Injury: Implications for Human Health,* Fowler, B.A., Ed., John Wiley & Sons, Chicester, 1987, 209.
99. Watson, W.P., Post-radiolabelling for detecting DNA damage, *Mutagenesis,* 2, 319, 1987.
100. Fong, A.T., Hendricks, J.D., Dashwood, R.H., Van Winkle, S., Lee, B.C., and Bailey, G.S., Modulation of diethylnitrosamine induced hepatocarcinogenesis and O^6-ethylguanine formation in rainbow trout by indole-3-carbinol, *Toxicol. Appl. Pharmacol.,* 96, 93, 1988.
101. Shugart, L., McCarthy, J., Jimenez, B., and Daniels, J., Analysis of adduct formation in the bluegill sunfish (Lepomis macrochirus) between BaP and DNA of the liver and hemoglobin of the erythrocytes, *Aquat. Toxicol.,* 9, 319, 1987.
102. Malins, D.C., Ostrander, G.K., Haimanot, R., and Williams, P., A novel DNA lesion in neoplastic livers of feral fish: 2,6-diamino-4-hydroxy-5-formamidopyrimidine, *Carcinogenesis,* 11, 1045, 1990.
103. Malins, D.C. and Haimanot, R., 4,6-diamino-5-formamidopyrimidine, 8-hydroxyguanine, and 8-hydroxyadenine in DNA from neoplastic liver of English sole exposed to carcinogens, *Biochem. Biophys. Res. Commun.,* 173, 614, 1990.
104. Malins, D.C. and Haimanot, R., The etiology of cancer: hydroxyl radical-induced DNA lesions in histologically normal livers of fish from a population with liver tumors, *Aquat. Toxicol.,* 20, 123, 1991.
105. Varanasi, U., Nishimoto, M., Reichert, W.L., and Le Eberhart, B.-T., Comparative metabolism of benzo(a)pyrene and covalent binding to hepatic DNA in English sole, starry flounder, and rat, *Cancer Res.,* 46, 3817, 1986.
106. Kulbacki-Garrity, S., Maccubbin, A.E., and Black, J.J., Unpublished data, 1990.
107. Dunn, B.P., Black, J.J., and Maccubbin, A.E., ^{32}P-postlabeling analysis of aromatic DNA adducts in fish from polluted areas, *Cancer Res.,* 47, 6583, 1987.
108. Dunn, B.P., Fitzsimons, J., Stalling, D., Maccubbin, A.E., and Black, J.J., Pollution-related aromatic adducts in liver from populations of wild fish, *Proc. Am. Assoc. Cancer Res.,* 31, 96, 1990.

109. Maccubbin, A.E., Black, J.J., and Dunn, B.P., ^{32}P-postlabeling detection of DNA adducts in fish from chemically contaminated waterways, *Sci. Total Environ.*, 94, 89, 1990.
110. Maccubbin, A.E. and Dunn, B.P., Unpublished data, 1990.
111. Varanasi, U., Reichert, W.L., and Stein, J.E., ^{32}P-postlabeling analysis of DNA adducts in liver of wild English sole (Parophrys vetulus) and winter flounder (Pseudopleuronectes americanus), *Cancer Res.*, 49, 1171, 1989.
112. Kurelec, B., Garg, A., Krca, S., Chacko, M., and Gupta, R.C., Natural environment surpasses polluted environment in inducing DNA damage in fish, *Carcinogenesis*, 10, 1337, 1989.
113. Nakatsuru, Y., Qin, X., Masahito, P. and Ishikawa, T., Immunological detection of in vivo aflatoxin B1-DNA adduct formation in rats, rainbow trout and coho salmon, *Carcinogenesis*, 11, 1523, 1990.
114. Montesanto, R., Ingram, A.J., and Magee, P.N., Metabolism of dimethylnitrosamine by amphibians and fish in vitro, *Experimentia*, 29, 599, 1973.
115. Ahokas, J.T., Saarni, H., Nebert, D.W., and Pelkonen, O., The in vitro metabolism and covalent binding of benzo[a]pyrene to DNA catalyzed by trout liver microsomes, *Chem.-Biol. Interact.*, 25, 103, 1979.
116. Varanasi, U., Gmur, D.J., and Krahn, M.M., Metabolism and subsequent binding of benzo[a]pyrene to DNA in pleuronectid and salmonid fish, in *Polynuclear Aromatic Hydrocarbons: Chemistry and Biological Effects*, Bjorseth, A. and Dennis, A.J., Eds. Battelle Press, Columbus, OH, 1980, 455.
117. Varanasi, U. and Gmur, D.J., Metabolic activation and covalent binding of benzo[a]pyrene to deoxyribonucleic acid catalyzed by liver enzymes of marine fish, *Biochem. Pharmacol.*, 29, 753, 1980.
118. Nishimoto, M. and Varanasi U., Benzo[a]pyrene metabolism and DNA adduct formation mediated by English sole liver enzymes, *Biochem. Pharmacol.*, 34, 263, 1985.
119. Nishimoto, M. and Varanasi, U., Metabolism and DNA adduct formation of benzo(a)pyrene and the 7,8-dihydrodiol of benzo(a)pyrene by fish liver enzymes, in *Polynuclear Aromatic Hydrocarbons: Fate and Effects*, Cooke, M.C. and Dennis, A.J., Eds., Battelle Press, Columbus, OH, 1986, 685.
120. Stein, J.E., Reichert, W.L., Nishimoto, M., and Varanasi, U., Overview of studies on liver carcinogenesis in English sole from Puget Sound; evidence for a xenobiotic chemical etiology. II. Biochemical studies, *Sci. Total Environ.*, 94, 51, 1990.
121. Zaleski, J., Steward, A.R., and Sikka, H.C., Metabolism of benzo(a)pyrene and (–)-*trans*-benzo(a)pyrene-7,8-dihydrodiol by freshly isolated hepatocytes from mirror carp, *Carcinogenesis*, 12, 167, 1991.
122. Steward, A.R., Zaleski, J., Gupta, R.C., and Sikka, H.C., Comparative metabolism of benzo(a)pyrene and (–)benzo(a)pyrene-7,8-dihydordiol by hepatocytes isolated from two bottom-dwelling fish, *Mar. Environ. Res.*, 28, 137, 1989.
123. Smolarek, T.A., Morgan, S.L., Moynihan, C.G., Lee, H., Harvey, R.G., and Baird, W.B., Metabolism and DNA adduct formation of benzo(a)pyrene and 7,12-dimethylbenz(a)anthracene in fish cell lines in culture, *Carcinogenesis*, 8, 1501, 1987.
124. Bailey, G.S., Taylor, M.J., and Selivonchick, D.P., Aflatoxin B1 metabolism and DNA binding in isolated hepatocytes from rainbow trout (Salmo gairdneri), *Carcinogenesis*, 3, 511, 1982.

125. Loveland, P.M., Wilcox, J.S., Pawlowski, N.E., and Bailey, G.S., Metabolism and DNA binding of aflatoxicol and aflatoxin B1 in vivo and in isolated hepatocytes from rainbow trout (Salmo gairdneri), *Carcinogenesis,* 8, 1065, 1987.
126. Loveland, P.M., Wilcox, J.S., Hendricks, J.D., and Bailey, G.S., Comparative metabolism and DNA binding of aflatoxin B1, aflatoxin M1, aflatoxicol, and aflatoxicol-M1 in hepatocytes from rainbow trout (Salmo gairdneri), *Carcinogenesis,* 9, 441, 1988.
127. Brookes, P., The binding to mouse skin mouse DNA of benzo(a)pyrene, its 7,8-diol, and 7,8-diol-9,10-epoxides in relation to the tumorigenicity of these compounds, *Cancer Lett.,* 6, 285, 1979.
128. Ashurst, S.W., Cohen, G.M., Nesnow, S., DiGiovanni, J., and Slaga, T.J., Formation of benzo(a)pyrene-DNA adducts and their relationship to tumor initiation in mouse epidermis, *Cancer Res.,* 43, 1024, 1983.
129. Bailey, G.S., Taylor, M.J., Loveland, P.M., Wilcox, J.S., Sinnhuber, R.O., and Selivonchick, D.P., Dietary modification of aflatoxin B1 carcinogenesis: mechanism studies with isolated hepatocytes from rainbow trout, *Natl. Cancer Inst. Monogr.,* 65, 379, 1984.
130. Fong, A.T., Swanson, H.I., Dashwood, R.H., Williams, D.E., Hendricks, J.D., and Bailey, G.S., Mechanisms of anti-carcinogenesis by indole-3-carbinol. Studies of enzyme induction, electrophile-scavenging, and inhibition of aflatoxin B1 activation, *Biochem. Pharmacol.,* 39, 19, 1990.
131. Goeger, D.E., Shelton, D.W., Hendricks, J.D., Pereira, C., and Bailey, G.S., Comparative effect of dietary butylated hydroxyanisole and beta-naphthoflavone on aflatoxin B1 metabolism, DNA adduct formation, and carcinogenesis in rainbow trout, *Carcinogenesis,* 9, 1783, 1988.
132. Baumann, P.C., Smith, W.D., and Parland, W.K., Tumor frequencies and contaminant concentrations in brown bullheads from an industrialized river and a recreational lake, *Trans. Am. Fish. Soc.,* 116, 79, 1987.
133. Maccubbin, A.E. and Ersing, N., Tumors in fish from the Detroit river, *Hydrobiologia,* 219, 301, 1990.
134. Brown, E.R., Hazdra, J.J., Keith, L., Greenspan, I., Kwapinski, J.B.G., and Beamer, P., Frequency of fish tumors in a polluted watershed as compared to non-polluted Canadian waters, *Cancer Res.,* 33, 189, 1973.
135. Sonstegard, R.A., Environmental carcinogenesis studies in fishes of the Great Lakes of North America, *Ann. N.Y. Acad. Sci.,* 298, 261, 1977.
136. Black, J.J. and Maccubbin, A.E., Unpublished data.
137. Croy, R.G., Nixon, J.E., Sinnhuber, R.O., and Wogan, G.N., Investigation of covalent aflatoxin B1-DNA adducts formed in vivo in rainbow trout (Salmo gairdneri) embryos and liver, *Carcinogenesis,* 1, 903, 1980.
138. Varanasi, U., Reichert, W.L., Le Eberhart, B.-T., and Stein, J.E., Formation and persistence of benzo(a)pyrene-diolepoxide-DNA adducts in liver of English sole (Parophrys vetulus), *Chem.-Biol. Interact.,* 69, 203, 1989.
139. James, M.O., Schell, J.D., Boyle, S.M., Altman, A.H., and Cromer, E.A., Southern flounder hepatic and intestinal metabolism and DNA binding of benzo(a)pyrene (BaP) metabolites following dietary administration of low doses of BaP, BaP-7,8-dihydrodiol or a BaP metabolite mixture, *Chem.-Biol. Interact.,* 79, 305, 1991.
140. Van Veld, P.A., Absorption and metabolism of dietary xenobiotics by the intestine of fish, *Rev., Aquat. Sci.,* 2, 185, 1990.

141. Bailey G., Selvinochick, D., and Hendricks, J., Initiation, promotion, and inhibition of carcinogenesis in rainbow trout, *Environ. Health Perspect.*, 71, 147, 1989.
142. Dashwood, R.H., Arbogast, D.N., Fong, A.H., Hendricks, J.D., and Bailey, G.S., Mechanism of anti-carcinogenesis by indole-3carbinol: detailed in vivo DNA binding dose response studies after dietary administration with aflatoxin, *Carcinogenesis*, 9, 427, 1988.
143. Dashwood, R.H., Breinholt, V., and Bailey, G.S., Chemopreventative properties of chlorophyllin: inhibition of aflatoxin B1 (AFB1)-DNA binding in vivo and antimutagenic activity against AFB1 and two heterocyclic amines in the Salmonella mutagenicity assay, *Carcinogenesis*, 12, 939, 1991.
144. Whitham, M., Nixon, J.E., and Sinnhuber, R.O., Liver DNA bound in vivo with aflatoxin B1 as a measure of hepatocarcinoma initiation in rainbow trout, *J. Natl. Cancer Inst.*, 68, 623, 1981.
145. Dashwood, R.H., Arbogast, D.N., Fong, A.H., Pereira, C., Hendricks, J.D., and Bailey, G.S., Quantitative inter-relationships between aflatoxin B1 carcinogen dose, indole-3-carbinol anti-carcinogen dose, target organ DNA adduction and final tumor response, *Carcinogenesis*, 10, 175, 1989.
146. Vasta, L.M., Hendricks, J.D., and Bailey, G.S., The significance of glutathione conjugation for aflatoxin B1 metabolism in rainbow trout and coho salmon, *Food Chem. Toxicol.*, 26, 129, 1989.
147. Reichert, W.L., French, B.L., Stein, J.E., and Varanasi, U., ^{32}P-postlabeling analysis of the persistence of bulky hydrophobic xenobiotic DNA adducts in liver of English sole (Parophrys vetulus), a marine fish, *Proc. Am. Assoc. Res.*, 32, 87, 1991.
148. Maccubbin, A.E., Kulbacki-Garrity, S., and Black, J.J., Unpublished data, 1989.
149. Hart, R.W., Setlow, R.B., and Woodhead, A.D., Evidence that pyrimidine dimers in DNA can give rise to tumors, *Proc. Natl. Acad. Sci. U.S.A.*, 74, 5574, 1977.
150. Dunn, B.P., DNA-carcinogen adducts in fish as a tool for measuring the biological dose of aquatic carcinogens, in *In Situ Evaluation of Biological Hazards*, Sandhu, S.S., Ed., Plenum Press, New York, 1990, 177.
151. Maccubbin, A.E. and Van Veld, P.A., Unpublished data, 1988.
152. Koreeda, M., Moore, P.D., Yagi, H., Yeh, H.J.C., and Jerina, D.M., Alkylation of polyguanylic acid at the 2-amino group and phosphate by the potent mutagen (±)-7β,8α-dihydroxy-9β,10β-7,8,9,10-tetrahydrobenzo(a)pyrene, *J. Am. Chem. Soc.*, 98, 6720, 1976.
153. Lesko, S.A., Chemical carcinogenesis: benzopyrene system, *Meth. Enzymol.*, 105, 539, 1984.
154. Cavalieri, E.L., Rogan, E.G., Devanesan, P.D., Cremonesi, P., Cerny, R.L., Gross, M.L., and Bodell, W.J., Binding of benzo(a)pyrene to DNA by cytochrome P-450 catalyzed one-electron oxidation in rat liver microsomes and nuclei, *Biochemistry*, 29, 4820, 1990.
155. Loeb, L.B. and Preston, B.D., Mutagenesis by apurinic/apyrimidinic sites, *Annu. Rev. Genet.*, 20, 201, 1986.
156. Cerutti, P.A., Prooxidant states and tumor promotion, *Science*, 227, 375, 1985.
157. Malins, D.C., Myers, M.S., and Roubal, W.T., Organic free radicals associated with idiopathic liver lesions of English sole (Parophrys vetulus) from polluted marine environments, *Environ. Sci. Technol.*, 17, 679, 1983.

158. Roubal, W.T. and Malins D.C., Free radical derivatives of nitrogen heterocycles in livers of English sole (Parophrys vetulus) with hepatic neoplasms and other liver lesions, *Aquat. Toxicol.,* 6, 87, 1985.
159. Wink, D.A., Kasprzak, K.S., Maragos, C.M., Elespuru, R.K., Misra, M., Dunams, T.M., Cebula, T.A., Koch, W.H., Andrews, A.W., Allen, J.S., and Keefer, L.K., DNA deaminating ability and genotoxicity of nitric oxide and its progenitors, *Science,* 254, 1001, 1991.
160. O'Connor, T.R., Boiteux, S., and Laval, J., Ring-opened 7-methylguanine residues in DNA are a block to *in vitro* DNA synthesis, *Nucl. Acids Res.,* 16, 5879, 1988.
161. Gorelick, N.J. and Wogan, G.N., Fluoranthene-DNA adducts: identification, and quantitation by an HPLC-^{32}P-postlabeling method, *Carcinogenesis,* 10, 1567, 1989.
162. Maccubbin, A.E., Caballes, L., Scappaticci, F., Struck, R.F., and Gurtoo, H.L., ^{32}P-postlabeling analysis of binding of the cyclophosphamide metabolite, acrolein, to DNA, *Cancer Commun.,* 2, 207, 1990.
163. Maccubbin, A.E., Caballes, L., Riordan, J.M., Huang, D.H., and Gurtoo, H.L., A cyclophosphamide/DNA phosphoester adduct formed *in vitro* and *in vivo, Cancer Res.,* 51, 886, 1991.
164. Lau, H.S.H. and Baird, W.M., Detection and identification of benzo(a)pyrene-DNA adducts by [^{35}S]phosphorothioate labeling and HPLC, *Carcinogenesis,* 12, 885, 1991.
165. Bohr, V.A., Gene specific DNA repair, *Carcinogenesis,* 12, 1983, 1991.
166. Loeb, L.A., Mutator phenotype may be required for multistage carcinogenesis, *Cancer Res.,* 51, 3075, 1991.
167. Lohman, P.H.M., Vijg, J., Uiterlinden, A.G., Slagboom, P., Gossen, J.A., and Berends, F., DNA methods for detecting and analyzing mutations in vivo, *Mutat. Res.,* 181, 227, 1987.
168. Frenkel, K., Donahue, J.M., and Banerjee, S.B., Benzo(a)pyrene-induced DNA damage: a possible mechanism for promotion by complete carcinogens, in *Oxy-Radicals in Molecular Biology and Pathology,* Cerutti, P., Fridovich, I., and McCord, J.M., Eds., Alan R. Liss, New York, 1988, 509.
169. Floyd., R.A., DNA-ferrous ion catalyzed hydroxyl free radical formation from hydrogen peroxide, *Biochem. Biophys. Res. Commun.,* 99, 1209, 1981.
170. Cavalieri, E.L., Rogan, E.G., Cremonesi, P., and Devanesan, P.D., Radical cations as precursors in the metabolic formation of quinones from benzo(a)pyrene and 6-fluorobenzo(a)pyrene. Fluoro substitution as a probe for one-electron oxidation in aromatic substrates, *Biochem. Pharmacol.,* 37, 2173, 1988.
171. Gelboin, H.V., Benzo(a)pyrene metabolism, activation, and carcinogenesis: role and regulation of mixed-function oxidases and related enzymes, *Physiol. Rev.,* 60, 1107, 1980.
172. Steeken, S., Purine bases, nucleosides and nucleotides: aqueous solution redox chemistry and transformation reactions of their radical cations and e$^-$ and OH adducts, *Chem. Rev.,* 89, 503, 1989.

Chapter 7

Expression of Oncogenes and Tumor Suppressor Genes in Teleost Fishes

Rebecca J. Van Beneden and Gary K. Ostrander

I. INTRODUCTION

The transformation of a normal cell into a cancer cell is a complex, multistep process. Two classes of genes are postulated to play important roles in this process: the oncogenes and the tumor suppressor genes. Both were named for their putative involvement in tumorigenesis; however, their true functions are key to the regulation of normal cellular growth. Moreover, it is their aberrant activity that has attracted so much attention. Though their presence was hypothesized much earlier,[1] oncogenes were first identified in 1976 as the transforming genes of retroviruses, which incorporated these cellular sequences from their host genomes.[2] The existence of suppressor genes, first implied by the familial genetic studies of Knudson,[3] was confirmed with the *in vitro* somatic cell hybrid experiments of Stanbridge and co-workers via the fusion of normal cells with malignant tumor cells.[4] The hybrid cells typically contained the normal chromosome complements and, most importantly, exhibited the phenotype of the normal parent cell. Conversely, when fusions were made using cells deficient in specific chromosomes (i.e., 11 and 14), the resulting hybrids exhibited the malignant phenotype. Thus, it was correctly interpreted that the malignant phenotype was produced by defects or losses of genes necessary for normal cell growth and function. Originally termed "recessive oncogenes," these are now more commonly referred to as "tumor suppressor genes."

Cellular oncogenes are important components of the mitogenic pathways of the cell. They can become activated by the incorporation of point mutations, deletions, insertions, or rearrangement, through chromosomal translocation, gene amplification, or proviral insertion.[5] Oncogenes have been classified into families based on the structure, function, and cellular location of their protein products. Over 60 oncogenes have now been detected. Their protein products function as growth factors, growth factor receptors, DNA-binding proteins, transcriptional activators, and other components of signal transduction pathways.

While the role of cellular oncogenes in normal cells is as dominant, positive regulators of growth and differentiation, the tumor suppressor genes act as negative regulators of cell proliferation. Although not as well understood, tumor suppressor genes appear to be similar to the cellular oncogenes in that their protein products are also components of intra- and extracellular signaling pathways. Tumor suppressor genes enable a cell to respond in the proper context to growth-inhibitory signals. Inactivation of these genes leads to the unregulated growth characteristic of a malignant neoplasm. Several tumor suppressor genes have already been isolated and characterized. These include the *Rb* (retinoblastoma), *WT* (Wilms' Tumor gene), *p53*, *NM23* (metastasis suppressor gene), *NF1* (neurofibromatosis type 1 gene), *DCC* (deleted-in-colon-carcinoma gene), *APC* (adenomatous polyposis coli), and *MCC* (mutated in colorectal cancer).

The *Rb* gene, first detected in retinoblastoma,[3] has also been implicated in the etiology of small-cell lung carcinoma,[6] bladder,[7] and mammary carcinomas,[8] as well as other human cancers.[9-13] Loss of the *WT* gene has been detected exclusively in Wilms' Tumor,[14] while inactivation of *p53* has been detected in a wide variety of human tumors.[15] The gene products of *Rb*, *WT*, and *p53* are all nuclear proteins and appear to be involved in transcriptional regulation. However, we know little of how these genes function at the molecular level or how they interact with other proteins (see discussion in Section IV).

The *NM23* gene was identified because of its differential expression in related murine melanoma cell lines of varying metastatic potential.[16] Higher levels of expression were detected in cell lines with lower metastatic potential via Northern blot analysis.

The *NF1* gene, which is responsible for the genesis of neurofibromatosis type I (i.e., von Recklinghausen's disease), also shows significant homology with the catalytic domains of mammalian GTPase-activating proteins.[17-19]

The protein product of *DCC* is a transmembrane receptor with homology to both immunoglobulins and fibronectins.[20] Loss of both alleles is diagnostic of progression to later stages of colon carcinoma. *APC* is inherited as an autosomal dominant trait and is characterized by the emergence of hundreds or even thousands of adenomatous polyps in the colon. If the polyps are not surgically removed, colon carcinoma(s) are likely to present at an earlier age than usual in sporadic cases.[21,22] Similarly, *MCC* encodes an 829 amino acid protein with a short region of similarity to the G-protein and is a strong candidate for the putative colorectal tumor suppressor gene located at 5q21.[23]

The existence of oncogenes and suppressor genes is not restricted to higher vertebrates. In fact, these genes are highly conserved in all taxonomic groups of recent Eumetazoa and even multicellular plants. Of the described oncogenes, one of the most widely distributed is the *ras* gene, which is found in the evolutionarily primitive yeast and slime mold as well as in all metazoan groups examined to date. Similarly, *src* is found in the earliest metazoans (parazoic sponges) and is distributed among all major taxa thus far examined, including humans and fish. For further detail of the evolutionary distribution of known oncogene systems in the animal kingdom, the interested reader is referred to Anders.[17] The known oncogenes and suppressor genes in fish are summarized in Table 1.

While the focus of oncogene research for the past decade has been on mammalian systems, a recent movement has been made toward the development of alternative model systems for the study of both human and environmental health. Of particular interest are the aquatic organisms. These animals are currently being developed as models to better understand normal and aberrant functioning of oncogene and suppressor gene products. Researchers are attempting to understand the response of these genes and its significance to anthropogenic perturbations in aquatic systems. To date, these studies have focused on the roles of oncogenes and suppressor genes in the onset and progression of neoplasia.

For over 20 years, tumors in fish and other lower vertebrates and invertebrates have been systematically documented by the Registry of Tumors in Lower Animals at the Smithsonian Institution in Washington, D.C.[25] While an overwhelming amount of data has been accumulated that documents the high incidence of neoplasia in fish residing in areas high in anthropogenic pollution[26-30] and numerous studies have reported the chemical induction of tumors in fish,[31] very little is known about the molecular basis of carcinogenesis in these animals. It was not until 1986 that the first oncogenes from fish were cloned and sequenced from the goldfish[32] and the rainbow trout.[33] Since that time, however, interest in fish as model systems for molecular studies has risen exponentially.

The development of aquatic models for studies in molecular carcinogenesis and toxicological studies has focused primarily on the teleosts (bony fishes). This group is the largest and most diverse class of vertebrates, with over 21,000 described species. Such research may provide more insight into basic mechanisms than those limited only to mammalian models.[25,31,34] Only recently have fish gained recognition as an integral part of the comparative approach to elucidating mechanisms of cancer common to different phylogenetic levels.[31] Their phylogenetic position relative to the other vertebrates makes them ideal for comparative carcinogenesis studies. This chapter summarizes the current knowledge of oncogenes and tumor suppressor genes in teleost fishes. It is outlined according to gene class:

1. Class I oncogenes — the tyrosine kinases
2. Class II oncogenes — the growth factors
3. Class III oncogenes — the nuclear oncoproteins with transactivating activity
4. Class IV oncogenes — the oncoproteins with GTP-binding ability
5. The tumor suppressor genes
6. Detection of unidentified transforming genes

A discussion of the *Xiphophorus* melanoma system, which represents some of the earliest research on cellular oncogenes and suppressor genes in any species, is also included. Finally, recommendations for future directions in the field are offered for consideration.

Table 1. Cellular Oncogenes and Tumor Suppressor Genes Reported in Fish

Oncogene	Species	Source	Ref.
myc	Rainbow trout (*Oncorhynchus mykiss*)	Liver, testes	33
ras	Goldfish (*Crassius auratus*)	Liver, erythrophoroma-derived cell line	32, 93
	Winter flounder (*Pseudopleuronectes americanus*)	Liver tumors	102
	Atlantic tomcod (*Microgadus tomcod*)	Liver tumors	107
	Rainbow trout (*O. mykiss*)	Normal, neoplastic liver	95, 96
src[a]	*Xiphophorus* sp.	Various tissues, melanoma	54, 68–70, 72
erb-A		Various tissues, melanoma	45
erb-B		Various tissues, melanoma	57, 59
Tu		Melanoma	57, 64
yes		Various tissues	78
fyn		Various tissues	78
abl	Atlantic tomcod (*M. tomcod*)	Liver	81
p53	Rainbow trout (*O. mykiss*)	Liver	118
Rb	Medaka (*Oryzias latipes*)	Liver, eye, brain, viscera	113
	Coelacanth (*Latimeria chalumnae*)	Liver	This paper
	Rainbow trout (*Oncorhynchus mykiss*)	Hepatocytes, brain	This paper
	English sole (*Parophyrs vetulus*)	Liver	This paper
wnt-1	Zebrafish (*Brachiodanio rerio*)	Various tissues	82
Unknown	Medaka (*Oryzias latipes*)	Cholangiocarcinoma	120, 121
Unknown	Northern pike (*Esox lucius*)	Lymphosarcoma	120

Source: Modified and updated from Van Beneden, R.J., in *Biochemistry and Molecular Biology of Fishes,* Hochachka, P.W. and Mommsen, T.P., Eds., Elsevier Press, Amsterdam, in press.

[a] pp60[c-src] activity has been detected in numerous other species; see Table 2.

II. CLASS I ONCOGENES: THE TYROSINE KINASES

The Class I oncogenes code for proteins that function as growth factor receptors. These proteins share two important properties: all are membrane bound, and all catalyze the phosphorylation of tyrosine residues. Before discovery of this class of genes, it was believed that phosphorylation of proteins occurred only on serine and threonine residues, which accounts for 99% of the amino acid phosphorylation in normal cells.[35] Since then, other growth factor receptors, which also phosphorylate tyrosine residues, have been described. This unusual phosphorylation pattern appears to be a critical feature of genes involved in control of cell proliferation. Tyrosine phosphorylation of intracellular target proteins is an important component in the initial step of signal transduction pathways, which control cell proliferation. Amplification of an extracellular signal initiates a complex series of events, which, under normal circumstances, is rapid, transient, and under strict control.

Approximately 30% of all known oncogenes encode protein kinases, which in addition to phosphorylation of tyrosine residues, share other similar structural features.[36] Ligand-binding domains are present in the extracellular amino terminal portion of these transmembrane proteins. A hydrophobic transmembrane section approximately 25 amino acids long typically spans the membrane. The cytoplasmic carboxy terminus of the protein contains the catalytic kinase domain. This class of oncogenes is subdivided into those that code for either receptor or nonreceptor proteins.

A. Receptor Tyrosine Kinases

1. Introduction — erb-B

The epidermal growth factor (EGF) receptor gene was the first growth factor receptor discovered to have protein kinase activity. The observation that protein kinase activity was enhanced by the binding of its ligand, EGF, first indicated that this activity may play a role in signal transduction. The c-*erb*-B proto-oncogene is the truncated, oncogenic form of the EGF receptor that lacks the extracellular binding domain.[37] It is amplified and overexpressed in a number of human tumors, notably squamous cell carcinoma and glioblastomas.[38] The absence of an extracellular ligand-binding domain is primarily responsible for its transforming ability and confers constitutive activity on the receptor.[39] Other structural modifications that have been noted in v-*erb*-B include a carboxy terminal deletion and point mutations, which, although not essential for neoplastic transformation, may contribute to the relative efficacy and host range of the virus that carries this oncogene.[39] An *erb*-B related gene, *erb*-B-2 *(her*-2/*neu)* has been frequently found amplified in human breast and ovarian cancers,[40] as well as in human salivary gland carcinoma.[41] *erb*-B-2 also codes for a cell surface receptor whose ligand has not yet been identified. Several oncogenes with tyrosine kinase activity have been described in the fish *Xiphophorus*. The *Xiphophorus* melanoma gene *(Tu/x-egfrB/ Xmrk/x-erb-B*)* and a related *erb*-B gene (x-*egfr*A) are discussed in the next section.

2. The Xiphophorus Melanoma System — Tu

The best-developed and longest-studied teleost tumor model is the hereditary melanoma of *Xiphophorus* interspecies hybrids.[42-51] Eighteen species of this small, freshwater viviparous fish are indigenous to Central America. They have evolved into numerous subspecies, which are distinguished by unique pigment patterns formed by three different types of pigment cells: pterinophores, purinophores, and melanin-containing melanophores.

Over 60 years ago, M. Gordon, G. Haussler, and C. Kosswig independently observed that interspecific F_1 hybrids of the platyfish *(X. maculatus)* and the swordtail *(X. helleri)* developed spontaneous melanomas, which were inherited in a Mendelian fashion.[42-44] These early researchers hypothesized that melanoma formation was related to increased expression of the melanin color genes. Later investigations by Anders et al.[52] led them to propose that hereditary melanomas in *Xiphophorus* hybrids were the result of the loss of a negative regulator of a specific tumor gene complex *(Tu)*. This complex is composed of three loci: (1) a pterninophore locus *(Ptr)*, which regulates differentiation of this class of pigment cells; (2) the compartment-specific loci (R_{co}), which restrict pigment cell differentiation to specific locations; and (3) the melanophore locus *(Mel-Tu)*, which if mutated or otherwise activated, transforms the melanin-containing pigment cells to neoplastic cells. Through classical genetic analysis, they determined that *Tu* expression in normal platyfish was regulated by a unlinked tumor suppressor gene *(Diff)*. The swordtail lacks both *Tu* and *Diff*. F_1 hybrids between these two species are hemizygous for *Tu* and *Diff*. Successive backcrosses to the swordtail parent result in progressive elimination of the regulatory genes, which in turn allows increased expression of *Tu* and development of malignant melanomas. For a more detailed discussion, reviews by Anders et al.[53] and Zechel et al.[45] are recommended.

The discovery of oncogenes in the mid-1970s ushered in a new direction in *Xiphophorus* research. Early Southern blot analysis detected DNA restriction fragments with homology to numerous vertebrate oncogenes: *abl, src, yes, fyn, erb*-B, *erb*-A, *fes, fgr, fms, fos, int, ras, myb, sis, mil, myc, hck, lck,* and *yes*.[47,54-56] Attention naturally focused on the identity of the melanoma tumor complex. Further analysis of *Tu* expression (as measured by the number and size of spots and the malignancy of the melanomas) showed that it was strongly correlated with elevated expression of the c-*src* gene.[54] Although their expression was highly correlated, further studies[56] indicated that *Tu* and *src* were not the same gene.

The search for the molecular identity of the *Tu* gene continued. In 1988, analysis of restriction fragment length polymorphisms (RFLPs) in a *Xiphophorus* v-*erb*-B homologue showed that this oncogene cosegregated with sex chromosome-linked melanoma in *Xiphophorus* hybrids.[57,58] Partial sequence analysis revealed not one but two distinct *Xiphophorus* genes (designated by Anders and colleagues as *x-egfr*A and *x-egfr*B) with homology to the human (c-*erb*-B) gene. The most highly conserved region was the tyrosine kinase domain. Additional studies of the *Tu* gene in M. Schartl's laboratory confirmed the observation that *Xiphophorus* melanoma DNA contained several v-*erb* related RFLPs.[59] A nearly full-length cDNA clone was isolated and sequenced from a melanoma-derived cell line (PSM). Its predicted amino acid sequence coded for a protein of about 130,000 Da with two strongly hydrophobic regions, one at the N-terminus, typical of a signal peptide, and a second flanked by acidic amino acids. These features suggested a transmembrane protein with an extracellular domain of 617 amino acids, a 23 amino acid transmembrane section, and a cytoplasmic domain of 501 amino acids. As with the *x-egfr*B and *x-egfr*A genes above, the PSM clone (designated *Xmrk*-*Xiphophorus* Melanoma Receptor Kinase gene) showed significant homology to the human EGF receptor. Comparison of *Xmrk* with the partial sequence of a genomic clone (the *Xiphophorus* EGF-receptor gene, *XER*) confirmed the observations of Anders et al. that they were two different genes. The difference in nomenclature adds to the confusion inherent in this system. The *Xiphophorus* melanoma-associated gene related to the human c-*erb*-B gene has been designated the *Tu* complex, *x-egfr*-B, *x-erb*-B*, and *Xmrk*.[58–60] To avoid further confusion, it will be referred to as *x-erb*-B* in the remaining discussion. For a detailed review of this gene, see Zechel et al.[60]

Northern blot analysis indicated that *x-erb*-B* was expressed in a *Xiphophorus* melanoma-derived cell line (PSM) at a 20-fold higher rate than in a fibroblast cell line (A_2).[57] Since the *x-erb*-B* gene itself was not amplified in the melanoma cells, overexpression may have occurred via changes in gene regulation. Further analysis of *x-erb*-B* expression[59] indicated the presence of a single 5.8-Kb transcript in embryos, which was also abundant in unfertilized eggs. Depressed levels were observed in most developmental stages. Elevated levels of *x-erb*-B* expression were found in both melanomas and the melanoma-derived PSM cell line, which confirms the results reported by Zechel et al.[57] Several smaller transcripts were also observed that were highly correlated with the degree of malignancy of the melanomas. Expression of the 5.8-Kb transcript was observed in all melanoma samples, regardless of the degree of malignancy. The data are consistent with the interpretation that *Xmrk* is the *Xiphophorus* tumor gene *Tu*, a novel receptor tyrosine kinase. Classic genetic analysis has indicated that the oncogenic potential of *Xmrk* is regulated by the *Diff* gene, which is postulated to act as a tumor suppressor gene.

Transforming genes have also been identified in melanomas in humans. Activated N-*ras* or H-*ras* oncogenes have been detected in 10% of the melanomas examined.[61] A *ras*-related gene, c-*MEL*, has also been identified.[62] One explanation for the low frequency of activated *ras* genes found in these tumors is that the majority of melanomas may develop via another mechanism.[63] As in *Xiphophorus*, the EGF receptor gene has been implicated. The EGF receptor gene is often duplicated or rearranged in human melanomas.[64] The available data suggest that this gene may be one of several growth factor receptor genes associated with melanoma progression. Studies of the homologous system in *Xiphophorus* may provide further insight into this system.

The *Xiphophorus* model remains perhaps the oldest and best-studied model of oncogene/suppressor gene interaction. Of the model's many unique aspects, the interaction of oncogenes (e.g., *x-erb*-B*) and suppressor gene(s) (e.g., *Diff*) is of prime importance. This is presently the only vertebrate system where intimate knowledge of oncogene/suppressor gene interactions appears readily accessible. Yet relatively few investigators interested in fundamental aspects of gene regulation and expression during oncogenesis have embraced this model. The obvious parallels to EGF receptor involvement in human melanoma appear to make this system an attractive and logical starting place for this basic research.

Xiphophorus has also failed to garner deserved attention among aquatic toxicologists. Of special interest is the fact that feral *Xiphophorus*, as well as inbred laboratory stocks, are nearly completely refractory to neoplasia (i.e., via exposure to strong mutagens/carcinogens such as X-rays and N-methyl-N-nitrosourea).[48] This is in sharp contrast to the sensitivity of hybrids, which readily develop tumors either spontaneously or after carcinogen exposure.[65] Cancerous neoplasms have been found in all neurogenic, epithelial, and mesenchymal tissues of these hybrid animals.[24] Thus, *Xiphophorus* could also be developed as a model for the study of the genetic basis of tumor susceptibility.

3. The Xiphophorus erb-B Gene

In contrast to X-*erb*-B*, the *x-egrf*A gene is not overexpressed in melanoma cells and was designated a ubiquitous, autosomal type 2 EGF gene. *x-egfr*A does not appear to be related to the *Tu* gene. Further studies of its function are warranted.

B. Nonreceptor Tyrosine Kinases

1. src

The v-*src* gene, the transforming gene of Rous sarcoma virus, was the first viral oncogene to be identified and the original model for both viral and cellular oncogenes.[67] *Src* was also the first oncogene identified as a protein tyrosine kinase gene that encoded a 60-Kd phosphoprotein product (pp60c-*src*). This kinase activity was specifically correlated with cell transformation.

The *src* protein was highly conserved in all organisms examined.[67] The presence of cellular-*src* genes in lower vertebrates was examined in several related studies.[68–70] Activity of pp60^{c-src} was assayed by the standard protocol of immunoprecipitation of the heavy chain of IgG from RSV-infected rabbit sera after incubation with ^{32}P-ATP.[71] The tyrosine kinase activity of pp60$^{c\text{-}src}$ was detected in all vertebrates examined, including fishes (Table 2). Confirmation of *src*-homologous sequences was demonstrated in subsequent studies by Southern blot analysis.[72]

Expression of the *src* gene was examined in *Xiphophorus* and other fish species by analysis of pp60$^{c\text{-}src}$-specific activity. The highest kinase activity was detected in extracts from brain and malignant melanoma of *Xiphophorus*.[69] Extracts from benign melanomas had lower levels of pp60$^{c\text{-}src}$ activity. Additional studies on *Xiphophorus* confirmed high expression of pp60$^{c\text{-}src}$ activity in hereditary melanomas of interspecies hybrids, as well as elevated activity in other tumors of various etiologies.[54] The relationship of elevated activity of the *src* protein to the etiology of the tumor remains to be determined. Tyrosine kinase activity was also detected in skin, liver, spleen, and testes of other fish species.[69] No activity was detected in muscle tissue.

Recent studies identified a second c-*src* gene product (pp60$^{c\text{-}src+}$), whose mRNA contained an 18-base insertion due to an alternative splicing event.[73] pp60$^{c\text{-}src+}$ has only been detected in cells of neural origin. Using an antibody prepared to the six amino acid insert of pp60$^{c\text{-}src+}$, Yang et al.[74] detected expression of the neuronal form in mammals, birds, and reptiles. They were unable to detect expression in brain extracts of primitive fish (catfish, lamprey) or crustaceans. However, using RT-PCR analysis, Raulf et al.[75] demonstrated two different *src* transcripts in RNA isolated from *Xiphophorus* brain tissue. Oligonucleotide primers (30 mers) were designed based on the known *Xiphophorus* c-*src* cDNA sequence, which represents the fibroblast form of c-*src*. These primers amplified a 682-base pair (bp) segment between exons 2 and 5, which included the expected insertion point at the junction of exons 3 and 4. The sequence of the 18-base insert region revealed that of the six amino acids it encoded, three were identical to the mammalian pp60$^{c\text{-}src+}$, and two were conservative changes that maintained the hydrophilic nature of this region. The conflicting results of these two studies may have been due to the inability of the antibody prepared to the mammalian form to recognize the protein in lower vertebrates.

The data summarized above suggest that the c-*src* gene is highly conserved in all vertebrates and primitive chordates examined based on Southern blot analysis. The *src* protein is also conserved based on antigenic specificity, biochemical properties, and tissue-specific expression.

pp60$^{c\text{-}src}$ activity has since been correlated with cell transformation in mammalian models and has also been reported in fish cell lines (e.g., PSM and A$_2$; melanoma-derived cell lines from *Xiphophorus*);[66,75] these should be amenable to further studies. In addition to studies with PSM and A$_2$, the melanized cell line (P15) from a goldfish tumor that can undergo reversible dedifferentiation and redifferentiation resulting in additional cell lines (P15D and P15D1)[76] should be examined. P15, which has stable phenotypic characteristics, was originally cloned

Table 2. Phylogenetic Occurrence of pp60^{src} Kinase Activity among Vertebrates

Taxonomic group	Species	Ref.
Mammals	Human *(Homo sapiens)*	128
	Rat *(Rattus rattus)*	70
	Mouse *(Mus domesticus)*	70
Birds	Chicken *(Gallus domesticus)*	70, 74, 129
	Quail *(Coturnix coturnix)*	70
Reptiles	Rattlesnake *(Crotalus atrox)*	74
	Turtle *(Terrapene carolina)*	74
Amphibians	Frog *(Rana catesbeiana),*	74
	(Xenopus laevis)	74
Teleosts	*Xiphophorus* sp.	70, 72
	Flatfish *(Pleuronectes platessa)*	70, 72
	Sea robin *(Trigla lucerna)*	70
	River grudgeon *(Gobio gobio)*	70
	Codfish *(Gadus morhua)*	70
	Flounder *(Platichtys flesus)*	70, 72
	European roach *(Leuciscus rutilus)*	70
	Blackstriped cichlid *(Cichlasoma nigrofasciata)*	70
	Eel *(Anguilla anguilla)*	72
	Catfish *(Ictalurus furcatus)*	74
Cartilaginous fish	Electric ray *(Torpedo marmorata)*	72
	Cat shark *(Scyliorhinus caniclus)*	70
Jawless fish	Western brook lamprey *(Lampetra planeri)*	70
	Lamprey *(Petromyzon marinus)*	74

Source: Modified and updated from Van Beneden, R.J., in *Biochemistry and Molecular Biology of Fishes,* Hochachka, P.W. and Mommsen, T.P., Eds., Elsevier Press, Amsterdam, in press.

from a goldfish erythrophoroma cell line (GEM 81).[77] Of additional interest is that the specific activity of tyrosinase in P15 cells decreases and increases during dedifferentiation and redifferentiation, respectively. How this may relate to tyrosine kinase activity is unknown. A closer examination of c-*src* expression in these cells is warranted.

2. yes, fyn

The c-*yes* and c-*fyn* genes have also been investigated in *Xiphophorus*.[78] These cellular oncogenes are members of the c-*src* subfamily. They encode for tyrosine kinases of around 60,000 Da; are highly conserved, especially in the carboxy terminal catalytic domain; and are associated with the inner surface of the cell membrane.[79]

The X-*yes* and X-*fyn* genes were cloned from a cDNA library derived from *X. helleri* brain mRNA.[78] A clone was also identified for the X-*yes* gene from a genomic library derived from *X. maculatus*. The X-*yes* gene intron/exon pattern of the kinase domain is identical to that of higher vertebrates. Sequence analyses of both X-*fyn* and X-*yes* cDNA clones specify two highly conserved regions: exons 3 to 6 and exons 7 to 12. Exons 3 to 6 contain the SH2 (*src*-homology-2) and SH3 (*src*-homology-3) regions, respectively, which are regulatory domains conserved in all *src*-related genes.[67] Exons 7 to 12 encode the carboxy terminal kinase domains. These domains are highly conserved at the predicted amino acid level, having 93% (X-*yes*) and 95% (X-*fyn*) homology to their human counterparts.

X-*yes*, X-*fyn*, and X-*src* all show similar patterns of tissue-specific expression, except for the higher amounts of X-*src* expression in gill and fin tissues. This is consistent with the speculation that all three genes may be under common regulatory control. Tissue expression patterns in *Xiphophorus* are similar to those described in higher vertebrates with two notable exceptions: (1) the absence of expression of X-*src* and X-*yes* in adult liver tissue; and (2) low levels of X-*fyn* expression in fin tissue fibroblasts, in contrast to high levels of c-*fyn* expression in human fibroblasts. The highest levels of expression of the three *src*-related genes were detected in melanoma tissue and a melanoma-derived cell line (PSM). The authors speculate that this may indicate a possible role for *src*-related tyrosine kinases in tumor progression.[78] Further studies in this area are certainly warranted.

3. abl

Members of the *abl*-related subgroup of nonreceptor tyrosine kinases have significant structural similarity to the *src* family described above.[80] Although the *abl* protein is substantially larger (150 Kd), both the SH2 and SH3 domains remain conserved. Activation of c-*abl* in chronic myelogenous leukemia (CML) involves a chromosomal translocation and subsequent structural alterations in the gene, resulting in elevated kinase activity. Similarly, v-*abl* attains transforming ability in the Abelson murine leukemia virus via fusion with the viral *gag* gene and loss of the SH3 domain, with a resulting increase in kinase activity.

Presently, very little is known about c-*abl* in fish species. The c-*abl* gene has been investigated in two fish species, the Atlantic tomcod *(Microgadus tomcod)* and *Xiphophorus*. A summary of these findings is presented below.

Atlantic tomcod residing in the Hudson River display an exceptionally high incidence of liver tumors (see Section V). Several studies have been made to address the molecular basis of this phenomenon. A recent report described significant allelic and genotypic frequency differences at the c-*abl* locus between Atlantic tomcod from the Hudson River (which have a high incidence of liver tumors) and a population from Maine (which has a low tumor incidence).[81] Three different c-*abl* domains were described, based on banding patterns of DNA restriction fragments hybridized at high stringency to a mouse v-*abl* probe. The authors did not correlate these domains with the homologous regions in mammalian *abl* oncogenes. Therefore, it is unknown at this time whether they have any functional significance. Allelic frequencies at the c-*abl* C domain were significantly different between Hudson River and Maine fish. Gene frequency data suggested that these were separate Mendelian populations. However, the Hudson River tomcod do not display a single c-*abl* allele, which is absent from fish in Maine. Gene expression analysis using Northern blots was unsuccessful. No polymorphisms were observed in several other oncogene loci examined (c-Ki-*ras*, c-Ha-*ras*, and c-*src),* nor were differences observed in mtDNA restriction patterns. The authors speculated that the Hudson River population may be genetically predisposed to develop hepatocellular carcinomas and that the observed c-*abl* polymorphisms may play a role. At this time, it is impossible to determine the role of c-*abl* in the etiology of this tumor. The polymorphism may simply be a marker that indicates that the Hudson River tomcod are a population distinct from the Maine population. They may be a genetically "more susceptible strain" for reasons entirely unrelated to the c-*abl* gene. On the other hand, the polymorphisms in the c-*abl* gene may be the result of a deletion of an important regulatory region (such as the SH3 domain), which would be predicted to activate the transforming ability of the c-*abl* gene, analogous to that in CML in humans and in the Abelson murine leukemia virus described previously. Substantial work remains to be done before the potential involvement of *abl* in tomcod neoplasia can be resolved.

The expression of *abl* has also been investigated in *Xiphophorus* embryonic, adult, and melanoma cells.[56] Although DNA sequences homologous to v-*abl* were identified by Southern blotting, RNA transcripts were not detected in melanoma cells or in any embryonic or adult tissue investigated.

While the *abl* gene has been conserved through evolutionary time, the studies with tomcod have shown that it remains susceptible to mutational events. Therefore, prior to final resolution of the potential involvement of *abl* in oncogenesis in aquatic species, it will be necessary to identify or develop suitable nucleic acid

III. CLASS II ONCOGENES: GROWTH FACTORS

The identification of the *sis* oncoprotein as platelet-derived growth factor (PDGF) confirmed the hypothesis proposed by Todaro in 1977 that overproduction of a growth factor by a responsive cell could result in uncontrolled cellular proliferation.[67] PDGF is the major growth factor required for the growth of fibroblast cells in culture. Further studies of this oncogene established that the *sis* oncoprotein binds to the PDGF receptor, thus initiating cellular transformation by continuous autocrine stimulation of a normal growth factor response pathway. Other members of this class of oncogenes related to fibroblast growth factors include *int-2, hst,* and *fgf-5,* as well as the acidic and basic fibroblast growth factors (FGFs).

Relatively little work has been done on these oncogenes in fish systems. Early studies of oncogenes in *Xiphophorus* identified sequences homologous to the *sis* and *int* oncogenes by Southern blot analysis.[47] A recent paper by Molven et al.[82] described the genomic structure and developmental expression pattern of the *wnt-1* (formerly named *int-1*) gene in the zebrafish *(Brachydanio rerio)*. The *wnt-1* gene was first identified as a retroviral integration site (for the mouse mammary tumor virus, MMTV) in mouse mammary carcinomas. Cell transformation was thought to occur due to the close proximity of the MMTV enhancer sequences to the *wnt-1* gene, which induced inappropriate expression of this gene. *wnt-1* is believed to play an important role in embryogenesis and is necessary for proper development of the midbrain and anterior hindbrain in the mouse. Studies of the zebrafish *wnt-1* gene indicated that the protein product is highly conserved. Furthermore, *in situ* hybridization revealed that it is expressed during the development of the neural tube and that the pattern of expression is nearly identical to that in the mouse. Additional studies of this gene may provide important clues to its role in developmental regulation. In addition to *wnt*, mRNA transcripts of other proto-oncogenes and growth factors are known to be present in high levels in early embryonic development (i.e., FGF, PDGF, TGF-β, *mos, myc, raf, rel,* and *src).*[67] This suggests that they function as maternal messages in early developmental stages. In *Xenopus,* maternal messages for basic FGF and a TGF-β-related message appear to function in mesoderm induction.

IV. CLASS III ONCOGENES: NUCLEAR ONCOPROTEINS

Eukaryotic transcription is controlled by the interaction of protein transcription factors with specific regulatory sequences of the gene. Transcription factors require two functional domains: a DNA-binding domain, which allows binding to the specific DNA regulatory sequence, and a *trans*-acting domain, which is involved in protein/protein interactions and enhances binding. Both of these domains can be provided through protein/protein interactions. Many of the oncoproteins that have been localized to the nucleus are known to bind DNA and act as transcriptional activators.[67] Members of this oncogene family include *erb*-A, *jun, fos, myc, ets, myb, ski,* and *rel.* Of these, only the *myc* and the *erb*-A genes have been cloned in teleosts.

A. *myc*

The activation of cellular *myc* oncogenes, c-*myc*, L-*myc*, and N-*myc*, has been described in a wide variety of neoplasms.[83,84] Activation of *myc* may occur via proviral insertion, chromosomal translocation, or gene amplification. In spite of extensive study of this oncogene, its precise mechanism of action is still a mystery. Data suggest that the *myc* protein is a transcription factor that interacts with several other nuclear proteins, including *max* and *mad*.[85]

The rainbow trout *myc* was one of the first fish oncogenes cloned and sequenced.[33] Alignment of the trout c-*myc* with the other *myc* sequences revealed that the third exon was most highly conserved overall. However, two domains within the second exon, *myc* boxes A and B, showed the highest degree of homology to known *myc* genes, 93 and 89%, respectively, at the predicted amino acid level. This suggests that these regions represent significant functional domains. Gene expression studies detected a 2.3-Kb RNA transcript in adult liver RNA by Northern blot analysis, slightly smaller than that reported for avian and mammalian c-*myc* (2.5 Kb). The first exon, the putative regulatory portion of the gene, was not cloned.

DNAs from several fish tumors were examined for amplification and rearrangement of this gene. DNA was isolated from feral northern pike *(Esox lucius)* lymphomas, feral white perch *(Morone americana)* liver tumors, and chemically induced liver tumors of the medaka *(Oryzias latipes)*.[86] Using Southern blots hybridized to trout c-*myc* probes, no amplification or rearrangement of c-*myc* was detected. A single exception was noted in DNA from one northern pike which had a unique *Eco*RI restriction pattern. However, this pattern was not repeated in any of the other northern pike tumors examined.

B. *erb* -A

In 1986, *erb*-A was identified as a thyroid hormone receptor-related oncogene, the first to be identified as a transcriptional regulator.[87] Mutations occurring primarily in the carboxy terminal end of *erb*-A prevent binding of the thyroid hormone. Recent studies revealed the unexpected finding that *erb*-A also represses transcription of thyroid hormone responsive genes. Current thought is that the normal thyroid hormone receptor induces transcription of genes involved in erythroid differentiation and therefore acts as a negative regulator of cell proliferation. *erb*-A expression blocks this action and thus promotes cell proliferation. Therefore, *erb*-A may be considered a suppressor gene.

Zechel et al.[45] recently reported species- and population-specific RFLPs for v-*erb*-A fragments in *Xiphophorus*. Partial sequence analysis of two *erb*-A homologous clones (x-th-r-1 and x-th-r-2) revealed two different *Xiphophorus* hormone receptors with homology to the human retinoic acid receptor and the human thyroid hormone receptor, respectively. As in humans, *Xiphophorus* genomes probably contain numerous related steroid receptor genes.

Both *Xiphophorus erb*-A-related genes were differentially expressed in *Xiphophorus* fibroblast and melanoma-derived cell lines. The role, if any, of *erb*-A gene expression in the development of genetic melanomas has not been established yet. As described previously for *src*, the melanized goldfish tumor cell line (P15)[58] is a promising system for initial studies of the role of *erb*-A in these melanoma cell lines. One possible avenue of research is the examination of *erb*-A message and protein product expression during dedifferentiation and redifferentiation. Subsequently, addition of steroid hormones to these cultures might provide further understanding of their potential role in cell transformation.

Other nuclear oncogenes that bear examination in fish species are *myb, fos, jun,* and *ets*. These genes have all been cloned in *Drosophila*. Studies of transcriptional activation in lower vertebrates might provide further clues to the molecular mechanisms of gene regulation. In addition, studies in fish would fill the evolutionary gap between mammals and invertebrates.

V. CLASS IV ONCOGENES: GTP-BINDING PROTEINS

A. *ras*

ras genes are highly conserved, and genes with *ras* homology have been identified in all eukaryotes studied, including mammals, fish, mollusks, *Drosophila,* plants, slime mold, and yeast.[88] The *ras* oncogenes are the most frequently detected oncogenes in human tumors. Approximately 20 to 30% of all human tumors examined to date contain activated *ras* genes. The high incidence of activated *ras* in various tumors of different etiologies suggests that the normal cellular protein performs a critical function in a wide variety of cell types. The *ras* gene superfamily now consists of over 40 related genes classified into three families: *ras, rho,* and *rab*.[89] Only studies of *ras* have been reported in fishes.

ras family genes encode for small (21 Kd), membrane-bound proteins involved in signal transduction.[88] Activated c-*ras* acts as a growth signal for the cell and is similar to other G proteins in that it binds GTP. Its native GTPase activity catalyzes the conversion of GTP to GDP and inactivates the protein, passing on the signal to an unknown receptor in the cytoplasm.

ras genes are activated to oncogenes by transformational changes that maintain the protein in its active form, which occurs primarily through point mutations at specific codons for amino acids that interact with guanine. Mutations at codons 12, 13, 59, and 61 occur at or near sites of interaction of the *ras* protein with the β- and/or α-phosphates of guanine nucleotides. Similarly, mutations at residues 116, 117, and 119 affect the interaction of the *ras* protein with the guanine base. *ras* mutations result in an increase in the amount of the active GTP-bound form of the *ras* protein by reducing GTPase activity and/or decreasing affinity for GDP. Not unlike studies of oncogenes in human cancer, *ras* has been the most frequently studied oncogene among fish species. Reviewed below are studies of *ras* in four different fish species, which have added to our understanding of the involvement of this gene in tumorigenesis.

B. Goldfish *ras*

ras was one of the first oncogenes to be cloned and sequenced in a fish. Studies with goldfish *(Carassius auratus)* in 1986[32] revealed extensive sequence homology between fish and mammalian *ras*. Not only were intron/exon boundaries identical to those in human *ras* genes, but the predicted amino acid sequence of the goldfish *ras* had 96% homology to K-*ras*, with the most deviation in the fourth exon. Furthermore, sequences at codons 12 and 61, reported hot spots for mutations in human tumors, were conserved. Sequence conservation in lower vertebrates further points out the importance of these regions.

Subsequently, Nemoto and co-workers[93] compared the sequence of the first exon of normal goldfish *ras* to a *ras* gene isolated from a cell line derived from a goldfish erythrophoroma.[94] Only a 5% variation at the nucleotide level was observed between the two clones. Interspecies identity at the nucleotide level for the goldfish clones relative to the mammalian *ras* clones was high and averaged 83% to human K-*ras*, 86% to rat H-*ras*, and 79% to human N-*ras*. No differences were found in the first exon region in DNA from normal goldfish and DNA from the erythrophoroma cell line, including the 12th codon. However, expression of *ras* in two erythrophoroma cell lines, as determined by dot blot analysis, was higher relative to normal goldfish liver. Unfortunately, regions involved in *ras* activation in other parts of the gene have not been investigated. Finally, conservation of the noncoding sequences upstream from the first exon in the two goldfish clones was noted. This suggests that this region may have some functional significance and should be examined further. This region near the initiation codon does not appear to be conserved between human c-H-*ras* and c-K-*ras* genes.

C. Rainbow trout *ras*

Two *ras* genes, designated trout *ras*-1 and *ras*-2, and a portion of a genomic *ras*-1 allele have been identified and partially (300 to 500 bp) sequenced from rainbow trout *(Oncorhynchus mykiss)* liver.[95] Reported DNA homology to human *ras* genes ranged from 76.8 to 87.1%. Although base changes have occurred among the trout, goldfish, and mammalian *ras* homologues, they are primarily neutral mutations. Minimal changes in base arrangement during higher vertebrate evolution have resulted in few changes in protein structure. Goldfish and trout *ras* genes appear to differ from mammalian *ras* genes to approximately the same extent. This is not surprising, given the minimal evolutionary distance between the goldfish (Cyprinidae) and trout (Salmonidae) families. It would be interesting to examine *ras* sequences in a phylogentically advanced fish such as English sole or winter flounder (family Pleuronectidae), both of which exhibit hepatocellular carcinomas under natural conditions.

The first report of experimentally induced *ras* mutations in any fish has been completed recently following exposure of rainbow trout to Aflatoxin B_1 (AFB_1).[96] AFB_1 is a potent mycotoxin that is metabolically activated to the AFB_1 8,9-epoxide capable of producing adducts in DNA at guanyl residues.[97] The DNA from 14 AFB_1-induced trout liver tumors of unknown pathology was examined for point mutations in exon 1 of the trout *ras*-1 and *ras*-2 genes. Ten of the fourteen tumor DNAs exhibited activating point mutations in the trout c-Ki-*ras* gene (trout *ras*-1). Seven of ten were codon 12 GGA→GTA transversions, two were codon 13 GGT→GTT transversions, and one was a codon 12 GGA→AGA transition. No mutations were detected in the first exon of the trout *ras*-2 gene or in DNA from control livers.

Previous studies suggest that AFB_1 may act to produce primarily G→T transversions, although the target sequence has not yet been identified. Recent studies of rat hepatocellular carcinoma DNA identified both GC→AT transitions and GC→TA transversions at codon 12 of c-K *ras*[98,99] in about 25% of tumors examined. Finally, although the pathology of the lesions in the trout studies was not reported, studies in humans have demonstrated that cholangiocarcinomas (also the most common lesion in trout liver following AFB_1 exposure) have activated *ras* sequences.[100,101]

The methodology described above for the rainbow trout was not designed to detect mutations in other exons or at other sites in exon 1, or did it detect multiple mutations. Also, changes in *ras* gene expression were not investigated. Hopefully, future studies will elucidate other mutations, if they exist, in the *ras* gene or message in neoplastic trout liver tissue. The fact that similar mutations in *ras* homologues occur in organisms that diverged evolutionarily over 400 million years ago strongly suggests that they may play a role in the etiology of these tumors. Moreover, *ras* proteins must be of fundamental importance to normal activities of the cell.

Investigation of normal *ras* function in fish would provide an important contribution to this area of research. *ras* proteins, like other G proteins, act to couple cell surface receptors to second messengers. The focus now is on the identity of the second messenger and other proteins in this signal transduction pathway. In the yeast, *Saccharomyces cerevisiae*, two *ras*-related genes have been detected, *RAS-1* and *RAS-2*.[90] In *S. cerevisiae*, one of the functions of the *RAS* proteins appears to be the regulation of adenylate cyclase activity.[91] However, there appear to be no interactions of *ras* with this second messenger in mammalian cells or other nonmammalian species. The GAP protein (GTPase-activating protein) is one protein which has been shown to interact directly with *ras* in mammals and *Xenopus*.[92] Fish *ras* genes, which appear to be highly conserved relative to mammalian genes, may also interact with GAP. This remains an intriguing area of investigation.

D. Flounder *ras*

Winter flounder *(Pseudopleuronectes americanus)* residing in Boston Harbor, in areas known to be contaminated with polycyclic aromatic hydrocarbons (PAHs), also exhibit a high incidence of both hepatocellular and cholangiocellular carcinomas.[29] Transfection analysis has be used by McMahon et al.[102] to detect activated c-Ki-*ras* oncogenes in tumor DNA from fish exhibiting either hepatocellular or cholangiocellular carcinomas. Briefly, DNA was isolated from the tumors and then transfected into NIH3T3 fibroblasts. Transfected cells were assayed for tumorigenicity by injection into nude mice. In a typical experiment, DNA from 7 of 13 of the diseased livers induced tumors in nude mice. Subsequently, DNA was isolated from the mouse tumors and subjected to a second round of transfection. These DNAs were also positive for transformation. Southern blot analysis of DNA digests from secondary transfectants revealed fragments, which hybridized to c-Ki-*ras* and appeared to be unique to flounder. DNA from primary liver samples, which tested positive in the nude mouse assay as well as nude mouse tumors showed GC→AT or GC→TA single base changes in the 12th codon of flounder c-Ki-*ras* when examined by DNA sequence analysis following PCR amplification. GC→AT changes were also detected in DNA from some diseased livers that were negative in the nude mouse assay. Finally, no mutations were detected at the 59th codon. Flounder detoxification enzymes have been reported to activate the PAH benzo(a)pyrene to mutagenic epoxides.[103] Similarly, induction of GC→TA or GC→AT base changes by BaP metabolites has been reported in both bacterial[104] and mammalian systems.[105] Thus, available data suggest that PAHs may be inducing similar activating mutations in the flounder *ras* gene. The authors conclude that environmental chemicals could play an important role in the neoplastic process, causing mutations at the level of the DNA. Future studies could be designed to investigate the correlation of liver pathology with precise genetic changes and thus define the role of these mutations(s) in the etiology of a naturally occurring epizootic in a feral vertebrate population.

E. Tomcod *ras*

The tomcod *(Microgadus tomcod)* is a bottom-dwelling fish found along the east coast of North America, including the Hudson River. The Hudson River is a heavily impacted area containing high levels of heavy metals, pesticides, polychlorinated biphenyls, and PAHs. The incidence of hepatocellular carcinomas approaches 90% among adult tomcod collected from the Hudson River.[106] Wirgin and co-workers analyzed genomic DNA isolated from tomcod tumors for transforming ability in the NIH3T3 transfection assay.[107] Six of nine tumor DNAs were positive in a primary transfection. Furthermore, transformed cells grew in soft agar, were tumorigenic in nude mice and DNA digests from both transformed cells and nude mouse tumors exhibited hybridization to a mammalian K-*ras* probe. These data suggest that an activated K-*ras* exists in tomcod liver tumors. Further confirmation, including DNA sequence analysis, is needed to support this hypothesis.

VI. SUPPRESSOR GENES

A. Retinoblastoma

The best-studied suppressor gene is the retinoblastoma *(Rb)* gene.[108,109] Retinoblastoma is the most common intraocular neoplasm of children appearing during the initial year of life, with a frequency of about 1 in 15,000 births. Nearly all retinoblastomas examined to date have been accompanied by either a loss or defect in the expression of the *Rb* gene product (p105Rb), a nuclear phosphoprotein with DNA-binding ability. p105Rb has been reported to complex with the transforming protein of DNA tumor viruses, leading to the suggestion that transformation by these viruses might require inactivation of tumor suppressor genes through protein/protein interactions.[110] Specifically, because of their cellular homology, the adenovirus E1A protein or SV40 large T antigen have been suggested as candidates to modulate the growth-limiting effects of the *Rb* gene. Subsequently, it was reported that p105RB was primarily underphosphorylated during the early G_1 phase and that it underwent three sequential phosphorylation steps as the cell entered S phase, progressed through G_2, and entered mitosis before becoming dephosphorylated.[111] Since the exact function of the *Rb* protein remains unknown, many important questions concerning the etiology of retinoblastoma development remain to be answered. Among these are the relationship of phosphorylation of the protein to its mode of action and the identity of the enzymes responsible for the regulation of its phosphorylation state.

Presently, examples of defective expression of the *Rb* gene have come only from human biopsy samples, tissue culture cell lines, and a single genetically altered mouse model.[112] A vertebrate model for examining chemically induced retinoblastoma, however, has not been developed. To study the consequences of punitive tumor promoters on suppressor gene activity, G.K. Ostrander has been developing a model system using the Japanese medaka. Retinoblastoma is observed following exposure of medaka to the carcinogen, methylazoxymethanol acetate.[113] The medaka is the only vertebrate model in which retinoblastoma can be chemically induced. In addition to medaka, we have recently demonstrated via Western blotting that the *Rb* gene product is widely distributed among fishes, including the very primitive coelacanth and the more evolutionarily advanced rainbow trout and English sole. We have detected its expression in nearly all tissues, with the highest levels in brain, gastrointestinal tract, and liver. Similar studies by Bernards et al.[114] and Destree et al.[115] have shown that this gene is widely distributed among all other major vertebrate groups, including amphibians, reptiles, birds, and several mammals. The wide distribution of the *Rb* protein among vertebrates provides evidence of its importance in a fundamental process related to growth control. Studies are currently underway to characterize potential alterations in *Rb* gene and gene product expression in this model. Our studies will continue to focus on the *Rb* suppressor gene and the influence of tumor promoters on the tumor suppressor gene product.

B. *p53*

The *p53* gene product was first identified as a 53-Kd nuclear phosphoprotein that complexed with the SV40 large T antigen in SV40-transformed murine cells.[116] Early studies established that *p53* cDNA cotransfected with *ras* could transform rat embryo fibroblasts. These initial results placed *p53* among the oncogenes. These clones, however, were actually mutant genes. Subsequently, numerous reports have established normal *p53* as a tumor suppressor gene. In contrast to the *ras* gene, mutations have been detected in more than 30 different codons within highly conserved regions.[117]

The structural evolution of the *p53* gene has recently been summarized.[118] Comparison of the predicted amino acid sequences of *p53* genes from human, monkey, mouse, rat, chicken, *Xenopus,* and rainbow trout revealed five highly conserved domains. Rainbow trout *p53* showed 90% homology at the predicted amino acid level to the human sequence within these five domains.

Northern blot analysis detected a 2.4-Kb transcript in rainbow trout. A 57-Kd protein was determined by SDS-PAGE. Binding of the rainbow trout *p53* and SV40 large T antigen was demonstrated *in vitro*.[118] Expression of *p53* has also been detected in the fish cell lines EPC (epithelioma papillosum cyprini) and CHSE 14 (chinook salmon embryo).[119] Southern blot analysis of genomic DNA from EPC cells confirmed the presence of sequences homologous to *p53* cDNA.

p53 is presently one of the most intensely studied genes among investigators utilizing mammalian carcinogenesis model systems. Studies are currently underway to investigate the role of *p53* in chemically induced tumors in the Japanese medaka in the laboratory of R.J. Van Beneden.

VII. DETECTION OF UNKNOWN ONCOGENES

Preliminary work in Van Beneden's laboratory has shown that DNAs isolated from tumors in feral northern pike from contaminated waters, as well as DNAs from chemically induced tumors in the laboratory in medaka, are able to transform NIH3T3 cells.[86,120,121]

Northern pike were collected from Ostego Lake, Gaylord, MI. Over 20% of the fish in this area showed the presence of large external lesions identified as lymphomas.

Preliminary studies on chemical induction of tumors were done with medaka treated with diethylnitrosamine. The medaka has been widely used in studies of chemical carcinogenesis. It is highly sensitive to a variety of chemicals, and tumors may be induced within a short period of time in nearly all tissues and organs.[120-126]

Transforming ability of fish tumor DNA was examined by DNA transfection into NIH3T3 cells. Tumorigenicity of the transformed cells was determined by the nude mouse assay.[127] A summary of results is presented in Tables 3 and 4. DNAs from both pike and medaka tumors were positive in these assays. DNA isolated from a cholangiocarcinoma, a very aggressive liver lesion, was very active by all criteria examined. DNA from diethylnitrosamine (DEN)-induced focal biliary hyperplasia, a preneoplastic liver condition, was marginally active in the transfection assay.

In order to confirm that the transformation of NIH3T3 cells was due to fish sequences, fish-specific sequences were demonstrated in transformed cells by Southern blot analysis. Efficiency of transformation has increased in secondary and tertiary transfections. Studies to identify these transforming genes are in progress. While transfection analysis is a classic, proven method for identification of activated oncogenes, it has several disadvantages. These include its ability to select for a limited set of oncogenes, as well as the expensive and labor-intensive cell culture involved. Alternative approaches include subtractive cDNA libraries and differential display PCR. Both of these methods measure differences in gene expression in normal vs. tumor tissue and may identify deregulated genes not detected by the transfection assay.

VIII. SUMMARY

Research in the area of molecular oncology in teleosts is nearly a decade behind that in mammalian models. However, the first oncogenes from fish were cloned only 7 years ago. The field is moving so rapidly now that the application of modern molecular techniques to aquatic systems awaits only the insight of the individual investigators. As summarized in this chapter, studies of teleost oncogenes and suppressor genes have provided major contributions to several important areas of research: (1) gene evolution, (2) mechanisms of carcinogenesis, (3) normal and aberrant oncogene/suppressor gene function, and (4) development of biomarkers sensitive to environmentally induced carcinogenesis. Potential problems for research in each of these areas are described in Sections VIII.A through VIII.D.

A. Gene Evolution

Although fish diverged from the higher vertebrates over 400 million years ago, certain domains in all the oncogenes and suppressor genes examined to date remain remarkably conserved. Studies on homologous genes in phylogenetically divergent organisms have provided important information on gene evolution and

Table 3. Secondary Transfection of Northern Pike DNA into NIH3T3 Cells

DNA[a] source	Standard focus assay (Ave. foci/μgDNA)	Nude mouse assay freq.	Growth in soft agar[b]
TR6-6Bc (normal liver)	2.6	2/16	0
TR6-1Cc, 1Da (lymphoma)	4.6	9/22	++

Source: Adapted from Van Beneden, R.J., et al., *Cancer Res.*, 50 (Suppl.), 5671s, 1990.

[a] DNA used in secondary transfection was isolated from cells or expanded foci from a primary transfection assay.
[b] Growth is relative to *mos*-transformed NIH3T3 cells. NIH3T3 cells were used as negative controls.

regulation. Currently, research has been focused on mammalian, *Drosophila*, and yeast systems. However, it is important to consider that the teleost fishes are uniquely placed to supply the missing information between the invertebrates and the higher vertebrates.

B. Mechanisms of Carcinogenesis

Investigation in rodent models has implicated the direct chemical activation of oncogenes by carcinogen treatments.[88] Evidence from flounder, trout, and tomcod *ras* studies and studies of chemically induced tumors in the medaka support this hypothesis. Fish are uniquely suited as models to assess the molecular mechanisms of aquatic toxicants; however, their use in such studies has often been overlooked.

C. Oncogene/Suppressor Gene Function

It is appropriate and, in fact, the responsibility of aquatic toxicologists to develop studies aimed at elucidating the normal and abnormal functions of oncogenes and suppressor genes. In doing so, it will not be enough simply to probe

Table 4. Summary of Medaka Secondary Transfection Analysis

DNA source	Focus DCF5	Assay DCF5+DEX	Colony QBSF	Selection QBSF+0.1%	Nude mouse assay[a]
Untreated medaka controls	0.57	0	—	—	2/16
DEN-induced focal biliary hyperplasia	0.77	0.53	—	—	10/16
DEN-induced cholangiocarcinoma	5.0	5.0	+	+	47/47

Source: Data summarized from Van Beneden, R.J., et al., *Cancer Res.*, 50 (Suppl.), 5671s, 1990.

[a] Tumors at <5 weeks postinjection.

fish tissues with mammalian probes and/or repeat other experiments first attempted in rodents. Instead, we must develop our own probes (i.e., from the fishes!) and experimental paradigms that exploit novel aspects of the biology of fishes. Only then will aquatic toxicologists be able to answer questions of fundamental importance and relevance to all species that cannot be addressed with the current "popular" systems.

D. Biomarkers

Finally, a number of feral fish populations exhibiting a high incidence of environmentally induced cancer (e.g., tomcod, English sole, winter flounder) appear to be suitable models for examination of the earliest molecular events associated with chemical carcinogenesis. Consequently, oncogene activation and/or mutation or deletion of suppressor genes in these species may serve as sensitive biomarkers of environmental carcinogenesis. Moreover, since many of these species appear amenable to laboratory culture and carcinogen exposure, meaningful experiments to corroborate field observations can be implemented.

IX. ACKNOWLEDGMENTS

We thank Dr. Keith Cheng and Dr. Maureen Krause for their critical reading of an earlier version of the manuscript. Studies of the northern pike lymphoma were supported by a special fellowship from the Leukemia Society of America to R.J. Van Beneden. Studies of transforming genes in chemically exposed medaka were supported by grants from the Environmental Protection Agency (R816277-01-0) and the U.S. Army Biomedical Research and Development Command (DAMD17091-C-1079) to R.J. Van Beneden. Studies of retinoblastoma gene expression were supported by grants from the National Institutes of Health (CA54590, CA58818, and RR07077) and the U.S. Army Biomedical Research and Development Command (DAMD17-93-J-3011) to G.K. Ostrander.

REFERENCES

1. Rous, P., A transmissible avian neoplasm. (Sarcoma of the common fowl), *Exp. Med.*, 12, 696, 1910.
2. Stehelin, D., Varmus, H.E., Bishop, J.M., and Vogt, P.K., DNA related to the transforming gene(s) of avian sarcoma viruses is present in normal avian DNA, *Nature (London)*, 260, 170, 1976.
3. Knudson, A.G., Jr., Mutation and cancer: statistical study of retinoblastoma, *Proc. Natl. Acad. Sci. U.S.A.*, 68, 820, 1971.

4. Stanbridge, E.J., Der, C.J., Doersen, C.J., Nishmi, R.Y., Peehl, D., Weissman, B.E., and Wilkinson, J.E., Human cell hybrids: analysis of transformation and tumorigenicity, *Science,* 215, 252, 1982.
5. Goldberg, Y.P., Parker, M.I., and Gevers, W., The genetic basis of cancer, *SAMJ,* 80, 99, 1990.
6. Rygaard, K., Sorenson, G.D., Pettengill, O.S., and Cate, C.C., An analysis of abnormalities of the retinoblastoma gene in human ovarian and endometrial carcinoma, *Cancer,* 66, 2150, 1970.
7. Ishikawa, J., Xu, H.-J., Hu, S.-X., Yandell, D.W., Maeda, S., Kamidoni, S., Benedict, W.F., and Takahashi, R., Inactivation of the retinoblastoma gene in human bladder and renal cell carcinomas, *Cancer Res.,* 51, 5736, 1991
8. Ong, G., Siora, K., and Gullick, W.J., Inactivation of the retinoblastoma gene does not lead to loss of TGF-β receptors or response to TGF-β in breast cancer cell lines, *Oncogene,* 6, 761, 1991.
9. Towatari, M., Adachi, K., Kato, H., and Saito, H., Absence of the human retinoblastoma gene product in the megakaryoblastic crisis of chronic myelogenous leukemia, *Blood,* 78, 2178, 1991.
10. Benedict, W.F., Xu, H.-J., Hu, S.-X., and Takahashi, R., Role of the retinoblastoma gene in the initiation and progression of human cancer, *J. Clin. Invest.,* 85, 988, 1990.
11. Gope, R., Christesen, M.A., Thorson, A., Lynch, H.T., Smyrk, T., Hodgson, C., Wildrick, D.M., Gope, M.L., and Boman, B.M., Increased expression of the retinoblastoma gene in human colorectal carcinomas relative to normal colonic mucosa, *J. Natl. Cancer Inst.,* 82, 310, 1990.
12. Sasano, H., Comerford, J., Silverberg, S.G., and Carrett, C.T., An analysis of abnormalities of the retinoblastoma gene in human ovarian and endometrial carcinoma, *Cancer,* 66, 2150, 1990.
13. Chen, Y.-C., Chen, P.-J., Yeh, S.-H., Tien, H.-F., Wang, C.-H., Tang, J.-L., and Hong, R.-L., Deletion of the human retinoblastoma gene in primary leukemias, *Blood,* 76, 2060, 1990.
14. Hastie, N.D., Bickmore, W.A., Maule, J.C., Porteous, D.J., and van Heyningen, V., Wilm's tumors locus at 11p13, in *Current Communications in Molecular Biology: Recessive Oncogenes and Tumor Suppression,* Cavenee, W., Hastie, N., and Stanbridge, E.J., Eds. Cold Spring Harbor Laboratory Press, Cold Spring Harbor, NY, 1989, 39.
15. Vogelstein, B., A deadly inheritance, *Nature (London),* 348, 681, 1990.
16. Steeg, P.S., Bevilacqua, G., Kapper, L., Thorgersson, V.P., Talmadge, J.E., Liotta, L.A., and Sobel, M.E., Evidence for a novel gene associated with low tumor metastatic potential, *J. Natl. Cancer Inst.,* 80, 200, 1988.
17. Wallace, M.R., Maarchuk, D.A., Andersen, L.B., Letcher, R., Odeh, H.M., Saulino, A.M., Fountain, J.W., Brereton, A., Nicholson, J., Mitchell, A.L., Brownstein, B.H., and Collins, F.S., Type 1 neurofibromatosis gene: identification of a large transcript disrupted in three NF1 patients, *Science,* 249, 181, 1990.
18. Cawthron, R.M., Weiss, R., Xu, G.F., Viskochil, D., Culver, M., Stevens, J., Robertson, M., Dunn, D., Gesteland, R., O'Connell, P., and White, R., A major segment of the neurofibromatosis type 1 gene: cDNA sequence, genomic structure, and point mutations, *Cell,* 62, 193, 1990.

19. Xu, G.F., O'Connell, P., Viskochil, D., Cawthon, R., Robertson, M., Culver, M., Dunn, D., Stevens, J., Gesteland, R., White, R., and Weiss, R., The neurofibromatosis type 1 gene encodes a protein related to GAP, *Cell*, 62, 599, 1990.
20. Fearon, E.R, Cho, K.R., Nigro, J.M., Kern, S.E., Simmons, J.W., Rupert, J.M., Hamilton, S.R., Preisinger, A.C., Thomas, G., Kinzler, K.W., and Vogelstein, B., Identification of a chromosome 18q that is altered in colorectal cancers, *Science*, 247, 49, 1990.
21. Bodmer, W., Bailey, C., Bodmer, J., Bussey, H., Ellis, A., Gorman, P., Lucibello, R., Murday, V., Rider, S., Scambler, P., Sheer, D., Solomon, E., and Spurr, N., Localization of the gene for familial adenomatous polyposis on chromosome 5, *Nature (London)*, 328, 614, 1987.
22. Leppert, M., Dobbs, M., Scambler, P., O'Connell, P., Nakamura, Y., Stauffer, D., Woodward, S., Burt, R., Hughes, J., Gardner, E., Lathrop, M., Wasmuth, J., Lalouel, J.-M., and White, R., The gene for familial polyposis coli maps to the long arm of chromosome 5, *Science*, 238, 1411, 1987.
23. Kinzler, K.W., Nilber, M.C., Vogelstein, B., Bryan, T.M., Levy, D.B., Smith, K.J., Preisinger, A.C., Hamilton, S.R., Hedge P., Markham, A., Carlson, M., Joslyn, G., Groden, J., White, R., Miki, Y., Miyoshi, Y., Nishisho, I., and Nakamura, Y., Identification of a gene located at chromosome 5q21 that is mutated in colorectal cancers, *Science*, 251, 1366, 19, 1991.
24. Anders, F., The Mildred Scheel 1988 Memorial Lecture: a biologist's view of human cancer, in *Modern Trends in Human Leukemia VIII*, Neth, R., Gallo, C.R., Greaves, M.F., Gaedicke, G., and Ritter, J., Eds., Springer-Verlag, Berlin, 1989, XXIV.
25. Harshbarger, J.C., Charles, A.M., and Spero, P.M., Collection and analysis of neoplasms in sub-homeothermic animals from a phyletic point of view, in *Phyletic Approaches to Cancer*, Dawe, C.J., Harshbarger, J.C., Kondo, S., Sugimura, T., and Takayama, S., Eds., Japan Scientific Society Press, Tokyo, 1981, 357.
26. Smith, C.E., Peck, T.H., Klauda, R.H., and McLaren, J.B., Hepatomas in Atlantic Tomcod *(Microgadus tomcod)* collected in the Hudson River Estuary, N.Y., *J. Fish Dis.*, 2, 313, 1979.
27. Baumann, P.C., Smith, W.D., and Ribick, M., Hepatic tumor levels and polynuclear aromatic hydrocarbon levels in two populations of brown bullheads *(Ictalurus nebulosus)*, in *Polynuclear Aromatic Hydrocarbons: Physical and Biological Chemistry*, Cooke, M., Dennis, A.J., and Fisher, G.L., Eds., Battelle Press, Columbus, OH, 1982, 93.
28. Black, J.J., Field and laboratory studies of environmental carcinogenesis in Niagara River fish, *J. Great Lakes Res.*, 9, 326, 1983.
29. Murchelano, R.A. and Wolke, R.E., Epizootic carcinoma in the winter flounder, *Pseudopleuronectes americanus*, *Science*, 220, 587, 1985.
30. Krahn, M.M., Rhodes, L.D., Myers, M.S., Moore, L.K., MacLeod, W.D., Jr., and Malins, D.C., Associations between metabolites of aromatic compounds in bile and occurrence of hepatic lesions in the English sole *(Parophyrs vetulus)* from Puget Sound, Washington, *Arch. Environ. Contam. Toxicol.*, 15, 61, 1986.
31. Hoover, K., Use of small fish species in carcinogenicity testing, *Natl. Cancer Inst. Monogr.*, 65, 1984.
32. Nemoto, N., Kodama, K., Tazawa, A., Prince Masahito, and Ishikawa, T., Extensive sequence homology of the goldfish *ras* gene to mammalian *ras* genes, *Differentiation*, 32, 17, 1986.

33. Van Beneden, R.J., Watson, D.K., Chen, T.T., Lautenberger, J.A., and Papas, T.S., Cellular *myc* (c-*myc*) in fish (rainbow trout): its relationship to other vertebrate *myc* genes and to the transforming genes of the MC29 family of viruses, *Proc. Natl. Acad. Sci. U.S.A.,* 83, 3698, 1986.
34. Prince Masahito, Ishikawa, T., and Sugano, H., Fish tumors and their importance in cancer research, *Jpn. J. Cancer Res. (Gann),* 79, 545, 1988.
35. Comoglio, P.M., DiRenzo, M.F., Gaudino, G., Ponzetto, C., and Prat, M., Tyrosine kinase and control of cell proliferation, *Am. Rev. Respir. Dis.,* 142, S16, 1990.
36. Stoler, A.B., Genes and cancer, *Br. Med. Bull.,* 47, 64, 1991.
37. Downward, J., Yarden, Y., Mayes, E., Scrace, G., Totty, N., Stockwell, P., Ullrich, A., Schlessinger, J., and Waterfield, M.D., Close similarity of epidermal growth factor receptor and v-*erb-B* oncogene protein sequences, *Nature (London),* 307, 521, 1984.
38. Aaronson, S.A., Growth factors and cancer, *Science,* 254, 1146, 1153, 1991.
39. Di Fiore, P.P., Pierce, J.H., Kraus, M.H., Segatto, O., King, C.R., and Aaronson, S.A., *erb*B-2 is a potent oncogene when overexpressed in NIH/3T3 cells, *Science,* 237, 178, 1987.
40. Slamon, D.J., Godolphin, W., Jones, L.A., Holt, J.A., Wong, S.G., Keth, D.E., Levin, W.J., Stuart, S.G., Udove, J., Ullrich, A., and Press, M.F., Studies of the HER-2/*neu* proto-oncogene in human breast and ovarian cancer, *Science,* 244, 707, 1989.
41. Semba, K., Kamata, N., Toyoshima, K., and Yamamoto, T., A v-*erb*-B related proto-oncogene, c-*erb*-B-2 is distinct from C-B-1/epidermal growth factor-receptor gene and is amplified in a human salivary gland adenocarcinoma, *Proc. Natl. Acad. Sci. U.S.A.,* 82, 6497, 1985.
42. Gordon, M., Pigment inheritance in the Mexican killifish. Interaction of factors in *Platypoecilus maculatus, J. Hered.,* 19, 551, 1928.
43. Haussler, G., Uber melanombildungen bei bastarden von *Xiphophorus helleri* and *Platypoecilus maculatus* var *rubra, Klin. Wochenschr.,* 7, 1561, 1928.
44. Kosswig, C., Uber kreuzungen zwischen den Teleostiern *Xiphophorus helleri* und *Platypoecilus maculatus, Z. Indukt. Abstamm. Vererbungsl.,* 47, 150, 1928.
45. Zechel, C., Schleenbecker, U., Anders, A., Pfütz, M., and Anders, F., Search for genes critical for the early and/or late events in carcinogenesis: studies in *Xiphophorus* (Pisces, Teleostei), in *Modern Trends in Human Leukemia VIII,* Neth, R., Gallo, C.R., Greaves, M.F., Gaedicke, G., and Ritter, J., Eds., Springer-Verlag, Berlin, 1989, 366.
46. Ozato, K. and Wakamatsu, Y., Multi-step regulation of oncogene expression in fish hereditary melanoma, *Differentiation,* 24, 181, 1983.
47. Anders, F., Schartl, M., Barnekow, A., and Anders, A., *Xiphophorus* as an *in vivo* model for studies on normal and defective control of oncogenes, *Adv. Cancer Res.,* 42, 191, 1984.
48. Anders, F., Schartl, M., Barnekow, A., Schmidt, C.R., Lüke, Jaenel-Dess, G.W., and Anders, A., The genes that carcinogens act upon, in *Haematology and Blood Transfusion 29: Modern Trends in Human Leukemia VI,* Neth, R., Gallo, C.R., Greaves, M.F., and Janka, Eds., Springer-Verlag, Berlin, 1985, 228.
49. Schartl, M., Homology of melanoma-inducing loci in the genus *Xiphophorus, Genetics,* 126, 1083, 1990.

50. Schwab, M., Genetic principles of tumor suppression, *Biochim. Biophys. Acta,* 989, 49, 1989.
51. Vielkind, J.R., Kallman, K.D., and Morizot, D.C., Genetics of melanomas in *Xiphophorus* fishes, *J. Aquat. Anim. Health,* 1, 69, 1989.
52. Anders, F., Tumor formation in platyfish-swordtail hybrids as a problem of gene regulation, *Experentia,* 23, 1, 1967.
53. Anders, A., Anders, F., and Klinke, K., Regulation of gene expression in the Gordon-Kosswig Melanoma System. I. The distribution of the controlling genes in the genome of the xiphophorin fish, *Platypoecilus maculatus* and the *Platypoecilus variatus,* in *Mutagenesis of Fish,* Schroder, H.J., Ed., Springer-Verlag, Berlin, 1973, 33.
54. Schartl, M., Schmidt, C., Anders, A., and Barnekow, A., Elevated expression of the cellular *src* gene in tumors of differing etiologies in *Xiphophorus, Int. J. Cancer,* 36, 199, 1985.
55. Anders, F., Gronau, T., Schartl, M., Barnekow, A., Jaenel-Dess, G., and Anders, A., Cellular oncogenes as ubiquitous constituents in the animal kingdom and as fundamentals in melanoma formation, in *Cutaneous Melanoma,* Veronesi, U., Cascinelli, N., and Santinami, N., Eds., Academic Press, New York, 1987, 351.
56. Maueler, W., Raulf, F., and Schartl, M., Expression of proto-oncogenes in embryonic, adult, and transformed tissue of *Xiphophorus* (Teleostei: Poeciliidae), *Oncogene,* 2, 421, 1988.
57. Zechel, C., Schleenbecker, U., Anders, A., and Anders, F., v-*erb*B related sequences in *Xiphophorus* that map to melanoma determining Mendelian loci and overexpress in a melanoma cell line, *Oncogene,* 3, 605, 1988.
58. Adam, D., Wittbrodt, J., Telling, A., and Schartl, M., RFLP for an EGF-receptor related gene associated with the melanoma oncogene locus of *Xiphophorus maculatus, Nucl. Acids Res.,* 16, 7212, 1988.
59. Wittbrodt, J., Adam, D., Malitschek, B., Mäueler, W., Raulf, F., Telling, A., Robertson, S.M., and Schartl, M., Novel putative receptor tyrosine kinase encoded by the melanoma-inducing *Tu* locus in *Xiphophorus, Nature (London),* 341, 415, 1989.
60. Zechel, C., Peter, H., Schleenbecker, U., Anders, A., and Anders, F., *Erb-B*[*a]: an "ignition spark" for the *Xiphophorus* melanoma machinery?, *Haematology and Blood Transfusion, Vol. 35, Modern Trends in Human Leukemia VIII,* Neth, R., Gallo, C.R., Greaves, M.F., Gaedicke, G., and Ritter, J., Eds, Springer-Verlag, Berlin, 1992, 213.
61. Albino, A.P., Le Strange, R., Oliff, A.I., Furth, M.E., and Old, L.J., Transforming *ras* genes from human melanoma: a manifestation of tumor heterogeneity?, *Nature (London),* 308, 69, 1984.
62. Padua, R.A., Barass, N.C., and Currie, G.A., A novel transforming gene in a human malignant cell line, *Nature (London),* 311, 671, 1984.
63. Fountain, J.W., Bale, S.J., Housman, D.E., and Dracopoli, N.C., Genetics of melanoma, *Cancer Surv.,* 9, 645, 1990.
64. Koprowski, H., Herlyn, M., Balaban, G., Parmiter, A., Ross, A., and Nowell, P., Expression of the receptor for epidermal growth factor correlates with increased dosage of chromosome 7 in malignant melanoma, *Somatic Cell Mol. Genet.,* 11, 297, 1985.

65. Montesano, R., Barth, H., Vainio, H., Wilborn, J., and Yamasaki, H., Long-term and short-term assays for carcinogenesis, in *IARC Scientific Publication No. 83*, International Agency for Research on Cancer, (IARC), Lyon, France, 1986, 103.
66. Mäuler, W., Barnekow, A., Eigenbrodt, E., Raulf, R., Falk, H.F., Telling, A., and Shartl, M., Different regulation of oncogene expression in tumor and embryonal cells of *Xiphophorus, Oncogene,* 3, 113, 1988.
67. Cooper, G.M., *Oncogenes,* Jones and Bartlett, Boston, MA, 1990.
68. Schartl, M., Barnekow, A., Bauer, H., and Anders, F., Correlations of inheritance and expression between a tumor gene and the cellular homolog of the Rous sarcoma virus-transforming gene in *Xiphophorus, Cancer Res.,* 42, 4222, 1982.
69. Barnekow, A., Schartl, M., Anders, F., and Bauer, H., Identification of a fish protein associated with a kinase activity and related to the Rous sarcoma virus transforming protein, *Cancer Res.,* 42, 2429, 1982.
70. Schartl, M. and Barnekow, A., The expression in eukaryotes of a tyrosine kinase which is reactive with pp60$^{v\text{-}src}$ antibodies, *Differentiation,* 23, 109, 1982.
71. Collett, M.S. and Erikson, R.L., Protein kinase activity associated with the avian sarcoma virus *src* gene product, *Proc. Natl. Acad. Sci. U.S.A.,* 75, 2021, 1978.
72. Barnekow, A. and Schartl, M., Comparative studies on the *src* proto-oncogene and its gene product pp60$^{c\text{-}src}$ in normal and neoplastic tissues of lower vertebrates, *Comp. Biochem. Physiol.,* 87B, 663, 1987.
73. Martinez, R., Mathey-Prevot, B., Bernards, A., and Baltimore, D., Neuronal pp60$^{c\text{-}src}$ contains a six-amino acid insert relative to its non-neuronal counterpart, *Science,* 237, 411, 1987.
74. Yang, X., Martinez, R., Le Beau, J., Wiestler, O., and Walter, G., Evolutionary expression of the neuronal form of the *src* protein in the brain, *Proc. Natl. Acad. Sci. U.S.A.,* 86, 4751, 1989.
75. Raulf, F., Robertson, S.M., and Schartl, M., Evolution of the neuron-specific alternative splicing product of the c-*src* proto-oncogene, *J. Neurosci. Res.,* 24, 81, 1989.
76. Chou, S.-C., Yang, C., Kimler, V.A., Taylor, J.D., and Tchen, T.T., Reversible dedifferentiation and redifferentiation of a melanized cell line from a goldfish tumor, *Cell Differentiation Dev.,* 28, 105, 1989.
77. Matsumoto, J., Ishikawa, T., Prince Masahito, and Takayama, S., Permanent cell lines from erthrophoromas in goldfish *(Carassius auratus), J. Natl. Cancer Inst.,* 64, 875, 1980.
78. Hannig, G., Ottilie, S., and Schartl, M., Conservation of structure and expression of the c-*yes* and *fyn* genes in lower vertebrates, *Oncogene,* 6, 361, 1991.
79. Semba, K., Yamanashi, Y., Nishizawa, M., Sukegawa, J., Yoshida, M., Sasaki, M., Yamamoto, T., and Toyoshima, K., *yes*-related protooncogene, *fyn,* belongs to the protein-tyrosine kinase family, *Proc. Natl. Acad. Sci. U.S.A.,* 83, 5459, 1986.
80. Cooper, G.M., Oncogenes as markers for early detection of cancer, *J. Cell. Biochem.,* 50(Suppl. 16G), 131, 1992.
81. Wirgin, I.I., D'Amore, M., Grunwald, C., Goldman, A., and Garte, S.J., Genetic diversity at an oncogene locus and in mitochondrial DNA between populations of cancer-prone Atlantic tomcod, *Biochem. Genet.,* 28, 459, 1990.
82. Molven, A., Njolstad, P.R., and Fjose, A., Genomic structure and restricted neural expression of the zebrafish *wnt-1 (int-1)* gene, *EMBO J.,* 10, 799, 1991.

83. Ascione, R., Sacchi, N., Watson, D.K., Fisher, R.J., Fujiwara, S., Seth, A., and Papas, T.S., Oncogenes: molecular probes for clinical application in malignant diseases, *Gene Anal. Tech.*, 3, 25, 1986.
84. Cole, M.D., The *myc* oncogene: its role in transformation and differentiation, *Annu. Rev. Genet.*, 20, 361, 1986.
85. Ayer, D. and Eisenman, R.N., *Mad:* a dimerization partner for *max*, Abstract, Eighth Annual Meeting of Oncogenes, 1992, 1.
86. Van Beneden, R.J., Watson, D.K., Chen, T.T., Lautenberger, J.A., and Papas, T.S., Teleost oncogenes: evolutionary comparison to other vertebrate oncogenes and possible roles in teleost neoplasms, *Mar. Environ. Res.*, 24, 339, 1988.
87. Weinberger, C., Thompson, C.C., Ong, E.S., Lebo, R., Gruol, D.J., and Evans, R.M., The c-*erb-A* gene encodes a thyroid hormone receptor, *Nature (London)*, 324, 641, 1986.
88. Barbacid, M., *ras* genes, *Annu. Rev. Biochem.*, 56, 779, 1987.
89. Downward, J., The *ras* superfamily of small GTP-binding proteins, *TIBS*, 15, 469, 1990.
90. De Feo-Jones, D., Scolnick, E.M., Koller, R., and Dhar, R., *ras*-related gene sequences identified and isolated from *Saccharomyces cerevisiae*, *Nature (London)*, 306, 707, 1983.
91. Toda, T., Uno, I., Ishikawa, T., Powers, S., Katoaoka, T., Broek, D., Cameron, S., Broach, J., Matsumoto, K., and Wigler, M., In yeast, *RAS* proteins are controlling elements of adenylate cyclase, *Cell*, 40, 27, 1985.
92. McCormick, F., *ras* GTPase activating protein: signal transmitter and signal terminator, *Cell*, 56, 5, 1989.
93. Nemoto, N., Kodama, K., Tazawa, A., Matsumoto, J., Prince Masahito, and Ishikawa, T., Nucleotide sequence comparison of the predicted first exonic region of goldfish *ras* gene between normal and neoplastic tissues, *J. Cancer Res. Clin. Oncol.*, 113, 56, 1987.
94. Masamoto, J., Ishikawa, T., Prince Masahito, and Takayama, S., Permanent cell lines from erthrophoromas in goldfish (*Carassius auratus*), *J. Natl. Cancer Inst.*, 64, 879, 1980.
95. Mangold, K., Chang, Y., Mathews, C., Marien, K., Hendricks, J., and Bailey, G., Expression of *ras* genes in rainbow trout liver, *Mol. Carcinogen.*, 4, 97, 1991.
96. Chang, Y., Mathews, C., Mangold, K., Marien, K., Hendricks, J., and Bailey, G., Analysis of *ras* gene mutations in rainbow trout liver tumors initiated by aflatoxin B1, *Mol. Carcinogen.*, 4, 112, 1991.
97. Muench, K.F., Misra, R.P., and Humayun, M.Z., Sequence specificity in aflatoxin B_1-DNA interactions, *Proc. Natl. Acad. Sci. U.S.A.*, 80, 6, 1983.
98. McMahon, G., Davis, E., and Wogan, G.N., Characterization of c-Ki-*ras* oncogene alleles by direct sequencing of enzymatically amplified DNA from carcinogen-induced tumors, *Proc. Natl. Acad. Sci. U.S.A.*, 84, 4974, 1987.
99. Sinha, S., Webber, C., and Marshall, C.J., Activation of *ras* oncogene in aflatoxin-induced rat liver carcinogenesis, *Proc. Natl. Acad. Sci. U.S.A.*, 85, 3673, 1988.
100. Gu, J.R., Hu, L.F., Cheng, Y.C., and Wan, D.F., Oncogenes in human primary hepatic cancer, *J. Cell Physiol.*, 4(Suppl.), 13, 1986.
101. Tada, M., Omata, M., and Ohto, M., Analysis of *ras* gene mutations in human hepatic malignant tumors by polymerase chain reaction and direct sequencing, *Cancer Res.*, 50, 1121, 1990.

102. McMahon, G., Huber, L.J., Moore, M.J., and Stegeman, J.J., Mutations in c-Ki-*ras* oncogenes in diseased livers of winter flounder from Boston Harbor, *Proc. Natl. Acad. Sci. U.S.A.,* 87, 841, 1990.
103. Stegeman, J.J., Skopek, T.R., and Thilly, W.G., Bioactivation of polycyclic aromatic hydrocarbons to cytotoxic and mutagenic products by marine fish, in Carcinogenic Polynuclear Aromatic Hydrocarbons in the Marine Environment, Richards, N., Ed., EPA-600/9-82-013, Environmental Protection Agency, Washington, D.C., 1982, 201.
104. Eisenstadt, E., Warren, A.J., Porter, J., Atkins, D., and Miller, J.H., Carcinogenic epoxides of benzo[*a*]pyrene and cyclopenta[*cd*]pyrene induce base substitutions via specific transversions, *Proc. Natl. Acad. Sci. U.S.A.,* 79, 1945, 1982.
105. Yang. J.-L., Maher, V.M., and McCormick, J.J., Kinds of mutations formed when a shuttle vector containing adducts of (±)-7β,8α-dihydroxy-9α,10α-epoxy-7,8,9,10-tetrahydrobenzo[*a*]pyrene replicates in human cells, *Proc. Natl. Acad. Sci. U.S.A.,* 84, 3787, 1987.
106. Dey, W., Peck, T, Smith, C., Cormier, S., and Kraemer, G.L., A Study of the Occurrence of Liver Cancer in Atlantic Tomcod *(Microgadus tomcod),* Final Report to the Hudson River Foundation, New York, 1986.
107. Wirgin, I., Currie, D., and Garte, S.J., Activation of the K-*ras* oncogene in liver tumors of Hudson River tomcod, *Carcinogenesis,* 10, 2311, 1989.
108. Den Otter, et al., Oncogene mutations in anti-oncogenes: a view, *Anticancer Res.,* 10, 475, 1990.
109. Weinberg, R.A., Finding the anti-oncogene, *Sci. Am.,* 259, 44, 1988.
110. Whyte, P., Buchkovitch, J.J., Horowitz, J.M., Friend, S.H., Raybuck, M., Weinberg, R.A., and Harlow, E., Association between an oncogene and anti-oncogene: the adenovirus E1A proteins bind to the retinoblastoma gene product, *Nature (London),* 334, 124, 1988.
111. DeCaprio, J.A., Ludlow, J.W., Figge, J., Shew, J.-Y., Huang, C.-M., Lee, W.-H., Marsilio, E., Paucha, E., and Livingston, D.M., SV40 large tumor antigen forms a specific complex with the product of the retinoblastoma susceptibility gene, *Cell,* 54, 275, 1988.
112. Windell, J.J., Albert, D.M., O'Brian, J.M., Marcus, D.M., Disteche, C.M., Bernards, R., and Mellon, P.L., Retinoblastoma in transgenic mice, *Nature (London),* 33, 665, 1990.
113. Ostrander, G.K., Shim, J.-K., Hawkins, W.E., and Walker, W.W., A vertebrate model for investigation of retinoblastoma, *Proc. 83rd Annu. Meet. Am. Assoc. Cancer Res.,* 33, 109, 1992.
114. Bernards, R., Schackleford, G.M., Gerber, M.R., Horowits, J.M., Friend, S.H., Schartl, M., Bogermann, E., Raport, J.M., McGee, T., Dryja, T.P., and Weinberg, R.A., Structure and expression of the murine retinoblastoma gene and characterization of its encoded protein, *Proc. Natl. Acad. Sci. U.S.A.,* 86, 6474, 1989.
115. Destree, O.H.J., Lam, K.T., Peterson-Maduro, L.J., Eizema, K., Diller, L., Gryka, M.A., Frebourg, T., Shibuya, E., and Friend, S.H., Structure and expression of the *Xenopus* retinoblastoma gene, *Dev. Biol.,* 153, 141, 1992.
116. Lane, O.P. and Crawford, L.V., T antigen is bound to a host protein in SV40-transformed cells, *Nature (London),* 278, 261, 1979.
117. Harris, A.L., Telling changes of base, *Nature (London),* 350, 377, 1991.

118. Soussi, T., Caron de Fromentel, C., and May, P., Structural aspects of the p53 protein in relation to gene evolution, *Oncogene*, 5, 945, 1990.
119. Smith, C.A.D, Louis, M.J., and Hetrick, F.M., A sequence homologous to the mammalian p53 oncogene in fish cell lines, *J. Fish Dis.*, 11, 525, 1988.
120. Van Beneden, R.J., Henderson, K.W., Blair, D.G., Papas, T.S., and Gardner, H.S., Oncogenes in hematopoietic and hepatic fish neoplasms, *Cancer Res.*, 50(Suppl.), 5671s, 1990.
121. Van Beneden, R.J., Henderson, K.W., Gardner, H.S., Blair, D.G., and Papas, T.S., New models for oncogene isolation in the study of carcinogenesis, Proceedings of Non-Mammalian Toxicity Assessment Research Rev., U.S. Army Biomedical Research and Development Laboratory, 1993, in press.
122. Ishikawa, T., Shimamine, T., and Takayama, S., Histologic and electron microscopy observations on dimethylnitrosamine-induced hepatomas in the small aquarium fish *(Oryzias latipes)*, *J. Natl. Cancer Inst.*, 55, 909, 1975.
123. Aoki, K. and Matsudaira, H., Induction of hepatic tumors in a teleost *(Oryzias latipes)* after treatment with methylazoxymethanol acetate, *J. Natl. Cancer Inst.*, 59, 1747, 1977.
124. Hawkins, W.E., Overstreet, R.M., and Walker, W.W., Carcinogenicity tests with small fish species, *Aquat. Toxicol.*, 11, 113, 1988.
125. Brittelli, M.R., Chen, H.H.C., and Muska, C.F., Introduction of bronchial (gill) neoplasms in the medaka fish *(Oryzias latipes)* by N-methyl-N'-nitro-N-nitrosoguanidine, *Cancer Res.*, 45, 3209, 1985.
126. Hyodo-Taguchi, Y. and Matsudaira, H., Induction of transplantable melanoma by treatment with N-methyl-N'-nitrosoguanidine in an inbred strain of the teleost *Oryzias latipes*, *J. Natl. Cancer Inst.*, 73, 1219, 1984.
127. Blair, D.G., Cooper, C.S., Oskarsson, M.K., Eader, L.A., and Van de Woude, G.F., New method for detecting cellular transforming genes, *Science*, 218, 1122, 1982.
128. Jacobs, C. and Rubsamen, H., The expression of the $pp60^{c\text{-}src}$ protein kinase in adult and fetal human tissue: high activities in some sarcomas and mammary carcinomas, *Cancer Res.*, 43, 1696, 1983.
129. Barnekow, A. and Bauer, H., The differential expression of the cellular *src*-gene product $pp^{c\text{-}src}$ and its phosphokinase activity in normal chicken cells and tissues, *Biochim. Biophys. Acta*, 782, 94, 1984.
130. Van Beneden, R.J., Oncogenes, in *Biochemistry and Molecular Biology of Fishes*, Hochachka, P.W. and Mommsen, T.P., Eds, Elsevier Press, Amsterdam, in press.

Chapter 8

Pathobiology of Chemical-Associated Neoplasia in Fish

Michael J. Moore and Mark S. Myers

"In the aggregate, neoplasms, like species, can be grouped according to major common characteristics, but as individuals none is exactly like any other."[1]

ABSTRACT

The study of neoplasia in aquatic animals has largely focused on morphology and epizootiology. Correlations have been sought between prevalences of specific lesion types and significant biological and environmental variables as potential etiological factors. These factors have included age, gender, and exposure to chemical contaminants and infectious agents, as well as predisposing genetic characteristics. Experiments have shown a number of chemicals to be hepatocarcinogens in fish. These include aflatoxin in rainbow trout and nitroso-compounds, azo compounds, methylazoxymethanol acetate, polynuclear aromatic hydrocarbons, and aromatic amines in trout, medaka, guppy, zebra danio, mangrove rivulus, sheepshead minnow, and topminnow. Furthermore, the carcinogenicity of extracts derived from chemically contaminated sediments has been demonstrated in the rainbow trout and the brown bullhead, and putatively preneoplastic focal hepatic lesions have been induced in English sole by exposure to similar sediment extracts. However, we know less of the significance of polychlorinated biphenyls (PCBs), chlorinated pesticides, and other nongenotoxic agents as promoters of the hepatocarcinogenic process in laboratory and feral species of fish. In this context, we discuss the mechanisms of chemical carcinogenesis in fish species exposed to environmentally important chemicals and the application of current and more sophisticated methods to elucidate these mechanisms.

I. INTRODUCTION

The biology of chemical-associated fish tumors is based on morphological descriptions of feral and experimental fish tumors and comparisons with tumors in other fish and mammalian species. Fish tumor biology is relevant to the study of tumors in mammals because fish anatomy, physiology, toxicology, and pathology often parallel those of mammals but at less complex levels. Understanding the histogenesis of fish tumors is necessary to develop mechanistic models of cancer biology in all vertebrates. Knowledge of fish tumor biology should also be central to the development and application of economic and predictive screening tests for potential carcinogens and promoters of carcinogenesis in higher vertebrates, including man. Furthermore, morphological markers of sublethal tumorigenic effects may be predictive of more cryptic biological changes. Such changes include reproductive impairment, teratogenicity, behavioral change, reduced competitiveness, and the resultant demise of specific populations. Appreciation of the utility of tumors and tumor-associated lesions as biomarkers of effects of chemical exposure in fish also permits better interpretation of efforts to monitor the status and trends of environmental damage in aquatic habitats.

This review will concentrate on the biology and histopathology of hepatic neoplasms and related lesions in experimental and feral fish species, the role of chemical contaminants in the induction of hepatic neoplasia, and the application of current and sophisticated methods to study the mechanisms of induction and the histogenesis of hepatic neoplasia in fish. The relationship between these hepatic lesions in feral fish and chemical exposure in the environment has received the greatest amount of attention and, as such, shows the strongest linkage between cause and effect. In contrast, while there have been numerous reports of skin neoplasms and related lesions in many fish species, including the development of epidermal papillomas in caged black bullheads,[2] the evidence for a strict chemical etiology for skin neoplasms in fish is less complete and convincing.[3-6]

II. TUMORS IN FISH: ASSOCIATIONS WITH EXPOSURE TO CHEMICAL CONTAMINANTS

A. An Historical Overview

Major reviews of liver neoplasia in native fish have been published in the past three decades.[6-10] A number of symposium proceedings have also been published.[11-15] Wellings[7] reviewed hepatic neoplasms in salmonids and other fish and cited early descriptions of liver tumors in salmonids.[16] In the 1960s, liver tumors in rainbow trout *(Oncoryhnchus mykiss)* were shown to be enzootic throughout hatcheries in the United States; aflatoxin-contaminated feed was clearly demonstrated as the causative agent.[17] Studies of aflatoxicosis in salmonids, liver tumors in native suckers and bullheads,[18] and experimental induction using diethylnitrosamine (DEN) in zebra danios[19] were the major advances in fish liver tumor biology in the 1960s. Harshbarger and Clark[6] cite liver tumor epizootics in 13 species, giving an overview of accessions to the Registry of Tumors in Lower Animals at the Museum of Natural History, Smithsonian Institution, Washington, D.C. Native species with notable liver neoplasm epizootics from around the world include suckers *(Catostomus commersoni)* and bullheads *(Ictalurus nebulosus)*,[3,5,18,20,21] hagfish *(Myxine glutinosa)*,[22,23] tomcod *(Microgadus tomcod)*,[24,25] English sole *(Pleuronectus vetulus, formerly Parophrys vetulus)*,[26-29] winter flounder *(Pleuronectes americanus, formerly Pseudopleuronectes americanus)*,[30-35] white croaker *(Genyonemus americanus)*,[29,36] dab *(Limanda limanda)*,[37] and mummichog *(Fundulus heteroclitus)*.[38]

B. Etiology

In 1985, Mix[9] reviewed the available evidence for an association between specific tumor epizootics in fish and shellfish and exposure to anthropogenic chemicals. Mix concluded that the evidence in the cases of multiple neoplasm types in several fish species from the Fox River in Illinois[39,40] and hepatic neoplasms and related lesions in English sole from Puget Sound[41,42] was convincing; in contrast, Mix felt that many of the other studies in fish and shellfish lacked adequate control sites, proper documentation of chemical contaminant exposure, and proper statistical verification of hypothesized associations. In this context, the high cost of chemical analysis has continued to limit the value of many epizootiological studies. Subsequent to this review, epizootics of neoplasia have been classed into two broad categories,[6] with respect to their relationship to xenobiotic exposure:

1. No obvious pollution association — lesions in this category include hemic, neural pigment cell, connective tissue, and gonadal neoplasms.
2. Pollution associated — this group includes epithelial neoplasms of the liver, pancreas, and gastrointestinal tract.

The second class shows strong circumstantial links to chemical carcinogen exposure, especially for hepatic neoplasms of hepatocellular or cholangiocellular origin. Historical, experimental, and physiological bases for the conclusion that liver neoplasms in fish are probably caused by chemical contaminants, especially those in sediments, are discussed by these authors.[6] They conclude that:

1. Epithelial tumors in fish, and specifically hepatic neoplasms, are strongly correlated with exposure to chemical contaminants.
2. Bottom-dwelling species are most commonly affected.
3. Certain species may be more sensitive to chemical carcinogenesis in the environment than others.

C. Skin Tumors

The link between chemical exposure and skin tumors is much weaker than for liver tumors, although chemical contaminants, especially as co-carcinogenic agents, may contribute to some epizootics.[43] Relevant studies include descriptions of epidermal papillomas in fishes of the Great Lakes,[44] skin painting[45,46] and cage experiments in bullheads,[2] and a global survey of papillomatosis in eels and purported epidermal papillomas in multiple flatfish species.[47] These last authors concluded that there were regional differences in prevalence that were not necessarily linked to pollution. This latter finding is not surprising, considering that the preponderance of evidence currently points to an infectious, but as yet unidentified, protozoan as the etiologic agent for the skin lesions in the flatfish species they surveyed.[48-52] These lesions are characterized by the presence of the so-called "X cell," which is the infectious agent composing the bulk of these lesions. The lesions are therefore not true neoplasms and are more appropriately referred to as pseudotumors. Viral influences have also been implicated in the etiology of epidermal tumors, as well as various mesenchymal tumors in fish.[53-60]

D. Nonepithelial Tumors

Many reports of nonepidermal and nonepithelial neoplasms, such as those described by Wellings,[7] precede the major era of chemical synthesis in this century. Furthermore, many tumors of diverse, mainly nonepithelial cell types have been suggested to be of a spontaneous nature.[61-65]

The remainder of this chapter discusses the tumor types most strongly linked to chemical contaminant exposure in native fish: hepatic neoplasms.

E. Comparative Morphology of Fish Liver Neoplasms

Reports of tumor epizootics from different geographic locations, types of chemical exposures, and species of fish reveal a broad morphological diversity, as illustrated by the examples shown in Figures 1 to 5. All of the tumors shown have been selected from a single species to illustrate the plasticity of cellular phenotype within a species. The interspecies variability in neoplasm types is shown in Table 1. Cholangiocellular neoplasms predominate in bullheads, winter flounder, and suckers, whereas the majority of neoplasms are hepatocellular in mummichog and English sole. Few studies have adequately considered the

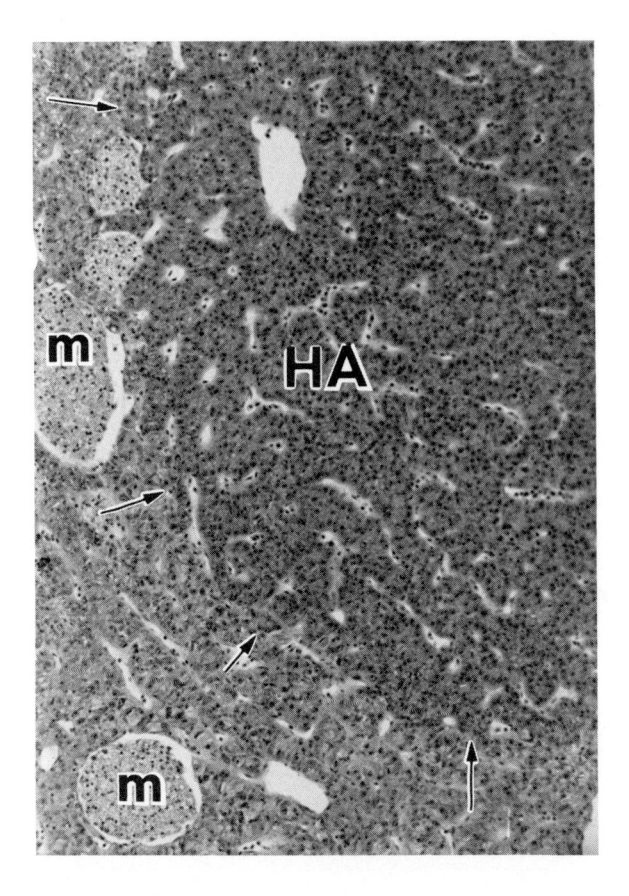

Figure 1. Hepatocellular adenoma (HA), basophilic type, in the liver of an adult English sole. The surrounding parenchyma is clearly separated from hepatocytes composing the nodule (arrows) and contains macrophage aggregates (m), structures rarely present within hepatic neoplasms or foci of cellular alteration in this species. The tubulosinusoidal architecture of the neoplasm is essentially normal. Hematoxylin and eosin stain (magnification × 156).

bases for these interspecific differences. Any or all of the following factors may be important:

1. The nature of the chemical(s) involved and the dose(s) and the duration of exposure are unique to each geographic site.
2. The life history, including spawning behavior, feeding strategy, and migratory behavior, is unique to a species and, to a degree, to each site.
3. The cellular and molecular responses of each species to chemical contaminant exposure depend on the unique genetic, biochemical, and morphological constitution of each species.

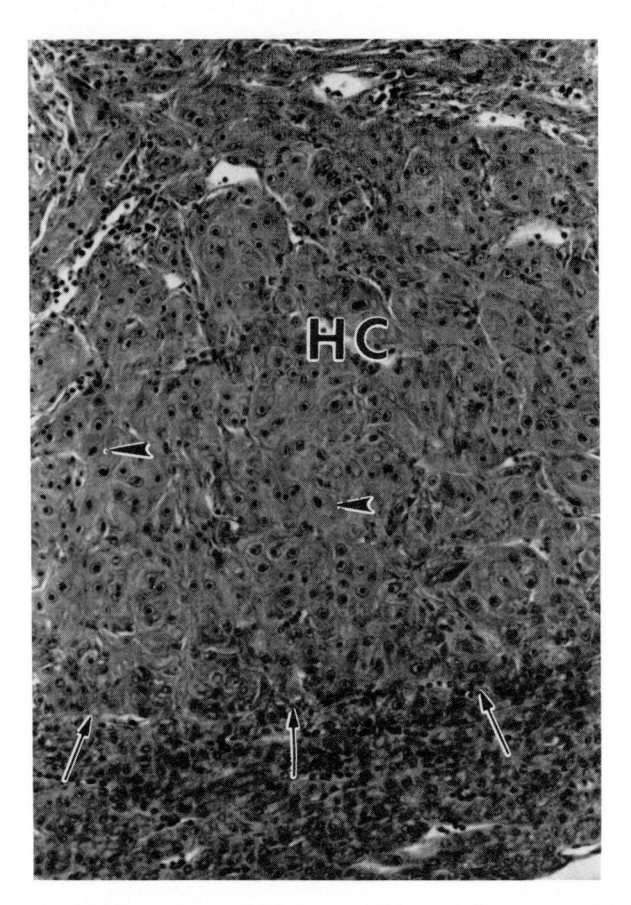

Figure 2. Hepatocellular carcinoma (HC), basophilic type, in liver of an adult English sole. The neoplasm (arrows point to border) is composed of enlarged, anaplastic, polygonal hepatocytes with enlarged, vesicular nuclei. Carcinomatous hepatocytes are arranged in disorganized tubulosinusoids, with loss of the typical polar orientation of the hepatocyte nuclei to the sinusoids. Horizontal arrowheads point to mitotic figures within the carcinoma. Compression of the sinusoids in the surrounding parenchyma is evident at the bottom of the micrograph. Hematoxylin and eosin stain (magnification ×256).

4. The relative stage in the life history at which the fish were examined varies in each field study and often among sites sampled for a species within a study.
5. Interpretation of histopathological features and subsequent morphological diagnosis is often unique to the examining histopathologist.

In this last regard, it is important to note that as research programs have progressed, some morphological descriptions have been updated. Aflatoxin-induced "hepatocellular" tumors[66] are now thought to be primarily mixed hepatocellular and cholangiocellular neoplasms.[67] Moreover, some hepatic neoplasm types induced by

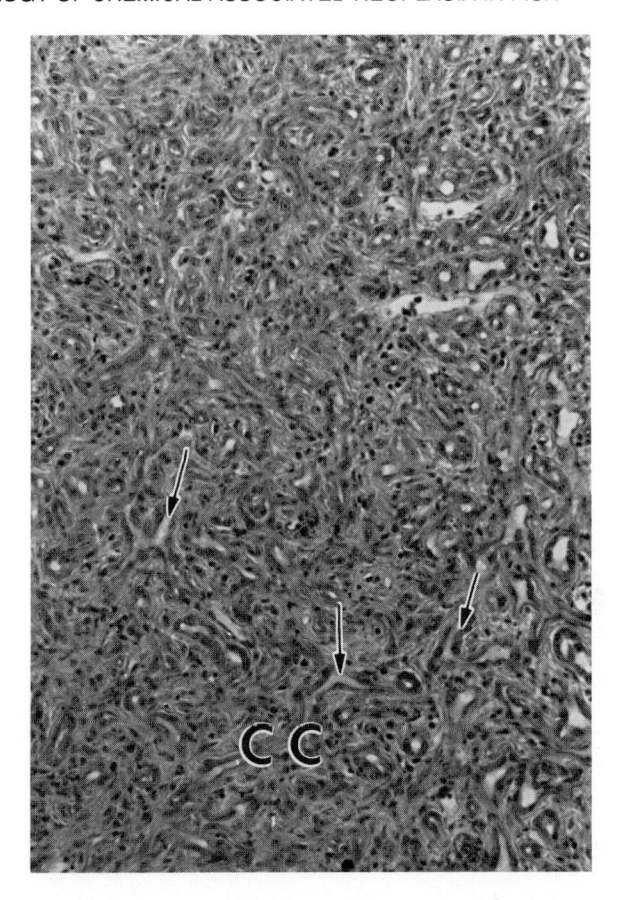

Figure 3. Cholangiocellular carcinoma (CC) in liver of an English sole. Neoplasm consists of poorly differentiated tubular structures lined by biliary epithelial cells within a slightly fibrous stroma. Arrows point to lumina of tubules. Normal hepatic parenchyma is not visible in this micrograph. Hematoxylin and eosin stain (magnification ×256).

DEN in *Poeciliopsis lucida* that were originally interpreted as hepatocellular carcinomas[68,69] are now appropriately diagnosed as cholangiocarcinomas.[70] These represent just two examples of differences in histopathological interpretation of hepatic neoplasms by individual investigators; in practice, these differences in subjective interpretation are quite common. Hopefully, these problems will diminish once a definitive atlas documenting diagnostic criteria for hepatic neoplasms and related lesions is published and made available.[71] However, considering the inherent subjectivity of histopathological diagnosis, diagnostic difficulties will certainly continue.

Fish pathologists are not alone in having difficulty standardizing diagnostic criteria. Edmondson and Craig[72] state that in the classification of human hepatic

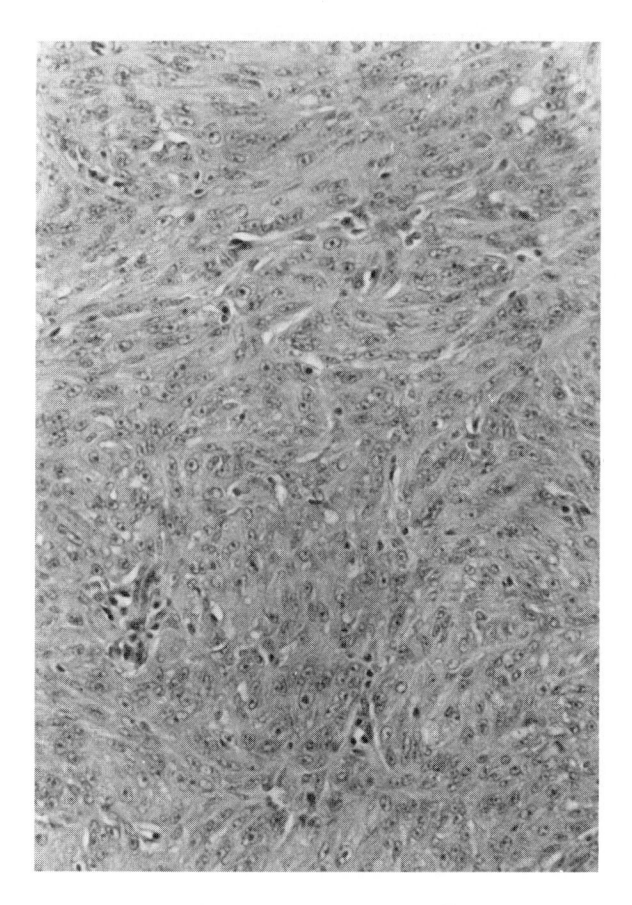

Figure 4. Probable cholangiocellular carcinoma, poorly differentiated, in liver of an English sole. Neoplasm consists of anaplastic, spindle-shaped cells, with oval nuclei arranged in disorganized, interlacing sheets. Tubular structures are not clearly evident, and this particular neoplasm was not confirmed as cholangiocytic by electron microscopy. However, neoplasms of similar histo- and cytomorphology have been ultrastructurally confirmed as cholangiocytic. Normal liver tissue is not visible. Hematoxylin and eosin stain (magnification × 312).

neoplasms sampled at autopsy, the terms cholangiocarcinoma, mixed cholangiocellular carcinoma, adenocarcinoma, and hepatocellular carcinoma with ductal transformation have all been used for the same lesion type by different institutions.

III. HISTOGENESIS OF HEPATIC NEOPLASIA

The question arises as to whether there is any fundamental difference between a "poorly differentiated hepatocellular carcinoma" and a "poorly differentiated cholangiocellular carcinoma" beyond subjective diagnostic interpretation. Are we

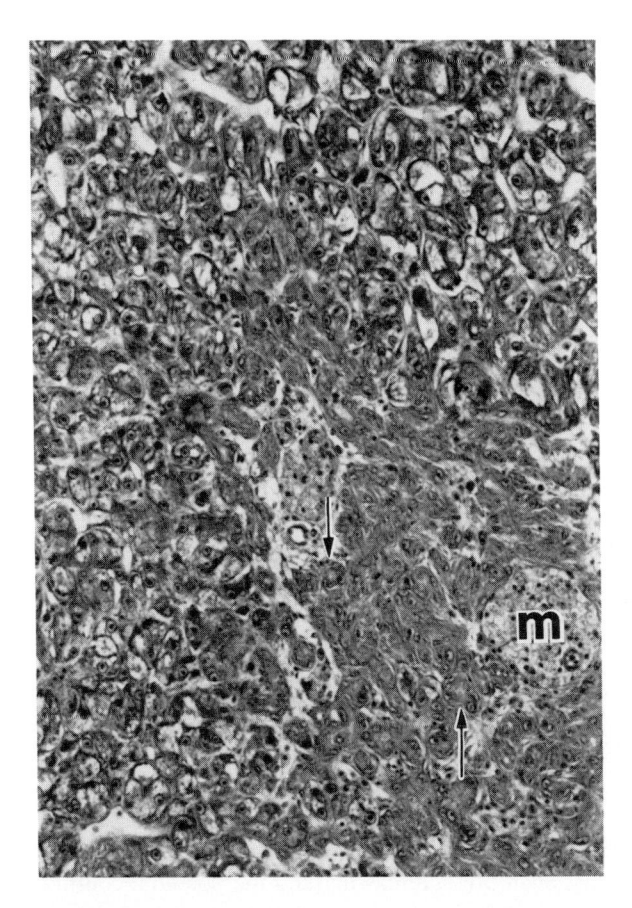

Figure 5. Mixed hepatocellular/cholangiocellular carcinoma in liver of an English sole, showing clearly separate components. Islands of larger, occasionally vacuolated, basophilic cells with large, vesicular nuclei represent the hepatocellular component, and the smaller, lighter-staining cells arranged in tubular-like structures (arrows), with smaller, oval nuclei are cholangiocytes. Macrophage aggregates (m) are also present. Hematoxylin and eosin stain (magnification ×256).

looking for two distinct oncogenetic mechanisms or just one? Is there a single epithelial stem cell population in liver, representing a common cell of origin of all hepatic epithelial neoplasms? These questions are still being actively debated in the field of mammalian hepatocarcinogenesis,[73,74] and perhaps a detailed analysis of the stepwise histogenesis of these lesions in fish species will add to our understanding of liver cancer in the broader context of comparative oncology.

There is a need for detailed experimental studies on the stepwise histogenesis of hepatic neoplasia in multiple fish species, given the differences in hepatic anatomy between teleosts and rodents.[75,76] The tubulosinusoidal system evident in fish liver[77-79] is probably a more primitive iteration of the mammalian system. The

Table 1. Relative Prevalence of Hepatocytic and Cholangiocytic Neoplasms in Fish from Contaminated Sites Where Normal and Grossly Abnormal Livers were Examined Histologically

Species	N	% HCC	% CCC	HCC/CCC	Ref.
Fundulus heteroclitus[a]	60	33	0	1	38
Parophrys vetulus[b]	151	15	5	.75	26
Ictalurus nebulosus[c]	456	17	27	.39	21
Pleuronectes americanus[d]	200	3	7	.30	30
Catostomus commersoni[e]	456	2[f]	11[f]	.15	5
P. americanus	236	1	9	.10	33
I. nebulosus	170	1[f]	10[f]	.09	5

Note: HCC = hepatocellular carcinoma; CCC = cholangiocellular carcinoma; HCC/CCC = index of neoplastic cell types (1 = all neoplasms were hepatocytic; 0 = all neoplasms were cholangiocytic).

[a] Mummichog.
[b] English sole.
[c] Brown bullhead.
[d] Winter flounder.
[e] White sucker.
[f] Estimated from published figure.

primary anatomical difference between mammalian and teleost liver is the lack of distinct lobules or acini in fish. Fish hepatic tubules can be regarded as randomly arranged unit components that, were they to be aligned in symmetrical units, could form mammalian-type lobules. The lack of a lobular or acinar architecture in fish and the resultant lack of portal and centrolobular landmarks have made description of hepatic abnormalities difficult. This has been exacerbated by the inherent tendency for fish histopathologists to use a frame of reference recognizable in mammalian terms. Such pressure has masked the potential contribution of study of the more primitive fish liver to understanding the histogenesis of liver neoplasia in higher vertebrates.

In contrast to experimental hepatocarcinogenesis research in rats and mice, studies of fish have infrequently employed serial sampling that specifically focuses on the stepwise development of lesions involved in hepatic neoplasia. Most experimental carcinogenicity studies in fish have typically examined fish only after a prolonged period after exposure and have assessed the development of frank neoplasms as the endpoint lesion of effect. In this regard, there is a diversity of lesions thought to precede the development of frank hepatic neoplasms in rodents (discussed in Section III.A).

A. Altered Hepatocellular Foci in Rodent Models of Hepatocarcinogenesis

Numerous studies in rats have shown the development of hepatocytic foci, referred to as tinctorially altered foci, enzyme-altered foci, and altered hepatocellular foci, or foci of cellular alteration. These foci develop subsequent to the cytotoxicity and compensatory proliferative phases following carcinogen exposure, are associated with hepatic neoplasms, and precede their development. These foci are clonal in origin,[80] are inducible by potent hepatocarcinogens, and increase in number and size following exposure to genotoxic hepatocarcinogens and various epigenetic promotional agents in a dose-dependent manner;[81] their incidence is correlated with subsequent incidence of adenomas and carcinomas. A temporal and morphological continuity between these foci and hepatocellular neoplasms has been established.[82] Altered hepatocellular foci are thus believed to be precursors to hepatocellular neoplasms. However, only a very small proportion of these focal lesions actually progress to such neoplasms; in fact, only one carcinoma develops for every 1000 to 10,000 foci observed prior to the development of the neoplasm.[82] Consequently, these focal lesions possess value as short- and medium-term predictors of hepatocarcinogenicity in rats,[83] especially in studies that employ quantitative stereological analysis of focal lesions and demonstrate the later development of actual hepatocellular neoplasms. However, in the absence of concurrent or subsequent development of hepatocellular neoplasms in long-term studies such as those carried out under the auspices of the National Toxicology Program, the induction of increased frequencies of foci of cellular alteration is generally considered as insufficient evidence for hepatocarcinogenicity. This same caveat should be recognized in fish hepatocarcinogenicity studies, assuming a similar rarity of progression from focus to neoplasm.

Foci of cellular alteration in rats have been described as basophilic, clear cell, and eosinophilic or acidophilic.[84] Two additional foci types, containing vacuolated cells and cells of mixed tinctorial phenotypes, have been described in mice.[85] More recent studies have defined and redefined further subtypes of hepatic foci.[83,86-88]

The precise biological role of the different types of altered hepatocellular foci in the development of hepatocellular neoplasia is not well defined even in rodents, although the generally accepted axiom is that the focal lesions progress to persistent neoplastic nodules or adenomas and thence to hepatocellular carcinoma.[89] However, despite the decades of research on the predictiveness of different types of altered hepatocellular foci for ultimate neoplasm development using multiple experimental models in rodents, the precise role of these highly heterogeneous focus phenotypes in stepwise hepatocarcinogenesis is still unresolved and controversial.[74,81,83,86-88,90-93] This is partially due to the continuing difficulty in establishing a full consensus on a standard set of consistent and objective criteria and a subsequent classification scheme for the diagnosis of focal hepatocellular lesions.

B. Oval Cells in Rodent Models of Hepatocarcinogenesis

Prior to and concurrent with the development of these focal lesions, there is also a proliferation of more primitive "oval" cells that arise from ductular cells in the periportal region of the liver acinus or lobule and are considered to be cytotoxicant-resistant candidates for the cellular origin of hepatocellular and biliary neoplasms in rats.[73,94,95] Most hepatocarcinogens also induce biliary ductular proliferation in the liver early in carcinogenesis.[96] Exposure to necrogenic doses of hepatocarcinogens results in the early proliferation of bipotent, "stem-like,"[97] periportal "oval cells," which can differentiate into either highly basophilic hepatocytes or bile ducts.[98-100] These oval cells are interpreted as the manifestation of the compensatory proliferative response following cytotoxicity in the early response to hepatocarcinogen exposure; in vivo and in vitro results in rodents are consistent with the conclusion that oval cells can undergo neoplastic transformation into either primary hepatocellular or cholangiocellular carcinomas.[73] Therefore, an alternative hypothesis currently being actively investigated is that hepatocellular and cholangiocellular tumors arise by aberrant differentiation of stem cells (e.g., oval cells) rather than by dedifferentiation of mature hepatocytes.

C. Histogenesis of Liver Neoplasia in Fish

What is known of the histogenesis of hepatic neoplasia in fish has been derived primarily from experimental studies in rainbow trout and medaka *(Oryzias latipes)*. With few exceptions, these studies have utilized a classification scheme for the diagnosis of hepatocellular focal lesions and neoplasms based on earlier studies in rats[84] and mice[85] and have followed the axiom that foci of hepatocellular alteration progress to hepatocellular neoplasms. However, species-specific deviations from the pattern observed in rodents are apparent in fish.

1. Rainbow Trout

In rainbow trout, although clear cell, eosinophilic, vacuolated, and basophilic foci have been induced by exposure to hepatocarcinogens,[101] the basophilic focus has been interpreted as the focal lesion most predictive of subsequent hepatocellular carcinoma development. In fact, cells composing the basophilic foci are morphologically indistinguishable from hepatocytes composing the carcinomas.[101] Most of these basophilic foci (interpreted as focal lesions <0.5 mm in diameter, with compressive lesions larger than this diagnosed as carcinomas) arise independently, but occasionally an eosinophilic/basophilic transformation is seen.[101] These eosinophilic foci correspond morphologically to the typical eosinophilic foci described in rats, but the fact that these foci in trout are usually infiltrated and destroyed by cells of the immune system has been interpreted as evidence against their role as precursor lesions to hepatocellular neoplasms.[101] However, recent studies with rainbow trout[102] have shown that in individuals exposed to N-methyl-N'-nitro-N-nitrosoguanidine (MNNG), hepatocytes composing this focus type are resistant to iron accumulation, a commonly utilized histochemical marker of preneoplastic hepatocytes in rodents.[103] The clear cell (glycogen storage) focus in trout is not considered functionally or morphologically analogous to the clear cell focus in rats, and these lesions are not associated with neoplasm development in this species. Vacuolated cell (lipid storing) foci occur prior to hepatic neoplasms, but are persistent and co-occur with neoplasms. These lesions are interpreted as manifestations of the toxic effects of carcinogens in trout, perhaps related to fatty change. Based on morphological evidence, vacuolated cell (lipid storing) foci in trout are not considered to be directly involved in the stepwise histogenesis of hepatocellular neoplasia.[101]

The basophilic focus in rainbow trout consists of fully transformed neoplastic hepatocytes and is therefore considered a microcarcinoma or carcinoma in situ which then progresses to hepatocellular carcinoma.[101,104] Earlier studies stated that no nodular lesion analogous to the neoplastic nodule or hepatocellular adenoma exists in this species. Therefore, the hepatocellular adenoma in trout was not diagnosed in those earlier studies. However, recent publications from the Oregon State University Department of Food Science and Technology[104-106] have utilized the term hepatocellular adenoma to describe larger (0.5 to 1.0 mm) basophilic foci with relatively normal tubulosinusoidal architecture. This research group has also adopted the paradigm that basophilic foci progress to hepatocellular adenomas with thickened, hyperplastic tubules and thence to compressive and/or invasive, trabecular hepatocellular carcinomas[105,106] that are well to poorly differentiated.

Purely cholangioproliferative lesions in the rainbow trout are less commonly observed than hepatocellular neoplasms,[101] although as stated previously, it is now recognized that the majority of carcinomas induced by carcinogen exposure in that species are a variable mixture of neoplastic hepatocytes and cholangiocytes. The histogenesis of these neoplasms is not known with certainty; however, simple bile duct proliferation may precede cholangiomas,[67] and recent studies in trout imply a role for putative oval cells in the genesis of biliary and mixed hepatobiliary neoplasms, as well as hepatocellular neoplasms. These studies also demonstrated pancreatic metaplasia in trout, further reinforcing the concept of a multipotent stem cell in liver.[67]

Until recently, the relationships among carcinogen dose, cell-specific cytotoxicity, regeneration, and the genesis of neoplasia had not been systematically studied in this species. In response to this need, Nunez and co-workers[107] have demonstrated, in the early phase of the cytotoxic and compensatory proliferative response following exposure to necrogenic doses of aflatoxin B_1 (AFB_1) that the cytotoxic phase is characterized by architectural disruption due to severe hepatocyte swelling and necrosis, with viable remaining hepatocytes showing pleomorphic, atypical nuclei, and foamy cytoplasm. Degenerative hepatocytes closely resemble those described in previous work on the early phases of AFB_1-induced hepatocarcinogenesis in trout,[108] as well as the megalocytic hepatocytes observed in feral English sole that are also interpreted as manifestations of cytotoxicity.[26] Of greatest importance in this study, however, was the emergence of small basophilic cells interspersed among the degenerative hepatocytes, which appeared to arise from the central region of the hepatic tubules, the location of biliary preductular cells or presumptive oval cells.[78] These oval cells occurred within hepatic tubules, often formed ductules, or were unorganized. Presumptive oval cells, and not mature hepatocytes, contained numerous mitotic figures and showed high mitotic indices from ^{3}H-thymidine labeling studies. These authors[107] hypothesized that these proliferative, regenerative oval cells in trout were analogous to cytotoxicant-resistant, γ-glutamyl transpeptidase positive and proliferative oval cells that emerge in the postcytotoxic, compensatory proliferative phase in rat hepatocarcinogenesis and have been proposed to be the critical target cells and cells of origin of hepatocytic and cholangiocytic neoplasms in rats.[73] Furthermore, because these cells differentiate into both hepatocytes and biliary epithelial cells in rats, in conjunction with the fact that the majority of chemically induced neoplasms in rainbow trout are mixed hepatobiliary, it is a reasonable hypothesis that these cells are the progenitors of hepatocytic, cholangiocytic, and mixed hepatobiliary tumors in the trout. This hypothesis deserves further attention in fish hepatocarcinogenesis research.

2. Medaka and Guppy

Other fish species commonly used in experimental hepatocarcinogenesis research have also provided information on the histogenesis of hepatic neoplasia. Studies in the medaka and guppy *(Poecilia reticulata)*[109] generally support the progression of altered hepatocellular foci (usually basophilic, less commonly eosinophilic) to well- or moderately well-differentiated nodular or trabecular, compressive lesions (hepatocellular adenomas) to poorly differentiated, invasive hepatocellular carcinomas.[110–113] This pattern appears to be consistent in other small fish species, such as zebra danio *(Danio rerio)*, mangrove rivulus *(Rivulus marmoratus)*, and sheepshead minnow *(Cyprinodon variegatus)*, as reviewed by Metcalfe.[109] Studies examining the early cytotoxic and compensatory proliferative phases of hepatocarcinogenicity in medaka following DEN exposure[113–115] have described initial degenerative changes, such as hepatocellular necrosis and vacuolization, "piecemeal" necrosis and apoptosis, and hepatocellular glycogen depletion. Separate studies in medaka and king cobra strain of guppy exposed to the direct-acting carcinogen methyazoxymethanol acetate documented hepatocytic megalocytosis in the early cytotoxic phase, within 3 weeks of exposure.[116] Cytotoxicity was not evident in biliary epithelial cells.[115] Spongiosis hepatis, characterized by cystic spaces lined by normal-appearing stellate, fat-storing perisinusoidal cells of Ito,[114] occurred subsequent to these changes, but was not strictly related to carcinogen exposure.[113] Later-appearing lesions following cytotoxicity, within what is characterized as the repopulation phase, were discrete foci of cells with oval nuclei, probably arising from proliferation of bile preductular or ductular epithelial cells corresponding to oval cells; other proliferative, basophilic hepatocellular lesions arose within or adjacent to the reticular network of perisinusoidal cells composing the spongiosis hepatis lesions.[114] Both of these lesions showed γ-glutamyl transpeptidase positivity,[114] often localized in pericanalicular, apical regions of hepatocytes.[115] Bunton et al.[113] described proliferative lesions characterized by cells with oval nuclei that sometimes formed tubular structures, which may have been precursors to neoplasms. The possibility that oval cells or biliary preductular/ductular epithelial cells may be progenitors of these early proliferative lesions and later-developing hepatocellular, cholangiocellular, and mixed hepatobiliary tumors induced in the medaka has already been raised.[114] Work that concentrates on identifying cell surface markers and histochemical and enzymatic features that can distinguish oval cells and biliary epithelial cells from mature hepatocytes is clearly needed to further elucidate the early histogenetic steps leading to hepatocytic, cholangiocytic, and mixed hepatobiliary neoplasms in small fish species such as medaka.

3. Histogenesis of Liver Neoplasia in Feral Fish Species

The following neoplastic and neoplasia-related lesions have been described in feral fish species: hepatocellular, cholangiocellular, and mixed heptobiliary neoplasms (Figures 1 to 5), foci of cellular alteration (Figures 6 to 8), hepatocellular regeneration (Figure 9), lesions interpreted as cytotoxic responses (e.g., megalocytic hepatosis in English sole)[26] (Figure 9), and other enigmatic neoplasia-associated lesions, such as hydropic vacuolation of hepatocytes/biliary epithelial cells (Figures 10 and 11).[33] However, experimental documentation of the complete histogenesis of hepatic neoplasia and the role of the full spectrum of lesions observed in field-captured specimens in the development of liver neoplasms has yet to be achieved in any feral species. In English sole, the histogenesis of hepatic neoplasia has been inferred to parallel that in the rodent, by a statistical analysis of hepatic lesion co-occurrence in field-captured specimens, in conjunction with age at earliest occurrence of particular lesion types.[26] Experimental studies in that species have shown chemical induction of hepatotoxic responses, including hepatocellular megalocytosis and nuclear pleomorphism, as well as hepatocellular regeneration, and foci of cellular alteration.[117] However, because serial samplings were not employed and frank neoplasms were not induced, the stepwise histogenesis of these lesions has yet to be experimentally determined in English sole. Studies focusing on this subject are currently in progress. In feral winter flounder, some studies also generally adopted the focus to adenoma to carcinoma and resistance-to-cytotoxicity paradigms in inferring the histogenesis of hepatic neoplasia,[30,34] with the uniquely vacuolated lesions (i.e., hydropic vacuolation) interpreted as a degenerative, hepatotoxic lesion.[118] Other studies of winter flounder[33] have suggested that the more likely histogenesis of hepatic neoplasms of biliary, hepatocellular, and mixed tumors may originate with the oval cell or preductular biliary epithelial cell, partially because the hydropically vacuolated

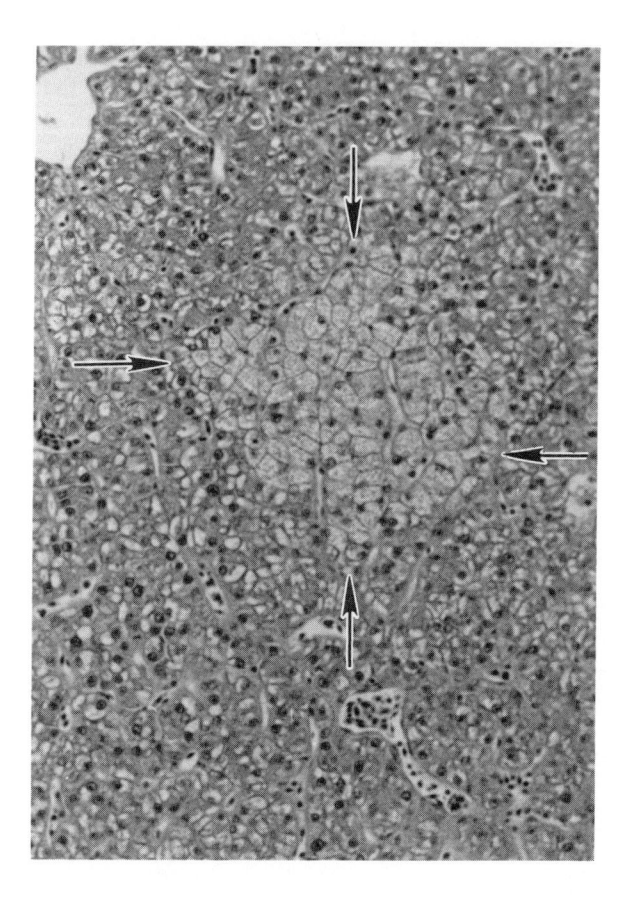

Figure 6. Clear cell focus of cellular alteration (arrows) in liver of an English sole, composed of essentially normal tubulosinusoids showing no compression of surrounding tissue and minimal cytologic alteration, aside from increased cytoplasmic glycogen. Hematoxylin and eosin stain (magnification ×256).

cells originate from preductular biliary epithelial cells and apparently are capable of replicative DNA synthesis (Figure 11).[152,201] However, the stepwise histogenesis of hepatic neoplasia in winter flounder also awaits experimental verification.

The report in mummichog from the Elizabeth River in Virginia of a spectrum of hepatic lesions, including multiple phenotypes of foci of cellular alteration, often containing more radically altered hepatocytes within them (i.e., "foci within foci"), hepatocellular adenomas, and well-to-poorly differentiated hepatocellular carcinomas, hepatoblastoma-like tumors, and cholangiocellular proliferative lesions[38] also conforms to the focus-adenoma-carcinoma scheme of histogenesis. Studies designed to experimentally verify the stepwise histogenesis of hepatic neoplasia in mummichog are in progress.[119] The nomenclature for foci of cellular alteration in this report also reflects the more recent tendency in rat

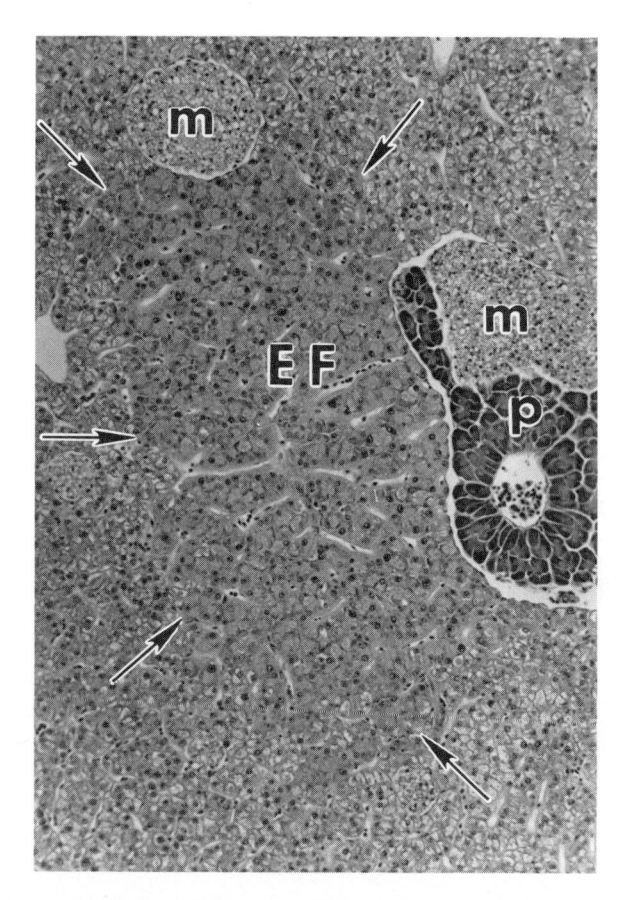

Figure 7. Eosinophilic focus of cellular alteration (EF, arrows show margin) in liver of an English sole. The tubulosinusoids blend with the surrounding parenchyma, and there is no compression. Some hepatocytes within the focus have somewhat enlarged nuclei, but cytologic alterations are otherwise minimal. Also shown are macrophage aggregates (m) exterior to the focus, and exocrine pancreas (p), a tissue normally present in English sole liver. Hematoxylin and eosin stain (magnification ×156).

hepatoproliferative lesion terminology to subclassify focal lesions of a particular tinctorial type. Moreover, the presence of atypical foci within representative examples of each of the clear cell, eosinophilic, and basophilic focus types was interpreted to suggest a preneoplastic role for these foci and that some foci could progress to carcinoma. Such atypical foci within foci of cellular alteration have generally not been reported in fish in the literature, although foci of highly anaplastic, carcinomatous hepatocytes within a benign-appearing hepatocellular adenoma were reported in English sole,[26] and subsequent examination of liver sections in this species has revealed similar "foci within foci."[120]

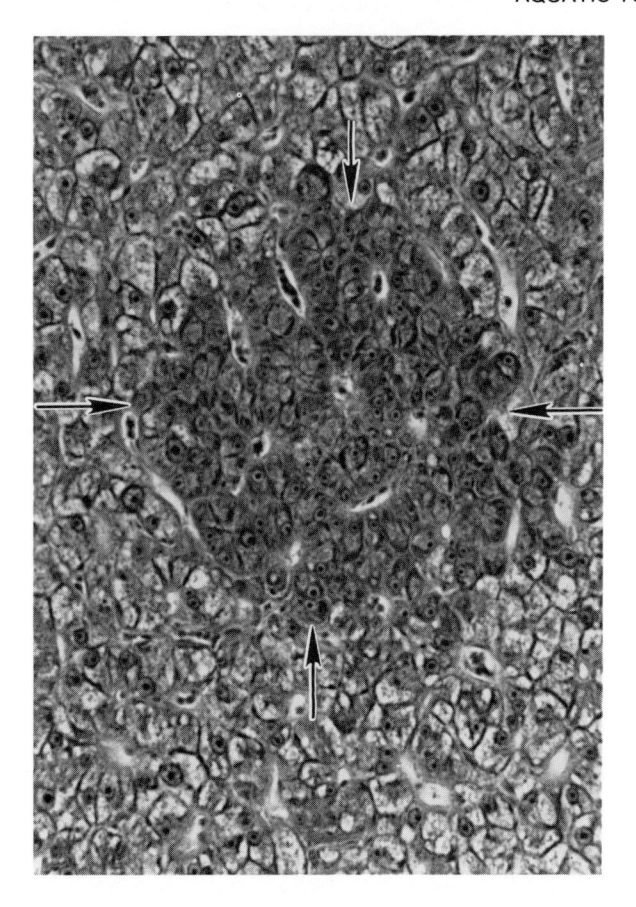

Figure 8. Basophilic focus of cellular alteration (arrows show margin) in liver of an English sole showing normal architecture and lack of compression. Note how the tubulosinusoids of the focus blend into those of the surrounding parenchyma. Hematoxylin and eosin stain (magnification ×256).

IV. HEPATIC NEOPLASM PREVALENCE IN FERAL FISH

It has been demonstrated that there are a number of important variables affecting the probability of finding a neoplastic or neoplasia-related lesion in a particular fish at a particular site. Therefore, it is imperative to consider the following parameters in the interpretation of epizootiological analyses of neoplasms and related lesions.

A. Age

In brown bullhead,[121] English sole,[122–124] winter flounder,[33,35] and white croaker,[110] the risk of occurrence of hepatic neoplasms increased significantly with fish age. Factors such as cumulative dose of xenobiotics absorbed,

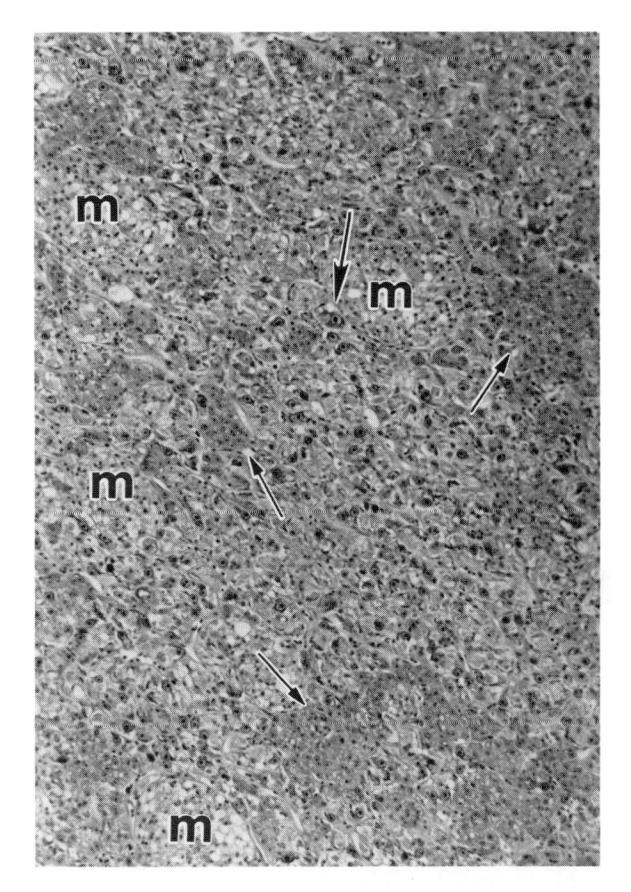

Figure 9. Liver of English sole affected with megalocytic hepatosis (hepatocellular cyto- and karyomegaly), with most of the hepatocytes in this section containing enlarged, hyperchromatic hepatocellular nuclei, with cellular hypertrophy (example shown by large arrow). Islands of smaller, more densely staining hepatocytes scattered throughout the section represent hepatocellular regeneration (small arrows). Also shown are several macrophage aggregates (m). Hematoxylin and eosin stain (magnification ×156).

bioaccumulation of xenobiotics in lipid stores, increasing levels of hepatic xenobiotic-DNA adducts,[125,126] or increasing failure of DNA repair may underlie this trend. A past practice has been to use fish length as an estimate of age.[34] However, fish length/age relationships can vary greatly among sampling sites in epizootiological surveys, and growth curves for male and female fish are typically quite different.[124] Therefore, it should be standard practice in epizootiological surveys of fish from multiple sampling sites to adjust the lesion (e.g., neoplasm) prevalence for both age and gender by multivariate statistical techniques such as logistic regression[127] or to strictly compare age- and gender-specific prevalences

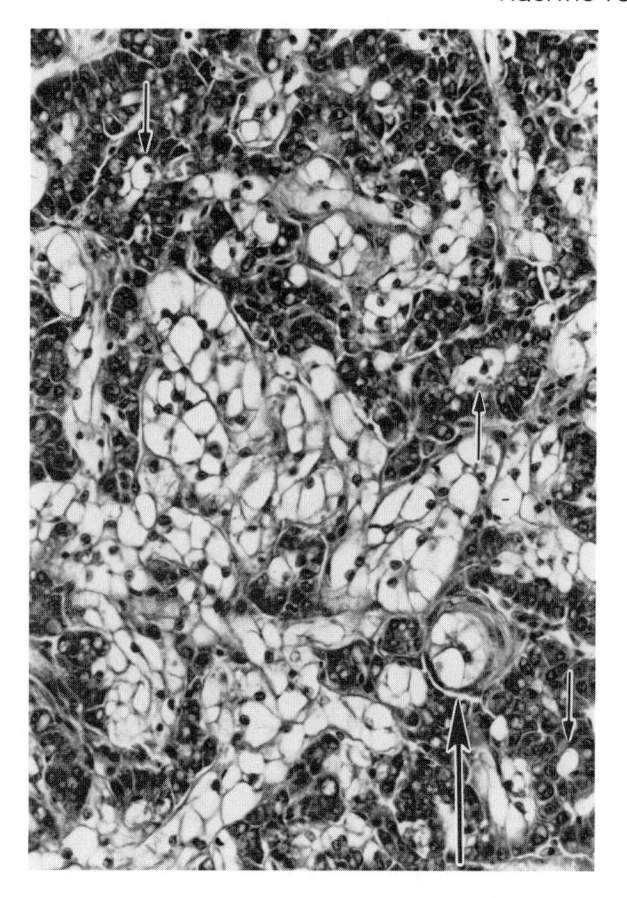

Figure 10. Hydropic vacuolation in a winter flounder from Deer Island Flats in Boston Harbor. Highly vacuolated cells are here primarily shown in tubular arrangement and represent affected hepatocytes as well as cholangiocytes within clearly defined bile ducts (large arrow). Highly basophilic cells are normal-appearing hepatocytes. Also visible (small arrow) are vacuolated preductular biliary cells in the central region of the hepatic tubules. Hematoxylin and eosin stain (magnification ×312).

among sites. This is especially true when the central hypothesis of the epizootiological survey is that tumor prevalences in populations captured at chemically contaminated sites should be higher than those at relatively uncontaminated reference or control sites. A further consideration is that many tumors in vertebrates develop spontaneously in old animals. This has been shown in goldfish

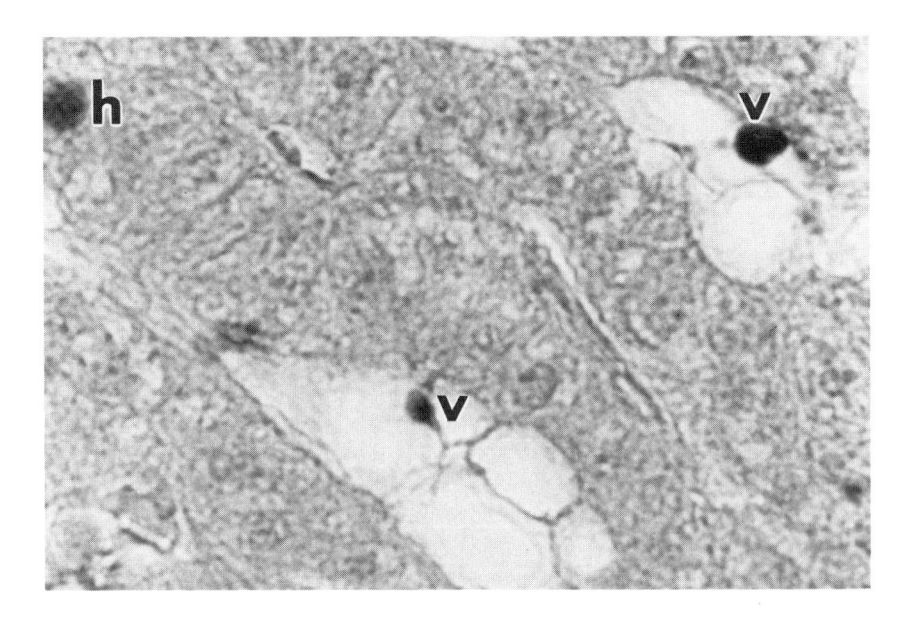

Figure 11. Detail of centrotubular and tubular hydropic vacuolation in winter flounder liver from Boston Harbor, stained with monoclonal antibody to bromodeoxyuridine (BrdU). The dark spheres represent anti-BrdU-labeled nuclei of nuclei undergoing replicative DNA synthesis. Multiple vacuolated cells nuclei (v) and one hepatocyte nucleus (h) stain for BrdU incorporation (magnification ×800).

(Carassius auratus)[128] and medaka.[64] However, with these experimental studies, it is important to consider the potential carcinogenicity of the diet. With regard to the relationship between fish age at sampling and neoplasm types detected, Hinton[129] has found that cholangiocellular neoplasms appear later than hepatocellular lesions in the experimental medaka system. In brown bullhead from the Black River, Baumann et al.[121] found a significantly higher prevalence of cholangiocarcinomas than hepatocellular carcinomas in 4-year-old brown bullhead (sexes combined) as compared to 3-year-old fish in which the prevalences of these neoplasm types were similar. These data could be interpreted to mean that cholangiocellular cancers generally develop later than hepatocellular types in this species, although histological examination of fish from more age classes would be necessary to confirm this. Aside from this study, to our knowledge there is no epizootiological survey in native fish that has comprehensively examined this relationship in species affected by both hepatocellular and cholangiocellular neoplasms.

B. Migration

Migratory patterns of fish are rarely adequately known. Consequently, in field studies examining liver neoplasms and related lesions in fish from multiple sites, there is a risk of detecting neoplasms in fish that are not normally resident at the site of capture, but have migrated from a distant site to the capture site. This confounding variable can sometimes be minimized by parallel collection of liver tissue from individuals for histopathology, organic chemistry, and/or determination of xenobiotic-DNA adducts, with subsequent analysis of hepatic levels of chemical compounds [e.g., chlorinated hydrocarbons, such as polychlorinated biphenyls (PCBs), p'-dichlorodiphenyl-trichloroethane (DDTs)], and/or xenobiotic-DNA adducts that are indicative of chronic exposure to contaminants. Although this protocol has been logistically difficult to apply and has involved considerable expenditure for chemical and biochemical analysis of individual liver samples, currently available screening methods for analysis of contaminants[130] and automated methods for DNA extraction have reduced the costs of such an approach considerably. The accepted strategy has been to select fairly territorial, demersal target species that are in continuous or intermittent contact with bottom sediments and are essentially nonmigratory or exhibit minimal seasonal movements for the purposes of reproduction. The best nonmigratory example is the mummichog. This species exhibits a home range of 30 to 40 m, with limited movements only in the fall and winter.[131] Examples of species with limited movement in the form of annual spawning migrations are English sole and winter flounder (north of Cape Cod). With the possible exception of the mummichog, fish migration must be considered in interpretation of the neoplasm prevalence data in any epizootiological survey in native fish, and highly migratory species should be avoided as target species whenever possible.

C. Tumor-Associated Mortality

There is a question as to whether there is a tumor-associated mortality factor in older fish. Baumann[121] compared tumor prevalence and age structure in brown bullheads from two sites in tributaries of Lake Erie. Liver tumor frequency was maximal in 4- to 5-year-old fish from the contaminated site. Older fish (6 to 7 years old) were absent from this site, but represented 18% of the catch from the uncontaminated site. The authors concluded that there was an age-selective mortality associated with high prevalences of liver carcinoma. Similarly, Vogelbein[119] finds the lesion prevalence in native mummichog resident at a highly polycyclic aromatic hydrocarbons (PAH)-contaminated site in the Elizabeth River, VA, to be lowest in the spring, suggesting a failure of the tumor-bearing animals to overwinter as successfully as the younger, tumor-free animals. Nonetheless, few field studies have addressed this issue of survival in tumor-bearing fish or the more general question as to the effects of hepatic neoplasm presence on individual fish health. With regard to the latter, abnormal values of several serum chemistry parameters (e.g, albumin, bilirubin, and total protein) that are characteristic of liver dysfunction and/or damage and impaired fish health (e.g., lowered hematocrits) have been demonstrated in Puget Sound English sole concurrently affected with hepatic neoplasms and related idiopathic hepatic lesions.[132] However, length/weight curves in tumor-bearing and tumor-free English sole are essentially the same,[133] and recent studies in English sole failed to show any clear effect of liver neoplasms or other toxicopathic hepatic lesions on mortality rate.[134] Because of the emphasis over the last decade on the use of hepatic neoplasms in native fish as biomarkers of chemical contaminant exposure and indicators of environmental degradation, rather than as predictors of declines in fish populations, the impact of hepatic neoplasia on survival and overall fish health status needs to be examined more closely. There are several major reasons why measures of population effects have not been included in epizootiological studies of disease in wild fish. First, it has proven very difficult to separate and estimate disease-related mortality from natural and fishing mortality, especially when epizootiologic studies on disease in fish do not utilize the tools of stock assessment. However, recent studies have begun to assess contaminant exposure effects at the population level, though not specifically related to the effects of neoplasms or other disease.[135] Second, most of the past studies have not emphasized collection of data necessary to assess potential population effects because population biologists have not been among the primary investigators working with fish pathologists. Consequently, these studies were not designed to assess the effects of disease states on mortality and rarely included collection of age frequency distribution data needed to calculate unbiased estimates of mortality rates.

V. EXPERIMENTAL CARCINOGENESIS IN AQUARIUM FISH SPECIES AND RAINBOW TROUT: RELEVANCE TO SIMILAR STUDIES IN NATIVE FISH SPECIES

Many direct- and indirect-acting chemical carcinogens have been tested in various aquarium fish species. Carcinogenicity tests in fish have recently been reviewed comprehensively.[109] It is not our intent to reiterate this synthesis, but to contrast the relative success of carcinogenicity testing in these aquarium species with attempts to experimentally induce neoplasms in feral species. Table 2 lists some of the more commonly tested chemicals whose carcinogenicity has been demonstrated in ten domesticated fish genera; the organs primarily affected are also listed. Most of the neoplasms described were in the liver; however, other organ systems were also involved. It should be noted that some of the studies, especially of larger fish species where multiple organs cannot be processed, embedded, and examined as easily as in small fish, may not have examined the histology of extrahepatic organs routinely. The diversity of tumors that may arise was illustrated

Table 2. Summary Studies of Selected Chemical Carcinogens and Their Target Organ(s) in Ten Genera of Fish

Species	DEN[a]	MNNG[b]	MAM[c]	BHA[d]	DMBA[e]	BAP[f]	AFB[g]	SED[h]
Danio	H[19]							
	H, EP[54]							
Medaka	H[154,159,160, 244]	G[246]	H[197,198,247]		H+[112]	H[111,112]		
		M[205]	H+[136]					
	H+[245]							
Rivulus marmoratus	H[248,249]		H+[136]	H[250]				
Guppy	H, EP[54]		H+[136]		H+[112,252]	H[111,112]		
	H[54,155]		P[251]					
Rainbow trout	H[67,182,188]	H, NB[102,253]			H[189,255]	H[138,256]	H[17,66,145,256,257]	H[139,140,258]
		H+[254]						
Fathead minnow			H+[136]					
Sheepshead minnow	H[259]		H+[136]					
Gulf killifish			H+[136]					
Inland silverside		P[260]	H+[136]					
Poeciliopsis spp.	H[68–70,261]				H[69,261,262]			

Note: H = liver; H+ = liver and other organs; P = pancreas; G = gill; E = eye; EP = esophageal papilloma; M = melanoma; HP = hemangiopericytoma; NB = nephroblastoma.

[a] Diethylnitroseamine.
[b] N-methyl-N′-nitro-N-nitrosoguanidine.
[c] Methylazoxymethanol acetate.
[d] Butylated hydroxyanisole.
[e] 7,12-dimethylbenz(a)anthracene.
[f] Benzo(a)pyrene.
[g] Aflatoxin B$_1$.
[h] Contaminated sediment.

by a study involving exposure of seven species to methylazoxymethanol acetate;[136] extrahepatic neoplasms occurred in retina, various mesenchymal tissues, exocrine pancreas, kidney, and nervous tissue of the medaka; in mesenchymal tissue, exocrine pancreas, and kidney of the guppy; and in choroid gland, mesenchymal tissues, and nervous tissue of the sheepshead minnow.

An important aspect of field epizootiological studies is that they represent a requisite step in identifying potential etiologic agents (e.g., classes of chemical contaminants), whose carcinogenic effects can then be tested in laboratory exposures. Kimura demonstrated the utility of this linkage in a survey of chromatophoromas in croaker *(Nibea mitsukurii)* in field and experimental studies.[137] There is only one other study that claims to have demonstrated an experimental causal link between exposure to specific chemicals present in and extracted from contaminated aquatic habitats and the development of actual hepatic neoplasms in a species that develops neoplasms in the wild. This involved brown bullheads exposed to Buffalo River sediment extract.[45,46,138] In this study, "incipient" neoplasms were described in one fish, and a well-differentiated cholangioma was described in another. While these experimental results suggested that hepatic lesions in feral fish could be due to exposure to sediment-associated contaminants, because of deficiencies in experimental design and technique, a very low sample size, and incomplete reporting/publishing of data, these studies have been regarded as inconclusive.[9] In the English sole, putatively preneoplastic focal lesions were induced in the liver by 13 monthly injections of an extract of an urban marine sediment, followed by a grow-out phase of 6 months, but hepatic neoplasms were not induced.[117] In the winter flounder, proliferative putative oval cells were induced in winter flounder following dietary exposure to technical grade chlordane and BaP.[33] Sediment extracts have been used to induce liver neoplasms in the rainbow trout by Metcalfe[139] and White,[140] cited by Manny and Kenaga.[141] These last studies are of great significance, but the rainbow trout is not known to develop contaminant-associated hepatic neoplasms in the wild. Given the well-differentiated nature of the neoplasms induced and the small sample size in the bullhead study and the fact that no hepatic neoplasms have been experimentally induced in the English sole and winter flounder studies, there is a need for further studies, with controlled experimental exposures to specific chemical(s) from relevant sites, in species known to develop contaminant-associated tumors in the wild. The relative ease with which *F. heteroclitus* can be adapted to laboratory experimentation makes this species perhaps the best candidate among the feral species for the detailed study of the temporal histogenesis of hepatic neoplasia. Such studies are in progress.[119]

The reasons for the difficulty in achieving successful experimental induction of frank hepatic neoplasms in feral fish species, in contrast to the success of hepatocarcinogenicity studies in laboratory-adapted fish species, are probably multiple. These are listed below as important factors to consider in future modeling efforts in feral fish species.

A. Differences in Genomic Susceptibility

We postulate that differences in the frequency of activated oncogenes may exist between inbred aquarium species and outbred wild-caught fish. This may result in a potentially greater number of genetic events and/or a longer duration of promotion being necessary for the development of neoplastic change in wild-caught fish. The role of such oncogenes and tumor suppressor genes has been investigated most completely in *Xiphophorus* species.[142,143] Xiphophorine fish from wild populations do not develop neoplasms; in contrast, certain backcrosses of *X. maculatus* (platyfish) with *X. helleri* (swordtail) as the recurrent parent produce offspring that develop neoplasms in a Mendelian pattern of inheritance.[144] Native species are presumably of a greater genetic diversity than highly inbred aquarium species, which although genetically viable, exhibit more limited genetic diversity.[70]

B. Life Stage-Dependent Cell Turnover

Failure to experimentally expose wild-caught fish species to known or suspected carcinogens at an early enough time point in their life history may be an additional factor explaining the relative lack of success in the experimental induction of hepatic neoplasms in these species. Rainbow trout are most sensitive to carcinogens at the late embryonic stage.[145] This may reflect the high rate of cell proliferation in the developing target organs and the consequently increased chance of fixation of chemically induced molecular lesions such as xenobiotic-DNA adducts, which may result in complete tumor initiation. Two of the studies in native fish mentioned in Section V[33,117] utilized wild-caught, juvenile flatfish of 1 to 2 years of age at the start of the experiments. Endogenous rates of cell proliferation in target organs such as the liver are relatively low in fish of this age.[120] A reasonable supposition that needs to be verified is that these experiments might be more successful if repeated with exposure of fish at the embryonic, larval, or early postlarval stages.

C. Husbandry

The successful induction of neoplasms in wild-caught species is also highly dependent upon development of effective husbandry methods, including those that will permit exposure of early life history stages to carcinogens, while allowing good survival through to the adult form. In these types of early life history exposure studies, as well as in longer-term exposures of juvenile or adults, it is also essential to be able to subsequently hold adults in a healthy state for periods of time that begin to approach the age at which hepatic neoplasms are commonly detected in wild fish residing in highly contaminated environments (at least several years). In English sole, experiments beyond an 18-month duration[117] have not been possible, but ongoing husbandry projects are addressing this critical research need. Metamorphosis is a critical period in the survival of flatfish in captivity, and attempts to rear large numbers of feral species such as English sole and winter flounder through this life history stage have been unsuccessful. Moreover, more basic research to determine the optimal physical holding facilities and the basic nutritional requirements of these native species needs to be done to assure the high survival rate and good fish health necessary in long-term carcinogenesis studies.

D. Partial Hepatectomy

Application of experimental procedures designed to dramatically increase cell proliferation in the liver subsequent to carcinogen exposure, such as a properly timed surgical or chemically induced partial hepatectomy, have been used commonly in rodent hepatocarcinogenesis studies.[146] These procedures have also been successfully used in small fish species[147] and rainbow trout[148,149] and could be utilized to accelerate the hepatocarcinogenic process in juvenile or adult individuals of feral fish species. These procedures are currently being applied to the study of hepatocarcinogenesis in English sole.

E. Route of Exposure

The choice of route of exposure and carrier of contaminant is complex. Studies have employed intraperitoneal, intramuscular, intragastric, and intravenous injections administered in a range of carriers, in addition to a diversity of dietary, water column, and sediment exposure methods. Each protocol is pharmacokinetically unique and, as a result, all studies are based to varying degrees on assumptions that may or may not be appropriate and valid.

VI. NEW RESEARCH DIRECTIONS

A. Histogenesis of Hepatic Neoplasia in Fish

The majority of the field and laboratory studies described above have focused on analysis of the end-stage lesion in the neoplastic process (i.e., frank neoplasms), with the primary goal(s) being the description of neoplasm type(s) and prevalences in field studies and their induction in laboratory studies. A greater focus on the stepwise histogenesis of each neoplasm type would allow greater comparison and contrast with the well-documented process of hepatocarcinogenesis in mammals, as well as expanding the number of experimentally proven histopathological biomarkers of carcinogen/toxicant exposure in wild fish. This approach would allow significant contribution to, and resolution of the current debates over the relative roles of lesion types such as putatively preneoplastic foci of altered hepatocytes,[74] areas of primitive biliary epithelial cells, such as oval cells,[150] hydropic vacuolation of hepatocytes and biliary epithelial cells,[33,118,151,152] and megalocytic hepatosis/hepatocellular karyo- and cytomegaly[26] within the context of histogenesis of hepatic neoplasia in fish. These experiments should include serial sample points, biochemical characterization of lesions by histochemistry and immunohistochemistry, and ultrastructural studies.

Studies that have included serial sampling to describe or infer lesion development and progression include experimental[110,113–115,153–155] and field analyses.[26,33,122,123] The experimental studies have shown that in small fish species, sequential development of putatively preneoplastic focal lesions can occur within weeks after carcinogen exposure. In contrast, field studies have shown that differences in lesion prevalence must be analyzed among a number of year classes or that the patterns of lesion co-occurrence must be statistically analyzed in numerous adult fish[26] to even indirectly infer the temporal histogenesis of hepatic neoplasia in wild fish. Moreover, in the one published experimental study where feral fish were chronically exposed by multiple intramuscular injections of an extract from a contaminated sediment[117] and foci of hepatocellular alteration were induced, histopathological examination was only done at the 18-month experiment termination point, thus precluding any consideration of the time frame of induction and histogenesis of the observed lesions. Furthermore, hepatic neoplasms are extremely rare in feral English sole[123] and winter flounder[33] less than 4 and 5 years old, respectively. These differences in time scale may, in part, reflect longer life spans of the feral fish, but this does not adequately explain the differences in time necessary for lesion induction. Differences in genetic susceptibility and life history stage at exposure, as discussed above, as well as level, mode, and duration of exposure are certainly also involved.

Ultrastructural analysis of cell morphology can add significant information to studies of histogenesis and can resolve inconsistencies in classification of abnormal cell types, as in distinguishing hepatocellular from cholangiocellular neoplasms or in identifying the cell types composing earlier progenitor lesions. This has significant logistic and cost considerations, but, nonetheless, in studies where ultrastructural analysis has been conducted, the quality and accuracy of the diagnostic descriptions is inevitably improved. Such studies include analysis of hepatic lesions in English sole,[156,157] rock sole,[158] winter flounder,[118,151] rainbow trout,[106] and medaka.[113,114,116,159]

B. Biochemical Characterization of Lesions

Studies of mammalian hepatocarcinogenesis and programs such as the National Toxicology Program responsible for testing the carcinogenic potential of chemicals in rodents have utilized enzymatic and histochemical markers (both positive/enhanced and negative/reduced) for foci of hepatocellular alteration. The more commonly assessed markers include enzymes such as γ-glutamyl transpeptidase(GGT, +), Mg^{+2}-dependent adenosine triphosphatase (ATPase, –), DT-diaphorase (DTD, +), glucose-6-phosphate dehydrogenase (G6PDH, +), glucose-6-phosphatase (G6P, –), alkaline phosphatase (ALKP, +), uridine diphosphoglucuronyl dehydrogenase (UDPGdH, +), cytochrome P450 isozymes (typically –, but variable, depending upon isozyme), glutathione S-transferases (GST, + for certain isozymes), and resistance to iron accumulation. The rationale for testing the activity of many of these enzymes in the focal lesions is their involvement in the metabolic activation (phase I) and inactivation (phase II) of hepatocarcinogens/hepatotoxicants, with decreases in phase I (e.g., P450 isozymes, DTD) and increases in phase II (e.g., GGT, GST, UDPGdH) enzymes serving as an adaptive response to the potential cytotoxicity/genotoxicity of these xenobiotics.[89]

Equivalent studies in fish are less common. Nakazawa and co-workers[160] described the extreme heterogeneity and sexual dimorphic characteristics, with respect to ATPase and G6P staining, of enzyme-altered and tinctorially altered foci and hepatic nodules in medaka induced by DEN. Hinton et al. and Teh et al.[110,161] have also described the histochemical features of enzyme-altered foci in *Cyprinodon* and medaka exposed to DEN. To summarize these studies, which are discussed in detail at the cellular level in Chapter 4, enhanced GGT activity in foci and variable GGT staining in neoplasms in both species have been shown; ATPase was reduced in foci and neoplasms of *Cyprinodon* and medaka; G6P was deficient in foci of both species; G6PDH was deficient in eosinophilic foci and an early hepatocellular carcinoma and positive in clear cell foci in medaka; acid phosphatase was positive in foci and enhanced in an early hepatocellular carcinoma in medaka; alkaline phosphatase was positive in eosinophilic and clear cell foci and an early hepatocellular carcinoma in medaka; and DTD and UDPGdH were enhanced in eosinophilic and clear cell foci and showed variable staining in early hepatocellular carcinomas in medaka. In contrast to Hinton's findings in these species, no GGT-positive foci of cellular alteration have been demonstrated in rainbow trout,[102] although recent studies on hepatic neoplasms in rainbow trout have biochemically and histochemically demonstrated induction of GGT, as well as induction of aldehyde dehydrogenase, DTD, and UDPGdH, and depression of P450IA1.[162] A particularly significant finding by Hinton's group[110,161] was the emergence of GGT-positive foci in medaka prior to the development of tinctorially altered foci detectable by standard hematoxylin and eosin (H&E) staining. These findings suggest that certain enzyme histochemical procedures, such as GGT, may provide more sensitivity than standard histopathological procedures in detecting important changes occurring early in the histogenesis of hepatic neoplasia. There is obviously a real need to further apply these methods in hepatocarcinogenesis experiments, as well as in field studies in fish, and to compare these histochemically altered foci and neoplasms in fish to their counterparts in rodents.

Other enzyme systems that have been studied immunohistochemically in fish include cytochrome P4501A in scup *(Stenotomus chrysops),* rainbow trout, winter flounder, mummichog, and English sole.[163-169] These studies, discussed in greater detail in Chapter 3, have shown significant induction of this enzyme in hepatocytes, and, in general, a reduction of expression in tinctorially altered hepatocellular foci and hepatic neoplasms in fish from contaminated sites.[167,168] These authors suggested this reduction to be an adaptive response in altered and neoplastic hepatocytes that confers upon these cells a resistance to cytotoxic and genotoxic compounds requiring activation by xenobiotic-metabolizing enzymes such as in the P450 system to exert their effects, with consequently increased survival in animals inhabiting a PAH-contaminated environment. These results essentially parallel P450-associated staining patterns in similar lesions in the rat.[170] Experimental exposure of scup to a tetrachlorobiphenyl congener, a tetrachlorodibenzofuran,[171] and to a mix of the same chemical classes in the field[166] resulted in induction of P4501A in multiple epithelia and in endothelia, but not in heart muscle, nerve cells, or smooth or skeletal muscle. This pattern of differential induction is reminiscent of the organ types discussed above that have putative chemical etiologies for tumors observed therein, namely epithelia, especially the liver, pancreas, and possibly skin. Thus, activation of P4501A may be a reliable biomarker of exposure to chemical carcinogens and one probable bioactivation route. The widespread induction of P4501A in endothelia suggests a potential role for these cells in detoxication of chemicals present in the vascular system.[172]

GST expression has also been examined immunohistochemically in rainbow trout[173] and suckers.[5,173,174] These studies showed that GST is less predictably induced and is more often reduced in liver foci and tumors in these fish species than in some rodent hepatocarcinogenesis models. These results suggest that hepatocarcinogenesis in these fish species, unlike in rodent models, is not associated with selection for GST-dependent resistance to cytotoxicity. These authors offered the alternative hypothesis that reduced GST expression could favor, by repeated damage to DNA by genotoxic metabolites, malignant progression of contaminant-associated liver neoplasms in fish exposed to carcinogens/toxicants detoxified by GSTs.[5] In contrast, a biochemical study in mummichog[175] showed GST levels in hepatic tumors and altered foci that were equivalent to normal liver; however, GST expression was not localized immunohistochemically. Recent results of a study using biochemical characterization of GST in AFB_1-induced mixed hepatobiliary carcinomas in rainbow trout showed clear enhancement of GST activity over control liver.[162] Clearly, GST expression in hepatocellular focal lesions and neoplasms in fish needs to be investigated more completely.

One diagnostically useful histochemical characteristic that appears to be consistent in rodents and fish is the resistance to iron accumulation in altered hepatocytes in all types of foci of hepatocellular alteration and neoplasms, as shown in the rainbow trout,[102] winter flounder,[33] and English sole.[26] In fact, in several species of adult feral fish that commonly have siderotic livers [e.g., English sole, rock sole *(Pleuronectes bilineata)*, hardhead catfish *(Arius felis)*, and black croaker

(Cheilotrema saturnum)], the absence of or reduction in stainable iron in individuals with siderotic livers is a useful and consistent diagnostic marker for more subtle tintorially altered foci.[120] This marker may precede the development of these lesions as typically detected in H&E-stained sections.[102] The mechanism of resistance of hepatocytes in putatively preneoplastic focal lesions and neoplasms to iron accumulation is not understood; however, as a marker in rats, this resistance indicates that the presence of exogenous and endogenous iron is abnormal in the hepatocytes composing these lesions.[176] More recent evidence in rats exposed to 2-acetylaminofluorene suggests that the mechanism of iron resistance is not related to the capacity of transferrin cell membrane receptors to bind iron, either by virtue of their number or binding affinity. They postulated that an alteration in the dissociation of iron from ferrotransferrin may explain the iron storage deficiency in hepatoctyes composing altered foci and neoplasms.[177]

To summarize the biochemical studies described above, there appear to be important differences in biochemical characterization of hepatic neoplasms and focal lesions between fish and rodents. Yet, too few fish species and too few biochemical markers have been examined in either experimental or field situations to know how much interspecies variability may exist for each of these markers. Because of the possibility that changes in one or several of these markers may precede the development of tinctorially altered foci (e.g., GGT in medaka[110] and iron resistance in rainbow trout),[102] their utility as earlier morphologic indicators of carcinogen/toxicant exposure in lab and field studies needs to be further explored. Furthermore, there is a need to examine multiple biochemical markers in the different types of tinctorially altered foci to gain a better understanding of the biochemical changes accounting for the various phenotypes.

C. Role of Epigenetic Carcinogens and Promoters in Carcinogenesis; Consideration of Other Promotional Stimuli and Assessment of Their Effects

1. Importance of Chemical Promoters/Modulators in Hepatocarcinogenesis

With some notable exceptions that have been reviewed recently,[178] the primary paradigm for experimental hepatocarcinogenesis in fish has involved a single, repeated, or continuous exposure of young fish to large doses of a genotoxic, initiating carcinogen, such as AFB_1, DEN, or an aromatic hydrocarbon, such as 7,12-dimethylbenz(a)anthracene or BaP, or to extracts of sediments containing a complex mixture of high levels of genotoxic PAHs, in addition to nitrogen-containing aromatic compounds and other uncharacterized compounds.[117] The complexity of the broad spectrum of contaminants usually present at much lower levels in the contaminated natural habitats contrasts significantly with this approach. It will, of course, be extremely difficult to conduct controlled modeling experiments that adequately address this complexity, but nonetheless it is important to test carcinogens and promoting agents other than genotoxicants, as well as other modulators of carcinogenesis. The research group that has studied epigenetic carcinogens in fish is that of Nunez et al.,[105] Bailey et al.,[179] and Dashwood et al.[180] These researchers have examined the role of enhancers and inhibitors of carcinogenesis, typically in AFB_1 or DEN-initiated rainbow trout. They have shown that 17-β-estradiol,[105] indole-3-carbinol, DDT,[181] and β-napthaflavone[182,183] can, if given after initiation, enhance tumor yield. Dieldrin was interpreted as a potential cocarcinogenic agent when fed together with AFB_1.[184] The role of a PCB mixture, Aroclor 1254, in rainbow trout hepatocarcinogenesis is complex; in trout cofed this mixture with AFB_1[185] or fed prior to AFB_1 initiation,[186] the PCB was profoundly inhibitory to tumor response. In AFB_1-initiated embryos subsequently fed PCBs after hatching, there was no effect on tumor response.[187] In contrast, PCBs dramatically enhanced tumor response when cofed with DEN;[188] furthermore, PCBs fed after dimethylbenz(a)anthracene also substantially enhanced tumor yield.[189] Therefore, the modulating activity of PCBs in rainbow trout is dependent upon the initiating carcinogen and timing of exposure. Likewise, these and other compounds, when given prior to or during initiation, may inhibit tumor formation.[179,181,183,185] In general, where reduced tumor response has been shown for a chemical modulator, the proposed mechanism has been an inhibition of the ultimate carcinogen in reaching its target site, mediated by induction of carcinogen detoxification pathways, as reflected by lower carcinogen-DNA adduct levels.[190,191]

In summary, these results show that the degree of inhibitory or promotional effect of such modulators is difficult to predict and may be substantially altered depending on the carcinogenic initiator used, the dose of the modulator, and the point in the exposure protocol at which the modulator is applied relative to the carcinogen.[181] The complex pattern of effects of these modulators, even in a single fish species, underscores the need for further research on promotional effects of environmental contaminants in additional aquarium and feral species.

2. Experimental Modulation of Hepatocarcinogenesis

A standard method of promoting and accelerating carcinogenesis in rodent models is to surgically remove 70% of the liver mass a short while after[146] or before[192] initiation (surgical partial hepatectomy) or to administer a hepatotoxic compound such as carbon tetrachloride after initiation[193] (chemical partial hepatectomy), serving to increase cell proliferation following or during carcinogen exposure and to both accelerate and enhance the development of focal lesions and neoplasms. Modifications of the surgical procedure have been used with partial success in medaka,[147] rainbow trout,[149,194] English sole,[120] and winter flounder,[195] with the chemical partial hepatectomy method used, with limited success, to enhance hepatocarcinogenesis in rainbow trout.[148] The promoting effects of temperature and consequently increased cell proliferation on tumorigenesis have also been explored as a modulator of carcinogenesis in fish.[54,147]

3. Assessment of Cell Proliferation and Other Promoting Effects

A common monitor of the potency of tumor promoters is the assessment of the amount of cell proliferation induced. This can be achieved by counting mitotic figures in tissue sections and determining a mitotic index, but these tend to be low in adult fish,[196] especially in the liver. Cell proliferation can also be estimated using tritiated thymidine uptake[107,147,197,198] and the exogenous, nonradioactive thymidine analogue bromodeoxyuridine (BrdU).[152,199] The latter method has been successfully applied to the assessment of cell proliferation in feral fish, including the determination of labeling indices in hepatic lesions such as cholangiocellular and hepatocellular carcinomas and others,[33,152,200] as illustrated in Figures 11 and 12. BrdU has also been used in parallel with an assay for ornithine decarboxylase (ODC) activity.[201] In this study, proliferative vacuolated and neoplastic cells in winter flounder were shown to have elevated ODC activity. The recent suggestion that ODC may be a proto-oncogene[202] means that the study of ODC as it relates to cell proliferation and transformation in fish should be a priority. Another method of determining mitotic activity is immunohistochemical localization of the endogenous proliferating cell nuclear antigen (PCNA), the expression of which is closely linked to cell replication.[203] A recent comparison between BrdU and PCNA methods in three species of fish has shown the relative equivalence of these methods.[204] The BrdU protocol and the PCNA method, especially, show potential for assessing cell proliferation and promotional effects in liver and hepatic lesions in wild-caught fish, as well as in carcinogenesis experiments with these species, and should be applied more frequently. One significant aspect of the interpretation of cell proliferation data in fish that has yet to be examined fully is the effect of environmental temperature and reproductive status on cell-labeling indices.

Another approach to investigating the effects of genotoxic vs. epigenetic carcinogens is described by Anders et al.[144] Tester strains were bred that either contained a tumor suppressor gene that could be overcome by a genotoxin or an oncostatic gene that could be overridden by tumor promoters.

An additional factor that should be considered in the analysis of tumor promotion in fish is chronic parasitism. Liver tumors in white suckers from industrially polluted areas of Lake Ontario are "consistently associated with chronic cholangiohepatitis and segmental cholangiofibrosis which may be associated with helmith parasitism in the liver."[5] Furthermore, fish from control sites also have biliary proliferation as a result of the presence of this parasite. Therefore, the interaction of chemicals with biological stimulants to cell proliferation needs to be considered.

D. Role of Genetics in Assessing the Effects of Carcinogens on Fish

Earlier, we discussed some genetic bases for why native fish may be less susceptible to carcinogens than aquarium-reared species. Dawe[62] discussed the

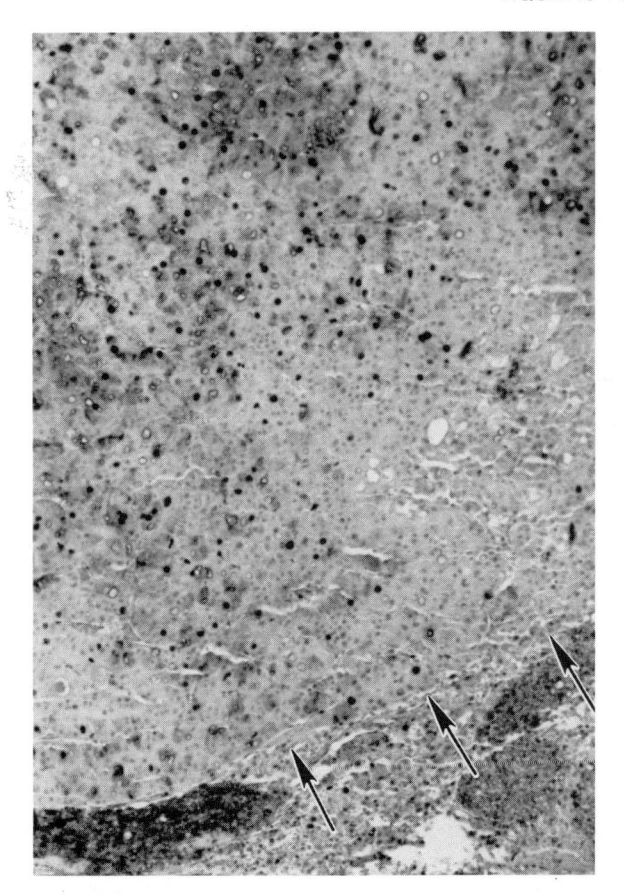

Figure 12. Section of a basophilic hepatocellular carcinoma in an English sole immunohistochemically stained with monoclonal antibody to bromodeoxyuridine (BrdU). The dark spheres represent anti-BrdU-labeled nuclei of hepatocytes undergoing replicative DNA synthesis, which are far more frequent within the carcinoma than in the surrounding hepatic tissue at the bottom of the figure, demonstrating a high rate of hepatocyte proliferation within the carcinoma. Arrows point to margin of the neoplasm; the large, dark, granular structures at bottom are macrophage aggregates that stain darkly due to hemosiderin content and do not represent positive anti-BrdU staining. BrdU monoclonal antibody, eosin counterstain (magnification ×156).

utility of the concept of an oncozoon. He suggested that hereditary factors such as oncogenes; tumor suppressor genes; genes associated with bioactivation, detoxication, immunologic responses, and their interactions have not been widely investigated, especially in tumor epizootics among feral fishes. In addition, there are suggestions that populations of fish living in chronically polluted habitats may be in various ways more resistant to toxic insults. Adult winter flounder exposed to technical grade chlordane showed a 100-fold lower maximum tolerable dose when collected from a clean offshore site than when

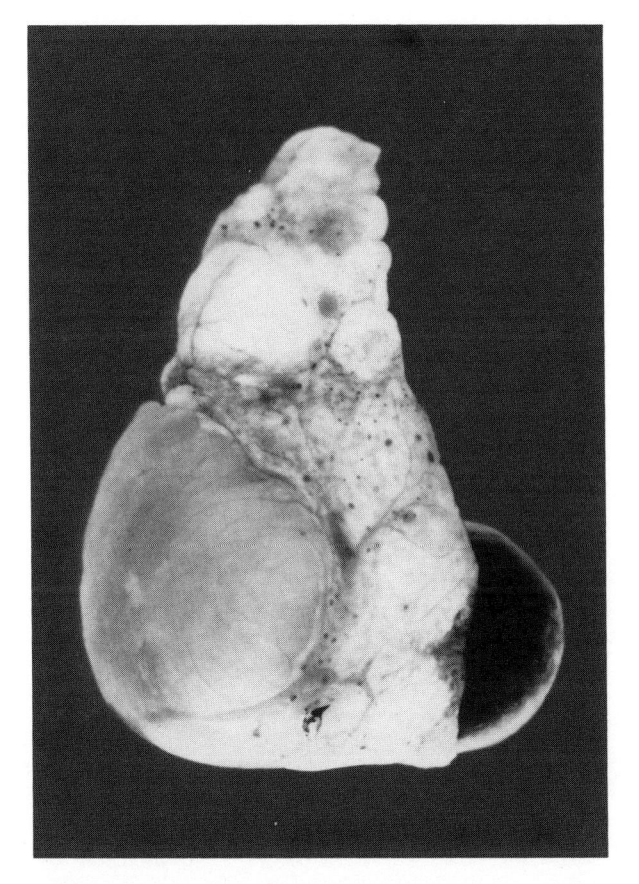

Figure 13. Multifocal, grossly visible lesions on the liver of a mummichog *(Fundulus heteroclitus)* from the Elizabeth River, VA.[119]

collected from a polluted site (Boston Harbor).[33] It was not established whether this was a heritable or acquired condition, but it is possible that populations of fish from chronically polluted sites have evolved mechanisms that result in increased tolerance to chronic chemical insult. Studies that have addressed the effects of altered genetics on tumorigenicity include those in *Poeciliopsis* clones and hybrids.[68,70] In the latter study, nine hybrid clones of nine different wild-type genotypes were compared for their response to DEN. A spectrum of tumor incidence was observed, suggesting there is a broad variability in the sensitivity of individuals in wild populations to carcinogens. In medaka,[205] where two inbred medaka strains were examined with very different sensitivities to MNNG, a transplantable melanoma was formed in only one strain. Until we have a much deeper understanding of the influence of genetics on carcinogenesis in fish, these issues will continue to blur our understanding of fish carcinogenesis. Recent advances in the genetics of oncogenesis are discussed in Chapter 7.

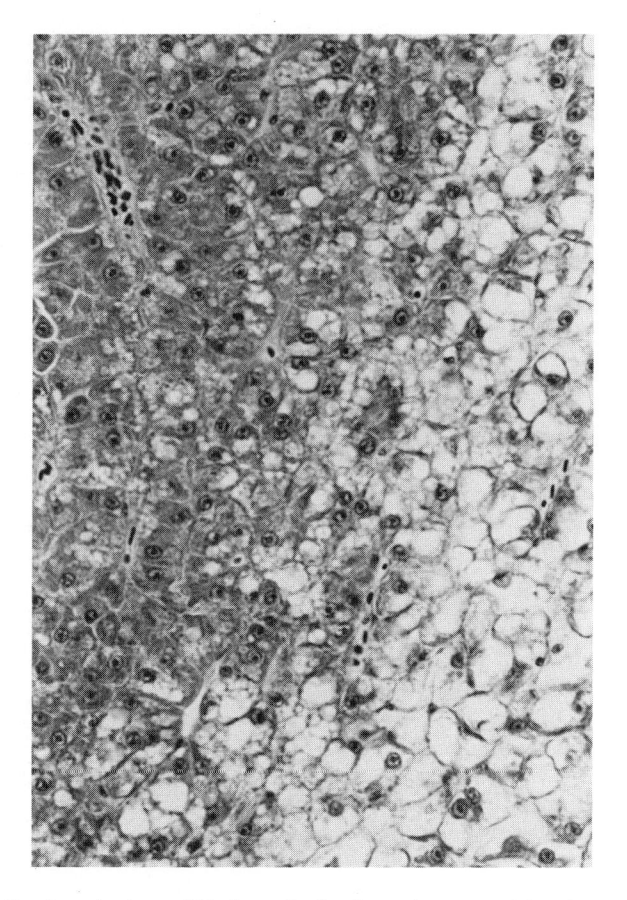

Figure 14. Section of a basophilic focus in the liver of a mummichog from the Elizabeth River, VA. Hematoxylin and eosin.[119]

E. Lesion Terminology and Diagnostic Criteria

The need for consistent diagnostic criteria and terminology for hepatic tumors and related lesions has been discussed previously in Section II.E. Differences in lesion criteria and terminology have made it difficult to directly compare lesion data generated in field and laboratory studies conducted by different investigators. Furthermore, there is a relative lack of knowledge regarding biological behavior of neoplastic lesions in fish as they relate to morphological features of lesion types, as compared to the relative prognostic accuracy of tumor diagnosis in mammalian systems where a mass of clinical information exists. This has led to a poor prognostic capability for morphological diagnosis in fish tumors. These problems are epitomized in the need to standardize the morphological criteria and degree of conservatism with which tumors are staged as benign or malignant in fish, recognizing the lack of

Figure 15. Section of an eosinophilic focus in the liver of a mummichog from the Elizabeth River, VA. Hematoxylin and eosin.[119]

information on biological behavior of these neoplasms. The extreme rarity of metastasis in fish vs. mammals makes it a relative nonissue, if one regards metastasis as a prerequisite for diagnosis of a malignancy. This position fails to acknowledge the broad variation in the degree of anaplasia and invasiveness seen in fish tumors. Perhaps, given the general lack of time series data on the ultimate fate of particular lesion types, it would be better to avoid all inferences of benign vs. malignant nature and simply grade neoplasms in a standard fashion, as suggested by Dawe[1] and Foulds.[206] Descriptive parameters should include degree of differentiation, extent of architectural disruption, cellular atypia, invasiveness, mitotic activity, and probable parent cell type. This last parameter is of particular concern in the liver, because hepatocytes and cholangiocytes may in fact arise from the same stem cell,[73] and any attempt to assign a poorly differentiated lesion to one or the other category is bound to raise controversy. Future developments in immunohistochemical characterization of

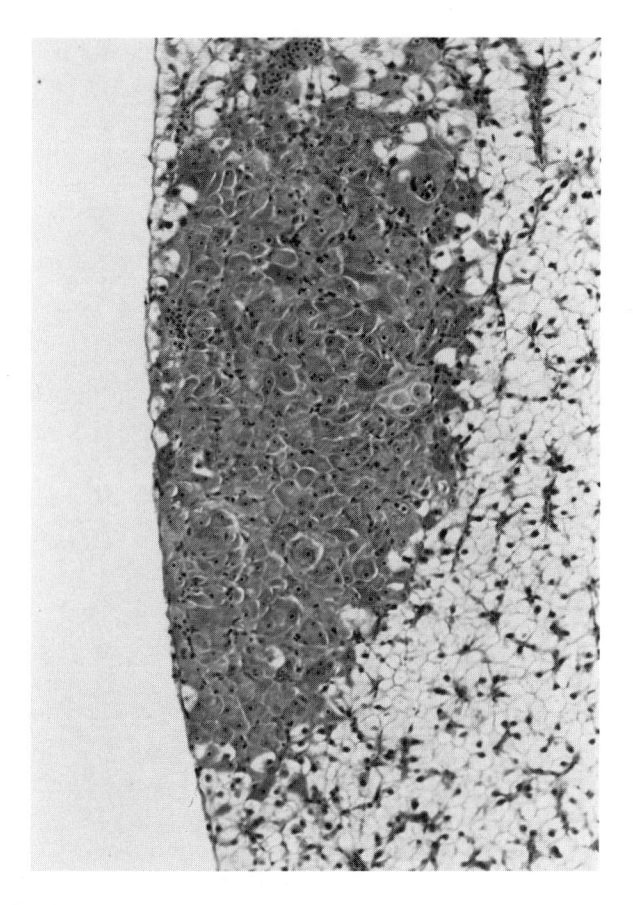

Figure 16. Section of a hepatocellular carcinoma in the liver of a mummichog from the Elizabeth River, VA. Hematoxylin and eosin.[119]

fish tumor cell types, such as via localization and identification of cytokeratins and cell surface glycoproteins, will be important in this regard.

F. Field Studies and Biomonitoring Programs Investigating Links Between Chemical Exposure and Neoplasms and Related Lesions in Fish; Strengths and Limitations

The studies showing the strongest and most consistent correlations between exposure to particular chemical contaminant classes and prevalences of toxicopathic hepatic lesions, including neoplasms, are those on the English sole in Puget Sound.[28,29,41,42,207,208] Sediment levels of PAHs correlate strongly with prevalences of liver lesions, including neoplasms, foci of cellular alteration, and the presumably degenerative/necrotic condition, megalocytic hepatosis. Sediment PCBs also correlate with liver neoplasm prevalence. Subsequent studies have further shown correlative links between sediment levels of PAHs, PCBs, and DDTs and hepatic lesions, including neoplasms, in English sole, as well as with other toxicopathic hepatic lesions in white croaker and starry flounder.[124] As measures of chemical bioaccumulation, hepatic levels of chlorinated compounds such as PCBs and DDTs have also been statistically associated with prevalences of hepatic neoplasms, altered foci, and other toxicopathic lesions in subsequent studies in English sole,[124,209] as well as with certain toxicopathic hepatic lesions (excluding neoplasms) in white croaker *(G. lineatus)* and starry flounder *(Platichthys stellatus).*[124] Similar field studies in winter flounder have shown statistical correlations between exposure to PAHs, DDTs, and chlordanes and prevalences of hydropic vacuolation of biliary epithelial cells and hepatocytes in the liver,[35] a lesion type thought to be involved in the histogenesis of hepatic neoplasia in this species.[33,151,201,210]

Other efforts to link chemical contamination to specific lesions, including hepatic and epidermal neoplasms, in freshwater species and ecosystems have focused on the Great Lakes Region, in brown bullheads,[3,21] suckers, and other species.[5,211] These studies are not comparable in scope to the extensive, multidisciplinary studies on the relationships between contaminant exposure and hepatic neoplasia in marine fish species;[35,124] they often do not examine all fish by histopathology,[21] and they lack rigorous statistical analyses of the relationship between chemical contaminants and observed effects. Nevertheless, the preponderance of the epizootiological and chemical data supports a positive influence of chemical contaminants, especially PAHs, on the etiology of neoplasms in these species.[212]

Such field studies on the hepatic histopathology in native fish exposed to chemical contaminants have provided information useful in formulating hypotheses that can be tested in laboratory exposure experiments. These contributions include (1) the correlational evidence and risk factor analyses showing a relationship between contaminant exposure and the presence of hepatic neoplasms and related lesions, which are therefore useful as histopathological biomarkers of contaminant exposure effects;[28,29,35,41,122,124,207–209,213] (2) data on hepatic lesions and age distribution of affected fish from which the probable histogenesis of hepatic neoplasia may be inferred;[33,50,123,124,151,152,201] (3) data on apparent species differences in susceptibility to contaminant-associated hepatic disease, including neoplasia;[41,124,214] and (4) data suggesting adverse health effects in fish affected by toxicopathic hepatic lesions.[132] Moreover, nationwide biomonitoring programs such as NOAA's National Benthic Surveillance Project and EPA's Environmental Monitoring and Assessment Program can provide consistent datasets of chemical, biochemical, and pathological variables.

However, the significant limitations of these and similar field studies must be realized. First, although certain chemical classes such as PAHs and organochlorines have been statistically identified as risk factors for hepatic neoplasia and other lesions, native fish are exposed to a complex mixture of contaminants. Several of these contaminant classes are known to covary in sediments and tissues,[35,41,124] and it is therefore difficult to clearly separate the relative influences of these variables on the risk of hepatic neoplasia or related lesions. Approaches to deal with this covariance of chemical classes have included the identification of covarying chemical classes by principal components analyses and subsequent correlational analysis between lesion types and summed levels of these classes.[41] However, this method still cannot separate the relative influence of each covarying class on disease risk. Moreover, epizootiological studies of fish disease have an inherent risk of identifying chemical risk factors that are simply covariant with another unmeasured chemical. Ultimately, hypotheses based on these types of relationships can only be tested, and cause and effect established, in long-term laboratory exposures. These experiments have been difficult to conduct on feral species because of the relative lack of background information on their husbandry.

Of additional concern is the fact that identified chemical risk factors for hepatic disease, including neoplasms, are often based on potential exposure (i.e., sediment levels), with actual exposure data as reflected by levels in liver tissue, bile, and stomach contents typically derived from composites, or means of multiple samples representing a site, rather than by the optimal method of analyzing chemicals in tissues of individual fish matched with histopathological and biological data. To date, the pragmatic considerations of cost of chemical analyses have dictated such an approach, but future improvements in analytical methodology that reduce cost and improve turnaround time of analytical results, such as rapid screening procedures for aromatic compounds, their metabolites in bile, and PCB congeners,[130] may permit more analyses of tissues from individual fish. Moreover, to date, there are no accurate measures of chronic exposure to PAHs, due to their extensive metabolism to compounds not detectable by routine methods. The most appropriate current method to estimate chronic PAH exposure, the quantitation of xenobiotic-DNA adducts by the P^{32} postlabeling method,[125,215,216] can be applied routinely to analyses in individual fish in field studies to address this need. These types of chemical exposure data in individual fish, as well as biological data on age, gender, and reproductive status, can then be incorporated into multivariate analyses (e.g., logistic regression) that determine the contribution of each variable to the risk of hepatic neoplasia or other lesions in individual fish. Such a strategy is currently being applied within the National Benthic Surveillance Project.

In summary, national biomonitoring programs have established the United States as one of the few countries in the world that has a strong commitment to and a reasonable understanding of the nature and extent of the chemical contamination of its coastal waters and the impact of such exposure on marine biota on a national and regional basis.

VII. POTENTIAL OF APPLYING CURRENT TECHNIQUES TO RESEARCH ON CHEMICAL INDUCTION OF NEOPLASIA IN FISH

A major focus of the preceding discussion has been progress and limitations in the study of the morphology of fish tumors and related lesions in the past 30 years. Future studies will continue to need a good descriptive basis, but there are a number of recently developed approaches that should be more fully integrated with classic histopathology. These include the following.

A. Examination of Potential Stem Cell Populations by Immunohistochemistry

We have discussed the need for more specific and accurate identification of cell types within preneoplastic and neoplastic lesions, especially in the liver. In mammalian cancer biology, there is an ever-growing range of reagent antibodies that allow the specific immunohistochemical identification of cell types. Important antigens include a series of cytokeratins, oncogenes, suppressor genes, α-fetoprotein, cell surface glycoproteins, and many others. In a recent study of 177 human hepatic tumors, 32 cases were reclassified once immunohistochemical data had been added to histopathological observations.[217] A few of these antigens have been successfully studied in the rainbow trout[218] and northern pike.[219] However, there is an underlying limitation; until the epitope of interest has been shown to be conserved in the fish species of interest, and the antibody has been shown by ELISA and Western blotting to be specific for the same antigen and to be functional as an immunohistochemical reagent, it is hard to have adequate confidence in the results generated using commercially available probes of mammalian origin. Furthermore, many antibodies are limited to use in frozen sections, whereas most archival material is paraffin embedded. Optimal method of fixation is also an important factor in successful immunohistochemical localization of many antibodies. An essential element in the validation and screening of any potential antibody is to have positive control tissue, preferably from the species of interest. Very few antigens are adequately characterized biochemically and genetically in fish to allow this to be a reality for fish tumor studies. Two studies that have applied immunohistochemical methods in fish successfully use antibodies generated to specific epitopes of the species of interest, those being to an oncofetal antigen in *Xiphophorus*[220] and to cytochrome P4501A in scup[166] and cod.[221] These latter reagents have also, through extensive biochemical characterization, been shown to specifically react with the same P450 isozyme in many fish and mammal species. They are, therefore, appropriate reagents for interspecies comparisons in fish. Other immunohistochemical characterizations in fish include use of mammalian antibodies to S-100 neuropeptide in coho salmon ependymoblastomas[222] and cytokeratins[224,225] in medaka and striped bass.[263] Mammalian antibodies that are available, but have yet to our knowledge to be adequately evaluated for use in the study of teleost hepatocarcinogenesis, include oval cell antibodies,[223] and connexins.[226-228]

The problem of epitope comparability is avoided in immunohistochemical studies using an exogenous antigen. An example of this is cell proliferation studies using the thymidine analogue, bromodeoxyuridine described above in Section VI.C.[152,201,229] The antibody is of a known specificity to the antigen that is administered to the animal before sampling. This technique will be of increasing significance as studies on histogenesis of neoplasia and the effects of tumor promoters on cell proliferation become more common.

B. Study of Cultured Cells

As the in vitro maintenance of epithelial cells becomes more routine, our ability to manipulate and understand neoplastic fish hepatic epithelia will increase. A cell line from a *Poeciliopsis* tumor has been described.[230] Blair and co-workers have described primary cultures of isolated trout liver cells.[231] Jenner and co-workers describe the increased DNA content of English sole hepatocytes from a contaminated site[232] using flow cytometry. Analysis and separation of specific cell types from a heterogeneous liver with a variety of lesion types will allow characterization and understanding of the cellular dynamics of the neoplastic process.

The remaining topics are mentioned here for completeness in consideration of chemical-associated neoplasia, but are discussed more fully elsewhere in this book.

C. Study of Macromolecular Adducts

Examining bulky chemical adducts to DNA and protein has become an important technique for "molecular dosimetry."[233] There have been a number of studies in fish[104,180,215,216,234-238] that have demonstrated the utility of these adducts as persistent bioindicators of contaminant exposure in native species and within the context of experimental hepatocarcinogenesis. This subject is discussed in greater detail in Chapter 6.

D. Transgenes

There has been little use of transgenic techniques in fish tumor research, but the possibilities are numerous. The recent applications of transgenic methods in fish have been reviewed by Maclean and Penman.[239] To our knowledge, to date there has been no transfer of oncogenic sequences in fish, comparable to studies done in rodents,[240] where single and paired oncogenes have been transferred into mice, with resultant synergism of two genes; such techniques could also be applicable to fish. However, there is an abstract of a meeting presentation[241] describing the insertion of reporter gene constructs into medaka and mummichogs to detect specific toxicants. Further details of this study were not available at the time of writing.

E. Athymic Mice

Immunodeficient athymic nude mice have been used to examine the tumorigenicity of Xiphophorine melanoma cells[242] and rodent fibroblasts transfected with DNA from winter flounder liver tumors.[243] The ability to use such murine models to test the tumorigenicity of transformed teleost cells should allow further characterization of specific oncogenic situations in many fish tumor models.

SUMMARY

In this chapter, we have shown that:

- There are strong and consistent statistical correlates between prevalences of hepatic neoplasms and neoplasia-related lesions in bottom-feeding fish and exposure to particular classes of environmental contaminants.
- There is a spectrum of morphological diversity of liver neoplasms in fish, with some species exhibiting primarily hepatocellular lesions and others being predominantly affected by cholangiocellular types. However, the accuracy of histopathologic diagnosis of hepatic tumor types is problematical, especially in the absence of ultrastructural confirmation and/or reliable immunohistochemical markers for the common target cell types.
- Elements of the histogenesis of teleost liver neoplasia mirror issues in rodent hepatocarcinogenesis, namely, the respective significance of cytotoxicity, compensatory cell proliferation, foci of altered hepatocytes, and primitive cholangiocytes or oval cells within that process.
- There are several laboratory-adapted fish species useful in the study of hepatocarcinogenesis, including the rainbow trout, medaka, guppy, *Poeciliopsis,* and the mummichog. In contrast to what is known about hepatocarcinogenesis in the rainbow trout, medaka, guppy, and *Poeciliopsis,* considerable experimental work remains to be done to document the biochemical mechanisms underlying and histogenesis of hepatic neoplasia in native species, including the mummichog, English sole, and winter flounder. This includes the evaluation of the effects of nongenotoxic modulators of this process that are common in aquatic environments where resident fish are affected by high prevalences of hepatic neoplasms and related lesions.
- There are a number of endogenous factors that influence the probability of neoplasm induction in native and laboratory fish species, in addition to exposure to chemicals. These include age, genome, and migratory habit. In field studies examining potential associations between contaminant exposure and disease prevalence among sampling sites, it is particularly important to include fish age as a variable in multivariate analyses assessing this relationship.

- In addition, there is a growing battery of diagnostic techniques and experimental interventions that have been and will be applied to furthering our understanding of carcinogenesis in fish. These include immuno- and enzyme histochemistry, biochemical methods such as quantitation of xenobiotic-DNA adducts, cell culture, partial hepatectomy by surgical or chemical means, and various technologies currently available within molecular biology.

ACKNOWLEDGMENTS

We would like to thank two anonymous reviewers and Tracy Collier, Carla Stehr, and Lyndal Johnson for their time in consideration of this manuscript and Wolfgang Vogelbein for providing the plates for Figures 13 to 16. M.J. Moore was partially supported by Grant #NA-90-AA-D-SG480 from the Coastal Ocean Program of the National Oceanic and Atmospheric Administration to the Woods Hole Oceanographic Institution Sea Grant Program. The views expressed herein are those of the authors and do not necessarily reflect the views of NOAA or any of its subagencies. Woods Hole Oceanographic Institution Contribution #8268.

REFERENCES

1. Dawe, C.J., Phylogeny and oconogeny, *Natl. Cancer Inst. Monogr.*, 31, 1, 1968.
2. Grizzle, J.M., Black bullhead: an indicator of the presence of chemical carcinogens, in *Water Chlorination: Chemistry, Environmental Impact and Health Effects,* Jolley, R.L., Ed., Lewis Publishers, Chelsea, MI, 1985, 451.
3. Baumann, P.C., Smith, W.D., and Parland, P.K., Tumor frequencies and contaminant concentrations in brown bullheads from an industrialized river and a recreational lake, *Trans. Am. Fish. Soc.*, 116, 79, 1987.
4. Smith, I.R. and Zajdlik, B.A., Regression and development of epidermal papillomas affecting white suckers, *Catostomus commersoni* (Lacepede), from Lake Ontario, Canada, *J. Fish Dis.*, 10, 487, 1987.
5. Hayes, M.A., Smith, I.R., Rushmore, T.H., Crane, T.L., Thorn, C., Kocal, T.E., and Ferguson, H.W., Pathogenesis of skin and liver neoplasms in white suckers from industrially polluted areas in Lake Ontario, *Sci. Total Environ.*, 94, 105, 1990.
6. Harshbarger, J.C. and Clark, J.B., Epizootiology of neoplasms in bony fish of North America, *Sci. Total Environ.*, 94, 1, 1990.
7. Wellings, S.R., Neoplasia and primitive vertebrate phylogeny: echinoderms, prevertebrates, and fishes — a review, *Natl. Cancer Inst. Monogr.*, 31, 59, 1968.
8. Sonstegard, R.A. and Leatherland, J.F., Comparative epidemiology: the use of fishes in assessing carcinogenic contaminants, *Contaminant Effects on Fisheries,* Cairns, V.W., Hodson, P.V., and Nriagu, J.O., Eds., *Adv. Environ. Sci. Technol.*, 16, 223, 1984.

9. Mix, M.C., Cancerous diseases in aquatic animals and their association with environmental pollutants: a critical review of the literature, *Mar. Environ. Res.*, 20, 1, 1986.
10. Vethaak, A.D. and ap Rheinallt, T., Fish disease as a monitor for marine pollution: the case for the North Sea, *Rev. Fish Biol. Fish.*, 2, 1, 1992.
11. Dawe, C.J. and Harshbarger, J.C., Neoplasms and related disorders of invertebrate and lower vertebrate animals, *Natl. Cancer Inst. Monogr.*, 31, 772, 1968.
12. Dawe, C., Scarpelli, D.G., and Wellings, S.R., Tumors in aquatic research, *Progr. Exp. Tumor Res.*, 20, 438, 1976.
13. Kraybill, H.F., Dawe, C.J., Harshbarger, J.C., and Tardiff, R.G., Aquatic pollutants and biologic effects with emphasis on neoplasia, *Ann. N.Y. Acad. Sci.*, 298, 604, 1977.
14. Hoover, K.L., Use of small fish species in carcinogenicity testing, *Natl. Cancer Inst. Monogr.*, 65, 1, 1984.
15. Metcalfe, C.D., Chemical contaminants and fish tumors, *Sci. Total Environ.*, 94, 1, 1990.
16. Plehn, M., Über einege bei fischen beobachtet geswülste und geswülstartige bildungen, *Ber Bayer Biol. Verssta*, 2, 55, 1909.
17. Halver, J.E., Crystalline aflatoxin and other vectors for trout hepatoma, in Trout Hepatoma Research Papers, Research Report #70 Washington,*U. S. Fish and Wildlife Service*, Washington, D.C., 1967, 78.
18. Dawe, C., Stanton, M., and Schwartz, F., Hepatic neoplasms in native bottom-feeding fish of Deep Creek Lake, Maryland, *Cancer Res.*, 24, 1194, 1964.
19. Stanton, M.F., Diethylnitroseamine induced hepatic degeneration and neoplasia in the aquarium fish, *Brachydanio rerio, J. Natl. Cancer Inst.*, 34, 117, 1965.
20. Black, J.J., Fish tumors as known field effects of contaminants, Chronic Effects of Toxic Contaminants on Large Lakes, Schmidtke, N.W., Ed., *Toxic Contam. Large Lakes*, 1, 55, 1988.
21. Baumann, P.C., Mac, M.J., Smith, S.B., and Harshbarger, J.C., Tumor frequencies in walleye *(Stizostedion vitreum)* and brown bullhead *(Ictalurus nebulosus)* and sediment contaminants in tributaries of the Laurentian Great Lakes, *Can. J. Fish. Aquat. Sci.*, 48, 1804, 1991.
22. Falkmer, S., Emdin, S.O., Ostberg, Y., Mattisson, Y.A., Johansson Sjobeck, M.-L., and Fange, R., Tumor pathology of the hagfish, *Myxine glutinosa*, and the river lamprey, *Lampetra fluviatilis, Progr. Exp. Tumor Res.*, 20, 217, 1976.
23. Falkmer, S., Marklund, S., Mattsson, P.E., and Rappe, C., Hepatomas and other neoplasms in the Atlantic hagfish *(Myxine glutinosa)*: a histopathologic and chemical study, *Ann. N.Y. Acad. Sci.*, 298, 342, 1977.
24. Cormier, S.M., Racine, R.N., Smith, C.E., Dey, W.P., and Peck, T.H., Hepatocellular carcinoma and fatty infiltration in the Atlantic tomcod, *Microgadus tomcod* (Walbaum), *J. Fish Dis.*, 12, 105, 1989.
25. Smith, C.E., Peck, T.H., Klauda, H., and McLaren, J.B., Hepatomas in Atlantic tomcod *(Microgadus tomcod)* collected in the Hudson River estuary, New York, *J. Fish Dis.*, 2, 313, 1979.
26. Myers, M.S., Rhodes, L.R., and McCain, B.B., Pathologic anatomy and patterns of occurrence of hepatic neoplasms, putative preneoplastic lesions and other idiopathic hepatic conditions in English Sole *(Parophrys vetulus)* from Puget Sound, Washington, *J. Natl. Cancer Inst.*, 78, 1987.

27. Becker, D.S., Ginn, T.C., Landolt, M.L., and Powell, D.B., Hepatic lesions in English sole *(Parophrys vetulus)* from Commencement Bay, Washington (USA), *Mar. Environ. Res.*, 23, 153, 1987.
28. Myers, M.S., Landahl, J.T., Krahn, M.M., Johnson, L.L., and McCain, B.B., Overview of studies on liver carcinogenesis in English sole from Puget Sound; evidence for a xenobiotic chemical etiology. 1. Pathology and epizootiology, *Sci. Total Environ.*, 94, 33, 1990.
29. Myers, M.S., Landahl, J.T., Krahn, M.M., and McCain, B.B., Relationships between hepatic neoplasms and related lesions and exposure to toxic chemicals in marine fish from the U.S. West Coast, *Environ. Health Perspect.*, 90, 7, 1991.
30. Murchelano, R.A. and Wolke, R.E., Epizootic carcinoma in the winter flounder, *Pseudopleuronectes americanus, Science,* 228, 587, 1985.
31. Gardner, G.R., Pruell, R.J., and Folmar, L.C., A comparison of both neoplastic and non-neoplastic disorders in winter flounder *(Pseudopleuronectes americanus)* from eight areas in New England, *Mar. Environ. Res.*, 28, 393, 1989.
32. Gardner, G.R. and Pruell, R.J., Quincy Bay. A Histopathological and Chemical Assessment of Winter Flounder, Lobster and Soft-Shelled Clam Indigenous to Quincy Bay, Boston Harbor and an *In Situ* Evaluation of Oysters Including Sediment (Surface and Cores) Chemistry, U.S. Environmental Protection Agency, Region 1, Boston, MA, 1988.
33. Moore, M.J., Vacuolation, proliferation and neoplasia in the liver of winter flounder, *Pseudopleuronectes americanus,* from Boston Harbor, Massachusetts, Technical Report, 91-28, 1991, Woods Hole Oceanographic Institution, Woods Hole, MA, 1991.
34. Murchelano, R.A. and Wolke, R.E., Neoplasms and nonneoplastic liver lesions in winter flounder, *Pseudopleuronectes americanus,* from Boston Harbor, Massachusetts, *Environ. Health Perspect.*, 90, 17, 1991.
35. Johnson, L.L., Stehr, C.M., Olson, O.P., Myers, M.S., Pierce, S.M., Wigren, C.A., McCain, B.B., and Varanasi, U., Chemical contaminants and hepatic lesions in winter flounder (Pleuronectes americanus) from the north east coast of the United States, *Environ. Sci., Technol.,* 1993, in press.
36. Malins, D.C., McCain, B.M., Brown, D.W., Myers, M.S., Krahn, M.M., and Chan, S.-L., Toxic chemicals, including aromatic and chlorinated hydrocarbons and their derivatives, and liver lesions in white croaker *(Genyonemus lineatus)* from the vicinity of Los Angeles, *Environ. Sci. Technol.,* 21, 765, 1987.
37. Kranz, H. and Dethlefsen, V., Liver anomalies in dab *Limanda limanda* from the southern North Sea with special consideration given to neoplastic lesions, *Dis. Aquat. Org.,* 9, 171, 1990.
38. Vogelbein, W.K., Fournie, J.W., Veld, P.A.V., and Huggett, R.J., Hepatic neoplasms in the mummichog *Fundulus heteroclitus* from a creosote-contaminated site, *Cancer Res.,* 50, 5978, 1990.
39. Brown, E.R., Hazdra, J.J., Keith, L., Greenspan, I., Kwapinski, I., and Beamer, P., Frequency of fish tumors found in a polluted watershed as compared to nonpolluted Canadian waters, *Cancer Res.,* 33, 189, 1973.
40. Brown, E.R., Sinclair, T., Keith, L., Beamer, P., Hazdra, J.J., Nair, V., and Callaghan, O., Chemical pollutants in relation to diseases in fish, *Ann. N.Y. Acad. Sci.,* 1977.

41. Malins, D.C., McCain, B.B., Brown, D.W., Chan, S.L., Myers, M.S., Landahl, J.T., Prohaska, P.G., Friedman, A.J., Rhodes, L.D., Burrows, D.G., Gronlund, W.D., and Hodgins, H.O., Chemical pollutants in sediments and diseases of bottom-dwelling fish in Puget Sound, Washington, *Environ. Sci. Technol.*, 18, 705, 1984.
42. Malins, D.C., Krahn, M.M., Brown, D.W., Rhodes, L.D., Myers, M.S., McCain, B.B., and Chan, S.-L., Toxic chemicals in marine sediment and biota from Mukilteo, Washington: relationships with hepatic neoplasms and other hepatic lesions in English sole *(Parophrys vetulus), J. Natl. Cancer Inst.*, 74, 487, 1985.
43. Smith, I.R., Ferguson, H.W., and Hayes, M.A., Histopathology and prevalence of epidermal papillomas epidemic in brown bullhead, *Ictalurus nebulosus* (Lesuer), and white sucker, *Catostomus commersoni* (Lacepede), populations from Ontario, Canada, *J. Fish Dis.*, 12, 373, 1989.
44. Baumann, P.C., Cancer in wild freshwater fish populations with emphasis on the Great Lakes, *J. Great Lakes Res.*, 10, 251, 1984.
45. Black, J.J., Field and laboratory studies of environmental carcinogenesis in Niagara River fish, *J. Great Lakes Res.*, 9, 326, 1983.
46. Black, J.J., Fox, H., Black, P., and Bock, F., Carcinogenic effects of river sediment extracts in fish and mice, *Water Chlorination: Chemistry, Environmental Impact, and Health Effects,* al., R.L.J.e., Ed., Plenum Press, New York, 1985, 415.
47. Stich, H.F., Acton, A.B., Dunn, B.P., Oishi, K., Yamazaki, F., Harada, T., Peters, G., and Peters, N., Geographic variations in tumor prevalence among marine fish populations, *Int. J. Cancer,* 20, 780, 1977.
48. Dawe, C.J., Bagshaw, and Poore, C.M., Amebic pseudotumors in pseudobranchs of Pacific cod, *Gadus macrocephalus, Proc. Am. Assoc. Cancer Res.*, March, 245, 1979.
49. Dawe, C.J., Polyoma tumors in mice and X cell tumors in fish, viewed through telescope and microscope. 11th Int. Symp. Princess Takematsu Cancer Res. Fund, Nakahara Memorial Lecture, *Phyletic Approaches to Cancer,* Dawe, C.J. et al., Eds., Japan Scientific Society Press, Tokyo, 1981, 19.
50. Myers, M.S., Pathologic Anatomy of Papilloma-Like Tumors in the Pacific Ocean Perch, *Sebastes alutus,* from the Gulf of Alaska, University of Washington, Seattle, 1981.
51. Harshbarger, J.C., Pseudoneoplasms in ectothermic animals, *Natl. Cancer Inst. Monogr.,* 65, 251, 1984.
52. Myers, M.S., X cell pseudotumors of skin, gill, and pseudobranch (parabranchial body) in teleost fishes, in *Pathobiology of Spontaneous and Induced Neoplasms in Fishes: Comparative Characterization, Nomenclature and Literature,* Dawe, C.J., Harshbarger, J., Wellings, S.R., and Strandberg, J.D., Eds., Academic Press, New York, in press.
53. Kimura, T., Yoshimizu, M., and Tanaka, M., Studies on a new virus (OMV) from *Onchorhynchus masou* — II. Oncogenic nature, *Fish. Pathol. Tokyo,* 15, 149, 1981.
54. Khudoley, V.V., Use of aquarium fish, *Danio rerio* and *Poecilia reticulata,* as test species for evaluation of nitrosamine carcinogenicity, *Natl. Cancer Inst. Monogr.*, 65, 65, 1984.
55. Bowser, P.R., Wolfe, M.J., Forney, J.L., and Wooster, G.A., Seasonal prevalence of skin tumors from walleye *(Stizostedion vitreum)* from Oneida Lake, New York, *J. Wildl. Dis.*, 24, 292, 1988.

56. Wolf, K., *Fish Viruses and Fish Virus Diseases,* Cornell University Press, Ithaca, New York, 1988.
57. Schmale, M.C., Transmissibility of a neurofibromatosis-like disease in bicolor damselfish, *Cancer Res.,* 48, 3828, 1988.
58. Bowser, P.R., Martineau, D., and Wooster, G.A., Effects of water temperature on experimental transmission of dermal sarcoma in fingerling walleyes, *J. Aquat. Anim. Health,* 2, 157, 1990.
59. Martineau, D., Bowser, P., Wooster, G., and Armstrong, L., Experimental transmission of a dermal sarcoma in fingerling walleyes *(Stizostedion vitreum), Vet. Pathol.,* 27, 230, 1990.
60. Martineau, D., Renshaw, R., Williams, J., Casey, J., and Bowser, P., A large unintegrated retrovirus DNA species present in a dermal tumor of walleye *Stizostedion vitreum, Dis. Aquat. Org.,* 10, 153, 1991.
61. Bernstein, J.W., Leukaemic lymphosarcoma in a hatchery-reared rainbow trout, *Salmo gairdneri,* Richardson, *J. Fish. Dis.,* 7, 83, 1984.
62. Dawe, C.J., Oncozoons and the search for carcinogen-indicator fishes, *Environ. Health Perspect.,* 71, 129, 1987.
63. Masahito, P., Ishikawa, T., and Takayama, S., Spontaneous spermatocytic seminoma in African lungfish, *Protopterus aethiopicus* Heckel, *J. Fish Dis.,* 7, 169, 1984.
64. Masahito, P., Aoki, K., Egami, N., Ishikawa, N., and Sugano, H., Life-span studies on spontaneous tumor development in the medaka *(Oryzias latipes), Jpn. J. Cancer Res.,* 80, 1058, 1989.
65. Torikata, C., Mukai, M., and Kageyama, K., Spontaneous olfactory neuroepithelioma in a domestic medaka *(Oryzias latipes), Cancer Res.,* 49, 2994, 1989.
66. Sinnhuber, R.O., Hendricks, J.D., Wales, J.H., and Putnam, G.B., Neoplasms in rainbow trout, a sensitive animal model for environmental carcinogenesis, *Ann. N.Y. Acad. Sci.,* 28, 398, 1976.
67. Lee, B.C., Hendricks, J.D., and Bailey, G.S., Metaplastic pancreatic cells in liver tumors induced by diethylnitroseamine, *Exp. Mol. Pathol.,* 50, 104, 1989.
68. Schultz, M.E. and Schultz, R.J., Diethylnitrosamine-induced hepatic tumors in wild vs. inbred strains of a viviparous fish, *J. Hered.,* 73, 43, 1982.
69. Schultz, R.J. and Schultz, M.E., Characteristics of a fish colony of *Poeciliopsis* and its use in carcinogenicity studies with 7,12-dimethylbenza[a]anthracene and diethylnitroseamine, *Natl. Cancer Inst. Monogr.,* 65, 5, 1984.
70. Schultz, M.E. and Schultz, R.J., Differences in response to a chemical carcinogen within species and clones of the livebearing fish, *Poeciliopsis, Carcinogenesis,* 9, 1029, 1988.
71. Ward, J.M., Hendricks, J.D., and Hinton, D.E., The liver, in *Pathobiology of Spontaneous and Induced Neoplasms in Fishes: Comparative Characterization, Nomenclature and Literature,* Dawe, C.J., Harshbarger, J., Wellings, S.R., and Strandberg, J.D., Eds., Academic Press, New York, in press.
72. Edmondson, H. and Craig, J., Neoplasms of the liver, in *Diseases of the Liver,* Lippincott, Philadelphia, PA, 1987, 1109.
73. Sell, S., Is there a liver stem cell?, *Cancer Res.,* 50, 3811, 1990.
74. Bannasch, P., Enzmann, H., Klimek, F., Weber, E., and Zerban, H., Significance of sequential cellular changes inside and outside foci of altered hepatocytes during hepatocarcinogenesis, *Toxicol. Pathol.,* 17, 617, 1989.

75. Elias, H. and Bengelsdorf, H., The structure of the liver of vertebrates, *Acta Anat.*, 14, 297, 1952.
76. Elias, H. and Sherrick, J.C., *Morphology of the Liver*, Academic Press, New York, 1969, 8.
77. Hampton, J.A., Lantz, R.C., and Hinton, D.E., Functional units in rainbow trout *(Salmo gairdneri,* Richardson) liver: III. Morphometric analysis of parenchyma, stroma, and component cell types, *Am. J. Anat.*, 185, 58, 1989.
78. Hampton, J.A., Lantz, R.C., Goldblatt, P.J., Lauren, D.J., and Hinton, D.E., Functional units in rainbow trout *(Salmo gairdneri)* liver: II. The biliary system, *Anat. Rec.*, 221, 619, 1988.
79. Hampton, J.A., McCuskey, P.A., McCuskey, R.S., and Hinton, D.E., Functional units in rainbow trout *(Salmo gairdneri)* liver: I. Arrangement and histochemical properties of hepatocytes, *Anat. Rec.*, 213, 166, 1985.
80. Rabes, H.M., Bucher, T., Hartmann, A., Linke, and Dunnwold, M., Clonal growth of carcinogen-induced enzyme-deficient preneoplastic cell populations in mouse liver, *Cancer Res.*, 42, 3220, 1982.
81. Williams, G.M., The significance of preneoplastic liver lesions in experimental animals, *Adv. Vet. Sci. Comp. Med.*, 31, 21, 1987.
82. Popp, J.A. and Goldsworthy, T.L., Defining foci of cellular alteration in short-term and medium-term rat liver tumor models, *Toxicol. Pathol.*, 17, 561, 1989.
83. Maronpot, R.R., Harada, T., Murthy, A.S.K., and Boorman, G., Documenting foci of hepatocellular alteration in two-year carcinogenicity studies: current practices of the National Toxicology Program, *Toxicol. Pathol.*, 17, 675, 1989.
84. Stewart, H.L., Williams, G.L., Keysser, C.H., Lombard, L.S., and Montali, R.J., Histologic typing of liver tumors in the rat, *J. Natl. Cancer Inst.*, 64, 179, 1980.
85. Frith, C.H. and Ward, J.L., A morphologic classification of proliferative and neoplastic hepatic lesions in mice, *J. Environ. Pathol. Toxicol.*, 3, 329, 1980.
86. Maronpot, R.R., Montgomery, C.A., Boorman, G.R., and McConnell, E.E., National Toxicology Program nomenclature for hepatoproliferative lesions of rats, *Toxicol. Pathol.*, 14, 263, 1986.
87. Harada, T., Maronpot, R.R., Morris, R.W., Stitzel, K.A., and A, B.G., Morphological and stereological characterization of hepatic foci of cellular alteration in control Fischer 344 rats, *Toxicol. Pathol.*, 17, 579, 1989.
88. Harada, T., Maronpot, R.R., Morris, R.W., and Boorman, G.A., Observations on altered hepatocellular foci in National Toxicology Program two-year carcinogenicity studies in rats, *Toxicol. Pathol.*, 17, 690, 1989.
89. Farber, E. and Sarma, D.S.R., Hepatocarcinogenesis: a dynamic cellular perspective, *Lab. Invest.*, 56, 4, 1987.
90. Maronpot, R.R. and Boorman, G.A., Interpretation of rodent hepatoproliferative alterations and hepatocellular tumors in chemical safety assessment, *Toxicol. Pathol.*, 10, 71, 1982.
91. Mori, H., Tanaka, T., Nishikawa, A., Takahashi, M., and Williams, G.M., DNA content of liver cell nuclei of N-2-fluorenylacetamide-induced altered foci in rats and human hyperplastic foci, *J. Natl. Cancer Inst.*, 69, 1277, 1982.
92. Weber, E., Moore, R.A., and Bannasch, P., Enzyme histochemical and morphological phenotype of amphophilic foci and amphophilic/tigroid cell adenomas in rat liver after combined treatment with dehydroepiandrosterone and N-nitrosomorpholine, *Carcinogenesis*, 9, 1049, 1988.

93. Zerban, H., Rabes, H.M., and Bannasch, P., Sequential changes in growth kinetics and cellular phenotype during hepatocarcinogenesis, *J. Cancer Clin. Oncol.*, 115, 329, 1989.
94. Farber, E., Similarities in the sequence of early histological changes induced in the liver of the rat by ethionine, 2-acetylaminofluorene, and 3′methyl-4-dimethylaminoazobenzene, *Cancer Res.*, 16, 142, 1956.
95. Tsao, M.-S. and Grisham, J.W., Hepatocarcinomas, cholangiocarcinomas, hepatoblastomas produced by chemically transformed cultured rat liver epithelial cells, *Am. J. Pathol.*, 127, 168, 1987.
96. Ward, J.M. and Vlahakis, G., Evaluation of hepatocellular neoplasms in mice, *J. Natl. Cancer Inst.*, 61, 807, 1978.
97. Shiojiri, N., Lemire, J.M., and Fausto, N., Cell lineages and oval cell progenitors in rat liver development, *Cancer Res.*, 51, 2611, 1991.
98. Grisham, J.W., Cell types in long-term propagable cultures of rat liver, *Ann N.Y. Acad. Sci.*, 349, 128, 1980.
99. Evarts, R.P., Nagy, P., Marsden, E., and Thorgeirsson, S., A precursor-product relationship exists between oval cells and hepatocytes in rat liver, *Carcinogenesis*, 8, 1737, 1987.
100. Germain, L., Blouin, M.-J., and Marceau, N., Biliary epithelial and hepatocytic cell lineage relationships in embryonic rat liver as determined by the differential expression of cytokeratins, alpha-fetoprotein, albumin and cell surface-exposed components, *Cancer Res.*, 48, 4909, 1988.
101. Hendricks, J.D., Meyers, T.R., and Shelton, D.W., Histological progression of hepatic neoplasia in rainbow trout *(Salmo gairdneri)*, *Natl. Cancer Inst. Monogr.*, 65, 321, 1984.
102. Lee, B.C., Hendricks, J.D., and Bailey, G.S., Iron resistance of hepatic lesions and nephroblastoma in rainbow trout (Salmo gairdneri) exposed to MNNG, *Toxicol. Pathol.*, 17, 474, 1989.
103. Williams, G.M., The pathogenesis of liver cancer caused by chemical carcinogens, *Biochim. Biophys. Acta*, 605, 167, 1980.
104. Goeger, D.E., Shelton, D.W., Hendricks, J.D., Pereira, C., and Bailey, G.S., Comparative effect of dietary butylated hydroxyanisole and beta-naphthoflavone on aflatoxin B1 metabolism, DNA adduct formation, and carcinogenesis in rainbow trout, *Carcinogenesis*, 9, 1793, 1988.
105. Nunez, O., Hendricks, J.D., Arbogast, D.N., Fong, A.T., Lee, B.C., and Bailey, G.S., Promotion of aflatoxin B_1 hepatocarcinogenesis in rainbow trout by 17-beta-estradiol, *Aquat. Toxicol.*, 15, 289, 1989.
106. Nunez, O., Hendricks, J.D., and Duimstra, J.R., Ultrastructure of hepatocellular neoplasms in aflatoxin B_1 (AFB_1)-initiated rainbow trout *(Oncorhynchus mykiss)*, *Toxicol. Pathol.*, 19, 11, 1991.
107. Nunez, O., Hendricks, J.D., and Fong, A.T., Inter-relationships among aflatoxin B_1 (AFB_1) metabolism, DNA-binding, cytotoxicity, and hepatocarcinogenesis, in rainbow trout *Oncorhynchus mykiss*, *Dis. Aquat. Org.*, 9, 15, 1990.
108. Wales, J.H., Degeneration and regeneration of liver parenchyma accompanying hepatomagenesis, Trout Hepatoma Research Conference Papers, Halver, J.E. and Mitchell, I.A., Eds., Research Report #70, Bureau of Sport Fisheries and Wildlife, Washington, D.C., 1967, 56.

109. Metcalfe, C.D., Tests for predicting carcinogenicity in fish, *Rev. Aquat. Sci.*, 1, 111, 1989.
110. Hinton, D.E., Couch, J.A., Teh, S.T., and Courtney, L.A., Cytological changes during progression of neoplasia in selected fish species, *Aquat. Toxicol.*, 11, 77, 1988.
111. Hawkins, W.E., Walker, W.W., Overstreet, R.M., Lytle, T.F., and Lytle, J.S., Dose-related carcinogenic effects of water-borne benzo[a]pyrene on livers of two small fish species, *Ecotoxicol. Environ. Safety*, 16, 219, 1988.
112. Hawkins, W.E., Walker, W.W., Overstreet, R.M., Lytle, J.S., and Lytle, T.F., Carcinogenic effects of some polycyclic aromatic hydrocarbons on the Japanese medaka and guppy in water borne exposures, *Sci. Total Environ.*, 94, 155, 1990.
113. Bunton, T., Hepatopathology of diethylnitrosamine in the medaka *(Oryzias latipes)* following short-term exposure, *Toxicol. Pathol.*, 18, 313, 1990.
114. Hinton, D.E., Lauren, D.J., Teh, S.J., and Giam, C.S., Cellular composition and ultrastructure of hepatic neoplasms induced by diethylnitrosamine in *Oryzias latipes*, *Mar. Environ. Res.*, 24, 307, 1988.
115. Lauren, D.J., Teh, S.J., and Hinton, D.E., Cytotoxicity phase of diethylnitrosamine-induced hepatic neoplasia in medaka, *Cancer Res.*, 50, 5504, 1990.
116. Hawkins, W.E., Overstreet, R.M., and Walker, W.W., Ultrastructual analysis of hepatic hepatocarcinogenesis in two small fish species, in *Proceedings of the 44th Annual Meeting of the Electron Microscopy Society of America,* San Francisco Press, Inc., San Francisco, CA, 1986.
117. Schiewe, M.H., Weber, D.D., Myers, M.S., Jacques, F.J., Reichert, W.L., Krone, C.A., Malins, D.C., McCain, B.B., Chan, S.-L., and Varanasi, U., Induction of foci of cellular alteration and other hepatic lesions in English sole *(Parophrys vetulus)* exposed to an extract of an urban marine sediment, *Can. J. Fish. Aquat. Sci.*, 48, 1750, 1991.
118. Bodammer, J.E. and Murchelano, R.A., Cytological study of vacuolated cells and other aberrant hepatocytes in winter flounder from Boston Harbor, *Cancer Res.*, 50, 6744, 1990.
119. Vogelbein, W.K., Personal communication.
120. Myers, M.S., Unpublished observations.
121. Baumann, P.C., Harshbarger, J.C., and Hartman, K.J., Relationship between liver tumors and age in brown bullhead populations from two Lake Erie tributaries, *Sci. Total Environ.*, 94, 71, 1990.
122. Landolt, M.L., Holmes, E.H., and Ostrander, G.K., Preneoplastic cellular changes associated with exposure to environmental contaminants in Puget Sound, Washington, *Mar. Environ. Res.*, 17, 334, 1985.
123. Rhodes, L.D., Myers, M.S., Gronlund, W.D., and McCain, B.B., Epizootic characteristics of hepatic and renal lesions in English sole, *Parophrys vetulus,* from Puget Sound, *J. Fish Biol.*, 31, 395, 1987.
124. Myers, M.S., Stehr, C.S., Olson, O.P., Johnson, L.L., McCain, B.B., Chan, S.-L., and Varanasi, U., Relationships between toxicopathic hepatic lesions and exposure to chemical contaminants in English sole *(Pleuronectes vetulus),* starry flounder *(Platichthys stellatus)* and white croaker *(Genyohemus lineatus)* from selected marine sites on the Pacific Coast, U.S.A., *Env. Health Perspect.*, 102, 1, 1994, in press.

125. Varanasi, U., Reichert, W.L., Le Eberhart, B.T., and Stein, J.E., Adducts in liver of wild English sole *(Parophrys vetulus)* and winter flounder *(Pseudopleuronectes americanus), Cancer Res.,* 49, 1171, 1989.
126. Varanasi, U., Reichert, W.L., Le Eberhart, B.T., and Stein, J.E., Formation and persistence of benzo[a]pyrene-diolepoxide-DNA adducts in liver of English sole *(Parophrys vetulus), Chem.-Biol. Interact.,* 69, 203, 1986.
127. Breslow, N.E. and Day, N.E., *Statistical Methods in Cancer Research,* 1st ed., International Agency for Research on Cancer (IARC), Lyon, France, 1980.
128. Etoh, H., Hyodo-Taguchi, Y., Aoki, K., Murata, M., and Matsudaira, H., Incidence of chromatoblastomas in aging goldfish *(Carassius auratus), J. Natl. Cancer Inst.,* 70, 523, 1983.
129. Hinton, D.E., Personal communication.
130. Krahn, M.M., Ylitalo, G.M., Buzitis, J., Chan, S.-L., and Varanasi, U., Rapid high-performance liquid chromatographic methods that screen for aromatic compounds in environmental samples, *J. Chromatogr.,* 1993, in press.
131. Lotrich, V.A., Summer home range and movements of *Fundulus heroclitus* (Family: Cyprinonodontidae) in a tidal creek, *Ecology,* 56, 191, 1975.
132. Casillas, E., Myers, M.S., Rhodes, L.D., and McCain, B.B., Serum chemistry of diseased English sole, *Parophrys vetulus* Girard, from polluted areas of Puget Sound, Washington, *J. Fish Dis.,* 8, 437, 1985.
133. McCain, B.B., Myers, M.S., Varanasi, U., Brown, D.W., Rhodes, L.D., Gronlund, W.D., Elliott, D.G., Palsson, W.A., Hodgins, H.O., and Malins, D.C., Pathology of Two Species of Flatfish from Urban Estuaries in Puget Sound, NOAA/EPA, EPA-600/7-82-00, 1982.
134. Johnson, L.L. and Landahl, J.T., Chemical contaminants, liver disease, and mortality rates in English sole *(Pleuronectes vetulus) Ecol. Appl.,* in press.
135. Barnthouse, L.W., Population-level effects, in *Ecological Risk Assessment,* G.W.S., Lewis Publishers, Chelsea, MI, 1993, 247.
136. Hawkins, W.E., Overstreet, R.M., Fournie, J.W., and Walker, W.W., Development of aquarium fish models for environmental carcinogenesis: tumor induction in seven species, *J. Appl. Toxicol.,* 5, 261, 1985.
137. Kimura, I., Taniguchi, N., Kumai, H., Tomita, I., Kinae, N., Yoshizaki, K., Ito, M., and Ishikawa, T., Correlation of epizootiological observations with experimental data: chemical induction of chromatophoromas in the croaker, *Nibea mitsukurii, Natl. Cancer Inst. Monogr.,* 65, 139, 1984.
138. Black, J.J., Maccubbin, A.E., and Schiffert, M., A reliable, efficient, microinjection apparatus and methodology for the *in vivo* exposure of rainbow trout and salmon embryos to chemical carcinogens, *J. Natl. Cancer Inst.,* 75, 1123, 1985.
139. Metcalfe, C.D., Cairns, V.W., and Fitzsimons, J.D., Experimental induction of liver tumors in rainbow trout *(Salmo gairdneri)* by contaminated sediment from Hamilton Harbour, Ontario, *Can. J. Fish. Aquat. Sci.,* 45, 2161, 1988.
140. White, D.S.J., Bowers, D., Jude, D., Moll, R., Hendricks, S., Mansfield, P., and Flexner, M., Exposure and Biological Effects of In-Place Pollutants (Bioassays), U.S. Environmental Protection Agency, Large Lakes Research Station, Grosse Ile, MI, 1987.
141. Manny, B.A. and Kenaga, D., The Detroit River: effects of contaminants and human activities on aquatic plants and animals and their habitats, *Hydrobiologia,* 219, 269, 1991.

142. Maueler, W., Raulf, F., and Schartl, M., Expression of proto-oncogenes in embryonic, adult, and transformed tissue of Xiphophorus (Teleostei: Poeciliidae), *Oncogene*, 2, 421, 1988.
143. Anders, F., Schartl, M., and Barnekow, A., Xiphophorus as an in vivo model for studies on oncogenes, *Natl. Cancer Inst. Monogr.*, 65, 1984.
144. Anders, A., Groger, H., Anders, F., Zechel, C., Smith, A., and Schlatterer, B., Discrimination of initiating and promoting carcinogens in fish, *Ann. Rech. Vet.*, 22, 273, 1991.
145. Hendricks, J.D., Wales, J.H., Sinnhuber, R.O., Nixon, J.E., Loveland, P.M., and R. A., S., Rainbow trout *(Salmo gairdneri)* embryos: a sensitive animal model for experimental carcinogenesis, *Fed. Proc.*, 39, 3222, 1980.
146. Solt, D.B. and Farber, E., New principle for the analysis of chemical carcinogenesis, *Nature (London)*, 263, 701, 1976.
147. Kyono-Hamaguchi, Y., Effects of temperature and partial hepatectomy on the induction of liver tumors in *Oryzias latipes, Natl. Cancer Inst. Monogr.*, 65, 337, 1984.
148. Kotsanis, N. and Metcalfe, C.D., Enhancement of hepatocarcinogenesis in rainbow trout with carbon tetrachloride, *Bull. Environ. Contam. Toxicol.*, 46, 879, 1991.
149. Ostrander, G.K., Blair, J.B., Hurst, J., and Stark, B., Response of rainbow trout to partial hepatectomy, *Aquat. Toxicol.*, 25, 31, 1993.
150. Fausto, N., Oval cells and liver carcinogenesis: an analysis of cell lineages in hepatic tumors using oncogene transfection techniques, in *Mouse Liver Carcinogenesis: Mechanisms and Species Comparisons,* Alan R. Liss, New York, 1990, 325.
151. Moore, M.J., Smolowitz, R., and Stegeman, J.J., Cellular alterations preceding neoplasia in *(Pseudopleuronectes americanus)* from Boston Harbor, *Mar. Environ. Res.*, 28, 425, 1989.
152. Moore, M.J. and Stegeman, J.J., Bromodeoxyuridine uptake in hydropic vacuolation and neoplasms in winter flounder liver, *Mar. Environ. Res.*, 34, 13, 1992.
153. Hinton, D.E. and Pool, C.R., Ultrastructure of the liver in channel catfish *Ictalurus punctatus* (Rafinesque), *J. Fish Biol.*, 8, 209, 1976.
154. Hinton, D.E., Lantz, R.C., and Hampton, J.A., Effect of age and exposure to a carcinogen on the medaka liver: a morphometric study, *Natl. Cancer Inst. Monogr.*, 65, 239, 1984.
155. Parland, W.K. and Baumann, P.C., Pathology and tumor development through time in guppies dosed with diethylnitrosamine (DEN), *J. Appl. Toxicol.*, 5, 265, 1985.
156. Stehr, C.M., Rhodes, L.D., and Myers, M.S., The ultrastructure and histology of hepatocellular carcinomas of English sole *(Parophrys vetulus)* from Puget Sound, Washington, *Toxicol. Pathol.*, 16, 418, 1988.
157. Stehr, C.M. and Myers, M.S., The ultrastructure and histology of cholangiocellular carcinomas in English sole *(Parophrys vetulus)* from Puget Sound, Washington, *Toxicol. Pathol.*, 18, 362, 1990.
158. Stehr, C.M., Ultrastructure of vacuolated cells in the liver of rock sole and winter flounder living in contaminated environments, in *Proceedings of the XII International Congress for Electron Microscopy,* San Francisco Press, San Francisco, CA, 1990, 522.
159. Bunton, T.E., Ultrastructure of hepatic hemangiopericytoma in the medaka *(Oryzias latipes), Exp. Mol. Pathol.*, 54, 87, 1991.

160. Nakazawa, T., Hamaguchi, S., and Kyono-Hamaguchi, Y., Histochemistry of liver tumors induced by diethylnitrosamine and differential sex susceptibility to carcinogenesis in *Oryzias latipes, J. Natl. Cancer Inst.*, 75, 567, 1985.
161. Teh, S.J. and Hinton, D.E., Detection of enzyme histochemical markers of hepatic preneoplasia and neoplasia in medaka (Oryzias-latipes), *Aquat. Toxicol.*, 24, 163, 1993.
162. Parker, L.M., Lauren, D.J., Hammock, B.D., Winder, B., and Hinton, D.E., Biochemical and histochemical properties of hepatic tumors of rainbow trout, Oncorhynchus mykiss, *Carcinogenesis*, 14, 211, 1993.
163. Miller, M.R., Hinton, D.E., Blair, J.J., and Stegeman, J.J., Immunohistochemical localization of cytochrome P450E in liver, gill and heart of scup *(Stenotomus chrysops)* and rainbow trout *(Salmo gairdneri), Mar. Environ. Res.*, 24, 37, 1988.
164. Miller, M.R., Hinton, D.E., and Stegeman, J.J., Cytochrome P-450E induction and localization in gill pillar (endothelial) cells of scup and rainbow trout, *Aquat. Toxicol.*, 14, 307, 1989.
165. Smolowitz, R.M., Moore, M.J., and Stegeman, J.J., Cellular distribution of cytochrome P-450E in winter flounder liver with degenerative and neoplastic disease, *Mar. Environ. Res.*, 28, 441, 1989.
166. Stegeman, J.J., Smolowitz, R.M., and Hahn, M.E., Immunohistochemical localization of environmentally induced cytochrome P450IA1 in multiple organs of the marine teleost *Stenotomus chrysops* (scup), *Toxicol. Appl. Pharmacol.*, 110, 486, 1991.
167. Van Veld, P.A., Vogelbein, W.K., Smolowitz, R., Woodin, B.R., and Stegeman, J.J., Cytochrome P4501A1 in hepatic lesions of a teleost fish *(Fundulus heteroclitus)* collected from a polyaromatic hydrocarbon-contaminated site, *Carcinogenesis*, 13, 505, 1992.
168. Myers, M.S., Willis, M.J., Husøy, A.M., Goksøyr, A., and Collier, T.K., Immunohistochemical localization of cytochrome P4501A1 in contaminant-associated hepatic lesions of English sole (Parophrys vetulus), *Abstr., 13th Annu. Meet. Soc. Environ. Toxicol. Chem.*, 231, 1992.
169. Myers, M.S., Willis, M.L., Husøy, A.M., Goksøyr, A., and Collier, T.K., Immunohistochemical localization of cytochrome P4501A1 in multiple types of contaminant-associated hepatic lesions in English sole (Pleuronectes vetulus), *Mar. Environ. Res.*, 1993, submitted.
170. Buchmann, A., Kuhlmann, W., Schwarz, M., Kunz, W., Wolf, C., Moll, E., Freidberg, T., and Oesch, F., Regulation of expression of four cytochrome P-450 isoenzymes, NADPH-cytochrome P-450 reductase, the glutathione transferases B and C and microsomal epoxide hydrolase in preneoplastic and neoplastic lesions in rat liver, *Carcinogenesis*, 6, 513, 1985.
171. Smolowitz, R.M., Hahn, M.E., and Stegeman, J.J., Immunohistochemical localization of cytochrome P450IA1 induced by 3,3',4,4'-tetrachlorobiphenyl and by 2,3,7,8-tetrachlorodibenzofuran in liver and extrahepatic tissues of the teleost *Stenotomus chrysops* (scup), *Drug Metab. Dispos.*, 19, 113, 1991.
172. Stegeman, J.J., Miller, M.R., and Hinton, D.E., Cytochrome P450IA1 induction and localization in endothelium of vertebrate (teleost) heart, *Mol. Pharmacol.*, 36, 723, 1989.
173. Kirby, G.M., Stalker, M., Metcalfe, C., Kocal, T., Ferguson, H., and Hayes, M.A., Expression of immunoreactive glutathione-S transferases in hepatic neoplasms induced by aflatoxin B1 or 1,2-dimethylbenzanthracene in rainbow trout *(Oncorhynchus mykiss), Carcinogenesis*, 11, 2255, 1990.

174. Kirby, G.M., Bend, J.R., Smith, I.R., and Hayes, M.A., The role of glutathione s-transferases in the hepatic metabolism of benzo(a)pyrene in white suckers *(Catostomus commersoni)* from polluted and reference sites in the Great Lakes, *Comp. Biochem. Physiol.*, 95C, 25, 1990.
175. Van Veld, P.A., Ko, U., Vogelbein, W.K., and Westbrook, D.J., Glutathione S-transferase in intestine, liver and hepatic lesions of mummichog *(Fundulus heteroclitus)* from a creosote-contaminated environment, *Fish Physiol. Biochem.*, 9, 369, 1991.
176. Williams, G.M., Hirota, N., and Rice, J.M., The resistance of spontaneous mouse hepatocellular neoplasms to iron accumulation during rapid iron loading by parenteral administration and their transplantability, *Am. J. Pathol.*, 94, 65, 1979.
177. Eriksson, L.C., Torndal, U.-B., and Andersson, G.N., The transferrin receptor in hepatocyte nodules: binding properties, subcellular distribution and endocytosis, *Carcinogenesis*, 7, 1467, 1986.
178. Calabrese, E.J., Baldwin, L.A., Scarano, L.J., and Kostecki, P.T., Epigenetic carcinogens in fish, *Rev. Aquat. Sci.*, 6, 89, 1992.
179. Bailey, G., Selivonchick, D., and Hendricks, J., Initiation, promotion, and inhibition of carcinogenesis in rainbow trout, *Environ. Health Perspect.*, 71, 147, 1987.
180. Dashwood, R.H., Arbogast, D.N., Fong, A.T., Pereira, C., Hendricks, J.D., and Bailey, G.S., Quantitative inter-relationships between aflatoxin B1 carcinogen dose, indole-3-carbinol anti-carcinogen dose, target organ DNA adduction and final tumor response, *Carcinogenesis*, 10, 175, 1989.
181. Bailey, G.S. and Hendricks, J.D., Environmental and dietary modulation of carcinogenesis in fish, *Aquat. Toxicol.*, 11, 69, 1988.
182. Fong, A.T., Hendricks, J.D., Dashwood, R.H., Van Winkle, S., Lee, B.C., and Bailey, G.S., Modulation of diethylnitrosamine-induced hepatocarcinogenesis and O^6-ethylguanine formation in rainbow trout by indole-3-carbinol, beta-naphthoflavone, and Aroclor 1254, *Toxicol. Appl. Pharmacol.*, 96, 93, 1988.
183. Fong, A.T., Swanson, H.I., Dashwood, R.H., Williams, D.E., Hendricks, J.D., and Bailey, G.S., Mechanisms of anti-carcinogenesis by indole-3-carbinol. Studies of enzyme induction, electrophile-scavenging, and inhibition of aflatoxin B_1 activation, *Biochem. Pharmacol.*, 39, 19, 1990.
184. Hendricks, J.D., Putnam, T.P., and Sinnhuber, R.O., Effect of dietary dieldrin on aflatoxin B_1 carcinogenesis in rainbow trout *(Salmo gairdneri)*, *J. Environ. Pathol. Toxicol.*, 2, 719, 1979.
185. Hendricks, J.D., Putnam, T.P., Bills, D.D., and Sinnhuber, R.O., Inhibitory effect of a polychlorinated biphenyl (Aroclor 1254) on Aflatoxin B_1 carcinogenesis in rainbow trout *(Salmo gairdneri)*, *J. Natl. Cancer Inst.*, 59, 1545, 1977.
186. Shelton, D.W., Coulombe, R.A., Pereira, C.B., Casteel, J.L., and Hendricks, J.D., Inhibitory effect of Aroclor 1254 on aflatoxin-initiated carcinogenesis in rainbow trout and mutagenesis using a *Salmonella*/trout hepatic activation system, *Aquat. Toxicol.*, 3, 229, 1983.
187. Hendricks, J.D., Putnam, T.P., and Sinnhuber, R.O., Null effect of dietary Aroclor 1254 on hepatocellular carcinoma incidence in rainbow trout (*Salmo gairdneri*) exposed to aflatoxin B1 as embryos, *J. Environ. Pathol. Toxicol.*, 4, 9, 1980.
188. Shelton, D.W., Hendricks, J.D., and Bailey, G.S., The hepatocarcinogenicity of diethylnitrosamine to rainbow trout and its enhancement by Arochlors 1242 and 1254, *Toxicol. Lett.*, 22, 27, 1984.

189. Hendricks, J.D., Arbogast, D.N., and Bailey, G.S., Arochlor 1254 (PCB) enhancement of 7,12-dimethylbenz[a]-anthracene (DMBA) hepatocarcinogenesis in rainbow trout, *Proc. Am. Assoc. Cancer Res.*, 31, 122, 1990.
190. Goeger, D.E., Shelton, D.W., Hendricks, J.D., and Bailey, G.S., Mechanisms of anticarcinogenesis by indole-3-carbinol: effect on the distribution and metabolism of aflatoxin B_1 in rainbow trout, *Carcinogenesis*, 7, 2025, 1986.
191. Shelton, D.W., Goeger, D.E., Hendricks, J.E., and Bailey, G.S., Mechanisms of anticarcinogenesis: the distribution and metabolism of aflatoxin B_1 in rainbow trout fed Arochlor 1254, *Carcinogenesis*, 7, 1065, 1986.
192. Tsuda, H. and Farber, E., Initiation of putative preneoplastic lesions by single doses of non-liver and liver carcinogens plus partial hepatectomy (PH), *Proc. Assoc. Cancer Res.*, 20, 15, 1979.
193. Solt, D.B., Cayama, E., Tsuda, H., Enomoto, K., Lee, G., and Farber, E., Promotion of liver cancer development by brief exposure to dietary 2-acetylaminofluorene plus partial hepatectomy or CCl4, *Cancer Res.*, 43, 188, 1983.
194. Metcalfe, C., Personal communication.
195. Moore, M.J., Unpublished observations.
196. Schultz, M.E., Kaplan, L.A.E., and Schultz, R.J., Initiation of cell proliferation in livers of viviparous fish *Poeciliopsis lucida* with 7,12-dimethylbenz(a)anthracene, *Environ. Res.*, 48, 248, 1989.
197. Aoki, K. and Matsudaira, H., Factors influencing tumorigenesis in the liver after treatment with methylazoxymethanol acetate in a teleost, *Oryzias latipes, Natl. Cancer Inst. Monogr.*, 65, 345, 1981.
198. Aoki, K. and Matsudaira, H., Factors influencing methylazoxymethanol acetate initiation of liver tumors in *Oryzias latipes*, carcinogen dosage and time of exposure, *Natl. Cancer Inst. Monogr.*, 65, 345, 1984.
199. Droy, B.F., Miller, M.R., Freeland, T.M., and Hinton, D.E., Immunohistochemical detection of CCl_4-induced, mitosis-related DNA synthesis in livers of trout and rat, *Aquat. Toxicol.*, 13, 155, 1988.
200. Moore, M.J. and Myers, M.S., Pathogenesis of liver neoplasia in English sole and winter flounder, in SETAC, 12th Annual Meeting, 1991.
201. Koza, R.A., Moore, M.J., and Stegeman, J.J., Elevated ornithine decarboxylase activity and cell proliferation in neoplastic and vacuolated liver cells of winter flounder *(Pleoronectes americanus), Carcinogenesis*, 14, 399, 1993.
202. Auvinen, M., Paasinen, A., Andersson, L.C., and E, H., Ornithine decarboxylase is critical for cell transformation, *Nature (London)*, 360, 355, 1992.
203. Hall, P.A., Levinson, D.A., Woods, A.L., Yu, C.C.W., Kellock, D.B., Watkins, J.A., Barnes, D.M., Gillett, C.E., Camplejohn, R., Dover, R., Waseem, N.H., and Lane, D.P., Proliferating nuclear cell antigen (PCNA) immunolocalization in paraffin sections — an index of cell proliferation with evidence of deregulated expression in some neoplasms, *J. Pathol.*, 162, 285, 1990.
204. Ortego, L.S., Moore, M.J., Myers, M.S., Vogelbein, W.K., Stegeman, J.J., and Hawkins, W.E., Cell proliferation in normal and neoplastic tissues of three species: comparison of bromodeoxyuridine and proliferating cell antigen immunohistochemistry, *Mar. Environ. Res.*, in press.
205. Hyodo-Taguchi, Y. and Matsudaira, H., Induction of transplantable melanoma by treatment with N-methyl-N′-nitro-N-nitrosoguanidine in an inbred strain of the teleost *Oryzias latipes, J. Natl. Cancer. Inst.*, 73, 1219, 1984.

206. Foulds, L., Tumor progression and neoplastic development, in *Cellular Control Mechanisms and Cancer,* Elsevier Press, Amsterdam, 1964, 242.
207. Malins, D.C., McCain, B.B., Myers, M.S., Brown, D.W., Krahn, M.M., Roubal, W.T., Schiewe, M.S., Landahl, J.T., and Chan, S.-L., Field and laboratory studies of the etiology of liver neoplasms in marine fish from Puget Sound, *Environ. Health Perspect.,* 71, 5, 1987.
208. Landahl, J., McCain, B.B., Myers, M.S., Rhodes, L.D., and Brown, D.W., Consistent associations between hepatic lesions in English sole *(Parophrys vetulus)* and polycyclic aromatic hydrocarbons in bottom sediment, *Environ. Health Perspect.,* 89, 195, 1990.
209. Myers, M.S., Olson, O.P., Johnson, L.L., Stehr, C.S., Hom, T., and Varanasi, U., Hepatic lesions other than neoplasms in subadult flatfish from Puget Sound, WA; relationships with indices of contaminant exposure, *Mar. Environ. Res.,* 34, 45, 1992.
210. Moore, M.J. and Stegeman, J.J., Bromodeoxyuridine uptake, cytochrome P4501A expression and presumptive oval cells in liver of winter flounder *(Pleuronectes americanus),* exposed to dietary benzo(a)pyrene and chlordane, *Proc. Am. Assoc. Cancer Res.,* 33, 126, 1992.
211. Maccubbin, A.E. and Ersing, N., Tumors in fish from the Detroit River, *Hydrobiologia,* 219, 301, 1991.
212. Black, J.J. and Baumann, P.C., Carcinogens and cancers in freshwater fishes, *Environ. Health Perspect.,* 90, 27, 1991.
213. Krahn, M.M., Rhodes, L.D., Myers, M.S., Moore, L.K., MacLeod, W.D., and Malins, D.C., Associations between metabolites of aromatic compounds in bile and the occurrence of hepatic lesions in English sole *(Parophrys vetulus)* from Puget Sound, Washington, *Arch. Environ. Contam. Toxicol.,* 15, 61, 1986.
214. Collier, T.K., Singh, S.V., Awasthi, Y.C., and Varanasi, U., Hepatic xenobiotic metabolizing enzymes in two species of benthic fish showing different prevalences of contaminant-associated liver lesions, *Toxicol. Appl. Pharmacol.,* 113, 319, 1992.
215. Maccubbin, A.E., Black, J.J., and Dunn, B.P., 32P-postlabeling detection of DNA adducts in fish from chemically contaminated waterways, *Sci. Total Environ.,* 94, 89, 1990.
216. Dunn, B.P., Carcinogen adducts as an indicator for the public health risks of consuming carcinogen-exposed fish and shellfish, *Environ. Health Perspect.,* 90, 111, 1991.
217. Hurlimann, M.D. and Gardiol, D., Immunohistochemistry in the differential diagnosis of liver carcinomas, *Am. J. Surg. Pathol.,* 15, 280, 1991.
218. Markl, J. and Franke, W.W., Localization of cytokeratins in tissues of the rainbow trout: fundamental differences in expression pattern between fish and higher vertebrates, *Differentiation,* 39, 97, 1988.
219. Thompson, J. and Kostiala, A., Immunological and ultrastructural characterization of true histiocytic lymphoma in the northern pike, *Esox lucius* L., *Cancer Res.,* 50, 5668, 1990.
220. Clauss, G., Winkler, C., Lohmeyer, J., Anders, F., and Schartl, M., Oncofetal antigen in *Xiphophorus* detected by monoclonal antibodies directed against melanoma-associated antigens, *Int. J. Cancer,* 45, 136, 1990.
221. Goksøyr, A., Andersson, T., Buhler, D.R., Stegeman, J.J., Williams, D.E., and Forlin, L., Immunochemical cross-reactivity of β-naphthoflavone-inducible cytochrome P450 (P450IA) in liver microsomes from different fish species and rat, *Fish Physiol. Biochem.,* 9, 1, 1991.

222. Masahito, P., Ishikawa, T., Yanagisawa, A., Sugano, H., and Ikeda, K., Neurogenic tumors in coho salmon *(Oncorhynchus kisutch)* reared in well water in Japan, *J. Natl. Cancer Inst.*, 75, 779, 1985.
223. Hixson, D.C. and Allison, J.P., Monoclonal antibodies recognizing oval cells induced in the liver of rats by N-2-fluorenylacetamide or ethionine in a choline-deficient diet, *Cancer Res.*, 45, 3750, 1985.
224. Germain, L., Goyette, R., and Marceau, N., Differential cytokeratin and alpha-fetoprotein expression in morphologically distinct epithelial cells emerging at the early stage of rat hepatocarcinogenesis, *Cancer Res.*, 45, 673, 1985.
225. Marceau, N., Germain, L., Goyette, R., Noel, M., and Gourdeau, H., Cell of origin of distinct cultured rat liver epithelial cells, as typed by cytokeratin and surface component selective expression, *Biochem. Cell Biol.*, 64, 788, 1986.
226. Oyamada, M., Krutovskikh, V.A., Mesnil, M., Partensky, C., Berger, F., and Yamasaki, H., Aberrant expression of gap junction gene in primary human hepatocellular carcinomas: increased expression of cardiac-type gap junction gene connexin 43, *Mol. Carcinogen.*, 3, 273, 1990.
227. Miyashita, T., Takeda, A., Iwai, M., and Shimazu, T., Single administration of hepatotoxic chemicals transiently decreases the gap-junction-protein levels of connexin 32 in rat liver, *Eur. J. Biochem.*, 196, 37, 1991.
228. Willecke, K., Hennemann, H., Dahl, E., Jungbluth, S., and Heynkes, R., The diversity of connexin genes encoding gap junctional proteins, *Eur. J .Cell Biol.*, 56, 1, 1991.
229. Miller, M.R., Blair, J.B., and Hinton, D.E., DNA repair synthesis in isolated rainbow trout liver cells, *Carcinogenesis*, 10, 995, 1989.
230. Hightower, L.E. and Renfro, J.L., Recent applications of fish cell culture to biomedical research, *J. Exp. Zool.*, 248, 290, 1988.
231. Blair, J.B., Miller, M.R., Pack, D., Barnes, R., Teh, S., and Hinton, D.E., Isolated trout liver cells: establishing short-term primary cultures exhibiting cell-to-cell interactions, *In Vitro Cell. Dev. Biol.*, 26, 237, 1990.
232. Jenner, N.K., Ostrander, G.K., Kavanagh, T.J., Livesey, J.C., Shen, M.W., Kim, S.C., and Holmes, E.H., A flow cytometric comparison of DNA content and glutathione levels in hepatocytes of English sole *(Parophrys vetulus)* from areas of differing water quality, *Arch. Environ. Contam. Toxicol.*, 19, 807, 1990.
233. Wogan, G.N. and Gorelick, N.J., Chemical and biochemical dosimetry of exposure to genotoxic chemicals, *Environ. Health Perspect.*, 62, 5, 1985.
234. Varanasi, U., Stein, J.E., Nishimoto, M., Reichert, W.L., and Collier, T.K., Chemical carcinogenesis in feral fish: uptake, activation, and detoxication of organic xenobiotics, *Environ. Health Perspect.*, 71, 155, 1987.
235. Varanasi, U., Reichert, W., and Stein, J., ^{32}P Postlabelling analysis of DNA adducts in liver of wild English sole *(Parophrys vetulus)* and winter flounder *(Pseudoleuronectes americanus)*, *Cancer Res.*, 49, 1171, 1989.
236. Varanasi, U., Reichert, W.L., Eberhart, B.-T.L., and Stein, J.E., Formation and persistence of benzo[a]pyrene-diolepoxide-DNA adducts in liver of English sole *(Parophrys vetulus)*, *Chem.-Biol. Interact.*, 69, 203, 1989.
237. Stein, J.E., Collier, T.K., Reichert, W.L., Casillas, E., Hom, T., and Varanasi, U., Bioindicators of contaminant exposure and sublethal effects — studies with benthic fish in Puget-Sound, Washington, *Environ. Toxicol. Chem.*, 11, 701, 1992.

238. Reichert, W.L., Stein, J.E., French, B., Goodwin, P., and Varanasi, U., Storage phosphor imaging technique for detection and quantitation of DNA adducts measured by the 32P-postlabeling assay, 13, 1475, 1992.
239. Maclean, N. and Penman, D., The application of gene manipulation to aquaculture, *Genet. Aquacult. III*, 85, 1, 1990.
240. Sinn, E., Muller, W., Pattengale, P., Tepler, I., Wallace, R., and Leder, P., Coexpression of MMTV/v-Ha-ras and MMTV/c-myc genes in transgenic mice: synergistic action of oncogenes in vivo, *Cell*, 49, 465, 1987.
241. Winn, R.N. and van Beneden, R.J. Development of transgenic fish as models for study of aquatic contaminants, in Second International Biotechnology Conference (IMBC '91), 1991.
242. Schartl, M. and Peter, R.U., Progressive growth of fish tumors after transplantation into thymus-aplastic (nu/nu) mice, *Cancer Res.*, 48, 741, 1988.
243. McMahon, G., Huber, L.J., Moore, M.J., Stegeman, J.J., and Wogan, G.N., Mutations in c-Ki-ras oncogenes in diseased livers of winter flounder from Boston Harbor, *Proc. Natl. Acad. Sci. U.S.A.*, 87, 841, 1990.
244. Ishikawa, T., Shimamine, T., and Takayama, S., Histologic and electron microscopy observations of diethylnitrosamine-induced hepatomas in small aquarium fish *(Oryzias latipes)*, *J. Natl. Cancer Inst.*, 55, 909, 1975.
245. Ishikawa, T., Masahito, P., and Takayama, S., Usefulness of the medaka, *Oryzias latipes*, as a test animal: DNA repair processes in medaka exposed to carcinogens, *Natl. Cancer Inst. Monogr.*, 65, 35, 1984.
246. Brittelli, M.R., Chen, H.H.C., and Muska, C.F., Induction of branchial (gill) neoplasms in the medaka fish *(Oryzias latipes)* by N-methyl-N'-nitro-N-nitrosoguanidine, *Cancer Res.*, 45, 3209, 1985.
247. Aoki, K. and Matsudaira, H., Induction of hepatic tumors after treatment with MAM acetate in *Oryzias latipes* and its inhibition by previous irradiation with X-rays, in *Radiation Effects on Aquatic Organisms*, 1980, 209.
248. Park, E.H. and Kim, D.S., Hepatocarcinogenicity of diethylnitrosamine to the self-fertilizing hermaphroditic fish Rivulus marmoratus (Teleostomi: Cyprinodontidae), *J. Natl. Cancer Inst.*, 73, 871, 1984.
249. Thiyagarajah, A. and Grizzle, J.M., Pathology of diethylnitrosamine toxicity in the fish *Rivulus marmoratus*, *J. Environ. Pathol. Toxicol. Oncol.*, 6, 219, 1985.
250. Eun-Ho, P., Hwa-Hyoung, C., and Young-Nam, C., Induction of hepatic tumors with butylated hydroxyanisole in the self-fertilizing hermaphroditic fish *Rivulus ocellatus marmoratus*, *Jpn. J. Cancer Res.*, 81, 738, 1990.
251. Fournie, J.W., Hawkins, W.E., Overstreet, R.M., and Walker, W.W., Exocrine pancreatic neoplasms induced by methylazoxymethanol acetate in the guppy *Poecilia reticulata*, *J. Natl. Cancer Inst.*, 78, 715, 1987.
252. Hawkins, W.E., Walker, W.W., Lytle, J.S., Lytle, T.F., and Overstreet, R.M., Carcinogenic effects of 7,12-dimethylbenz[a]anthracene on the guppy, *(Poecilia reticulata)*, *Aquat. Toxicol.*, 15, 63, 1989.
253. Hendricks, J.D., Scanlan, R.A., Williams, J.L., Sinnhuber, R.O., and Grieco, M.P., Carcinogenicity of N-methyl-N-nitroso-N'-nitrosoguanidine to the livers and kidneys of rainbow trout *(Salmo gairdneri)* exposed as embryos, *J. Natl. Cancer Inst.*, 64, 1511, 1980.

254. Kimura, I., Miyaki, T., and Yoshizaki, K., Induction of tumors of the stomach, of the liver, and of the kidney in rainbow trout by intrastomach administration of N-methyl-N-nitroso-N'-nitrosoguanidine (MNNG), *Proc. Jpn. Cancer Assoc.*, 35, 16, 1976.
255. Metcalfe, C.D. and Sonstegard, R.A., Microinjection of carcinogens into rainbow trout embryos: an in vivo carcinogenesis assay, *J. Natl. Cancer Inst.*, 73, 1125, 1984.
256. Hendricks, J.D., Meyers, T.R., Shelton, D.W., Casteel, J.L., and Bailey, G.S., Hepatocarcinogenicity of benzo(a)pyrene to rainbow trout by dietary exposure and intraperitoneal injection, *J. Natl. Cancer Inst.*, 74, 839, 1985.
257. Hendricks, J.D., The use of rainbow trout *(Salmo giardneri)* in carcinogen bioassay, with special emphasis on embryonic exposure, in *Phyletic Approaches to Cancer*, Dawe, C.J. et al., Eds., *Japan Scientific Society Press, Tokyo*, 1981, 227.
258. Metcalfe, C.D., Balch, G.C., Cairns, V.W., Fitzsimons, J.D., and Dunn, B.P., Carcinogenic and genotoxic activity of extracts from contaminated sediments in western Lake Ontario, *Sci. Total Environ.*, 94, 125, 1990.
259. Couch, J.A. and Harshbarger, J.C., Effects of carcinogenic agents on aquatic animals: an environmental and experimental overview, *Environ. Carcinogen. Rev.*, 3, 63, 1985.
260. Grizzle, J.M., Putnam, M.R., Fournie, J.W., and Couch, J.A., Microinjection of chemical carcinogens into small fish embryos: exocrine pancreatic neoplasm in Fundulus grandis exposed to N-methyl-N'-nitro-N-nitrosoguanidine, *Dis. Aquat. Org.*, 5, 101, 1988.
261. Schultz, M.E. and Schultz, R.J., Transplantable chemically-induced liver tumors in the viviparous fish *Poeciliopsis, Exp. Mol. Pathol.*, 42, 320, 1985.
262. Schultz, M.E. and Schultz, R.J., Induction of hepatic tumors with 7,12-dimethylbenz[a]anthracene in two species of viviparous fishes (genus *Poeciliopsis), Environ. Res.*, 27, 337, 1982.
263. Bunton, T.E., The immunocytochemistry of cytokeratin in fish tissues, *Vet. Pathol.*, 30, 418, 1993.

Chapter 9

Metal Regulation in Aquatic Animals: Mechanisms of Uptake, Accumulation, and Release

G. Roesijadi and W.E. Robinson

I. INTRODUCTION

Aquatic animals are naturally exposed to a variety of metals whose chemical forms and concentrations are governed by natural geochemical processes and anthropogenic activities. These metals include both essential elements required to support biological processes and nonessential metals with no known biological function. Cellular functions are critical to processes involved in metal uptake, regulation, utilization, and release. Toxicity can be attributed to their dysfunction and the resultant interaction of metals with inappropriate cellular structures. Investigations at the cellular level will advance our understanding of the mechanisms by which aquatic animals respond to metal exposure.

Attempts to discern general patterns of response to metals in aquatic animals are complicated by the wide diversity of species, most of whose fundamental biology is incompletely understood. The large body of work on responses of aquatic animals to metals has been the subject of recent volumes in which the extent of this diversity is amply documented.[1,2] In this chapter, we focus on mechanisms associated with metal uptake, internal transport, intracellular storage, and release and attempt to derive unifying principles for cellular and biochemical responses that underlie current observations. When necessary, we have relied on data from other animal systems, particularly mammals, to fill in the gaps in knowledge of aquatic species; we have also identified topics of potentially fruitful research.

A generalized organismal model for metal uptake, translocation, storage, and release in aquatic animals is shown in Figure 1. The gills, digestive tract, and integument represent the sites of metal uptake. Metals are subsequently transported to internal organs for utilization, storage, and release. Components of the blood and hemolymph are the vehicles for this transport. Cellular mechanisms associated with various organ systems operate in a highly integrated fashion to coordinate metal uptake, transport, and release.

II. BIOAVAILABILITY OF METALS AND IMPLICATIONS FOR UPTAKE BY ORGANISMS

The "Borderline" and "Class B" metal and metalloid ions (classification of Nieboer and Richardson[3]) are of specific interest here. Often referred to as "heavy" or "trace" metals in general nomenclature, they include ions such as As^{3+}, Cd^{2+}, Co^{2+}, Cr^{2+}, Cu^{2+}, Fe^{2+}, Fe^{3+}, Ga^{3+}, In^{3+}, Mn^{2+}, Ni^{2+}, Pb^{2+}, Sb^{3+}, Sn^{2+}, Sn^{4+}, Ti^{2+}, and V^{2+} as borderline ions and Ag^+, Au^+, Bi^{3+}, Cu^+, Hg^{2+}, Pb^{4+}, Pd^{2+}, Pt^{2+}, and Tl^{3+} as Class B ions. Their nitrogen- and sulfur-seeking properties contribute to their ability to form complexes with such centers in biologically important ligands.[3] Each is potentially toxic, if present at sufficiently high concentrations, and can occur in the aquatic environment as anthropogenic contaminants.

The chemical speciation of metals in aquatic systems is dependent on the specific physical/chemical factors that prevail in local environments. Factors such as salinity, dissolved organics, pH, hardness, and sedimentary load all influence the prevailing chemical forms of metals in aquatic systems. These, in turn, influence metal bioavailability and toxicity.

Particulate forms include those sorbed onto suspended mineral and organic particles and metals present in food organisms. Soluble chemical forms include simple aquated metal ions, metal ion complexes with inorganic anions, and metal ion complexes with organic ligands such as amino, fulvic, and humic acids.[4] Environmental conditions that cause a shift of the dominant metal species from one form to another not only affect the bioavailability of the metal, but can also direct uptake to different pathways. Thus, the effective exposure of individuals is not determined by the total metal concentration in a particular environment, but by the concentrations of individual chemical forms. These are not equally bioavailable, and the effect of a single metal species may have greater significance in explaining the biological response than the total metal concentration.

Particle-bound metals are often ingested by aquatic organisms in association with food. In vertebrates, the lowered pH of the stomach helps to solubilize

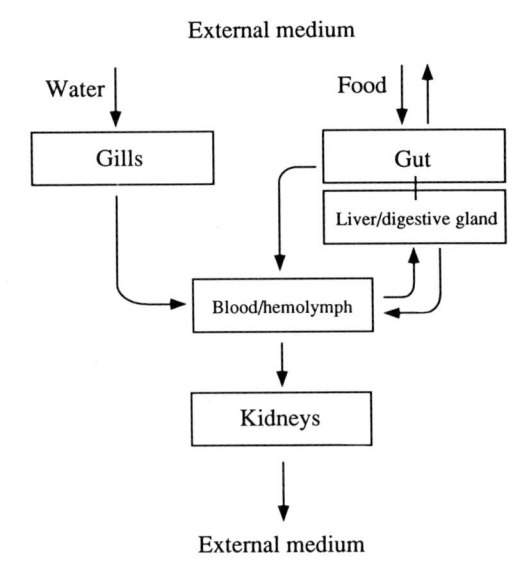

```
                        External medium
      Water │                Food │ ↑
            ▼                     ▼ │
      ┌──────────┐          ┌──────────┐
      │  Gills   │          │   Gut    │
      └────┬─────┘          └────┬─────┘
           │                ┌────┴──────────────┐
           │                │Liver/digestive gland│
           │                └────┬──────────────┘
           │     ┌──────────────┐│
           └────▶│Blood/hemolymph│◀┘
                 └───────┬──────┘
                         ▼
                 ┌──────────────┐
                 │   Kidneys    │
                 └───────┬──────┘
                         ▼
                   External medium
```

Figure 1. Generalized scheme for metal absorption, distribution, and release in aquatic animals. Once absorbed across the epithelial barriers of the gill or the digestive system, metals are distributed to internal organs. Following redistribution to sites for excretion in the liver/digestive gland or kidneys, metals are released from the organism by excretory mechanisms. (From Roesijadi,[59] with permission.)

metals, which are then absorbed in the intestine under alkaline conditions. The digestive tracts of invertebrates, on the other hand, generally lack highly acidic extracellular conditions. Endocytosis of food particles and digestion by lysosomes play prominent roles in the processing of metal-associated particles, and metals may be solubilized by lysosomal activity.

For dissolved metals, there is considerable evidence for the view that the ionic form in solution is the major bioavailable form in aquatic environments.[5-7] Many studies in which free metal ion concentrations have been controlled with chelating agents or measured with electrochemical procedures implicate the free metal ion concentration rather than the total metal concentration as the major determinant of metal accumulation or toxicity. With dissolved Cu, Cd, and Zn, for example, the ionic form or analytically labile form, in some cases, correlates best with metal accumulation or toxicity.[8-10]

The relative importance of the bioavailability of other chemical forms of dissolved metals, particularly organically chelated forms, is not as clearly understood.[11,12] The presence of organic ligands in the surrounding water is reported to increase Cd bioavailability in the mussel *Mytilus edulis*.[13] In fish, complexation with various hydrophobic ligands such as xanthates, diethyldithiocarbamates, and dithiophosphates can also increase Cd bioavailability.[14-16] The likely explanation of these observations is related to the enhanced hydrophobic characteristics imparted by the ligand and the resultant increased potential for dissolution of the complex into membrane lipids.[12] Organometallic compounds, such as the organomercurials and organotins, are derived from covalent binding of the metal and organic moiety and represent the best known examples of such behavior.[17] Generalizations regarding the influence of ligands on bioavailability of metals must take into consideration the chemical nature of specific metal/ligand complexes. The extensive data for mammals indicate that a number of naturally occurring organic metal-binding substances (L-amino acids, citrate, phosphate, gluconate, oxalate) and synthetic metal chelators (nitroloacetate, EDTA) enhance intestinal uptake of Cu and Zn.[18] The question of whether these metals are absorbed in a form bound to absorbable ligands or whether these ligands present the metals to the intestinal cells in a manner that facilitates the subsequent uptake of the uncomplexed metal has yet to be resolved.[18]

With Hg and Cd, diffusion of the neutrally charged chloro-complexes $HgCl_2$ and $CdCl_2$ across an artificial lipid bilayer exceeds that of the ionic Hg^{2+} and Cd^{2+}.[19,20] Thus, even in the absence of organic chelation, the ionic form cannot be categorically stated as the sole biologically significant form of metals. The magnitude of difference between diffusion of the charged and neutral forms is much less with Cd than with Hg, however, and uptake of Cd^{2+} rather than $CdCl_2$ is generally considered to be more biologically significant on the basis of empirical observations in vivo.[20] These observations have ecological implications because aquatic environments are characterized by differences in the chloride concentration as one moves from freshwater to strictly marine environments. Mercury speciation is chloride dominated in both seawater and freshwater of pH 6, while, with Cd, the chloride dominance observed in seawater diminishes as the salinity decreases.[21] Thus, differences in the external chloride concentrations in freshwater and marine environments are expected to influence Hg speciation to a lesser degree in comparison with that of Cd, so long as other factors remain constant.

Microenvironments at the sites of metal uptake can differ from the surrounding water and have important consequences on speciation, availability, and biological effects of metals.[22] In rainbow trout exposed to Al at low pH, for example, an increase in pH in the branchial microenvironment is attributable to the effects of expired CO_2 and ammonia and results in reduced solubility of Al and its precipitation on the gill surfaces.[23] This is associated with both respiratory and ionoregulatory impairment.

In order to achieve a better understanding of the significance of metal speciation on chemical/biological interactions, the ensuing effects of chelation on both metal speciation and membrane permeability need to be understood.[12] Some general principles (e.g., the ready bioavailability of ionic forms of numerous metals) have emerged from the research conducted to date. Routine analytical or numerical approaches adequate for estimating the concentrations of the various chemical forms of metals in natural waters are not available.[24] Thus, in most biological studies, the nature of the external environment with respect to the chemical forms of the metals, whether natural or experimental, usually remains a variable that is neither clearly understood nor controlled experimentally. Specific data on speciation are generally lacking for most of the published studies on metal uptake or accumulation. Dissolved and particulate forms follow fundamentally different pathways that need to be more clearly understood. The unique pathways for uptake of ionic and organically chelated forms of dissolved metals need to be identified and characterized.[12] Whether separate pathways also exist for different valence states and inorganic ionic complexes of a single metal needs to be delineated as well.

III. MECHANISMS OF METAL UPTAKE

A. General Considerations

The absorption of metals by aquatic animals involves transfer of metals to the circulatory system across the epithelial barrier of gills, digestive system, or integument. This transfer across epithelial cells includes three basic elements: (1) uptake by the apical membrane, the interface with the external environment; (2) movement through the cell and interaction with intracellular ligands; and (3) efflux across the basolateral membrane, the interface with the circulatory system.[25] Organs that serve as the sites for uptake (e.g. gills, intestine, and digestive gland) also tend to concentrate metals and, therefore, exhibit relatively high potentials for bioaccumulation.

The cellular uptake of metals is a membrane-based phenomenon that can be assigned to one of two general schemes, depending on whether the uptake is based on membrane transport or the endocytotic processes of phagocytosis or pinocytosis. Dissolved metals would be expected to be taken up by exposed body surfaces such as the gills. Particulate metals are most commonly ingested and then taken up after solubilization in the gut. They can also be phagocytosed and solubilized in endocytotic vesicles.

Fundamentally different capabilities for the uptake of metals associated with food exist in the invertebrates and vertebrates. The invertebrates possess the capability for both extracellular and intracellular digestion, while the vertebrates are considered to rely on the former.[26] Thus, metal uptake by endocytosis would be expected to be a process of greater significance in the various invertebrate species. Once endocytosed, the particulate metal complexes can be broken down and the metal redistributed to other intracellular ligands. In vertebrates, metal complexes must be broken down to more simple structures in the intestinal lumen before uptake via various membrane-dependent transport pathways can occur.

B. Epithelial Uptake of Dissolved Metals from the Environment

The gill[27-30] and intestine[31] are the primary sites for uptake of soluble metals from the aquatic environment. In soft-bodied invertebrate species, the body wall may also be an important site of soluble metal uptake.

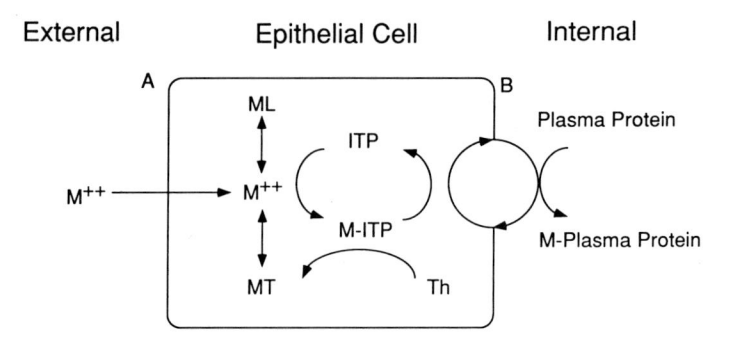

Figure 2. Model for metal absorption across the gill and intestine/digestive gland based on the reported behaviors of CRIP and metallothionein during the intestinal absorption of zinc in mammals.[52] CRIP is depicted here as a more generic intracellular transporter protein (ITP) that is involved in the normal absorption of metals across the epithelium, transferring them to plasma proteins via a transmembrane transporter system at the basolateral membrane. Induction of metallothionein by high concentrations of metals will result in an increase in a high-affinity ligand pool that competes with the ITP and retards metal absorption into the circulatory system. Metal absorption is regulated by the coordinated expression of ITP and metallothionein. (A, apical membrane of cell; B, basolateral membrane of cell; ITP, intracellular transport protein; M, metal; ML, metal and nonthionein-ligand complexes, exclusive of ITP; MT, metallothionein; and Th, apothionein).

trout gills is proposed to be mediated by Ca-ATPase.[38] In the tilapia intestine, Cd transport across the basolateral membrane follows pathways for Ca that involve Ca-ATPase and Cd-Ca exchange.[31] Because little unbound metal is expected to occur within cells, transporter molecules such as CRIP would be expected to present metal ions to efflux mechanisms that would, in turn, supply metals to transport molecules (e.g., albumins in mammalian plasma) in the circulatory system.

C. Endocytosis as a Mechanism for Metal Uptake

It is well known that endocytosis is used by invertebrates for uptake of materials from the environment.[55] The digestive cells of most invertebrates are capable of phagocytosis and intracellular digestion of food particles. Additionally, external organs, such as the gills, are also capable of phagocytosis of particles from the environment. Metals associated with particles have been shown to be accumulated by endocytotic mechanisms in various invertebrate species.[55] Once metals are brought into a cell by endocytosis, they are incorporated into lysosomal vesicles and subsequently follow intracellular pathways that are only partially understood.[56] Pathways for absorption into the blood may include transfer of the particles or vesicles to blood cells or release of vesicle contents directly into the blood plasma. However, such pathways remain to be elucidated. Although considered to be a viable mechanism, there are only a few examples implicating endocytosis in the uptake of metals from the environment.[12] Colloidal ferric hydroxide[57] and Pb,[55] for example, are taken up by endocytosis by mussel gills. More likely, endocytosis is the mechanism that is used in assimilating metals from food in the digestive tract of invertebrates, although specific examples to support such a hypothesis are lacking. Quantitative assessments of the contribution of endocytosis as a pathway for metal uptake are currently lacking.

IV. INTRACELLULAR STORAGE AND SEQUESTRATION

A. General Considerations

Sequestration of metals in an immobilized form occurs throughout the various tissues and organs involved in pathways for metal uptake, transport, utilization, and release. Sequestration by any one compartment in the chain may be temporary, a single step in the sequence of events that starts at the sites of uptake (gills, intestine, integument) and proceeds to the sites of detoxification and either long-term storage or excretion (i.e., liver or equivalent organ and kidney). Two of the best-studied intracellular structures involved in metal sequestration and storage are the metallothioneins and the intracellular vesicle-bound granules. General properties of these two systems are addressed here. Undoubtedly, they interact in coordinated functions in those tissues where both are present and where each has been shown to play a significant role in metal binding and sequestration. The possible roles of metallothionein in transmural metal transport and endocytotic vesicles in metal uptake have been discussed in preceding sections. These structures are also relevant to the subsequent sections on metal transport and release in aquatic animals.

B. Metallothioneins

Metallothioneins are a family of low-molecular-mass, metal-binding proteins believed to function in the regulation of the essential metals Cu and Zn and in the detoxification of these and nonessential metals such as Cd and Hg.[58,59] Their role in sequestering metals is well established, and their induction by metal exposure is associated with conferring protection against metal toxicity. Studies on the regulation of metallothionein gene expression have firmly established that induction by metals is a direct response to increases in the intracellular metal concentration and mediated through the action of transacting metal-binding regulatory factors.[60]

The capacity for metallothionein induction is greatest in tissues that are active in metal uptake, storage, and excretion. In aquatic animals, metallothioneins have been identified in the small intestine,[54] liver,[61–63] and gills[64] of fish and in the digestive gland[65,66] and gills[30,67–69] of mollusks and crustaceans. This induction results in relatively high concentrations of metallothionein-bound metals in these organs and, in cases such as Cd, results in a slower turnover of the metal. Thus, in the mammalian intestine, binding to metallothionein is now considered to retard transfer of metals to the blood,[49,52,70] with the intestine serving as a filter for the subsequent translocation to internal organs. Binding of metals to induced metallothioneins in the gut and gills of aquatic animals probably has an effect on metal absorption similar to that noted above for the mammalian intestine. Binding of metals to metallothionein enhances bioaccumulation in these organs, as well as in the liver and kidney. Whether a protein analogous to CRIP may co-occur with metallothionein and function as a metal transporter in the intestines and gills of aquatic animals remains to be determined.

Although a detailed understanding of the specific intracellular interactions associated with metallothionein function has yet to be deduced, it is clear that the protein has a central role in regulating the availability of intracellular Cu and Zn for essential cell function. Metallothionein can donate Zn or Cu to metalloproteins, such as carbonic anhydrase, pyridoxal kinase, and hemocyanin,[71-73] and activate them. Thionein, the apoprotein form, can remove metals from Zn-finger proteins.[74,75] Inactivation of these transacting regulatory factors through this latter mechanism has been proposed to function in the regulation of gene expression. In mammals, fish, and invertebrates, metallothionein synthesis is developmentally regulated[76-78] and, possibly, involved in the regulation of gene expression by controlling Zn availability to regulatory factors. Its synthesis is also regulated in higher vertebrates by a number of factors apart from metals.[60] The normal functions of metallothionein are critical to the cell and are superimposed on any toxicological response.

Increased levels of metallothionein and metals bound to metallothionein appear to be characteristic of exposure to Cd, Cu, and, possibly, Zn in aquatic animals. The ability of metallothionein to sequester metals has unequivocally been linked to a metal-detoxification function in studies on yeast whose metallothionein genes have been genetically altered.[79] Although yet to be confirmed, the cellular interactions involving metallothioneins can be expected to follow two general lines, the first being the interception and binding of metal ions that are initially taken up by the cell and the second being the removal of metals from nonthionein ligands that include cellular targets of toxicity. The latter may represent a "rescue" function for structures that have been reversibly impaired by inappropriate metal binding.[80] It still remains to be determined whether metallothionein induction may interfere with the normal regulation of metal-dependent systems. In this context, metal-induced thionein synthesis may result in an increased pool of ligands that can inappropriately remove essential metals from active sites of other molecules and possibly interfere with cellular function.

Measurement of metallothionein induction has been proposed as a cellular indicator of metal exposure and toxicity in aquatic animals (see Reference 59 for discussion). This induction confers enhanced metal tolerance to both cells[81-83] and intact individuals.[84-88] Because the organism is protected, it does not succumb as readily to metal toxicity, and, coupled with the relatively long turnover time for metallothionein-bound metals, higher burdens of metals can be accumulated than would otherwise occur. Thus, one of the correlates of metallothionein induction is that an increased metal burden can be tolerated by the individual. A possible consequence of this increased capacity for sequestration is an increase in the potential for trophic transfer of metals. Such compensatory mechanisms for responding to metal exposure may have as yet undetermined ecological or public health consequences, the latter when the species affected is a human food source.

The ready manipulation of the metallothionein gene has made possible prospects for constructing transgenic organisms possessing a metallothionein promoter and an appropriate reporter for easy quantification of gene expression. For medaka, transgenic fry with a trout metallothionein-A or mouse metallothionein-I promoter and chloramphenicol acetyltransferase (CAT) reporter have been developed through recombinant techniques.[89] Metal exposure of these individuals results in a high CAT activity that is mediated initially through the metallothionein promoter. Further development and refinement of this and similar transgenic systems is expected to lead to a greater understanding of metallothionein function and the possibility for metallothionein-based molecular monitors for metals in the aquatic environment. Of greater interest in the context of this review is that easy detection of metallothionein promoter activity in different tissues and organs under various physiological conditions and life history stages may provide revealing information on the relationship between metallothionein gene expression and the physiological function of the proteins in metal regulation and transport.

C. Intracellular Deposits

All aquatic animals contain a wide variety of membrane-bound intracellular deposits, many of which bind metals. Classified as either "granules" or "concretions," these structures are generally associated with the digestive or excretory tissues of invertebrates (i.e., midgut, digestive gland, hepatopancreas, Malpighian tubules, and kidney). They are also found in the connective tissues of both vertebrates and invertebrates (e.g., as lipofuschin granules), as well as in specialized cells of some organisms (e.g., Cu-containing pore cells in gastropods). The metal content of these various granule types varies considerably.

Brown[90] identified three types of Cu-, Fe-, and Ca-containing granules. Copper-containing granules are composed of phosphorus and Cu, but little additional metal besides some Fe. These rather homogeneous granules are limited to arthropods and may be involved in Cu regulation. Similarly, ferric phosphate-containing granules found only in some echinoderms may function in Fe regulation.

Calcium-containing concretions can be subdivided into the Ca carbonate and Ca phosphate types.[56,90] Calcium carbonate granules are often found in the connective tissue of arthropods and gastropods and, apparently, serve as an easily mobilized store of Ca. These granules may be formed by the mineralization of an organic template produced by the Golgi.[91] Calcium carbonate granules are probably not important for the physiological regulation of heavy metals, although some metals such as Pb may substitute for Ca ions to a limited degree.[56]

Calcium phosphate granules are ubiquitous throughout the invertebrate phyla and are thought to play a major role in metal detoxification (and possibly elimination).[56,91] These granules are primarily composed of Ca and Mg phosphates, but they can incorporate high concentrations of such metals as Al, Ag, Ba, Co, Fe, Mn, Pb, Sn, and Zn, as well as lower concentrations of Cu, Cd, Cr, Hg, and Ni. The ratio of (Ca + Mg):P varies considerably between species and individuals.[92] The amount incorporated of each metal is highly variable, even within the same species, and probably relates to variable environmental metal concentrations.[56,93] These granules generally have a low percentage (<10%) of organic matter,[93] although granules in the kidney of *M. edulis* are an exception to this pattern, with 46% carbon and only 5 to 6% metal.[94]

Metal-rich Ca phosphate granules are generally associated with digestive and excretory tissues.[90] The granules are usually 0.2 to 3 µm in diameter in most species, although 40-nm diameter particles have been identified in the hemocytes of a number of marine bivalves.[95] Exceptionally large granules of 10 to 15 µm and up to 1.9 mm diameter have been reported in some bivalve species.[93] These large granules are extracellular and can continue to grow after exocytosis or holocrine or apocrine secretion. The granules can be composed of pyrophosphate or orthophosphate, depending on species.[96] Both types are virtually insoluble in saline, making them excellent "sinks" for immobilizing metals in a nontoxic, unavailable form.

The initiation of intracellular granule formation is not well understood. No single mechanism may account for the diversity of granule types observed in aquatic organisms.[97] In some species, an organic matrix or template may first be produced by the endoplasmic reticulum or Golgi and housed within a membrane-limited vacuole. This template can then be mineralized.[97,98]

Current evidence supports the view that the cellular lysosomal system is involved in most instances of granule formation.[56,90] Lysosomal systems are present in virtually all invertebrate cells and are particularly well developed in digestive and excretory cells such as those of the molluscan digestive gland and kidney and arthropod hepatopancreas. Particulate material, such as food, foreign particles, membranes, cellular organelles, and cytoplasmic proteins, are endocytosed and incorporated into secondary and tertiary lysosomes. Metals associated with these materials would be similarly incorporated. In the mussel kidney, the breakdown of metallothionein/metal complexes results in incorporation of Cu, Cd, Hg, and Zn into lysosomal granules.[99] Similarly, ferritin complexes are broken down in the lysosomal system of mammalian liver cells and produce hemosiderin and residual bodies rich in Fe.[100]

In lysosomally formed Ca phosphate granules, lipofuscin is generally the organic material left after membrane decomposition. Lipofuscin may prove to be the primary organic constituent of all granules that are produced via the lysosomal system, regardless of the type of tissue where the granules are produced or the metal content of the granules. In the bivalve *Mercenaria mercenaria,* lipofuscin was identified histochemically in both kidney granules with high metal concentrations and digestive gland granules with much lower metal contents.[93] Any metals bound to the membranes would be incorporated into the residual body during lipofuscin formation. Both Cd and Zn bind weakly and reversibly to lipofuscin by a passive adsorptive process.[101]

Subsequent intracellular or extracellular growth of Ca phosphate granules may occur by a mechanism different from the granule's initial formation. Very little is known about the accretion of organic material by existing granules, although incorporation of organic substances is evidently reduced as granules attain a larger size.[93] Accretion of metals to the granule periphery has been reported both in vivo and in vitro.[99,101,102] The concentric rings observed in granule sections by electron microscopy have been interpreted as alternate periods of surface accretion and resorption.[103] Known changes in pH inside the secondary and tertiary lysosomes may also favor metal deposition onto the granule surface.[104] Active transport of vesicles from the endoplasmic reticulum across the vacuolar membrane and precipitation of vesicular contents onto the growing granule surface have also been proposed as a mechanism of granule accretion.[105] Accretion of metal ions onto vesicle-bound granules may also occur by a surface absorption/corrosion process that involves the binding of metals to sites where Ca has been dissociated during a H^+-induced dissolution of the granule surface[106] or where Mg has been released from the granules.[107]

Additional work is needed to further characterize the process of initial stages of granule formation and the subsequent growth of granules and incorporation of metals.

D. Other Mechanisms

The only other well-characterized system for the intracellular storage and sequestration of metals involves ferritin, a 450-kDa protein that is ubiquitous in vertebrate (and probably invertebrate) tissues.[108] In the invertebrates, ferritin has been isolated from body tissues of crustaceans, chitons, gastropods, cephalopods, and bivalves.[109-111] Circulating ferritin has been measured in the blood plasma of both chitons and limpets.[112,113]

V. INTERNAL TRANSPORT OF METALS

A. General Considerations

The delivery of metals to internal organs and the subsequent redistribution among organs[114-117] occur as a result of metal transport via the circulatory system. There are differences in the developmental origin of the circulatory systems of invertebrates, such as mollusks and arthropods, which possess an open circulatory system, and other invertebrates and vertebrates, which possess a closed circulatory system. The nomenclature used to describe analogous components of the respective circulatory systems is specific for each. However, for convenience and in recognition of functional similarities, we use a general terminology when describing structures that comprise the circulatory system.[118] Thus, for the invertebrates, the terms "blood" and "hemolymph" are used synonymously, as are "blood cells" and "hemocytes." Vertebrate circulatory structures are referred to as "blood" and "blood cells." "Plasma" is the cell-free blood or hemolymph.

It has been known for some time that the plasma proteins of vertebrates play the dominant role in metal transport.[119-121] The blood plasma contains diverse proteins that transport a wide range of metals. Binding of metals to plasma proteins can be either specific (e.g., Fe by transferrin and Cu by ceruloplasmin) or nonspecific (e.g., Ca, Ni, and Zn by serum albumin). Although metals can be taken up by mammalian erythrocytes and leukocytes, these types of blood cells are not considered carriers of metal.

In invertebrates, blood cells and hemocytes have traditionally been considered the primary vehicles for transporting metals. Many types of these cells are highly mobile, phagocytic, and capable of passing between cells of tissue layers.[118] Certain cell types can harbor high concentrations of metals.[122-125] It is believed that these cells are capable of carrying metals and transferring their metal loads to other tissues, presumably through exocytosis. However, recent evidence indicates that invertebrate plasma proteins may play a much greater role in metal transport than was previously thought, one that is analogous to the situation in vertebrates. Invertebrate plasma proteins bind metals[126-128] and, as discussed in Section V.C. may be responsible for the bulk of the metal transport that occurs in the circulatory system of these animals.

B. Blood Cells and Hemocytes

Studies on the role of blood in metal transport in fish are scarce. Thus, in lieu of specific information, mammalian systems currently serve as models for deducing the role of blood cells in metal transport in fish. Mammalian blood cells can accumulate metals. Erythrocytes, for example, readily take up Pb, Zn, and Cd.[129,130] With Cd, 90% of the total Cd loading by blood cells may be bound by metallothionein.[131]

Vertebrate blood cells do not appear to be involved in metal transport. Metals in these cells are not readily available for exchange with external receptors. Intravenous injection of Ca-EDTA, for example, does not mobilize either Zn or Pb from blood cells.[129] The turnover of metals in vertebrate blood cells is very slow and not consistent with a transporter function associated with delivery of metals to sites of utilization.

Considerably more data are available on metals in blood cells of aquatic invertebrates, especially in tunicates and mollusks, in comparison with fish. In tunicates, V, Fe, or a mixture of both metals, depending on species, are concentrated in blood cells.[123,124] These metals are localized in membrane-limited vacuoles in signet ring, compartment, granulocyte, and, at much lower concentrations, morula cells.[118,132]

In mollusks, hemocytes are known to concentrate a variety of metals,[56,127,133] which are sequestered by several intracellular ligands. Metal concentrations in hemocytes are much higher than in the surrounding plasma. However, because hemocytes comprise only a small percentage of both the weight and the volume of bivalve blood (e.g., 2.1% of wet weight for *Mytilus edulis*),[134] the actual metal burden contained in the blood cells, expressed as mass of metal, is usually a small fraction of the total body burden and is often much less than the load carried by the plasma.[92,125]

Special mechanisms for concentrating exceptionally high levels of Cu and Zn have been observed in hemocytes of some species of oysters. As a result of this specialization, a higher percentage of the whole blood Cu and Zn load is localized in these cells. In *Ostrea edulis*, 70 to 75% of the hemolymph Cu and 42 to 77% of the hemolymph Zn occur in hemocytes.[122,135] In the extreme case, "green-sick" oysters accumulate such high levels of Cu in their hemocytes and tissues that they have acquired a greenish coloration.[122] Crassostreid oysters contain hemocytes that concentrate either Cu or Zn.[136] *O. edulis* also possesses separate Cu- or Zn-containing hemocytes[122] with concentrations up to 400 mM Cu and up to 1.2 M Zn.[137] A third type of hemocyte that sequesters both Cu and Zn also exists in *O. edulis, O. angasi,* and *Crassostrea gigas*.[137]

Metals in invertebrate hemocytes are often localized in membrane-limited vesicles or vacuoles. These vesicles appear to be part of the cellular lysosomal system in hemocytes of mollusks and arthropods and so-called "leukocytes" of tunicates.[56,92,118] However, in the colored, vacuolated blood cells of tunicates (i.e., in the signet ring, morula, and compartment cells), the vacuoles appear to be specialized repositories of either V or Fe, rather than part of the lysosomal system.[118]

In addition to vacuolar storage, metals in hemocytes may be bound to cytoplasmic proteins. Copper, zinc, and cadmium are bound to cytoplasmic molecules <3 kDa and metallothioneins in the hemocytes of mussels[138,139] and oysters.[140]

Turnover rates of metals in invertebrate blood cells vary by species and metal. In the crab *Scylla serrata* injected with Cu chloride, much of the Cu accumulated by hemocytes 2 hr after injection is released within the next 2-hr period.[141] In contrast, in hemocytes of the hard clam *Mercenaria mercenaria*, elimination of Cd is very slow and that of Ag is negligible.[134] Slow release of metals from hemocytes would argue against a role in metal transport.

Currently, there is no direct evidence to support the hypothesis that metals are transferred from hemocytes to other cells. The prevailing view that metals are transported by bivalve hemocytes and transferred to tissues such as the kidney and digestive gland for eventual elimination (Figure 3) is based on the proximity of hemocytes to renal and hepatic tissues in histological sections rather than on direct observations of such transfer. Particulate metals, such as ferric hydroxide precipitates, are reported to be taken up by hemocytes of *Mytilus edulis* through endocytosis.[57] How this Fe is supposed to be released by the hemocytes and passed on to the other tissues remains to be determined. Hypothetically, exocytosis of residual bodies, with an accompanying endocytosis by the target tissue cell, may be envisioned for this transfer. Soluble cytoplasmic metals may be transferred from hemocytes by a similar process. In either case, fine coordination would be required between the hemocyte and target cell. It would seem equally likely that hemocytes may simply sequester metal over their cellular lifetimes, releasing the metal into the blood plasma once they die and lyse (Figure 3). In addition, hemocytes may help to reduce the body burden of metals through diapedesis, which is the one-way migration of hemocytes out of the tissues into the gut lumen or surrounding water (Figure 3). Short-term, in vivo studies using radioisotopic metals, coupled with the use of pharmacological agents to block endo- and exocytosis, would be very helpful in determining mechanisms of both uptake and release in hemocytes.

C. Plasma

Plasma proteins play the dominant role in metal transport in mammals.[39,119–121] This is probably the case for fish, where even the most primitive of species contain plasma proteins that are homologous to those of mammals.[142] In plasma of winter flounder, binding of Cu and Zn to plasma proteins is both metal and gender related.[143] Copper is bound to proteins of 170 kDa in both sexes. With Zn, about 95% is bound to proteins of 76 kDa in both sexes. In males, the remainder of the plasma Zn is bound to proteins of 186 kDA, while, in females, the remainder is distributed between the 186-kDa proteins and others of 340 to 370 kDa. The larger proteins in females are hypothesized to be circulating vitellogenin. A novel 66-kDa Zn-binding protein has been reported recently in plasma of albacore tuna.[144]

There is increasing evidence that plasma proteins in invertebrates may also have an important function in metal transport. In the marine bivalves, plasma usually carries a much higher percentage of the total metals in whole blood in comparison with hemocytes. In *Mercenaria mercenaria*, for example, the blood plasma contains 99% of the whole blood load of Zn, 95% of the Cr and Cu, 80% of the Ba, and 77% of the Fe.[127] This situation in marine bivalves may also prove to be the case for other aquatic invertebrate species.

Relatively specific and strong binding is considered characteristic of metal-binding proteins (e.g., ferritin, transferrin, metallothionein), but is not a requisite of proteins involved in metal transport. Metal binding to transport proteins may either be nonspecific, with weak binding of a variety of metals and low affinity

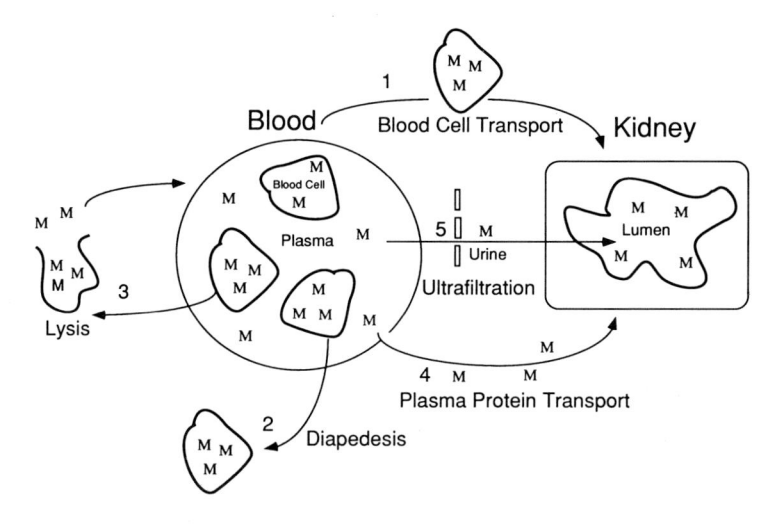

Figure 3. Pathways for metal transport to the kidney or other target organ in invertebrates such as a marine bivalve mollusk. Metals (M) can be measured in both hemocytes and blood plasma. The traditional view (1) purported that hemocytes transported metals in intracellular vacuoles to the kidney or other tissues and then transferred the metals to these tissues by as yet to be determined mechanisms. Hemocytes may also reduce the body burden of metals via diapedesis (2). Alternatively, metals may simply be sequestered over the life of the hemocytes (3) and released back into the plasma upon cell lysis. Recent evidence indicates that blood plasma (4) may play a prominent role in metal transport. With respect to the kidney, primary urine, produced by ultrafiltration, may transport a small fraction of the total plasma metal concentration for uptake by kidney cells.

constants (K_a), or specific, with strong binding of particular metal ions and high K_a. Metal transport proteins usually bind only one or two metal atoms per molecule.[56] Serum albumin of mammals, a 66.5-kDa plasma protein, is probably the best-known example of a metal transport protein with low affinity and low carrying capacity for metals. One of the functions of serum albumin is transport of Ca, Cu, Zn, Ni, and Cd.[120,121] As an indication of its variable carrying capacity for different metals, a single site on the human serum albumin molecule binds and transports Ni, whereas 16 sites are involved for Ca and Zn.[119,145] The ability of human serum albumin to bind most metals is weak, with relatively low K_a's of 10^2 M^{-1} for Ca, 10^2 M^{-1} for Zn, 10^5 M^{-1} for Ni, and 10^2 M^{-1} for Cd.[119] Yet human serum albumin is the primary transport protein for Ca, Zn, and Ni in humans.[121,146] The low affinities facilitate mobilization and transfer of metals to enzymes, metallothioneins, and other metalloproteins in cells of target tissues.

The role of plasma proteins in Fe transport in mammals is well characterized. Iron must be rapidly bound to proteins in order to protect the organism from the Fe^{3+}-catalyzed Haber–Weiss reaction, which reduces Fe^{3+} to Fe^{2+} and produces free OH* radicals via the Fenton reaction.[147] In vertebrates, transferrin binds and transports Fe to tissues requiring the metal or to the liver for storage as a ferritin-bound complex.[148] Transferrin can also bind other metals. For example, Cd binds to transferrin at two sites with K_a's of 10^5 to 10^6 M^{-1}.[149] Iron in the Fe^{3+} form is strongly bound by both transferrin and ferritin with K_a's $\geq 10^{22}$ M^{-1}. Thus, a strong redox environment with pH below 5.5 is required to dissociate the Fe from transferrin and subsequently transfer it to ferritin for storage.[150] Transferrin has yet to be identified in any aquatic invertebrate, although a "transferrin-like" protein of 41 kDa has been identified in an ascidian.[151]

Ceruloplasmin, a 150-kDa μ_2-globulin, is a specific Cu transport protein in mammals.[18,39,119] Ceruloplasmin is secreted by the liver and tightly binds approximately 90 to 95% of the Cu present in the plasma.[18,145] The remainder is primarily bound to serum albumin. Although the relative importance of ceruloplasmin and serum albumin in Cu transport has been the subject of controversy, the primary role of serum albumin appears to be transport of newly absorbed Cu from the gut to the liver. The lifetime of any Cu-serum albumin association is thus relatively short. Copper is subsequently released from the liver tightly bound to ceruloplasmin.[39,120]

Additional vertebrate plasma proteins that may be involved in metal transport currently include the Cu- and Zn-binding proteins in winter flounder;[143] the 66-kDa protein that binds three molecules of Zn per molecule of protein in albacore tuna plasma;[144] a histidine-rich glycoprotein in humans that binds Zn, Ni, and Cd with a higher carrying capacity and slightly higher affinity than serum albumin;[152] and a mammalian high molecular weight Cu-containing plasma protein of 270 kDa.[153] Nonprotein transport ligands include free amino acids.[56]

The examples presented above for vertebrates alert us to the possibility that both specific and nonspecific metal-transport proteins may be present in the plasma of aquatic invertebrates. To date, evidence that plasma proteins play a major role in metal transport in invertebrates is still limited. Serum albumins, and possibly other plasma proteins identified in mammals, do not occur in invertebrates.[119,120,142] However, it is likely that analogous proteins exist in these organisms, and future efforts may demonstrate this.

Metal binding to respiratory pigments, apart from that needed to activate the proteins, appears to be a general phenomenon in aquatic invertebrates.[91] Zinc, cadmium, and mercury bind to arthropod hemocyanin,[126,154–157] which has been proposed as the primary Zn transport proteins in the decapod crustaceans.[126,157] Other high molecular weight plasma proteins that bind Cd, Pu, and Fe in crabs[158] and Ca, Cd, Co, Cu, Fe, Mn, Sr, and Zn in crayfish[159] may prove to be subunits of hemocyanin.[109] Hemoglobin contained in coelomocytes of the polychaete *Glycera dibranchiata* binds Cd.[160]

Other uncharacterized invertebrate plasma proteins include a 150-kDa protein that binds two atoms of Fe per molecule in the crab *Cancer magister*[161] and high molecular weight plasma proteins that bind Cd and Zn in gastropod[125] and bivalve mollusks.[68,92,127]

Cadmium binds weakly and nonspecifically to the entire suite of plasma proteins of *M. mercenaria* and *Mytilus edulis*, with K_a's of about 10^4 M^{-1} and binding capacities (C_L) of about 1.5 mM.[127,134] At low metal concentrations, 5% or less of the total Cd in the blood plasma of *Mercenaria mercenaria* is present as either free Cd, soluble inorganic Cd complexes, or complexes with soluble organic molecules <1 kDa in mass. More than 40 polypeptides occur in the plasma of these bivalves. At least six of these bind Fe, and over 13 bind Cd in *Mytilus edulis*. The identities and characteristics of the individual proteins remain to be determined.

The evidence presented above indicates that plasma proteins may play a central role in metal transport in aquatic invertebrates and reflects a situation analogous to that in mammals. In view of the greater carrying capacity usually observed in invertebrate plasma, in comparison with blood cells or hemocytes, the plasma may be a far more important component of the metal transport mechanism in invertebrates than has previously been recognized.

VI. RELEASE OF METALS

A. Conceptualization of Metal Release

Aquatic organisms utilize a variety of mechanisms to eliminate metals from their bodies. The overall process is a species-specific, organ- and tissue-specific, and metal- and ligand-specific process. In a single individual, the kinetics of metal release are expected to be very complex and to reflect the diverse compartments from which metals must be mobilized. Additionally, physical and chemical parameters, such as temperature and salinity, may affect the rate of release in aquatic animals.[162-165] However, due to the inherent variability in biological systems and the fact that whole body release rates are usually reported, the observed kinetics of metal release in aquatic species are usually described by a two-compartment model (Figure 4).[162,166] Although additional complexity is recognized for each compartment, fitting data to more complex models is usually discouraged by statistical ambiguity and the difficulty of discriminating among compartments at fine levels of resolution. The two-compartment model provides a useful framework for the development of hypotheses based on more complex interactions. In the two-compartment model, metals are accumulated into a rapidly exchanging compartment, from which metals are easily mobilized, and a slowly exchanging compartment, in which metals are tightly bound. The amount of metal in each of these depends on a variety of factors, the most important of which is the length of time that has been available for sequestration into slowly exchanging compartments. The easily mobilized metals include those that are adsorbed to external surfaces, complexed to external mucus, or complexed to low-affinity intracellular and extracellular

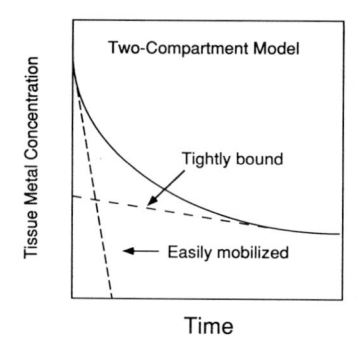

Figure 4. A theoretical metal depuration curve based on a two-compartment model of metal loss over time. Metal is rapidly released during the early stages of depuration from the easily mobilized compartment (e.g., surface absorption, mucus, low-affinity ligands). Subsequent metal loss from the tightly bound compartment (e.g., Ca concretions, metallothioneins, high-affinity ligands) is considerably slower. The size of the easily mobilized and tightly bound compartments varies appreciably for different metals, species of organism, and environmental conditions.

ligands. Release from these binding sites may take hours (e.g., Cd),[167] days (e.g., Zn),[162] or weeks (e.g., Ni, Cu, Pb).[168–170] The tightly bound metals include those that are sequestered by calcified concretions, metallothionein, ferritin, and ceruloplasm. Release from these sites depends on the turnover rate of the metal and the ligand and can take months, seasons, or years.

The two-compartment model can be used to explain apparent discrepancies in data on metal release in aquatic organisms. The lack of metal release observed in some cases indicates that the metals can be sequestered predominantly in the tightly bound compartment.[127,171,172] Observations that release can be curvilinear in some species,[169] exponential in others,[165,167,170] and biphasic in still others[162,168] can also be explained on this basis. In the cases of curvilinear and exponential release, the initial rapid loss would correspond to release from the easily mobilized compartment, whereas the later, slower release would be from the tightly bound compartment. The rate at which metals are released appears to be directly related to the rate of accumulation.[173,174] Exposure to high metal concentrations for short periods of time is expected to result in rapid accumulation by the easily mobilized, labile compartment. Low-level, chronic exposure seen in the natural environment favors slow filling of the tightly bound compartment, and release rates are correspondingly slower.

B. Pathways of Metal Release

1. Renal Pathways

Due to the minimal information on metal binding to fish plasma proteins, the potential for metals to be either retained or passed through the renal ultrafilter of fishes cannot be assessed presently. Thus, excretion of metals in mammals must serve as a model for urinary metal loss in fish.

The renal pathway is the primary route for the excretion of a variety of metals, such as Co, Cd, Sn, Ni, Cr, Mg, Zn, and Cu in mammals.[175] In rat urine, Cu, Zn, Ni, and Cd are bound by low-molecular-mass compounds of 0.5 to 5 kDa.[176] Urinary monitoring for free metal ion is used for the clinical diagnosis of Pb, Cd, and Hg poisoning.[175] Similarly, the presence of elevated urinary levels of metallothionein, metallothionein-bound metals, and proteinuria are indicative of toxic-metal-induced renal damage.[177-179]

Unlike mammals, whose excretory systems are reasonably well characterized, a consensus on the sites of ultrafiltration in aquatic invertebrates does not exist. The antennal and maxillary glands of crustaceans, the branchial heart appendages of cephalopods, and the ventricle wall in prosobranch gastropods have all been identified as sites of ultrafiltration on the basis of morphological evidence.[180] In bivalves, identification of this site remains controversial. Ultrastructural studies, using low-molecular-weight markers such as colloidal gold, have identified the pericardial glands as potential sites of ultrafiltration.[181] It is generally recognized that the primary urine of bivalves travels from the pericardial cavity, which surrounds the heart, to the kidney proper, via the renopericardial ducts. Ultrafiltration in bivalve mollusks allows passage of molecules smaller than 45 to 83 kDa.[134,182] Because plasma proteins that are smaller than this range are not prevalent, the vast majority of protein-bound metals will probably not pass into the primary urine. Additionally, experiments on *Mercenaria mercenaria* using ^{109}Cd have shown that 5% or less of the plasma-borne radionuclide is associated with very low-mass substances (<1 kDa), such as amino acids. Thus, the transport of metals to the kidney of bivalves via ultrafiltration (Figure 3) may not be a major pathway for metal excretion.

As mentioned previously, a number of aquatic invertebrate species sequester metals in Ca phosphate concretions.[90] Molluscan kidney concretions form a major subset of these invertebrate concretions. They are best studied in the marine bivalves. Although primarily intracellular,[90,135] very large extracellular concretions >1 mm can also be formed in some species of bivalves.[93] Whether these concretions can be excreted in the urine under natural conditions has yet to be determined. Conflicting data do not allow definitive conclusions.[92,98,103,183,184]

2. Digestive System Pathways

A second route for release of metals is through elimination with the feces. In terrestrial vertebrates, metals initially accumulated in the liver can be excreted in the bile. In humans, for example, the majority of the Pb, As, and Fe is secreted with the bile rather than in the urine; and Co, Cd, Sn, Ni, Cr, Mg, Zn, and Cu are also present in bile but at lower concentrations than in the urine.[175] Under normal circumstances, most of the Cd in rats is reported to be excreted with the feces, via the bile.[185,186] However, Cd and Cd-metallothionein are elevated in urine when the extent of Cd accumulation in the liver is associated with hepatic disorders. The source of this Cd is believed to be leakage from the liver to the circulatory system and a consequence of hepatic Cd toxicity. To date, no studies have specifically examined fish. Thus, the importance of biliary excretion of metals has not been determined for these aquatic vertebrates. Direct measurement of biliary and urinary output of metals is needed for estimation of the relative quantitative importance of the two routes.

In aquatic invertebrates, metals may be released by the discharge of digestive tissue residual bodies. In mollusks and crustaceans, these membrane-limited granules occur as a product of intracellular digestion by lysosomes in digestive gland and hepatopancreas cells. They are prominent features of these cells. The residual bodies can contain a variety of metals, including Ca and Mg, an organic matrix, and phosphate counterions. Thus, they may be considered to be one of the group of Ca concretions discussed previously. They are released from digestive gland cells by exocytosis during the later part of the normal digestive cycle[187] and are eliminated as part of the feces. It is unlikely that metals accumulated in these residual bodies would be reabsorbed during passage through the gut.[188]

Lipofuscin granules are also present in a wide range of aquatic invertebrates and vertebrates. These fluorescent granules are thought to be formed by the lysosomal degradation of membranes and are often found dispersed in pockets within connective tissues or intercalated between epithelial cells. Because they are formed by lysosomal action, they have characteristics similar to the residual bodies formed during intracellular digestion of ingested material. A variety of metals have been shown to be associated with such granules in the digestive gland. For example, Ca, Cd, Cr, Cu, Fe, Mn, Ni, and Zn occur in digestive gland granules of *M. mercenaria*.[93] Since more lipofuscin granules are found in older animals, these granules are thought to accumulate with age rather than be eliminated.

3. Diapedesis

Based primarily on histological evidence, several investigators have described a process referred to as "diapedisis," which is defined as a one-way migration of molluscan hemocytes from internal tissues, through epithelial layers, and into either the gut lumen or the surrounding water.[189,190] This process is considered to be involved with elimination. Metals sequestered by hemocytes would be eliminated from the body during this process. Unfortunately, it is very difficult to determine the importance of this route of metal release, and, as noted earlier, hemocytes do not as a general rule accumulate a significant portion of the body burden of metals. The oyster, for which diapedisis has been described, is an exception. More information on the turnover rate of hemocytes and the contribution of diapedesis to this turnover is required before definitive conclusions regarding the significance of diapedesis in regulating metal levels can be made.

VII. SUMMARY AND CONCLUSIONS

Processes associated with metal uptake, transport, and release are integrally linked to both environmental conditions and the intrinsic biological functions of organisms. Studies, to date, have identified the importance of the chemical speciation of metals in controlling metal uptake and a number of the biological mechanisms involved in metal regulation and metabolism. However, much of the work has been of a phenomenological nature, and the specific cellular and biochemical mechanisms underlying the generalized responses remain to be elucidated.

It is clear that an understanding of metal absorption will require understanding of membrane-dependent phenomena associated with the influx and efflux of metals across epithelial barriers. For this to be realized will also require a better understanding of membrane transport in aquatic organisms. For example, metals often are taken up via pathways that exist for essential nutrients. Thus, while it is known, for example, that Cd can be taken up through Ca channels,[41] the existence of such channels has only recently been reported in uptake organs of both fish[25] and invertebrates.[43] Similarly, cellular mechanisms associated with transmural transport and the function of blood and hemolymph in metal transport by the circulatory system have yet to be studied in sufficient detail to provide an accurate view of the actual processes that are involved. The basic understanding of the cellular physiology and biochemistry underlying such processes is currently lacking. Thus, while understanding toxicologically related responses is dependent on understanding basic biological function, the latter is often not available. Perhaps, the need for such information for designing toxicological studies will stimulate greater efforts in elucidating the basic biology of underlying fundamental processes.

Much less is known about the mechanisms of metal release from aquatic animals than is known about metal uptake and accumulation. Additionally, because few investigations have been conducted on fish, we must currently rely on mammalian models for deducing the responses of this major group. Of the numerous studies on aquatic invertebrates that have documented metal release, few have examined the actual mechanisms involved in this process. Of those that have, few provide even qualitative estimates of the contributions of the various pathways. Renal pathways, digestive system pathways, and diapedesis have received the most attention, although much of the information is still anecdotal. Some metals may also be released with gametes during spawning.[191] No investigators have examined the possibility that metals are released in a soluble form. Thus, it is not clear which of the proposed pathways of metal loss in invertebrates are of greatest significance.

While it is clear that much remains to be learned, it is equally clear that knowledge of the various mechanisms involved in metal uptake, accumulation, and release in aquatic organisms has advanced considerably over the past decade. This knowledge is fragmentary, however, concentrating on a few groups of organisms (e.g., bivalves, crustaceans), a limited number of metals (e.g., Cd, Cu, Zn), and a handful of metal ligands. Much is known, for example, about vertebrate and invertebrate metallothioneins, yet the presence and role of other intracellular metal-binding proteins and transmural transporters have only recently received attention. Many additional metal regulatory steps are expected to be identified in the years to come. The most exciting areas of future research will deal with the integration and coordination of these separately described steps, not only to elucidate toxicological responses to metals, but ultimately to understand the routine regulation of physiologically important elements.

REFERENCES

1. Furness, R.W. and Rainbow, P.S., *Heavy Metals in the Environment,* CRC Press, Boca Raton, FL, 1990, 256.
2. Newman, M.C. and McIntosh, A.W., *Metal Ecotoxicology: Concepts and Applications,* Lewis Publishers, Chelsea, MI, 1991, 399.
3. Niebohr, E. and Richardson, D.H.S., The replacement of the nondescript term 'heavy metals' by a biologically and chemically significant classification of metal ions, *Environ. Pollut. Ser. B,* 1, 3, 1980.
4. Förstner, U. and Wittman, G.T.W., *Metal Pollution in the Aquatic Environment,* Springer-Verlag, New York, 1979, 486.
5. Borgman, U., Metal speciation and toxicity of free metal ions to aquatic biota, in *Aquatic Toxicology,* Nriagu, J.O., Eds., John Wiley & Sons, New York, 1983, 47.
6. Luoma, S.N., Bioavailability of trace metals — a review, *Sci. Total Environ.,* 28, 1, 1983.
7. Brezonik, P.L., King, S.O., and Mach, C.E., The influence of water chemistry on trace metal bioavailability and toxicity to aquatic organisms, in *Metal Ecotoxicology: Concepts & Applications,* Newman, M.C. and McIntosh, A.W., Eds., Lewis Publishers, Chelsea, MI, 1991, 1.
8. Sunda, W.G., Engel, D.W., and Thuotte, R.M., Effect of chemical speciation on toxicity of cadmium to grass shrimp, *Palaemonetes pugio:* importance of free cadmium ion, *Environ. Sci. Technol.,* 12, 409, 1978.
9. Zamuda, C.D. and Sunda, W.G., Bioavailability of dissolved copper to the American oyster *Crassostrea virginica.* I. Importance of chemical speciation, *Mar. Biol.,* 66, 77, 1982.
10. O'Brien, P., Rainbow, P.S., and Nugegoda, D., The effect of chelating agent EDTA on the rate of uptake of zinc by *Palaemon elegans* (Crustacea: Decapoda), *Mar. Environ. Res.,* 30, 155, 1990.
11. Pärt, P. and Wikmark, G., The influence of some complexing agents (EDTA and citrate) on the uptake of cadmium in perfused rainbow trout gills, *Aquat. Toxicol.,* 5, 277, 1984.
12. Simkiss, K. and Taylor, M.G., Metal fluxes across the membranes of aquatic organisms, *CRC Crit. Rev. Aquat. Sci.,* 1, 173, 1989.

13. George, S.G. and Coombs, T.L., The effects of chelating agents on the uptake and accumulation of cadmium by *Mytilus edulis, Mar. Biol.*, 39, 261, 1977.
14. Block, M. and Pärt, P., Increased availability of cadmium to perfused rainbow trout *(Salmo gairdneri,* Rich.) gills in the presence of the complexing agents diethyl dithiocarbamate, ethyl xanthate and isopropyl xanthate, *Aquat. Toxicol.*, 8, 295, 1986.
15. Block, M., Glynn, A.W., and Pärt, P., Xanthate effects on cadmium uptake and intracellular distribution in rainbow trout *(Oncorhynchus mykiss)* gills, *Aquat. Toxicol.*, 20, 267, 1991.
16. Block, M. and Pärt, P., Uptake of ^{109}Cd by cultured gill epithelial cells from rainbow trout *(Oncorhynchus mykiss), Aquat. Toxicol.*, 23, 137, 1992.
17. Rainbow, P.S., Heavy metal levels in marine invertebrates, in *Heavy Metals in the Marine Environment,* Furness, R.W. and Rainbow, P.S., Eds., CRC Press, Boca Raton, FL, 1990, 67.
18. Cousins, R.J., Absorption, transport, and hepatic metabolism of copper and zinc: special reference to metallothionein and ceruloplasmin, *Physiol. Rev.*, 65, 238, 1985.
19. Gutnecht, J., Inorganic mercury (Hg^{2+}) transport through lipid bilayer membranes, *J. Membr. Biol.*, 61, 61, 1981.
20. Gutnecht, J., Cadmium and thallous ion permeabilities through lipid bilayer membranes, *Biochim. Biophys. Acta*, 735, 185, 1983.
21. Turner, D.R., Whitfield, M., and Dickson, A.G., The equilibrium speciation of dissolved components in freshwater and seawater at 25°C, *Geochim. Cosmochim. Acta*, 45, 855, 1981.
22. Playle, R.C. and Wood, C.M., Water chemistry changes in the gill micro-environment of rainbow trout: experimental observations and theory, *J. Comp. Physiol. B,* 159, 527, 1989.
23. Playle, R.C. and Wood, C.M., Water pH and aluminum chemistry in the gill microenvironment of rainbow trout during acid and aluminum exposures, *J. Comp. Physiol. B*, 159, 539, 1989.
24. Luoma, S.N. and Carter, J.L., Effects of trace metals on aquatic benthos, in *Metal Ecotoxicology: Concepts & Applications,* Newman, M.C. and McIntosh, A.W., Eds., Lewis Publishers, Chelsea, MI, 1991, 1.
25. Verbost, P.M., van Rooij, J., Flik, G., Lock, R.A.C., and Wendelaar Bonga, S.E., The movement of cadmium through freshwater trout branchial epithelium and its interference with calcium transport, *J. Exp. Biol.*, 145, 185, 1989.
26. Morton, J.E., *Guts,* 2nd ed., Edward Arnold Publishers, London, 1979, 59.
27. Williams, D.R. and Giesy, J.P., Relative importance of food and water sources to cadmium uptake by *Gambusia affinis, Environ. Res.*, 16, 326, 1978.
28. Carpene, E. and George, S.G., Absorption of cadmium by gills of *Mytilus edulis* (L.), *Mol. Physiol.*, 1, 23, 1981.
29. Roesijadi, G., Uptake and incorporation of mercury into mercury-binding proteins of gills of *Mytilus edulis* as a function of time, *Mar. Biol.*, 66, 151, 1982.
30. Roesijadi, G. and Klerks, P., A kinetic analysis of Cd-binding to metallothionein and other intracellular ligands in oyster gills, *J. Exp. Zool.*, 251, 1, 1989.
31. Schoenmakers, T.J.M., Klaren, P.H.M., Flik, G., Lock, R.A.C., Pang, P.K.T., and Wendelaar Bonga, S.E., Actions of cadmium on basolateral plasma membrane proteins involved in calcium uptake by fish intestine, *J. Membr. Biol.*, 127, 161, 1992.

32. Stacey, N.H. and Klaassen, C.D., Cadmium uptake by isolated rat hepatocytes, *Toxicol. Appl. Pharmacol.*, 55, 448, 1980.
33. Foulkes, E.C., On the mechanism of cellular cadmium uptake, *Biol. Trace Elem. Res.*, 21, 195, 1989.
34. Foulkes, E.C., Further findings on the mechanism of cadmium uptake by intestinal mucosal cells (step 1 of Cd absorption), *Toxicology*, 70, 261, 1991.
35. Blazka, M.E. and Shaikh, Z.A., Cadmium and mercury accumulation in rat hepatocytes: interactions with other metal ions, *Toxicol. Appl. Pharmacol.*, 113, 118, 1992.
36. Bobilya, D.J., Briske-Anderson, M., and Reeves, P.G., Zinc transport into endothelial cells is a facilitated process, *J. Cell. Physiol.*, 151, 1, 1992.
37. Percival, S.S. and Harris, E.D., Copper transport from ceruloplasmin: characterization of the cellular uptake mechanism, *Am. J. Physiol.*, 258, C140, 1990.
38. Spry, D.J. and Wood, C.M., A kinetic method for the measurement of zinc influx *in vivo* in the rainbow trout and the effects of waterborne calcium on flux rates, *J. Exp. Biol.*, 142, 425, 1989.
39. Harris, E.D., Copper transport: an overview, *Proc. Soc. Exp. Biol. Med.*, 196, 130, 1991.
40. Spry, D.J. and Wood, C.M., Zinc influx across the isolated, perfused head preparation of the rainbow trout *(Salmo gairdneri)* in hard and soft water, *Can. J. Fish. Aquat. Sci.*, 45, 2206, 1988.
41. Hinkle, P.M., Kinsella, P.A., and Osterhoudt, K.C., Cadmium uptake and toxicity via voltage-sensitive calcium channels, *J. Biol. Chem.*, 262, 16333, 1987.
42. Blazka, M.E. and Shaikh, Z.A., Differences in cadmium and mercury uptakes by hepatocytes: role of calcium channels, *Toxicol. Appl. Pharmacol.*, 110, 355, 1991.
43. Roesijadi, G. and Unger, M.E., Cadmium uptake in the gills of the mollusc *Crassostrea virginica* and inhibition by calcium channel blockers, *Aquat. Toxicol.*, 195, 1993.
44. Simkiss, K., Lipid solubility of heavy metals in saline solutions, *J. Mar. Biol. Assoc. U.K.*, 63, 1, 1983.
45. Wrench, J.J., Biochemical correlates of dissolved mercury uptake by the oyster *Ostrea edulis*, *Mar. Biol.*, 47, 79, 1978.
46. Hosey, M.M. and Lazdunski, M., Calcium channels: molecular pharmacology, structure and regulation, *J. Membr. Biol.*, 104, 81, 1988.
47. Scheuhammer, A.M., The dose-dependent deposition of cadmium into organs of Japanese quail following oral administration, *Toxicol. Appl. Pharmacol.*, 95, 153, 1988.
48. Min, K.-S., Nakatsubo, T., Kawamura, S., Fujita, Y., Onosaka, S., and Tanaka, K., Effects of mucosal metallothionein in small intestine on tissue distribution after oral administration of cadmium compounds, *Toxicol. Appl. Pharmacol.*, 113, 306, 1992.
49. Ohta, H. and Cherian, M.G., Gastrointestinal absorption of cadmium and metallothionein, *Toxicol. Appl. Pharmacol.*, 107, 63, 1991.
50. Birkenmeier, E.H. and Gordon, J.I., Developmental regulation of a gene that encodes a cysteine-rich intestinal protein and maps near the murine immunoglobulin heavy chain locus, *Proc. Natl. Acad. Sci. U.S.A.*, 83, 2516, 1986.
51. Hempe, J.M. and Cousins, R.J., Cysteine-rich intestinal protein binds zinc during transmucosal zinc transport, *Proc. Natl. Acad. Sci. U.S.A.*, 88, 9671, 1991.

52. Hempe, J.M. and Cousins, R.J., Cysteine-rich intestinal protein and intestinal metallothionein: an inverse relationship as a conceptual model for zinc absorption in rats, *J. Nutr.*, 122, 89, 1992.
53. Shears, M.A. and Fletcher, G.L., The binding of zinc to the soluble proteins of intestinal mucosa in winter flounder *(Pseudopleuronectes americanus), Comp. Biochem. Physiol.*, 64B, 297, 1979.
54. Shears, M.A. and Fletcher, G.L., The relationship between metallothionein and intestinal zinc absorption in the winter flounder, *Can. J. Zool.*, 62, 2211, 1984.
55. Coombs, J. and George, S.G., Mechanisms of immobilization and detoxification of metals in marine organisms, in *Physiology and Behavior of Marine Organism,* McLusky, D.S. and Berry, A.J., Eds., Pergamon Press, Oxford, 1978, 179.
56. George, S.G., Subcellular accumulation and detoxication of metals in aquatic animals, in *Physiological Mechanisms of Marine Pollutant Toxicity,* Vernberg, W.B., Calabrese, A., Thurberg, F.P., and Vernberg, F.J., Eds., Academic Press, New York, 1982, 3.
57. George, S.G., Pirie, B.J.S., and Coombs, T.L., The kinetics of accumulation and excretion of ferric hydroxide in *Mytilus edulis* (L.) and its distribution in the tissues, *J. Exp. Mar. Biol. Ecol.*, 23, 71, 1976.
58. Kagi, J.H.R. and Kojima, Y., Chemistry and biochemistry of metallothioneins, in *Metallothionein II,* Kagi, J.H.R. and Kojima, Y., Eds., Birkhauser-Verlag, Basel, 1987, 25.
59. Roesijadi, G., Metallothioneins in metal regulation and toxicity in aquatic animals, *Aquat. Toxicol.*, 22, 81, 1992.
60. Thiele, D.J., Metal-regulated transcription in eukaryotes, *Nucl. Acids Res.*, 20, 1183, 1992.
61. Overnell, J., Berger, C., and Wilson, K., Partial amino acid sequence of metallothionein from plaice *(Pleuronectes platessa), Biochem. Soc. Trans.*, 9, 217, 1981.
62. Olsson, P.-E. and Haux, C., Rainbow trout metallothionein, *Inorg. Chim. Acta,* 107, 67, 1985.
63. Shears, M.A. and Fletcher, G.L., Hepatic metallothionein in the winter flounder *(Pseudopleuronectes americanus), Can. J. Zool.,* 1602, 1985.
64. Olsson, P.-E. and Hogstrand, C., Subcellular distribution and binding of cadmium to metallothionein in tissues of rainbow trout after exposure to ^{109}Cd in water, *Environ. Toxicol. Chem.*, 6, 867, 1987.
65. Viarengo, A., Pertica, M., Mancinelli, G., Zanicchi, G., Bouquegneau, J.M., and Orunesu, M., Biochemical characterization of copper-thioneins isolated from the tissues of mussels exposed to the metal, *Mol. Physiol.*, 5, 41, 1984.
66. Olafson, R.W., Sim, R.G., and Boto, K.G., Isolation and characterization of the heavy metal-binding protein metallothionein from marine invertebrates, *Comp. Biochem. Physiol.*, 62B, 407, 1979.
67. Viarengo, A., Pertica, M., Mancinelli, M., Zanicchi, G., and Orunesu, M., Rapid induction of copper-binding proteins in the gills of metal-exposed mussels, *Comp. Biochem. Physiol.,* 67C, 215, 1980.
68. Nolan, C.V. and Duke, E.J., Cadmium accumulation and toxicity in *Mytilus edulis:* involvement of metallothioneins and heavy-molecular weight protein, *Aquat. Toxicol.*, 4, 153, 1983.

69. Engel, D.W., Brouwer, M., and Thurberg, F.P., Comparison of metal metabolism and metal-binding proteins in the blue crab and the American lobster, in *Marine Pollution and Physiology: Recent Advances*, Vernberg, F.J. Thurberg, F.P., Calabrese, A., and Vernberg, W., Eds., University of South Carolina, Columbia, 1985.
70. Petering, D.H., Goodrich, W.H., Krezoski, S., Weber, D., Shaw, C.F., III, Spieler, R., and Zettergren, L., Metal-binding proteins and peptides for the detection of heavy metals in aquatic organisms, in *Biomarkers of Environmental Contamination*, McCarthy, J.F. and Shugart, L.R., Eds., Lewis Publishers, Chelsea, MI, 1990, 239.
71. Udom, U.O. and Brady, F.O., Reactivation in vitro of zinc-requiring apo-enzymes by rat liver zinc-thionein, *Biochem. J.*, 187, 329, 1980.
72. Churchich, J.E., Scholz, G., and Kwok, F., Activation of pyridoxal kinase by metallothionein, *Biochim. Biophys. Acta*, 996, 181, 1989.
73. Brouwer, M. and Brouwer-Hoexum, T., Glutathione-mediated transfer of copper (I) into American lobster apohemocyanin, *Biochemistry*, 31, 4096, 1992.
74. Zeng, J., Heuchel, R., Schaffner, W., and Kägi, J.H.R., Thionein (apometallothionein) can modulate DNA binding and transcription activation by zinc finger containing Sp1, *FEBS Lett.*, 279, 310, 1991.
75. Zeng, J., Vallee, B.L., and Kägi, J.H.R., Zinc transfer from transcription factor IIIA fingers to thionein clusters, *Proc. Natl. Acad. Sci. U.S.A.*, 88, 9984, 1991.
76. Nemer, M., Travaglini, E.C., Rondinelli, E., and D'Alonzo, J., Developmental regulation, induction, and embryonic tissue specificity of sea urchin metallothionein gene expression, *Dev. Biol.*, 102, 471, 1984.
77. Olsson, P.-E., Zafarullah, M., Foster, R., Hamor, T., and Gedamu, L., Developmental regulation of metallothionein mRNA, zinc and copper levels in rainbow trout, *Salmo gairdneri, Eur. J. Biochem.*, 193, 229, 1990.
78. Andrews, G.K., Huet-Hudson, Y.M., Paria, B.C., McMaster, M.T., De, S.K., and Dey, S.K., Metallothionein gene expression and metal regulation during preimplantation mouse embryo development (MT mRNA during early development), *Dev. Biol.*, 145, 13, 1991.
79. Hamer, D.H., Thiele, D.J., and Lemontt, J.E., Function and autoregulation of yeast copper-thionein, *Science*, 228, 685, 1985.
80. Huang, P.C., Personal communication.
81. Hildebrand, C.E., Tobey, R.A., and Campbell, E.W., A cadmium-resistant variety of the Chinese hamster (CHO) cell with increased metallothionein induction capacity, *Exp. Cell Res.*, 124, 237, 1979.
82. Beach, L.R. and Palmiter, R.D., Amplification of the metallothionein-I gene in cadmium resistant mouse cells, *Proc. Natl. Acad. Sci. U.S.A.*, 78, 2110, 1981.
83. Gick, G.G. and McCarty, K.S., Amplification of the metallothionein-I gene in cadmium- and zinc-resistant Chinese hamster ovary cell, *J. Biol. Chem.*, 15, 9048, 1982.
84. Pruell, R.J. and Engelhardt, F.R., Liver cadmium uptake, catalase inhibition and cadmium thionein production in the killifish *(Fundulus heteroclitus)* induced by experimental exposure, *Mar. Environ. Res.*, 3, 101, 1980.
85. Kito, H., Tazawa, T., Ose, Y., Sato, T., and Ishikawa, T., Protection by metallothionein against cadmium toxicity, *Comp. Biochem. Physiol.*, 73C, 135, 1982.
86. Roesijadi, G., Drum, A.S., Thomas, J.M., and Fellingham, G.W., Enhanced mercury tolerance in marine mussels and relationship to low molecular weight, mercury-binding proteins, *Mar. Pollut. Bull.*, 13, 250, 1982.

87. Roesijadi, G. and Fellingham, G.W., Influence of Cu, Cd, and Zn preexposure on Hg toxicity in the mussel *Mytilus edulis, Can. J. Fish. Aquat. Sci.*, 44, 680, 1987.
88. Aoki, Y., Hatakeyama, S., Kobayashi, N., Sumi, Y., Suzuki, T., and Suzuki, K.T., Comparison of cadmium-binding protein induction among mayfly larvae of heavy metal resistant *(Baetis thermicus)* and susceptible species *(B. yoshinensis and B. sahoensis), Comp. Biochem. Physiol.*, 93C, 345, 1989.
89. Inoue, K., Akita, N., Shiba, T., Satake, M., and Yamashita, S., Metal-inducible activities of metallothionein promoters in fish cells and fry, *Biochem. Biophys. Res. Commun.*, 185, 1108, 1992.
90. Brown, B.E., The form and function of metal-containing granules in invertebrate tissues, *Biol. Rev.*, 57, 621, 1982.
91. Simkiss, K. and Mason, A.Z., Metal ions: metabolic and toxic effects, in *The Mollusca*, Hochachka, P.W., Ed., Academic Press, New York, 1983, 362.
92. George, S.G. and Pirie, B.J.S., Metabolism of zinc in the mussel, *Mytilus edulis* (L.): a combined ultrastructural and biochemical study, *J. Mar. Biol. Assoc. U.K.*, 60, 575, 1980.
93. Sullivan, P.A., Robinson, W.E., and Morse, M.P., Isolation and characterization of granules from the kidney of the bivalve *Mercenaria mercenaria, Mar. Biol.*, 99, 359, 1988.
94. George, S.G., Coombs, T.L., and Pirie, B.J.S., Characterization of metal-containing granules from the kidney of the common mussel, *Mytilus edulis, Biochim. Biophys. Acta*, 716, 61, 1982.
95. Marsh, M.E. and Sass, R.L., Distribution and characterization of mineral-binding phosphoprotein particles in Bivalvia, *J. Exp. Zool.*, 234, 237, 1985.
96. Simkiss, K. and Taylor, M.G., Convergence of cellular systems of metal detoxification, *Mar. Environ. Res.*, 28, 211, 1989.
97. George, S.G. and Pirie, B.J.S., The occurrence of sub-cellular particles in the kidney of the marine mussel, *Mytilus edulis*, exposed to cadmium, *Biochim. Biophys. Acta*, 580, 234, 1979.
98. Fowler, B.A. and Gould, D., Ultrastructural and biochemical studies of intracellular metal-binding patterns in kidney tubule cells of the scallop Placopecten magellanicus following prolonged exposure to cadmium or copper, *Mar. Biol.*, 97, 207, 1988.
99. George, S.G., Heavy metal detoxication in the mussel *Mytilus edulis* — composition of Cd-containing kidney granules (tertiary lysosomes), *Comp. Biochem. Physiol.*, 76C, 53, 1983.
100. Sternlieb, I. and Goldfischer, S., Heavy metals and lysosomes, in *Lysosomes in Biology and Pathology*, Dingle, J.T. and Dean, R.T., Eds., American Elsevier Publishers, New York, 1976, 185.
101. George, S.G., Heavy metal detoxication in *Mytilus* kidney — an in vitro study of Cd- and Zn-binding to isolated tertiary lysosomes, *Comp. Biochem. Physiol.*, 76C, 59, 1983.
102. Simkiss, K., Cellular discrimination processes in metal accumulating cells, *J. Exp. Biol.*, 94, 317, 1981.
103. Carmichael, N.G., Squibb, K.S., and Fowler, B.A., Metals in the molluscan kidney: a comparison of two closely related bivalve species *(Argopecten)*, using X-ray microanalysis and atomic absorption spectroscopy, *J. Fish. Res. Board Can.*, 36, 1149, 1979.

104. DeDuve, C. and Wattiaux, R., Functions of lysosomes, *Annu. Rev. Physiol.*, 28, 435, 1966.
105. Wessing, A. and Eichelberg, D., Ultrastructural aspects of transport and accumulation of substances in the Malpighian tubules, in *Excretion. Fortschritte der Zoology*, Wessing, A., Eds., Gustav Fischer Verlag, Stuttgart, 1975, 148.
106. Taylor, M.G., Simkiss, K., Greaves, G.N., and Haries, J., Corrosion of intracellular granules and cell death, *Proc. R. Soc. Lond.*, 234B, 463, 1988.
107. Nott, J.A. and Nicolaidou, A., The cytology of heavy metal accumulations in the digestive glands of three marine gastropods, *Proc. R. Soc. Lond.*, 237B, 347, 1989.
108. Bezkorovainy, A., Biochemistry of nonheme iron in man. I. Iron proteins and cellular iron metabolism, *Clin. Physiol. Biochem.*, 7, 1, 1989.
109. Guary, J.C. and Negrel, R., Plutonium and iron association with metal-binding proteins in the crab *Cancer pagurus* (L.), *J. Exp. Mar. Biol. Ecol.*, 42, 87, 1980.
110. Bottke, W., Electrophoretic and immunologic studies on the structure of a mollusc ferritin, *Comp. Biochem. Physiol.*, 81B, 325, 1985.
111. Webb, J., Macey, D.J., and Talbot, V., Identification of ferritin as a major high molecular weight zinc-binding protein in the tropical rock oyster, Saccostrea cuccullata, *Arch. Environ. Contam. Toxicol.*, 14, 403, 1985.
112. Burford, M.A., Macey, D.J., and Webb, J., Hemolymph ferritin and radula structure in the limpets *Patelloida alticostata* and *Patella peronii* (Mollusca: Gastropoda), *Comp. Biochem. Physiol.*, 83A, 353, 1986.
113. Kim, K.-S., Webb, J., and Macey, D.J., Properties and role of ferritin in the hemolymph of the chiton *Clavarizona hirotosa*, *Biochim. Biophys. Acta*, 884, 387, 1986.
114. Bjerregaard, P., Influence of physiological condition on cadmium transport from haemolymph to hepatopancreas in *Carcinus maenas*, *Mar. Biol.*, 106, 199, 1990.
115. Cunningham, P.A. and Tripp, M.R., Factors affecting the accumulation and removal of mercury from tissues of the American oyster *Crassostrea virginica, Mar. Biol.*, 31, 311, 1975.
116. Betzer, S.B. and Pilson, M.E.Q., Copper uptake and excretion by *Busycon canaliculatum* L., *Biol. Bull.*, 148, 1, 1975.
117. Holwerda, D.A., Hemelraad, J., Veenhof, P.R., and Zandee, D.I., Cadmium accumulation and depuration in *Anodonta anatina* exposed to cadmium chloride or cadmium-EDTA complex, *Bull. Environ. Contam. Toxicol.*, 40, 373, 1988.
118. Ratcliffe, N.A. and Rowley, A.F., *Invertebrate Blood Cells*, Vols. 1 and 2, Academic Press, New York, 1981, 641.
119. Putnam, F.W., *The Plasma Proteins*, Vols. 1 and 2, Academic Press, New York, 1975.
120. Peters, T. and Reed, R.G., Serum albumin as a transport protein, in *Transport of Protein*, Blauer, G. and Sund, H., Eds., de Gruyter & Co., Berlin, 1978, 57.
121. Scott, B.J. and Bradwell, P.R., Metal binding to serum protein, in *Protides of the Biological Fluids*, Peeters, N., Ed., Pergamon Press, Oxford, 1984, 15.
122. George, S.G., Pirie, B.J.S., Cheyne, A.R., Coombs, T.L., and Grant, M.B., Detoxication of metal by marine bivalves: an ultrastructural study of the compartmentation of copper and zinc in the oyster *Ostrea edulis, Mar. Biol.*, 45, 147, 1978.
123. Hawkins, C.J., Kott, P., Parry, D.L., and Swinehart, J.H., Vanadium content and oxidation state related to ascidian phylogeny, *Comp. Biochem. Physiol.*, 76B, 555, 1983.

124. Michibata, H., Terada, T., Anada, N., Yamakawa, K., and Numakunai, T., The accumulation and distribution of vanadium, iron, and manganese in some solitary ascidians, *Biol. Bull.*, 171, 672, 1986.
125. Langston, W.J. and Zhou, M., Cadmium accumulation, distribution, and metabolism in the gastropod *Littorina littorea:* the role of metal-binding proteins, *J. Mar. Biol. Assoc. U.K.*, 67, 585, 1987.
126. Zatta, P., Zinc transport in the haemolymph of *Carcinus maenas* (Crustacea, Decapoda), *J. Mar. Biol. Assoc. U.K.*, 64, 801, 1984.
127. Robinson, W.E. and Ryan, D.K., Transport of cadmium and other metals in the blood of the bivalve molluscs Mercenaria mercenaria, *Mar. Biol.*, 97, 101, 1988.
128. Depledge, M.H., Studies on copper and iron concentrations, distributions and uptake in the brachyuran *Carcinus maenas* (L.) following starvation, *Ophelia*, 30, 187, 1989.
129. Araki, S. and Oono, H., Behavior of lead and zinc in plasma, erythrocytes, and urine and ALAD in erythrocytes following intravenous infusion of Ca EDTA in lead workers, *Arch. Environ. Health,* 39, 363, 1984.
130. Kunimoto, M., Miyasaka, K., and Miura, T., Changes in membrane properties of rat red blood cells induces by cadmium accumulating in the membrane fraction, *J. Biochem.*, 99, 397, 1986.
131. Duval, G. and Grubb, B.R., Tissue accumulation of cadmium as a function of blood concentration, *Biol. Trace Elem. Res.*, 9, 101, 1986.
132. Anderson, D.H. amd Anderson, J.H.S., The distribution of vanadium and sulfer in the blood cells, and the nature of vanadium in the blood cells and plasma of the ascidian, *Ascidia ceratodes, Comp. Biochem. Physiol.*, 99A, 585, 1991.
133. George, S.G., Pirie, B.J.S., and Coombs, T.L., Isolation and elemental analysis of metal-rich granules from the kidney of the scallop *Pecten maximus* (L.), *Exp. Mar. Biol. Ecol.*, 42, 143, 1980.
134. Robinson, W.E., Unpublished data.
135. George, S.G., Pirie, B.J.S., and Frazier, J.M., Effects of cadmium exposure on metal-containing amoebocytes of the oyster *Ostrea edulis, Mar. Biol.*, 76, 63, 1983.
136. Ruddell, C.L. and Rains, D.W., The relationship between zinc, copper and the basophils of two crassostreid oysters, *C. gigas* and *C. virginica, Comp. Biochem. Physiol.*, 51A, 585, 1975.
137. Pirie, B.J.S., George, S.G., Lytton, D.G., and Thomson, J.D., Metal-containing blood cells of oysters: ultrastructure, histochemistry and X-ray microanalysis, *J. Mar. Biol. Assoc. U.K.*, 64, 115, 1984.
138. George, S.G., Heavy metal detoxication in the mussel Mytilus edulis — composition of Cd-containing kidney granules (tertiary lysosomes), *Comp. Biochem. Physiol.*, 76C, 53, 1983.
139. Steinert, S.A. and Pickwell, G.V., Expression of heat shock proteins and metallothionein in mussels exposed to heat stress and metal ion challenge, *Mar. Environ. Res.*, 24, 211, 1988.
140. Martoja, M. and Martin, J.-L., Detoxification of cadmium by the oyster *Crassostrea gigas* (Mollusc, Bivalve). Characterization of a cadmium-binding protein in the zinc and copper-containing amoebocytes, *C.R. Acad. Sci. Paris*, 300, 549, 1985.
141. Balaji, R., Mullainadhan, P., and Arumugam, M., *In vivo* binding of exogenous copper to haemolymph fractions of estuarine crab *Scylla serrrata* (Forskal), *J. Exp. Mar. Biol. Ecol.*, 128, 241, 1989.

142. Doolittle, R.F., Evolution of the vertebrate plasma proteins, in *The Plasma Proteins,* 2nd ed., Putnam, F.W., Ed., Academic Press, Orlando, FL, 1984, 317.
143. Fletcher, P.E. and Fletcher, G.L., Zinc- and copper-binding proteins in the plasma of winter flounder *(Pseudopleuronectes americanus), Can. J. Zool.,* 58, 609, 1980.
144. Dyke, B., Hegenauer, J., and Saltman, P., Isolation and characterization of a new zinc-binding protein from albacore tuna plasma, *Biochemistry,* 26, 3228, 1987.
145. Laurell, C.B., Metal-binding plasma proteins and cation transport, in *The Plasma Proteins,* Putnam, F.W., Ed., Academic Press, New York, 1960, 349.
146. Chilvers, D.C., Dawson, J.B., Bahreyni-Toosi, M.-H., and Hodgkinson, A., Identification and determination of copper- and zinc-protein complexes in blood plasma after chromatographic separation on DEAE-Sepharose Cl-6B, *Analyst,* 109, 871, 1984.
147. Halliwell, B. and Gutteridge, J.M.C., Oxygen toxicity, oxygen radicals, transition metals and disease, *Biochem. J.,* 219, 1, 1984.
148. Aisen, P. and Listowsky, I., Iron transport and storage proteins, *Annu. Rev. Biochem.,* 49, 357, 1980.
149. Harris, W.R. and Madsen, L.J., Equilibrium studies on the binding of cadmium(II) to human serum transferrin, *Biochemistry,* 27, 284, 1988.
150. Chrichton, R.R. and Charloteaux-Wauters, M., Iron transport and storage, *Eur. J. Biochem.,* 164, 485, 1987.
151. Martin, A.W., Huebers, E., Huebers, H., Webb, J., and Finch, C.A., A mono-cited transferrin from a representative deuterstome: the ascidian *Pyura stolonifera, Blood,* 64, 1047, 1984.
152. Guthans, S.L. and Morgan, W.T., The interaction of zinc, nickel and cadmium with serum albumen and histidine-rich glycoprotein assessed by equilibrium dialysis and immunoadsorbent chromatography, *Arch. Biochem. Biophys.,* 218, 320, 1982.
153. Weiss, K.C. and Linder, M.C., Copper transport in rats involving a new plasma protein, *Am. J. Physiol.,* 218, 377, 1985.
154. Martin, J.-L.M., Wormhoudt, A.V., and Ceccoldi, H.J., Zinc-hemocyanin binding in the hemolymph of *Carcinus maenas* (Crustacea, Decapoda), *Comp. Biochem. Physiol.,* 58A, 193, 1977.
155. Wright, D.A., The uptake of cadmium into the haemolymph of the shore crab *Carcinus maenas:* the relationship with copper and other divalent cations, *J. Exp. Biol.,* 67, 147, 1977.
156. Brouwer, M. and Engel, D.W., Stoichiometry and functional consequences of Hg(II) and Cd(II) binding to arthropod hemocyanins, in *Physiological Mechanisms of Marine Pollutant Toxicity,* Vernberg, W.B., Calabrese, A., Thurberg, I.P., and Vernberg, F.J., Eds., Academic Press, New York, 1982, 289.
157. Engel, D.W., Metal regulation and molting in the blue crab, *Callinectes sapidus:* copper, zinc, and metallothionein, *Biol. Bull.,* 172, 69, 1987.
158. Bjerregaard, P. and Vislie, T., Effects of cadmiun on hemolymph composition in the shore crab *Carcinus maenas, J. Ecol. Physiol. Ser.,* 27, 135, 1985.
159. Lyon, R., Taylor, M., and Simkiss, K., Ligand activity in the clearance of metals from the blood of the crayfish *(Austropotamobius pallipes), J. Exp. Biol.,* 113, 19, 1984.
160. Rice, M.A. and Chien, P.K., Uptake, binding and clearance of divalent cadmium in *Glycera dibranchiata* (Annelida: Polychaeta), *Mar. Biol.,* 53, 33, 1979.

161. Huebers, H.A., Finch, C.A., and Martin, A.W., Characterization of an invertebrate transferrin from the crab *Cancer magister* (Arthropoda), *J. Comp. Physiol.*, 148B, 101, 1982.
162. Keckes, S., Ozretić, B. and Krajnović, M., Loss of ^{65}Zn in the mussel *Mytilus galloprovincialis, Malacologia*, 7, 1, 1968.
163. Denton, G.R.W. and Burdon-Jones, C., Influence of temperature and salinity on the uptake, distribution and depuration of mercury, cadmium and lead by the black-lip oyster *Saccostrea echinata, Mar. Biol.*, 64, 317, 1981.
164. Dahlgaard, H., Effects of season and temperature on long-term in situ loss rates of Pu, Am, Np, Eu, Ce, Ag, Tc, Zn, Co, and Mn in a Baltic *Mytilus edulis* population, *Mar. Ecol. Prog. Ser.*, 33, 157, 1986.
165. Van Dolah, F.M., Siewicki, T.C., Collins, G.W., and Logan, J.S., Effects of environmental parameters on the elimination of cadmium by eastern oysters, *Crassostrea virginica, Arch. Environ. Contam. Toxicol.*, 16, 733, 1987.
166. Ruzic, I., Two-compartment model of radionuclide accumulation into marine organisms. I. Accumulation from a medium of constant activity, *Mar. Biol.*, 15, 105, 1972.
167. Scholz, N., Accumulation, loss and molecular distribution of cadmium in *Mytilus edulis, Helgolander Meeresunters*, 33, 68, 1980.
168. Zaroogian, G. and Johnson, M., Nickel uptake and loss in the bivalves *Crassostrea virginica* and *Mytilus edulis, Arch. Environ. Contam. Toxicol.*, 13, 411, 1984.
169. Zaroogian, G., Studies on the depuration of cadmium and copper by the American oyster Crassostrea virginica, *Bull. Environ. Contam. Toxicol.*, 23, 117, 1979.
170. Roesijadi, G., Young, J.S., Drum, A.S., and Gurtisen, J.M., Behavior of trace metals in *Mytilus edulis* during a reciprocal transplant field study, *Mar. Ecol. Prog. Ser.*, 18, 155, 1984.
171. Behrens, W.J. and Duedall, I.W., The behavior of heavy metals in transplanted hard clams, *Mercenaria mercenaria, J. Cons. Int. Explor. Mer*, 39, 223, 1981.
172. Luten, J.B., Bouquet, W., Burggraaf, M.M., Rouchbaar, A.B., and Rus, J., Trace metals in mussels *(Mytilus edulis)* from the Waddenzee, coastal North Sea and the estuaries of Ems, Western and Eastern Scheldt, *Bull. Environ. Contam. Toxicol.*, 36, 770, 1986.
173. Riisgård, H.U., Bjørnestad, E., and Møhlenberg, F., Accumulation of cadmium in the mussel *Mytilus edulis:* kinetics and importance of uptake via food and sea water, *Mar. Biol.*, 96, 349, 1987.
174. Salanki, J. and V.-Bologh, K., Uptake and release of mercury and cadmium in various organs of mussels *(Anodonta cygrea* L.), *Symp. Biol. Hung.*, 29, 325, 1985.
175. Ishihara, N. and Matsushiro, T., Biliary and urinary excretion of metals in humans, *Arch. Environ. Health*, 41, 324, 1986.
176. Sarkar, B., Metal-protein interactions in transport, accumulation, and excretion of metals, *Biol. Trace Elem. Res.*, 21, 137, 1989.
177. Kjellström, T., Elinder, C.-G., and Friberg, L., Conceptual problems in establishing the critical concentration of cadmium in human kidney cortex, *Environ. Res.*, 33, 284, 1984.
178. Bremner, I., Mehra, R.K., Morrison, J.N., and Wood, A.M., Effects of dietary copper supplementation of rats on the occurrence of metallothionein-I in liver and its secretion into blood, bile and urine, *Biochem. J.*, 235, 735, 1986.

179. Mitane, Y., Tohyama, C., and Saito, H., The role of metallothionein in the elevated excretion of copper in urine from people living in a cadmium-polluted area, *Fund. Appl. Toxicol.*, 6, 285, 1986.
180. Kümmel, G., Filtration structures in excretory systems: a comparison, in *Comparative Physiology,* Bolis, L., Schmidt-Nielsen, K., and Maddrell, S.H.P., Eds., North-Holland, Amsterdam, 1973, 221.
181. Morse, M. P., Comparative functional morphology of the bivalve excretory system, *Am. Zool.*, 27, 737, 1987.
182. Hevert, F., Urine formation in the lamellibranchs: evidence for ultrafiltration and quantitative description, *J. Exp. Biol.*, 111, 1, 1984.
183. Rheinberger, R., Hoffman, G.L., and Yevich, P.P., The kidney of the quahog *(Mercenaria mercenaria)* as a pollution indicator, in *Animals as Monitors of Environmental Pollutants,* National Academy of Science, Washington, D.C., 1979, 119.
184. Regoli, F., Nigro, M., and Orlando, E., Effects of copper and cadmium on the presence of renal concretions in the bivalve *Donacila cornea, Comp. Biochem. Physiol.*, 102C, 189, 1992.
185. Cherian, M.G., Onosaka, S., Carson, G.K., and Dean, P.A.W., Biliary excretion of cadmium in rat. V. Effects of structurally related mercaptans on chelation of cadmium from metallothionein, *J. Toxicol. Environ. Health*, 9, 389, 1982.
186. Tanaka, K., Effects of hepatic disorder on the fate of cadmium in rats, in *Biological Roles of Metallothionein,* Foulkes, E.C., Ed., Elsevier North-Holland, New York, 1982, 237.
187. Morton, B.S., Feeding and digestion in Bivalvia, in *The Mollusca*, Salenddin, A.S.M. and Wilbur, K.M., Eds., Academic Press, New York, 1983, 65.
188. Arumugam, M. and Ravindranath, M.H., Copper toxicity in the crab, *Scylla serrata,* copper levels in tissues and regulation after exposure to a copper-rich medium, *Bull. Environ. Contam. Toxicol.*, 39, 708, 1987.
189. Tripp, M.R., Disposal by the oyster of intracardially injected red blood cells of vertebrates, *Proc. Natl. Shellfish Assoc.*, 48, 143, 1957.
190. Brown, A.C. and Brown, R.J., The fate of thorium dioxide injected into the pedal sinus of Bullia *(Gastropoda prosobranchiata), J. Exp. Biol.*, 42, 509, 1965.
191. Lowe, D.M. and Moore, M.N., The cytochemical distribution of zinc (ZnII) and iron (FeIII) in the common mussel, *Mytilus edulis,* and their relationship with lysosomes, *J. Mar. Biol. Assoc. U.K.*, 59, 851, 1979.

Chapter 10

Behavioral Mechanisms of Metal Toxicity in Fishes

Daniel N. Weber and Richard E. Spieler

I. A SYNTHESIS OF PROXIMATE MECHANISMS OF BEHAVIOR, ETHOLOGICAL PARADIGMS, AND METAL TOXICITY

At present, two major divisions, physiological and behavioral, exist within aquatic toxicology. With admitted bias, it appears to us that, in general, physiological toxicologists have presented a wealth of data on the effects of toxicants on a vast number of physiological variables. With the exception of morbidity, however, seldom do these data give insight into organismal effects (i.e., what do these physiological changes mean for the animal in terms of social interaction or behavioral performance?). In contrast, today's ethologists are primarily concerned with evolutionary adaptations and processes. This approach is not as useful in polluted waters, due to the frequent inability of organisms to adapt to such massive disturbance of their habitat. With this in mind, behavioral toxicologists have presented data on the effects of some toxicants on specific behaviors. A cursory review of the literature would suggest that investigators seldom address the physiological mechanisms that explain how that particular behavioral effect was produced. Yet, it is clear to us that these two approaches, physiological and behavioral, are not and should not be mutually exclusive; they clarify and reinforce one another.[1] In this chapter, we seek to synthesize these two fields by examining the full toxicant-physiological-behavioral axis and, in so doing, evaluate the effects of environmental contamination on the physiological mechanisms of behavior. Clearly, this is a nontraditional ethological approach.

To accomplish this goal, we divide this chapter into five sections: one section on mechanistic models of behavior; three on hormonal, neurological, and energetic and respiratory mechanisms of behavior; and one on future research directions, although additional ideas are also included throughout each section. Regrettably, there is considerable overlap among these rather arbitrarily defined sections, yet they serve to categorize the diverse studies under discussion. As much as we might have preferred to restrict our synthetic view of behavioral toxicology to research with fishes, there are significant gaps in the literature on fish toxicology; thus, a number of mammalian citations are included where necessary to make our point. Clearly, the assumption we will make is that fish and mammalian physiological mechanisms are fundamentally similar. A quick review of the literature would indicate that there is sufficient data to support such a concept.

Before beginning, however, it is important to say a few words about those transitional elements often labeled as "heavy metals," which have the potential to either enhance or upset normal physiological functions. The concentrations of waterborne metals are an important aspect of their toxicity. More critical, however, is the bioavailability of metals, which is correlated to a variety of environmental factors independent of concentration, e.g., the rate of metal mineralization, pH (and, therefore, oxidation state), organic acid concentration, adsorption to sediment, water hardness, temperature, dietary exposure, anthropogenic input, and level of sewage treatment.[2] Yet bioavailability does not necessarily correlate with biological effects. The kinetics of metal metabolism may affect the degree of toxicity. Metals such as cadmium, copper, or zinc may be bound to cysteine-rich proteins (e.g., metallothionein)[3] or incorporated into nonlabile pools (e.g., lead in bone).[4] Thus, only a small portion of the total metal intake may actually be biologically active in affecting such parameters as behavior. It is with that small amount that this chapter concerns itself.

Many models have been suggested to understand both the proximate (physiological) and ultimate (evolutionary) mechanisms of behavior. A few of these are useful as paradigms for behavioral toxicity. In this section, we will examine the utility of these models. Because the mechanisms of behavioral ontogeny can be altered by metals, we will also evaluate one model that examines the long-term effects of life stage-specific perturbations.

Under normal conditions in uncontaminated environments, changes in fish responsiveness to stimuli are nonrandom processes; specific behavioral sequences are initiated by specific stimuli at specific stages of an animal's life. These "prewired" or "hard-wired" behavioral performances have been labeled "fixed action patterns" (FAP). Such behaviors are triggered by an external stimulus (releaser), which activates a neural network (innate releasing mechanism) that first filters the stimulus response and then, if a specific threshold is reached, facilitates specific behavioral sequences. This will be detailed in Section III. Each

FAP, as species specific as any morphological characteristic, maintains four fundamental qualities: (1) it is under genetic control; (2) a feedback mechanism is not required or is needed only minimally to begin or stop the sequence; (3) the action, even if the releaser is removed or manipulated experimentally, goes to completion once the brain is stimulated by a releaser; and (4) the behavior is not a simple reflexive motion, rather, it is an intricate coordination of several muscles or muscle groups. In general, ethologists assume that sequences of behaviors are adaptive and will, if uninterrupted, be carried to completion before the individual begins a second set of directed actions. But what happens if a FAP is interrupted or fails to be completed due to the physiological interaction of a toxic metal? Obviously, integration of the fish with its environment is affected, presumably in a nonadaptive fashion.

The potential for disruption of behavioral integration (i.e., individual movements that occur sequentially to accomplish a specific task) is even more apparent when one considers that these behaviors are dependent upon both the intensity of the stimuli and the seasonal physiological state of the animal (hereafter referred to by us as Huntingford's first paradigm).[5] Because the internal physiological state of the fish varies throughout the day and year, identical stimuli may result in very different effects (Figure 1A). Depending on the seasons (I to IV), the fish may be differentially sensitive to some specific stimulus (S_1). In turn, the physiological effects (E_1 to E_4) may range from none to very intense, with resultant behaviors ranging similarly. Likewise, if more than one stimulus (e.g., S_1 and S_2) simultaneously affects the internal state (Figure 1B), then again, dependent upon temporal relationships, there may be no or multiple physiological changes and one might observe no or multiple behaviors (sequential patterns, guided by attendant physiological shifts). One can observe that, dependent on the photoperiod regime, adult fathead minnows may or may not form schools (personal observation). Thus, dependent upon the physiological changes induced by one stimulus (S_1) (i.e., photoperiod), a second stimulus (S_2) (i.e., food) may release either aggressive displays in nonschooling fish or socially facilitated feeding among schooling groups. de Bruin[6] cites examples of other fish species demonstrating similar reactions to two or more simultaneous environmental stimuli. This can be extended to situations involving changes in behavior due to environmental contamination. One can hypothesize that, depending on various temporal relationships, the intensity and, perhaps, direction of contaminant effect will vary. We will examine this possibility further in Section III.C.

The integration of both individual behaviors and behavioral sequences form specific, recognizable, stereotypic patterns for a species. In general, the degree of linkage between the change in internal state due to a given stimulus and the overt response determines the time interval between a behavioral response due to that stimulus.[5] An example of a strong linkage yielding short time intervals would be the visual and olfactory stimulus of the presence of food inducing feeding behaviors. Other stimuli, such as changes in photoperiod, may begin a slower change in physiological state that eventually, but not immediately, alters

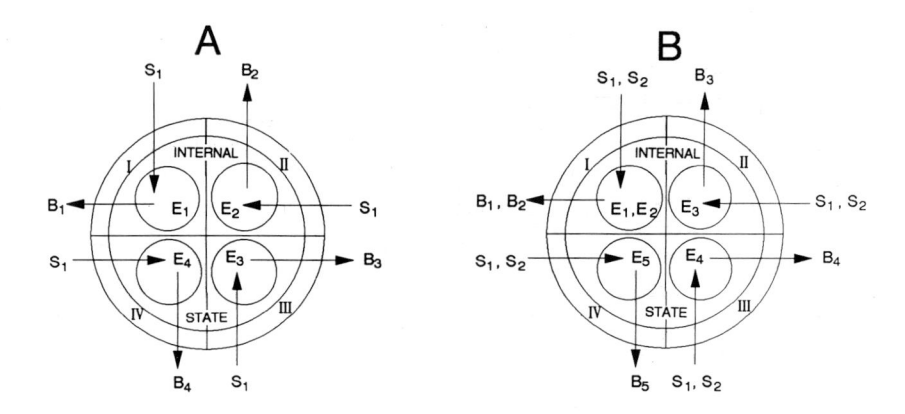

Figure 1. Temporal interactions between the external environment and internal state of the individual fish. I to IV = different seasons; S_1 and S_2 = different environmental stimuli; B_1 to B_5 = different behavioral responses to specific environmental stimuli; E_1 to E_5 = different internal states upon which environmental stimuli act and which provide the fundamental mechanisms that direct seasonal-specific behaviors (based upon Huntingford[5]). (A) A single environmental stimulus inducing a single behavioral response. (B) Multiple environmental stimuli inducing multiple behavioral responses.

reproductive behavior. If two separate behavioral patterns, e.g., B_1 and B_2, (Figure 1B) are inducible by similar changes in internal states, they will tend to co-occur when the causal stimuli are strong. For example, reproductive and aggressive behaviors are, at least in part, controlled by the anterior dorsal telencephalon and the preoptic area.[6] Changes in generalized neural activity in either of these regions could result in altering two distinct behavioral patterns. Unfortunately, this does not take into account feedback mechanisms that could also be affected by metal contaminants.

From a toxicological viewpoint, these are important concepts, especially when examining subtle, sublethal effects. Different patterns of behavior (feeding, reproduction, and defense) accomplish specific goals within the context of ever-changing daily and seasonal environmental requirements. Contaminants may alter specific physiological characteristics of fishes that link overt responses to specific external stimuli. Thus, by inducing these inappropriate responses to time-based environmental requirements, metal-induced alterations may be detrimental to long-term survival of a fish population, even at levels sublethal for the individual. One possible example would be metal binding to a hormone receptor site, resulting in reduced synthesis of specific metabolic intermediates required for the expression of a particular behavior, even though sufficient hormone has been synthesized in response to an environmental stimulus. Although there are sufficient data in the mammalian literature demonstrating the capacity of toxic metals to bind at hormone receptor sites,[7,8] we have found no studies that correlate this particular phenomenon to actual behavioral effects in any vertebrate species, especially fishes. Although such work may be difficult to conduct due to myriad biochemical paths affected by metals, we submit that such correlations are critical.

One can also analyze behavior in terms of mechanisms animals use to "decide" whether to display a specific pattern of behaviors. Huntingford[5] visualized a space-state model of interconnected action groups with specific motivational bases, each related to the complex interplay between the individual's internal state and its external stimuli (hereafter referred to by us as Huntingford's second paradigm). Given a particular point in time, the myriad independent physiological variables that control behavior can be placed on axes of a multidimensional-space graph. Motivational space is the subset of those physiological variables that can be actually altered by behavior. A second subset, cue space, is the mechanism used to translate a given set of stimuli into a proper sequence of behaviors. When these two subsets interact, the range of *possible* observed behaviors is found within the causal factor space. The individual must then give priority values to this "list." The method of ranking this range of possible behavioral responses to a given situation (i.e., how to time share between competing behaviors to achieve a specific goal) is found in a region called candidate space. Whether the fish should eat or rest, defend a nest or attract a mate, migrate or stay are decisions that need the immediate attention of that individual. As will be suggested later, this paradigm is useful in describing some behavioral observations of metal-exposed fishes.

Another perspective used by many behavioral toxicologists is the examination of selected motions that reflect simple reflex-arc behaviors.[9,10] The value of this type of analysis is that these behaviors are easy to quantify, are easily manipulated by toxic substances, and thus are useful in bioassay tests. In general, such tests do not necessarily further an understanding of the alterations in mechanisms of complex, sequential behavioral patterns (FAPs) and will not be examined in this chapter.

The examination of differential sensitivity to metals during various periods of an individual's life history is another research method for behavioral toxicologists. Behavioral ontogeny occurs in both the young of all species and the adults of those species that display parental care[11-13] and is directly dependent on the physiological pathways involved in ontogenetic change. As a result, one could hypothesize that toxic metals may induce very different age-specific results. On one level, Huntingford's[5] first paradigm (Figure 1) could be used with seasons I to IV being changed to distinct life history stages; similar stimuli at different ages cause unequal responses. Alternatively, one may design experiments and analyze the data using Chalmers'[14] (Figure 2) concept of developmental pathways of behavior. In this view, an individual researcher or animal must assess behavioral performance against some internal threshold that determines if behavioral competency has been reached, i.e., a set point. For example, feeding is either initiated or ended when some threshold of hunger or satiation is reached. Whether the set point is achieved is dependent on age, the method of assessing performance, development of the required physiological, as well as anatomical and morphological, machinery, and, central to our thesis, the frequency and intensity of perturbations. When the set point has been reached, the behavior either is initiated or terminated. For example, during the lengthening springtime photoperiod, an adult

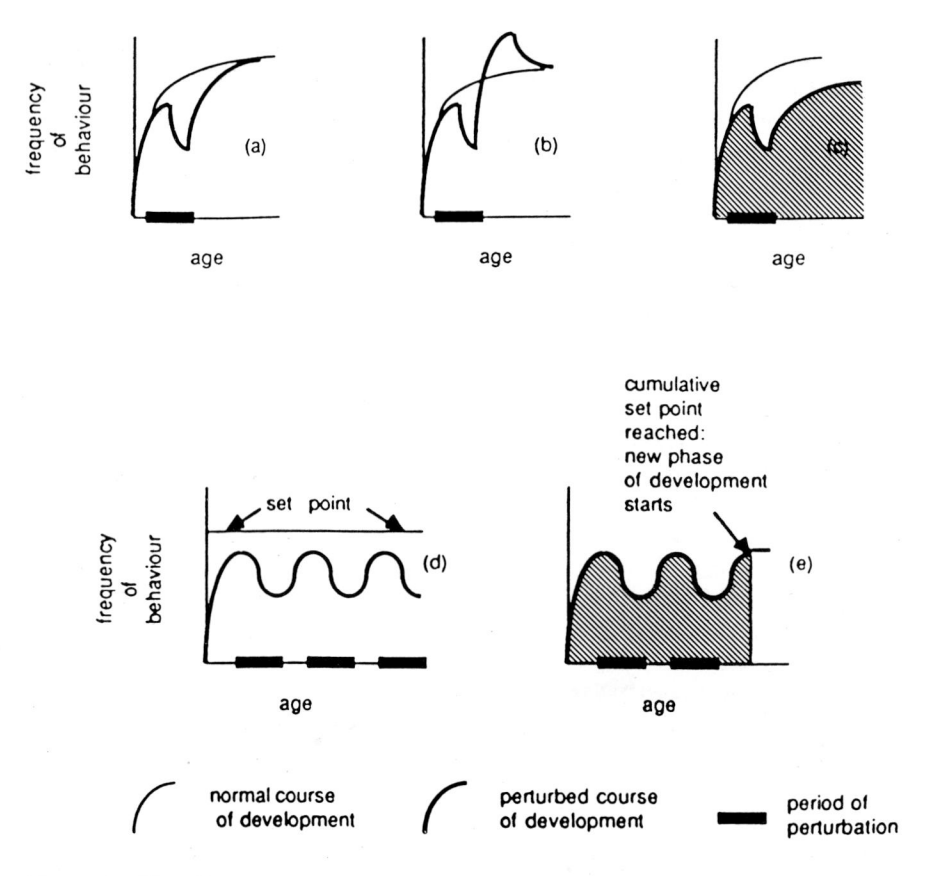

Figure 2. The effects of environmental disturbances on the expression and development of specific behaviors. Asymptote on curves depicting normal course of development = set point at which, once attained, a new phase of development may begin. X-axis (age) = increasing age during life cycle of individual. (a) Instantaneous assessment of performance using an age-dependent set point; (b) cumulative assessment of performance using an age-dependent set point; (c) and (d) instantaneous assessment of performance with an age-independent set point; and (e) cumulative assessment of performance with an age-independent set point. (Reprinted with permission from Chalmers, 1987.[14] Copyright © 1987 by Academic Press.)

fish increasingly shows signs of both secondary sexual characteristic development and changing hormonal concentrations. Until the set point of full maturity is reached, the full range of reproductive behaviors (FAPs) may not be expressed. Certainly, this simplistic view is complicated by those factors (temperature, photoperiod, or feeding regime) that can alter the rate and direction of developmental processes in fishes. Yet this model does have utility for initial hypothesis testing.

Some behaviors, such as feeding patterns or the determination of social hierarchies, have an age-dependent set point. Under such conditions, styles of eating; the amount, size, and type of food ingested; or the nature and intensity of agonistic displays are stereotypic for each age class and therefore could be estimated by some mathematical function. Successful performance assessment may be either an instantaneous or cumulative procedure. If instantaneous, then the consequences of not eating or not being aggressive, for example, are immediately matched to some physiological need. Cumulative assessments imply that the matching between a behavior and successful performance occurs over a longer period of time; it may require social conditioning, learning, or advanced physiological development.

Other behaviors, such as escape behavior, possess age-independent set points. Again, the assessment of performance, in this case escape, may be either instantaneous (determination of risk in the face of a potential predator) or cumulative. In the latter case, age-independent decisions (e.g., foraging when a potential predator is present) are related to cumulative physiological effects of hunger and the conflicting motivations to feed in a dangerous situation.

If behaviors are altered by an environmental contaminant during a critical developmental period, the full expression of these behaviors may be suppressed either permanently or temporarily. Permanent changes occur if the physiological, anatomical, or morphological machinery that drives or permits that behavior is irreversibly altered, as in nerve damage or diminution of important color patterns. If the behavior is suppressed only as long as the stress is applied or, alternatively, if the length of time a stress is applied is insufficient to cause permanent damage, then Chalmers[14] predicted different postperturbation outcomes dependent on the method of assessing performance and the nature of the set point. Bear in mind that these predictions do not exclude other possible outcomes; rather, if these expected behaviors are observed after a perturbation, the hypothesis is supported. The predictions are outlined below and diagrammed in Figure 2:

1. Instantaneous assessment of performance with age-dependent set points. Prediction after perturbations are removed: an eventual return to the original behavioral set point (Figure 2A). Example: psychomotor control during predation where prey-size choice is age dependent.
2. Cumulative assessment of performance with age-dependent set points. Prediction after perturbations are removed: initially, behavior will be expressed more frequently than with controls until the cumulative amount of behavior has matched the age-dependent set point (Figure 2B). Example: satiation level and feeding rate, where amount eaten and quality of food chosen are age dependent.
3. Instantaneous assessment of performance with age-independent set points. Prediction after perturbations are removed: a lateral displacement from the normal behavioral development pathway and a permanent suppression of the set point (Figure 2C). Example: escape behavior.
4. Cumulative assessment of performance with age-independent set points. Prediction after perturbations are removed: a lateral displacement from the normal behavioral development pathway, but original set point reached eventually due to cumulative behavior patterns. Example: motivational decisions to forage or escape.

The few examples shown suggest that, from a toxicological viewpoint, these predictions may prove useful to investigators attempting to model results from life history experiments of pulse exposure to toxic metals at critical periods of behavioral development.

From the perspective of behavioral mechanisms, behaviors are the result of integrating biochemical and physiological functions in response to environmental stimuli. It is possible, therefore, that the overt behaviors that result from these proximate mechanisms are readily measurable and attributable to environmental changes. For example, a single stimulus that induces the FAP for escape behavior requires the proper integration of bioenergetic, nervous, hormonal, and muscle functions. After the introduction of a toxic substance, a change in behavior may occur that exceeds the normal range of variability (e.g., over- or understimulation of key neural networks). The remainder of this chapter will, in part, explore how the paradigms described by Huntingford[5] and Chalmers[14] can be integrated into an overall understanding of the mechanisms that are altered by a sublethal level of toxic heavy metals, which may ultimately manifest themselves by changing specific behavioral patterns.

II. HORMONAL MECHANISMS OF BEHAVIOR

There are two integrating systems that control behavior: nervous and endocrine. Hormones affect behavioral displays by affecting gene expression, influencing signal reception and sending, activating neuron clusters, determining future effects on brain structure and function, and producing messenger substances, especially in the brain. In turn, changes in behavior may influence the synthesis, concentration, and receptivity of specific hormones.[15] Table 1 provides an overview of some of the possible behavioral changes that occur in fishes due to hormonal influence. Note that we are not including pheromone-induced behavioral responses, as little is known about interactions of metals with pheromonal communication. We do admit, however, that this topic may yet prove to be a highly fruitful area of research.

A. Metal Effects on Hormone Action

Of particular interest here are the effects of toxic heavy metals on the long-lasting, long-term changes associated with hormonal balance. Behaviors induced by endocrine activity are dependent upon hormone synthesis, release, circulating levels, and receptivity. Before one can apply the various behavioral paradigms to metal/behavior interactions, the metal/hormone interactions need examination. Only then can a proper perspective on metal-induced alterations in the mechanism of hormone-based behaviors be evaluated.

Table 1. Hormonal Influence on Some Representative Behavioral Mechanisms in Fishes

Hormone	Physiological/behavioral effects	References
Thyroid hormone system (TSH = thyroid stimulating hormone; T_3 = triiodothyronine; T_4 = thyroxine)	High in migration to freshwater, low in migration to saltwater	26
	T_3: no effect alone on estradiol levels; with gonadotropin induces estradiol	124
	T_4: increase visual and olfactory sensitivity; extremely important in neural growth; increase aggression via TSH-inducing T_4; decrease estradiol levels, high in migration to freshwater, low in migration to saltwater	26, 125
	T_{3+4}: antagonistic to seasonal migration in trout; enhanced movement to unshaded areas in salmonids; enhanced olfactory learning in salmon	126–128
Cortisol	Induced under stress of crowding: if low, fish aggressive; if high, submissive (self-defense-based agonism); migration; increased saltwater preference in stickleback	66, 129, 130
Prolactin	Increases parental behaviors: nest building, fanning, buccal incubation; with steroids: increases parental behaviors, induces brood pouch development, increases fat metabolism; important in osmoregulation; decreased aggression to predator; schooling	131–133
Melatonin	Secretion by pineal during dark hours, influences circadian/circannual rhythms; gonadal growth; schooling behavior	66, 133, 134
Androgens	Primarily 11-ketotestosterone; transforms kidney tubule cells to serous and mucous gland cells for glue secretion and nest construction in sticklebacks	135
	Sex-specific discharge frequencies for male/female recognition in weakly electric fish	136
	Increases swimming movements, not essential for migration	26
	Increases territory, nest building, courtship behaviors; appears to act centrally (anterior-ventral telencephalon) via aromatized telencephalon; increases after successfully defeating opponent; increases aggression (resource-based agonism)	66, 137
	With FSH + LH: increases prebreeding aggression and territory establishment, nest building	129, 138
	Increases olfactory sensitivity	139
Estrogen	Primarily estradiol; increases olfactory sensitivity	66
	Decreases agonistic behavior	
	For internal fertilizing species: increases sexual activity — cycle of E_2 mirrors cycle of sexual activity	137
Progesterone	Increases olfactory sensitivity	66
Gonadotropins	Promote sexual maturation and reproductive behavior	140

1. Hormone Synthesis

Few studies of fishes have examined metal effects specifically on hormone synthesis (as opposed to measuring circulating concentrations), the first biochemical step required to induce specific behavioral patterns at a specific time of day, season, or year (Figure 1). We will briefly highlight two studies on steroidogenesis and the metal-specific differences in hormone synthesis response. Atlantic croaker *(Micropogonias undulatus)* exposed to either 1.34 mg dietary lead per 70 g/day or 0.2 mg dietary lead per kilogram per day for 30 days had both 17β-estradiol and testosterone levels lowered, although steroidogenic activity of the ovary remained unaffected.[16,17] In contrast, *M. undulatus* exposed to 1 mg cadmium per liter seawater for 40 days during ovarian recrudescence were found to have elevated ovarian growth and increased plasma estrogen, implying increased vitellogenic activity.[16] Unfortunately, although there are clear changes in the hypothalamo/pituitary/gonad axis, neither study directly implicates toxic metal exposure to specific alterations in hormone synthesis, let alone in behavioral patterns. However, such correlations are complicated by the potential number of sites for endocrine toxicity, as outlined in the female reproductive axis of teleost fishes (Figure 3).

The work on the thyroid axis gives further appreciation of the multiplicity of sites where metals can affect hormone synthesis. Impaired thyroid function in the lead-exposed catfish *(Clarias batrachus)* was correlated to decreased iodine uptake in fish exposed to waterborne lead.[18] Using either mercury- or cadmium-exposed *Channa punctatus*, Bhattacharya et al.[19] found that iodide peroxidase activity was depressed, resulting in lower circulating thyroxine titers. Suppression of 5′-monoiodinase by aluminum also caused lowered triiodothyronine concentrations in *Oncorhynchus mykiss*.[20] Although the thyroid hormones do have specific behavioral effects (Table 1), data are required to correlate metal-induced changes in thyroid activity to altered behaviors.

2. Hormone Release

Metal-induced changes in behavior cannot be fully appreciated without an understanding of the potential role metals have in upsetting the process of hormone release from specific endocrine glands. Unfortunately, very few studies have examined the effects of metal toxicity specifically on hormone release in fishes. In one of these investigations, Thomas[16,17a] noted increased pituitary secretion of gonadotropin hormones, followed by increases in plasma estradiol and ovarian growth in cadmium-exposed (1 mg/L for 40 days) *M. undulatus*.

Despite the lack of concrete data on effects of metals on hormone release, we believe this is probably a significant area of hormone/metal interaction. This supposition is supported by the fact that many of the neurotransmitters and ions that are affected by heavy metals are also involved in hormone release mechanisms (e.g., catecholamines, indolamines, and calcium; see Section III).

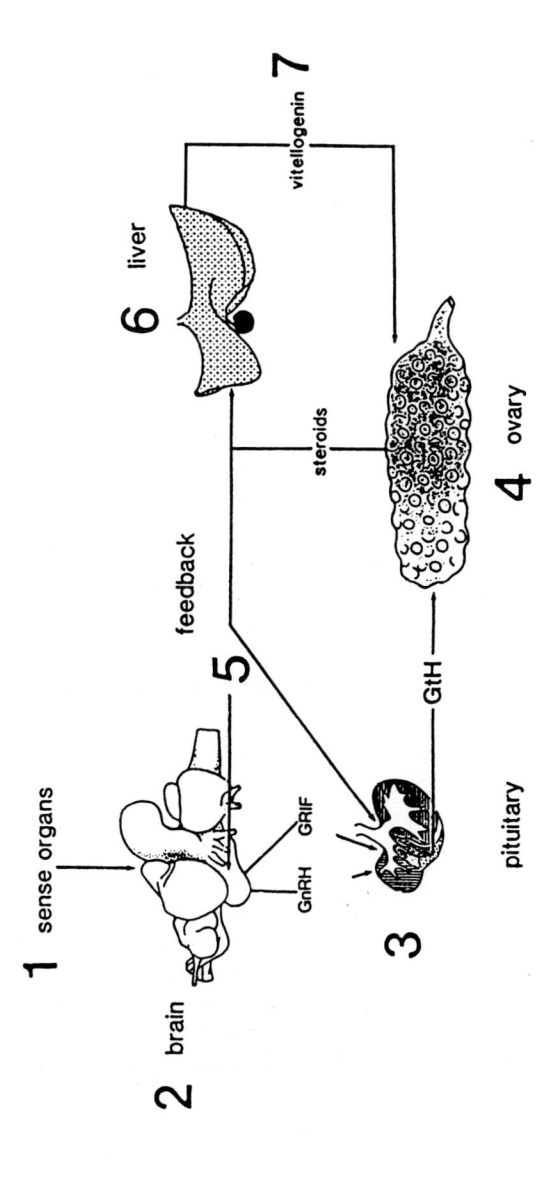

Figure 3. Schematic representation of a hypothalamic-pituitary-gonadal axis in a female teleost fish. Numbers refer to possible sites of chemical interference. GnRH = gonadotropin-releasing hormone; GRIF = gonadotropin release inhibiting factor; GtH = gonadotropin hormone. (Reprinted with permission from Thomas, 1991.[17] Copyright © 1991 by Wiley-Liss, Inc.)

3. Circulating Hormone Concentrations

There has been a considerable amount of work done on effects of metals on circulating hormone levels in fishes. Without extensively reviewing this topic, we would note that the thyroid hormones or steroid levels have been mainly assayed, due to the low availability of homologous assays for the protein hormones of fishes.

It must be pointed out that often metals have no *immediate* effect on specific hormone levels. Responses over time, both daily and longer, need to be evaluated. In Section III.C, we will further discuss the dyschronogenic effects of metals. Hormone levels do show circadian variations,[21] and metal interactions with hormone regulation may show unequal temporal effects, as suggested in Huntingford's first paradigm.

Neither dietary nor waterborne cadmium affected cortisol levels in largemouth bass *(Micropterus salmoides)*[3] or coho salmon *(O. kisutch)*,[22] respectively. In contrast, sublethal, waterborne copper did induce a stress response of increased cortisol in *O. kisutch*.[22] Apparently, only certain metals have the capacity to interact with the hypothalamic/pituitary/interrenal tissue axis and its regulation of cortisol levels in fishes. What this might mean behaviorally is, at present, unclear. Cortisol in fishes, however, can have a dual function. It can regulate metabolic activity and, consequently, the associated ability for movement; that is, it can behave as a glucocorticoid (see Section IV). It can also have a major role in plasma ionic and osmotic homeostasis; it can behave as a mineralocorticoid. Seawater adaptation in sockeye salmon *(O. nerka)* is hampered by prolonged, elevated levels of cortisol.[23] If exposure to specific metals induces an extended stress response at critical life history stages (smoltification and migration to the sea), the linkage between physiological adaptation and behavioral response could be weakened, and the individual could either not migrate or be physiologically compromised during the attempt.

Sublethal levels of lead suppress levels of serum thyroxine and probably triiodothyronine but not cortisol in rainbow trout *(O. mykiss)*.[24] Lower thyroid hormone levels may be detrimental for fish metabolism[25] and migration behavior.[26] Additionally, thyroid hormones are important in neural development during embryogenesis.[27] Not only would lowered thyroid hormone levels be adverse for central nervous system (CNS) development, but they it could alter neural development in the peripheral nervous system (PNS) as well. During development, the axonal distribution in white muscle of larval and juvenile catfish *(Ictalurus nebulosus)* changes.[28] By altering the patterns of nerve connections in muscle at specific stages of the fish's life history, the ability to express specific movements critical to future behaviors may be changed. This is an area of research that needs further testing; it has direct application to Chalmers'[14] hypotheses. The effects of lead on circulating thyroid hormones may be mediated by the pituitary. Decreases in thyroid-stimulating hormone (TSH) release in children exposed to lead have also been observed.[29] Effects of lead on TSH release in fish, which apparently also stimulates freshwater preference in diadromous migratory species,[26] are not yet known. A critical examination of metal exposure, both to embryos and to premigratory individuals, is needed; specifically, the correlation to changes in both immediate and future behavioral patterns should be examined. Again note that none of these toxicological studies evaluates changes related to the species' chronobiology.

Reproductive, parental, and migratory behaviors are likely candidates to be used in examining the interactions of metals with the mechanisms of circannual rhythms, as in Figure 1. To date, few data exist to correlate changes in yearly behavioral patterns with mechanisms of metal toxicity. Specific changes in behavioral and endocrine physiology do occur, as shall be shown, but experiments to clearly demonstrate the connection are few, especially in fishes.

A significant negative correlation between blood lead of male rats and intratesticular testosterone levels, as well as decreased intratesticular sperm counts, was observed by Sokol et al.[30] These authors did not find changes in serum LH values but did find suppressed FSH levels. The suppression of testosterone, along with a failure to show elevated LH and FSH concentrations, suggests a site of lead toxic action at the hypothalamic/pituitary/gonadal axis.[30] Circulating testosterone and 5α-dihydrotestosterone were suppressed in adult male rats injected with 5 mg cadmium for 1 and 2 weeks, although androstenedione levels were unaltered. These reduced levels were correlated to lowered sexual drive of the experimental animals.[31] In most fishes, 11-ketotestosterone is the predominant androgen. Nonetheless, since reduced sexual behavior in fishes has been noted with a variety of metal exposure regimes, it is likely that similar mechanisms of toxicity occur.

Increases in serum prolactin of lead-exposed rats were correlated with decreased hypothalamic concentrations of dihydroxyphenylacetic acid, due to decreased dopamine receptor density in the pituitary;[32] this is consistent with the putative role of dopamine as the prolactin-inhibiting factor in mammals. Increases in serum prolactin can cause delayed sexual development and reproductive failure.[31] Nickel-exposed rats[31] and cadmium-exposed cichlids *(Oreochromis mossambicus)*[33] were found to have elevated prolactin concentrations. No correlations with the mechanisms of hormonal control of behavior were made in those studies.

From a behavioral perspective, prolactin appears to be an important hormone in parental and migratory activity in some fishes. The movement from freshwater toward saltwater is perhaps aided by increased levels of circulating prolactin, which alter serum osmolality.[26,33] Epidermal mucus secretions from discus-fish *(Symphysodon* spp.) parents are used by the fry as food.[34] Because these secretions are induced by fish prolactin (and, to a smaller extent, by mammalian prolactin),[35] this genus could be developed as a tool to examine cadmium, lead, or nickel toxicity in fishes. An exciting and major benefit of this model is the quantifiable alterations in maternal-offspring behavior that could result from measurable, metal-induced changes in prolactin synthesis, release, circulating levels, and receptivity at critical periods of the life history. Investigators searching for an appropriate animal model to examine toxic metal/hormone/behavior interactions should give serious consideration to this genus.

4. Hormone Reception

In order to fully evaluate hormonal mechanisms of behavior, it is necessary also to examine metal-induced effects on hormone binding and the cascade of cellular metabolic events that are initiated by hormone receptions. These biochemical events are critical to organizing proper behavioral responses to environmental cues (Figure 1). The degree of permanence in receptor interference by toxic metals may also provide a mechanism for evaluating behavioral observations that test Chalmars'[14] predictions (Figure 2). Regrettably there is too little data on fishes, or other vertebrates for that matter, to make such an evaluation at present.

There is one study with fish that indicates that the reception to the genetic-expression axis of steroid hormones may be affected by metals. Cadmium-exposed, estradiol-injected male and female flounder *(Platichthys flesus)* displayed significantly depressed serum levels of the yolk-precursor protein, vitellogenin. Changes in the amount of yolk may be particularly critical, not only during embryogenesis but after hatching when the yolk-sac fry are still dependent on the yolk for food. In the females but not in estradiol-injected males, the decrease in vitellogenin occurred with reduced hepatic RNA to DNA ratios,[36] suggesting a potential for low estrogen receptivity in the liver. Other examples of receptor function interference in metal-exposed fishes are still needed, especially those that bear directly on behavior, e.g., metal interference with calmodulin-related protein activation[36a] and the possible effects on neuronal activation.[25]

In mammals, it appears that the binding of some but not all neuropeptide-releasing hormones can be affected by metals. In a dose-dependent fashion, lead displaced the binding of growth hormone-releasing factor in rat anterior pituitary receptors without any affect on thyroid-releasing hormone binding capacity.[7] In rats specifically, metals have been shown to affect binding of a protein hormone, FSH. Reduced testicular steroidogenesis in lead-exposed rats at the onset of puberty appears to result from lowered binding of FSH to Sertoli cells and a resultant decrease in cAMP production, and reduction in 3β-hydroxysteroid dehydrogenase activity.[8]

B. Behavioral Correlates of Metal-Induced Endocrine Toxicity

Alterations in the proximate mechanisms of behavior, the physiological basis for behavioral response to environmental stimuli (Figure 1), due to sublethal metal exposure help to explain the dramatic changes observed in fish reproductive behavior, even after sublethal, short-term exposure. Several workers have noted decreased spawning activity in a variety of freshwater teleosts, fathead minnows *(Pimephales promelas),* exposed to lead or copper immediately prior to spawning.[37–39]

For *P. promelas,* male nest preparation and maintenance activities are required for successful spawning. Using Huntingford's[5] first paradigm (Figure 1), it is possible to create testable hypotheses that may help biologists appreciate why fish at different stages of sexual development (approximating different times of year) were apparently affected unequally by lead exposure.[37,38] Specific behavioral pathways may be "primed" by sex steroids in the brain (E_n, where n = 1, 2, etc.; Figure 1). Concurrent or later introduction of lead would inhibit further behavioral development or expression (decrease in B_n), either by suppressing further gonadal development, causing gonadal regression, or blocking receptor sites for various neurotransmitters or hormones (decrease in E_n).[8,40]

Studies aimed at evaluating toxic effects of metals during behavioral ontogeny in fish are few. Pickering et al.[39] demonstrated that when fish were exposed to varying concentrations of copper at either 6, 3, or 0 months (I to IV, Figure 1) before spawning, only the higher, albeit sublethal, levels of copper reduced spawning activity (B_n) and then only when introduced immediately prior to ovipositioning. The authors did not make any linkage to the underlying physiological mechanisms (E_n) that would explain these behavioral aberrations. Although more work is needed to examine this possibility, it is probably fair to state that when it comes to metal exposure of fish "timing is everything."

The studies above were under conditions of constant exposure, albeit at different times of development, and data were collected while the fish were still exposed. To explore another area that is ecologically relevant, pulsed-exposure experiments that mimic spring snow melts, storm runoffs, etc. at various critical periods of behavioral development and which can be correlated to specific physiological changes that alter behavior, are needed. Then investigators will be able to explore the possible utility of Chalmers'[14] hypotheses in terms of the effects of short-term, sublethal environmental contamination on age-specific behaviors.

Metals may have additional effects on reproductive behavior by altering sensitivity to environmental variables, if the exposure precedes the development of increased brain receptivity to specific hormones. Such data fit well with the paradigm in Figure 1. This will be examined further in Section III.

Previous workers[41] have examined how low-level contaminants in aquatic systems affect specific displays within the overall fish courtship ritual, but apparently, other than studies using lead,[37,38] no reports of pollutants altering the sequence of these behaviors exist; none of these studies made the toxic metal/hormone/behavior linkage. Alterations and interruptions in the proper sequence of behaviors may prevent the completion of a FAP, even though the requisite environmental stimuli are present. Investigators need to be attentive to such metal-induced alterations in the integration of behavioral patterns. If it is assumed that the internal state has been changed due to interactions between metals and physiological processes, then one may hypothesize a decrease in the size of the motivational and cue space subsets. Additionally, if metal exposure suppresses neuroendocrine controls for feeding motivation and/or the ability of the fish to use environmental cues correctly to analyze feeding situations, it is likely that the range of potential responses to prey presence (causal factor space) will be smaller and less effective than for those individuals in uncontaminated water. Furthermore, metal exposure may shift individual ability to give appropriate priority to significant behavioral responses (candidate space), especially sequential actions. The result is significantly different behavioral responses to given stimuli in clean vs. contaminated systems. Thus, an understanding of physiological changes in metal-exposed fish may provide a basis for appreciating ethological models of toxicity. Potential neurobiological changes in feeding motivation, as well as ability, are discussed in Section III.E.4.

Although only examined as two-event sequences, Weber[37,38] observed an increase in the number of statistically significant differences between reproductive display sequences in control vs. lead-exposed male *P. promelas*. Some sequences rarely occurred in lead-exposed males. Furthermore, the intensification of behavioral expression over time (behavioral ontogeny) was also suppressed by lead.

The importance of changes in the integration of specific behavioral acts into an overall pattern that ensures reproductive success is clear. Lead, for example, may be altering the mechanisms by which specific actions are coordinated into an integrated whole. It is insufficient for a male *P. promelas* to merely hover under or to hover and then touch the substrate ceiling with its dorsal pad. The sequence must develop to circling, cleaning, etc. to ensure quality nest site preparation for attracting a female and, subsequently, egg protection. That this behavioral development occurred less in lead-exposed fish than controls may indicate lowered integrative capacity, due in part to suppression of hormone synthesis, release, or receptivity. Such analyses of metal-induced alterations in behavioral sequences during fish reproductive behavior after exposure to heavy metals have not been reported previously in the literature. Additional validation of these findings is required.

III. NEUROBIOLOGICAL MECHANISMS OF BEHAVIOR

When considering neurobiological mechanisms of neurotoxicity, two general points should be kept in mind:

1. In the most simplistic view, neural control of behavior can follow a direct path: stimulus → motor neurons → target muscle. In most cases, this simple pathway is expanded to stimulus → sense organs → sensory interneurons → motor neurons → specific target muscles. Often, however, the final behavior, in turn, affects brain function, in which case the path should be illustrated as stimulus → sense organs → interneurons → brain region(s) → descending pathways → interaction with motor neurons → target organ(s) → specific behavior → ascending pathway → interaction with many neurons → brain region(s). Thus, our first point is that the complex nervous pathway from stimulus to behavior has multiple sites for metal/neurotoxic interactions; one needs to be cautious in ascribing the mechanism of behavioral toxicity to a specific site.
2. Behavioral responses to neural control are often quick and of short duration. Further, neural activity within and between brain regions can differ on a temporal basis.[42] Regional ablation studies in fishes[43] have made it clear that more than one

brain region may control a specific behavior, and alterations of any of the regions capable of control may affect the behavior. Thus, our second point is that metal accumulation in different regions of the brain may alter specific behavioral patterns at some specific times of day or seasons and not at others; temporal aspects of a toxic response need to be considered.

Silbergeld[4] postulated three principles of lead neurotoxicity. We believe her principles have broad application to neurotoxic effects of other metals as well. First, changes in neurochemical function can be detected before pathological damage becomes evident. Second, within the CNS and PNS, observable neurotoxic effects of a metal appear to be specific for different neurotransmitter pathways. Third, the effects on neurotransmitter pathways are a result of the specific effects of a metal on the ionic neurotransmission mechanisms, e.g., enzyme regulation, transport processes, exocytosis, or phosphoregulation. We will begin our discussion first by addressing how metals alter neural transmission, second by examining metal-induced changes in information processing by the central and peripheral nervous systems, and third by suggesting what such changes mean in terms of behavior.

A. Presynaptic Metal Toxicity

Although there are various reports both from laboratory experiments and field trials of heavy metal-induced effects on neurotransmitter levels in teleost brains,[44-47] most of the work on neurotoxicity has been done with mammals. From work with mammals, using the motor endplate as an example, different metals may induce a range of both amplitude reductions ($Pb^{+2} \ggg Mg^{+2} \gg Mn^{+2} = Co^{+2} > Cd^{+2}$) and miniature endplate potential frequency increases ($Pb^{+2} = Hg^{+2} > Cd^{+2} > Zn^{+2} > Cr^{+3}$). At this point, the fundamental presynaptic mechanism of metal-induced changes in both the frequency and amplitude of neural transmission seems to be similar for all metals; there is a competition with calcium for receptors located on the external presynaptic nerve terminal membrane, as well as intracellular effects on calcium buffering and energy metabolism.[48]

Cooper et al.[48] reviewed the complex interactions of a variety of heavy metals on synaptic transmission (Figure 4). The underlying themes of toxic metal interaction at the synapse appear to be not only interference with the voltage-gated entry of calcium into the nerve terminal, a requisite step for evoked transmitter release, but metal entry into the nerve terminal as well, where further alterations in neurotransmitter release can occur. Specifically, three primary effects on neurotransmitter release exist: alteration of membrane calcium receptors; competition with calcium within calcium channels; and chemical effects within the cell that alter exocytosis of neurotransmitters either directly or indirectly. For lead, at least, these effects are reversible once external lead is removed, although the residual level of intracellular lead may still affect nerve transmission. From an environmental toxicity perspective, the issue of reversibility is crucial. Yet little data exist on the amelioration of metal-induced behavioral effects in fishes once metal exposure ceases. The degree to which the effects of metal exposure on nerve transmission is reversible would certainly be critical to a testing of Chalmers'[14] predictions of the behavioral effects of pulsed perturbations on either age- dependent or ageindependent behavior (i.e., does the animal attain its pre-exposure behavioral set point once the toxic metal is removed from the environment, and if not, what alternative behavioral patterns would be observed?).

In general, the process of metal neurotoxicity at the nerve terminal appears to be one of an initially very rapid blockage of external calcium receptors, followed by metal entry into the terminal region. Once the metal is internalized, several biochemical effects interference with calcium buffering systems (some of which may be irreversible) occur and are seen in an increase in miniature endplate potentials.[49] It must be noted here that while *acute* exposures to any metal that increases internal calcium concentrations will increase the quantal release of neurotransmitter,[50] *chronic* lead and mercury exposure can actually inhibit transmitter release.[49] Implications of such acute vs. chronic neurotoxic effects for long-term behavioral changes are unclear. Again, the issue of reversibility needs to be addressed.

B. Postsynaptic and Synaptic Cleft Metal Toxicity

Activity within the synaptic cleft is also important in the proper coordination of neural transmission and its translation into behavior. Thus, an examination of the possible connection between this site of neurotoxicity and behavioral change would be useful. Alteration of synaptic cleft enzyme activity, as well as binding to postsynaptic receptor sites, such as from lead and cadmium exposure,[51] has been shown in some "lower" vertebrates. Data from fish studies, however, are few.

Although neither fish nor mammalian studies have adequately connected neurobiological changes to specific metal-induced changes in behavior, a brief examination of some of the data is useful. In fishes, changes in steady-state levels of neurotransmitters and their metabolites have been observed by Katti and Sathyanesan.[46,47] This indicates potential changes in synthesis, release, or catabolism. Catfish *(Clarias batrachus)* were exposed for 150 days to 5 ppm of lead (a concentration, excluding sediment levels, not usually found in aquatic systems) and increases in brain serotonin (5-HT) and decreases in γ-amino butyric acid (GABA) were observed. For juvenile *P. promelas* exposed for only 28 days at 1 ppm lead, similar increases in brain 5-HT were observed.[44] In addition, norepinephrine (NE) levels increased. No changes in dopamine (DA) levels occurred in *P. promelas*, and the DA levels that were observed were comparable to levels found in other fish species.[45-47]

In both the CNS and PNS, acetylcholinesterase (AChE) activity in fishes is suppressed by the presence of heavy metals. Nemcsók and Hughes[52] and Nemcsók

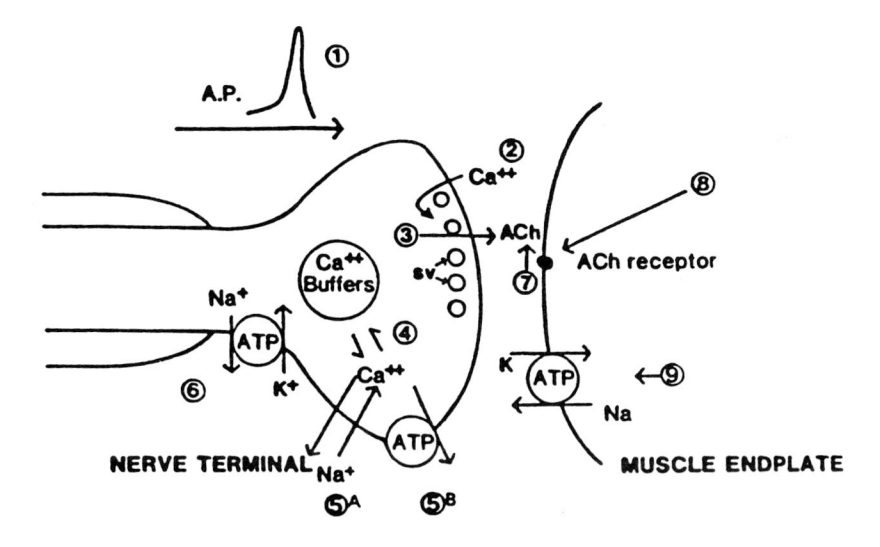

Figure 4. A simplified representation of a generalized neuromuscular junction. Possible sites of action and processes affected by heavy metals are (1) action potential propagation; (2) voltage-gated entry of calcium into the nerve terminal; (3) site-specific interaction of calcium with vesicles and plasmalemma altering exocytosis; (4) buffering of calcium by intracellular organelles; (5A, 5B) calcium extrusion by sodium-calcium exchange and ATP-driven calcium pump; (6) maintenance of membrane potential by ATP-dependent sodium-potassium pump; (7) diffusion of acetylcholine within the synaptic cleft and hydrolysis of acetylcholine by acetylcholinesterase; (8) interaction of acetylcholine with endplate receptors; and (9) maintenance of endplate potential and other membrane properties. Equivalent situations exist for other neurotransmitters. (Reprinted by permission from Cooper et al., 1986.[48] Copyright © 1986 by Raven Press, Ltd.)

et al.[53] demonstrated the differential effects of zinc and copper on brain, heart, muscle, and serum AChE activity in *Oncorhynchus mykiss* and carp *(Cyprinus carpio)*. Whereas zinc had no effect on any organ, copper inhibited AChE activity in all tissues tested. Lower AChE levels in the heart muscle potentiates vagal tone, which, in turn, causes adverse effects on muscle metabolism and blood circulation. Presumably, these effects could, in turn, affect behavioral performance, although no data exist to our knowledge to support this. The topic of metal-induced metabolic dysfunction and its potential effects on behavior will be further discussed in Section IV. Shaw and Panigrahi[54] observed similar AChE-inhibited results in the brains of mercury-exposed fishes. Because of decreased catabolism of acetylcholine in the synaptic cleft,[55] postsynaptic neurons are hyperpolarized for a longer-than-normal period of time. Behaviorally, this may be seen as "hyperactive" or uncoordinated movements due to muscle contractions extending over a greater time period for a given stimulus or longer stimulation of neuronal projections from CNS nuclei. Alternatively, it may change the refractory times between successive above-threshold stimulations of the postsynaptic neuron. This, in turn, may result in uncoordinated integration of behavioral sequences, i.e., FAPs. Studies on *P. promelas* (see Section III.C) demonstrate a decrease in mean daily levels of brain AChE in response to lead intoxication,[37] although no connection to changes in behavior was suggested.

C. Dyschronogenic Effects

Specific differences among various toxicological studies have been attributed to such variables as species, age, dosage, and duration or route of administration. Brain distribution of both neurotransmitters and their synthesis and degradation enzyme systems in fish, for example, appear to show wide species-specific variations.[56,57] A significant cause of variation among studies could, however, be an interaction among dosage, sampling time, and endogenous rhythms. According to the paradigm presented in Figure 1, temporal relationships between environmental stimuli and physiological states result in differential behavioral responses. A metal-produced phase shift or disruption in endogenous neurotransmitter rhythms could result in a host of varying results and interpretations, depending on the time of sampling. This possibility was supported by experiments with lead-exposed minnows.[37] Changes in brain neurotransmitter levels between control and lead-exposed fish occurred at night for NE, vanillylmandellic acid (VMA), homovanillic acid (HVA), and hydroxindole acetic acid (5-HIAA), but during the morning hours for AChE. Furthermore, the rhythms of neurotransmitter systems may have been altered to a point that they no longer had the same phase relationship to each other. These data may suggest temporal, lead-induced alterations in neurotransmitter synthesis and turnover and may be a potential basis for observed changes in behavior in lead-exposed individuals. Depending upon the specific brain regions where such shifts occurred, the lack of coordination in synthesis, release, and metabolism of various neurotransmitters could cause significant, maladaptive behavioral alterations. Temporal synchrony among endogenous and exogenous rhythms is a requisite aspect of normal vertebrate physiology. A disruption of this synchrony may affect a wide-ranging panoply of variables, including health, reproduction, migration, and behavior. We are unaware of any other data on dyschronogenic effects of metals on brain neurotransmitters and their metabolites in fishes or other animals. Metal-induced dyschronogenicity, if demonstrated to be a widespread phenomenon, would provide significant support for Huntingford's[5] first paradigm as an important organizing principle in defining underlying mechanisms of behavioral toxicity. This should prove a fertile area for research.

Simultaneous examination at smaller time intervals and within specific brain regions is now needed. In fishes, the lateral hypothalamus (LH) has been implicated as a feeding control center. In addition, the following areas have, when stimulated, been shown to be important in inducing feeding behaviors in fishes: nucleus recessus lateralis (NRL), nucleus diffusus lobi inferioris (NDLI), and the nucleus glomerulosus (NG), which has fiber connections to the NDLI. Stimulation of various regions of the telencephalon (e.g., olfactory) also have induced feeding behavior, possibly via the medial forebrain bundle (MFB) to the inferior lobe of the hypothalamus. Lesions of the LH have caused aphagia, and lesions of the NRL have caused decreased growth. The NRL is noradrenergic.[58] Presumably, a depression of NE at entrained feeding times could result in immediate satiation, while a phase shift of NE patterns in the ventro-medial hypothalamus (VMH) or other noradrenergic feeding centers could allow the animal to eventually finish its allotment of food within a 24-hr period. In fact, if, due to toxic metal exposure, NE is sufficiently increased within feeding control centers, the fish might eat more than control fish during a 24-hr period. If this hypothesis was proven to be correct, it would be a strong argument against Smith's[59] conclusion that environmental contaminants affect feeding behavior by primarily altering the PNS rather than the CNS. However, as he rightly points out, feeding responses are the result of an integration of olfactory, visceral-gustatory, aggressive, locomotor, visual, and auditory (including lateral line) inputs, all of which can be altered by metal exposure.

D. Brain Regionalization

We now return to our second point regarding neurotoxicity, the role of regional brain effects. Compartmentalization of lead neurotoxicity has been noted in rat brains. Govoni et al.,[40] using their own results as well as reviewing the data of other workers, found that specific neurotransmitter levels, synthesis, and turnover rates vary with brain region. Compound these findings with potential effects on circadian rhythms of these areas, and one quickly realizes that the toxic effects of lead, and most likely other metals as well, within the CNS are very complex.

The roles of the telencephalon in fish reproduction and aggression have been reviewed by de Bruin[6] and Smith[59] (Table 2). After lesions or ablations of the telencephalon, various deficiencies in reproductive behavior were noted. Noble[43] found that telencephalic lesions in the cichlid *Hemichromis bimaculatus* resulted in decreased coordination and synchronization of sexual behavior; a decrease in the series of stereotypic sequences of mating behavior was observed in two poecillids, *Lebistes reticulatus* and *Xiphophorus helleri*. What is interesting is the similarity of these findings to data from experiments conducted on lead-exposed, reproductively active *P. promelas*,[37,38] in which the normal sequential patterns of male nesting and courtship activity were disrupted by lead intoxication, as described in Section II.B. Presumably, these observations indicate an intersection of endocrine and CNS effects due to short-term lead exposure.

Locomotor activity is the result of the integration of several sources of neural information (Table 2). Affecting any of the efferent or afferent neurons or CNS nuclei may cause changes in the locomotor pattern (cue and causal factor spaces).[5] Additionally, changes in the circadian control of

Table 2. Brain Regions and Neurotransmitters Possibly Involved in Different Fish Behaviors

Behavior unit	Brain region	Neurotransmitters
Locomotor	Medulla oblongata	Serotonin, acetylcholine
	Mesencephalon	CA
	Diencephalon	GABA
		Peptides
Aggression	Telencephalon	Serotonin
	Diencephalon	CA
Schooling	Mesencephalon	Serotonin
	Telencephalon	Acetylcholine
	Medulla oblongata	CA
	Diencephalon	
	Cerebellum	
Conditioned responses	Telencephalon	Serotonin, peptides
	Diencephalon	CA
Temperature selection	Diencephalon	Serotonin
	Mesencephalon	Acetylcholine
Respiration	Medulla oblongata	Acetylcholine, peptides
		CA
Feeding	Telencephalon	Serotonin
	Diencephalon	Acetylcholine
	Mesencephalon	CA
	Medulla oblongata	
	Cerebellum	
Fear	Medulla oblongata	Glycine, glutamate
		GABA
Rheotropism	Mesencephalon	Serotonin
	Medulla oblongata	Acetylcholine
	Cerebellum	CA
Electrical activity	Medulla oblongata	Acetylcholine
	Cerebellum	
	Mesencephalon	
Salinity preference	Diencephalon	CA
Reproduction	Diencephalon	Serotonin
	Telencephalon	CA

Source: Reprinted with permission from Smith, J.R., *Aquatic Toxicology,* Vol. 2, Weber, L.J., Ed., Raven Press, New York, 1984. Copyright© 1984 by Raven Press, Ltd..

locomotor rhythms may also occur. Lead-induced changes in locomotor activity rhythms in goldfish *(Carassius auratus)* were observed by Spieler.[60] The lead-exposed goldfish displayed increased activity (as percent daily activity) during the dark period (in comparison to controls) and a much flatter and broader acrophase of activity during the light period (Figure 5). A similar flattening and elevation of circadian locomotor activity was observed in copper-exposed sea catfish *(Arius felis).*[61] Whether this response to metal intoxication is a change in the circadian distribution of activity or a generalized hyperactivity needs to be determined. Missing are further data correlating metal-induced changes in specific brain regions to particular behaviors or behavioral sequences using the pathways described at the beginning of this section.

E. Interactions of the PNS and CNS with Toxic Metals

1. Sensitivity to Environmental Stimuli

If the toxic metal/neurophysiological/behavioral axis is as in Figure 1, then it is likely that the threshold for reception of environmental cues will be either increased or decreased, the direction of change being dependent on the metal and its oxidation state, fish species, site of metal interaction, and temporal physiological rhythms. Metals have been shown to suppress neuron sensitivity to normal stimuli in fishes.[62,63] Heiligenberg[64] suggested that the specific behaviors of the cichlid *Haplochromis burtoni* occur only when environmental stimulation crosses some threshold long enough for that behavior to be expressed. If, in fact, environmental cues are not being appropriately analyzed due to such disruptions in physiological function as desensitized receptors, it is possible for observed behavioral changes to occur without any gross morphological damage to key brain regions that either receive environmental cues or translate them into specific adaptive behaviors. For example, short-term, sublethal levels of waterborne lead did not cause any gross morphological damage in the olfactory bulb, retina, hypothalamus, anterior dorsal telencephalon (an important region of testosterone receptivity), or pituitary.[64a] Such exposure, however, did cause numerous behavioral disorders.[37-39] These studies do not rule out the possibility that there is a change in neurotransmitter dynamics within the CNS regions, but do indicate that sensory reception and PNS transmission may be important and overlooked sites of metal toxicity. As an aside, the studies noted above support the hypothesis that behavioral changes due to environmental contamination can occur long before histopathological lesions become obvious and are thus a sensitive measure of the presence of pollutants.

To be adaptive, a specific behavior must not only be completed, but it must demonstrate a proper orientation to critical aspects of a fish's habitat and be adjusted to changing conditions by the proper analysis of incoming information. For example, lack of proper breeding substrate orientation by lead-exposed male *P. promelas* also reduces reproductive success, since the female may not be attracted to a site that is not properly prepared or maintained. In nature, *P. promelas* eggs are oviposited on the underside of objects, presumably for protection against predation. Orienting nest preparation behaviors to the proper site is therefore critical for success. Sublethal levels of metal contamination can change this orientation and therefore may be highly detrimental to long-term fish population survival.[37,38] Orientation within a substrate should not be confused with substrate selection, which may involve a different set of environmental cues and neural pathways. Lead-exposed smallmouth bass *(M. dolumieui)* did not demonstrate any changes in substrate selection; lead did not alter their preference for a shelter.[65] This was also observed by Weber[37,38] in that *P. promelas* exposed to lead could be found under the breeding substrate for equally long periods as control fish. The difference was that control fish spent more of their time preparing the

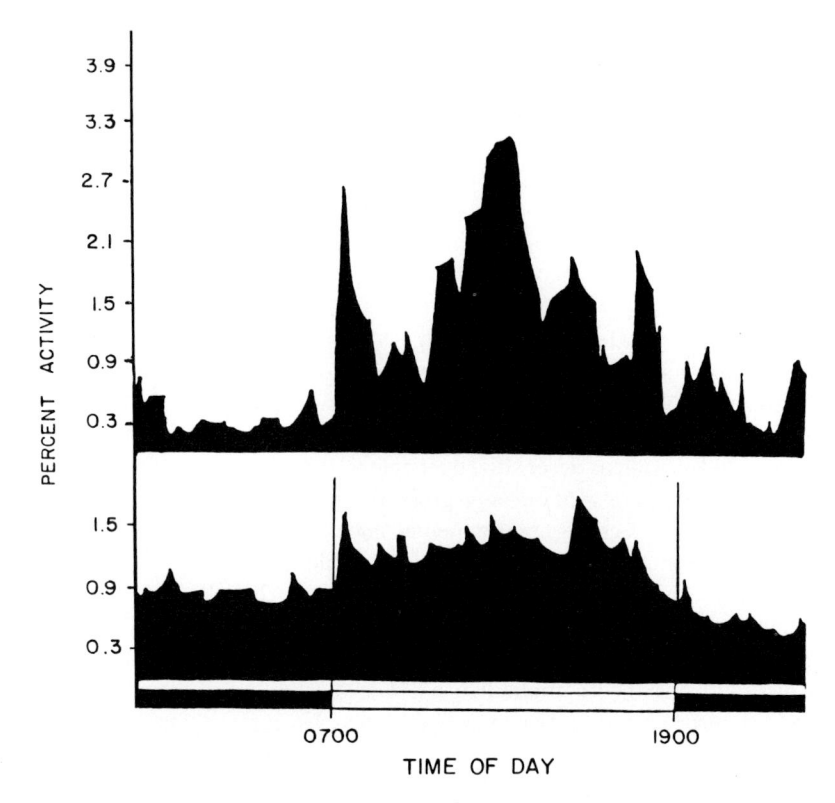

Figure 5. Daily activity rhythms of goldfish *(C. auratus)* after receiving either 0.0 (upper graph) or 0.4 ppm lead, as lead acetate, for 28 days. White bar on X-axis represents 12 hr of lights on (0700 h to 1900 h); black bar represents 12 hr of lights off (1900 h to 0700 h). Percent activity is based upon level of locomotor activity during a specific recording period (15 min) in comparison to total recorded activity for 24 hr. (Reprinted with permission from Spieler.[60] Copyright © 1987 by University of Wisconsin, Milwaukee.)

underside of the substrate for egg deposition by the female, whereas the lead-exposed fish spent significantly less time at the ceiling of the substrate. Further demonstrations of metal-induced effects of site orientation vs. selection are needed.

Because different aspects of a behavior can be affected, these observations may indicate that lead selectively disrupts specific regions of the brain. Orientation to a breeding substrate by a sexually active male may be influenced by many environmental variables, such as O_2 concentration gradients, water velocity, light, substrate texture, etc. Further, sensitivity of either CNS or PNS sites to these variables may be under the influence of either androgens or thyroid hormones.[25,66] Thus, the interaction of altered levels of sex steroids, as seen in Section II, may have a profound effect on neurobiological aspects of behavior during critical periods of the fish's life history. This is precisely what is implied by Huntingford's[5] first paradigm; temporal changes in physiological states are differentially affected by exposure to environmental contaminants. Some catfishes, such as *I. nebulosus*, use electro-orientation. Cadmium accumulates in the receptor cells and doubles the threshold necessary for electrical stimulation. The effect is reversible once cadmium is removed.[63] The authors did not quantify whether pre-exposure electro-orientation was similar to postexposure levels. Such data would be useful in evaluating Chalmers'[14] hypotheses. In copper-exposed *Torpedo marmorata*, the electric lobe neurons display mitochondrial and cytoplasmic pathologies characteristic of cellular stress.[67] Whether or not such changes could be repaired by the fish was not reported. The reversibility of such anatomical damage may be critical to the proper function of the neuron and the expression of the associated behaviors.

2. Sexual Dichromism

In some fishes, the control of color change is predominantly under neuroendocrine control.[68] Pigment aggregating neurons pass through a medullary center and finally to the melanophores in the skin. Stimulation of the optic tectum and the centro-median telencephalon results in darkening, whereas stimulation of areas in the anterior dorsal and posterior dorso-lateral telencephalon cause paling. These nerves are apparently under sympathetic control. Coloration changes in many species of fishes are also under sex steroid control. It would be of interest to correlate metal-induced suppression of sexual dichromism[37,38] with the possible interaction of the PNS neurons and sex steroid sensitive brain regions. It is possible that there may be critical developmental periods, in terms of both neural growth and steroid synthesis and receptivity, when the ability to communicate present or future reproductive status would be most compromised by metal exposure. We know of no studies that examine the effects of toxic metals on fish coloration and the behavioral implications of such alterations, especially if the degree of color change can be modeled as a dose-response curve. The ability to associate a specific, quantifiable morphological change with altered behavior would be useful in defining toxic metal effects on proximate mechanisms of behavior.

3. Group Behavior

Group activity involves behavior over a variety of time frames (e.g., the milliseconds involved in the changing patterns of schooling movements; daily cycles of feeding aggregation; reproductive aggregations; and daily, as well as seasonal, cycles of migration). Schooling behavior, the highly polarized movement within a shoal of fish, is the result of three sensory modalities: mechanoreception (lateral line system), olfaction (telencephalon), and vision (mesencephalon or optic tectum)[59] (see Table 2 for neurotransmitters associated with those regions). Immediately upon contact with adverse stimuli, schools of fish either display the flash expansion (similar to a fireworks display) or fountain effect (movement away from followed by regrouping behind the adversive stimuli). Amazingly, during either maneuver, no collsions have ever been observed in fishes with their sensory apparati intact.[69] We believe that this trait will prove to be useful in developing a clear picture of the toxic metal/neurobiological/behavioral pathway.

Because the neural endings are in intimate contact with the water, the lateral line system is a likely candidate for being affected by waterborne metals, although at present this system has not been used as a model of the interplay between neurotoxicity and behavior. This system is sensitive to transitory displacements of water. The lateral line canal runs along various lengths of the head and trunk, depending upon the species, and connects to the vestibular system in the brain. Visual stimuli are also required to complement information supplied by the lateral line to adequately assess the individual's position within the school.[69]

By either changing the threshold level to induce an electrical stimulus along the axon or suppressing neurotransmitter activity at specific synapses, information about interindividual distances and velocity could be altered. The result would be differential responses in terms of group shape, size, or velocity during either stimulated or nonstimulated situations. Experiments with Atlantic silversides *(Menidia menidia)* demonstrated changes in swimming speed, distance to nearest neighbor, mean direction of travel, rate of change of direction, and depth of school in tank that occurred after only 1 day of exposure to sublethal levels of copper.[70] Cadmium caused individual *P. promelas* to orient perpendicularly to the rest of the group and to change directions more frequently.[71] Such activity may reduce the benefits of schooling behavior. Apparently, distance to the second nearest neighbor is also important for the fish to consider,[69] but no data are available concerning the effects of heavy metals on this variable. Neither of the above studies suggested changes in the lateral line system as an explanation of the metal-induced behavioral changes in school activity. Schooling responses to toxic levels of particular metals may prove to be a highly sensitive method for determining levels of toxicity, although, as yet, attempts to standardize these experiments or determine the underlying physiological mechanisms have not been made.

The structure of fish groups may change throughout the day. Circadian rhythms of shoaling behavior in goldfish *(C. auratus)* are influenced by opioid neuropeptide β-endorphin stimulation of the pineal gland.[72] Although we know of no evidence of toxic levels of metals interfering with neuropeptide synthesis, release, or receptivity, this also may be a promising avenue of future research.

Related to schooling behavior is the phenomenon of socially facilitated feeding among some taxonomic groups of fishes, e.g., cyprinids.[73] Visual, lateral line, and olfactory cues are critical to the coordination of this form of feeding activity. Social status, alterable by metals,[10] and sex may also play a role; the first fish to investigate a potential food source may always be one particular fish. This behavioral category may be best analyzed using the second paradigm of Huntingford,[5] because each level of "decision-making" requires a specific capacity to integrate environmental cues with appropriate physiological and behavioral responses. To our knowledge, no data exist on the effects of heavy metals on the physiological and behavioral dynamics of socially facilitated feeding. Researchers investigating the behavioral toxicity of metals may wish to frame their studies in terms of how motivational, cue, causal, and candidate spaces are affected. For example, one could ask this set of questions as it relates to feeding behavior. How does metal exposure alter the physiology of hunger, various environmental variables used by the fish to detect food, range of possible reactions to the presence of food, and interactions of the lead fish's movement toward food and the ability of other group members to detect that behavior?

4. Feeding Behavior

Feeding abilities are a species-specific function of any or all of the following physiological variables: locomotor coordination, mouth structure, visual acuity, smell, taste, touch, hearing, learning, and dietary requirements. One can analyze metal-induced changes in feeding behavior using the second paradigm of Huntingford.[5] In this view, the observed daily feeding patterns are a function of CNS (see Section III.C) and PNS (especially the gut) activity that control, modify, and organize motivational, cue, and candidate space subsets. Several studies, using a wide variety of metals, have indicated that feeding abilities[37,74,75,75a] as well as the appetite,[37,76] (i.e., motivational space) of fishes are adversely affected by metal intoxication. Studies on feeding abilities indicate that lead, mercury, copper, or aluminum decreased psychomotor coordination and, thereby, reduced the level of successful predation on microcrustacean prey items. Suppression of appetite (motivational space) may indicate changes in neuronal control at higher "satiety" centers in the brain, olfactory receptors, PNS sites in the gastrointestinal tract, or some combination of these factors.

Predation ability is a highly accurate estimate of the animal's visual acuity. Nyman,[77] using zebra fish *(Brachydanio rerio)*, hypothesized that if given a simultaneous choice between small vs. large prey items, lead-exposed fish should, due to a loss of psychomotor control, prefer smaller, easier-to-catch prey items. Weber[37] examined the possibility of changes in the size or shape of the candidate space, but found that the opposite occurred; unexposed juvenile fathead minnows switched much earlier (within 2 vs. >7 days for lead-exposed fish) to the easier-to-catch, smaller prey. It was suggested that this may reflect a change in learning ability. Additionally, age-specific responses may be altered by metal exposure in either instantaneous or cumulative patterns similar to those described by Chalmers'[14] concept of developmental pathways of behavior. Cleveland et al.[78] found that feeding behavior of brook trout *(Salvelinus fontinalis)* at older life stages was less sensitive to various combinations of pH and aluminum concentrations than of those at younger life stages. Examinations of age-specific alterations in feeding success that correlate psychomotor coordination with specific affected developing neural pathways are needed.

Peripheral control of feeding behavior is, in part, under the influence of gut and stomach extension. In fishes, the control of peristaltic activity and gut-brain communication are similar to mammals[79] and affects the fish's motivational space. Yet a potential internal conflict arises in the integration (cue space) of CNS and PNS control of feeding behavior. One of the toxic effects of metals is decreased nutrient absorption. By suppressing mechanisms of nutrient absorption through the gut epithelium (see Section IV.A),[80] as well as increased mucous cell production, disruption of intestinal brush border, or edematous tissue,[37,81] peripheral or central receptors sensitive to concentrations of substances that control satiety responses will remain at subthreshold levels. To adjust to a perceived energy and nutrient imbalance by glucostatic and other receptors in the brain, the fish, theoretically, should increase feeding. Yet numerous authors have observed a decrease in feeding. Thus, apparently a synergistic effect of the CNS and PNS in lowering growth rates after metal-metal intoxication includes decreases in absorption and feeding. Possibly, CNS control is more critical to feeding regulation (candidate space), decreasing both the size and frequency of feeding bouts, and the PNS simultaneously affects greater clearance rates and reduced absorption.

The possible dual effects of CNS feeding regulation and PNS gut activity during metal exposure were examined by Weber.[37] It has been noted previously in Section III.C that CNS neurotransmitters are altered. Weber[37] additionally observed that approximately 33 to 50% of the intestine from lead-exposed fish was clear of any feces, whereas intestines of control fish, even after a 24-hr fast, were packed with material. Furthermore, it was demonstrated that lead-exposed fish passed the food faster than controls — by 3 to 4 days (Figure 6). This at least suggests that lead increased peristaltic activity. In fishes, a decrease in NE from the sympathetic nerves decreases gut motility, an increase of acetylcholine (ACh) from the parasympathetic nerves increases motility, or possibly, an increase in 5-HT in some fishes increases gastrointestinal activity.[82] However, such changes were not observed in lead-exposed adult *P. promelas*[37] (Table 3). It may be that other factors need to be considered to evaluate neurochemical mechanisms of metal-induced feeding behavior alterations. Changes in levels or activity of specific intestinal neuropeptides (see Section III.E.3), e.g., cholecystokinin or vasoactive intestinal polypeptide, may enhance or depress gut motility and thus affect feeding motivation. Alterations in either the ratio of ACh to NE or actual ACh and NE levels may have a similar effect. Finally, different times of day of sampling may yield conflicting results as they relate to the levels of critical neurotransmitters and neuropeptides.

It is possible that these changes in the perceived or actual level of hunger due to metal exposure are critical factors in the expression of feeding behaviors that may not be very beneficial to the individual. Fish that are infected with parasites will

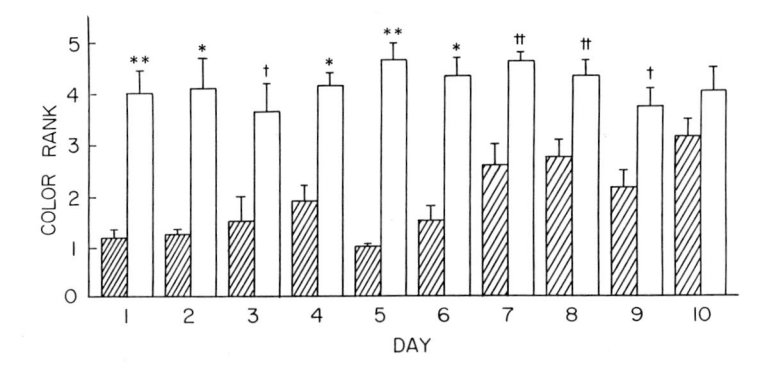

Figure 6. Mean rate of passage of colored flake food through the intestine of control (▨) and 1.0 ppm lead-exposed (☐) fathead minnows *(P. promelas)* (from Weber.[34]) * $p < 0.05$, ** $p < 0.01$, † $p < 0.005$, †† $p < 0.001$. Bars = standard error of the mean.

forage more quickly and actively after a frightening stimulus than unparasitized individuals.[83] If a predator is the adversive stimulus, this rapid recovery could be disadvantageous. Although toxic metal-exposed fishes tend to be preyed upon more often than control individuals,[71] no conclusive test has been conducted to evaluate the roles that hunger and predation risk may have played in these experiments. Changes in behavioral motivation, as in Huntingford's[5] second paradigm, may be traceable to neuroendocrine alterations in either the CNS, the PNS, or both.

IV. ENERGETIC AND RESPIRATORY MECHANISMS OF BEHAVIOR

A. Energetic Considerations in Behavioral Analyses

Specific locomotor, feeding, or reproductive activity movements and morphological development of color, shape, or size during sexual maturity in response to environmental changes or social interactions require energy. With only a finite store of energy-containing lipids and proteins, any alteration in the availability, level, metabolism, or physiological compartmentalization of energy ultimately may change the capacity for appropriate behavioral reponses.[84] Thus, after a brief, generalized summary of behavioral energetics, we will discuss the nature of metal interference with normal metabolism and the possible resultant effects on behavior. It must be noted that there is very little research at present on the energetics of behavioral toxicology in fishes. Specifically, there is almost no information on the energy required for specific parts of behavioral sequences, let alone about what happens when toxic metals are introduced into a system. Thus, many of our comments will be speculative.

Pandian,[85] reviewing a number of important relationships in fish energetics, noted the following:

Table 3. Neurotransmitter Analysis of Intestines from Control and Lead-Exposed Fathead Minnows (*P. promelas*)

	Lead concentration	
	0.0 ppm	1.0 ppm
ng NE/mg intestine	0.13 ± 0.06	0.28 ± 0.4[a]
ng MHPG/mg intestine	0.07 ± 0.02	0.24 ± 0.08[a]
ng 5-HT/mg intestine	0.23 ± 0.03	0.17 ± 0.02[a]
U AChE/mg intestine	0.98 ± 0.14[b]	0.43 ± 0.24[a]

Source: From Weber, D.N., Physiological and behavioral effects on waterborne lead on fathead minnows, Ph.D. Dissertation, University of Wisconsin, Milwaukee, 1991.

Note: NE = norepinephrine, MHPG = 3-methoxy,4-hydroxyphenylglycan, 5-HT = 5-hydroxytryptophan (serotonin), and AChE = acetylcholinesterase.

[a] *t*-test, $p<0.05$.
[b] One unit = a µmole of acetic acid from acetylcholine in 30 min at 25°C at pH 7.8.

1. Fishes, as ectotherms, are efficient in utilizing protein as their primary source of energy because they do not need to expend large amounts of energy on body temperature maintenance, they expend low levels of energy countering gravity, and they can remove ammonia waste rapidly through the gills; lipids can also be used for energy.
2. The energy content of the food is important in determining the rate of stomach or gut (in species with no stomach, e.g., some cyprinids) evacuation.
3. The cost of biochemical transformation of nutrients, specific dynamic action (SDA), is positively correlated to ration and temperature.
4. Those species that display daily migratory patterns tend to feed at higher temperatures and expend less energy, particularly SDA, during digestion at lower temperatures.
5. Reproductive energetics, including breeding and parental care activities, vary with age, life span, and number of breeding seasons in the fish's life time.

That toxic levels of metals in aquatic systems alter feeding activity and gut motility has already been described in this chapter. We now will suggest that the role of metals in adversely affecting other components of the energy balance equation (see below) could also potentially impinge on behavior.

Models have been developed to predict feeding behavior patterns based on the interaction of sets of asymptotically increasing rate curves that relate food density, energy per unit food, and energy required to obtain food.[86] These curves delineate the potential scope of activity for an individual fish. This scope includes locomotor capacity, type and density of prey preferred, water temperature preference, and growth rates. It should also include ability to perform specific behaviors within a given FAP critical to the successful accomplishment of the particular task. Changes in these curves have clear applications to the energetics-based behavioral alterations observed in metal-exposed fishes in that a decrease in scope for activity limits the behavioral options the individual or group may have (causal factor space). It is also possible that the ordering of which behaviors should be elicited (candidate space) may be changed. We have seen no data to test this idea.

The balance of caloric input vs. output is correlated with degree of hunger; activity of the CNS and PNS; storage of adipose tissue; use of carbohydrates, lipids, and proteins in both the growth and repair of tissues; SDA and standard metabolism; and level and pattern of whole animal activity.[87] Using a generalized energetics equation [88] modified to include behavior, it is possible to compartmentalize the various components of bioenergetic relationships in a fish. This is by no means the most recent or sophisticated energetics model, but, precisely because it is a simplified version of more complex equations, it facilitates easier initial correlations to the mechanisms of behavioral toxicology and more easily directs researchers toward testable hypotheses.

$$C = (R_s + aR_{r-s} + bR_{f-s} + cR_{a-s}) + (P_g + P_{rg} + P_{fs} + P_b) + (F + U + S)$$

where C = consumed food
R_s = standard metabolism
R_r = routine metabolism
R_f = SDA
R_a = active metabolism
P_g = somatic growth
P_{rg} = reproductive growth
P_{fs} = fat stores
P_b = energy for specific behaviors
F = feces
U = urine
S = secretions
a,b,c = constants related to fraction of time used for that variable

This equation offers a glimpse into the dynamic nature of relationships and the available energy a fish may have for specific activities. Additionally, one can appreciate that directed motion and behavioral patterns can also be analyzed energetically and that toxic exposure to metals can, by altering the allocation of energy, potentially change these patterns. This is an exciting area of research, and its potential to give insights into proximate mechanisms of behavior has not yet been fully realized.

Energetic evaluations of reproductive behavior in fishes under natural conditions have been conducted.[89,90] These workers analyzed energy utilization for specific components of reproductive tissue, fecundity, parental investment, somatic growth, and respiration (P_g, P_{rg}, R). Reproductive success, then, is in part a result of the individual fish's proper allocation of energetic resources to the morphological and physiological development required for proper behavioral expression. How a metal-stressed fish deals with this is not well understood.

It has already been stated that specific hormones are decreased by metals. This is important because the distribution of energy to reproductively important sites is dependent, to a large extent, on seasonal variations in sex steroid levels.[91,92] Estrogens create a negative energy balance ($P_{fs\text{-initial}} > P_{fs\text{-final}}$) by decreasing white adipose tissue lipid stores (P_{fs}; less fatty acid uptake and more fatty acid release), increasing voluntary (R_a) exercising, and increasing heat loss (R). Testosterone reduces lipid content ($P_{fs\text{-initial}} > P_{fs\text{-final}}$), increases muscle mass (P_g), and increases voluntary exercise (R_a). On top of this is the circannual and circadian variation of thyroid hormones and their role in lipolysis.[21,25,91] From Table 1, it is evident that changes in these energy-related hormones are also critical to specific behaviors, as explained in Section II.A. At present, the role lower levels of these hormones may have in behavioral energetics after exposure to toxic metals is unclear.

B. Behavioral Effects Potentially Related to Metal-Altered Metabolism

Directed movement, that is, motion toward or away from an environmental response such as light (phototaxis) or water current (rheotaxis), requires energy (R_a) and a sensory system. Inhibition or stimulation of any of the integrated steps of metabolism (Figure 1) could alter the capacity for sustained, directed activity. In other words, changes in diet choice, feeding ability, energy required to metabolize food (SDA), energetically important hormones, or conversion of consumed food into behaviorally important structures may seriously compromise the fish's ability to display a full complement of critically useful behaviors due to a decreased scope for activity.

Many workers have examined a fish's sustained swimming capacity in order to evaluate the effect of a contaminant on fish behavior as a function of changes in biochemical activity. This easily measured and quantifiable behavioral parameter is a function of qualitative and quantitative changes in energy stores ($P_{fs\text{-initial}} \neq P_{fs\text{-final}}$), capacity for oxygen utilization (R), degree of dependence on anaerobic metabolism, size and age of fish, and mitochondrial function. These factors may act singly or in combination to effect a behavioral change.

Weber[37] examined the effects of short-term, sublethal lead exposure on swimming capacity of adult fathead minnows *(P. promelas)*. In comparison to unexposed controls, lead-exposed fish displayed varying levels of decreased ability for sustained activity; larger fish showed greater decreases after lead exposure. During exposure to aluminum at different pH levels, brook trout *(S. fontinalis)* displayed a complex, age-specific swimming capacity response. Newly hatched fry were most sensitive to aluminum at pH 5.5 and older juveniles only at pH 7.2.[78] Although neither study examined responses of metal-exposed fishes returned to uncontaminated water, such an examination of reversibility would be critical to testing Chalmers'[14] hypotheses. Interestingly, due to different oxidation states and species of aluminum at varying pH levels, the mechanism of sensitivity may be closely related to aluminum chemistry. Other metals also have several pH-dependent oxidation states; this is an issue that needs further study, although it is certainly complicated by the behavioral alterations induced by pH changes alone.[93]

Swimming capacity is also affected by the oxygen-carrying capacity of erythrocytes. Lead and zinc inhibit aminolevulinic acid dehydratase (ALAD) activity in fishes.[94] Decreases in ALAD activity result in decreased heme synthesis and, therefore, decreased hemoglobin production, as in tilapia *(Oreochromis hornorum)*.[95] Additionally, at least in mammals[96] and presumably in fishes, lead induces concomitant increases in aminolevulinic acid synthetase (ALAS) and decreases in ferrochelatase activity, both required steps in hemoglobin synthesis. Thus, it might possibly be argued that, at least on one level, the decrease in oxygen-carrying capacity in the blood of lead- or zinc-exposed fishes is a fundamental reason for decreased sustained swimming capacity. Decreased oxygen utilization capacity of copper-exposed fishes further complicates swimming capacity.[52] The capacity for sustained activity during heavy metal exposure, therefore, may be decreased simply due to an inability to use lower levels of available food stores (P_{fs}). If so, then these behavioral changes may be reversible once the toxic insult is removed. The degree, pattern, and rate of recovery may be an age-dependent response, which is assessed either instantaneously or cumulatively.[14]

To what extent changes in hemoglobin synthesis may affect burst activity (escape behavior or aggressive displays) is, at present, unclear. Fishes, as a group, have a much larger proportion of white muscle than other vertebrates. Since, in most fishes, red muscle mass (i.e., the site of maximum aerobic capacity) composes only a small percentage of total muscle mass, most of the aerobic capacity in fishes will occur in white muscle.[97] In male carp, for example, 40 to 65% of whole-body cytochrome c oxidase activity is found in the white muscle. Furthermore, most of the blood flow in both arctic grayling *(Thymallus arcticus)* and rainbow trout *(Oncorhynchus mykiss)* is found in the white muscle. The potential of both burst and sustained motion being compromised by metal exposure could be particularly critical to such behavioral sequences as courtship. In this sequence, prolonged activity is punctuated by quick movements. Therefore, we suggest that improper energy allocation or insufficient energy stores resulting from metal intoxication might be responsible for decreased activity or interrupted sequential patterns.

Decreases in overall aerobic capacity and changes in energy reserves or metabolism may affect both stamina and burst activity. Fishes experiencing decreased oxygen-carrying capacity, as well as changes in serum glucose (P_{fs}),[98,99] liver glycogen stores (P_{fs}),[100,101] key metabolic enzymes (R),[102,103] lipid and cholesterol content (P_{fs}),[104,105] and fatty acid metabolism (R)[106] during metal exposure, demonstrate an inability to maintain prolonged exercise. Fishes under stress may catabolize lipids differently. It has been noted for salmon *(O. nerka)* that the catabolism of lipids differs at various swimming speeds. Salmon swimming for short distances at high water velocities (R_a) tend to metabolize a greater percentage of unsaturated fatty acids[86] (i.e., fatty acids that have more double bonds and thus a greater energy requirement for catabolism). This extra metabolic demand (R_f) to obtain maximum usable energy may be a critical factor in reducing the level of available energy for sustained swimming. Studies using larval herring *(Clupea harengus)* indicated that the level of total body lipids, especially those fatty acids crucial to visual and neural development, have a dramatic effect on larval swimming performance. Both orientation and ability to coordinate muscle movements are altered at low levels of specific fatty acids.[84] Although it is not yet clear whether this effect is due to initial stores vs. utilization of fatty acids, changes in larval lipid metabolism could account for some behavioral aberrations in metal-exposed fishes.

Capacity for sustained activity in fishes should also be sensitive to changes in oxygen diffusion across the gill lamellae. This assumes, of course, a correlation between oxygen consumption and whole-body metabolism. Correlations with gill damage and toxicant exposure have been shown using a variety of environmental organic and inorganic contaminants.[107-109] Clubbing of secondary gill lamellae, cell death, and interlamellar hyperplasia after metal exposure may be sufficient to reduce oxygen diffusion (R_s).[107]

Although several studies highlight the apparent conflict between gill damage by metal exposure and an increase in respiratory need, some of these difficulties may be compensated for behaviorally (i.e., increased respiratory rate, movement to oxygen-rich water). Changes in oxygen uptake, as well as increases in energy used to obtain oxygen, may affect energy available for allocation to specific behaviors, especially if they demonstrate temporal relationships, e.g., swimming, feeding, or reproduction — behaviors not initiated as a direct response to metabolic changes but as a consequence of energy availability. During examination of circadian rhythms of oxygen uptake in *P. promelas* exposed to 1.0 ppm lead for 2 weeks, it was noted that during periods of greatest activity (dawn, feeding time, and dusk), lead-exposed fish nearly doubled their oxygen uptake[37] (Figure 7). Thompson et al.[108] demonstrated overall ventilatory rate increase with increased gill purge frequencies and shallower ventilatory movements in zinc- and copper-exposed *L. macrochirus*. Diamond et al.[109] confirmed these results in zinc- and cadmium-exposed *Lepomis macrochirus*. Unfortunately, the data at present are insufficient for answering the question of whether disruptions in behaviors are a consequence of metabolic changes caused by metal-induced effects on energetics. If increased stress leads to increased oxygen uptake and increased metabolic activity for maintenance (R_f or R_s), then with less available energy for directed movement (e.g., nest preparation, courtship, or migration), critical social interactions fail to develop properly. Thus, a clear connection between physiological effects of metal exposure with behavioral alterations could be demonstrated.

Presently, no data exist on the changes in whole-animal reproductive energy allocations, specifically including allocations for behavior, due to toxic metal exposure. Experiments need to be conducted to assess energetic-based mechanisms of altered behavior and reproductive success. Areas that still need investigation include direct vs. indirect effects on the gill's ability for oxygen uptake;

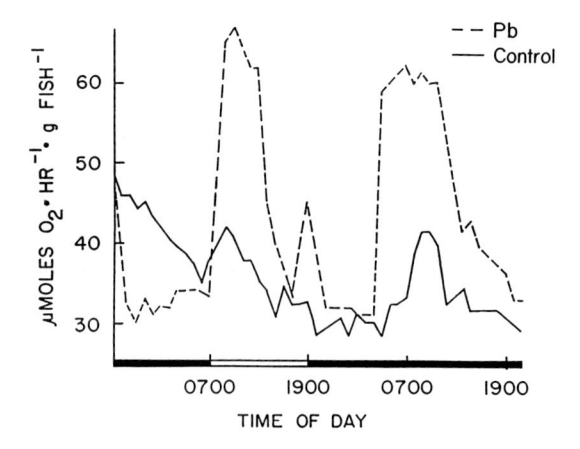

Figure 7. Selected example of the effects of lead on circadian oxygen consumption uptake rhythms in fathead minnows, *P. promelas*. (—) Control; (- - -) 1.0 ppm lead. Light/dark cycle indicated by white and black bars. (from Weber)[37]

reversibility of the effects on prolonged or burst activity; evaluation of metal exposure time dependency (i.e., length of exposure) on activity level; effects due to either continuous exposure or single pulses at varying times of day followed by behavioral tests at varying times of day; and differential effects of organic (e.g., alkyl or acetate salt) vs. inorganic metals and age-, sex-, circadian-, circannual-, and size-specific dependencies (all of which can be tested against the behavioral paradigms outlined in Figures 1 and 2).

V. DIRECTIONS FOR FUTURE RESEARCH

We have attempted to point out deficits in our knowledge of the mechanisms of behavioral metal toxicity throughout this chapter. In this section, we highlight some particularly promising, or needed, areas of research.

We believe that toxicity testing is most meaningful if it is integrated into a mechanistic approach of behavioral toxicology. Again, we admit that this demands an approach that is not necessarily within the mainstream of current ethological studies. Because human interaction with nature is frequently on a scale that often precludes a species' ability to adapt, evolutionary questions are probably less important than mechanistic questions to evaluate behavioral change. To do this, future research should be aimed at cause-and-effect relationships between observed changes in fish physiology and altered behavior. Our sections on neuro- and endocrine bases of behavior have suggested how the toxic metal/physiological/behavioral axis can be investigated. Detailed analysis on cascading physiological events, critical sites of interaction, and temporal sensitivity (Figure 1) are now needed to evaluate behavioral changes after metal exposure.

neurochemistry not only affects behavior, but that the reverse is equally true.[11, 118a] The role of previous exposure experience in ameliorating or exacerbating the behavioral effects of future exposure to heavy metals is central to understanding critical periods in the life history. It has been suggested by Weber[37,38] that if a male fish is exposed to lead after some level of reproductive maturity and the associated "priming" or "imprinting" of specific regions of the brain for increased sensitivity to sex steroids has been achieved, then the toxic effects, related to reproduction, may be lessened in comparison to the effects in a less reproductively mature male. Hypothetically, previous behavioral experience by altering brain structure may, to some degree, confer a "behavioral resistance" to metal toxicity for other behaviors as well. This may be of great importance to migratory species, for example, that imprint a home range smell to facilitate a return to the same location for breeding.[12] If exposure was before this imprinting, the effects may be greater than if it occurred afterward.

A genetic basis for fish behavior has been demonstrated.[119] Metals, such as lead, mercury, and aluminum, have been shown to affect specific gene products or processes.[120,120a] These metals may also interact with genes and gene products that control behavior in fishes, although we know of no data to support this hypothesis. If such interactions alter specific behaviors, a fundamental mechanism of behavioral toxicity could be identified. This should prove to be an exciting avenue of collaborative research for ethologists, toxicologists, and molecular biologists.

Individuals involved in risk assessment and regulatory policy need baseline data to determine dose-response curves for specific behavioral alterations, such as for calculating a lowest observable effect concentration (LOEC) under different environmental conditions or maximum acceptable toxicant concentration (MATC) for life history tests. It is possible that such curves are not linear overall concentrations but show biphasic or other patterns.[121] Such information will enhance the usefulness of behavioral toxicity in evaluating the relative harm of different environmental contaminants.

No metal ever exists either as an isolated contaminant or in a single oxidation state in any natural aquatic system. Thus, in nature, the level of toxicity will vary depending upon the specific chemical milieu. Metal toxicity in fishes is intimately related to the particular metal species, binding capacity to organic ligands, and presence of other metals.[122] Basic laboratory research is required on the relationship of various mixes and chemical forms of contaminants on behavioral toxicity. Such data will allow investigators to then conduct field validation studies of laboratory experiments, which to date are rare.[123]

One of the great problems in behavioral toxicology is the lack of a standardization of tests. Behavioral tests are highly sensitive to the detection of environmental contaminants,[122] but the techniques used between laboratories vary significantly, as do methods of data analysis. Differences may occur between and within species that further compound the problem of comparative analyses.

We now come full circle. Beginning with an analysis of two behavioral paradigms that might be useful in studying the toxic effects of metals and continuing with separate analyses of various physiological mechanisms that might explain behavioral toxicity, we offer the following concluding statement both as a challenge and an underlying philosophy of behavioral toxicology. Ultimately, the most critical problem facing behavioral toxicologists is the multifaceted dimension of the many physiological mechanisms involved in eliciting behavior, many of which are altered by even the lowest detectable concentrations of single- or multiple-metal exposure regimes. Sorting out the myriad endocrine, neural, energetic, and respiratory pathways that control behavior, both directly and indirectly, and determining which are affected by toxic levels of heavy metals will indeed be a herculean task. If researchers struggle with it, however, the rewards will be great, for it will not only allow scientists to understand why metals are toxic but to perhaps grasp even the fundamental mechanisms of normal fish behavior.

ACKNOWLEDGMENTS

Preparation of this chapter was supported by National Institutes of Environmental Health Sciences grant ES 04184.

REFERENCES

1. Scherer, E., Behavioral responses as indicators of environmental alterations: approaches, results, developments, *J. Appl. Icthyol.*, 8, 122, 1992.
2. O'Dell, B.L., Bioavailability of essential and toxic trace elements, in Symposium programmed by the American Institute of Nutrition and the Society for Environmental Geochemistry and Health at the 66th Annual Meeting of the Federation of American Societies for Experimental Biology, New Orleans, LA, 1982, 1714.
3. Weber, D.N., Eisch, S., Spieler, R.E., and Petering, D.H., Metal redistribution in largemouth bass *(Micropterus salmoides)* in response to restrainment stress and dietary cadmium: role of metallothionein and other metal-binding proteins, *Comp. Biochem. Physiol.*, 101C, 255, 1992.
4. Silbergeld, E.K., Experimental studies of lead neurotoxicity: implications for mechanisms, dose-reponse and reversibility, in *Lead Versus Health*, Rutter, M. and Rutter Jones, R., Eds., John Wiley & Sons, New York, 1983, 191.
5. Huntingford, F., *The Study of Animal Behaviour,* Chapman and Hall, London, 1984, chap. 3.
6. de Bruin, J.P.C., Telencephalon and behavior in teleost fish: a neuroethological approach, in *Comparative Neurology of the Telencephalon*, Ebbesson, S.O.E., Ed., Plenum Press, New York, 1980, chap. 7.
7. Lau, Y.-S., Camoratto, A.M., White, L.M., and Moriarty, C.M. Effect of lead on TRH and GRF binding in rat anterior pituitary membranes, *Toxicology*, 68, 169, 1991.

8. Wiebe, J.P., Salhanick, A.I., and Myers, K.I., On the mechanism of action of lead in the testis: *in vitro* suppression of FSH receptors, cyclic AMP and steroidogenesis, *Life Sci.*, 32, 1997, 1983.
9. Alkahem, H.F., Effect of sublethal copper concentrations on the behavior of cichlid fish, *Oreochromis niloticus, Z. Angew. Zool.*, 76, 93, 1989.
10. Henry, M.G. and Atchison, G.J., Behavioral changes in bluegill *(Lepomis macrochirus)* as indicators of sublethal effects of metals, *Environ. Biol. Fish*, 4, 37, 1979.
11. Browman, H.I., Embryology, ethology and ecology of ontogenetic critical periods in fish, *Brain Behav. Evol.*, 34, 5, 1989.
12. Dodson, J.J., The nature and role of learning in the orientation and migratory behavior of fishes, *Environ. Biol. Fish*, 23, 161, 1988.
13. Ridgway, M.S., Developmental stage of offspring and brood defense in smallmouth bass *(Micropterus dolomieui), Can. J. Zool.*, 66, 1722, 1988.
14. Chalmers, N.R., Developmental pathways in behaviour, *Anim. Behav.*, 35, 659, 1987.
15. Barchas, J.D., Akil, H., Elliott, G.R., Holman, R.B., and Watson, S.J., Behavioral neurochemistry: neuroregulators and behavioral states, *Science*, 200, 964, 1978.
16. Thomas, P., Effects of Aroclor 1254 and cadmium on reproductive endocrine function and ovarian growth in Atlantic croaker, *Mar. Environ. Res.*, 28, 499, 1989.
17. Thomas, P., Teleost model for studying the effects of chemicals on female reproductive endocrine function, *J. Exp. Zool.*, Suppl. 4, 126, 1991.
17a. Thomas, P., Effects of cadmium on gonadotropin secretion from Atlantic croaker pituitaries incubated *in vivo*, *Mar. Environ. Res.*, 35, 141, 1993.
18. Katti, S.R. and Sathyanesan, A.G., Lead nitrate induced changes in the thyroid physiology of the catfish *Clarias batrachus* (L.), *Ecotoxicol. Environ. Safety*, 13, 1, 1987.
19. Bhattacharya, T., Bhattacharya, S., Ray, A.K., and Dey, S., Influence of industrial pollutant on thyroid function in *Channa punctatus* Bloch, *Ind. J. Exp. Biol.*, 27, 65, 1989.
20. Brown, S.B., Maclatchy, D.L., Hara, T.J., and Eales, J.G., Effects of low ambient pH and aluminum on plasma kinetics of cortisol, T_3 and T_4 in rainbow trout, *Oncorhynchus mykiss, Can. J. Zool.*, 68, 1537, 1990.
21. Spieler, R.E. and Noeske, T.A., Diel variations in circulating levels of triiodothyronine and thyroxine in goldfish, *Carassius auratus, Can. J. Zool.*, 57, 665, 1979.
22. Schreck, C.B. and Lorz, H.W., Stress response of coho salmon *(Oncorhynchus kisutch)* elicited by cadmium and copper and potential use of cortisol as an indicator of stress, *J. Fish. Res. Board Can.*, 35, 1124, 1978.
23. Franklin, C.E., Forster, M.E., and Davison, W., Plasma cortisol and osmoregulatory changes in sockeye salmon transferred to sea water — comparison between successful and unsuccessful adaptation, *J. Fish Biol.*, 41, 113, 1992.
24. Spieler, R.E. and Weber, D.N., Effects of waterborne lead on circulating thyroid hormones and cortisol in rainbow trout, *Med. Sci. Res.*, 19, 477, 1991.
25. Hadley, M.E., *Endocrinology*, Prentice Hall, Englewood Cliffs, NJ, 1992.
26. Baggerman, B., Some endocrine aspects of fish migration, *Gen. Comp. Endocrinol.*, Suppl. 1, 188, 1962.
27. Gorbman, A., Thyroid function and its control in fishes, in *Fish Physiology, Vol. 2: The Endocrine System*, Hoar, W.S. and Randall, D.J., Eds., Academic Press, New York, 1969, 241–274.

28. Raso, D.S., The changing axonal distribution in white muscle of the developing catfish, *Ictalurus nebulosus* (Leseur), *Neurosci. Lett.*, 139, 50, 1992.
29. Huseman, C.A., Moriarty, C.M., and Angle, C.R., Childhood lead toxicity and impaired releases of thyrotropin-stimulating hormone, *Environ. Res.*, 42, 524, 1987.
30. Sokol, R.Z., Madding, C.E., and Swerdlow, R.S., Lead toxicity and hypothalamic-pituitary-testicular axis, *Biol. Reprod.*, 33, 722, 1985.
31. Cooper, R.L., Goldman, J.M., and Rehnberg, G.L., Pituitary function following treatment with reproductive toxins, *Environ. Health Perspect.*, 70, 177, 1986.
32. Govoni, S., Lucchi, L., Battaini, F., Spano, P.F., and Trabucchi, T., Chronic lead treatment affects dopaminergic control of prolactin secretion in rat pituitary, *Toxicol. Lett.*, 20, 237, 1984.
33. Fu, H., Lock, R.A.C., and Bonga, S.E.W., Effect of cadmium on prolactin cell activity and plasma electrolytes in the freshwater teleost *Oreochromis mossambicus*, *Aquat. Toxicol.*, 14, 295, 1989.
34. Hildemann, W.H., A cichlid fish, *Symphysodon discus*, with unique nutritive habits, *Cichlidae, Lond.*, 2, 32, 1975.
35. Blüm, V., Experimente mit Teleosteer-Prolaktin, *Zool. Jb. (Allg. Zool.)*, 77, 335, 1972.
36. Povlsen, A.F., Korsgaard, B., and Bjerregaard, P., The effect of cadmium on vitellogenin metabolism in estradiol-induced flounder, *Platichthys flesus* L., males and females, *Aquat. Toxicol.*, 17, 253, 1990.
36a. Behra, R., *In vitro* effects of cadmium, zinc and lead on calmodulin-dependent actions in *Oncorhynchus mykiss*, *Mytilus* sp., and *Chlamydomonas reinhardtii*, *Arch. Environ. Contam. Toxicol.*, 24, 21, 1993.
37. Weber, D.N., Physiological and behavioral effects of waterborne lead on fathead minnows (*Pimephales promelas*), Ph.D. dissertation, University of Wisconsin, Milwaukee, 1991.
38. Weber, D.N., Exposure to sublethal levels of waterborne lead alters reproductive behavior patterns in fathead minnows *(Pimephales promelas)*, *Neurotoxicology*, 14, 347, 1993.
39. Pickering, Q., Brungs, W., and Gast, M., Effect of exposure time and copper concentration on reproduction of the fathead minnow *(Pimephales promelas)*, *Water Res.*, 11, 1079, 1977.
40. Govoni, S., Memo, M., Lucchi, L., Spano, P.F., and Trabucchi, T., Brain neurotransmitter systems and chronic lead intoxication, *Pharmacol. Res. Commun.*, 12, 447, 1980.
41. Schröder, J.H. and Peters, K., Differential courtship activity of competing guppy males *(Poecilia reticulata* Peters; Pisces: Poeciliidae) as an indicator for low concentrations of aquatic pollutants, *Bull. Environ. Contam. Toxicol.*, 40, 396, 1988.
42. Popek, W., Seasonal variations in circadian rhythm of hypothalamic catecholamine content in the eel *(Anguilla anguilla* L.), *Comp. Biochem. Physiol.*, 75C, 193, 1983.
43. Noble, G.K., The function of the corpus striatum in the social behavior of fishes, *Anat. Rec.*, 64, 34, 1936.
44. Weber, D.N., Russo, A.C., Seale, D.B., and Spieler, R.E., Waterborne lead affects feeding abilities and neurotransmitter levels of juvenile fathead minnows *(Pimephales promelas)*, *Aquat. Toxicol.*, 21, 71, 1991.
45. Munkittrick, K.R., Martin, R.J., and Dixon, D.G., Seasonal changes in whole brain amine levels in white sucker exposed to elevated levels of copper and zinc, *Can. J. Zool.*, 68, 869, 1990.

46. Katti, S.R. and Sathyanesan, A.G., Lead nitrate induced changes in the brain constituents of the freshwater fish *Clarias batrachus* (L.), *Neurotoxicology*, 7, 47, 1986a.
47. Katti, S.R. and Sathyanesan, A.G., Changes in the hypothalamoneurohypophysial complex of lead treated teleostean fish *Clarias batrachus* (L.), *Z. Mikrosk. Anat. Forsch., Leipzig*, 100, 347, 1986b.
48. Cooper, G.P., Suszkiw, J.B., and Manalis, R.S., Presynaptic effects of heavy metals, in *Cellular and Molecular Neurotoxicology*, Narahashi, T., Ed., Raven Press, New York, 1986, 1.
49. Manalis, R.S., Cooper, G.P., and Pomeroy, S.L., Effects of lead on neuromuscular transmission in frog, *Brain Res.*, 294, 95, 1983.
50. Rahamimhoff, R., Lev-Tov, A., Meiri, H., Rahamimhoff, H., and Nussonovitch, I., I. Regulation of acetylcholine liberation from presynaptic nerve terminals, *Monogr. Neural Sci.*, 7, 3, 1980.
51. Cooper, G.P. and Manalis, R.S., Interactions of lead and cadmium on acetylcholine release at the frog neuromuscular junction, *Toxicol. Appl. Pharmacol.*, 74, 411, 1984.
52. Nemcsók, J.G. and Hughes, G.M., The effect of copper sulphate on some biochemical parameters in rainbow trout, *Environ. Pollut.*, 49, 77, 1988.
53. Nemcsók, J., Németh, A., Buzás, Z.S., and Boross, L., Effects of copper, zinc and paraquat on acetylcholinesterase activity in carp *(Cyprinus carpio* L.), *Aquat. Toxicol.*, 5, 23, 1984.
54. Shaw, B.P. and Panigrahi, A.K., Brain acetylcholinesterase activity studies in some fish species collected from a mercury contaminated estuary, *Water Air Soil Pollut.*, 53, 327, 1990.
55. Suresh, A., Sivaramakrishna, B., Victoriamma, P.C., and Radhakrishnaiah, K., Comparative study on the inhibition of AChE activity in the freshwater fish, *Cyprinus carpio*, by mercury and zinc, *Biochem. Int.*, 26, 367, 1992.
56. Hornby, P.J. and Piekut, D.T., Distribution of catecholamine-synthesizing enzymes in goldfish brains: presumptive dopamine and norepinephrine neuronal organization, *Brain Behav. Evol.*, 35, 49, 1990.
57. Sloley, B.D. and Rehnberg, B.G., Noradrenaline, dopamine, 5-hydroxytryptamine and tryptophan concentrations in the brains of four cohabiting species of fish, *Comp. Biochem. Physiol.*, 89C, 197, 1988.
58. Peter, R.E., The brain and feeding behavior, in *Fish Physiology, Vol. 8: Bioenergetics and Growth*, Hoar, W.S., Randall, D.J., and Brett, J.R., Eds., Academic Press, New York, 1979, 280.
59. Smith, J.R., Fish neurotoxicology, in *Aquatic Toxicology*, Vol. 2, Weber, L.J., Ed., Raven Press, New York, 1984, 107.
60. Spieler, R.E., Dyschronogenic effects of environmental contaminants, in *Annual Report of the Marine and Freshwater Biomedical Core Center*, University of Wisconsin, Milwaukee, 1987, 62.
61. Steele, C.W., Effects of sublethal exposure to copper on diel activity of sea catfish, *Arius felis, Hydrobiology*, 178, 135, 1989.
62. Baatrup, E., Structural and functional effects of heavy metals on the nervous system including sense organs of fish, *Comp. Biochem. Physiol.*, 100C, 253, 1991.
63. Neuman, I.S.A., van Rossum, C., Bretschneider, F., Teunis, P.F.M., and Peters, R.C., Biomonitoring cadmium deteriorates electro-orientation performance in catfish, *Comp. Biochem. Physiol.*, 100C, 259, 1991.

64. Heiligenberg, W., Processes governing behavioral states of readiness, *Adv. St. Behav.*, 2, 173, 1974.
64a. Weber, D.N., Unpublished data.
65. Coughlan, D.J., Gloss, S.P., and Kubota, J., Acute and sub-chronic toxicity of lead to the early life stages of smallmouth bass *(Micropterus dolomieui), Water Air Soil Pollut.*, 28, 265, 1986.
66. Munro, A.D. and Pitcher, T.J., Hormones and agonistic behaviour in teleosts, in *Control Processes in Fish Physiology*, Rankin, J.C., Pitcher, T.J., and Duggan, R.T., Eds., Wiley Interscience, New York, 1983, chap. 9.
67. Enesco, H.E., Pisanti, F.A., and Alojtotaro, E., The effect of copper on the ultrastructure of *Torpedo marmorata* neurons, *Mar. Pollut. Bull.*, 20, 232, 1989.
68. Iwata, K.S. and Fukuda, H., Central control of color changes in fish, in *Responses of Fish to Environmental Changes*, Chavin, W., Ed., Charles C Thomas, Springfield, 1973, chap. 11.
69. Partridge, B.L., The structure and function of fish schools, *Sci. Am.*, 246, 114, 1982.
70. Koltes, K.H., Effects of sublethal copper concentrations on the structure and activity of Atlantic silverside schools, *Trans. Am. Fish. Soc.*, 114, 413, 1985.
71. Sullivan, J.F., Atchison, G.F., Kolar, P.J., and McIntosh, A.W., Changes in the predator-prey behavior of fathead minnows *(Pimephales promelas)* and largemouth bass *(Micropterus salmoides)* caused by cadmium, *J. Fish. Res. Board Can.*, 35, 446, 1978.
72. Kavaliers, M., Day-night rhythms of shoaling behavior in goldfish: opioid and pineal involvement, *Physiol. Behav.*, 46, 167, 1989.
73. Morgan, M.J. and Colgan, P.M., The role of experience in foraging shoals of bluntnose minnows *(Pimephales notatus), Behav. Process.*, 16, 87, 1988.
74. Weis, J.S. and Khan, A.A., Effects of mercury on the feeding behavior of the mummichog, *Fundulus heteroclitus* from a polluted habitat, *Mar. Environ. Res.*, 30, 243, 1990.
75. Gunn, J.M. and Noakes, D.L.G., Latent effects of pulse exposure to aluminum and low pH on size, ionic composition and feeding efficiency of lake trout *(Salvelinus namaycush)* alevins, *Can. J. Fish. Aquat. Sci.*, 44, 1418, 1987.
75a. Deloney, A.J., Little, E.E., Woodward, D.F., Brumbaugh, W.G., Farag, A.M., and Rabeni, C.F., Sensitivity of early-life stage golden trout to low pH and elevated aluminum, *Environ. Toxicol. Chem.*, 12, 1223, 1993.
76. Beitinger, T.L., Behavioral reactions for the assessment of stress in fishes, *J. Great Lakes Res.*, 16, 495, 1990.
77. Nyman, H.-G., Sublethal effects of lead (Pb) on size selective predation by fish — applications on the ecosystem level, *Verh. Int. Verein. Limnol.*, 21, 1126, 1981.
78. Cleveland, L., Little, E.E., Hamilton, S.J., Buckler, D.R., and Hunn, J.B., Interactive toxicity of aluminum and acidity in early life stages of brook trout, *Trans. Am. Fish. Soc.*, 115, 610, 1986.
79. Morley, J.E., Neuropeptide regulation of appetite and weight, *Endocrine Rev.*, 8, 256, 1987.
80. Farmanfarmaian, A., Pugliese, K.A., and Sun, L.-Z., Mercury inhibits the transport of D-glucose by the intestinal brush border membrane vesicles of fish, in Fifth Int. Symp. Responses of Marine Organisms to Pollutants, Plymouth, England, U.K., 1989, 247.

81. Crespo, S., Nonnotte, G., Colin, D.A., Leray, C, Nonnotte, L., and Aubree, A., Morphological and functional alterations induced in trout intestine by dietary cadmium and lead, *J. Fish Biol.*, 28, 69, 1986.
82. Fänge, R. and Grove, D., Digestion, in *Fish Physiology, Vol. 8: Bioenergetics and Growth*, Hoar, W.S., Randall, D.J., and Brett, J.R., Eds., Academic Press, New York, 1979, 162.
83. Giles, N., Predation risk and reduced foraging activity in fish: experiments with parasitized and non-parasitized three spine sticklebacks, *Gasterosteus aculeatus* L., *J. Fish Biol.*, 31, 37, 1987.
84. Navarro, J.C. and Sargent, J.R., Behavioral differences in starving herring *(Clupea harengus* L.) larvae correlate with body levels of essential fatty acids, *J. Fish Biol.*, 41, 509, 1992.
85. Pandian, T.J., Fish, in *Animal Energetics*, Vol. 2, Pandian, T.J., Ed., Academic Press, New York, 1986, chap. 7.
86. Dabrowski, K., Takashima, F., and Law, Y.K., Bioenergetic model of planktivorous fish feeding, growth and metabolism: theoretical optimum swimming speed of fish larvae, *J. Fish Biol.*, 32, 443, 1988.
87. Krueger, H.M., Saddler, J.B., Chapman, G.A., Tinsley, I.J., and Lowry, R.R., Bioenergetics, exercise and fatty acids of fish, *Am. Zool.*, 8, 119, 1968.
88. Calow, P., Adaptive aspects of energy allocation, in *Fish Energetics: New Perspectives*, Tytler, P. and Calow, P., Eds., Croom and Helm, London, 1985, 13.
89. Mrowka, W. and Schierwater, B., Energy expenditure for mouthbrooding in a cichlid fish, *Behav. Ecol. Sociobiol.*, 22, 161, 1988.
90. Hirshfield, M.F., An experimental analysis of reproductive effort and cost in the Japanese medaka, *Oryzias latipes, Ecology*, 61, 282, 1980.
91. Wade, G.N., Sex steroids and energy balance: sites and mechanisms of action, in *Reproduction: A Behavioral and Neuroendocrine Perspective*, Komisarak, B.R., Siegel, H.I., Cheng, M.-F., and Feder, H.H., Eds., The New York Academy of Science, New York, 1986, 389.
92. Scott, A.P., Bye, V.J., and Baynes, S.M., Seasonal variations in sex steroids of female rainbow trout *(Salmo gairdneri* Richardson), *J. Fish Biol.*, 17, 587, 1980.
93. Lorenz, J.J. and Taylor, D.H., Effects of a chemical stressor on the parental behavior of convict cichlids with offspring in early life stages of development, *Trans. Am. Fish. Soc.*, 121, 315, 1992.
94. Rodrigues, A.L., Bellinaso, M.L., and Dick, T., Effect of some metal ions on blood and liver delta-aminolevulinate dehydratase of *Pimelodus maculata* (Pisces, Pimelodidae), *Comp. Biochem. Physiol.*, 94B, 65, 1989.
95. Tabache, L.M., Martinez, C.M., and Sanchex-Hidalgo, E., Comparative study of the toxic effect on gill and haemoglobin of tilapia fish, *J. Appl. Toxicol.*, 10, 193, 1990.
96. GESAMP, Cadmium, lead and tin in the marine environment, UNEP Regional Seas Reports and Studies No. 56, United Nations Environmental Programme, United Nations, 1985.
97. Goolish, E.M., Aerobic and anaerobic scaling in fish, *Biol. Rev.*, 66, 33, 1991b.
98. Pratap, H.B. and Bonga, S.E.W., Effects of water-borne cadmium on plasma cortisol and glucose in the cichlid fish *Oreochromis mossambicus, Comp. Biochem. Physiol.*, 95, 313, 1990.

99. Hernandez-Pascual, M.D. and Tort, L., Metabolic effects after short-term sublethal cadmium exposure to dogfish, *Scyliorhinus canicula, Comp. Biochem. Physiol.*, 94C, 261, 1989.
100. Wester, P.W., Canton, J.H., van Iersel, A.A.J., Kranjnc, E.I., and Vaessen, H.A.M.G., The toxicity of bistri-n-buthyltin-oxide, TBTO, and di-n-butyltin dichloride, DBTC, in the small fish species *Oryzias latipes* medaka and *Poecilia reticulata* guppy, *Aquat. Toxicol.*, 16, 53, 1990.
101. Lowe-Jinde, L. and Niimi, A.J., Short-term and long-term effects of cadmium on glycogen reserves and liver size in rainbow trout *(Salmo gairdneri* Richardson), *Arch. Environ. Contam. Toxicol.*, 13, 759, 1984.
102. Gill, T.S., Tewari, H., and Pande, J., Use of the fish enzyme system in monitoring water quality effects of mercury on tissue enzymes, *Comp. Biochem. Physiol.*, 97C, 287, 1990.
103. Kramer, V.J., Newman, M.C., and Ultsch, G.R., Changes in concentrations of glycolysis and Krebs cycle metabolites in mosquitofish, *Gambusia holbrooki*, induced by mercuric chloride, *Environ. Biol. Fish.*, 34, 315, 1992.
104. Katti, S.R. and Sathyanesan, A.G., Changes in tissue lipid and cholesterol content in the catfish *Clarias batrachus* (L.) exposed to cadmium chloride, *Bull. Environ. Contam. Toxicol.*, 32, 486, 1984.
105. Tulasi, S.J., Reddy, P.V.M., and Rao, J.V.R., Accumulation of lead and effects on total lipids and lipid derivatives in the freshwater fish *Anabasz testudineus* (Bloch), *Ecotoxicol. Environ. Safety*, 23, 33, 1992.
106. Holm, G., Norrgren, L., and Lindén, O., Reproductive and histopathological effects of long-term experimental exposure to bistributyltinoxide, TBTO, on the three-spined stickleback, *Gasterosteus aculeatus* Linnaeus, *J. Fish Biol.*, 38, 373, 1991.
107. Evans, D.H., The fish gill: site of action and model for toxic effects of environmental pollutants, *Environ. Health Perspect.*, 71, 47, 1987.
108. Thompson, K.W., Hendricks, A.C., Nunn, G.L., and Cairns, J., Jr., Ventilatory response of bluegill to heavy metals (Zn^{++} and Cu^{++}), *Water Res. Bull.*, No. 5, 719, 1983.
109. Diamond, J.M., Parson, M.J., and Gruber, D., Rapid detection of sublethal toxicity using fish ventilatory behavior, *Environ. Toxicol. Chem.*, 9, 3, 1990.
110. Sandheinrich, M.B. and Atchison, G.J., Sublethal toxicant effects on fish foraging behavior: empirical vs. mechanistic approaches, *Environ. Toxicol. Chem.*, 9, 107, 1990.
111. Whitehead, C. and Brown, J.A., Endocrine responses of brown trout, *Salmo trutta* L., to acid, aluminum and lime dosing in a welsch hill stream, U.K., *J. Fish Biol.*, 35, 59, 1989.
112. Larsson, Å., Haux, C., and Sjöbeck, M.-L., Fish physiology and metal pollution: results and experiences from laboratory and field studies, *Ecotoxicol. Environ. Safety*, 9, 250, 1985.
113. Spieler, R.E., Noeske, T.A., and Seegert, G.L., Diel variations in sensitivity of fishes to potentially lethal stimuli, *Progr. Fish Cult.*, 39, 144, 1977.
114. Weis, J.S. and Weis, P., Effects of heavy metals on development of the killifish, *Fundulus heteroclitus, J. Fish Biol.*, 11, 49, 1977.

115. Regan, C.M., Neural cell adhesion molecules, neuronal development and lead toxicity, *Neurotoxicology*, 14, 69, 1993.
116. Choi, B.H., Effects of methylmercury in the developing brain, *Progr. Neurobiol.*, 32, 447, 1989.
117. Akberali, H.B. and Earnshaw, M.J., Copper-stimulated respiration in the unfertilized egg of the eurasian perch, *Perca fluviatilis* (L)., *Comp. Biochem. Physiol.*, 78C, 349, 1984.
118. Sparks, R.E., Waller, W.T., and Cairns, J., Jr., Effect of shelters on the resistance of dominant and submissive bluegills *(Lepomis macrochirus)* to a lethal concentration of zinc, *J. Fish. Res. Board Can.*, 29, 1356, 1972.
118a. Matt, K.S., Neuroendocrine mechanisms of environmental integration, *Amer. Zool.*, 33, 266, 1993.
119. Bakker, T.C.M., Aggressiveness in sticklebacks *(Gasterosteus aculeatus* L.): a behaviour-genetic study, *Behaviour*, 98, 1, 1986.
120. Misra, S., Zafarullah, M., Price-Haughey, J., and Gedamu, L., Analysis of stress-induced gene expression in fish cell lines exposed to heavy metals and heat shock, *Biochim. Biophys. Acta*, 1007, 325, 1989.
120a. Heagler, M.G., Newman, M.C., Medvey, M., and Dixon, P.M., Allozyme genotype in mosquito fish, *Gambusia holbrooki*, during mercury exposure: temporal stability, concentration effects and field verification, *Environ. Toxicol. Chem.*, 12, 385, 1993.
121. McNicol, R.E. and Scherer, E., Behavioral responses of lake whitefish, *Coregonus clupeaformis*, to cadmium during preference-avoidance testing, *Environ. Toxicol. Chem.*, 10, 225, 1991.
122. Rand, G.M. and Petrocelli, S.R., *Fundamentals of Aquatic Toxicology*, Hemisphere Publishing Corp., Washington, D.C., 1985.
123. Hartwell, S.I., Cherry, D.S., and Cairns, J., Jr., Field validation of avoidance of elevated metals by fathead minnows *(Pimephales promelas)* following *in situ* acclimation, *Environ. Toxicol. Chem.*, 6, 189, 1987b.
124. Cyr, D.G. and Eales, J.G., *In vitro* effects of thyroid hormones on gonadotropin-induced estradiol-17β secretion by ovarian follicles of rainbow trout, *Salmo gairdneri*, *Gen. Comp. Endocrinol.*, 69, 80, 1988.
125. Majumdar, A.C., Shende, D.D., and Pandey, O.P., Effect of thyroxine on the incorporation of super(3)H-lysine in the brain, ovary, liver, and muscle proteins of Lata fish *(Ophiocephalus punctatus)*, *J. Nucl. Agric. Biol.*, 12, 46, 1983.
126. Birks, E.K., Ewing, R.D., and Hemmingsen, A.R., Migration tendency in juvenile steelhead trout, *Salmo gairdneri* Richardson, injected with thyroxine and thiourea, *J. Fish Biol.*, 26, 291, 1985.
127. Iwata, M., Yamanome, T., Tagawa, M., Ida, H., and Hirano, T., Effects of thyroid hormones on phototaxis of chum and coho salmon juveniles, in *Salmonid Smoltification III. Proceedings of a Workshop Sponsored by the Directorate for Nature Management*, Hansen, L.P., Clarke, W.C., Saunders, R.L., and Thorpe, J.E., Eds., Norwegian Fisheries Research Council, Norwegian Smolt Producers Association and Statkraft, Trondheim, Norway, 1989, 329–338.
128. Morin, P.-P., Dodson, J.J., and Dore, F.Y., Thyroid activity concomitant with olfactory learning and heart rate changes in Atlantic salmon, *Salmo salar*, during smoltification, *Can. J. Fish. Aquat. Sci.*, 46, 131, 1989.

129. Munro, A.D. and Pitcher, T.J., Steroid hormones and agonistic behavior in a cichlid teleost, *Aequidens pulcher, Horm. Behav.*, 19, 353, 1985.
130. Audet, C., Fitzgerald, G.J., and Guderly, H., Prolactin and cortisol control of salinity preferences in *Gasterosteus aculeatus* and *Apeletes quadracus, Behaviour,* 93, 36, 1985.
131. de Ruiter, A.J.H., Wendelaar Bonga, S.E., Slijkhuis, H., and Baggerman, B., The effect of prolactin on fanning behavior in the male three-spined stickleback, *Gasterosteus aculeatus* L., *Gen. Comp. Endocrinol.*, 64, 273, 1986.
132. Hirano, T., The spectrum of prolactin action in teleosts, in *Comparative Endocrinology: Developments and Directions*, Ralph, C.L., Ed., Alan R. Liss, New York, 1986, 53.
133. Sparwasser, K., The influence of metoclopramide and melatonin on activity and schooling behavior in *Chromis viridis* (Cuvier, 1830; Pomacentridae, Teleostei), *Mar. Ecol.*, 8, 297, 1987.
134. Falcon, J. and Collin, J.-P., Photoreceptors in the pineal of lower vertebrates: functional aspects, *Experientia*, 45, 909, 1989.
134a. Bittman, E.L., The sites and consequences of melatonin binding in mammals, *Amer. Zool.,* 33, 200, 1993.
135. de Ruiter, A.J.H. and Mein, C.G., Testosterone-dependent transformation of nephronic tubule cells into serous mucous gland cells in stickleback kidneys *in vivo* and *in vitro, Gen. Comp. Endocrinol.,* 47, 70, 1982.
136. Freedman, E.G., Olyarchuk, J., Marchaterre, M.A., and Bass, A.H., A temporal analysis of testosterone-induced changes in electric organs and organ discharges of mormyrid fishes, *J. Neurobiol.*, 20, 619, 1989.
137. Stacey, N.E., Hormones and reproductive behaviour in teleosts, in *Control Processes in Fish Physiology*, Rankin, J.C., Pitcher T.J., and Duggan R.T., Eds., Wiley Interscience, New York, 1983, chap. 7.
138. Hoar, W.S., Reproductive behavior of fish, *Gen. Comp. Endocrinol.*, Suppl. 1, 206, 1962.
139. Demski, L.S. and Hornby, P.J., Hormonal control of fish reproductive behavior: brain-gonadal steroid interaction, *Can. J. Fish. Aquat. Sci.*, 39, 36, 1982.
140. Sorenson, P.W. and Winn, H.E., The induction of maturation and ovulation in American eels, *Anguilla rostrata* (LeSuer), and the relevance of chemical and visual cues to male spawning behavior, *J. Fish Biol.*, 25, 261, 1984.

Chapter 11

Use of Physiologically Based Toxicokinetic Models in a Mechanistic Approach to Aquatic Toxicology

James M. McKim and John W. Nichols

I. INTRODUCTION

Molecular toxicologists, biochemists, and cancer experts all have the capabilities at their fingertips to investigate the responses of organ and cellular systems to chemical stressors, but they are frequently unable to link these responses to whole animal exposures. This problem is particularly evident in the field of aquatic toxicology, because fish and other aquatic organisms are often exposed simultaneously to chemicals in water and to chemicals complexed with food, particulates, and dissolved organic material. One promising approach to dealing with this problem involves the use of physiologically-based toxicokinetic (PB-TK) models. In this chapter, we describe recent progress toward the development and validation of PB-TK models for fish and show how these models can be used to understand the dynamic relationships between applied chemical dose and target organ dose.[1-8]

In aquatic toxicology, mechanistic studies involving histopathology, organ function, macromolecular binding, depletion of specific cellular constituents, carcinogenesis, cell death, etc., are usually related to either a water exposure concentration or to an applied single or multiple bolus dose. Standard toxicity tests with waterborne chemicals provide no information on the chemical dose received internally, while tests with single or multiple bolus doses provide an applied dose, but no insight on the dose received at the site of damage. This situation is further complicated by not knowing whether changes seen in the target organ are caused by the parent chemical or a reactive metabolite. PB-TK models are designed to address these concerns by providing estimates of the time course of both parent chemical and metabolites at the site of toxic action. Because these models are based on an animal's physiological and biochemical attributes, they can be used to extrapolate dosimetry estimates across species, regardless of exposure route.

Through the use of PB-TK models, dose-response studies at all levels of biological organization can be more directly related to target organ dose under a variety of exposure scenarios. In turn, mechanistic toxicology can provide a deeper understanding of the cellular and subcellular processes leading to toxicological responses. An understanding of these processes is essential to determine accurately the quantitative inputs needed for development of a second type of physiologically based model, termed a physiologically-based tissue response (PB-TR) model. Once the general mechanism of action for a chemical is understood, PB-TK models that describe tissue dosimetry can be linked with PB-TR models that describe the mechanistic processes responsible for a toxicological response. Combining these linked models with a knowledge of applied dose or exposure concentration will then allow accurate predictions of target organ dose and expected tissue response. Finally, research on the basic biochemical and molecular mechanisms of specific classes of chemicals can form, through the future use of these two types of models, the basis for health and environmental risk assessments vital to the safe use of chemicals and to the preservation of our environment.

The principal limitations of a PB-TK modeling approach relate to acquisition of the physiological, biochemical, and physical-chemical information essential for model development and parameterization. Generally, PB-TK models require the following inputs for accurate predictions of tissue dosimetry: (1) a complete set of physiological, biochemical, and morphological information for the species of interest, (2) an understanding of the physiological mechanisms that control chemical flux across the gills, skin, or gastrointestinal (GI) tract, (3) a quantitative description of the equilibrium distribution (partitioning) of toxicant (parent and metabolite) between the blood and tissues, and (4) knowledge of the site of toxic action.

Detailed physiological information is lacking for most aquatic vertebrates; however, many of the required inputs have been measured in certain large species (e.g., rainbow trout and channel catfish) and can be extrapolated to other species using allometric scaling techniques. In addition, it can be shown that, in many cases, model behavior is largely controlled by a restricted subset of parameter inputs. In such cases, it may only be necessary to obtain accurate estimates for this limited parameter set, while relying on approximate values for other model inputs. The basic mechanisms (physical and biological) controlling chemical uptake and elimination by fish are presently being investigated and in many cases are well enough understood to allow parameterization of prototype models for neutral organic compounds. In contrast, efforts to understand the uptake and elimination of ionized compounds (organic and inorganic) are just beginning. Similarly, current methods for in vitro determination of equilibrium chemical partition coefficients are limited to use with volatile compounds, although research is now under way to develop methods capable of providing these critical inputs for other classes of chemicals. Finally, the site of action of many toxic compounds remains to be determined. In the interim, however, surrogates for target organ concentration, such as chemical concentration in arterial blood, total chemical body burden, or area under the arterial blood curve (AUC), can be used effectively as modeling endpoints.

The intent of this chapter is to review the development of PB-TK models for fish, with special emphasis on (1) dose concepts in aquatic toxicology, (2) general concepts of physiologically based dosimetry and tissue response models, (3) conceptualization and parameterization of fish PB-TK models, (4) route of exposure considerations, (5) examples and present status of fish PB-TK models, and (6) current needs and future applications in the development and evaluation of fish PB-TK and PB-TR models. The application of PB-TK models to aquatic invertebrates has not yet been investigated and will not be discussed; however, it is our opinion that a modified modeling approach (eg., adapted for use with an open circulatory system) could be used with invertebrates, depending upon the availability of required physiological, biochemical, and morphological information.

II. DOSE CONCEPTS IN AQUATIC TOXICOLOGY

A. Applied and Target Organ Dose

One of the most important principles in toxicology is that the magnitude of toxic response depends on both the amount of chemical (or metabolite) at the site of toxic action and the length of time that it remains there. The applied dose in a toxicity test controls only the maximal concentration that can be attained. The total amount of the applied dose that actually reaches the target organ is the net result of many internal processes, including (1) absorption, (2) tissue distribution, (3) biotransformation to active and/or inactive metabolites, and (4) excretion. The impact that these processes have on the amount of parent chemical reaching the site of action is summarized in Figure 1[9] for inhalation, oral, and dermal exposure routes. Toxicokinetics deals with the movement of chemical mass within an organism and therefore encompasses all of the internal processes listed above.

If we are to link mechanistic toxicology to effects seen in the whole animal, the difference between applied dose, absorbed dose, effective dose, and dose actually

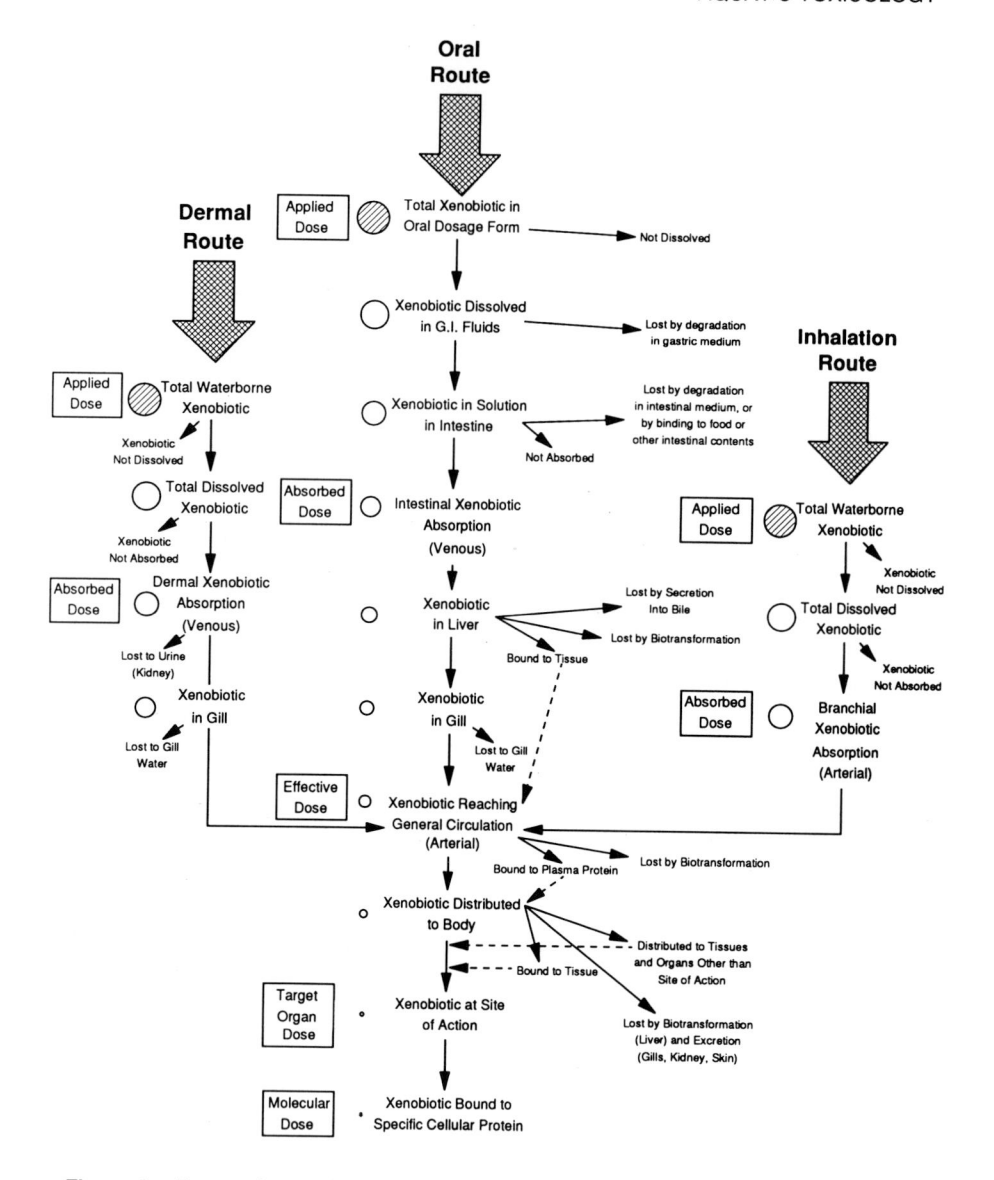

Figure 1. Factors that modify the amount of xenobiotic reaching the site of action following inhalation, oral, and dermal exposure (◉ — applied dose, ○ — progressive reduction in applied dose; modified from Levine)[9].

reaching the target organ must be understood (Figure 1). In addition, the route of exposure can strongly impact the effective dose and the disposition of a chemical within an animal. For example, oral exposures provide direct venous blood inputs of chemical that are subject to first-pass elimination in the liver and gills, while chemicals taken up across the dermal surface are subject to first-pass elimination in the kidney and gills. Each of these elimination pathways reduces the amount of parent chemical and/or metabolite reaching the systemic arterial circulation. In contrast, xenobiotics absorbed across the gills enter the systemic circulation directly, with no mitigating first-pass effects in the liver, kidney, or gills.

B. Aquatic Dose

Those involved with aquatic toxicity testing have generally not dealt with any kind of internal chemical dose, but instead with the concentration of chemical in the exposure water (LC_{50}, MATC, NOEC).[10] This approach allows a relative comparison of effects between chemicals and species, but does not provide an understanding of the absorbed dose received by the animal. The importance of internal dose as a determinant of toxicity in fish was suggested by McCarty,[11-13] who multiplied LC_{50} values by bioconcentration factors to estimate lethal chemical body burdens. However, this research did not address the mechanisms controlling internal dose/response relationships. More importantly, these descriptive studies provided no information on the dose received at the site of action (target organ dose and molecular dose).

Obtaining environmentally relevant measures of chemical absorption is of critical importance to the development of predictive dosimetry models. Work on the quantitation of chemical absorption across the gills of fishes has provided the tools necessary to monitor the uptake of chemicals from inspired water.[14-16] These studies have yielded direct measurements of total absorbed dose (e.g., milligram per kilogram per hour) received across the gills and have made available techniques to explore the mechanisms controlling both gill and dermal flux of waterborne xenobiotics. However, if fish PB-TK models are to achieve their maximum potential, they must also incorporate mechanistic descriptions of chemical flux across the GI tract, so that all relevant uptake and elimination routes are included.

III. PHYSIOLOGICALLY BASED DOSIMETRY AND TISSUE RESPONSE MODELS

A. Data-Based vs. Physiologically Based Models

Toxicokinetic models for fish can be classified generally as data-based[17] or physiologically based.[18-21] Both types of models employ one or more compartments that act as repositories for the chemical of interest. A set of kinetic rate expressions is used to relate chemical flux into and out of these compartments (or between a compartment and the environment) to chemical concentrations within the compartments. In data-based modeling, the total number of compartments and the values of kinetic rate constants are determined by fitting model simulations to experimental observations. The identity of these compartments may or may not relate to actual tissues and organs. Data-based models have been used extensively to describe the uptake and elimination of chemicals by fish and have proven useful as tools for understanding the processes that control chemical accumulation.[22-25] In practice, however, the limitations of curve fitting restrict such models to two or three compartments. More importantly, the use of data-based models for extrapolation is limited by the need to fit a new set of parameters empirically for each new compound, species, and route of exposure.

PB-TK models are based on the physiology, biochemistry, and anatomy of the animal under investigation, and as a result can be used to predict the quantitative behavior of the experimental time course without being based on it.[18] In a typical PB-TK model, the animal is divided up into major tissue compartments (Figure 2)[26] and includes the following information on each compartment: (1) tissue (organ) volume, (2) arterial blood flow as percentage of cardiac output, (3) equilibrium chemical partitioning between the tissue compartment and blood, and (4) biotransformation rate and capacity parameters. In addition, a description of chemical flux across exchange surfaces, such as the lungs of a mammal or the gills of a fish, must be included for inhalation exposures, while similar information is required for dermal and GI tract exposures. These model inputs are either available from the literature or are acquired through experimentation with the species to be modeled. Once all of the model inputs are available and the model is parameterized, simulations are generated by solving mass-balance differential equations for each compartment.[27,28] Kinetic experiments are then designed to evaluate the model. Because physiologically based models have biological and physical-chemical integrity, they can be used to extend predictions well beyond the range of experimental conditions. The superior extrapolative capabilities of these models derive from (1) the ability to change physiological parameter values,[20] (2) the capability of altering exposure route by changing model input equations, and (3) the capacity to model new chemicals by changing equilibrium chemical partition coefficients and metabolic rate and capacity constants. Finally, one of the great strengths of this modeling approach is that it provides a conceptual framework from which hypotheses can be described based on biological processes, predictions made on the basis of these descriptions, and hypotheses revised based on comparisons to experimental data.[18] This provides a truly iterative process by which tissue dosimetry can be described along with the primary mechanisms that control it. For those readers wishing a more historical perspective on PB-TK models, we suggest a review by Bischoff.[29]

B. Tissue Response Models

In order to effectively use PB-TK models to extrapolate target tissue dosimetry across species, two general assumptions must be made. The first is that one species is a true surrogate of another. This means that, qualitatively, the target tissue in one species is the same in all species being studied, and that the effects

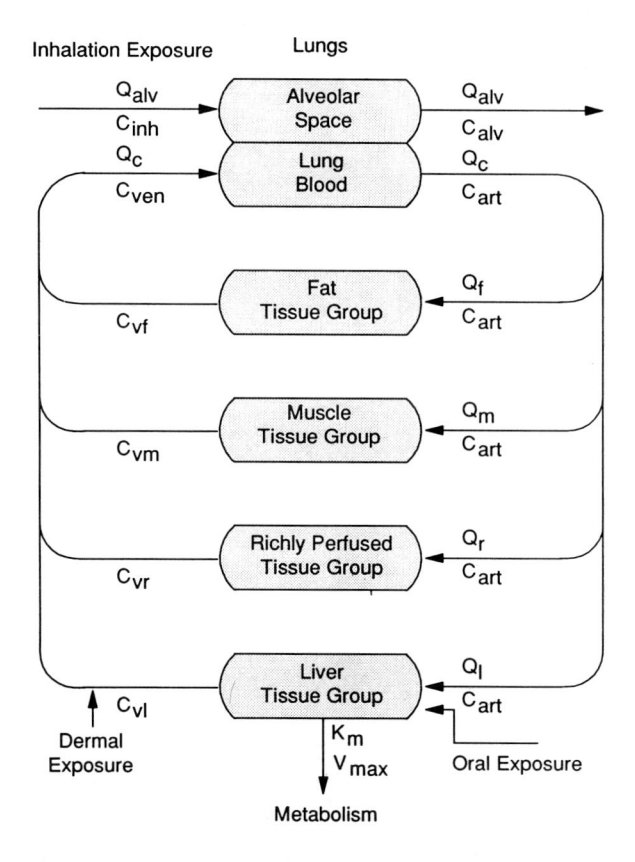

Figure 2. Schematic representation of a mammalian PB-TK model for gases and vapors. Symbols and abbreviations are as follows: Q_{alv} — alveolar ventilation; Q_c — cardiac output; C_{art} — arterial blood concentration; C_{ven} — mixed venous blood concentration; Q_i — tissue blood flow; C_{vi} — concentration in venous blood exiting the tissue (subscripts (i): r — richly perfused tissue; m — poorly perfused tissue; f — fat; l — liver); and V_{max} and K_m — rate and capacity parameters for Michaelis-Menten metabolism. Routes of exposure (inhalation, oral, and dermal) are shown in relation to their input to the bloodstream (modified from Andersen[26]).

in one species are the same as in others. The second assumption is that of dose equivalence. Under this assumption, the target organ (tissue) dose responsible for causing a specific toxicological response in one species is quantitatively the same as in other species being tested. These assumptions are suspect in some situations, but in general seem to hold up over a wide range of species and chemicals.[30]

PB-TK models provide estimates of the target organ dose (parent or metabolite) that corresponds to an applied dose or exposure condition. This dose interacts with cellular constituents of the target organ, either reacting with one or a few classes of critical macromolecules or interacting in a less specific manner with a variety of cellular receptors. In either case, these events have the potential to be toxic to the cell, and the amount of chemical involved in these interactions is termed the molecular dose.[26,31] A cascade of molecular events then follows these interactions and becomes the toxicodynamic portion of the process, which involves the direct molecular mechanism of toxic action that produces the observed toxic response (Figure 3).

To complete the chain of events leading from target organ dose to a toxic response requires a PB-TR model that sequentially describes the biology of the mechanism of toxic action (Figure 3). Research efforts that link tissue dosimetry models with biological effects models are just beginning and are probably hindered most by the lack of knowledge of biochemical toxicity mechanisms.[32,33] Once these mechanisms are well understood and tissue response models are available to describe them, it will be possible to use linked PB-TK and PB-TR models to mechanistically relate whole-animal exposures to target organ responses and resulting effects upon the organism.

IV. DEVELOPMENT OF PHYSIOLOGICALLY BASED TOXICOKINETIC MODELS FOR FISH

A. General Considerations

Like all living organisms, fish are composed of an array of complex chemical and biological systems. Therefore, it is tempting to conclude that any model whose structure is based upon this complexity must be so cumbersome as to be of little or no utility. Fortunately, the uptake of chemical contaminants by fish appears to be driven by relatively few characteristics of their anatomy and physiology. This, in turn, has made it possible to employ a physiologically-based approach to modeling these processes. In this and the following two sections, we will identify those considerations that have been shown to be important in the development of PB-TK models for fish. The mathematical treatment of these topics will be kept as simple as possible. Instead, emphasis will be placed on techniques by which to select and assess various simplifying assumptions. Readers who are interested in a more detailed description of the mathematical aspects of physiologically based kinetic modeling are referred to several good reviews of the subject.[29,34-36]

B. Compartments and Blood Flow Relationships

A physiologically based model is made up of several compartments, each of which corresponds to tissues and organs with similar kinetic characteristics (Figure 4 and Table 1). The transfer of material within the model is mediated strictly by blood flow, and the relationships between compartments reflect as closely as possible their anatomical relationships within the organism. The total number of

```
Chemical          Target Tissue Dose           Molecular              Toxicological
Exposure          of Active Chemical           Dose                   Response

Physiologically                   Chemical or              Physiologically
Based                             Biochemical              Based
Tissue Dosimetry                  Interactions             Tissue Response
Models                                                     Models
```

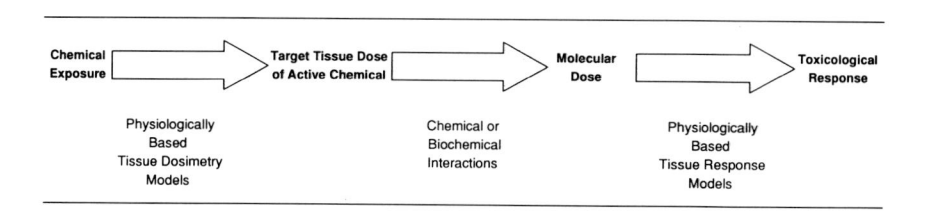

Figure 3. Linkage of physiologically based tissue dosimetry models (PB-TK) with physiologically based tissue response models (PB-TR) for extrapolation of dose/response relationships (from Andersen[26]).

compartments within a model is determined by the desire to reproduce the observed kinetic behavior of the compound and by the need to model chemical concentrations in specific tissues of interest. In general, muscle plays a dominant role in the distribution of water soluble (hydrophilic) compounds because of its large size relative to that of other tissues. In contrast, the distribution of fat soluble (hydrophobic) compounds tends to be dominated by uptake into adipose tissue. Additional compartments that are commonly identified in both fish and mammalian models include the liver, kidney, and gastrointestinal tract.

The fish model schematic presented in Figure 4 differs from that of published models for mammals (see Figure 2) in two important respects: (1) an emphasis on portal blood flows, both to the liver and to the kidney, and (2) incorporation of a gill description to account for chemical flux between the fish and ventilated water. In fish, the liver receives venous blood via the hepatic portal vein, while the kidney receives venous blood from the caudal vein.[37,38] The toxicokinetic significance of these portal blood supplies has not yet been investigated. However, studies with rainbow trout suggest that venous blood constitutes greater than 80% of the total blood supply to both organs.[39]

When a tissue receives only arterial blood, chemical delivery rate to the tissue may be calculated as the product of chemical concentration in arterial blood and tissue blood flow rate (usually expressed as a fraction of cardiac output). A special case exists, however, when an organ receives both arterial and venous (portal) blood. Under these circumstances, a fraction of the compound originally contained in portal blood has been removed during prior perfusion of another (upstream) tissue. Chemical concentration in mixed blood is then calculated by summing the flow-weighted contributions of both arterial and portal blood (see Section VI.A).

C. Mass-Balance Descriptions of Chemical Uptake and Distribution

At the core of a physiologically based dosimetry model is a set of simultaneous mass-balance differential equations that describe the rate of change of the amount of chemical within each of the tissue compartments. Considering at this time just

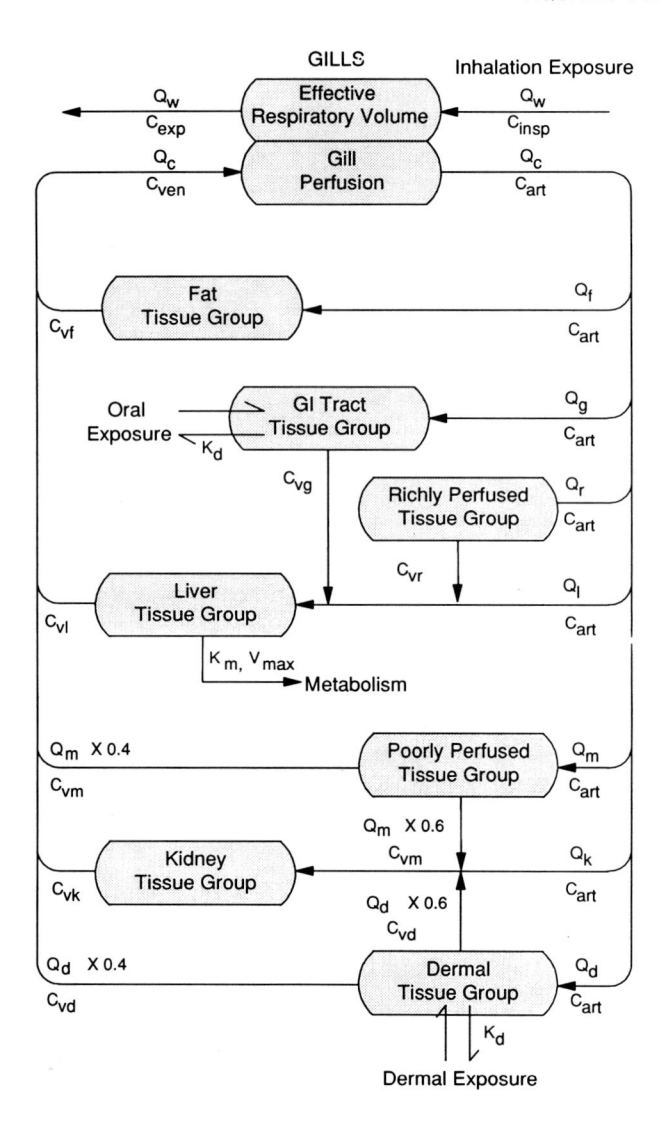

Figure 4. Schematic representation of a PB-TK model for the uptake and disposition of waterborne organic chemicals by fish (from Nichols et al.[1]). Symbols and abbreviations used in the model schematic are presented in Table 1.

the transport of chemical by blood, the general form of the equation for each tissue (i) may be written as:*

$$V_i \, dC_i/dt = Q_i \, (C_{art} - C_{vi}) \qquad (1)$$

---
* Symbols and abbreviations used throughout the text are defined in Table 1.

Table 1. Abbreviations and Symbols Used to Describe a Physiologically Based Toxicokinetic Model for Fish[a]

Q_v	Ventilation volume (L/hr)
VO_2	Oxygen consumption rate (mg O_2/hr)
Q_w	Effective respiratory volume (L/hr)
C_{insp}	Concentration in inspired water (µg/L)
C_{exp}	Concentration in expired water (µg/L)
EE	Chemical extraction efficiency (% reduction of C_{insp})
R_{bw}	Blood:water partition coefficient (µg/L in blood/µg/L in water)
Q_c	Cardiac output (L/hr)
C_{art}	Concentration in arterial blood (µg/L)
C_{ven}	Concentration in mixed venous blood (µg/L)
k_{fc}	First order elimination rate constant (/hr)
K_m	Michaelis-Menten constant for enzymatic reaction (µg/L blood)
V_{max}	Maximum enzymatic reaction rate (µg/hr)

Subscripts (i) for compartments

l	Liver compartment
f	Fat compartment
m	Poorly perfused compartment
r	Richly perfused compartment
k	Kidney compartment

Q_i	Blood flow to the compartment (L/hr)
V_i	Compartment volume (L)
C_i	Concentration in the compartment (µg/L)
A_i	Amount of chemical in the compartment (µg)
$A_{met\,i}$	Amount of chemical metabolized within the compartment (µg)
C_{vi}	Concentration in venous blood exiting the compartment (µg/L)
R_i	Tissue:blood partition coefficient (µg/L in tissue/µg/L in blood)

For compartments receiving mixed, arterial, and portal blood

Q_{mi}	Summed blood flow to the compartment (L/hr)
C_{mi}	Concentration in mixed blood (µg/L)
Q_{pi}	Portal blood flow to the compartment (L/hr)
C_{pi}	Concentration in portal blood (µg/L)

Source: Nichols, J.W., et al., *Toxicol. Appl. Pharmacol.*, 106, 433, 1990.

[a] This model was developed to describe inhalation exposure studies with rainbow trout. Chemical distribution to tissues was assumed to be flow-limited.

or equivalently,

$$dA_i/dt = Q_i\,(C_{art} - C_{vi}) \tag{2}$$

The specific form that this equation takes depends, in turn, upon the factors that limit chemical transfer from blood to tissues. For this discussion, two possibilities will be examined: flow and diffusion limitations. It is important to realize, however, that other factors may come into play, including active transport and high-affinity binding to macromolecular constituents in tissues or in plasma.[29,34–36]

Diffusion-limited chemical distribution occurs when chemical diffusion across the capillary wall limits mass transfer to the rest of the tissue compartment. Under these conditions, the tissue compartment must be divided into two subcompartments, representing the blood and extravascular spaces, between which is interposed a diffusion barrier.* Diffusive flux may then be modeled using the Fick relationship:

$$\text{Flux} = \text{permeability (P)} \cdot \text{surface area (SA)} \cdot \text{concentration gradient} \quad (3)$$

In practice, chemical permeability and the surface area for diffusion are difficult to estimate. The permeability-area product is therefore frequently treated as a single-transport parameter k_s, the value of which must be determined from exposure data. Each subcompartment is considered to be a well-mixed phase; therefore, venous blood exiting the tissue has the same chemical concentration as that of the vascular subcompartment. The concentration gradient is usually referenced to the extravascular tissue space, in which case the magnitude and direction of the gradient are determined by the difference between chemical concentration in venous blood leaving the tissue (C_{vi}) and that in the extravascular space (C_{ei}) divided by a tissue-to-blood equilibrium partition coefficient (R_i). With the foregoing assumptions, a mass-balance equation for the vascular subcompartment (bi) may be written as:

$$dA_{bi}/dt = Q_i(C_{art} - C_{vi}) - k_s (C_{vi} - C_{ei}/R_i) \quad (4)$$

and for the extravascular subcompartment (ei),

$$dA_{ei}/dt = k_s (C_{vi} - C_{ei}/R_i) \quad (5)$$

Chemical concentration in the whole tissue (i) is then calculated from the volume-weighted contributions of the blood and extravascular spaces:

$$C_i = (V_{bi}C_{vi} + V_{ei}C_{ei})/V_i \quad (6)$$

Chemical distribution to tissues is said to be flow-limited when the rate-limiting step for mass transfer is the blood perfusion rate of the tissue. This assumption implies that chemical transfer across the capillary wall and cell membranes is rapid, with the result that equilibrium occurs between the chemical in blood and in the tissues that the blood is perfusing. Under these conditions, the chemical concentration in venous blood exiting the tissue equals that of the tissue itself divided by a tissue-to-blood equilibrium partition coefficient (R_i). The two-subcompartment diffusion model then reduces to one compartment, and a mass-balance on the compartment can be written as:

$$dA_i/dt = Q_i (C_{art} - C_i/R_i) \quad (7)$$

---

* Alternatively, diffusion limitation can occur at the cell membrane. In this case, the mathematical treatment is identical, but different volumes must be assigned to represent the intracellular and extracellular spaces.

Compared to the diffusion-limited model, the flow-limited model is both simpler mathematically and requires a reduced set of parameter estimates. It is, therefore, frequently used as a default assumption when information on the distribution of a new chemical is lacking. At some point, however, it is important to test the validity of this assumption. This is most easily accomplished by infusing a bolus of the material intravenously and then measuring chemical residues in tissues during the initial, distributive phase of the blood concentration–time curve. Residue data are then compared with levels predicted by a flow-limited model. Concordance between observed and predicted values suggests that the flow-limited assumption is valid. A diffusion-limited model is indicated when tissue residues are consistently less than those predicted.

D. Metabolism, Urinary, and Biliary Elimination

Fish metabolically transform most of the potentially toxic chemicals to which they are exposed. This metabolism, although largely protective, has the potential to increase toxicity by creating chemically reactive metabolites. In general, these reactions may be classified as "phase I"- or "phase II"-type reactions. Phase I reactions usually increase the polarity of a molecule through oxidative, reductive, and hydrolytic processes. The most important of these reactions are carried out by cytochrome P450-dependent mixed function oxidase (MFO) enzymes located in the endoplasmic reticulum of cells. Phase II reactions are synthetic and involve conjugation of phase I metabolites with polar cellular substances, such as glucuronic acid, sulfate, and glutathione. The products of these reactions are then excreted in urine, bile, or gill water.[40-43]

When modeling these reactions, it is implied that one is modeling the rate-limiting step in the metabolic chain of events. If the metabolite is assumed to be less toxic than the parent compound, the important question is whether metabolic clearance reduces parent chemical concentrations at sensitive target tissues. Alternatively, the metabolites formed may themselves be highly toxic. In this case, it is important to determine whether the reaction products leave the site of formation, thereby exerting toxic effects distant from the metabolizing tissue. For some compounds, a complete description of distribution and metabolism may require linked models for the parent chemical and its reaction products.

Xenobiotic metabolism probably occurs to a limited extent in all tissues, but it is usually modeled as a process localized to one compartment. The choice of this compartment and of the kinetics of the reaction (zero order, first order, or nonlinear) varies with the species and chemical in question. In general, however, elimination may be modeled by incorporating an expression for the rate of metabolism ($dA_{met\ i}/dt$) into the mass-balance equation for the eliminating compartment (here assuming flow-limited distribution):[21,44]

$$dA_i/dt = Q_i\ (C_{art} - C_i/R_i) - dA_{met\ i}/dt \tag{8}$$

Referenced to the chemical concentration in venous blood exiting the eliminating compartment, first-order elimination may be described by the equation:

$$dA_{met\ i}/dt = k_{fc}\ C_{vi}\ V_i \tag{9}$$

Saturable pathways can be described by a Michaelis-Menten-type of equation, where V_{max} is the maximum rate of reaction and K_m equals the chemical concentration in venous blood leaving the compartment at one half V_{max}:

$$dA_{met\ i}/dt = V_{max}\ C_{vi}/(K_m + C_{vi}) \tag{10}$$

A variety of compounds are actively secreted into urine or bile without prior biotransformation. This is particularly true of weak acids and bases and for compounds that are structurally similar to biologically active materials (see for example, the PB-TK model for methotrexate, Section VIII). The kinetic description of this process is identical to that for metabolism, except that rate and capacity parameters are used to characterize the concentration dependence of the active transport system.

In the absence of any metabolism or active secretion, a parent compound can be eliminated in urine and bile, due simply to chemical partitioning. For most compounds, the contribution of these processes to the overall kinetics of the parent chemical will be unimportant. Freshwater fish produce substantial quantities of urine (2.0 to 5.0 mL/kg/hr in rainbow trout); however, this urine is extremely dilute (<20 mM) and has a limited capacity for hydrophobic substances.[2,45] In contrast, fish bile exhibits a moderately high affinity for hydrophobic compounds, but bile flow rates are quite low (0.05 mL/kg/hr in rainbow trout).[46] Nevertheless, for some very hydrophobic compounds, partitioning of the parent chemical to bile and subsequent expulsion in feces may represent an important route of elimination.

E. Scaling

A wide range of anatomical, physiological, and metabolic parameters have been shown to vary with body weight (BW) according to the allometric equation:[47]

$$Y = a\ (BW)^b \tag{11}$$

where Y is the parameter of interest and a and b are constants.
Most of these data have been collected from mammals; however, the available information suggests that generalizations made for mammals also tend to hold for fish. Thus, organ weights tend to scale within and across species with the first power of body weight, while oxygen consumption rate, ventilation volume, and gill surface area scale to an exponent of between 0.7 and 0.9. The total surface area of fish scales to the 0.67 power of body weight, in accordance with the surface law.[48]

Data obtained from rainbow trout suggest that cardiac output scales with the first power of body weight;[39] however, these data were collected across a relatively narrow range of animal sizes (100 to 1000 g) and may not be representative. In mammals, there is considerable evidence that cardiac output scales to the 0.75 power of body weight, while percent of cardiac output to different organs remains approximately constant. Also in mammals, the Michaelis-Menten rate parameter V_{max} appears to scale with the 0.75 power of body weight, while the capacity constant K_m tends to remain constant.[49] Unfortunately, the available biotransformation data for fish are too limited to permit similar generalizations.

F. Growth

The need to incorporate growth into a PB-TK model for fish will depend largely upon the chemical of interest. Compounds that exhibit rapid uptake and elimination kinetics tend to approach steady-state within a short period of time. Assuming constant exposure, tissue concentrations of these materials will then remain relatively constant throughout the life of the organism (provided that equilibrium partitioning behavior does not change). Growth becomes an important consideration when the time required for a chemical to approach steady-state is similar to or greater than the time that it takes for the exposed organism to grow from a juvenile into an adult. In such cases, growth has the effect of reducing tissue concentrations below those that would have been predicted for the animal had it not increased in size. Termed "growth dilution," the mathematical basis for this outcome is given by rearrangement of Equation 1 and incorporation of an expression to describe the rate of increase of the volume of the accumulating compartment:

$$dC_i/dt = Q_i(C_{art} - C_{ven})/(dV_i/dt) \qquad (12)$$

The simplest description of growth assumes a fixed rate of volume increase (i.e., dV_i/dt = a constant value). Unfortunately, data for most fish species suggest more complicated patterns of growth, frequently characterized by two or more "growth phases." Under these conditions, it will be necessary for the modeler to collect information for the species of interest and develop a mathematical description of these data.

V. ROUTES OF EXPOSURE FOR AQUATIC VERTEBRATES

A. Importance of Chemical Hydrophobicity

The principal route of exposure of fish to a given compound is likely to depend upon a variety of factors, including water chemistry, the physical and chemical characteristics of the compound, and the biology of the species in question. In general, very hydrophobic (log K_{ow} >6.0) compounds are believed to be accumulated by fish primarily from food, while compounds exhibiting low to moderate hydrophobicity (log K_{ow} <4.0) are thought be accumulated by absorption from water, occurring primarily at the gills.[50–52] Consistent with this suggestion is the observation that hydrophobic compounds exhibit a high affinity for organic substrates (dissolved organic material and suspended sediment) and, as a result, appear to be relatively unavailable for uptake across fish gills.[53a] Moreover, fish prey items, including smaller fishes, tend to accumulate hydrophobic compounds to very high levels relative to the background water concentration, providing a rich source of chemical for uptake across the gut. For compounds in the intermediate (>4.0 but <6.0) log K_{ow} range, exposure routes are not as well characterized, and it is reasonable to expect that both food and water contribute significantly to chemical uptake. Finally, the skin may represent an important route of chemical uptake and elimination for very small fish or for juveniles of larger species.[53b]

B. Branchial Uptake

Using respirometer-metabolism chambers, McKim and co-workers[54] directly measured the uptake of various organic chemicals across the gills of rainbow trout. When these data were plotted against chemical log K_{ow}, a striking relationship was observed between branchial uptake and chemical hydrophobicity (Figure 5). Briefly, uptake rates were low when log K_{ow} was <1.0, increased sharply between log K_{ow} 1.0 and 3.0, leveled off between log K_{ow} 3.0 and 6.0, and then declined when log K_{ow} exceeded 6.0. This general pattern of uptake was subsequently observed in free-swimming exposures with guppies[55] and now serves to guide efforts to determine the physiological and chemical factors that control chemical uptake at fish gills. In the discussion that follows, attention will be focused on the efforts of researchers in our laboratory to develop a model for chemical uptake at fish gills.[56–58] It should be noted, however, that alternative approaches exist and have been used to model to the same general set of observations.[59,60]

C. Flow-Limited Gill Model

The structure of fish gills reflects their primary function as a gas exchange and osmoregulatory organ.[61] Gill surface area is maximized by an intricate arrangement of plate-like structures, termed secondary lamellae, each of which is oriented longitudinally to the direction of water flow (Figure 6).[62] Deoxygenated blood flows through these lamellae in a posterior-to-anterior direction, providing for countercurrent exchange with water.

In the absence of significant diffusion limitation, it can be shown that equilibrium between blood and water will occur at either the upstream or downstream (with respect to water) end of gill lamellae. When equilibrium occurs at the

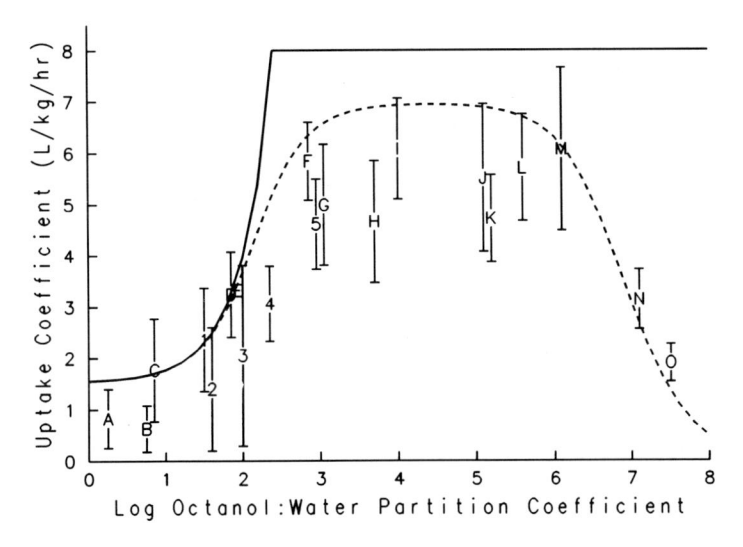

Figure 5. Observed and predicted relationship of gill uptake rate to octanol/water partition coefficient (modified from Erickson and McKim[57]). Sublethal exposures are denoted by A — ethyl formate; B — ethyl acetate; C — 1-butanol; D — nitrobenzene; E — p-cresol; F — chlorobenzene; G — 2,4-dichlorophenol; H — 2,4,5-trichlorophenol; I — 1-decanol; J — 1-dodecanol; K — pentachlorophenol; L — hexachlorobenzene; M — 2,5,2′,5′-tetrachlorobiphenyl; N — fenvalerate; and O — mirex. Lethal exposures are denoted by 1 — benzaldehyde; 2 — 2,4-dinitrophenol; 3 — MS-222; 4 — malathion; 5 — 1-octanol. The solid line denotes the prediction of a flow-limited gill model (Equation 19). The dashed line denotes the prediction of a gill model that incorporates both diffusion limitations and adsorption to dissolved organic material.

upstream end (afferent water with efferent blood, termed blood flow-limited exchange), a mass-balance equation for chemical flux (F_g) may be written as:

$$F_g = Q_c (C_{art} - C_{ven}) \tag{13}$$

From these assumptions, it also follows that the chemical concentration in arterial blood will be equal to the product of inspired water concentration and a constant (R_{bw}) that describes equilibrium chemical partitioning between the two phases:

$$C_{art} = C_{insp} R_{bw} \tag{14}$$

Substituting Equation 14 into Equation 13 and then rearranging yields an equation for chemical flux, referenced to incoming water concentration:

$$F_g = Q_c R_{bw} [C_{insp} - (C_{ven}/R_{bw})] \tag{15}$$

Alternatively, when equilibrium occurs at the downstream end of the lamellae

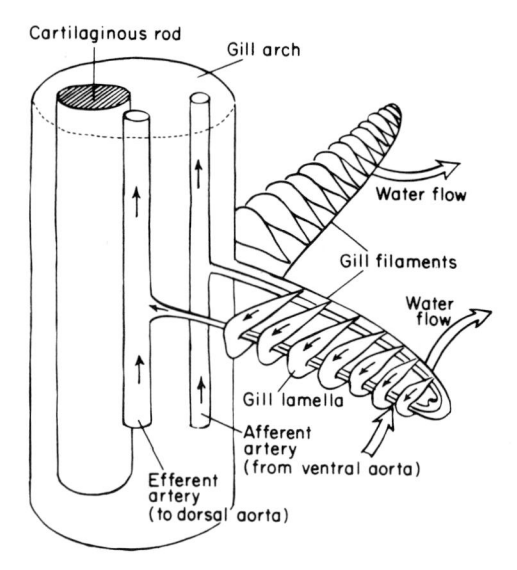

Figure 6. Diagram of fish gills showing countercurrent blood and water flows (from Wedemeyer et al.,[62] by permission of TFH publications).

(efferent water with afferent blood, termed ventilation-limited exchange), a mass balance on the gill may be written as:

$$F_g = Q_w \, (C_{insp} - C_{exp}) \tag{16}$$

and the chemical concentration in expired water will be related to that in venous blood according to the relationship:

$$C_{exp} = C_{ven}/R_{bw} \tag{17}$$

Substituting Equation 17 into Equation 16 and then rearranging yields a second possible description of chemical flux:

$$F_g = Q_w \, [C_{insp} - (C_{ven}/R_{bw})] \tag{18}$$

The final equation for chemical flux at fish gills may be derived by noting that the terms $Q_c \, R_{bw}$ in Equation 15 and Q_w in Equation 18 represent exchange coefficients that relate chemical flux to the difference in chemical activities in inspired water and venous blood. Chemical flux under a given set of circumstances will therefore be determined by whichever of these coefficients limits exchange:

$$F_g = \text{Min} \, [Q_w, \, Q_c \, R_{bw}] \, [C_{insp} - (C_{ven}/R_{bw})] \tag{19}$$

D. Use of the Flow-Limited Gill Model to Predict Chemical Uptake

Since its initial development, the simple flow-limited gill model described above has been expanded to include the effects of both diffusion limitation and chemical adsorption to dissolved organic material. The resulting model satisfactorily describes all of the important features of the data presented in Figure 5, including the decline of uptake of very hydrophobic compounds.[57] Nevertheless, it is important to recognize that uptake rates for many chemicals can be adequately predicted simply from flow limitations. Thus, for chemicals of intermediate hydrophobicity (log K_{ow} >3.0 but <6.0), uptake is predicted to be dependent almost entirely on water flow rate (Figure 7). In contrast, when the amount of chemical presented to the fish in water approaches or exceeds the capacity of blood to carry it away (log K_{ow} <3.0), the model predicts that cardiac output will become an important determinant of uptake rate. Recent experimental evidence obtained by independent manipulation of gill ventilation and cardiac output tends to support both of these predictions.[63]

In Equations 16, 18, and 19, the water flow term Q_w corresponds to the volume of water that achieves equilibrium with blood in perfused gill lamellae. This flow rate is less than the total ventilation volume due to shunting of water around the gills and to the fact that not all lamellae are perfused with blood at any given time. An estimate of this flow rate can be obtained if it is assumed that the volume of water from which chemicals are taken up is the same as that available for uptake of oxygen (termed the effective respiratory volume, and abbreviated ERV, Figure 7). Q_w may then be calculated from a fish's oxygen consumption rate and the inspired-to-expired difference in oxygen concentration in respiratory water:[57]

$$Q_w = ERV = VO_2/(O_{2\ insp} - O_{2\ exp}) \qquad (20)$$

The value of this approach is that it may be used to estimate the initial uptake rate of moderately hydrophobic (ventilation-limited) compounds in very small fish or in fish that do not otherwise lend themselves to experimental manipulation. An estimate of $O_{2\ exp}$ can be obtained if the partial pressure of expired water is assumed to be the same as that of venous blood (P_VO_2). Data from a variety of fish species suggest that P_VO_2 averages between 20.0 and 30.0 torr (range = 7.0 to 42.9 torr) or about 15% of saturation with the atmosphere.[64-72] Thus, $O_{2\ exp} \sim (0.15)\ O_{2\ insp}$. In an air-saturated system, $O_{2\ insp}$ simply equals the aqueous solubility of oxygen at the temperature of interest. With these assumptions, Q_w can be estimated by measuring VO_2 for the species of interest. Initial uptake rate (i.e., when $C_{ven} = 0$) may then be calculated by using this value as an input to Equation 19.

E. Dermal Uptake

In general, when fish of different sizes are compared to one another, smaller fish are characterized by: (1) relatively large ratios of total body surface area to

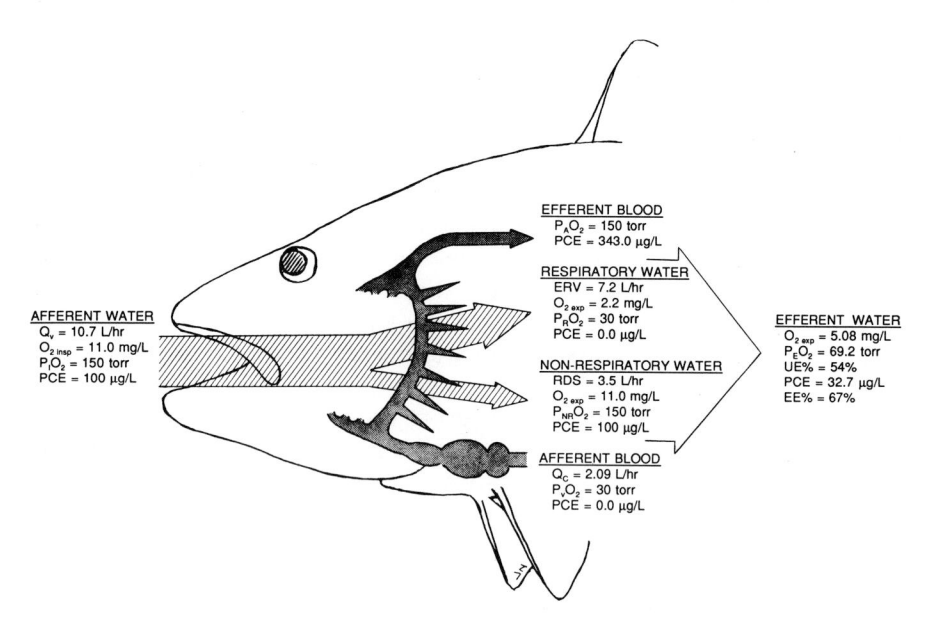

Figure 7. Branchial uptake of oxygen and pentachloroethane (PCE) at fish gills, illustrating the relationship between uptake of oxygen and chemical when chemical flux is ventilation-limited. The values shown approximate those obtained in studies with 1 kg rainbow trout exposed at 12°C.[1] For simplicity, the chemical concentration in venous blood is set equal to zero (as would be true at the beginning of an exposure). Symbols and abbreviations are as follows: Q_V — ventilation volume; ERV — effective respiratory volume; RDS — respiratory deadspace; Q_C — cardiac output; $O_{2\,insp}$ — oxygen concentration in inspired water; $O_{2\,exp}$ — oxygen concentration in expired water; P_iO_2 — oxygen partial pressure [subscripts (i): I — inspired water; E — expired water; R — respiratory water; NR nonrespiratory water; A — arterial blood; V — venous blood]; and UE% — oxygen uptake efficiency; EE% — chemical extraction efficiency.

body weight, (2) decreased ratios of gill surface area to total body surface area, and (3) decreased skin thickness. Collectively, these factors can be expected to result in an increase in chemical flux across the skin relative to that across the gills. Physiological support for this suggestion is provided by the observation that 80% of total oxygen uptake by posthatch chinook salmon takes place across the skin.[73] The potential importance of this dosing route in aquatic toxicology is underscored by the fact that most of the toxicity data for fish have been gathered using small fish (e.g., guppies, fathead minnows, and Japanese medaka) or the young of larger species (in embryo-larval testing).

Dermal flux can be described by treating the skin as a discrete compartment into which chemicals move according to their permeability, the surface area for diffusion, and the magnitude of the diffusive gradient (Fick diffusion). Assuming that chemical transport from the tissues to blood is flow-limited, a mass balance on the skin compartment may be written by referencing chemical concentrations in blood and water to that in the dermal tissues using tissue-to-blood (R_d) and tissue-to-water (R_{dw}) partition coefficients:[74]

$$dA_d/dt = Q_d(C_{art} - C_{ven}/R_d) + P_d\ SA_d\ (C_{water} - C_d/R_{dw}) \quad (21)$$

An important feature of this description is that chemical permeability (P_d) and total surface area (SA_d) are treated as independent terms (in contrast to the equation for diffusion-limited flux within tissues, which utilizes a lumped permeability-area product term). The total dermal surface area of fish may be estimated from established allometric relationships. By modeling to dermal exposure data (preferably venous blood concentration), it thus becomes possible to solve for the apparent value of P_d. Using this approach, McDougal and co-workers characterized the permeation of several organic vapors across the skin of rats.[74] Studies with rainbow trout and channel catfish suggest that the same approach can be applied to modeling dermal uptake of chemicals by fish.[3]

F. Dietary Uptake

PB-TK models for dietary exposure have not yet been fully developed for fish. Guidance for the development of such models, therefore, must come from existing models for mammals.[75] Briefly, the factors thought to control uptake of chemicals across the gut include:

1. rate of food digestion in the stomach
2. rate of gastric evacuation to the intestinal tract
3. rate of food passage through the different intestinal regions
4. rate of fecal egestion
5. changes in meal volume with the absorption of water and nutrients
6. changes in the relative affinity of chemicals for food and for the tissues of the digestive tract (due primarily to uptake of dietary lipid)
7. rate of chemical diffusion across the gastrointestinal epithelium

The number and identity of model compartments will depend upon the species of interest. In general, the morphology of the fish gastrointestinal tract is simpler than that of mammals; however, marked species differences are known to exist. For example, in some fish, the surface area of the small intestine is greatly increased by the presence of blind diverticula, called pyloric ceca, while in other species, the small intestine is a short, simple tube. Similarly, some species possess a discrete stomach, while others do not.[76,77]

Each gut compartment may consist of two (tissues and lumen) or three (blood, tissues, and lumen) subcompartments, depending on the factors thought to limit chemical transport. Chemical flux from the lumen to the tissues is generally modeled as a diffusion-limited process, characterized by a permeability-area product term and the difference in chemical activity between the tissues and the lumenal contents. Chemical flux from the tissues to blood may then be modeled as a flow or diffusion-limited process, as previously described.

Chemicals taken up across the gut are considered to enter the general circulation via the hepatic portal vein.[38] The liver, therefore, has an opportunity to act on chemicals before they perfuse other tissues in the body. For certain compounds, the gut tissues themselves may also represent an important site of metabolism.[78] Blood exiting the liver flows directly to the heart and from there to the gills. Thus, during dietary exposures, the gills act principally as an eliminating organ, and the overall rate of accumulation represents the net result of dietary input, metabolic transformation, branchial elimination, and fecal egestion.

The required complexity of a gut model may depend upon the chemical of interest. Studies with rainbow trout suggest that dietary uptake of chemicals with log K_{ow} values <4.0 can be adequately simulated using a gut description consisting only of a stomach compartment, from which uptake occurs in a diffusion-limited manner.[4] Under these conditions, chemical elimination at the gills continually reestablishes the chemical gradient that favors chemical diffusion across the stomach epithelium, and uptake is essentially complete before the stomach contents are evacuated to the lower intestine. With increasing log K_{ow}, or alternatively when a compound is tightly bound to the food matrix, we can expect the upper and lower intestines to play important roles in chemical uptake, requiring the development of full PB-TK gut models. Finally, chemical flux from the fish to the gut contents may represent, in some cases, a significant route of elimination if, for example, the level of chemical contamination in the diet is reduced, resulting in a net outward diffusion gradient.

G. Multiple Exposure Routes

It should be emphasized that as progress is made toward the description of oral and dermal exposure routes, this information can be combined with existing descriptions of chemical flux at fish gills to provide models capable of handling more than one exchange surface. Such models could be used to simulate the results of combined, oral, dermal, and inhalation exposures, as might be expected to occur in an environmental setting, or alternatively to predict the relative contribution of each exposure route for different chemicals, species, and environmental conditions.

VI. MODEL PARAMETERIZATION

A. Physiological Inputs

Parameterization of a physiologically based model for fish requires: (1) estimates of cardiovascular and respiratory function, (2) compartment volumes and blood perfusion rates, (3) metabolic and excretory rates, and (4) equilibrium partition coefficients for blood and tissues. Depending upon the fish species and chemical of interest, some or all of this information may be available in the literature. In most cases, however, various of these parameters must be measured in the laboratory or estimated from other information using established physiological relationships or allometric scaling methods.

The physiology of rainbow trout *(Oncorhynchus mykiss)* is the best characterized of all fish species, and information is available for major organ and tissue volumes, cardiac output, organ blood flows as a percentage of cardiac output, ventilation volume, oxygen consumption rate, and urine and bile flow rates (Table 2).[39,45,46,65,79–85] In addition, the physiology of trout has been examined with respect to the effects of various environmental parameters, including temperature[39] and dissolved oxygen content.[86] Significant cardiovascular and respiratory information also exists for several other teleosts, including the channel catfish *(Ictalurus punctatus)*,[71,87,88] starry flounder *(Platichthys stellatus)*,[67,70] common carp *(Cyprinus carpio)*,[72,89] tench *(Tinca tinca)*,[68] and winter flounder *(Pseudopleuronectes americanus)*,[69] and for the dogfish shark *(Squalus acanthias)*,[64,90–92] an elasmobranch. Additional sources review the cardiovascular and respiratory physiology of fishes generally.[93,94]

Of the compartment-specific parameters, those which apply to the fat compartment are frequently the most difficult to obtain, due to the diffuse nature of this tissue. One means of estimating the volume of this compartment is to obtain lipid content data for the whole fish and for the various tissues in the model. The volume of the fat compartment can then be calculated by subtracting from the total lipid content of the fish the amount of lipid that is attributable to "lean" tissues.[1]

Arterial blood flows to individual organs can be estimated using radiolabeled microspheres.[39,95] When injected into the bloodstream, these spheres lodge in tissue capillary beds. The total radioactivity found in a tissue is then considered to be proportional to the amount of blood perfusing the tissue. The microsphere method is based on the assumption that spheres distribute homogeneously in the bloodstream and that the obstruction of some capillaries has no effect on the overall pattern of blood distribution. In addition, it is critical that the spheres be sized appropriately for the capillary diameter of the species of interest.

The microsphere method works well when a tissue receives only arterial blood, but it tends to underestimate total blood flow to organs (e.g., liver and kidney) that receive portal blood, due to the series arrangement of capillary beds. Total blood flow to these organs, therefore, must be estimated by summing both arterial and portal inputs:

$$Q_{mi} = Q_i + Q_{pi} \tag{22}$$

where Q_{pi} is defined as equal to all or a fraction of arterial flow to the upstream compartment.

Table 2. Physiological Parameters Used in a Physiologically Based Toxicokinetic Model for Adult Rainbow Trout[a]

Ventilation volume (L water/hr)	Q_v	10.60
Effective respiratory volume (L water/hr)	Q_w	7.20
Oxygen consumption rate (mg O_2/hr)	VO_2	63.0
Cardiac output (L blood/hr)	Q_c	2.07
Arterial blood flow to tissues (L blood/hr)[b]	Q_l	0.060
	Q_f	0.176
	Q_m	1.242
	Q_r	0.476
	Q_k	0.116
Tissue group volumes (L)	V_l	0.012
	V_f	0.098
	V_m	0.818
	V_r	0.063
	V_k	0.009

Source: Nichols, J.W. et al., *Toxicol. Appl. Pharmacol.*, 106, 433, 1990.

[a] Values given are for a 1 kg trout at 12°C.
[b] Calculated by multiplying blood flow as a percent of Q_c times 2.07 L blood per hour.

With C_{pi} set equal to the chemical concentration in venous blood exiting the upstream compartment, chemical concentration in mixed blood can be calculated by summing the flow-weighted contributions of arterial and portal blood:

$$C_{mi} = (Q_i C_{art} + Q_{pi} C_{pi})/Q_{mi} \tag{23}$$

A complete mass balance for the compartment may then be written as:

$$dA_i/dt = Q_{mi}(C_{mi} - C_{vi}) \tag{24}$$

B. Chemical Partitioning

Equilibrium chemical partition coefficients describe the chemical partitioning that would occur between any two phases (e.g., blood-to-water or tissue-to-blood), if these phases were allowed to come to thermodynamic equilibrium. In reality, thermodynamic equilibrium seldom, if ever, occurs in living organisms; however, the difference between thermodynamic equilibrium and actual partitioning can be used to estimate the direction and rate at which chemicals will move from one phase to another.

In vitro methods have been developed to estimate chemical partition coefficients for volatile compounds. Although initially applied in mammalian PB-TK modeling efforts,[96,97] these methods have since been modified for use with fish tissues.[98] Briefly, samples (e.g., water, blood, tissue homogenates, urine, and bile) are placed in closed containers, and a known quantity of the compound is introduced as a saturated vapor. Allowing sufficient time for equilibrium to be achieved, the relative affinity of the chemical for each sample matrix is estimated by the extent to which the headspace vapor concentration declines. Sample-to-sample partition coefficients are then calculated by dividing respective sample-to-air partitioning values, using appropriate blanks to compensate for the use of sample diluents and chemical adsorption to the assay system.

The advantage of using an in vitro partitioning system is that equilibrium partitioning estimates can be obtained without having to expose fish to the chemical. Model simulations can then be generated a priori, providing a true test of the model's predictive power. Additional methods have been developed to obtain in vitro partitioning estimates for nonvolatile compounds.[99] To date, however, they have not been applied in PB-TK modeling efforts with fish.

When in vitro partitioning information is not available, equilibrium partition coefficients must be estimated from in vivo exposure data. Ideally, these data should be collected from animals exposed continuously until they are at or very near steady-state. Alternatively, when a compound approaches steady-state slowly but on terminating the exposure remains within the organism for an extended period of time, it is possible to obtain partitioning estimates by infusing a bolus of the material intravenously. Blood and tissues are then sampled after several days or weeks. In either case, it must be understood that estimates obtained from in vivo exposure data are likely to underestimate true equilibrium partitioning values when a tissue is involved in the elimination of a compound (as by metabolism or excretion in urine or bile).

C. Kinetic Rate and Capacity Parameters

Chemical elimination in mammalian PB-TK models is usually assumed to take place in the liver. There is no reason, however, not to incorporate metabolism from some other compartment if in vivo data support this possibility. Following identification of the eliminating compartment(s), a judgment must be made concerning the kinetics of the process over the range of concentrations to which the tissue is exposed. Accumulating in vivo evidence from mammals suggests that phase I metabolic pathways tend to exhibit saturable (Michaelis-Menten) kinetics, while phase II processes often follow first-order kinetics.[100] Unfortunately, little of this type of information is presently available for fish.

In vitro methods have been employed to estimate kinetic rate and capacity parameters for incorporation into PB-TK models for mammals[101] and fish.[8] As with collection of chemical partitioning information, the advantage of using an in vitro system is that model simulations can be obtained without the prior need to expose fish. This type of information is difficult to interpret, however, when the relationship between in vivo and in vitro chemical concentrations is not well understood or when experimental manipulations affect the activity of the enzyme system.

Lacking in vitro parameter estimates, in vivo methods can be used to determine whether appreciable elimination of a compound occurs by some route other than simple diffusion across the gills or skin. One such method involves the use of fish respirometer-metabolism chambers. Employed as described by McKim and Goeden,[14] these chambers allow direct calculation of chemical extraction efficiency from inspired water. During an inhalation exposure, extraction efficiency (EE) is related to net chemical uptake according to the relationship:

$$\text{net uptake } (\mu g/hr) = EE\ (\%) \times Q_v\ (L/hr) \times C_{insp}\ (\mu g/L) \qquad (25)$$

and reflects both chemical distribution to tissues and the extent to which the chemical is eliminated by systemic (extrabranchial) routes. At steady-state, net uptake at the gills is determined by the amount of chemical taken up by eliminating tissues during each pass of blood through the fish. An estimate of total systemic elimination, therefore, may be obtained by measuring extraction efficiency when fish are at or near steady-state.[1] The disadvantage of this approach is that it is limited to use with compounds that approach steady-state rapidly and for which metabolism has a measurable effect on the kinetics of the parent compound.

Alternatively, as experience is gained with a model, and assuming a high degree of confidence in chemical partitioning estimates, it is possible to use the model itself to estimate rate and capacity parameters for a chemical of interest. In practice, metabolism parameters are first set equal to zero, and the experimental data are examined for evidence to the contrary. When such evidence is obtained, these parameters are then adjusted until model outputs simulate the data. Thus, the model itself becomes a tool for investigation of xenobiotic metabolism. This approach has been developed extensively in studies with mammals and has been used to investigate the effects of metabolic inducers and inhibitors.[44] The advantages of this approach are its lack of reliance upon steady-state data and its utility for compounds that undergo limited metabolism (assuming that a method exists for analysis of the metabolic product). It should be noted, however, that this approach yields *apparent* kinetic parameter values which may represent the net result of several biological processes. Therefore, until such time as in vivo information for the compound can be collected to test model assumptions (e.g., primary involvement of the liver), direct comparison of these values to estimates obtained in vitro is problematic.

VII. MODEL VALIDATION AND APPLICATION

In the strict sense of the word, a model cannot be proven to be valid; it can only be shown to be wrong. By definition, therefore, model validation is a subjective process, the objective of which is to determine how accurately the model represents the system of interest. Validation consists of two processes:

verification and evaluation. Verification is defined as the process of determining whether the model structure faithfully represents underlying theory. Evaluation is the process of examining and appraising model performance by comparing model outputs to measured data.

To evaluate model performance, the modeler must choose appropriate performance measures, usually one or more endpoints to which one is modeling. In physiologically based modeling of fish, typical endpoints include chemical concentration in blood and exhaled water, and chemical extraction efficiency from inspired water. It is important that performance measures be easily attainable, with a high degree of precision and accuracy, so that deviation between model outputs and observed values can be characterized.

The development and validation of a PB-TK model must be viewed as an iterative process (Figure 8). Following the identification of a problem, relevant physiological, biochemical, and physical-chemical information are combined using mass-balance principles to create a model for the system of interest. As experience with the model is gained, a sensitivity analysis may be undertaken to assess the relative contribution of individual parameters to model performance. This information can then be combined with the modeler's confidence in the various parameter estimates to determine whether and how the model should be amended and which of the parameters require further investigation.

The value of a PB-TK model derives, in part, from the ability to change input variables in a systematic fashion, thereby addressing specific questions concerning the system under study. Several examples of this strategy have already been described, including determination of metabolic rate and capacity parameters and skin permeation coefficients. These efforts are constrained, however, by the need to assign physiologically reasonable values to all parameters estimates. In addition, as experience is gained with a particular fish species, it becomes necessary to use the same physiological and morphological inputs for all chemicals.

Finally, it should be noted that natural systems tend to be highly variable. Therefore, it follows that when a model is parameterized using average values for a species, it may not accurately predict the outcome of a single experiment utilizing a limited number of animals. A detailed presentation of methods by which to assess and incorporate the effects of parameter uncertainty is outside the scope of this chapter. Various methods exist, however, and have been applied to PB-TK models for mammals.[102,103]

VIII. EXAMPLES OF PB-TK MODELS FOR FISH

A. Methotrexate

The first physiologically based model for fish was published by Zaharko and co-workers in 1972.[6] Drawing upon previous experience with a PB-TK model for mice,[104] these authors developed an intravenous infusion model to simulate the

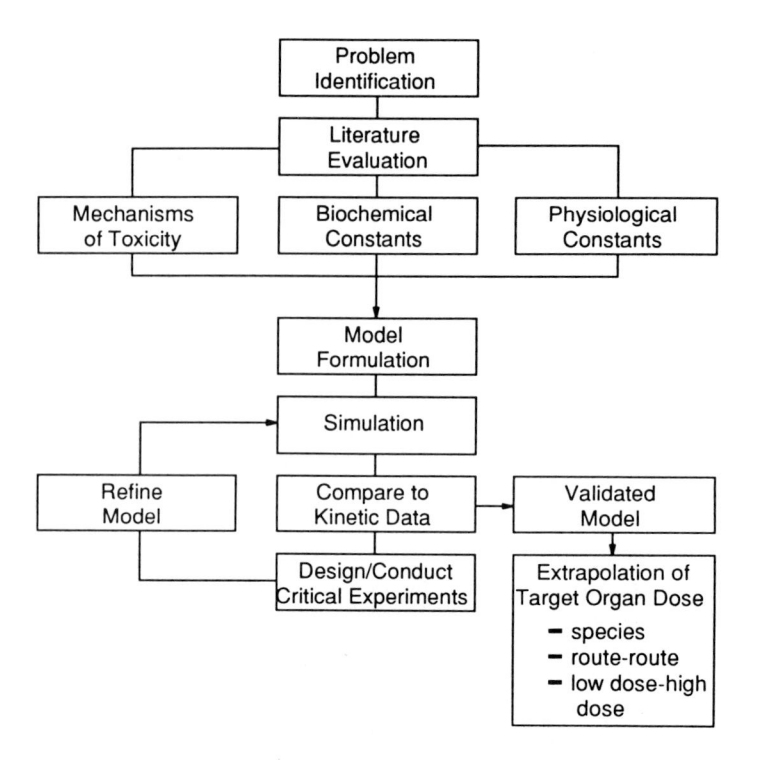

Figure 8. Simplified flow chart for the development and validation of a PB-TK model (modified from Andersen[30]).

distribution and elimination of methotrexate in the sting rays, *Dasyatidae sabina* and *D. sayi*. Methotrexate is a folic acid analogue that has been used both as a chemotherapeutic agent and as an immunosuppressant. As with all antimetabolites, an important factor in the use of methotrexate is the balance between therapeutic and toxic dose levels.

When methotrexate was injected into sting rays, it disappeared from plasma in a multiphasic fashion, becoming most concentrated in the liver. The kinetic behavior of methotrexate in rays was accurately described using a flow-limited model consisting of four tissue compartments: the liver, kidney, gastrointestinal tract, and muscle. Earlier studies in mammals suggested that methotrexate was poorly metabolized. This was confirmed for rays both in vitro and in vivo.

Two important observations in this study illustrate the potential utility of PB-TK models for fish. First, by using circulation velocity as an equivalent time-scaling factor, kinetic data from rays were found to be similar to data collected previously from mice. This result suggested that the same physiological processes govern the distribution and elimination of methotrexate in both species. More importantly, it shows how these models may be used to extrapolate kinetic information between fish and mammals.

Second, by changing just one physiological parameter (blood flow rate to the muscle compartment), improvements were made in the agreement between observed and predicted residues in muscle, liver, kidney, and plasma. Although it does not constitute proof of such, this observation suggests the possibility that initial blood-flow estimates were incorrect. As information for a given species accumulates, it should be possible to use PB-TK models to provide physiological insights where information does not exist or is contradictory.

B. Phenol Red

Bungay and co-workers[7] developed a PB-TK model to describe the distribution and elimination of phenol red (phenolsulfonphthalein or PSP) in the dogfish shark. When administered to healthy humans, PSP is rapidly cleared from blood and appears in the urine, causing it to turn bright red. Therefore, it is routinely used in clinical practice to assess renal function. Preliminary studies with dogfish suggested that PSP was eliminated by this species in both the urine and bile; however, the appearance of PSP in gall bladder bile was delayed following intravenous injection by as much as 4 hr.[105]

The kinetic behavior of PSP was adequately described using a flow-limited intravenous infusion model, despite the fact that nearly 70% of the material was bound to plasma protein after 4 hr. Elimination by the liver and kidney was modeled using first-order rate constants, each representing the sum of all processes responsible for clearance of both the parent compound and its primary metabolite, a glucuronide conjugate. Conjugation was assumed to be followed immediately by secretion (i.e., no back-diffusion into blood); therefore, mass-balance equations were required only for the parent compound. The delayed appearance of PSP in bile was described by incorporating two "bile duct" subcompartments. The movement of material through these subcompartments was then calculated as the product of a residence time factor and the rate of transport to each compartment. Residence time factors and rate constants for both urinary and biliary elimination were estimated by fitting model simulations to the experimental data.

C. Pyrene

The distribution and elimination of pyrene in rainbow trout were modeled by Law and co-workers[8] using a PB-TK intra-arterial infusion model. Pyrene is one of a large number of polycyclic aromatic hydrocarbons (PAHs) that result from incomplete combustion of organic material. Concern for these compounds has arisen due to their ubiquitous distribution in the environment and because several (notably, benzo(a)pyrene) have been shown to be potent carcinogens. With the exception of muscle, the distribution of pyrene to tissues was assumed to be flow-limited. Uptake by muscle was modeled as a diffusion-limited process, with the barrier to diffusion assumed to be the cell membrane. Chemical flux across this barrier was calculated as the product of the diffusion gradient and a lumped transport parameter (permeability area product term), the value of which was determined from the exposure data.

Previous work indicated that trout eliminate pyrene primarily as water-soluble metabolites in both urine and bile.[107] The model was designed, therefore, to incorporate both first-order (kidney and liver) and saturable (liver only) clearance pathways. Rate and capacity parameters for saturable metabolism were estimated using an in vitro isolated hepatocyte assay. First-order rate constants for clearance from the liver and kidney were determined from the exposure data. As with earlier PB-TK models for the sting ray[6] and shark,[7] elimination of pyrene across the gills was assumed to represent an insignificant contribution to the overall kinetics of distribution and elimination. Predicted pyrene concentrations in the gills, liver, gut, kidney, muscle, and blood were consistent with experimental observations for up to 6 days after dosing.

D. Hexachloroethane, Pentachloroethane, and 1,1,2,2-Tetrachloroethane

The pioneering work of Zaharko, Bungay, and co-workers[6,7] illustrated the potential utility of PB-TK models for fish; however, the use of these models in aquatic toxicology was limited by the fact that neither included an environmentally relevant route of exposure (branchial, oral, or dermal). Nichols and co-workers[1] addressed this problem by incorporating a countercurrent description of chemical flux at fish gills[56] into a PB-TK model for rainbow trout. The resulting model was then evaluated by exposing trout to pentachloroethane (PCE) in fish respirometer-metabolism chambers.

The selection of PCE for this investigation was based upon three considerations. First, the kinetics of PCE in rats had been described previously using a PB-TK model.[106] Second, it was anticipated that PCE would exhibit rapid uptake and accumulation kinetics, making it possible to obtain data from fish at or near steady-state. Finally, because PCE is volatile, it was possible to obtain chemical partitioning estimates using an in vitro vial equilibration method.[98]

Parameterized with physiological information from the literature, the model accurately predicted PCE accumulation in blood (Figure 9), fat, kidney, liver, and muscle and its extraction from inspired water. Extraction efficiency declined to near 0% when fish were exposed to PCE for 264 hr, suggesting that fish were at or near steady-state and that systemic elimination of PCE was small. This conclusion was further supported by the correspondence between in vitro chemical partitioning estimates and PCE residues in blood and tissues, expressed as blood-to-water and tissue-to-blood concentration ratios.

Extending their earlier work with PCE, Nichols and co-workers[2] developed a PB-TK model to describe the uptake of three chlorinated ethanes by rainbow trout. Hexachloroethane (HCE) and 1,1,2,2-tetrachloroethane (TCE) were chosen along

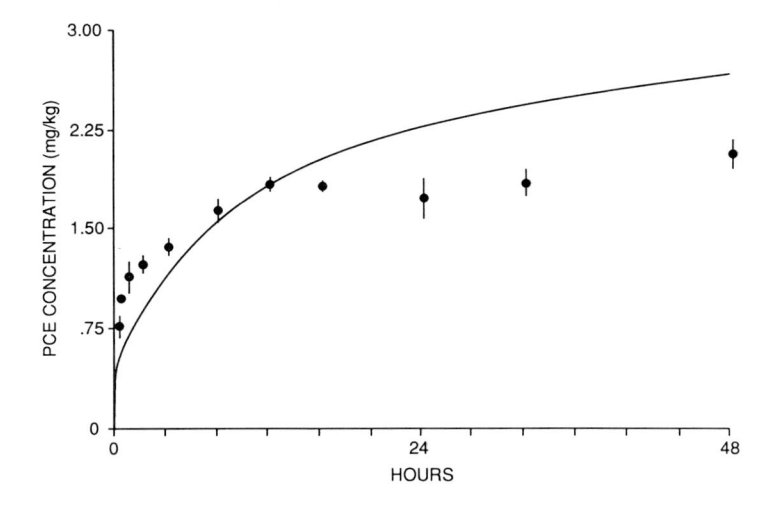

Figure 9. Time course for PCE concentration in arterial blood (from Nichols et al.[1]). The model simulation (solid line) was prepared by setting the input concentration to that used in the corresponding validation study (mean = 140 µg PCE/per liter). Observed values are shown as individual points. Each point represents the mean ± SD; N = 4 fish.

with PCE to provide data for a homologous series of compounds of similar molecular weight but different relative hydrophobicity. As before, chemical uptake from water was described using a countercurrent gill model.[57] Equilibrium chemical partition coefficients for all three compounds were again estimated using an in vitro vial equilibration technique.[98]

For each chemical and test, simulations were obtained by entering chemical-specific partitioning information and the water exposure concentration. With metabolic rate and capacity inputs set at zero, the model accurately described the extraction of all three compounds from inspired water (Figure 10) and their accumulation in blood and tissues (Figure 11 and Table 3). Subsequent analysis of extraction efficiency data confirmed that the gills of trout represent the primary route of uptake and elimination of these compounds.

Differences in chemical hydrophobicity had a marked effect on both the uptake and distribution of the three compounds. TCE (log K_{ow} = 2.64) was close to steady-state in trout after 48 hr, while PCE (log K_{ow} = 3.63) approached steady-state in 264 hr. In contrast, HCE (log K_{ow} = 4.61) continued to accumulate in fish even after 600 hr of continuous exposure. With increasing hydrophobicity, the fat compartment in the model assumed greater importance, both as a repository for the compound and as a determinant of the overall kinetics of distribution.

Inhalation studies with rainbow trout were followed by a similar set of investigations utilizing channel catfish as the test species.[5] The objective of these studies was to determine whether the same model structure, when parameterized

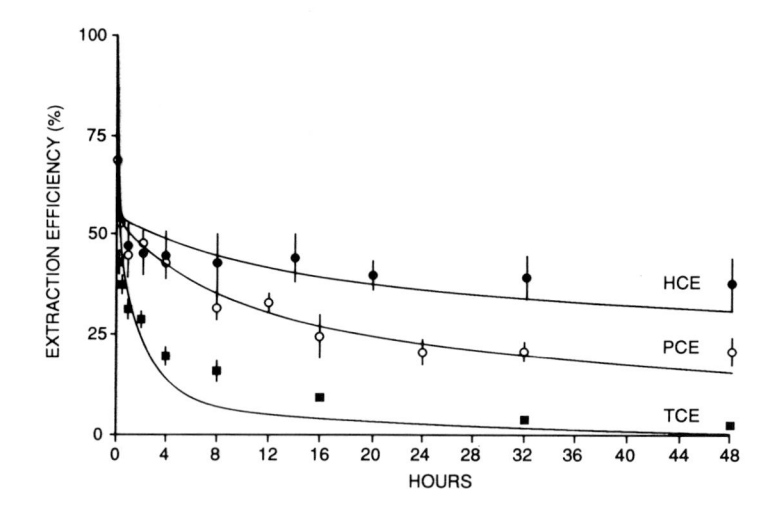

Figure 10. Time course of 48 hr for chloroethane uptake from inspired water (from Nichols et al.[2]). Chemical uptake is expressed as percentage extraction efficiency, calculated as $(C_{insp} - C_{exp})/C_{insp}$. Model simulations were prepared by setting input concentrations to exposure levels measured in the corresponding validation studies. Model simulations are shown as solid lines, the observed values as individual points. The data in this figure were obtained from fish exposed to (mean) 1060 µg TCE/per liter, 140 µg PCE/per liter, and 62 µg HCE/per liter. Each point represents the mean ± SE; N = 4 (TCE and HCE) or 5 (PCE) fish.

for a new species, would continue to describe the observed kinetics of chemical accumulation in blood and tissues. Physiological parameters were obtained from the literature or were measured during the experiments. Model predictions were again evaluated by exposing channel catfish to HCE, PCE, and TCE in fish respirometer-metabolism chambers.

In general, the experimental time course information from channel catfish closely resembled that obtained from trout, despite the fact that channel catfish were tested at a higher temperature (20°C vs. 12°C for trout). Model simulations tended to confirm this result and suggested that the correspondence between species was due to a similarity in the physiological parameters (ERV, P_{bw}) that control chemical uptake at fish gills (see Section V.C). As with trout, the model accurately predicted chemical residues in various tissues and organs (Table 4). Extraction efficiency information again indicated that catfish eliminate very little of these compounds by metabolism or other extrabranchial routes.

Inhalation studies with both trout and catfish provided support for the basic model structure and demonstrated the utility of the countercurrent gill description. This information was then used to investigate the dermal route of exposure.[3] Rainbow trout were again exposed to all three chloroethanes in fish respirometer-metabolism chambers; however, the exposure was limited to the trunk portion of the fish extending posteriorly from the head region. Under these conditions, the

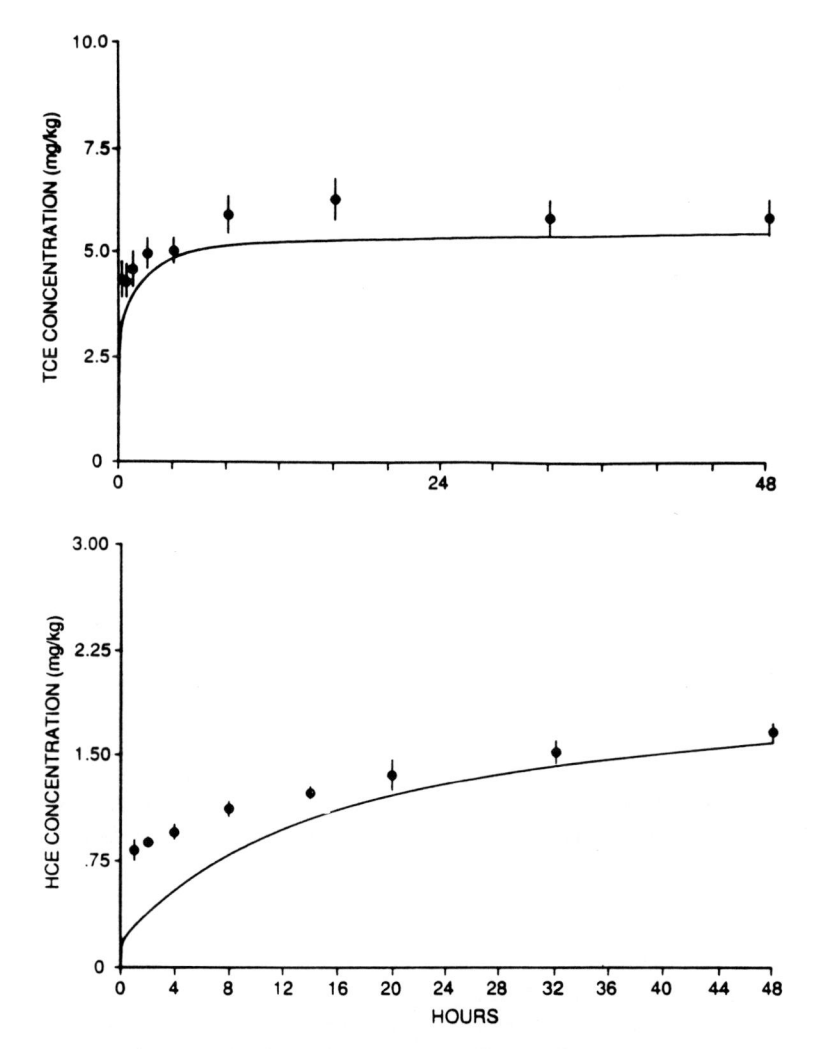

Figure 11. Time course of 48 hr for hexachloroethane (HCE) and 1,1,2,2-tetrachloroethane (TCE) concentration in arterial blood (from Nichols et al.[2]). Model simulations are shown as solid lines, with the observed values as individual points. The data in this figure are from the same fish from which extraction efficiency data in Figure 10 were obtained.

gills served exclusively as an eliminating organ. By cannulating the fish from both the dorsal aorta and ventral aortal, it was also possible to obtain a complete chemical mass-balance with which to directly assess gill function.

Chemical uptake across the skin was modeled as a Fick diffusion process, as previously described (see Section V.E). Dermal permeability coefficients (P_d) were obtained by fitting model simulations to observed arterial blood data. All

Table 3. Observed and Predicted Residues of HCE, PCE, and TCE in Blood and Tissues from Rainbow Trout, Expressed as Tissue to Blood Concentration Ratios[a]

	HCE Obs.	HCE Pred.	PCE Obs.	PCE Pred.	TCE Obs.	TCE Pred.
Blood to water	61.36	55.17	23.54	25.58	6.60	5.14
Tissue to blood						
Fat	175.07	158.04	119.86	84.42	53.74	38.55
Kidney	7.04	3.27	6.33	3.15	2.90	3.07
Liver	3.85	3.50	3.13	2.83	2.62	2.55
Muscle	3.09	3.16	3.35	3.21	1.54	2.46

Source: Nichols, J.W., et al., *Toxicol. Appl. Pharmacol.*, 110, 374, 1991. With permission.

[a] Residue information is reported as tissue to blood concentration ratios to normalize for exposure concentration. Model simulations were obtained by setting input concentrations and exposure durations to those used in the corresponding validation studies. Observed values were obtained from fish exposed to (mean) 6.8 µg HCE per liter for 600 hr, 159.1 µg PCE per liter for 264 hr, and 102.4 µg TCE per liter for 48 hr. Each value represents the mean of data from four fish.

other model inputs were unchanged from previous inhalation studies. Parameterized in this manner, the model accurately described the kinetics of accumulation of all three chemicals in both venous and arterial blood and their elimination in expired water (Figure 12). The permeability coefficient for each chemical appeared to remain constant with exposure concentration; however, a comparison between chemicals suggested a general increase in permeability with increasing chemical hydrophobicity (mean P_d values for TCE, PCE, and HCE were 0.11, 0.13, and 0.23 cm/hr, respectively).

Table 4. Observed and Predicted Residues of HCE, PCE, and TCE in Blood and Tissues from Channel Catfish, Expressed as Tissue to Blood Concentration Ratios[a]

	HCE Obs.	HCE Pred.	PCE Obs.	PCE Pred.	TCE Obs.	TCE Pred.
Blood to water	58.78	44.66	22.12	18.15	5.46	5.56
Tissue to blood						
Fat	44.25	68.33	47.14	70.05	53.37	53.66
Kidney	3.91	1.90	4.16	3.71	3.88	2.16
Liver	1.34	1.50	1.93	2.67	2.16	2.18
Muscle	0.28	0.36	0.39	0.61	0.62	0.92

Source: Nichols, J.W., et al., *Aquat. Toxicol.*, 27, 83, 1993. With permission.

[a] Residue information is reported as tissue to blood concentration ratios to normalize for exposure concentration. Model simulations were obtained by setting input concentrations and exposure durations to those used in the corresponding validation studies. Observed values were obtained from fish exposed to (mean) 6.48 µg HCE per liter for 48 hr, 196.9 µg PCE per liter for 48 hr, and 140.7 µg TCE per liter for 48 hr. Each value represents the mean of data from five fish.

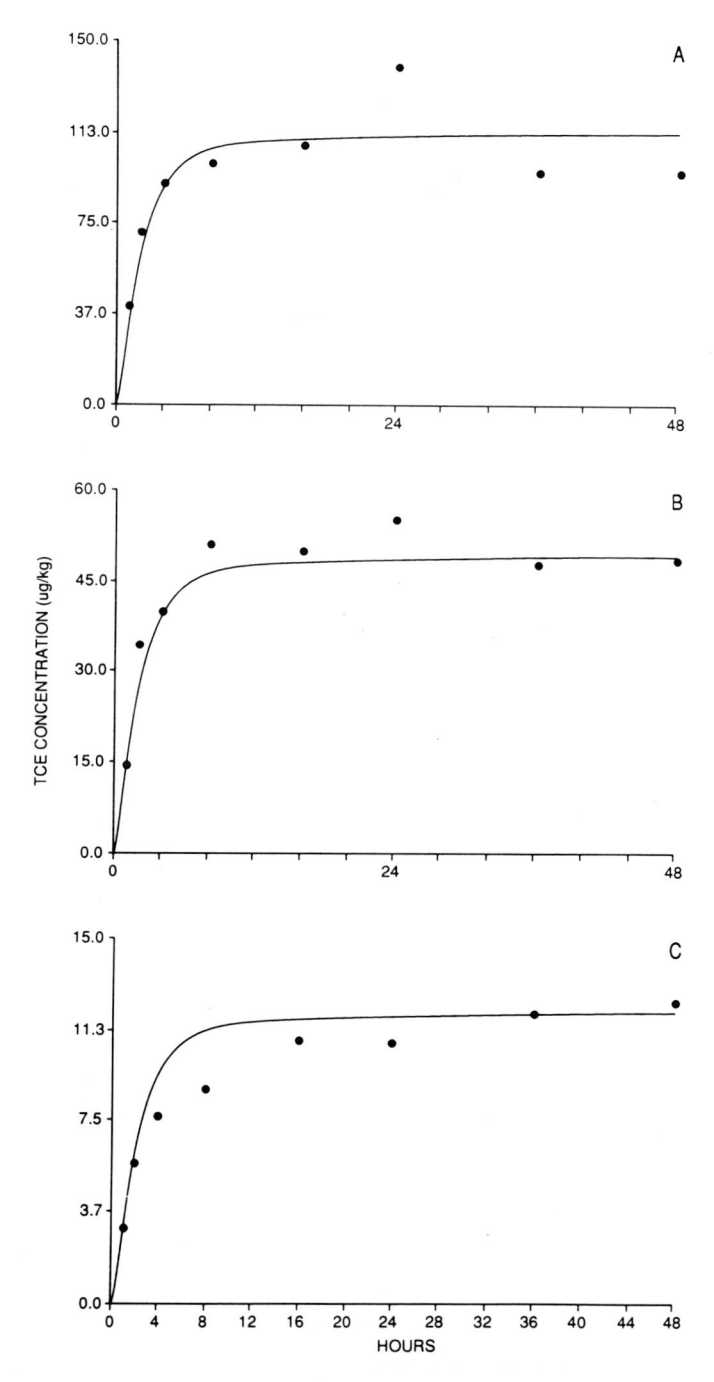

Figure 12. Time course of 48 hr for 1,1,2,2-tetrachloroethane (TCE) concentration in (A) venous blood, (B) arterial blood, and (C) expired water (from Nichols et al.[3]). Exposures were limited to the trunk portion of the fish (dermal only; mean concentration = 1370 µg TCE/per liter). Model simulations are represented as solid lines, with the observed values as individual points. Each point represents the mean of data from five fish.

IX. CURRENT RESEARCH NEEDS

PB-TK models incorporating environmentally relevant exposure routes (inhalation, dermal, and oral) have been used to describe the uptake and disposition of a homologous series of chloroethanes in both rainbow trout and channel catfish.[1-5] Additional fish models have been developed to describe the tissue disposition and biotransformation of selected chemicals that were introduced directly into the general circulation.[6-8] These prototype models demonstrate the feasibility of using a PB-TK modeling approach with aquatic vertebrates. However, continued progress toward the development of predictive models for use in aquatic toxicology and environmental risk assessment will require significant advances in our understanding of *biotransformation reactions, chemical partitioning, toxicity mechanisms (toxicodynamics),* and *oral uptake dynamics.* In addition, research efforts must be directed toward understanding the uptake and disposition of ionized chemicals (organic and inorganic) and to improving our knowledge of the physiology and biochemistry of as yet untested vertebrate and invertebrate species.

A. Biotransformation

If PB-TK models are to accurately predict tissue dose metrics for a wide variety of chemicals and provide for meaningful species extrapolations, research emphasis must be placed on the development and validation of rapid, in vivo and in vitro methods for predicting biotransformation products and rates. One approach discussed earlier in this chapter is to fit model simulations to the observed experimental data by solving for apparent in vivo rate and capacity parameters. This technique has been used successfully in gas uptake studies with rats,[44] and its value lies in its relative simplicity and in its ability to represent complex patterns of metabolism.[100,101,106] A similar approach was used in PB-TK modeling efforts with rainbow trout to quantitate systemic elimination of chloroethanes through the analysis of gill extraction efficiency data.[1,2] It must be noted, however, that because these in vivo approaches rely on a process of curve fitting, they are not predictive. Moreover, their implementation is critically dependent upon the accuracy of the other parameters in the model.

Early investigators suggested that in vitro experimental systems could be used to estimate metabolic rate and capacity parameters for incorporation into PB-TK models,[109,110] but the utility of this approach has yet to be demonstrated conclusively. In vitro methods generally require incubation of the chemical in question with liver microsomal preparations, hepatocyte cultures, or whole-liver homogenates.[111-115] Many of the problems encountered with these approaches stem from not having the same inter- and intracellular environment present in both the in vivo and in vitro systems. However, recent efforts reviewed by Wilkinson[116] are encouraging, as they provide a dataset for several classes of chemicals, which seems to indicate reasonably good agreement between in vitro values generated with both hepatocyte cultures and microsomal preparations and metabolism values measured in vivo (Table 5). In vitro incubation studies, utilizing both isolated hepatocytes and microsomal preparations, have also been conducted with fish,[117-119] but there are presently no studies to assess the potential of using these data to predict in vivo metabolism.

Two new approaches hold considerable promise for future progress in animal metabolism studies. The first involves the use of in vivo microdialysis. With this technique, it is possible to follow the kinetics of both parent compounds and their major metabolites in the blood and tissues of intact animals.[120] In a recent study, in vivo microdialysis was employed to monitor the concentrations of phenol and phenylglucuronide (a primary metabolite of phenol) in the bloodstream of rainbow trout, demonstrating the utility of this method for applications in aquatic toxicology.[121]

The second approach involves the use of precision-cut organ slices. The potential benefits of an organ slice preparation relate to the maintenance of cellular integrity. Technical problems associated with the isolation and culture of isolated hepatocytes have the potential to damage cells, and, once isolated, these cells tend to lose metabolic activity and become dedifferentiated. Smith and co-workers[122] reviewed the potential for application of cultured liver slices in biotransformation and toxicity studies. More recently, research with a variety of substituted bromobenzenes demonstrated that cultured liver slices were more representative of in vivo toxicity than were isolated hepatocytes.[123] To date, however, this technique has not been applied in studies involving aquatic organisms.

B. Partitioning of Nonvolatiles

A second critical research need to advance the use of PB-TK models in aquatic as well as mammalian toxicology is the development of reliable techniques for the determination of chemical-specific, equilibrium partition coefficients for nonvolatile hydrophobic (log K_{ow} >4.0) compounds. Partition coefficients for volatile compounds can be readily obtained using established vial equilibration methods,[96-98] while the partitioning of chemicals that reach steady-state rapidly can be measured in vivo, as demonstrated for a series of drugs by Schuhmann and co-workers.[124] Presently, however, there is no easy way to obtain partition coefficients for nonvolatile chemicals that come to steady-state only after a long exposure period. Since these compounds are often the ones causing the most environmental concern, it is essential that we continue to search for a reliable way to estimate their equilibrium-partitioning behavior. To date, some success has been achieved in the determination of in vitro partition coefficients for nonvolatiles through the application of equilibrium dialysis[99,125,126] and ultrafiltration of tissue homogenates.[127] However, further studies on these and other approaches are clearly required.

Table 5. Comparison Between the In Vivo Hepatic Extraction Ratio in the Rat and Other Species and That Predicted by the Venous Equilibration Model, Based on In Vitro Estimates of V_{max} and K_m[a]

Drug	Preparation	In Vitro prediction	In Vivo measurement	Metabolic pathway
Acetaminophen	Hepatocytes	0.7–0.72	0.67	Sulfation/gluronidation
Alprenolol	9000 × g supernate	0.92	>0.90	Oxidation
	9000 × g supernate	0.90	0.98	
	Microsomes	0.60	0.98	
	Hepatocytes	0.75	0.98	
Antipyrine	9000 × g supernate	0.04	0.01	Oxidation
Carbamazepine	9000 × g supernate	0.05	0.04	Oxidation
Chlorpheniramine	600 × g supernate[b]	0.89	0.67–0.84	Oxidation
Felodipine	Microsomes	0.62	0.65	Oxidation
	Microsomes	0.91	0.80	
	Microsomes[c]	0.80	0.83	
	Microsomes[d]	0.80	0.84	
Hexobarbital	9000 × g supernate	0.44	0.33	Oxidation
5-Hydroxytryptamine	Homogenate	0.68	0.59	MAO
Lidocaine	9000 × g supernate	0.80	>0.90	Oxidation
Morphine	Microsomes[e]	0.09–0.14	0.61–0.71	Glucuronidation
Phenacetin	Hepatocytes	0.54	0.87–0.95	Oxidation
Phenytoin	9000 × g supernate	0.50	0.53	Oxidation
	Microsomes	0.03	0.05	
Propranolol	9000 × g supernate	0.83	>0.90	Oxidation

Source: Wilkinson, G.R., *Pharmacokinetics in Risk Assessment, Vol. 8,* Drinking Water and Health, National Academic Press, Washington, D.C., 1987, 80. With permission.

[a] Extraction ratios (E) were calculated from in vitro information according to the relationship: $E = CL_{int} \cdot f_b/(Q + CL_{int} \cdot f_b)$; where CL_{int} = free intrinsic clearance rate = V_{max}/K_m, f_b = fraction of substrate that is unbound in blood, and Q = blood flow to the organ. This treatment assumes that reaction conditions are first order. The symbols V_{max} and K_m represent the sum of all metabolic activity responsible for the disappearance of substrate.
[b] Rabbit.
[c] Dog.
[d] Human.
[e] Monkey.

C. Toxicodynamics

PB-TK models provide the necessary toxicokinetic dose metrics for tissue exposure and relate administered or applied dose to actual target tissue or target organ dose. The next step and a critical future need involves the development of toxicodynamic (PB-TR) models to describe the chemical/biological interactions (toxic mechanisms) at the target organ that ultimately lead to a toxic response. Such models do not have to include a complete description of the cellular events or biochemical mechanisms leading to toxicity, but should address general aspects of the nature and causes of a particular toxic response.

In order to link PB-TK and PB-TR models, it will be necessary to determine the dose metric that is most appropriate for a given mode of action. Accordingly, some of the more important questions that must be considered in choosing the correct expression of tissue dose are:[30]

1. Is the effect related to generalized chemical reactivity or is it due to binding to a specific cellular receptor?
2. Is toxicity caused by the parent chemical or a metabolite?
3. If a metabolite, is it highly reactive with a short half-life or does it have a long half-life and circulate throughout the body?
4. Are the effects reversible cytotoxic effects or irreversible carcinogenic changes?
5. If the effect is cancer induction, is it genotoxic or epigenetic in origin?

D. Oral Uptake Dynamics

Until recently, a major portion of the research on the uptake of xenobiotic chemicals by fish and other aquatic animals was concentrated on the diffusion of dissolved chemical across the respiratory surface.[58,108] However, present information suggests that for very hydrophobic chemicals, (log K_{ow} >6.0) chemical ingestion in the diet presents a major exposure route.[50-52] In the water column, these hydrophobic chemicals rapidly adsorb to particulates (including plankton) and dissolved organic material, making them unavailable for diffusion across the respiratory surface.[53a,58] Nevertheless, these chemicals can still get into the organism through consumption of contaminated food or ingestion of sediments. The mechanisms controlling intestinal absorption of xenobiotic chemicals by fish are just beginning to be understood and must be explored further. In addition, research on the capacity of the fish intestine to directly metabolize xenobiotics as they are absorbed from the gut lumen is highly noteworthy and should be continued. A recent review by Van Veld[78] provides an excellent overview of research progress through the last decade on the anatomical, physiological, biochemical, and absorptive attributes of the fish gut.

Dietary uptake of xenobiotic chemicals by fish is frequently characterized by introducing the chemical via gavage into the stomach and determining an apparent first-order rate constant for absorption. This technique is somewhat controversial, however, because of the important influence of the carrier (organic solvent, corn oil, synthetic diet, etc.) on chemical bioavailability.[78] With regard to future PB-TK model development, a major research need is to complete the basic research necessary to formulate and parameterize detailed descriptions of the fish gut exchange surface. Once this is accomplished, fish PB-TK models for extrapolation across all routes of environmental exposure (inhalation, dermal, and oral) will become a reality.

Finally, it should be reemphasized here that once a PB-TK model is established for a given chemical and species, the route of exposure can be changed by inserting into the model a mathematical description for the new exchange surface (Figures 2 and 5). The rest of the inputs to the model (physiology, metabolism, organ volumes, etc.) remain fixed.

X. FUTURE APPLICATIONS

PB-TK models are rapidly presently being used in human health risk assessments for noncarcinogenic as well as carcinogenic compounds, and their utility is destined to grow in the future. Similarly, there is increased interest in the development of PB-TK models for fish, and we anticipate that in the future these models will become important tools for protecting the aquatic environment.

A. Fish Models

The major future challenges for PB-TK models in aquatic toxicology are: (1) accurate predictions of parent chemical and/or metabolite dose metrics associated with specific toxic events, (2) interspecies extrapolation of dosimetry information regardless of whether the route of exposure was inhalation, dermal, or oral, and (3) development of linkages between tissue dosimetry (PB-TK) models and tissue response (PB-TR) models. These capabilities would contribute to the development of highly accurate, scientifically defensible, environmental risk assessments and provide the information necessary to compare tissue and cellular sensitivity of aquatic animals exposed to specific chemicals on a dose basis (milligram per kilogram). In addition, meeting these challenges would reduce the costs of testing chemicals on a variety of species, curtail the need for large numbers of test animals, and allow aquatic toxicologists to expand, through extrapolation, their useful toxicity database by including toxicity data on mammalian species as well as a variety of lower vertebrates.

The future routine use of fish PB-TK models in environmental risk assessment will depend in large part on the ability to develop the necessary databases required to supply critical model inputs for aquatic vertebrates. A schematic that describes the information essential for fish PB-TK models is presented in Figure 13. It emphasizes three major database inputs for a complete model.[128] These are (1) information on toxicity mechanisms and dose-response, (2) species-specific biological information, and (3) species-specific information on special systems capable of altering sensitivity to certain chemicals. By combining information contained within these databases with models such as those described earlier in this chapter, dose-response relationships can be predicted for tissues and animals, as outlined in Figure 13.

The most complex and critical database required for effective PB-TK modeling contains the available tissue-response and dose-response data sets for different species and is referred to in Figure 13 as the Critical Toxicity Reference System.[128] Such a database is available for mammalian species;[129] however, a similar one for aquatic species must be developed from the present Aquatic Toxicity Information Retrieval (AQUIRE) System.[130] This will be a costly effort, but one that is critical for the future use of PB-TK models in aquatic toxicology and environmental risk assessment.

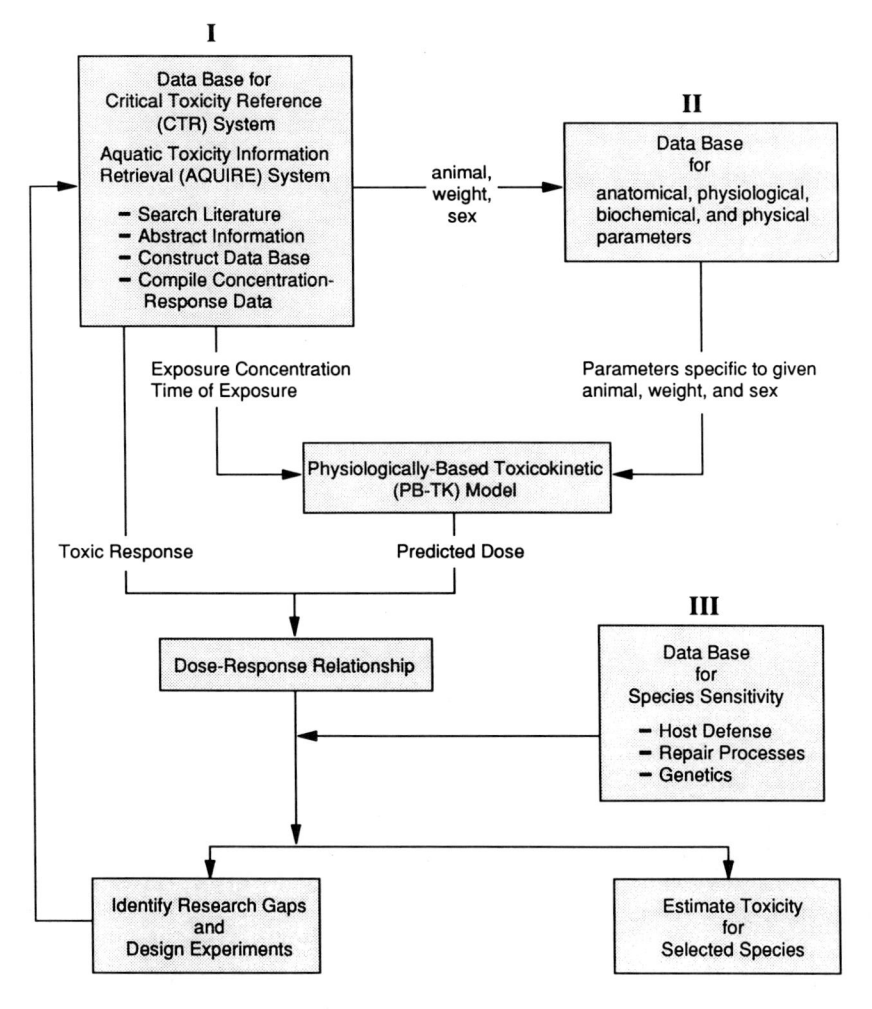

Figure 13. Elements of an integrated PB-TK modeling approach, emphasizing the major database inputs (modified from Miller et al.[126]).

B. Predictive Toxicology

At present, aquatic toxicology and environmental risk assessment are driven primarily by descriptive studies involving sets of toxicity tests designed to answer specific questions about the responses of selected aquatic animals to chemicals known to be present in the environment. This means that, in most cases, laboratory toxicity tests (acute, subacute, subchronic, and chronic) with single chemicals and selected species still provide the major source of dose-response information for toxicity databases. This descriptive approach is an essential part of toxicological research, but it is slow and costly and cannot hope to keep abreast of an environmental chemical inventory that now numbers 50,000 industrial chemicals, and the 1000 new chemical structures cleared for production each year that may eventually find their way into aquatic ecosystems.[131,132]

If aquatic toxicologists are to make any progress on a toxicological evaluation of this large and growing chemical inventory, a predictive capability must be developed. Two scientific approaches that can provide this predictive capability are quantitative structure activity relationships (QSAR) and PB-TK models.

QSAR equations relate the physical and chemical attributes of a compound to its effects on biological material. The utility of QSAR methods in aquatic toxicology was first established by Koneman, Hermens, Veith, and co-workers,[133-138] and these equations are now routinely used in environmental risk assessment to predict the toxicity of structurally similar chemicals. The value of the QSAR approach is that once equations are developed for a given mode of action, toxic chemical concentrations in the exposure water can be accurately predicted for specific classes of chemicals without the necessity of conducting toxicity tests. The general knowledge of toxic mechanisms required for development of QSAR equations is provided by studying the differences between toxic response sets [fish acute toxicity syndromes (FATS)] elicited by specific classes of chemicals.[139-143] These predictions are presently limited primarily to acute responses to nonreactive chemicals with nonspecific modes of action (e.g., narcosis). However, current research efforts are aimed at formulating acute and chronic QSAR equations for reactive chemicals, that take into account the effects of both parent chemicals and metabolites.

PB-TK models can be used to build upon QSAR efforts by extending chemical extrapolations based on chemical structure to biological extrapolations across species and routes of exposure. Thus, empirically derived QSAR equations allow toxicologists to predict water concentrations, and in some cases whole body residues, that are likely to produce a toxic response. However, once the correct dose metric is selected for a particular mode of toxic action represented by a specific QSAR equation, then the toxicity of a whole class of chemicals can be extrapolated to a second species simply by adjusting appropriate physiological-biochemical inputs to a PB-TK model.

Finally, we anticipate that, in the future, scientists will use QSAR methods to link structural attributes of a chemical to: (1) in vivo biotransformation, (2) specific tissue and molecular toxicological responses, and (3) tissue partitioning and cellular binding affinities. At this point, QSAR will extend PB-TK models by providing, from chemical structure alone, the major inputs required to parameterize PB-TK models and to link PB-TK and PB-TR models.

XI. CONCLUSIONS

Fish dosimetry (PB-TK) models have been developed, parameterized, and validated for use in aquatic toxicology and risk assessment. The importance of these models to basic mechanistic toxicology lies in their ability to provide tissue dose metrics that can be related to responses monitored at the target organ and molecular levels of organization, while in environmental risk assessment, their value derives from their use for extrapolation across dosing regimes, species, and routes of exposure.

Biological response (PB-TR) models describe the key processes and their rates that result in an observed toxicological response. Our understanding of these processes is in the formative stages of development. In time, however, the linkage of PB-TK and PB-TR models will provide complete quantitative descriptions of the events that underlie toxic responses to chemical stressors and, in so doing, will bring together the two major areas of toxicology: toxicokinetics and toxicodynamics.

REFERENCES

1. Nichols, J.W., McKim, J.M., Andersen, M.E., Gargas, M.L., Clewell, H.J., III, and Erickson, R.J., A physiologically based toxicokinetic model for the uptake and disposition of waterborne organic chemicals in fish, *Toxicol. Appl. Pharmacol.*, 106, 433, 1990.
2. Nichols, J.W., McKim, J.M., Lien, G.J., Hoffman, A.D., Bertelsen, S.L., and Elonen, C.M., A physiologically-based toxicokinetic model for dermal absorption of organic chemicals by fish, *Environ. Toxicol. Chem.*, 1994, submitted.
3. McKim, J.M. and Nichols, J.W., Physiologically-based toxicokinetic modeling of the dermal uptake of three waterborne chloroethanes in rainbow trout *(Oncorhynchus mykiss), Toxicologist*, 11(Abstract), 35, 1991.
4. Nichols, J.W., McKim, J.M., and Hoffman, A., Physiologically-based modeling of the oral uptake of two chloroethanes in the rainbow trout *(Oncorhynchus mykiss)*, in Development and Validation of a Fish Physiologically-Based Toxicokinetic Model for Use in Environmental Risk Assessment, final report submitted to Air Force Office of Scientific Research, Air Force Systems Command, United States Air Force, Grant AFOSR-ISSA-89-0060, 1992.
5. Nichols, J.W., McKim, J.M., Lien, G.J., Hoffman, A.D., and Bertelsen, S.L., Physiologically-based toxicokinetic modeling of three waterborne chloroethanes in channel catfish, *Ictaturus punctatus, Aquat. Toxicol.*, 27, 83, 1993.
6. Zaharko, D.S., Dedrick, R.L., and Oliverio, V.T., Prediction of the distribution of methotrexate in the sting rays *Dasyatidae sabina* and *sayi* by use of a model developed in mice, *Comp. Biochem. Physiol.*, 42, 183, 1972.7.
7. Bungay, P.M., Dedrick, R.L., and Guarino, A.M., Pharmacokinetic modeling of the dogfish shark *(Squalus acanthus):* distribution and urinary and biliary excretion of phenol red and its glucuronide, *J. Pharmacokin. Biopharm.*, 4, 377, 1976.
8. Law, F.C.P., Abedini, S., and Kennedy, C.J., A biologically based toxicokinetic model for pyrene in rainbow trout, *Toxicol. Appl. Pharmacol.*, 110, 390, 1991.
9. Levine, R.R., *Pharmacology: Drug Actions and Reactions*, Little, Brown and Co., Boston, MA, 1987, 513.
10. Rand, G.M. and Petrocelli, S.R., Eds., Introduction, in *Fundamentals of Aquatic Toxicology: Methods and Applications*, Hemisphere, Washington, D.C., 1985, chap. 1.

11. McCarty, L.S., The relationship between aquatic toxicity QSARs and bioconcentration for some organic chemicals, *Environ. Toxicol. Chem.*, 5, 1071, 1986.
12. McCarty, L.S., Relationship between toxicity and bioconcentration for some organic chemicals. I. Examination of the relationship, in *QSAR in Environmental Toxicology — II,* Kaiser, K.L.E., Ed., D. Riedel, Dordrecht, Holland, 1987, 207.
13. McCarty, L.S., Relationship between toxicity and bioconcentration for some organic chemicals. II. Application of the relationship, in *QSAR in Environmental Toxicology — II,* Kaiser, K.L.E., Ed., D. Riedel, Dordrecht, Holland, 1987, 221.
14. McKim, J.M. and Goeden, H.M., A direct measure of the uptake efficiency of a xenobiotic chemical across the gills of brook trout *(Salvelinus fontinalis)* under normoxic and hypoxic conditions, *Comp. Biochem. Physiol.*, 72, 65, 1982.
15. McKim, J.M. and Heath, E.M., Dose determinations for waterborne 2,5,2′,5′-[^{14}C]tetrachlorobiphenyl and related pharmacokinetics in two species of trout *(Salmo gairdneri* and *Salvelinus fontinalis):* a mass-balance approach, *Toxicol. Appl. Pharmacol.*, 68, 177, 1983.
16. Black, M.C., Millsop, D.S., and McCarthy, J.F., Effects of acute temperature change on respiration and toxicant uptake by rainbow trout *(Salmo gairdneri), Physiol. Zool.*, 64, 145, 1991.
17. Gibaldi, M. and Perrier, D., *Pharmacokinetics, Vol. 1, Drugs and the Pharmaceutical Sciences,* Marcel Dekker, New York, 1975, 329.
18. Clewell, H.J., III and Andersen, M.E., Risk assessment extrapolations and physiological modeling, *Toxicol. Ind. Health*, 1, 111, 1985.
19. Bischoff, K.B. and Brown, R.H., Drug distribution in mammals, *Chem. Eng. Prog. Symp. Ser.*, 62, 33, 1966.
20. Dedrick, R.L., Animal scale-up, *J. Pharmacokinet. Biopharm.*, 1, 435, 1973.
21. Ramsey, J.C. and Andersen, M.E., A physiologically based description of the inhalation pharmacokinetics of styrene in rats and humans, *Toxicol. Appl. Pharmocol.*, 73, 159, 1984.
22. Karara, A.H. and Hayton, W.L., Pharmacokinetic model for the uptake and disposition of di-2-ethylhexyl phthalate in sheepshead minnow *(Cyprinodon variegatus), Aquat. Toxicol.*, 5, 181, 1984.
23. Stehly, G.R. and Hayton, W.L., Disposition of pentachlorophenol in rainbow trout *(Salmo gairdneri):* effect of inhibition of metabolism, *Aquat. Toxicol.*, 14, 1, 131, 1989.
24. Barron, M.G., Schultz, I.R., and Hayton, W.L., Presystemic branchial metabolism limits di-2-ethylhexyl phthalate accumulation in fish, *Toxicol. Appl. Pharmacol.*, 98, 49, 1990.
25. Barron, M.G., Stehly, G.R., and Hayton, W.L., Pharmacokinetic modeling in aquatic animals. I. Models and concepts, *Aquat. Toxicol.*, 17, 187, 1990.
26. Andersen, M.E., Physiological modeling of tissue dosimetry, *CIIT Activ.*, 9, 5–6, 1, 1989.
27. Clewell, H.J., III and Andersen, M.E., A multiple dose-route physiological pharmacokinetic model for volatile chemicals using ACSL/PC, in *Languages for Continuous System Simulation*, Cellier, F.D., Ed., Society for Computer Simulation, San Diego, CA, 1986, 95.
28. Clewell, H.J., III and Andersen, M.E., Improving toxicology testing protocols using computer simulations, *Toxicol. Lett.*, 49, 139, 1989.

29. Bischoff, K.B. Physiologically Based Pharmacokinetic Modeling, in *Pharmacokinetics in Risk Assessment, Vol. 8, Drinking Water and Health,* National Academic Press, Washington, D.C., 1987, 36.
30. Anderson, M.E., Tissue dosimetry in risk assessment or What's the Problem here anyway?, in *Pharmacokinetics in Risk Assessment, Vol. 8,* Drinking Water and Health, National Academic Press, Washington, D.C., 1987, 8.
31. Conolly, R.B., Biologically-based models for toxic effects: tools for hypothesis testing and improving health risk assessments, *CIIT Activ.,* 10, 5, 1990.
32. Conolly, R.B., Reitz, R.H., Clewell, H.J., III, and Andersen M.E., Pharmacokinetics, biochemical mechanism and mutation accumulation: a comprehensive model of chemical carcinogensis, *Toxicol. Lett.,* 43, 189, 1988.
33. Conolly, R.B., Reitz, R.H., Clewell, H.J., III, and Anderson, M.E., Biologically-structured models and computer simulation: application to chemical carcinogenesis, *Comm. Toxicol.,* 2, 305, 1988.
34. Tuey, D.B., Toxicokinetics, in *Introduction to Biochemical Toxicology,* Hodgson, E. and Guthrie, F.E., Eds., Elsevier, New York, 1980, 40.
35. Gerlowski, L.E. and Jain, R.K., Physiologically based pharmacokinetic modeling: principles and applications, *J. Pharm. Sci.,* 72, 1103, 1983.
36. Rowland, M., Physiologic pharmacokinetic models and interanimal species scaling, *Pharmac. Ther.,* 29, 49, 1985.
37. Smith, L.S. and Bell, G.R., A Practical Guide to the Anatomy and Physiology of Pacific Salmon, Miscellaneous Special Publication 27, Department of the Environment, Fisheries and Marine Service, Ottawa, Canada, 1975.
38. Thorarensen, H., McLean, E., Donaldson, E.M., and Farrell, A.P., The blood vasculature of the gastrointestinal tract in chinook, *Oncorhynchus tshawytscha* (Walbaum), and coho, *O. kisutch* (Walbaum), salmon, *J. Fish Biol.,* 38, 525, 1991.
39. Barron, M.G., Tarr, B.D., and Hayton, W.L., Temperature-dependence of cardiac output and regional blood flows in rainbow trout, *(Salmo gairdneri)* Richardson, *J. Fish Biol.,* 31, 735, 1987.
40. Sipes, G.I. and Gandolphi, A.J., Biotransformation of toxicants, in *Cassarett and Doull's Toxicology,* 3rd ed., Klaassen, C.D., Amdur, M.O., and Doull, J.D., Eds., MacMillan, New York, 1986, 64.
41. James, M.O., Conjugation of organic pollutants in aquatic species, *Environ. Health Perspect.,* 71, 97, 1987.
42. Kleinow, K.M., Melancon, M.J., and Lech J.J., Biotransformation and induction: implications for toxicity, bioaccumulation and monitoring of environmental xenobiotics in fish, *Environ. Health Perspect.,* 71, 105, 1987.
43. Buhler, D.R. and Williams, D.E., The role of biotransformation in the toxicity of chemicals, *Aquat. Toxicol.,* 11, 19, 1988.
44. Gargas, M.L., Andersen, M.E., and Clewell, H.J., III, A physiologically based simulation approach for determining metabolic constants from gas uptake data, *Toxicol. Appl. Pharmacol.,* 86, 341, 1986.
45. James Curtis, B. and Wood, C.M., The function of the urinary bladder *in vivo* in the freshwater rainbow trout, *J. Exp. Biol.,* 155, 567, 1991.
46. Schmidt, D.C. and Weber, L.J., Metabolism and biliary excretion of sulfobromophthalein by rainbow trout *(Salmo gairdneri), J. Fish. Res. Board Can.,* 30, 1301, 1973.

47. Adolph, E.F., Quantitative relations in the physiological constitutions of mammals, *Science*, 109, 579, 1949.
48. Schmidt-Nielsen, K., *Scaling: Why is Animal Size So Important?*, Cambridge University Press, Cambridge, England, 1984.
49. Travis, C.C., White, R.K., and Ward, R.C., Interspecies extrapolation of pharmacokinetics, *J. Theor. Biol.*, 142, 285, 1990.
50. Bruggeman, W.A., Opperhuizen, A., Wijbenga, A., and Hutzinger, O., Bioaccumulation of superlipophilic chemicals in fish, *Toxicol. Environ. Chem.*, 7, 173, 1984.
51. Thomann, R.V. and Connolly, J.P., Model of PCB in the Lake Michigan lake trout food chain, *Environ. Sci. Technol.*, 18, 65, 1984.
52. Oliver, B.G. and Niimi, A.J., Bioconcentration factors of some halogenated organics for rainbow trout: limitations in their use for prediction of environmental residues, *Environ. Sci. Technol.*, 19, 842, 1985.
53a. Black, M.C. and McCarthy, J.F., Dissolved organic macromolecules reduce the uptake of hydrophobic organic contaminants by the gills of rainbow trout *(Salmo gairdneri)*, *Environ. Toxicol. Chem.*, 7, 593, 1988.
53b. Lien, G.J. and McKim, J.M. Predicting branchial and cutaneous uptake of 2,2',5,5'-tetrachlorobiphenyl in fathead minnows (*Pimephales promelas*) and Japanese medaka (*Oryzias latipes*): rate limiting factor, *Aquat. Toxicol.*, 27, 15, 1993.
54. McKim, J., Schmieder, P., and Veith, G., Absorption dynamics of organic chemical transport across trout gills as related to octanol-water partition coefficient, *Toxicol. Appl. Pharmacol.*, 77, 1, 1985.
55. Saarikoski, J., Lindstrom, M., Tyynila, M., and Viluksela, M., Factors affecting the absorption of phenolics and carboxylic acids in the guppy *(Poecilia reticulata)*, *Ecotoxicol. Environ. Safety*, 11, 158, 1986.
56. Erickson, R.J. and McKim, J.M., A simple flow-limited model for exchange of organic chemicals at fish gills, *Environ. Toxicol. Chem.*, 9, 159, 1990.
57. Erickson, R.J. and McKim, J.M., A model for exchange of organic chemicals at fish gills: flow and diffusion limitations, *Aquat. Toxicol.*, 18, 175, 1990.
58. McKim, J.M. and Erickson, R.J., Environmental impacts on the physiological mechanisms controlling xenobiotic transfer across fish gills, *Physiol. Zool.*, 64, 39, 1991.
59. Gobas, F.A.P.C. and MacKay, D., Dynamics of hydrophobic organic chemical bioconcentration in fish, *Environ. Toxicol. Chem.*, 6, 495, 1987.
60. Barber, M.C., Suarez, L.A., and Lassiter, R.R., Bioconcentration of nonpolar organic pollutants by fish, *Environ. Toxicol. Chem.*, 7, 545, 1988.
61. Hughes, G.M., General anatomy of the gills, in *Fish Physiology*, Vol. 10, Hoar, W.S. and Randall, D.J., Eds., Academic Press, New York, 1984, 1.
62. Wedemeyer, G.A., Meyer, F.P., and Smith, L., *Environmental Stress and Fish Diseases*, T.F.H. Publications, Neptune, NJ, 1982, 102.
63. Schmeider, P.K. and Weber, L.J., Blood flow and water flow limitations on gill uptake of organic chemicals in the rainbow trout (*Oncorhynchus mykiss*), *Aquat. Toxicol.*, 24, 103, 1992.
64. Taylor, E.W., Short, S., and Butler, P.J., The role of the cardiac vagus in the response of the dogfish *Scyliorhinus canicula* to hypoxia, *J. Exp. Biol.*, 70, 57, 1977.
65. Cameron, J.N. and Davis, J.C., Gas exchange in rainbow trout *(Salmo gairdneri)* with varying blood oxygen capacity, *J. Fish. Res. Board Can.*, 27, 1069, 1970.
66. Stevens, E.D., Some aspects of gas exchange in tuna, *J. Exp. Biol.*, 56, 809, 1972.

67. Watters, K.W., Jr. and Smith, L.S., Respiratory dynamics of the starry flounder *Platichthys stellatus* in response to low oxygen and high temperature, *Mar. Biol.*, 19, 133, 1973.
68. Eddy, F.B., Blood gases of the tench *(Tinca tinca)* in well aerated and oxygen-deficient waters, *J. Exp. Biol.*, 60, 71, 1974.
69. Cech, J.J., Rowell, D.M., and Glasgow, J.S., Cardiovascular responses of the winter flounder *Pseudopleuronectes americanus* to hypoxia, *Comp. Biochem. Physiol.*, 57, 123, 1977.
70. Wood, C.M., McMahon, B.R., and McDonald, D.G., Respiratory gas exchange in the resting starry flounder, *Platichthys stellatus*: a comparison with other teleosts, *J. Exp. Biol.*, 78, 167, 1979.
71. Burggren, W.W. and Cameron, J.N., Anaerobic metabolism, gas exchange, and acid-base balance during hypoxic exposure in the channel catfish *(Ictalurus punctatus)*, *J. Exp. Zool.*, 213, 405, 1980.
72. Takeda, T., Ventilation, cardiac output and blood respiratory parameters in the carp *(Cyprinus carpio)* during hyperoxia, *Respir. Physiol.*, 81, 227, 1990.
73. Ure, D. and Rombough, P.J., Cutaneous and branchial contributions to oxygen uptake by alevins of chinook salmon *(Oncorhynchus tshawytscha)*, Proceedings of the 23rd Annual Prairie Universities Biological Seminars, cited by Rombough, P.J. and Moroz, B.M., *J. Exp. Biol.*, 154, 1, 1990.
74. McDougal, J.N., Jepson, G.W., Clewell H.J., III, MacNaughton, M.G., and Andersen, M.E., A physiological pharmacokinetic model for dermal absorption of vapors in the rat, *Toxicol. Appl. Pharmacol.*, 85, 286, 1986.
75. Bungay, P.M., Dedrick, R.L., and Matthews, H.B., Enteric transport of chlordecone (Kepone®) in the rat, *J. Phamacokinet. Biopharm.*, 9, 309, 1981.
76. Kapoor, B.G., Smith, H., and Verighina, I.A., The alimentary canal and digestion in teleosts, *Adv. Mar. Biol.*, 13, 109, 1975.
77. Buddington, R.K. and Diamond, J.M., Pyloric ceca of fish: a "new" absorptive organ, *Am. J. Physiol.*, 252, G65, 1987.
78. Van Veld, P.A., Absorption and metabolism of dietary xenobiotics by the intestine of fish, *Aquat. Sci.*, 2, 185, 1990.
79. Holeton, G.F. and Randall, D.J., The effect of hypoxia upon the partial pressure of gases in the blood and water afferent and efferent to the gills of the rainbow trout, *J. Exp. Biol.*, 317, 1967.
80. Stevens, E.D., The effect of exercise on the distribution of blood to various organs in rainbow trout, *Comp. Biochem. Physiol.*, 25, 615, 1968.
81. Davis, J.C. and Cameron, J.N., Water flow and gas exchange at the gills of rainbow trout *(Salmo gairdneri)*, *J. Exp. Biol.*, 54, 1, 1971.
82. Denton, J.E. and Yousef, M.K., Body composition and organ weights of rainbow trout *(Salmo gairdneri)*, *J. Fish Biol.*, 8, 489, 1976.
83. Kiceniuk, J.W. and Jones, D.R., The oxygen transport system in trout *(Salmo gairdneri)* during sustained exercise, *J. Exp. Biol.*, 69, 247, 1977.
84. Wood, C.M. and Shelton, G., The reflex control of heart rate and cardiac output in the rainbow trout: interactive influences of hypoxia, haemmorhage, and systemic vasomotor tone, *J. Exp. Biol.*, 87, 271, 1980.
85. Neumann, P., Holeton, G.F., and Heisler, N., Cardiac output and regional blood flow in gills and muscles after exhaustive exercise in rainbow trout *(Salmo gairdneri)*, *J. Exp. Biol.*, 105, 1, 1983.

86. Randall, D.J., The control of respiration and circulation in fish during exercise and hypoxia, *J. Exp. Biol.*, 100, 275, 1982.
87. Burggren, W.W. and Cameron, J.N., Anaerobic metabolism, gas exchange, and acid-base balance during hypoxic exposure in the channel catfish *(Ictalurus punctatus)*, *J. Exp. Zool.*, 213, 405, 1980.
88. McKim, J.M., Nichols, J.W., Lien, G.J., and Bertelsen, S.L., Respiratory-cardiovascular physiology and chloroethane gill flux in the channel catfish *(Ictalurus punctatus)* (Rafinesque), *J. Fish Biol.*, 1993, in press..
89. Glass, M.L., Andersen, N.A., Kruhoffer, M., Williams, E.M., and Heisler, N., Combined effects of environmental PO_2 and temperature on ventilation and blood gases in the carp *Cyprinus carpio* L., *J. Exp. Biol.*, 148, 1, 1990.
90. Piiper, J., Baumgarten, D., and Meyer, M., Effects of hypoxia upon respiration and circulation in the dogfish *Scyliorhinus stellaris*, *Comp. Biochem. Physiol.*, 36, 513, 1970.
91. Kent, B., Pierce, M., and Pierce, E.C., III, Blood flow distribution in *Squalus acanthias:* a sequel, *Bull. Mt. Desert Isl. Biol. Lab.*, 13, 64, 1973.
92. Metcalfe, J.D. and Butler, P.J., Differences between directly measured and calculated values for cardiac output in the dogfish: a criticism of the Fick method, *J. Exp. Biol.*, 99, 255, 1982.
93a. Hoar, W.S and Randall, D.J., Eds., *Fish Physiology*, Vols. 1–11, Academic Press, New York.
93b. Hoar, W.S., Randall, D.J., and Farrell, A.P., Eds., *Fish Physiology*, Vol. 12, Academic Press, New York, 1992.
94. Satchell, G.H., *Physiology and Form of Fish Circulation*, Cambridge University Press, Cambridge, England, 1991.
95. Cameron, J.N., Blood flow distribution as indicated by tracer microspheres in resting and hypoxic arctic grayling *(Thymallus arcticus)*, *Comp. Biochem. Physiol.*, 52, 441, 1975.
96. Sato, A. and Nakajima, T., A vail-equilibration method to evaluate the drug-metabolizing enzyme activity for volatile hydrocarbons, *Toxicol. Appl. Pharmacol.*, 47, 41, 1979.
97. Gargas, M.L., Burgess, R.J., Voisard, D.E., Cason, G.H., and Andersen, M.E., Partition coefficients of low-molecular-weight volatile chemicals in various liquids and tissues, *Toxicol. Appl. Pharmacol.*, 98, 87, 1989.
98. Hoffman, A.D., Bertelsen, S.L., and Gargas, M.L., An *in vitro* gas equilibration method for determination of chemical partition coefficients in fish, *Comp. Biochem. Physiol.*, 101, 47, 1992.
99. Sultatos, L.G., Kim, B., and Woods, L., Evaluation of estimations *in vitro* of tissue/blood distribution coefficients for organothiophosphate insecticides, *Toxicol. Appl. Pharmacol.*, 103, 52, 1990.
100. Gargas, M.L., Clewell, H.J., III, and Andersen, M.E., Metabolism of inhaled dihalomethanes *in vivo:* differentiation of kinetic constants for two independent pathways, *Toxicol. Appl. Toxicol.*, 82, 211, 1986.
101. Reitz, R.H., Mendrala, A.L., Park, C.N., Andersen, M.E., and Guengerich, F.P., Incorporation of *in vitro* enzyme data into the physiologically-based pharmacokinetic (PB-PK) model for methylene chloride: implications for risk assessment, *Toxicol. Lett.*, 43, 97, 1988.

102. Farrar, D., Allen, B., Crump, K., and Shipp, A., Evaluation of uncertainty in input parameters to pharmacokinetic models and the resulting uncertainty in output, *Toxicol. Lett.*, 49, 371, 1989.
103. Bois, F.Y., Zeise, L., and Tozer, T.N., Precision and sensitivity of pharmacokinetic models for cancer risk assessment: tetrachloroethylene in mice, rats, and humans, *Toxicol. Appl. Pharmacol.*, 102, 300, 1990.
104. Bischoff, K.B., Dedrick, R.L., Zaharko, D.S., and Longstreth, J.A., Methotrexate pharmacokinetics, *J. Pharm. Sci.*, 60, 1128, 1971.
105. Guarino, A.M. and Anderson, J.B., Biliary and urinary excretion of phenol red and its glucuronide in the dogfish shark *(Squalus acanthias), Xenobiotica*, 6, 1, 1976.
106. Gargas, M.L. and Andersen, M.E., Determining kinetic constants of chlorinated ethane metabolism in the rat from rates of exhalation, *Toxicol. Appl. Pharmacol.*, 99, 344, 1989.
107. Kennedy, C.J. and Law, F.C.P., Toxicokinetics of selected polycyclic aromatic hydrocarbons in rainbow trout following different routes of exposure, *Environ. Toxicol.*, 9, 133, 1990.
108. Hunn, J.B. and Allen, J.L., Movement of drugs across the gills of fishes, *Annu. Rev. Pharmacol.*, 14, 1, 1974.
109. Bischoff, K. and Dedrick, R.L., Thiopental pharmacokinetics, *J. Pharm. Sci.*, 57, 1346, 1968.
110. Dedrick, R.L., Forrester, D.D., Cannon, J.N., El Dareer, S.M., and Mellett, L.B., Pharmacokinetics of 1-β-D-arabinofuranosylcytosine (Ara-C) deamination in several species, *Biochem. Pharmacol.*, 22, 2405, 1973.
111. Pang, K.S., Kong, P., Terrell, J.A., and Billings, R.E., Metabolism of acetaminophen and phenacetin by isolated rat hepatocytes: a system in which spatial organization inherent in the liver is disrupted, *Drug Metab. Dispos.*, 13, 42, 1985.
112. Rane, A., Sawe, J., Lindberg, B., Svensson J.-O, Garle, M., Erwald, R., and Jorulf, H., Morphine glucuronidation in the rhesus monkey. A comparative *in vivo* and *in vitro* study, *J. Pharmacol. Exp. Ther.*, 229, 571, 1984.
113. Hilderbrand, R.L., Andersen, M.E., and Jenkings, L.J., Jr., Prediction of *in vivo* kinetic constants for metabolism of inhaled vapors from kinetic constants measured *in vitro, Fund. Appl. Toxicol.*, 1, 403, 1981.
114. Sato, A. and Nakajima, T., A vial equilibration method to evaluate the drug metabolizing enzyme activity for volatile hydrocarbons, *Toxicol. Appl. Pharmacol.*, 47, 41, 1979.
115. Collins, J.M., Blake, D.A., and Enger, P.G., Phenytoin metabolism in the rate. Pharmacokinetic correlations between *in vitro* hepatic microsomal enzyme activity and *in vivo* elimination kinetics, *Drug Metab. Dispos.*, 6, 251, 1978.
116. Wilkinson, G.R., Prediction of *in vitro* parameters in drug metabolism and distribution from *in vitro* studies, in *Pharmacokinetics in Risk Assessment, Vol. 8,* Drinking Water and Health, National Academic Press, Washington, D.C., 1987, 80.
117. Baski, S.M. and Frazier, J.M., Isolated fish hepatocytes — model systems for toxicology research, *Aquat. Toxicol.*, 16, 229, 1990.
118. Dady, J.M., Bradbury, S.P., Hoffman, A.D., Voit, M.M., and Olson, D.L., Hepatic microsomal *N*-hydroxylation of aniline and 4-chloroaniline by rainbow trout *(Onchorhyncus mykiss), Xenobiotica*, 12, 1605, 1991.

119. Funari, E., Zoppini, A., Verdina, A., De Angelis, G., and Vittozzi, L., Xenobiotic-metabolizing enzyme systems in test fish. I. Comparative studies of liver microsomal monooxygenases, *Ecotoxicol. Environ. Safety*, 13, 24, 1987.
120. Lunte, C.E. and Scott, D.O., Sampling living systems using microdialysis probes, *Anal. Chem.*, 63, 773, 1991.
121. McKim, J.M., Jr., McKim, J.M., Sr., Naumann, S., Hammermeister, D.E., Hoffman, A.D., and Klaassen, C.D., *In vivo* microdialysis sampling of phenol and phenyl glucuronide in blood of unanesthetized rainbow trout: implications for toxicokinetic studies, *Fund. Appl. Toxicol.*, 20, 190, 1993.
122. Smith, P.F., Fisher, R., McKee, R., Gandolfi, A.J., Krumdieck, C.L., and Brendel, K., Precision cut liver slices: a new *in vitro* tool in toxicology, in *In Vitro Toxicology: Model Systems and Methods*, McQueen, C.A., Ed., Telford Press, New Jersey, 1988, 93.
123. Fisher, R., Hanzlik, R.P., Gandolfi, A.J., and Brendel, K., Toxicity of ortho-substituted bromobenzenes in rat liver slices: a comparison to isolated hepatocytes and the whole animal, *In Vitro Toxicol.*, 4, 173, 1991.
124. Schuhmann, G., Fichtl, B., and Kurz, H., Prediction of drug distribution *in vivo* on the basis of *in vitro* binding data, *Biopharm. Drug Dispos.*, 8, 73, 1987.
125. Lin, J.H., Sugiyama, Y., Awazu, S., and Hanano, M., *In vitro* and *in vivo* evaluation of the tissue-to-blood partition coefficient for physiological pharmacokinetic models, *J. Pharmacokinet. Biopharm.*, 10, 637, 1982.
126. Igari, Y., Sugiyama, Y., Awazu, S., and Hanano, M., Comparative physiologically based pharmacokinetics of hexobarbital, phenobarbital and thiopental in the rat, *J. Pharmacokinet. Biopharm.*, 10, 53, 1982.
127. Kurz, H. and Fichtl, B., Binding of drugs to tissues, *Drug Metab. Rev.*, 14, 467, 1983.
128. Miller, F.J., Overton, J.H., Jr., Smolko, E.D., Graham, R.C., and Menzel, D.B., Hazard assessment using an integrated physiologically based dosimetry modeling approach: ozone, in *Pharmacokinetics in Risk Assessment, Vol. 8,* Drinking Water and Health, National Academic Press, Washington, D.C., 1987, 353.
129. Smolko, E.D., McKee, D.J., and Menzel, D.B., Critical toxicity reference system. I. An approach for managing quantitative toxicity data, *J. Am. Coll. Toxicol.*, 5, 589, 1986.
130. Russo, R.C. and Pilli, A., AQUIRE: Aquatic Information Retrieval Toxicity Data Base. Project Description, Guidelines and Procedures, EPA 600/8-84-021, Environmental Research Laboratory, Duluth, MN, 1984.
131. DiCarlo, F.J., Bickart, P., and Auer, C.M., Role of the structure-activity team (SAT) in the premanufacture notification (PMN) process, in *QSAR in Toxicology and Xenobiochemistry,* Richy, M., Ed., Elsevier, Amsterdam, 1985, 433.
132. McCutcheon, R.S., Toxicology and the law, in *Casarett and Doull's Toxicology: The Basic Science of Poisons*, 2nd ed., Doull, J., Klaassen, C.D., and Amdur, M.O., Eds., Macmillan, New York, 1980, 727.
133. Konemann, H., Quantitative structure-activity relationships in fish toxicity studies. Part 1: relationship for 50 industrial pollutants, *Toxicology*, 19, 209, 1981a.
134. Konemann, H., Fish toxicity tests with mixtures of more than two chemicals: a proposal for quantitative approach and experimental results, *Toxicology*, 19, 229, 1981b.

135. Veith, G.D., Call, D.J., and Brooke, L.T., Structure-toxicity relationships for the fathead minnow: narcotic industrial chemicals, *Can. J. Fish. Aquat. Sci.*, 40, 743, 1983.
136. Veith, G.D., Defoe, D., and Knuth, M., Structure-activity relationships for screening organic chemicals for potential ecotoxicity effects, *Drug Metab. Rev.*, 15, 1295, 1985.
137. Veith, G.D. and Broderius, S.J., Structure-toxicity relationships for industrial chemicals causing type (II) narcosis syndrome, in *QSAR in Environmental Toxicology — II,* Kaiser, K.L.E., Ed., D. Reidel, Dordrecht, The Netherlands, 1987, 385.
138. Hermens, J.L.M., Quantitative structure-activity relationships of environmental pollutants, in *Handbook of Environmental Chemistry, Vol. 2, Reactions and Processes,* Hutzinger, O., Ed., Springer-Verlag, Berlin, 1989.
139. McKim, J.M., Schmieder, P.K., Carlson, R.W., Hunt, E.P., and Niemi, G.J., Use of respiratory-cardiovascular responses of rainbow trout *Salmo gairdneri* in identifying acute toxicity syndromes in fish. Part 1. Pentachlorophenol, 2,4,-dinitrophenol, tricaine methanesulfonate and 1-octanol, *Environ. Toxicol. Chem.*, 6, 295, 1987.
140. McKim, J.M., Schmieder, P.K., Niemi, G.J., Carlson, R.W., and Henry, T.R., Use of respiratory-cardiovascular responses of rainbow trout *Salmo gairdneri* in identifying acute toxicity syndromes in fish. Part 2. Malathion, carbaryl, acrolein and benzaldehyde, *Environ. Toxicol. Chem.*, 6, 313, 1987.
141. McKim, J.M., Bradbury, S.P., and Niemi, G.J., Fish acute toxicity syndromes and their use in the QSAR approach to hazard assessment, *Environ. Health Perspect.*, 71, 171, 1987.
142. Bradbury, S.P., Henry, T.R., Niemi, G.J., Carlson, R.W., and Snarski, V.M., Use of respiratory-cardiovascular responses of rainbow trout *(Salmo gairdneri)* in identifying acute toxicity syndromes in fish. Part 3. Polar narcotics, *Environ. Toxicol. Chem.*, 8, 247, 1989.
143. Bradbury, S.P., Carlson, R.W., Niemi, G.J., and Henry, T.R., Use of respiratory-cardiovascular responses of rainbow trout *(Oncorhynchus mykiss)* in identifying acute toxicity syndromes in fish. Part 4. Central nervous system seizure agents, *Environ. Toxicol. Chem.*, 10, 115, 1991.

Index

A

AAF, see 2-Acetylaminofluorene
Abl gene, 305–307
ABT, see Aminobenzotriazole
Acetaminophen, 51, 222
Acetanilide, 109
2-Acetylaminofluorene (AAF), 57, 95, 96, 357
Acetylcholinesterase (AChE), 438–440, 448
AChE, see Acetylcholinesterase
Acid phosphatase (ACP), 226
ACP, see Acid phosphatase
Acrolein, 58
Acrylamide, 222
Active transport, 5
Adenomas, 73, 296, 331, 337, 339, 342, see also Tumors
Affinity chromatography, 66
Aflatoxicol, 274
Aflatoxin B, 43, 52, 53, 72, 114, 143
  detoxification of, 143
  DNA adducts and, 268, 271, 272, 274, 275, 277
  oncogenes and, 311
  tumors and, 339, 356, 357, 358
Aflatoxin M, 271, 274
Aflatoxins, 56, 91, 327, 332, see also specific types
Aglycones, 39, see also specific types
AHH, see Arylhydrocarbon hydroxylase

Ah receptor nuclear translocator (ARNT), 173
Ah receptors, 149, 156
  in aquatic species, 135–141
  evolution of, 172–174
ALAD, see Aminolevulinic acid dehydratase
ALAS, see Aminolevulinic acid synthetase
Alcohols, 38, 39, 50, 110, see also specific types
Aldehyde dehydrogenase (ALDH), 38, 227
ALDH, see Aldehyde dehydrogenase
Aldrin, 128
Aldrin epoxidase, 128
Aliphatic alcohols, 50, see also specific types
Alkaline phosphatase (ALKP), 226, 355
Alkanes, 110, see also specific types
Alkenals, 60, see also specific types
Alkoxybiphenyls, 91, see also specific types
ALKP, see Alkaline phosphatase
Alkylins, 113, see also specific types
Alkyl phenoxazones, 126, see also specific types
Allyl formate, 222
α-Naphthoflavone (ANF), 109
Aluminum, 398, 453, 458
Amidases, 38, see also specific types
Amides, 39, see also specific types
Amines, 16, 38, 39, 50, 327, see also specific types
Amino acids, 5, 38, 56, see also specific types

free, 404
  metals and, 388, 390, 404, 407
  oncogenes and, 301
Aminobenzoic acid, 38
Aminobenzotriazole (ABT), 111
Aminolevulinic acid dehydratase (ALAD), 453
Aminolevulinic acid synthetase (ALAS), 453
2-Aminophenol, 44
Aminopyrine, 128, 129
Androgens, 98, see also specific types
Androstenedione, 128
ANF, see α-Naphthoflavone
Aniline, 7, 128
1-Anilino-8-naphthalene-sulfonate, 65
Antibodies, 119, see also specific types
  to CYP1A, 126, 143, 215, 217
  mammalian, 367
  monoclonal, 18, 114, 119–121, 215
  oval cell, 368
  polyclonal, 114, 129, 130
  protein cross-reactions with, 122
  specific, 114
  tumors and, 367
Antioxidant responsive elements (AREs), 53–54, 72
Antioxidants, 245–246, see also specific types
Arachidonic acid, 98, 147, 242
Arachidonic acid metabolites, 242
AREs, see Antioxidant responsive elements
ARNT, see Ah receptor nuclear translocator
Aroclor, 18, 47
Aromatase, 114, 125
Aromatic amines, 50, 327, see also specific types
Aromatic nitrogens, 7, see also specific types
Arsenic, 4, 407
Arylhydrocarbon hydroxylase (AHH), 123, 126, 128, 131, 149, 150
  biomarkers and, 167
  DNA adducts and, 270
  extrahepatic cytochrome P450 and, 153
  induction of, 151–152, 170
  in kidney, 154–155
  SARs and, 107, 111, 113
Aryl mercaptans, 39, see also specific types
Aryl phenols, 41, see also specific types
Aryltransferase, 56
Athymic mice, 369
Autofluorescence, 219

Autoradiography, 22
Azo compounds, 5, 327, see also specific types

B

Back-diffusion, 497
Bacterial enzymes, 16, see also specific types
BaP, see Benzo(a)pyrene
Barium, 398
Behavioral mechanisms, 421–459
  description of, 421–428
  energetic, 449–455
  feeding, 447–449
  future research on, 455–459
  group, 445–447
  hormones and, 428–436
  neurobiological, 436–449
  reproductive, 426, 433
  research on, 455–459
  respiratory, 449–455
  sequential, 425
Benzaldehyde, 227
Benzene, 110
Benzophetamine, 129
Benzo(a)pyrene (BaP), 10, 12, 13, 18–20, 22, 43, 51
  clearance of, 22
  as CYP1A inducer, 134
  cytochrome P450 and, 130, 143
  cytotoxicity of, 37
  DNA adducts and, 267, 270–272, 274, 280
  hybridizations and, 123
  metabolic activation of, 143, 145
  metabolism of, 19, 107, 109, 143, 146, 154
  PAH metabolism and, 105
  transformation of, 128
  tumors and, 20, 21, 357
Benzo(a)pyrene (BaP)-7,8-dihydriols, 154
Benzo(a)pyrene (BaP)-7,8-dihydrodiol, 12, 13, 21, 143
Benzo(a)pyrene (BaP)-7,8-dihydrodiol-9,10-epoxide, 12
Benzo(a)pyrene (BaP)-7,8-diol, 13
Benzo(a)pyrene (BaP) diol epoxide, 1056
Benzo(a)pyrene (BaP)-7,8-diol-9,10-epoxide, 280
Benzo(a)pyrene (BaP) hydroxylase, 118
Benzo(a)pyrene (BaP) monooxygenase, 16
Benzo(a)pyrene (BaP)-4,5-oxide, 61, 62, 66
Benzo(a)pyrene (BaP)-7,8-oxide, 13

Benzphetamine, 114
β-Naphthoflavone (BNF), 19, 47, 54, 71, 72, 123, 132, 140
  biological organization and, 210
  CYP1A and, 152, 156
  cytochrome P450 and, 111, 118, 130, 164, 165
  DNA adducts and, 274, 277
  hormones and, 150
  tumors and, 358
BHA, see Butylated hydroxyanisole
Bile acids, 5, 11, 52, 53, 98, see also specific types
Bile salts, 50
Biliary epithelial cells, 216–217, 223, 341, 354
Bilirubin, 40–43, 47, 53
Bioaccumulation, 3
Bioactivation, 89, 143–146, 356
Bioavailability
  biotransformation as modulator of, 12–20
  intrinsic factors in, 4–12
  of metals, 388–391
  modulation of, 12–20
Biological organization, 210–212
Biological specificity, 86
Biomagnification factor, 3
Biomarkers, 166–169, 207–231
  altered hepatocytes and, 227–228
  biliary epithelial cells and, 216–217, 223
  biological organization and, 210–212
  cytochemistry and, 223–227
  cytochrome P450 as, 166–169, 210, 211, 217, 218, 227
  endothelial cells and, 217–218, 223
  epithelial cells and, 216–217, 223
  fat-storing cells of Ito and, 218–219
  hepatocytes and, 220, 227–228
  lectins as, 227
  liver biology and, 214
  liver embryology and, 213
  liver histopathologic, 222–226
  liver intercellular matrix and, 220–221
  liver as site of responses of, 212
  liver tumors and, 229
  oncogenes and, 317
  requirements for, 221–227
  targets of toxicants and, 210–212
  tumors and, 229, 328, 356, 357
Biotransformation, 1–2, 4
  as bioavailability modulator, 12–20

  as disposition modulator, 12–20
  enzymatic, 12
  of enzymes, 12
  in gastrointestinal tract, 16–20
  individual variation in, 151–152
  microfloral, 16
  of monooxygenases, 88–92
    individual variation in, 151–152
    oxidative, 92–103
  pathways for, 37–39
  PB-TK models and, 471, 504–505
  preconsumptive, 12–16
Biotransformation enzymes, 16, 19, 59, 89, 90, 91, see also specific types
Biphenyls, 91, 128, see also specific types
Bladder carcinoma, 296
Blood flow relationships, 476–477
BNF, see β-Naphthoflavone
Body burdens, 1, 2
Brain regionalization, 441–442
Brass, 250
Breast carcinoma, 296, 307
Bromobenzene, 222
Bromodeoxyuridine, 368
Bromosulfothalein, 65
Butylated hydroxyanisole (BHA), 274
Butylhydroquinone, 55

C

Cadmium, 5, 113, 389, 390, 396
  accumulation of, 408
  behavioral mechanisms and, 433
  binding of, 404, 405, 406, 422
  in blood cells, 400, 401
  calcium exchange with, 394
  in erythrocytes, 400
  immunosuppressive effects of, 249, 250
  intracellular deposits of, 399
  metallothioneins and, 400, 401, 422
  plasma proteins and, 403
  presynaptic toxicity of, 437
  release of, 406, 407, 408
  toxicity of of, 445
  transferrin and, 404
  transport of, 404, 405
  uptake of, 392
Calcium, 437
  binding of, 404
  cadmium exchange with, 394
  concretions of, 408

free, 242
  intracellular deposits of, 397–399
  plasma proteins and, 403
  release of, 393, 408
  transport of, 400
  uptake of, 392
Calcium carbonate, 397, 398
Calcium channels, 392
Calcium phosphate, 397–399, 407
Calmodulin-related protein, 434
Carbamates, 39, 43, see also specific types
Carbazole, 7
Carbonic anhydrase, 396
Carbon monoxide, 113
Carbon tetrachloride, 221
Carboxylic acids, 38, 39, see also specific types
Carcinogenesis, 327, see also Carcinogens; Tumors
  epigenetic carcinogens and, 357–359
  experimental, 349–353
  experimental modulation of, 358
  in liver, 222, 268, 335–338, 355, see also Liver tumors
    experimental modulation of, 358
    promoters/modulators in, 357–358
  mechanisms of, 316
  promoters/modulators in, 357–358
Carcinogens, 143–146, 330, 359–361, see also Carcinogenesis; specific types
Carcinomas, 73, 337, 338, 342, see also Tumors; specific types
  bladder, 296
  cholangiocellular, 334, 338
  colon, 296
  liver cell, 333, 334, 339
  mammary, 296, 307
  mouse mammary, 307
  small-cell lung, 296
Carrier-mediated diffusion, 5
CAT, see Chloramphenicol acetyltransferase
Catalase, 242, 245, 253
Catechols, 50, see also specific types
CDNB, see 1-Chloro-2,4-dinitrobenzene
Cell surface glycoproteins, 367, see also specific types
Cell turnover, 352
Cellular stress, 445
Central nervous system (CNS), 432, 437–439, 441, 443–449
Ceruloplasmin, 400, 404

CGD, see Chronic granulonatous disease
Chelation, 390, 391
Chemical specificity, 86
Chemiluminescence, 243, 246, 248–250, 251, 252–254
Chemoprotectants, 59, see also specific types
Chloramphenicol acetyltransferase (CAT), 397
Chlorinated compounds, 327, 348, see also specific types
Chlorinated phenoxyacetic acids, 38, see also specific types
Chlorine substitution, 9
p-Chlorobenzyl N,N-diethylthiocarbamate S-oxidase, 94
Chloro-complexes, 390, see also specific types
Chlorodinitrobenzene, 56
1-Chloro-2,4-dinitrobenzene (CDNB), 14, 52, 59, 60, 62, 63, 70, 71
  3-MC-induced, 72
  in salmon, 67–69
  in shank, 64
  in skate, 66
1-Chloro-2,4-dinitrobenzene (CDNB)-conjugating activity, 73
Chlorophenolics, 43, see also specific types
Cholangiocellular tumors, 332, 334, 338, 339, 354, 363
Cholesterol, 98, 453
Cholesterol side-chain cleavage, 124
Chondroitin 4-sulfate, 220
Chondroitin 6-sulfate, 220
Chondronectin, 220
Chromatography, 47, 66, 69, 278, see also specific types
Chromium, 402, 407, 408, 437
Chronic granulonatous disease (CGD), 244, 245
Chronic toxicity, 221–227
Chylomicrons, 10
Cinnamic acid, 98
Citrate, 390
CL, see Chemiluminescence
Cleavage, 124
Cloning, 41, 123–125, 131
CNS, see Central nervous system
Cobalt, 5, 398, 404, 407, 437
Co-carcinogens, 330
Collagen, 220
Colon carcinoma, 296

INDEX

Colony stimulating factors, 242
Colorectal tumors, 296
Colorimetry, 246
Complement receptors, 242
Concretions, 397, 408
Conjugation, 89
Connexins, 368
Copper, 250, 389, 390
  binding of, 404, 422
  in blood cells, 401
  intracellular, 396, 397
  plasma proteins and, 402, 403
  release of, 406–408
  transport of, 400, 404
  uptake of, 392
Copper-binding proteins, 404, see also specific types
Copper-containing plasma proteins, 404, see also specific types
Corticosteroids, 98, see also specific types
Covalent binding, 20–21
CRIP, see Cysteine-rich intestinal protein
Cross hybridization, 55
Cross-immunoreactivity, 68
Cross-reactivity, 53, 114, 122
Cultured cells, 368
Cumene hydroperoxide, 53, 69
CYP1A, 71, 74, 103, 118
  allelic variants of, 123
  antibodies to, 126, 143, 215, 217
  benzo(a)pyrene metabolism and, 143, 146
  biomarkers and, 218
  catalytic activities and, 140
  cellular localization of expression of, 155–160
  in developmental stages of fish, 152–153
  expression of, 105
  extrahepatic cytochrome P450 and, 153
  functions of, 143
  inducers of, 107, 109, 111, 113
    Ah receptors and, 139–141
    structure-activity relationships of, 134
    temporal aspects of, 132
    toxicity of, 141
  induction of, 168–169, 170
  N-terminus of, 124
  recognition of, 120–121
  steroid metabolism and, 147
  suppression of, 150
  synthesis of, 111
  tumors and, 162–164

Cysteine-rich intestinal protein (CRIP), 393, 394, 396
Cysteine-rich proteins, 393, 394, 396, 422, see also specific types
Cysteinyl, 14
Cysteinyl-glycine, 14
Cytochemistry, 223–227
Cytochrome b, 97, 104, 244, 245
Cytochrome c, 246
Cytochrome c oxidase, 453
Cytochrome P450, 17–20, 92, 94, 97–103
  in amphibians, 125–127
  in aquatic species, 103–113, 125–127
  catalytic functions of, 105–106
  cloned, 115–117
  diversity of, 113–131
    cloning and, 123–125, 131
    hybridizations and, 123–125, 131
    immunological relationships and, 118–123, 130
  microsomal, 104
  purified, 115–117, 129–130
  SARs and, 106–113
  substrate SARs and, 106–110
  as biomarker, 166–169, 210, 211, 217, 218, 227
  biotransformation of, 151–152
  carcinogen bioactivation and, 143–146
  catalytic functions of, 105–106
  cloned, 115–117
  CYP1A and, 152–153, 155–160
  distribution of, 157–168
  diversity of functions of, 98–101
  efficiency of, 105
  in endothelium, 160–162
  evolutionary aspects of, 171–174
  expression of, 148–153
  extrahepatic, 153–164
  functions of, 98–101
    toxicity associated with, 141–147
  gene regulation in, 131–134
  gene superfamily of, 101–103
  in heart, 155
  hormones and, 150–151
  induction of as biomarker, 166–169
  inhibitors of, 129
  in invertebrates, 106, 129–131
  in isolated cells, 164–166
  in kidney, 154–155
  in liver, 212, 223
  in mammals, 127–129

metabolism and, 97, 99, 105
microsomal, 104
natural products and, 169–171
nomenclature for, 101–103
PB-TK models and, 481
peroxidase localization of, 210
purified, 115–117
reaction mechanism of, 97–98
in reptiles, 125–127
sex differences and, 150–151
stereoselectivity of, 146
substrate metabolism and, 146–147
substrate SARs and, 106–110
temperature acclimation and, 149
toxicity associated with functions of, 141–147
tumors and, 355, 356, 367
variables in expression of, 148–153
in vertebrates, 105–106
Cytochrome P450-dependent monooxygenases, 18, 38
Cytochrome P450 reductase, 129
Cytokeratins, 367, see also specific types
Cytoplasmic enzymes, 241, see also specific types
Cytoplasmic proteins, 401, see also specific types
Cytosolic enzymes, 50, 59, see also specific types
Cytosols, 70, 244, see also specific types

D

2,4-D, see 2,4-Dinitrophenol
DA, see Dopamine
DAB, see Diaminobenzidine
Data-based models, 473–474
DCFH, see 2′,7′-Dichlorofluorescin
DCFH-DA, see 2′,7′-Dichlorofluorescin-diacetate
DCNB, see 1,2-Dichloro-4-nitrobenzene
DDT, see p′-Dichlorodiphenyltrichloroethane
Dealkylation, 91, 98
Degradation enzymes, 440, see also specific types
DEN, see Diethylnitrosamine
Dermal exposure routes, 2, 487–489
Dermatin sulfate, 220
Detoxification, 89, 143, 396, 398
Diagnostic substrates, 106–110, see also specific types

Diaminobenzidine (DAB), 243, 246, 253
Diapedesis, 408
Dibenzanthracene, 118
Dicarbonyls, 39, see also specific types
p′-Dichlorodiphenyltrichloroethane (DDT), 4, 348, 358, 365
2′,7′-Dichlorofluorescin (DCFH), 244
2′,7′-Dichlorofluorescin-diacetate (DCFH-DA), 244
Dichloronitrobenzene, 66, 68
1,2-Dichloro-4-nitrobenzene (DCNB), 56, 64–65
Dieldrin, 4
Dietary exposure routes, 3
  bioavailability and, 4–12
  in PB-TK models, 489–490
Dietary-inducing agents, 18–19, see also specific types
Di-2-ethyl hexylphthalate, 43
Diethylnitrosamine (DEN), 57, 91, 227, 277, 315
  tumors and, 329, 333, 340, 355, 357, 358, 361
Diffusion, 5, 6, 390, 392, see also specific types
  carrier-mediated, 5
  passive, 5, 6, 13
  in PB-TK models, 480, 481, 488, 497
Digestibility, 8
Digestive enzymes, 8, see also specific types
Digitalis, 37
Digitoxigenins, 43, see also specific types
Diglycerides, 10, see also specific types
Dihydrodiol-epoxide, 143
Dihydrodiols, 152, see also specific types
5α-Dihydrotestosterone, 433
N,N-Dimethylaniline-N-oxidase, 94
Dimethylbenzanthracene, 134, 358
1,2-Dimethylbenzanthracene, 72
7,12-Dimethylbenzanthracene, 10, 357
Dimethylsulfonioproprionate (DMSP), 96
2,4-Dinitrophenol, 38, 392
Diolepoxide, 13, 152
Dioxins, 105, 137, see also specific types
Disposition modulation, 12–20
DMSP, see Dimethylsulfonioproprionate
DNA adducts, 267–284
  field studies of, 278–280
  in fish, 271–280
  formation of, 269–271
  in vitro studies of, 272–274

INDEX 527

in vivo studies of, 275–277
laboratory studies of, 272–277
metabolism and, 270–271
tumors and, 345, 348, 352, 358, 366
DNA binding, 20–21
DNA-binding proteins, 296, see also specific types
DNA hydroperoxides, 60, see also specific types
DNA repair, 271, 345
DNA synthesis, 342
Dopamine, 438
Dose concepts, 471–476
Drug metabolizing enzymes, 89, 90, see also specific types
Dyes, 5, 107, 244, see also specific types

E

EB, see Ethidium bromide
Ecdysteroids, 98
ECOD, see Ethoxycoumarin O-deethylase
EDTA, 390
EETs, see Epoxyeicosatrienoic acids
EGF, see Epidermal growth factor
Elastin, 220
Electrolytes, 5
Electron acceptors, 92
Electron transfers, 104
Endocrine toxicity of metals, 434–436
Endocytosis, 389, 394–395, 402
Endothelial cells, 217–218, 223
Endothelium, 160–162
Energetic behavioral mechanisms, 449–455
Energy availability, 454
Energy metabolism, 437
Enterohepatic cycling, 13
Enzymes, 89, 92, see also specific types
 bacterial, 16
 biochemistry of, 91
 biotransformation, 16, 19, 59, 89–91
 biotransformation of, 12
 biotransformation pathways and, 37–39
 characterization of, 46–50
 cytochemistry of, 226
 cytoplasmic, 241
 cytosolic, 50, 59
 degradation, 440
 digestive, 8
 diversity of, 91
 drug metabolizing, 89, 90

efficiency of, 105
homodimeric, 58
hydrolytic, 16, 38
inducibility of, 18–19
metabolic, 453
molecular biology of, 91
oxidative, 38
phase I, 38, 74
phase II, see Phase II enzymes
phase-II-conjugating, 211
purification of, 46–50
redox, 38
regulation of, 437
Ependymoblastomas, 367
Epidermal growth factor (EGF) receptor gene, 299, 301, 302
Epigenetic carcinogens, 357–359
Epithelial cells, 214
 biliary, 216–217, 223, 341, 354
 differentiating features of, 216–217
 metal transfer across, 391
 metal uptake by, 391–394
Epoxidation, 98
Epoxide, 211
Epoxide hydrolases, 20, 38
9,10-Epoxy-7,8-dihydroxybenzo(a)pyrene, 51
Epoxyeicosatrienoic acids (EETs), 59, 147
1,2-Epoxy-3-(p-nitrophenoxy)propane, 59
1,2-Epoxy-3-(p-phenoxy)propane, 63, 67, 68
Erb-A gene, 308–309
Erb-B gene, 302
EREs, see Estradiol response elements
EROD, see Ethoxyresorufin O-deethylase
Erythrocytes, 400, 453
Esterases, 38
Estradiol, 114, 128, 133, 147, 151, 358
17β-Estradiol, 43, 358
Estradiol receptors, 151
Estradiol response elements (EREs), 151
Estrogens, 98, 434, see also specific types
Estrone, 50
Ethacrynic acid, 63
Ethanol, 122
Ethers, 110, see also specific types
Ethidium bromide (EB), 244
7-Ethoxycoumarin, 51, 128, 130
Ethoxycoumarin O-deethylase (ECOD), 113
7-Ethoxyresorufin, 128
Ethoxyresorufin O-deethylase (EROD), 109, 111, 113, 123, 126, 128, 134, 150, 227
N-Ethylmaleimide, 59

Evolution, 171–174, 315–316, 421
Exocytosis, 400, 402, 437
Exposure routes, 1, see also specific types
 dermal, 2, 487–489
 dietary, 3, 4
  bioavailability and, 4–12
  in PB-TK models, 489–490
 multiple, 490
 oral, 2, 507
 in PB-TK models, 483–490
 respiratory, 2
 uptake and, 2
Extrahepatic cytochrome P450, 153–164

F

FAPs, see Fixed action patterns
FATS, see Fish acute toxicity syndromes
Fat-soluble drugs, 9, see also specific types
Fat-storing cells of Ito, 218–219
Fatty acid derivatives, 98, see also specific types
Fatty acids, 5, 9, 98, 173, 454, see also specific types
 digestion of, 212
 free, 10
 hydroxylation of, 147
 metabolism of, 453
 utilization of, 454
FCA, see Foci of cellular alteration
Fc receptors, 242
Feeding behavior, 447–449
Ferric hydroxide, 394–395
Ferricytochrome c, 245
Ferritin, 402
Fetoprotein, 367
FGFs, see Fibroblast growth factors
Fibroblast growth factors (FGFs), 307
Fibronectin, 220
Fibronectin receptors, 242
Filtration, 5
Fish acute toxicity syndromes (FATS), 510
Fixed action patterns (FAPs), 422–223, 425, 426, 428, 435, 440
Flavine monooxygenases, 38
Flavonoids, 19, 20, 98, see also specific types
Flavoprotein monooxygenases (FMOs), 92–96, see also specific types
Flavoproteins, 104, see also specific types
Flow cytometry, 368
Flow-limited gill model, 484–487

Fluorescence, 219, 246
FMOs, see Flavoprotein monooxygenases
Foci of cellular alteration (FCA), 223–227
Follicle-stimulating hormone (FSH), 433, 434
Food chain components, 3–4
Free fatty acids, 10
FSH, see Follicle-stimulating hormone
Fulvic acid, 388
Functionalization, 89
Fyn gene, 305

G

GABA, see Gamma-amino butyric acid
GAGs, see Glycosaminoglycans
Gamma-amino butyric acid (GABA), 438
Gas chromatography (GC), 278
Gastric evacuation rates, 9
Gastrointestinal pH, 9
Gastrointestinal tract, 16–22
GC, see Gas chromatography
Gene evolution, 315–316
Geochemical transformations, 87
GGT, see γ-Glutamyl transpeptidase
Glucan, 242
Glucocorticoid response elements (GREs), 151
Glucocorticoids, 173, see also specific types
Gluconate, 390
Glucose-6-phosphate dehydrogenase, 60, 152, 215, 217, 219, 226, 355
β-Glucosidases, 14, 16, see also specific types
β-Glucosides, 14, see also specific types
Glucuronic acid, 38
β-Glucuronidases, 16, 20, see also specific types
Glucuronidation, 38, 40, 43
Glucuronides, 13, 39, 43, see also specific types
γ-Glutamyl transpeptidase (GGT), 14, 217, 226, 227, 340, 355, 357
Glutathione, 14, 16
Glutathione peroxidase, 53, 71, 242, 245
Glutathione reductase, 60
Glutathione S-transferases (GSTs), 17–19, 52–73, 211, 227, see also specific types
 activities of, 60–62
 α-class, 53–56
 in fish, 60–73

INDEX

activities of, 60–62
  immunological investigations and, 70–71
  inducibility of, 71–73
  occurrence of, 60–62
  purification of, 64–70
  tissue distribution of, 62–64
gene structure of, 54–58
immunological investigations and, 70–71
inducibility of, 71–73
mammalian, 53–60
microsomal, 59
μ-class, 56–57
occurrence of, 60–62
π-class, 57–58
purification of, 64–70
θ-class, 58–59
tissue distribution of, 62–64
tumors and, 355, 356
Glycerol, 10
Glycogen, 215, 338
Glycolmethacrylate (GMA), 220
Glycoproteins, 220, 367, 404, see also specific types
Glycosaminoglycans (GAGs), 219–221, see also specific types
GMA, see Glycolmethacrylate
Golgi apparatus, 10
G6PDH, see Glucose-6-phosphate dehydrogenase
G protein, 244
Granules, 397, see also specific types
GREs, see Glucocorticoid response elements
Growth factor receptors, 296, see also specific types
Growth factors, 296, 307, see also specific types
GSTs, see Glutathione S-transferases
GTP, see Guanosine triphosphate
GTPase, see Guanosine triphosphatase
Guanosine triphosphatase-activating proteins, 296, see also specific types
Guanosine triphosphate (GTP)-binding proteins, 309–313, see also specific types

H

Haber-Weiss reaction, 403
HAHs, see Halogenated aromatic hydrocarbons
Halide, 241

Halogenated alkanes, 110, see also specific types
Halogenated aromatic hydrocarbons (HAHs), 105, 134, 139, 147, 154, see also specific types
Halogenated phenols, 41, see also specific types
HCE, see Hexachloroethane
HDLs, see High density lipoproteins
Heat shock proteins, 211
Heme oxygenase, 113
Heme-porphyrin system, 211
Heme protein monooxygenases, 92–93, see also Cytochrome P450
Hemocyanin, 396, 404
Hemocytes, 249–252, 400–402, 408
Heparin, 220
Heparin sulfate, 220
Hepatoblastomas, 342
Hepatocarcinogens, 268, see also specific types
Hepatocytes, 215–216, 223, 363, see also Liver; Liver cells
  altered, 221, 227–228, 354
  biomarkers and, 220
  degenerative, 340
  differentiating features of, 215–216
  fate of, 227–228
  tumors and, 338, 339
Hepatomas, 137, 165
Heterocyclics, 37, see also specific types
HETEs, see Hydroxyeicosatrienoic acids
Hexachlorobiphenyl, 119
Hexachloroethane (HCE), 498–503
Hexanal, 227
Hexose monophosphate shunt, 242
High density lipoproteins (HDLs), 10
Histidine-rich glycoproteins, 404
Homodimeric enzymes, 58, see also specific types
Homovanillic acid (HVA), 243, 246, 440
Hormones, 91, see also specific types
  behavioral mechanisms and, 428–436
  circulating, 432–433
  cytochrome P450 and, 150–151
  follicle-stimulating, 433, 434
  luteinizing, 433
  maturation, 133
  metal effects on actions of, 428–433
  molting, 131
  pituitary, 151

reception of, 434
release of, 430
retinoid, 173
steroid, see Steroids
synthesis of, 430
thyroid, 173, 432
thyroid-releasing, 434
thyroid-stimulating, 432
Humic acid, 388
HVA, see Homovanillic acid
Hyaluronic acid, 220
Hybridizations, 123–126, 131
Hydrocarbons, 87, see also specific types
   chlorinated, 348
   halogenated aromatic, 105, 134, 139, 147, 154
   polycyclic aromatic, see Polycyclic aromatic hydrocarbons (PAHs)
   solubilization of, 10
   tumors and, 348
Hydroethidine, 244
Hydrogen peroxide, 241, 243, 246, 253, 280
Hydrolases, 242
Hydrolysis, 13, 38, 242
Hydrolysis products, 10, see also specific types
Hydrolytic enzymes, 16, 38, see also specific types
Hydroperoxides, 56, 98
Hydrophilicity, 6
Hydrophobicity, 483–484
4-Hydroxy alkenals, 60, see also specific types
Hydroxyanisole, 56, 59, 72
Hydroxybenzo(a)pyrene, 13, 16
4-Hydroxy biphenyl, 43
Hydroxyeicosatrienoic acids (HETEs), 147
Hydroxylamines, 50, see also specific types
16α-Hydroxylase, 155
17α-Hydroxylase, 114, 125
Hydroxylation, 38, 91, 97, 98, 109, 133, 147
16α-Hydroxylation, 129
Hydroxyl radicals, 241
4-Hydroxynon-2-enal, 59
4-Hydroxynonenals, 56
3β-Hydroxysteroid dehydrogenase, 434
3α-Hydroxysteroids, 43, 47, see also specific types
3β-Hydroxysteroids, 47, see also specific types
17β-Hydroxysteroids, 40, 43, 47, see also specific types

4-Hydroxytryptamine, 40
Hydruxindole acetic acid, 440

I

Immonological cross-reactivity, 53
Immunoblotting, 50, 63, 69, 70
Immunochemical relatedness, 123, 126
Immunocytochemistry, 63
Immunoglobulin E, 242
Immunoglobulin G, 130, 217, 242
Immunoglobulin M, 242
Immunohistochemistry, 356, 367–368
Immunological cross-reactivities, 114
Immunological relationships, 118–123, 130
Immunoreactivity, 71
Immunotoxicants, 250, see also specific types
Indole-3-carbinol, 274, 358
Inhibition/inactivation SARs, 110–113
Inhibitor SARs, 110–113
Insecticides, 38, 43, see also Pesticides; specific types
Intercellular matrix of liver, 220–221
Interleukins, 242
Ion exchange chromatography, 47, 69
Ionizable hydrogens, 6
Iron, 5
   accumulation of, 355, 357
   binding of, 404
   in blood cells, 401
   intracellular deposits of, 397, 398
   plasma proteins and, 403, 404
   release of, 407, 408
   transport of, 400, 403, 404
   uptake of, 392
Isosafrole, 123
Ito cells, 218–219

K

Keratin sulfate, 220
Ketones, 110, see also specific types
11-Ketotestosterone, 155, 433
Kidney, 392, 393, 396, 406–407
Kupffer cells, 214

L

Laminin, 220
Lampreycides, 43, 44, see also specific types
Lauric acid, 98, 114, 147

INDEX

LC, see Lethal concentrations
LDLs, see Low density lipoproteins
Lead, 5, 395
  binding of, 404
  in erythrocytes, 400
  intracellular deposits of, 398
  neurotoxicity of, 437
  presynaptic toxicity of, 437
  release of, 406, 407
  toxicity of of, 452, 453, 458
Lectins, 227, see also specific types
Lethal concentrations, 473
Leukemia, 251
Leukotrienes, 56
LH, see Luteinizing hormone
Lipid peroxidation, 53, 58, 59
Lipofuscin, 398, 408
Lipophilic compounds, 92, see also specific types
Lipophilicity, 6
Lipoproteins, 10, see also specific types
Lithocholic acid, 53
Liver
  biology of, 214
  as biomarker response site, 212
  carcinogenesis in, 222, 268, 335–338, 355, see also Liver tumors
    experimental modulation of, 358
    promoters/modulators in, 357–358
  cytochrome P450 in, 212, 223
  embryology of, 213
  estrogen receptivity in, 434
  intercellular matrix of, 220–221
  lectins as biomarkers of cells in, 227
  metallothioneins in, 395, 396
  metals in, 392
  organogenesis and, 213
Liver cells, 215–220, see also Hepatocytes; specific types
  biomarkers and, 223
  carcinomas of, 333, 334, 339
Liver histopathologic biomarkers, 222–226
Liver tumors, 229, 328, 330–334, 369
  in fish, 344–349
  histogenesis of, 334–343, 353–354
LOEC, see Lowest observable effect concentration
Low density lipoproteins (LDLs), 10
Lowest observable effect concentration (LOEC), 458
Luminol, 243, 250, 253

Luteinizing hormone (LH), 433
Lysosomal granules, 398
Lysosomes, 215, 389, 408

M

Macromolecular adducts, 368
Macromolecules, 20–22
Macrophage aggregates, 219–220
Macrophages, 219–220
Magnesium, 44, 217, 355, 399, 407, 408, 437
Magnesium phosphate, 398
Mammary carcinoma, 296, 307
Manganese, 5, 398, 404, 408, 437
Mannose, 242
Mass-balance descriptions, 477–481
Mass spectrometry (MS), 278
MATC, see Maximum acceptable toxicant concentration
Matrix parameters, 7–12
Maturation hormones, 133, see also specific types
Maximum acceptable toxicant concentration (MATC), 458, 473
3-MC, see 3-Methylcholanthrene
MDP, see Methylenedioxyphenyl
Mechanistic models, 3
Melanomas, 300–302, 309, 369
Membrane-bound proteins, 309, see also specific types
Mercapturic acid, 14
Mercury, 390, 392, 404, 407, 458
Metabolic enzymes, 453, see also specific types
Metabolic transformation, see Biotransformation
Metabolism, 89
  of benzo(a)pyrene, 19, 107, 109, 143, 146, 154
  of benzo(a)pyrene-7,8-dihydrodiol, 143
  cytochrome P450 and, 97, 99, 105
  DNA adducts and, 270–271
  of drugs, 91
  energy, 437
  of estradiol, 147
  of fatty acids, 453
  of lipids, 454
  metal-altered, 452–455
  oxidative, 89
  in PB-TK models, 481–482, 505
  of polychlorinated biphenyls, 105

of polycyclic aromatic hydrocarbons, 38, 105, 146
of procarcinogens, 95
of steroids, 147, 172
of substrates, 146–147
of testosterone, 147
Metal-binding proteins, 249, 407, see also specific types
Metal-binding substances, 390, see also specific types
Metal chelators, 390, see also specific types
Metallothioneins, 207, 211, 392, 393, 395–397, 422, see also specific types
  cadmium and, 400, 401, 422
  measurement of, 396
  plasma proteins and, 402
  release of, 407
Metals, 4, 250, 387–410, see also specific types
  absorption of, 391
  accumulation of, 392, 408
  binding of, 404, 422
  bioavailability of, 388–391
  in blood cells, 400–402
  detoxification of, 396, 398
  digestive system release of, 407–408
  dyschronogenic effects of, 440–441
  elimination of, 398
  endocrine toxicity of, 434–436
  endocytosis and, 394–395
  in epithelial cells, 391–394
  excretion of, 407
  in hemocytes, 400–402
  hormone action and, 428–433
  internal transport of, 399–405
  intracellular deposits of, 397–399
  intracellular storage of, 395–399
  in invertebrates, 391
  in kidney, 392, 393
  in liver, 392
  metabolism alteration and, 452–455
  neurotoxicity of, 437, 438
  nitrogen-seeking properties of, 388
  plasma proteins and, 400, 402–405
  postsynaptic toxicity of, 438–440
  presynaptic toxicity of, 437–438
  regulation of, 207, 387
  release of, 387, 388, 405–408
  sequestration of, 395–399
  storage of, 388, 395–399
  sulfur-seeking properties of, 388
  toxicity of, 207, 387, 397

behavioral mechanisms of, 421–459
description of, 421–428
energetic, 449–455
future research on, 455–459
hormones and, 428–436
neurobiological, 436–449
research on, 455–459
respiratory, 449–455
  central nervous system and, 432, 437, 438, 441, 443–449
  endocrine, 434–436
  peripheral nervous system and, 432, 437, 438, 443–449
  postsynaptic, 438–440
  presynaptic, 437–438
  synaptic cleft, 438–440
translocation of, 388
transport of, 392, 399–405
trophic transfer of, 397
turnover rates of, 401
uptake of, 387, 388–391, 390
  endocytosis and, 394–395
  mechanisms of, 391–395
utilization of, 387
in vertebrates, 391
Metal transport proteins, 403, see also specific types
Methimazole N-oxidase, 94
Methotrexate, 495–497
7-Methoxycoumarin, 128
p-Methoxyphenyl-1,3-dithiolane, 146
Methylazoxymethanol acetate, 327
3-Methylcholanthrene (3-MC), 10, 18, 19, 41, 47, 49
  as CYP1A inducer, 134
  cytochrome P450 and, 126
  GSTs and, 54, 56, 71, 72
Methylenedioxyphenyl (MDP), 111
1-Methyl-2-nitro-1-nitrosoguanidine, 56
N-Methyl-N'-nitro-N-nitrosoguanidine (MNNG), 338, 361
N-Methyl-N-nitrosourea, 302
Methyltransferase, 271
4-Methylumbelliferone, 40
Metyrapone, 111
MFOs, see Mixed-function oxidases
Michaelis-Menten kinetics, 493
Microfloral biotransformation, 16
Mixed-function oxidases (MFOs), 38, 92, 212, 245, 270, 481, see also specific types

MMTV, see Mouse mammary tumor virus
MNNG, see N-Methyl-N'-nitro-N-nitrosoquanidine
Molecular cloning, 41
Molecular oxygen, 92, 93
Molting hormones, 131, see also specific types
Monoclonal antibodies, 18, 114, 119–121, 215
Monodigitoxide, 43
Monoglycerides, 10, see also specific types
Monooxygenases, 19, 87–175, see also specific types
 in amphibians, 125–127
 in aquatic species, 94–96, 103–113, 125–127, 135–141
  catalytic functions of, 105–106
  cloned, 115–117
  diversity of, 113–131
   cloning and, 123–125, 131
   hybridizations and, 123–125, 131
   immunological relationships and, 118–123, 130
  microsomal, 104
  purified, 115–117, 129–130
  SARs and, 106–113
  substrate SARs and, 106–110
 benzo(a)pyrene, 16
 as biomarkers, 166–169, 210, 211
 biotransformation of, 88–92
  individual variation in, 151–152
  oxidative, 92–103
 carcinogen bioactivation and, 143–146
 catalytic functions of, 105–106
 cloned, 115–117
 cytochrome P450-dependent, 18, 38
 distribution of, 157–168
 diversification of, 93
 diversity of functions of, 98–101
 efficiency of, 105
 in endothelium, 160–162
 evolutionary aspects of, 171–174
 expression of, 148–153
 extrahepatic, 153–164
 FAD-, 14
 flavine, 38
 flavoprotein, 92–96
 functions of, 98–101
  toxicity associated with, 141–147
 gene regulation in, 131–134
 heme protein, 92–93
 hormones and, 150–151
 inactivation of, 112
 inhibition of, 112
 in invertebrates, 106, 129–131
 in isolated cells, 164–166
 in mammals, 127–129
 microsomal, 104
 natural products and, 169–171
 oxidative biotransformation of, 92–103
 perspective on, 87–88
 purified, 115–117
 questions regarding, 96
 reaction mechanism of, 93–94, 97–98
 in reptiles, 125–127
 sex differences and, 150–151
 substrate SARs and, 106–110
 toxicity associated with functions of, 141–147
 types of, 92
 variables in expression of, 148–153
 in vertebrates, 105–106
Monooxygenation, 13
Morphine, 37, 43
Mouse mammary carcinoma, 307
Mouse mammary tumor virus (MMTV), 307
MPO, see Myeloperoxidase
MS, see Mass spectrometry
Myc gene, 308
Myeloperoxidase (MPO), 241

N

Naphthalene, 2, 43, 51
Naphthalene aminotriazole (NAT), 111
Naphthoflavone, see α-Naphthoflavone (ANF); β-Naphthoflavone (BNF)
α-Naphthoflavone (ANF), 20, 109
1-Naphthol, 40, 44, 50, 51
2-Naphthol, 52
1-Naphthol-uridine diphosphate-glucuronosyltransferase, 18
NAT, see Naphthalene aminotriazole
NBT, see Nitroblue tetrazolium
NE, see Norepinephrine
Neoplasia, see Tumors
Neurobiological behavioral mechanisms, 436–449
Neurofibromatosis, 296
Neurotoxicity of metals, 438
Neurotransmitters, 438, 440, see also specific types

Nickel, 400, 403, 404, 406, 407, 408, 433
Nitric oxide, 113
p-Nitrobenzylchloride, 56, 59, 63, 64, 65, 68, 70
Nitroblue tetrazolium (NBT), 243, 245–247, 253
Nitrogen heterocyclic compounds, 7, 111, see also specific types
Nitrogen-seeking properties of metals, 388
Nitroloacetate, 390
4-Nitrophenol, 40, 47
Nitropyrene oxide, 56
Nitrosamines, 110, see also specific types
Nitroso-compounds, 327, see also specific types
Nonepithelial tumors, 330
Nonreceptor tyrosine kinases, 302–307
Nonvolatile partitioning, 505–506
Norepinephrine, 438, 440, 441, 448
Northern blot analysis, 49, 211, 296, 306
Nuclear oncoproteins, 307–309, see also specific types
Nucleic acid hybridizations, 123–125, 131

O

Octanol-water partition coefficient, 3
Octene-1,2-oxide, 62
ODC, see Ornithine decarboxylase
Oncogenes, 295–317, see also specific types
 class I, 299–307
 class II, 307
 class III, 307–309
 class IV, 309–313
 detection of unknown, 314–315
 discovery of, 300
 evolution of, 315–316
 functions of, 316–317
 tumors and, 367
Oncoproteins, 307–309, see also specific types
Oral exposure routes, 2, 507
Organochlorines, 3, 212, see also specific types
Organogenesis, 213
Organometallics, 113, see also specific types
Ornithine decarboxylase (ODC), 359
Osteonectin, 220
Oval cells, 337–338, 354, 368
Oxalate, 390

Oxidants, 211, 245–246, see also specific types
Oxidation, 38, 89, 94, 146, 243
Oxidative biotransformation, 92–103
Oxidative enzymes, 38, see also specific types
Oxidative metabolism, 89
Oxidative stress, 246
Oxolinic acid, 11

P

p53 gene, 314
PAHs, see Polycyclic aromatic hydrocarbons
PAPS, see 3'-Phosphoadenosine 5' phosphosulfate
Parasitism, 252–254
Partition coefficients, 3
Passive diffusion, 5, 6, 13
PBO, see Pieronyl butoxide; Piperonyl butoxide
PB-TK, see Physiologically based toxicokinetic
PB-TR, see Physiologically based tissue response
PCBs, see Polychlorinated biphenyls
PCE, see Pentachloroethane
PCN, see Pregnenolone 16α-carbonitrile
PCNA, see Proliferating cell nuclear antigen
PCP, see Pentachlorophenol
PCR, see Polymerase chain reaction
PDGF, see Platelet-derived growth factor
2,3',4,4',5-Pentachlorobiphenyl, 134
Pentachloroethane (PCE), 498–503
Pentachlorophenol (PCP), 43, 51, 250, 251
7-Pentoxyresorufin, 128
Pentoxyresorufin O-deethylase (PROD), 109, 126, 128
Peripheral nervous system (PNS), 432, 437, 438, 443–449
Peroxidase-catalyzed oxidation, 243
Peroxidases, 69, 98, 243, see also specific types
 glutathione, 242, 245
 localization of, 210
 phagosomal, 244
 reactive oxyten intermediates and, 252, 253
Peroxides, 93, 97, see also specific types
Peroxide shunt, 98
Persistence of drugs, 12

INDEX

Pesticides, 37, 38, 52, 327, see also specific types
Phagocytosis, 243, 253, 394
Phagosomal peroxidases, 244
Phase I enzymes, 38, 74, see also specific types
Phase-II-conjugating enzymes, 211, see also specific types
Phase II enzymes, 37–75, see also specific types
  activities of, 44–45, 60–62
  biotransformation pathways and, 37–39
  characterization of, 46–50
  in fish, 43–50
  future directions for study of, 73–74
  gene structures of, 40–43, 54–58, 74
  glutathione S-transferases, see Glutathione S-transferases (GSTs)
  immunological investigations and, 70–71
  inducibility of, 71–73
  microsomal, 59
  occurence of, 52, 60–62
  properties of, 39–40, 44–45
  purification of, 46–50, 64–70, 69
  reactions of, 39–40, 43–44, 50, 52
  sulfotransferases, 50–52
  tissue distribution of, 62–64
  tissue expression of, 46
  UDP-glucuronosyl transferases, see Uridine diphosphate-glucuronosyl transferases
Phase I reactions, 89, see also specific types
Phase II reactions, 89, see also specific types
Phenanthrene, 43, 128
Phenobarbital, 19, 43, 55, 56, 119
Phenolic steroids, 50, see also specific types
Phenolphthalein, 44
Phenol red, 246, 497
Phenols, 39–43, 47, 49–52, see also specific types
  aryl, 41
  halogenated, 41
  oxidation of, 243
  in PB-TK models, 505
Phenolsulfonphthalein (PSP), 497
Phenoxyacetic acids, 38
4-Phenyl-3-butene-2-one, 68
Phenylglucuronide, 505
Phenylthiazole, 43–44
Phorbol 12-O-tetradecanoate 13-acetate-responsive elements (TREs), 57

Phosphates, 390, 397–399, 407, see also specific types
Phosphatidyl choline, 47
3′-Phosphoadenosine 5′ phosphosulfate (PAPS), 50
Phosphoinositide hydrolysis, 242
Phospholipid hydroperoxides, 59, 60
Phospholipids, 5, 10, see also specific types
Phosphoregulation, 437
Phototaxis, 452
Physicochemical properties, 1, 6–7, see also specific types
Physiologically based tissue response (PB-TR) models, 470, 471, 473–476, 510
Physiologically based toxicokinetic (PB-TK) models, 469–511
  applications of, 494–495, 508–510
  biotransformation in, 471, 504–505
  blood flow relationships and, 476–477
  capacity parameters in, 493–494
  chemical partitioning in, 492–493
  dermal uptake in, 487–489
  development of, 476–483
  dietary uptake in, 489–490
  dose concepts in, 471–476
  examples of, 495–503
  exposure routes in, 483–490
  for fish, 476–483
    examples of, 495–503
    future applications of, 508
  future applications of, 508–510
  growth in, 483
  hexachloroethane in, 498–503
  hydrophobicity in, 483–484
  kinetic rate in, 493–494
  limitations of, 470
  mass-balance descriptions in, 477–481
  methotrexate in, 495–497
  oral uptake dynamics in, 507
  parmeterization in, 490–494
  pentachloroethane in, 498–503
  phenol red in, 497
  physiological inputs into, 490–492
  in predictive toxicology, 509–510
  pyrene in, 497–498
  research needs on, 504–507
  scaling in, 482–483
  1,1,2,2,-tetrachloroethane in, 498–503
  validation of, 494–495
Phytoallexins, 99

Pinocytosis, 5
Piperonyl butoxide (PBO), 111, 113
Pituitary hormones, 151
Plasma proteins, 400, 402–405, 406, see also specific types
Platelet-derived growth factor (PDGF), 307
Platelet-derived growth factor (PDGF) receptor, 307
PLHC-1 cells, 166
PMA, 248, 251, 252
PNS, see Peripheral nervous system
Polychlorinated biphenyls (PCBs), 3, 4, 10, 47, 111, 126, 127, 139
  AHH induction and, 152
  biomarkers and, 212
  cogeners of, 366
  CYP1A and, 168
  degree of contamination with, 129
  hormones and, 150
  metabolism of, 105
  tumors and, 327, 348, 358, 364–366
Polyclonal antibodies, 114, 129, 130
Polycyclic aromatic hydrocarbons (PAHs), 4, 10, 13, 19, 42–43, 47, 51, 71, 109, see also specific types
  accumulation of, 113
  activation of, 145
  Ah receptors and, 139
  cancer and, 20
  CYP1A and, 134, 168, 169
  cytochrome P450 and, 105
  dietary, 18
  diolepoxide formation and activation of, 152
  diol-epoxides for, 143
  gene responsiveness to, 54
  glucuronidation of metabolites of, 38
  intraperitoneal injection of, 72
  metabolism of, 38, 105, 146
  reactive oxyten intermediates and, 245, 250
  tumors and, 327, 348, 356, 357, 364, 365
Polymerase chain reaction (PCR), 315
Polymorphisms, 59
Postsynaptic metal toxicity, 438–440
Preconsumptive biotransformation, 12–16
Predictive toxicology, 509–510
Pregnenolone 16α-carbonitrile (PCN), 43, 123
Presynaptic metal toxicity, 437–438
Procarcinogens, 95, see also specific types
PROD, see Pentoxyresorufin O-deethylase
Progesterones, 114, 129, 133, see also specific types
Prolactin, 433

Proliferating cell nuclear antigen (PCNA), 228, 359
Prostaglandins, 53, see also specific types
Protein kinase C, 242
Protein kinases, 299, see also specific types
Proteins, see also specific types
  calmodulin-related, 434
  copper-binding, 404
  copper-containing plasma, 404
  cross-reactions of with antibodies, 122
  cysteine-rich, 393, 394, 396, 422
  cytoplasmic, 401
  DNA-binding, 296
  G, 244
  guanosine triphosphatase-activating, 296
  guanosine triphosphate-binding, 309–313
  heat shock, 211
  intestinal, 393, 394, 396
  membrane-bound, 309
  metal-binding, 249, 407
  metal transport, 403
  plasma, 400, 402–405, 406
  purification of, 118, 129–130
  stress, 211
  tumors and, 368
  zinc-binding, 404
  zinc-finger, 396
  zinc-transporting, 393
Proteinuria, 407
Proteoglycans, 220, 221, see also specific types
PSP, see Phenolsulfonphthalein
Pyrene, 497–498
Pyrethrins, 37, see also specific types
Pyrethroids, 38, see also specific types
Pyridoxal kinase, 396
Pyrimidine, 5
Pyrocatalytic transformations, 87
Pyrolysis, 37

Q

QSARs, see Quantitative structure-activity relationships
Quantitative structure-activity relationships (QSARs), 510

R

Radionuclides, 407
Ras gene, 310–313
Reactive oxygen intermediates (ROIs), 241–257

aquatic species production of, 245–249
assays of in clinical setting, 244–245
background on, 241–245
fish production of, 247–249, 250–251
hemocyte-mediated production of, 249–251
invertebrate production of, 246–247
of mammalian blood cells, 241–242
methods of, 242–244
molluscan disease production of, 251–254
parasitism and, 252–254
problems with, 254–257
production of, 245–249
  hemocyte-mediated, 249–251
  by molluscan diseases, 251–254
Receptor tyrosine kinases, 299–302
Recklinghausen's disease, 296
Redox enzymes, 38, see also specific types
Redox potential, 97
Reductases, 16, 104
Reduction, 38
Regulatory factors, 242
Resistance to drugs, 52
Resorufins, 107, 126
Respiratory behavioral mechanisms, 449–455
Respiratory exposure routes, 2
Respiratory pigments, 404
Restriction fragment length polymorphisms (RFLPs), 300, 301
Retinoblastoma, 296, 313–314
Retinoid hormones, 173
RFLPs, see Restriction fragment length polymorphisms
Rheotaxis, 452
Rhodamine, 244
RNA binding, 21
ROIs, see Reactive oxygen intermediates
Routes of exposure, see Exposure routes

S

SARs, see Structure-activity relationships
SDM, see S-Sulfadimethoxine
Selenium, 2, 53
Serotonin, 438
Serum albumin, 400, 403, 404
Sexual dichromism, 445
Signal transduction, 242, 296
Silver, 4
Singlet oxygen, 241
Skin tumors, 330
Small-cell lung carcinoma, 296

SOD, see Superoxide dismutase
Southern blot analysis, 54, 55, 124
Spectrometry, 278, see also specific types
Src gene, 302–305
Stem cells, 366–368
Stereoselectivity, 146
Steroid-derived compounds, 98, see also specific types
Steroids, 40, 43, 109, 173, see also specific types
  cytochrome P450 gene regulation and, 133
  3α-hydroxy, 43, 47
  3β-hydroxy, 47
  17β-hydroxy, 43, 47
  hydroxylation of, 97
  metabolism of, 147, 172
  phenolic, 50
Stilbene oxide, 56, 72, 211
Stress proteins, 211, see also specific types
Structure-activity relationships (SARs)
  of CYP1A inducers, 134
  inhibition/inactivation, 110–113
  inhibitor, 110–113
  quantitative, 510
  of substrates, 106–110
Styrene oxide, 14, 56, 66
Substrates, 129, see also specific types
  diagnostic, 106–110
  hydroxylation of, 98
  low concentrations of, 38
  metabolism of, 146–147
  oxidation of, 243
  selectivity of, 17
  specificity of, 45–46, 53–54, 56, 57, 59
  structure-activity relationships of, 106–110
Sugars, 5, see also specific types
Sulfadimethoxine, 6
S-Sulfadimethoxine (SDM), 11
Sulfatases, 16
Sulfation, 38, 89
Sulfides, 38
Sulfones, 14
Sulfotransferases, 16, 17, 50–52, see also specific types
Sulfoxides, 14, see also specific types
Sulfur-seeking properties of metals, 388
Superoxide, 97, 241, 243
Superoxide dismutase (SOD), 241–243, 245, 253
Synaptic cleft metal toxicity, 438–440

T

2,4,5-T, 38
Tannins, 98
Target organ dose, 471–473
Taurolithochloate, 52
TBT, see Tributyltin
TCB, 156
TCDD, 139, 147, 153, 162, 164, 165
TCDF, 156
TCE, see 1,1,2,2-Tetrachloroethane
TEF, see Toxic equivalency factor
Temperature acclimation, 149
Testosterone, 114, 129, 147, 433
2,2',4,4'-Tetrachlorobiphenyl, 105
3,3',4,4'-Tetrachlorobiphenyl, 140
1,1,2,2,-Tetrachloroethane (TCE), 498–503
Tetracycline, 37
Tetrafluoromethylnitrophenol, 44
Thiocarbamates, 39, see also specific types
Thiols, 14, 16, 38, 61, see also specific types
Thionein, 396
Thiones, 38, see also specific types
Thymidine, 368
Thyroid hormones, 173, 432
Thyroid-releasing hormone, 434
Thyroid-stimulating hormone (TSH), 432
Tin, 398
Tissue response models, 470, 471, 473–476, 510
TMA, see Trimethylamine
TMAO, see Trimethylamine oxide
Toxic equivalency factor (TEF), 165, 166
Transcriptional activators, 296
Transferrin, 400, 402–404
Transformations, 87, 128, see also specific types
　bio-, see Biotransformation
　metabolic, see Biotransformation
Transgenes, 368–369
TREs, see Phorbol 12-O-tetradecanoate 13-acetate-responsive elements
Tributyltin (TBT), 113, 249, 250, 251
Trichloroethane, 110
Trifluoromethylnitrophenol, 43
Triglycerides, 9, 10, see also specific types
Trihydroxyprogesterone, 133
Trimethylamine (TMA), 94, 96
Trimethylamine N-oxidase, 94
Trimethylamine oxide (TMAO), 94
Trophic transfer, 1–23
　food chain components and potential for, 3–4
　gastrointestinal tract biotransformation and, 16–20
　macromolecule binding and, 20–22
　matrix properties and, 7–12
　of metals, 397
　microfloral biotransformation and, 16
　physicochemical properties and, 1, 6–7
　physiological properties and, 7–12
　potential for, 3–4
　preconsumptive biotransformation and, 12–16
TSH, see Thyroid-stimulating hormone
Tumors, 327–370, see also specific types
　age and, 344–347
　biochemical characterization of, 354–357
　biomonitoring of, 364–366
　cholangiocellular, 332, 334, 338, 339, 354, 363
　colorectal, 296
　comparative morphology of, 330–334
　diagnosis of, 362–364
　etiology of, 329–330
　experimental, 349–353
　field studies of, 364–366
　in fish, 328–334
　　age and, 344–347
　　biomonitoring of, 364–366
　　experimental, 349–353
　　field studies of, 364–366
　　genetics and, 359–361
　　histogenesis of, 338–343, 353–354
　　in liver, 344–349
　　migration and, 348
　　mortality and, 348–349
　　research on, 367–369
　gastrointestinal tract, 20–21
　genetics and, 359–361
　genomic susceptibility to, 352
　histogenesis of, 334–343, 353–354
　history of, 328–329
　liver, see Liver tumors
　migration and, 348
　morphology of, 330–334
　mortality and, 348–349
　nonepithelial, 330
　reactive oxyten intermediates and, 251–252
　research on, 367–369
　skin, 330
　susceptibility to, 352
　terminology for, 362–364

INDEX

Tumor suppressor genes, 296, 309, 313–314, 316–317, 367, see also specific types
Two-compartment model, 405, 406
Tyrosine kinases, 299–307, see also specific types
   nonreceptor, 302–307
   receptor, 299–302

U

UDP, see Uridine diphosphate
UDPGA, see Uridine diphosphate-glucuronic acid
UDPGDH, see Uridine diphosphate glucose dehydrogenase
UDPGdH, see Uridine diphosphoglucuronyl dehydrogenase
UDPGTs, see UDP-glucuronosyl transferases
UDS, see Unscheduled DNA synthesis
Ultraviolet radiation, 277
Unscheduled DNA synthesis (UDS), 271
Uridine diphosphate glucose dehydrogenase (UDPGDH), 226, 227
Uridine diphosphate-glucuronic acid (UDPGA), 39
Uridine diphosphate glucuronosyl dehydrogenase, 215
Uridine diphosphate-glucuronosyl transferases (UDPGTs), 16–18, 39–50, 74, see also specific types
   activities of, 44–45
   biomarkers and, 227
   characterization of, 46–50
   in fish, 43–50
   gene structure of, 40–43
   mammalian, 39–43
   properties of, 39–40, 44–45
   purification of, 46–50
   reactions of, 39–40
   tissue expression of, 46
Uridine diphosphate-glucuronyltransferase, 19
Uridine diphosphoglucuronyl dehydrogenase (UDPGdH), 355

V

Vanillylmandellic acid (VMA), 440
Very low density lipoproteins (VLDLs), 10, see also specific types

Vesicle-bound granules, 395
Vitamin A, 218, 219
Vitamin D, 98
Vitamins, 10, 242, see also specific types
Vitellogenesis, 133
Vitellogenin, 212
VLDLs, see Very low density lipoproteins
VMA, see Vanillylmandellic acid

W

Western blot analysis, 18, 95, 130, 367
Western immunoblotting, 70
Wilms' tumor gene, 296

X

X cell, 330
Xiphophorus erb-B gene, 302
Xiphorphorine melanoma cells, 369
X-rays, 302

Y

Yes gene, 305

Z

Zinc, 389, 390
   binding of, 393, 404, 422
   in blood cells, 400, 401
   in erythrocytes, 400
   intracellular, 396
   intracellular deposits of, 398, 399
   plasma proteins and, 402, 403
   presynaptic toxicity of, 437
   release of, 406, 407, 408
   toxicity of of, 453
   transport of, 393–394, 400, 404
   uptake of, 392
Zinc-binding proteins, 404, see also specific types
Zinc-finger proteins, 396, see also specific types
Zinc-transporting protein, 393
Zymosan, 246, 248, 250, 252, 254, see Toxic equivalency